Geophysical Monograph Series

Including

IUGG Volumes

Maurice Ewing Volumes

Mineral Physics Volumes

GEOPHYSICAL MONOGRAPH SERIES

Geophysical Monograph Volumes

1 **Antarctica in the International Geophysical Year** *A. P. Crary, L. M. Gould, E. O. Hulburt, Hugh Odishaw, and Waldo E. Smith (Eds.)*
2 **Geophysics and the IGY** *Hugh Odishaw and Stanley Ruttenberg (Eds.)*
3 **Atmospheric Chemistry of Chlorine and Sulfur Compounds** *James P. Lodge, Jr. (Ed.)*
4 **Contemporary Geodesy** *Charles A. Whitten and Kenneth H. Drummond (Eds.)*
5 **Physics of Precipitation** *Helmut Weickmann (Ed.)*
6 **The Crust of the Pacific Basin** *Gordon A. Macdonald and Hisashi Kuno (Eds.)*
7 **Antarctic Research: The Matthew Fontaine Maury Memorial Symposium** *H. Wexler, M. J. Rubin, and J. E. Caskey, Jr. (Eds.)*
8 **Terrestrial Heat Flow** *William H. K. Lee (Ed.)*
9 **Gravity Anomalies: Unsurveyed Areas** *Hyman Orlin (Ed.)*
10 **The Earth Beneath the Continents: A Volume of Geophysical Studies in Honor of Merle A. Tuve** *John S. Steinhart and T. Jefferson Smith (Eds.)*
11 **Isotope Techniques in the Hydrologic Cycle** *Glenn E. Stout (Ed.)*
12 **The Crust and Upper Mantle of the Pacific Area** *Leon Knopoff, Charles L. Drake, and Pembroke J. Hart (Eds.)*
13 **The Earth's Crust and Upper Mantle** *Pembroke J. Hart (Ed.)*
14 **The Structure and Physical Properties of the Earth's Crust** *John G. Heacock (Ed.)*
15 **The Use of Artificial Satellites for Geodesy** *Soren W. Henricksen, Armando Mancini, and Bernard H. Chovitz (Eds.)*
16 **Flow and Fracture of Rocks** *H. C. Heard, I. Y. Borg, N. L. Carter, and C. B. Raleigh (Eds.)*
17 **Man-Made Lakes: Their Problems and Environmental Effects** *William C. Ackermann, Gilbert F. White, and E. B. Worthington (Eds.)*
18 **The Upper Atmosphere in Motion: A Selection of Papers With Annotation** *C. O. Hines and Colleagues*
19 **The Geophysics of the Pacific Ocean Basin and Its Margin: A Volume in Honor of George P. Woollard** *George H. Sutton, Murli H. Manghnani, and Ralph Moberly (Eds.)*
20 **The Earth's Crust: Its Nature and Physical Properties** *John C. Heacock (Ed.)*
21 **Quantitative Modeling of Magnetospheric Processes** *W. P. Olson (Ed.)*
22 **Derivation, Meaning, and Use of Geomagnetic Indices** *P. N. Mayaud*
23 **The Tectonic and Geologic Evolution of Southeast Asian Seas and Islands** *Dennis E. Hayes (Ed.)*
24 **Mechanical Behavior of Crustal Rocks: The Handin Volume** *N. L. Carter, M. Friedman, J. M. Logan, and D. W. Stearns (Eds.)*
25 **Physics of Auroral Arc Formation** *S.-I. Akasofu and J. R. Kan (Eds.)*
26 **Heterogeneous Atmospheric Chemistry** *David R. Schryer (Ed.)*
27 **The Tectonic and Geologic Evolution of Southeast Asian Seas and Islands: Part 2** *Dennis E. Hayes (Ed.)*
28 **Magnetospheric Currents** *Thomas A. Potemra (Ed.)*
29 **Climate Processes and Climate Sensitivity (Maurice Ewing Volume 5)** *James E. Hansen and Taro Takahashi (Eds.)*
30 **Magnetic Reconnection in Space and Laboratory Plasmas** *Edward W. Hones, Jr. (Ed.)*
31 **Point Defects in Minerals (Mineral Physics Volume 1)** *Robert N. Schock (Ed.)*
32 **The Carbon Cycle and Atmospheric CO_2: Natural Variations Archean to Present** *E. T. Sundquist and W. S. Broecker (Eds.)*
33 **Greenland Ice Core: Geophysics, Geochemistry, and the Environment** *C. C. Langway, Jr., H. Oeschger, and W. Dansgaard (Eds.)*
34 **Collisionless Shocks in the Heliosphere: A Tutorial Review** *Robert G. Stone and Bruce T. Tsurutani (Eds.)*
35 **Collisionless Shocks in the Heliosphere: Reviews of Current Research** *Bruce T. Tsurutani and Robert G. Stone (Eds.)*
36 **Mineral and Rock Deformation: Laboratory Studies —The Paterson Volume** *B. E. Hobbs and H. C. Heard (Eds.)*
37 **Earthquake Source Mechanics (Maurice Ewing Volume 6)** *Shamita Das, John Boatwright, and Christopher H. Scholz (Eds.)*
38 **Ion Acceleration in the Magnetosphere and Ionosphere** *Tom Chang (Ed.)*
39 **High Pressure Research in Mineral Physics (Mineral Physics Volume 2)** *Murli H. Manghnani and Yasuhiko Syono (Eds.)*
40 **Gondwana Six: Structure, Tectonics, and Geophysics** *Gary D. McKenzie (Ed.)*
41 **Gondwana Six: Stratigraphy, Sedimentology, and Paleontology** *Garry D. McKenzie (Ed.)*
42 **Flow and Transport Through Unsaturated Fractured Rock** *Daniel D. Evans and Thomas J. Nicholson (Eds.)*
43 **Seamounts, Islands, and Atolls** *Barbara H. Keating, Patricia Fryer, Rodey Batiza, and George W. Boehlert (Eds.)*

44 **Modeling Magnetospheric Plasma** *T. E. Moore and J. H. Waite, Jr. (Eds.)*
45 **Perovskite: A Structure of Great Interest to Geophysics and Materials Science** *Alexandra Navrotsky and Donald J. Weidner (Eds.)*
46 **Structure and Dynamics of Earth's Deep Interior (IUGG Volume 1)** *D. E. Smylie and Raymond Hide (Eds.)*
47 **Hydrological Regimes and Their Subsurface Thermal Effects (IUGG Volume 2)** *Alan E. Beck, Grant Garven, and Lajos Stegena (Eds.)*
48 **Origin and Evolution of Sedimentary Basins and Their Energy and Mineral Resources (IUGG Volume 3)** *Raymond A. Price (Ed.)*
49 **Slow Deformation and Transmission of Stress in the Earth (IUGG Volume 4)** *Steven C. Cohen and Petr Vaníček (Eds.)*
50 **Deep Structure and Past Kinematics of Accreted Terranes (IUGG Volume 5)** *John W. Hillhouse (Ed.)*
51 **Properties and Processes of Earth's Lower Crust (IUGG Volume 6)** *Robert F. Mereu, Stephan Mueller, and David M. Fountain (Eds.)*
52 **Understanding Climate Change (IUGG Volume 7)** *Andre L. Berger, Robert E. Dickinson, and J. Kidson (Eds.)*
53 **Plasma Waves and Instabilities at Comets and in Magnetospheres** *Bruce T. Tsurutani and Hiroshi Oya (Eds.)*
54 **Solar System Plasma Physics** *J. H. Waite, Jr., J. L. Burch, and R. L. Moore (Eds.)*
55 **Aspects of Climate Variability in the Pacific and Western Americas** *David H. Peterson (Ed.)*
56 **The Brittle-Ductile Transition in Rocks** *A. G. Duba, W. B. Durham, J. W. Handin, and H. F. Wang (Eds.)*
57 **Evolution of Mid Ocean Ridges (IUGG Volume 8)** *John M. Sinton (Ed.)*
58 **Physics of Magnetic Flux Ropes** *C. T. Russell, E. R. Priest, and L. C. Lee (Eds.)*
59 **Variations in Earth Rotation (IUGG Volume 9)** *Dennis D. McCarthy and Williams E. Carter (Eds.)*
60 **Quo Vadimus Geophysics for the Next Generation (IUGG Volume 10)** *George D. Garland and John R. Apel (Eds.)*
61 **Cometary Plasma Processes** *Alan D. Johnstone (Ed.)*
62 **Modeling Magnetospheric Plasma Processes** *Gordon R. Wilson (Ed.)*
63 **Marine Particles: Analysis and Characterization** *David C. Hurd and Derek W. Spencer (Eds.)*
64 **Magnetospheric Substorms** *Joseph R. Kan, Thomas A. Potemra, Susumu Kokubun, and Takesi Iijima (Eds.)*
65 **Explosion Source Phenomenology** *Steven R. Taylor, Howard J. Patton, and Paul G. Richards (Eds.)*
66 **Venus and Mars: Atmospheres, Ionospheres, and Solar Wind Interactions** *Janet G. Luhmann, Mariella Tatrallyay, and Robert O. Pepin (Eds.)*
67 **High-Pressure Research: Application to Earth and Planetary Sciences (Mineral Physics Volume 3)** *Yasuhiko Syono and Murli H. Manghnani (Eds.)*
68 **Microwave Remote Sensing of Sea Ice** *Frank Carsey, Roger Barry, Josefino Comiso, D. Andrew Rothrock, Robert Shuchman, W. Terry Tucker, Wilford Weeks, and Dale Winebrenner*
69 **Sea Level Changes: Determination and Effects (IUGG Volume 11)** *P. L. Woodworth, D. T. Pugh, J. G. DeRonde, R. G. Warrick, and J. Hannah*
70 **Synthesis of Results from Scientific Drilling in the Indian Ocean** *Robert A. Duncan, David K. Rea, Robert B. Kidd, Ulrich von Rad, and Jeffrey K. Weissel (Eds.)*
71 **Mantle Flow and Melt Generation at Mid-Ocean Ridges** *Jason Phipps Morgan, Donna K. Blackman, and John M. Sinton (Eds.)*
72 **Dynamics of Earth's Deep Interior and Earth Rotation (IUGG Volume 12)** *Jean-Louis Le Mouël, D.E. Smylie, and Thomas Herring (Eds.)*
73 **Environmental Effects on Spacecraft Positioning and Trajectories (IUGG Volume 13)** *A. Vallance Jones (Ed.)*
74 **Evolution of the Earth and Planets (IUGG Volume 14)** *E. Takahashi, Raymond Jeanloz, and David Rubie (Eds.)*
75 **Interactions Between Global Climate Subsystems: The Legacy of Hann (IUGG Volume 15)** *G. A. McBean and M. Hantel (Eds.)*
76 **Relating Geophysical Structures and Processes: The Jeffreys Volume (IUGG Volume 16)** *K. Aki and R. Dmowska (Eds.)*
77 **The Mesozoic Pacific: Geology, Tectonics, and Volcanism—A Volume in Memory of Sy Schlanger** *Malcolm S. Pringle, William W. Sager, William V. Sliter, and Seth Stein (Eds.)*
78 **Climate Change in Continental Isotopic Records** *P. K. Swart, K. C. Lohmann, J. McKenzie, and S. Savin (Eds.)*
79 **The Tornado: Its Structure, Dynamics, Prediction, and Hazards** *C. Church, D. Burgess, C. Doswell, R. Davies-Jones (Eds.)*
80 **Auroral Plasma Dynamics** *R. L. Lysak (Ed.)*
81 **Solar Wind Sources of Magnetospheric Ultra-Low Frequency Waves** *M. J. Engebretson, K. Takahashi, and M. Scholer (Eds.)*

82 **Gravimetry and Space Techniques Applied to Geodynamics and Ocean Dynamics** *Bob E. Schutz, Allen Anderson, Claude Froidevaux, and Michael Parke (Eds.)*
83 **Nonlinear Dynamics and Predictability of Geophysical Phenomena** *William I. Newman, Andrei Gabrielov, and Donald L. Turcotte (Eds.)*
84 **Solar System Plasmas in Space and Time** *J. Burch, J. H. Waite, Jr. (Eds.)*
85 **The Polar Oceans and Their Role in Shaping the Global Environment** *O. M. Johannessen, R. D. Muench, and J. E. Overland (Eds.)*
86 **Space Plasmas: Coupling Between Small and Medium Scale Processes** *Maha Ashour-Abdalla, Tom Chang, and Paul Dusenbery (Eds.)*
87 **The Upper Mesosphere and Lower Thermosphere: A Review of Experiment and Theory** *R. M. Johnson and T. L. Killeen (Eds.)*

Maurice Ewing Volumes

1 **Island Arcs, Deep Sea Trenches, and Back-Arc Basins** *Manik Talwani and Walter C. Pitman III (Eds.)*
2 **Deep Drilling Results in the Atlantic Ocean: Ocean Crust** *Manik Talwani, Christopher G. Harrison, and Dennis E. Hayes (Eds.)*
3 **Deep Drilling Results in the Atlantic Ocean: Continental Margins and Paleoenvironment** *Manik Talwani, William Hay, and William B. F. Ryan (Eds.)*
4 **Earthquake Prediction—An International Review** *David W. Simpson and Paul G. Richards (Eds.)*
5 **Climate Processes and Climate Sensitivity** *James E. Hansen and Taro Takahashi (Eds.)*
6 **Earthquake Source Mechanics** *Shamita Das, John Boatwright, and Christopher H. Scholz (Eds.)*

IUGG Volumes

1 **Structure and Dynamics of Earth's Deep Interior** *D. E. Smylie and Raymond Hide (Eds.)*
2 **Hydrological Regimes and Their Subsurface Thermal Effects** *Alan E. Beck, Grant Garven, and Lajos Stegena (Eds.)*
3 **Origin and Evolution of Sedimentary Basins and Their Energy and Mineral Resources** *Raymond A. Price (Ed.)*
4 **Slow Deformation and Transmission of Stress in the Earth** *Steven C. Cohen and Petr Vaníček (Eds.)*
5 **Deep Structure and Past Kinematics of Accreted Terrances** *John W. Hillhouse (Ed.)*
6 **Properties and Processes of Earth's Lower Crust** *Robert F. Mereu, Stephan Mueller, and David M. Fountain (Eds.)*
7 **Understanding Climate Change** *Andre L. Berger, Robert E. Dickinson, and J. Kidson (Eds.)*
8 **Evolution of Mid Ocean Ridges** *John M. Sinton (Ed.)*
9 **Variations in Earth Rotation** *Dennis D. McCarthy and William E. Carter (Eds.)*
10 **Quo Vadimus Geophysics for the Next Generation** *George D. Garland and John R. Apel (Eds.)*
11 **Sea Level Changes: Determinations and Effects** *Philip L. Woodworth, David T. Pugh, John G. DeRonde, Richard G. Warrick, and John Hannah (Eds.)*
12 **Dynamics of Earth's Deep Interior and Earth Rotation** *Jean-Louis Le Mouël, D.E. Smylie, and Thomas Herring (Eds.)*
13 **Environmental Effects on Spacecraft Positioning and Trajectories** *A. Vallance Jones (Ed.)*
14 **Evolution of the Earth and Planets** *E. Takahashi, Raymond Jeanloz, and David Rubie (Eds.)*
15 **Interactions Between Global Climate Subsystems: The Legacy of Hann** *G. A. McBean and M. Hantel (Eds.)*
16 **Relating Geophysical Structures and Processes: The Jeffreys Volume** *K. Aki and R. Dmowska (Eds.)*
17 **Gravimetry and Space Techniques Applied to Geodynamics and Ocean Dynamics** *Bob E. Schutz, Allen Anderson, Claude Froidevaux, and Michael Parke (Eds.)*
18 **Nonlinear Dynamics and Predictability of Geophysical Phenomena** *William I. Newman, Andrei Gabrielov, and Donald L. Turcotte (Eds.)*

Mineral Physics Volumes

1 **Point Defects in Minerals** *Robert N. Schock (Ed.)*
2 **High Pressure Research in Mineral Physics** *Murli H. Manghnani and Yasuhiko Syona (Eds.)*
3 **High Pressure Research: Application to Earth and Planetary Sciences** *Yasuhiko Syono and Murli H. Manghnani (Eds.)*

Geophysical Monograph 88

Active Margins and Marginal Basins of the Western Pacific

Brian Taylor
James Natland
Editors

American Geophysical Union

Published under the aegis of the AGU Books Board.

Library of Congress Cataloging-in-Publication Data

Active margins and marginal basins of the western Pacific / Brian Taylor. James Natland, editors.
p. cm. — (Geophysical monograph : 88)
Includes bibliographical references.
ISBN 0-87590-045-3
1. Geology—Pacific Ocean. 2. Continental margins—East Asia.
I. Taylor, Brian, 1953– . II. Natland, James P. (James Preston)
III. Series.
QE350.4.A28 1995
551.46'2'08'094—dc20 95-9944
CIP

ISSN 0065-8448

ISBN 0-87590-045-3

This book is printed on acid-free paper.

Printed in the United States of America.

CONTENTS

CONTENTS

PREFACE

This volume focuses on the volcanic, fluid, sedimentary, and tectonic processes occurring in the trench-arc-backarc systems of the western Pacific—a natural regional focus for studies of these themes. The results of ocean drilling and associated site surveys in the western Pacific have brought fundamental changes to our understanding of volcanism, crustal deformation, fluid circulation, and sedimentation in active margins and marginal basins. Our goal here is to synthesize the results of ocean drilling in a multi-disciplinary manner, including a comparison of the findings from drilling legs having similar themes, and to emphasize the significance of these results to the broader geoscience community.

Intense land and marine studies of the island arcs and collision zones of the western Pacific were conducted in preparation for, and accompanying, Ocean Drilling Program (ODP) activities in the region from late 1988 to early 1991. When the results of these individual studies became available, 70 of the more than 250 scientists who participated in this effort met to discuss them at a conference in Monterey in January 1993. Collaboration was encouraged, and the papers in this volume were submitted after a year of further data integration and interpretation. (Not all of the topics addressed by drilling or discussed at the conference are represented here.)

Volcanism

The middle Eocene initiation of intra-oceanic subduction in the western Pacific was characterized by depleted boninitic and tholeiitic series volcanism that in ~10 m.y. formed an extensional terrane, ~300 km wide and thousands of kilometers long, that is the best modern analog to supra-subduction zone ophiolites (see the first paper, by Bloomer et al.).

Neogene or Quaternary arc tholeiitic sills occur halfway to the trench in all the forearcs drilled and have compositions geochemically closest to the adjacent Quaternary arc volcanoes (R. Taylor et al.). From a source beneath the volcanic front, the magmas must have travelled laterally up to 100 km via dikes transgressing the lower forearc. This implies trench-perpendicular compression in the forearc synchronous with extension in the backarc.

The spatial and temporal evolution of island arcs is discussed by Arculus et al., Clift, and Cambray et al. The arcs do not evidence a consistent and monotonic increase in crustal thickness or alkalinity, but rather a variability associated with state of stress (cycles of arc rifting, backarc spreading, transpression or collision), changes in mantle source composition, and a lithospheric filter effect. Explosive, high-silica volcanism reached peak intensities during the initial rifting of the Japan, Izu-Bonin-Mariana, Sulu and Lau arcs. In several but not all (Mariana) systems, arc activity declined markedly (or even terminated in the Izu-Bonins) at the start of subsequent backarc spreading.

The formation of backarc basins involves extensive arc rifting (either in front of and/or behind the volcanic front), lateral rift migration and longitudinal spreading propagation, synchronous but chemically distinct rift and (frontal and/or remnant) arc volcanism, with slightly wet but otherwise MORB-like lavas erupting in the earliest rifts, together with subsequent arc-like lavas, and N-MORB dominance only after substantial spreading (see the papers by Clift and by Hawkins).

Basalts from all the western Pacific marginal basins have isotopic characteristics similar to those of Indian rather than Pacific MORB, indicating that the Pacific-Indian asthenoshere isotopic boundary has been located along the Pacific subduction boundary for at least the last 50 m.y. (Hickey-Vargas et al.). In detail, these two mantle isotopic provinces have interfingered in space and time during the evolution of the boundary arcs and backarc basins (New Hebrides: Crawford et al.; Lau: Hawkins).

Fluids, Diagenesis, Metamorphism

Closed-system, low-temperature alteration of rapidly deposited volcanogenic sediments in forearc and intra-arc basins to hydrous clay and zeolite minerals, principally smectite, produces saline pore fluids ($CaCl_2$ brines), low permeabilities, and large concentration gradients (Martin et al., Torres et al.). In contrast, biogenic productivity controls the formation of bedded cherts, authigenic carbonates, and barites (from remobilized barium concentrated at the base of the sulfate reduction zone).

Serpentine (hydrated peridotite) seamounts in the Izu-Bonin-Mariana forearc vary between two end-member types: mud volcanoes (diapirs from the mantle) and horst blocks (metamorphosed ultramafics emplaced in the crust) (Fryer et al.). All types are associated with large-scale faulting. Their similarities to "sedimentary" serpentinite terranes on land suggest that many of the latter may have formed prior rather than syn/post obduction. The venting fluids have subducted slab components and may have exotic compositions (e.g., pH=12.5).

The low-temperature (150–250° Celsius), high-pressure (5–6 kb), mineral assemblage lawsonite + pumpellyite + hematite found in metabasite clasts entrained in Conical seamount provides direct evidence of incipient blueschist-facies metamorphism beneath the Mariana forearc (Maekawa et al.).

Sedimentation

Volcanic input, vertical tectonics, and eustacy are primary variables that control the style of sedimentation in backarc and forearc basins (Marsaglia et al., Underwood et al.). Submarine gravity flows distribute felsic pumice and volcaniclastic debris, episodically erupted from submarine arc calderas, throughout the Sumisu backarc rift, where sedimentation rates are 1–4 km/m.y. Similar, but more distal, deposits were drilled in the lower Miocene Sulu backarc basin. Rather than inundation beneath a thick volcaniclastic apron, the stratigraphy of isolated sub-basins in the Lau Basin reflects an eastward migration of extension and intra-rift volcanism. Eustacy and tectonically controlled sill depths modulated pelagic/hemipelagic sedimentation in the Japan, Sulu, and Celebes seas. The Izu-Bonin forearc basin turbidites of mixed pyroclastic and epiclastic detritus were deposited by a noncyclic basin-plain style of sedimentation rather than by a channelized submarine fan or slope apron. Drilling penetrated through the decollement and into basement at the toe of the Nankai accretionary prism: the sedimentary facies pass upwards from abyssal-plain hemipelagites, through a sandy trench wedge, to a thin slope apron sequence. Paleoflow directions require that the turbidites experienced a phase of upslope movement onto the outer wall and then back toward the Nankai trench axis. In the New Hebrides ridge-trench collision, accreted strata include trench turbidites and pelagic deposits delaminated from the subducting basement along kaolinite-rich horizons.

Tectonics

Improved tectonic reconstructions of the western Pacific, in part facilitated by paleomagnetic studies of FMS-oriented core and Ar-Ar dating, indicate middle Eocene subduction initiation at ~50 Ma (prior to the Hawaiian-Emperor bend) and multiphase rotations at both basin-wide and forearc scales. In the case of the Japan Sea opening, Jolivet et al. show that a model including distributed continental strain as well as propagating oceanic spreading in a pull-apart geometry between meridional dextral shear zones can explain the principal paleomagnetic and structural observations. For the Philippine Sea Plate, Hall et al. incorporate land and ODP paleomagnetic data to derive an evolutionary model that produces 90° of clockwise rotation since 50 Ma but with no rotation between 25 and 40 Ma other than possibly at forearc sites.

Several of the inner marginal basins of the western Pacific, including those of the Sulu, Celebes, and Japan seas, started to close and subduct soon after they were formed (Rangin et al.). In this process, crustal sections from basin margins may be thrust into young accretionary wedges and subsequently form ophiolites.

Acknowledgments

The Ocean Drilling Program is supported financially by the governments of 19 nations. All but one western Pacific island nation generously granted permission to drill in their waters. The western Pacific drilling program was designed largely by the JOIDES Western Pacific Panel, in conjunction with the Tectonics, Lithosphere, and Sediments and Ocean History Panels, from proposals submitted by the international scientific community, and with advice from the whole JOIDES structure during 5 years of planning (1984-1988). The success of the program is a direct consequence of the excellent site surveys and the dedication of all who served on the drillship and who supported the operation from onshore. The conference in Monterey and the preparation of material for this volume were funded by the U.S. Science Advisory Committee of the Joint Oceanographic Institutions Inc. Diane Henderson and the SOEST Publications Group, University of Hawaii, copy-edited all manuscripts, reformatted "camera-ready" submissions, and made our job as scientific editors manageable. The rigorous manuscript reviews by those listed below significantly improved the volume.

Reviewers

James Allan
William Bryan
David Christie
Robert Coleman
John Compton
Brian Cousens
Keith Crook
Darrel Cowan
Hugh Davies
Charles DeMets
William Dickinson
Patrick Dobson
Trevor Falloon
James Hein
Richard Hiscott
Fred Hochstaedter
Ray Ingersoll
Jill Karsten
Barbara Keating
Masato Koyama
John Mahoney
Shigenori Maruyama
Russell McDuff
Ralph Moberly
Gregory Moore
James Natland
Renzo Sartori
Tetsuro Seno
Jon Stephenson
Brian Taylor
Tracy Vallier
Jon Woodhead
Geoff Wheat
Seiya Uyeda

Brian Taylor
University of Hawaii
Honolulu, Hawaii

James Natland
RSMAS
Miami, Florida

Editors

Early Arc Volcanism and the Ophiolite Problem: A Perspective From Drilling in the Western Pacific

Sherman H. Bloomer[1], Brian Taylor[2], Christopher J. MacLeod[3], Robert J. Stern[4], Patricia Fryer[5], James W. Hawkins[6], and Lynn Johnson[7]

Marine geologic studies and Deep Sea Drilling Project (DSDP) and Ocean Drilling Program (ODP) transects of the intraoceanic Izu-Bonin-Mariana (IBM) forearcs have generated several important new ideas about the origin and evolution of these terranes. The first, and most striking, is the recognition that the initial phases of volcanism in these subduction zones developed nearly synchronously in the middle to late Eocene over a zone up to 300 km wide and thousands of kilometers long . This early, or "infant," arc volcanism was characterized by the eruption of very depleted boninitic and arc tholeiitic lava compositions and occurred in extensional environments (as evidenced by dikes on Chichi-jima in the Bonin Islands). This infant arc volcanism was built on, or displaced, the pre-existing oceanic crust and had igneous production rates much higher than those of mature arcs—eruption rates on the order of those in slow-spreading ridges. This initial arc volcanism is unlike that developed during the "normal" or mature phases of arc activity and is a plausible mechanism for developing supra-subduction zone ophiolites. In fact, the duration of volcanism, volcanic and plutonic rock compositions, and structural setting in the IBM forearc are virtually identical to those in the Troodos ophiolite. The exposure of Eocene arc basement immediately adjacent to the axis of the Mariana-Bonin Trench requires some subduction erosion since the Eocene. However, if the Eocene basement did form by unusual, voluminous volcanism in an extensional environment, the required amount of erosion may only be 20 to 50 km; the vertical tectonics of the forearc indicate there has been little erosion of the forearc since the Oligocene. There is also clear evidence of minor, episodic, post-Eocene incorporation of older Pacific plate sediments and crustal fragments into the outer forearc. These forearcs are structurally and compositionally complex, and are clearly direct products of subduction, rather than just trapped pieces of oceanic crust. They have geological and geochemical similarities to supra-subduction zone ophiolites like those in Troodos and Oman and represent a previously unappreciated type of crustal construction. They are, in fact, the closest crustal analog to supra-subduction zone ophiolites that has been found in *any* modern geologic environment.

[1]Department of Earth Sciences, Boston University, Boston, Massachusetts
[2]Department of Geology and Geophysics, University of Hawaii, Honolulu Hawaii
[3]Institute of Oceanographic Sciences, Surrey, United Kingdom
[4]Center for Lithospheric Studies, University of Texas at Dallas, Richardson, Texas
[5]School of Ocean and Earth Science and Technology, University of Hawaii, Honolulu, Hawaii
[6]Geological Research Division, Scripps Institution of Oceanography, La Jolla, California
[7]Marine Physics Branch, Naval Research Laboratory, Washington, DC

Active Margins and Marginal Basins of the Western Pacific
Geophysical Monograph 88

1. INTRODUCTION

Forearcs are an integral part of intraoceanic convergent margins. Their igneous basement contains a record of the early evolution of the subduction system [e.g., *Hussong and Uyeda, 1981*; *Bloomer*, 1983; *Ishii*, 1985; *Taylor*, 1992] and the volcaniclastic sediments covering them document the tectonic and geochemical evolution of the arc. Deep Sea Drilling Project (DSDP) and Ocean Drilling Program (ODP) transects of the intraoceanic Izu-Bonin-Mariana and Tonga forearcs, in combination with marine geological and geophysical studies, have shed light on two critical geological problems—the nature of the earliest arc volcanism associated with subduction and the origin of so-called "supra-subduction zone" ophiolites. These results are, in fact, among the most exciting to emerge from the first phase of the Ocean Drilling Program.

The nature of the arc volcanism associated with the onset of subduction has rarely been considered. We are used to thinking about arcs as long lines of central volcanoes, relatively fixed in space, which are associated with deep, and essentially steady-state, subducting slabs. However, the rupture of oceanic lithosphere during the initiation of a subduction zone must produce transient thermal and mechanical states that yield magmatism of a sort very different from the point-source volcanism we see in mature subduction zones. The record of that earliest arc volcanism should, conversely, tell us something about the thermal and mechanical conditions under which subduction began and therefore something about how subduction begins. In this paper we review evidence that the earliest record of arc volcanism lies in the present-day forearcs of intraoceanic arcs.

Study of the western Pacific forearcs has made an important contribution to the debate about the origin of ophiolites. It has been known for a number of years that several of the large, relatively intact Tethyan ophiolites such as the Troodos ophiolite in Cyprus or the Semail ophiolite in Oman differ in significant respects from the mafic crust formed at mid-oceanic spreading centers [e.g., *Rautenschlein et al.*, 1985; *Serri,* 1981; *Bloomer and Hawkins*, 1983]. The sheeted dikes in those ophiolites and the abundant structural evidence of normal faulting provided compelling arguments for the formation of these ophiolites by seafloor spreading. However, the geochemistry of the lavas in the ophiolites equally clearly requires that they formed in some setting above a subduction zone [e.g., *Pearce et al.*, 1984]. Appropriate modern analogs for these supra-subduction zone (SSZ) ophiolites have been difficult to find. They are certainly not like modern island arcs, with lines of central edifices and thick volcaniclastic aprons. Backarc basins have provided the most appealing analog for the origin of SSZ ophiolites, but these ophiolites rarely have an associated remnant or active arc, nor are they covered by volcaniclastic sediments shed from those arcs, as is the case for most crust formed in a backarc basin. It is also difficult to understand how backarc basin crust can be emplaced as an ophiolite without the involvement of adjacent arc and forearc crust.

We shall argue that the best analog in an intraoceanic subduction zone for these SSZ ophiolites actually lies in the forearc crust formed in the earliest stages of subduction [*Natland and Tarney*, 1981; *Ishii*, 1985]. Note that we refer here specifically to the large, complete, Tethyan-type ophiolites, which have a marked subduction zone geochemical signature. There are a number of ophiolites, commonly partially dismembered, (for example in the Alps, the Appenines, and the Alaskan accretionary prisms) which do appear to have structural, petrologic, and geochemical affinities with crust formed at mid-oceanic spreading centers.

Scientific ocean drilling has been an absolutely essential tool in efforts to understand the geology of the forearc regions of the western Pacific intraoceanic island arcs. Drilling provides the only way to determine the age or composition of the forearc basement, which is covered by a thick volcaniclastic section. Boreholes also provide a means to constrain the post-formation tectonic history of the forearcs, as it is recorded in sediment compositions, structures, and paleomagnetic signatures.

We review here our present understanding of the geology of the forearcs of the Izu-Mariana-Bonin and Tonga systems and examine some of the inferences that can be drawn about volcanism during the initiation of subduction. We discuss some of the similarities between supra-subduction zone ophiolites and the crust constructed during that early arc volcanism. We shall show that the initial phases of volcanism in these subduction zones developed nearly synchronously during the middle to late Eocene over a zone up to 300 km wide [*Natland and Tarney*, 1981; *Hussong and Uyeda, 1981*; *Taylor*, 1992] and thousands of kilometers long [*Stern and Bloomer*, 1992; *Taylor*, 1992]. This early, or "infant," arc volcanism was characterized by the eruption of very depleted boninitic and arc tholeiitic lava compositions [e.g., *Meijer*, 1980; *Ishii*, 1985; *Bloomer*, 1987; *Murton et al.*, 1992] and developed in an extensional setting in places possibly to the point of true seafloor spreading [*Stern and Bloomer*, 1992]. This infant arc crust was built on, or displaced, the pre-existing oceanic crust and had igneous production rates much higher than those of mature arcs—eruption rates on the order of those in slow-spreading ridges. This mode of arc volcanism is unlike that developed during the "normal" or mature phases of arc activity and is a plausible mechanism for developing supra-subduction zone ophiolites [what *Pearce et al.*, 1984, termed "pre-arc" ophiolites; *Hawkins et al.*, 1984; *Robertson*, 1990; *Stern and Bloomer*, 1992]. In fact, the duration of volcanism, volcanic and plutonic rock compositions, and structural setting of the IBM forearc are virtually identical to those of the Troodos ophiolite.

This synthesis draws heavily on recent reviews by *Taylor* [1992] and *Stern and Bloomer* [1992] and the results from Legs 125 and 126 in the Izu-Bonin-Mariana forearcs and Leg 135 in the Tonga forearc [*Bloomer et al.*, 1994]. It also builds on the syntheses from DSDP drilling in the Mariana forearc during Leg 60 [*Hussong and Uyeda, 1981*; *Natland and Tarney, 1981*].

2. BACKGROUND

The models for the evolution of subduction zones discussed here are built around studies of the intraoceanic subduction zones of the western Pacific. Most details of the discussion are drawn from the Izu-Bonin-Mariana (IBM) forearc, as it has the greatest density of geological, geophysical, and drilling data. The existing information from the Tonga forearc generally corroborates the picture constructed in the IBM system. The study of forearc geology and structure has concentrated on these two areas because they are where it is easiest to constrain the geologic hisotries of the forearc igneous basement. Both the Tonga and the IBM systems have had relatively simple histories since their initiation around 45 Ma. There have been no reversals of subduction, no collisions with arc or continental fragments, and there is no evidence that either is floored by older arc or continental fragments. Both are systems built entirely within older oceanic crust. Both lack significant accretionary prisms, because of the thin sedimentary sections on the subducting plates and in the trenches, so the foreac basement remains accessible, in places by dredging and in others by drilling through relatively thin sedimentary sequences. Models for the origin of these intraoceanic forearcs may not be applicable in all regards to subduction zones formed in different settings (e.g., accretionary and continental margin subduction zones), but they are end-member cases and provide an important starting point for understanding the mechanics of one mode of subduction initiation and consequent early arc volcanism.

2.1. *The Izu-Bonin-Mariana Forearc*

The Izu-Bonin-Mariana intraoceanic arc-trench system is the type example with which other older or less well-studied non-accretionary systems are compared (Figure 1). Subduction along the edge of the Philippine Sea began sometime in the mid-Eocene. Marine geophysical studies, together with drilling on Deep Sea Drilling Project (DSDP) legs 6, 31, 58, 59, and 60, have confirmed the essential tenets of *Karig* [1971] and *Packham and Falvey* [1971] that backarc basins, such as the Shikoku Basin, Parece Vela Basin, and Mariana Trough (Figure 1), formed by seafloor spreading between the formerly contiguous remnant arcs (Palau-Kyushu Ridge and West Mariana Ridge) and the active IBM arc. The forearc of the IBM system has existed since the initiation of subduction, and was formed either by trapping of older oceanic crust or by early arc volcanism. It has been rafted eastward (relative to the West Philippine Basin) during successive episodes of arc rifting and backarc basin spreading. Oligocene rifting produced significant stretching in the Bonin part of the forearc, with accompanying volcanism, but most subsequent rifting has been confined to the backarc regions [*Taylor,* 1992]. The forearc platform has been modified since the initiation of subduction by the accumulation of thick volcaniclastic sediments [*LaTraille and Hussong*, 1980; *Mrozowski et al.*, 1981; *Arculus et al.*, 1992], the development of serpentinite diapirs piercing the outer forearc and inner trench slope [*Hussong and Fryer*, 1981; *Bloomer* 1983; *Fryer et al.*, 1985; *Fryer and Fryer*, 1985; *Fryer et al.*, 1990], and the incorporation of minor amounts of mid-ocean ridge basalts, intraplate basalts, and associated sediments from the subducting plate into the outer forearc [*Azema and Blauchet*, 1981; *Bloomer*, 1983; *Fryer et al.*, 1985; *Johnson and Fryer*, 1990; *Johnson et al.*, 1991].

The Palau-Kyushu Ridge, the oldest line of arc volcanoes in the IBM system, was well established by the early Oligocene [*Klein et al.*, 1980]. We are interested in the

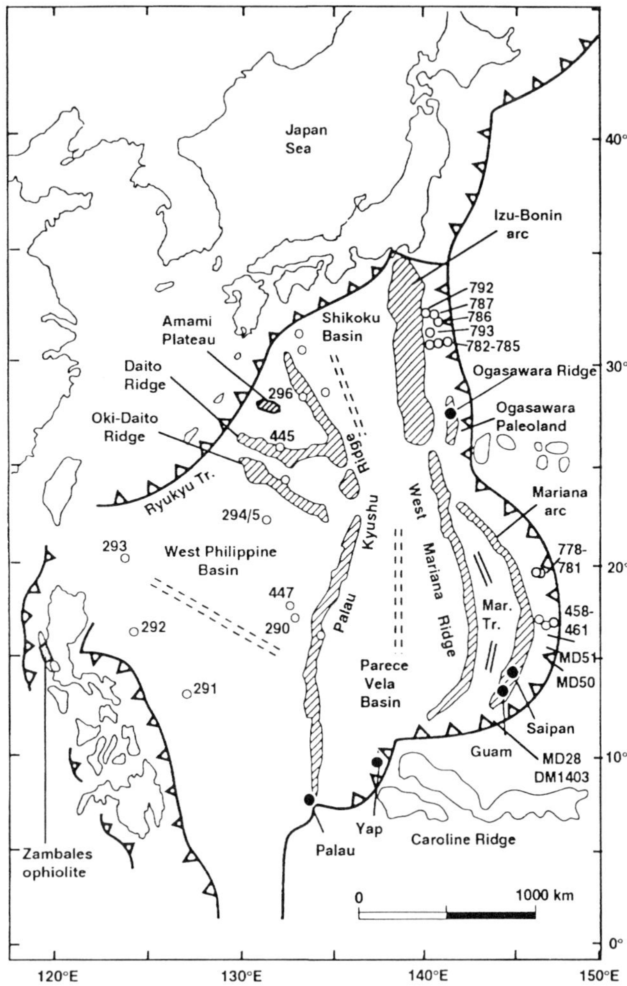

Fig. 1. Map of the Philippine Sea, showing principal dredge, island, and borehole sample sites. MD and DM indicate dredge sites, open circles are DSDP or ODP boreholes, solid circles are islands with basement exposures. Locations with Eocene to early Oligocene rocks are underlined. From *Pearce et al.* [1992b].

history of volcanism prior to that time, which requires samples of the older (Eocene) parts of the IBM forearc. These samples are available from the western edge of the forearc, in uplifted basement on the frontal arc islands of Guam, Tinian, and Saipan [Figure 1; *Ingle*, 1975; *Reagan and Meijer,* 1984], from dredged quartz diorites, andesites and tonalites along the northernmost Palau-Kyushu Ridge [*Shiki*, 1985], and from the island of Palau [*Pearce et al.*, 1992b]. Eocene to early Oligocene arc volcanics have been sampled from the outer edge of the forearc, along the outer arc high, at DSDP sites 458 and 459, ODP sites 782, and 786 and on the Bonin Islands [*Hussong et al.*, 1981; *Natland and Tarney*, 1981; *Umino,* 1985; *Dobson*, 1986; *Fryer et al.*, 1990]. Geochemically similar lavas have been dredged

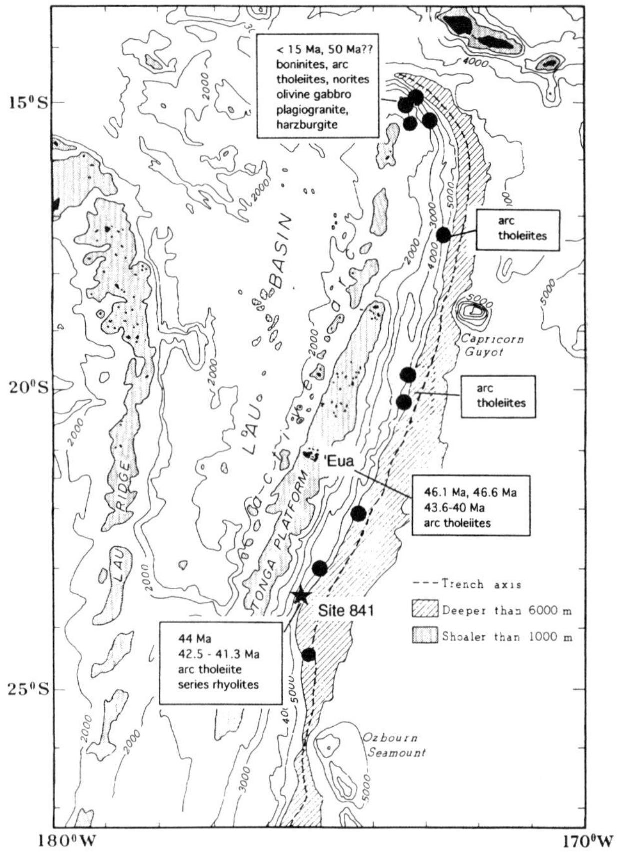

Fig. 2. Map of the Tonga Trench and Lau Basin region showing principal sample sites with rock compositions and ages, where available. Dredge locations shown by circles [from *Bloomer and Fisher*, 1987; *Fisher and Engel*, 1969; *Vallier et al.*, 1985; *Gnibdenko et al.,* 1985; *Sharaskin et al.*, 1983; *Falloon et al.*, 1987], ODP site 841 shown by diamond; 'Eua data and ages from *Duncan et al.* [1985] and *Cunningham and Anscombe* [1985].

from the inner wall of the IBM trench [*Dietrich et al.*, 1978; *Bloomer* 1983; *Ishii*, 1985, *Bloomer and Hawkins*, 1983].

2.2. The Tonga Forearc

The Tonga forearc is a part of a complex array of arcs and backarcs that have developed as a result of the subduction of the Pacific plate below the Indo-Australia plate since at least middle Eocene time [Figure 2; e.g., *Herzer and Exon*, 1985]. As in the IBM system, marine geological and geophysical work has documented a complex post-Eocene history of arc volcanism, rifting, and backarc basin spreading, leaving a festoon of ridges and basins including the Tonga platform (extinct, coralline-capped Eocene to Miocene volcanic islands), the active Tofua arc [e.g., *Ewart et al.*, 1977], the actively spreading Lau Basin, and the inactive Lau Ridge, a remnant arc [*Karig*, 1971; *Cole et al.*, 1990].

The Tonga forearc and platform, like their analogs in the IBM region, have existed since Eocene time, when they formed as part of the ancient Melanesian arc, which included fragments of what are now the Fiji, Lau, Tonga, Kermadec, and New Hebrides arcs. Since the forearc first formed (by trapping of older ocean crust or by arc volcanism) it, too, has accumulated a thick clastic sedimentary cover during its post-Eocene evolution [*Herzer and Exon*, 1985]. The forearc has been faulted and fractured during arc and backarc rifting [*Greene and Wong*, 1983; *Scholl et al.*, 1985] but there is little evidence in the forearc for either the extensive serpentine diapirism seen in the IBM forearc or for significant post-Eocene rifting. *Bloomer and Fisher* [1987] documented minor accretion of seamount fragments along the inner trench slope and *Gnibdenko et al.* [1985] and *Ballance et al.* [1989] have documented accretion of seamount fragments to the trench slope near the intersection of the trench with the Louisville Ridge seamount chain. Geophysical surveys of the trench and the exposure of arc rocks adjacent to the subducting Pacific plate have been interpreted to require ongoing subduction erosion, rather than long-term accretion [*Lonsdale*, 1986; *Bloomer and Fisher*, 1987]. It appears that the Tonga forearc has undergone less post-Eocene modification than is the case for much of the IBM forearc.

Samples of the forearc basement in Tonga are limited. The basement of the western Tonga forearc is exposed only on the island of 'Eua (Figure 2). That basement comprises depleted island-arc-tholeiitic basalts, andesites, and dacites [*Ewart and Bryan*, 1972; *Hawkins and Falvey*, 1985; *Cunningham and Anscombe*, 1985]. Seismic reflection profiles suggest that the Eocene basement on 'Eua can be traced across the forearc and that it crops out on the landward trench slope, just outboard of the trench-slope break [Figure 3; *Greene and Wong*, 1983; *Scholl et al.*, 1985; *Gnibidenko et al.*, 1985]. Indeed, fragments of arc tholeiitic basalts and andesites, highly depleted serpentinites, and gabbros have been dredged from a number of places on the landward slope (Figure 2) supporting the suggestion that the entire forearc is floored by arc crust [*Fisher and Engel*, 1969; *Vallier et al.*, 1985; *Bloomer and Fisher*, 1987]. The only samples from the outer forearc with good age control are from ODP site 841, on the trench-slope break southeast of 'Eua (Figure 2).

An enigmatic dredge collection from the northern Tonga Trench includes boninites, arc tholeiitic basalts and andesites, highly depleted serpentinites, and gabbros [Figure 2; *Falloon et al.*, 1987; *Sharaskin et al.*, 1983]. Boninite glasses from these dredge hauls have yielded ages of < 10 Ma (A. Crawford, pers. comm., 1992), but gabbroic rocks from the same suite give a 50 Ma age [*Acland et al.*, 1990]. The crustal exposures at the northern trench slope are in a complicated region at the terminus of the active arc and Lau Basin, along a trench-trench transform. They may be a combination of older forearc basement overprinted by interaction with a northeast-propagating arm of the Lau

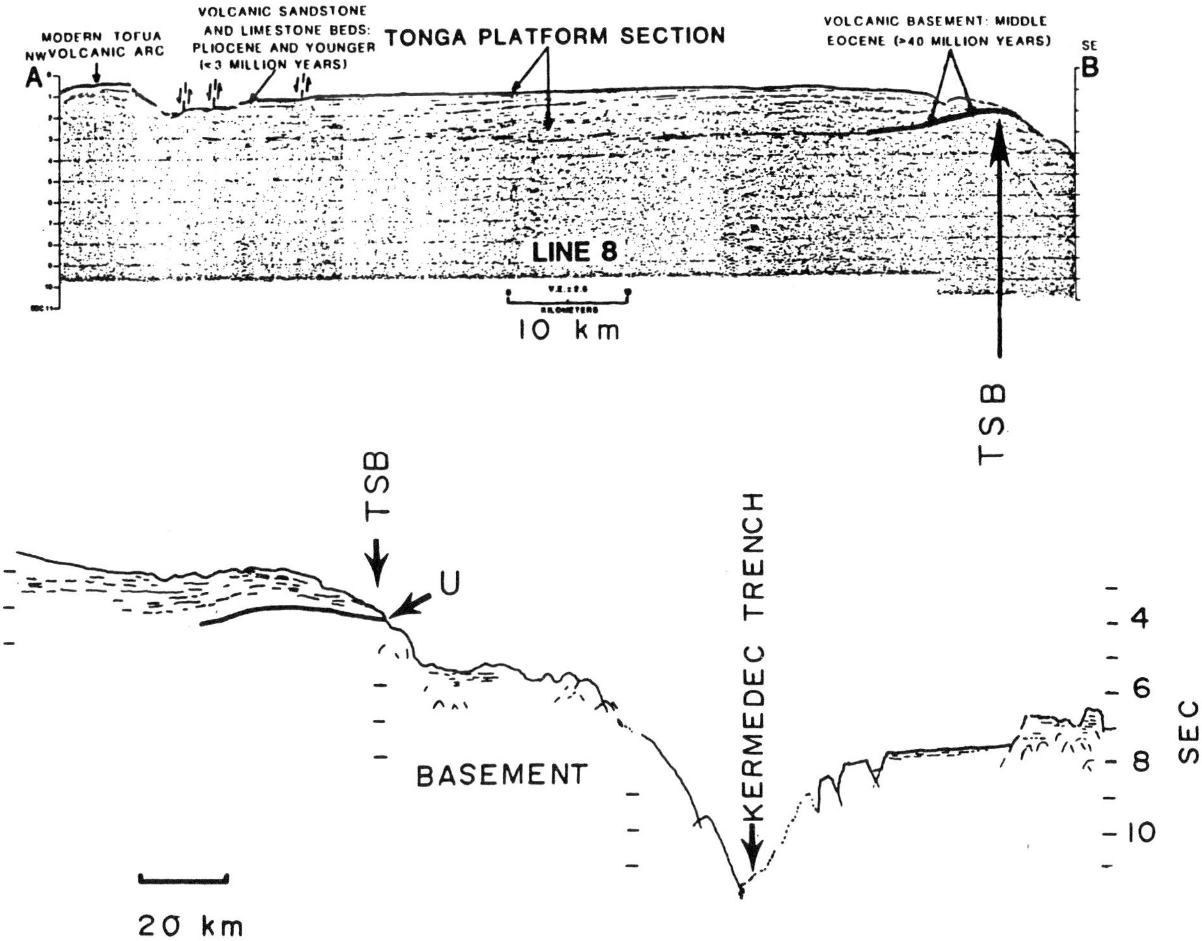

Fig. 3. A. Seismic section across the Tonga forearc interpreted to show volcanic basement, equivalent to Eocene basement rocks exposed on 'Eua on the Tonga Ridge, outcropping near the trench slope break, from *Scholl et al.*, [1985], *Greene and Wong* [1983]; B. Seismic profile across northern Kermedec trench showing similar outcropping of igneous basement at the trench-slope break [*Gnibidenko et al.,* 1985].

Basin. Because of the present ambiguities surrounding the interpretation of these northernmost exposures, we exclude them from our subsequent discussions of the geology of the forearc. More information is required on the ages and stratigraphy of the crustal section in the north before it can be reliably placed in a regional context.

3. COMPOSITION OF THE FOREARC BASEMENT

3.1. *Izu-Bonin-Mariana*

3.1.1. *Volcanic compositions.* The various forearc sample sites (Figure 4), reconstructed to their approximate relative positions in the early Oligocene. This reconstruction does not take into account rotations of the Philippine Sea plate, and is used here only to allow discussion of the widths and lengths of the forearc when it formed. The distance from the trench to the Oligocene volcanic front was at least 200-300 km, assuming no significant subduction erosion of the outer forearc [*Taylor*, 1992].

The composition of the forearc basement is shown schematically in Figure 5. *All* of the sections with stratigraphic control—boreholes or island sites—comprise only volcanics of island-arc tholeiite and boninite series volcanics and their derivatives. Volcanic facies include sheet flows, pillow lavas, dikes, breccias and hyaloclastite. Pillowed flows on Guam [*Reagan and Meijer*, 1984], the Bonin Islands [*Dobson*, 1986), and in site 458 (*Natland and Tarney*, 1981] all document submarine eruptions. Many of the forearc samples are from basement highs, which are interpreted to be volcanic constructs (Figure 6), and there is evidence that at site 786 [*Lagabrielle et al.*, 1992], on Hahajima in the Bonin Islands [*Hanzawa*, 1947], and on Guam and Saipan [*Cloud et al.*, 1956; *Tracey et al.*, 1964] those constructs were at or near sea-level in the mid- to late

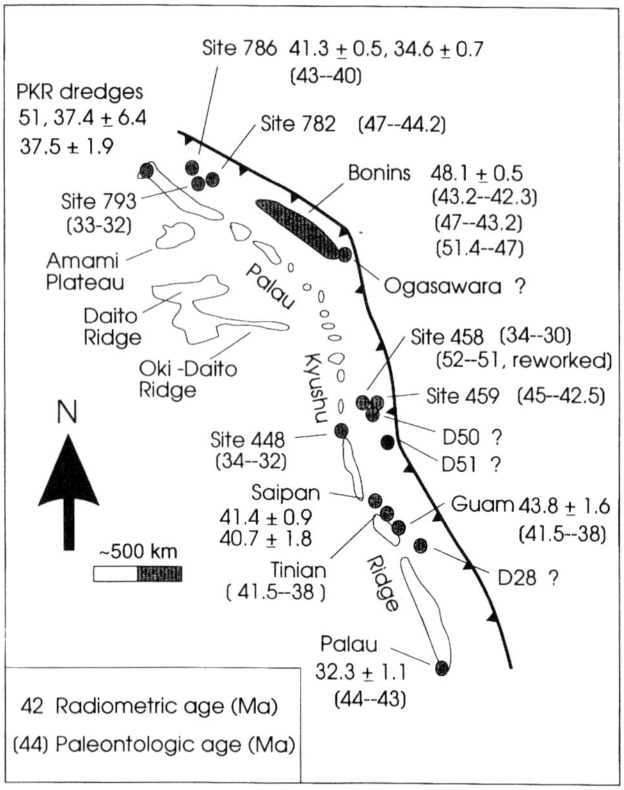

Fig. 4. Reconstructed geometry of the IBM subduction system in the early Oligocene, showing the approximate relative location of sample sites with Eocene to early Oligocene ages. The reconstruction has simply collapsed arc/backarc units younger than early Oligocene and has ignored subsequent rotations of the Philippine Sea plate. The figure is intended to illustrate the relative positions of sites to each other, not their absolute geographic location. Palau-Kyushu dredge ages from *Ozima et al.*, [1977] and *Shiki* [1985]; Saipan, Tinian, Guam and Palau from *Meijer et al.* [1983], *Ingle* [1975], *Tracey et al.* [1964], and *Cloud et al.* [1956]; Bonin Islands from *Dobson* [1986], *Hanazawa* [1947], and *Kaneoka et al.* [1970]; site 448, *Martini et al.* [1980]; sites 458 and 459, *Ellis* [1981] and *Kling* [1981]; sites 782, 786 and 793 from *Taylor et al.* [1990].

Eocene. Fractionated lava compositions (tholeiitic andesites and dacites) are common on the emergent islands and in the drilled forearc sequences; this may reflect a bias in sampling these small to large volcanic constructs. However, dredges along the inner trench slope recover similar distributions of mafic to felsic tholeiitic and boninitic derivatives throughout the trench slope, indicating this to be the typical composition of the basement.

Dredges along the inner trench slope (Figures 1, 5) show that this province of tholeiitic/boninitic crust extends all the way to the boundary with the Pacific plate. Some of those dredges also recovered the only "oceanic" lithologies found in the forearc. Metavolcanics of intraplate and mid-ocean ridge composition have been found in a few hauls from the lower trench slope and outer forearc [*Bloomer*, 1983; *Johnson et al.*, 1991; *Ishii,* 1985]. These Pacific plate compositions occur with Mesozoic-age cherts [*Bloomer*, 1983; *Johnson et al.*, 1991] and represent fragments of older crust faulted into the forearc basement or engulfed by forearc magmatism. Reworked Cretaceous fauna were also found at DSDP leg 60 sites 460 and 461 on the landward slope of the trench [*Ellis*, 1981; *Kling*, 1982]. MORB-like clasts have also been found within serpentine diapirs piercing the IBM forearc [*Johnson*, 1992]

The widespread similarity in composition of the forearc basement is striking. All of the apparently autochthonous lavas in the basement are clearly of arc affinity, as evidenced

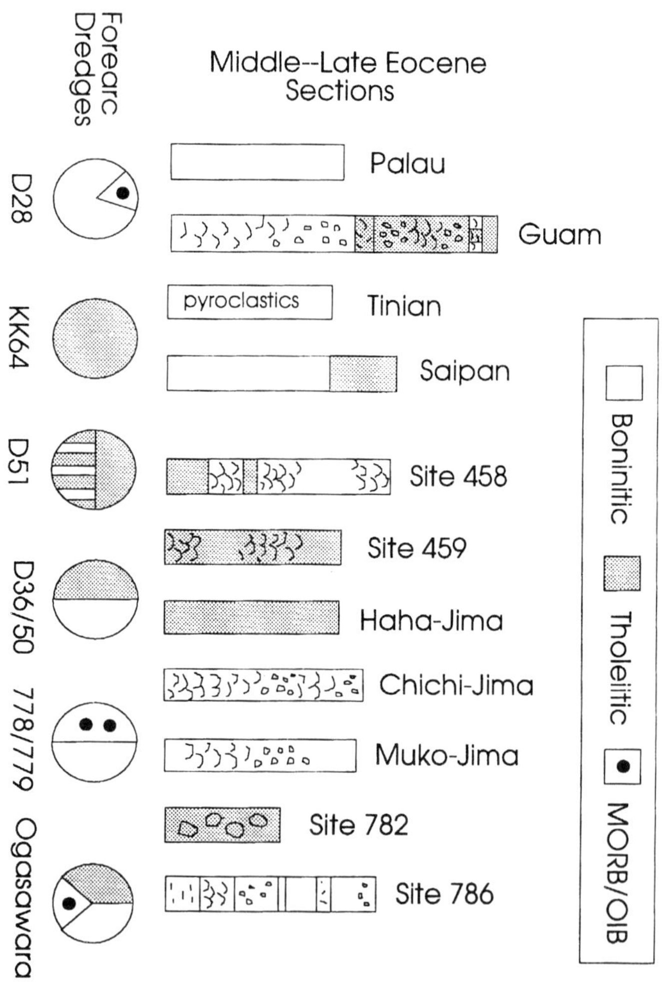

Fig. 5. Summary of the compositions of IBM forearc basement in-situ sections (on islands or in boreholes) and as inferred from dredges on the easternmost edge of the forearc, adjacent to the subducting Pacific plate. Data sources as in Figure 4 and *Bloomer* [1983], *Ishii* [1985], *Johnson et al.* [1991], *Umino* [1985].

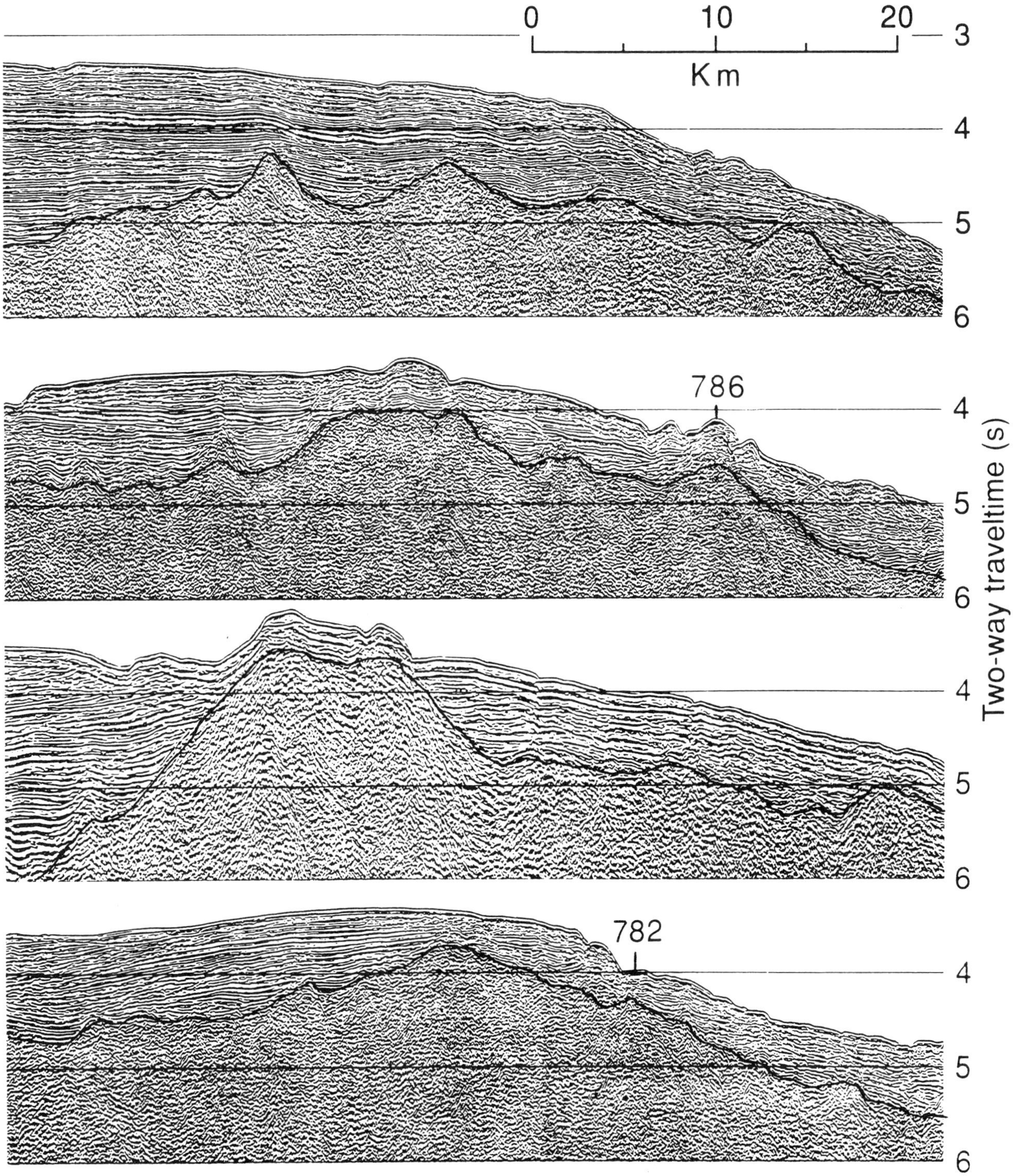

Fig. 6. Seismic profiles across the outer-arc basement high [from *Taylor*, 1992] of the Bonin forearc showing onlap of forearc sediments onto basement irregularities interpreted to be volcanic constructions.

by their trace element signatures, including alkali and alkaline earth enrichments, Ta and Nb depletions, and low abundances of heavy rare earth elements (Table 1; Figure 7). Equally striking is that the forearc crust contains two geochemically distinct, but intermingled lava stems (Figure 8). Both the tholeiitic (high TiO_2) and boninitic (low TiO_2)

TABLE 1. Representative Chemical Analyses of Eocene Arc Volcanic Rocks from the Forearcs of the IBM and Tonga Systems.

Area:	Site 786 IBM	Site 786 IBM	Site 786 IBM	Mariana Trench	Mariana Trench	Mariana Trench	Mariana Trench	Mariana Trench	'Eua, Tonga	'Eua, Tonga	'Eua, Tonga	'Eua, Tonga
SiO_2	53.47	58.44	64.80	53.60	61.90	50.67	51.70	61.79	48.40	50.60	53.80	58.25
TiO_2	0.20	0.23	0.28	0.18	0.17	0.48	0.57	0.81	0.67	0.65	0.77	1.04
Al_2O_3	12.05	11.43	14.72	10.77	11.66	14.98	15.94	14.13	20.90	16.10	16.90	15.87
Fe_2O_3	8.47	6.15	5.39	10.88	8.00	11.23	12.25	9.60	12.49	12.63	12.79	11.42
MnO	0.13	0.13	0.04	0.15	0.11	0.19	0.15	0.14	0.15	0.24	0.22	0.20
MgO	13.04	7.40	2.36	15.53	8.92	9.18	8.06	1.64	5.25	7.87	4.86	3.32
CaO	5.27	4.95	5.00	6.98	4.13	9.00	8.07	5.88	9.40	10.30	6.95	4.43
Na_2O	3.19	3.23	3.82	1.00	3.48	3.42	2.19	3.62	2.28	1.54	2.51	4.30
K_2O	0.67	0.20	0.69	0.66	0.70	1.11	1.11	0.47	0.49	0.10	0.50	0.17
P_2O_5	0.03	0.04	0.08	0.05	0.05	0.06	0.07	0.12	<0.05	<0.05	0.05	0.09
Sum	100.85	96.86	98.41	99.80	99.12	100.32	100.11	98.20	100.03	100.03	99.35	99.09
LOI	4.33	4.66	1.23	-	-	-	-	-	-	-	-	-
Zn	63	77	48	-	-	-	-	-	100	78	81	-
Ni	306	202	26	384	150	89	87	<8	11	58	12	29
Cr	1036	670	51	876	486	205	70	<20	19.7	117.0	47.6	-
V	166	148	152	161	100	203	258	131	560	490	330	324
Cu	60	70	34	-	-	-	-	-	130	140	4	-
Zr	35	45	62	32	63	36	38	85	-	-	-	62
Sc	29.0	27.6	20.4	-	-	-	-	-	37.0	48.4	40.0	-
Y	7.2	8.9	10.1	7	-	11	15	35	9	10	20	36
Sr	107	144	222	71	220	116	135	108	180	110	140	128
Ba	47	50	62	17	44	35	12	27	44	20	71	38
Rb	6.4	1.5	9.1	10	10	<6	<6	8	-	-	-	<1
Nb	0.53	0.65	1.03	<5	-	8	<5	<5	-	-	-	5
Cs	-	-	-	-	-	-	-	-	-	--	-	-
U	0.18	0.26	0.27	-	-	-	-	-	-	50-	50	-
Th	0.27	0.31	0.55	-	-	-	-	-	-	--	-	-
Pb	1.50	4.66	2.24	-	-	-	-	-	8	6	6	-
Co	45.0	32.0	13.9	-	-	-	-	-	28.1	40.5	30.8	-
Ta	0.03	0.05	0.06	-	-	-	-	-	-	-	-	-
Hf	0.90	1.15	1.67	-	-	-	-	-	0.64	0.61	1.10	-
La	1.81	2.62	3.02	-	-	-	-	-	0.68	0.57	1.12	-
Ce	3.86	5.52	6.46	2.90	-	-	4.24	8.90	-	2.27	-	-
Pr	0.59	0.80	0.96	-	-	-	-	-	-	-	-	-
Nd	2.84	3.65	4.64	1.59	-	-	3.36	7.41	-	-	-	-
Sm	0.81	0.97	1.24	0.46	-	-	1.14	2.64	1.05	1.02	1.71	-
Eu	0.27	0.34	0.39	0.15	-	-	0.41	0.90	-	0.45	0.66	-
Gd	0.97	1.20	1.59	0.58	-	-	1.60	3.55	-	-	-	-
Tb	0.16	0.21	0.24	-	-	-	-	-	0.30	-	0.48	-
Dy	1.05	1.21	1.49	0.68	-	-	1.79	4.74	2.01	-	3.16	-
Ho	0.23	0.27	0.32	-	-	-	-	-	-	-	-	-
Er	0.69	0.77	1.03	0.49	-	-	1.28	3.53	-	-	-	-
Tm	0.11	0.14	0.16	-	-	-	-	-	0.22	0.24	0.41	-
Yb	0.80	0.92	1.04	-	-	-	-	3.50	1.40	1.71	2.33	-
Lu	0.14	0.15	0.15	-	-	-	-	-	0.22	0.27	0.34	-

from *Pearce et al.*, 1992b; *Murton et al.*, 1992; *Bloomer*, 1987; *Hawkins and Falvey*, 1985; *Cunningham and Anscombe*, 1985; *Dobson*, 1986, *Stern et al.*, 1991; *Meijer, 1983*; *Shiraki et al.*, 1980;
all iron as Fe_2O_3, oxides in weight percent, elements in ppm, - indicates a value not determined.

TABLE 1. (continued)

Area:	Site 841 Tonga	Guam IBM	Guam IBM	Guam IBM	Site 458 IBM	Palau Is.-IBM	Saipan IBM	Bonin Is.-IBM	Bonin Is.-IBM	Bonin Is.-IBM	Bonin Is.-IBM	Site 786 IBM
SiO_2	77.58	55.44	54.95	51.80	54.23	53.65	78.20	52.80	53.40	57.00	58.01	47.23
TiO_2	0.22	0.34	0.38	0.36	0.34	0.47	0.17	0.10	0.14	0.13	0.14	0.31
Al_2O_3	12.04	15.05	14.84	16.83	15.95	12.01	10.58	9.16	11.27	10.21	11.64	12.06
Fe_2O_3	2.19	6.81	7.60	8.26	10.28	8.12	1.47	8.55	8.30	8.66	10.21	8.05
MnO	0.07	0.19	0.14	0.14	0.11	0.16	0.04	0.16	0.10	0.12	0.15	0.14
MgO	0.26	9.71	9.95	9.81	8.25	11.95	0.22	12.81	10.03	10.86	9.94	11.40
CaO	1.71	9.32	8.65	10.85	7.53	12.50	1.28	9.17	9.25	6.23	6.82	11.90
Na_2O	4.46	2.36	2.20	1.64	2.12	0.89	3.54	1.17	1.06	1.56	2.42	2.12
K_2O	1.17	0.62	0.64	0.12	1.15	0.05	1.69	0.47	0.47	0.51	0.66	0.26
P_2O_5	0.03	0.05	0.07	0.05	0.04	0.21	0.12	0.01	0.02	0.01	0.01	0.04
Sum	99.68	99.89	99.42	99.86	100.00	100.01	97.25	94.35	94.03	95.31	100.00	98.97
LOI	5.42	4.77	3.81	5.18	-	4.01	3.00	-		-	4.43	5.46
Zn	39	-	-	-	86	54	-	-	-	-	66	57
Ni	3	164	166	113	71	205	1	240	218	258	103	374
Cr	6	-	-	-	276	659	4	854	556	815	481	762
V	4	-	-	-	199	256	-	115	149	169	176	193
Cu	16	-	-	-	45	157	-	-	-	-	89	27
Zr	173	46	53	30	37	20	125	30	33	37	27	26
Sc	12	-	-	-	-	-	-	-	-	-	30.3	27.1
Y	61	10	61	10	5.6	15.7	-	8	8	9	4.0	8.2
Sr	56	144	119	114	101	133	92	87	87	79	86	139
Ba	190	121	82	37	16	13.2	175	23	24	26	35	15
Rb	12	11	6	2	14.9	1.0	15	9	14	15	13.0	7.6
Nb	1.4	-	1	1	2.20	0.23	-	-	-	-	0.66	0.53
Cs	0.31	-	-	-	-	-	-	-	-	-	-	-
U	0.27	-	-	-	0.16	-	0.85	-	-	-	0.13	0.14
Th	0.41	-	-	-	0.13	-	1.6	-	-	-	0.16	0.18
Pb	4.00	-	-	-	1.79	-	-	-	-	-	1.51	0.84
Co	4	-	-	-	39.0	-	-	-	-	-	35.0	44.6
Ta	0.10	-	-	-	0.06	0.01	-	-	-	-	0.06	0.02
Hf	4.55	-	-	-	0.80	0.60	-	-	-	-	0.75	0.70
La	4.28	-	-	-	0.52	0.99	5.50	-	-	-	0.86	1.50
Ce	13.32	-	-	-	2.00	2.18	9.96	-	-	-	1.80	3.05
Pr	2.36	-	-	-	0.40	0.16	-	-	-	-	0.24	0.50
Nd	13.26	-	-	-	1.54	2.46	7.71	-	-	-	1.24	2.61
Sm	4.94	-	-	-	0.53	0.99	2.48	-	-	-	0.47	0.80
Eu	1.14	-	-	-	0.20	0.37	0.54	-	-	-	0.14	0.31
Gd	6.96	-	-	-	1.35	0.87	3.31	-	-	-	0.46	1.06
Tb	1.18	-	-	-	0.14	0.26	-	-	-	-	0.08	0.20
Dy	8.46	-	-	-	1.62	1.40	3.42	-	-	-	0.62	1.30
Ho	1.87	-	-	-	0.37	0.33	-	-	-	-	0.13	0.30
Er	5.53	-	-	-	0.99	1.07	2.13	-	-	-	0.48	0.80
Tm	0.91	-	-	-	0.23	0.17	-	-	-	-	0.07	0.10
Yb	5.79	-	-	-	0.68	1.14	2.40	-	-	-	0.68	0.87
Lu	1.00	-	-	-	0.11	0.19	0.43	-	-	-	0.12	0.13

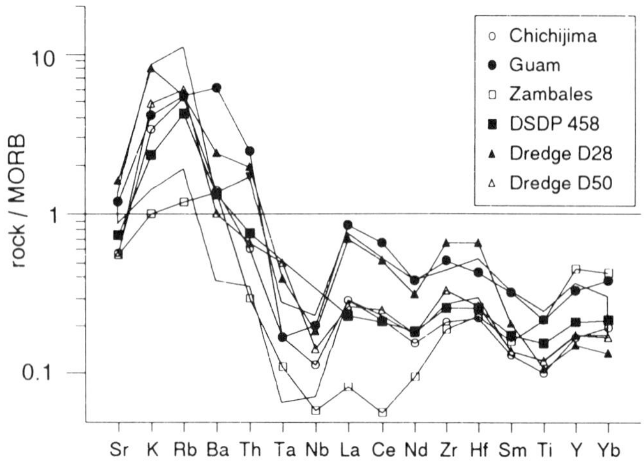

Fig. 7. Chemical compositions of several Eocene forearc basement samples (and one sample from the Eocene Zambales ophiolite in the Philippines) showing the distinctive arc geochemical signature in these volcanics [from *Pearce et al.*, 1992b].

series fractionate from mafic to very silicic compositions [*Bloomer* 1983; *Pearce et al.*, 1992b; *Umino*, 1985; *Meijer*, 1983]. There is no clear evolutionary sequence to the lava stems either. Boninitic series lavas are very abundant, but both precede and succeed tholeiitic series (Figure 5); some sections are dominantly tholeiitic, others dominantly boninitic. However, boninitic lavas are absent in all arc and backarc volcanic sections formed in the IBM region after the late Eocene [Figure 5; *Taylor* , 1992; *Taylor et al.*, 1992a].

3.1.2. *Plutonic rocks.* Some samples of the lower crustal components of the forearc basement have been recovered by dredging on serpentine seamounts (*Johnson*, 1992), on the outer trench-slope break (*Johnson and Fryer*, 1990) and along the inner trench slope and as clasts incorporated in serpentinite seamounts drilled on leg 125 (*Fryer et al.*, 1990). The gabbros and ultramafic rocks in these dredges have affinities with arc plutonic rocks, and are compositionally distinct from the gabbros and peridotites dredged from mid-ocean ridges and backarc basins (*Bloomer and Hawkins*, 1983; *Ishii*, 1985).

Gabbros from the outer forearc are typically orthopyroxene-clinopyroxene-plagioclase accumulates, in which orthopyroxene and clinopyroxene are early appearing cumulate phases (*Bloomer and Hawkins*, 1983; *Ishii,* 1985). This in contrast to most oceanic gabbroic suites, in which orthopyroxene appears only as a late intercumulus phase or as a cumulus mineral only in fractionated ferrogabbros (e.g., *Hodges and Papike*, 1976; *Bloomer et al.*, 1991). The forearc gabbros also have more calcic plagioclase coexisting with given clinopyroxene composition than do their mid-ocean or backarc counterparts (Figure 9). Crystallization sequences and mineral chemistries in the gabbros are consistent with a genetic relationship to the associated arc tholeiitic and boninitic lavas.

The ultramafic rocks from the forearc are clinopyroxene-poor harzburgites, with residual spinels having very high Cr/Cr+Al (*Bloomer and Hawkins*, 1983; *Pearce et al.*, 1992b), indicating degrees of melt depletion higher than are typical of peridotites from mid-ocean ridges and fracture zones [Figure 10; *Dick and Bullen*, 1984; *Dick et al.*, 1984]. Mineralogic and trace element characteristics of the harzburgites dredged and drilled from the forearc indicate them to be the mantle residue left after the extraction of the boninite and arc tholeiite magmas [*Bloomer and Hawkins*, 1983; *Pearce et al.*, 1992b].

3.2. *Tonga Forearc*

3.2.1 *Volcanic rocks.* The best described forearc samples in Tonga are from the island of 'Eua on the inboard side of the forearc and from site 841 on the outboard side (Figure 2). The basement on 'Eua comprises depleted island-arc-tholeiitic basalts, andesites, and dacites [*Ewart and Bryan*, 1972; *Hawkins and Falvey*, 1985; *Cunningham and Anscombe*, 1985]. The basement rocks yield radiometric ages of 46 to 40 Ma [*Ewart et al.*, 1977; *Duncan et al.*, 1985] and, locally, are unconformably overlain by upper middle Eocene calcareous conglomerates and breccias [*Cunningham and Anscombe*, 1985]. Oligocene lava flows (33-31 Ma) and Miocene dikes (19-17 Ma) indicate later episodes of arc activity [*Duncan et al.*, 1985]. The 211 meters of igneous basement drilled (in 4810 m of water) at site 841 are composed of low-K rhyolites, rhyolitic tuffs,

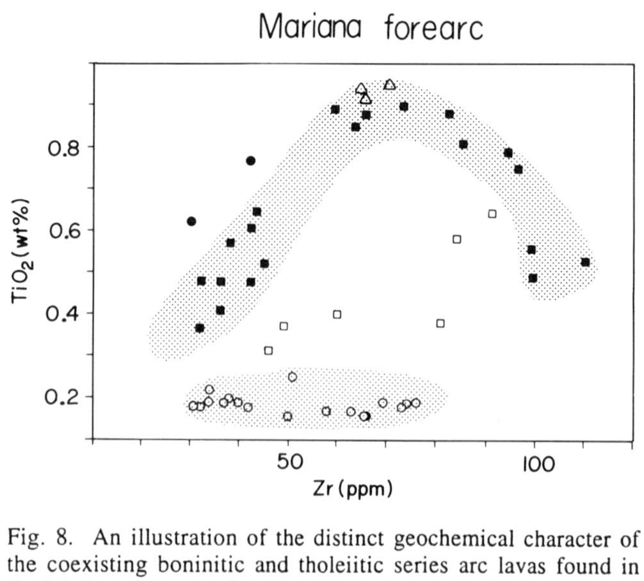

Fig. 8. An illustration of the distinct geochemical character of the coexisting boninitic and tholeiitic series arc lavas found in the Mariana forearc. Data for dredges D28 and D50 in the outer forearc, from *Bloomer* [1987]. Filled symbols are tholeiitic series lavas, open triangles are tholeiitic andesites, open circles and open squares are boninite series lavas.

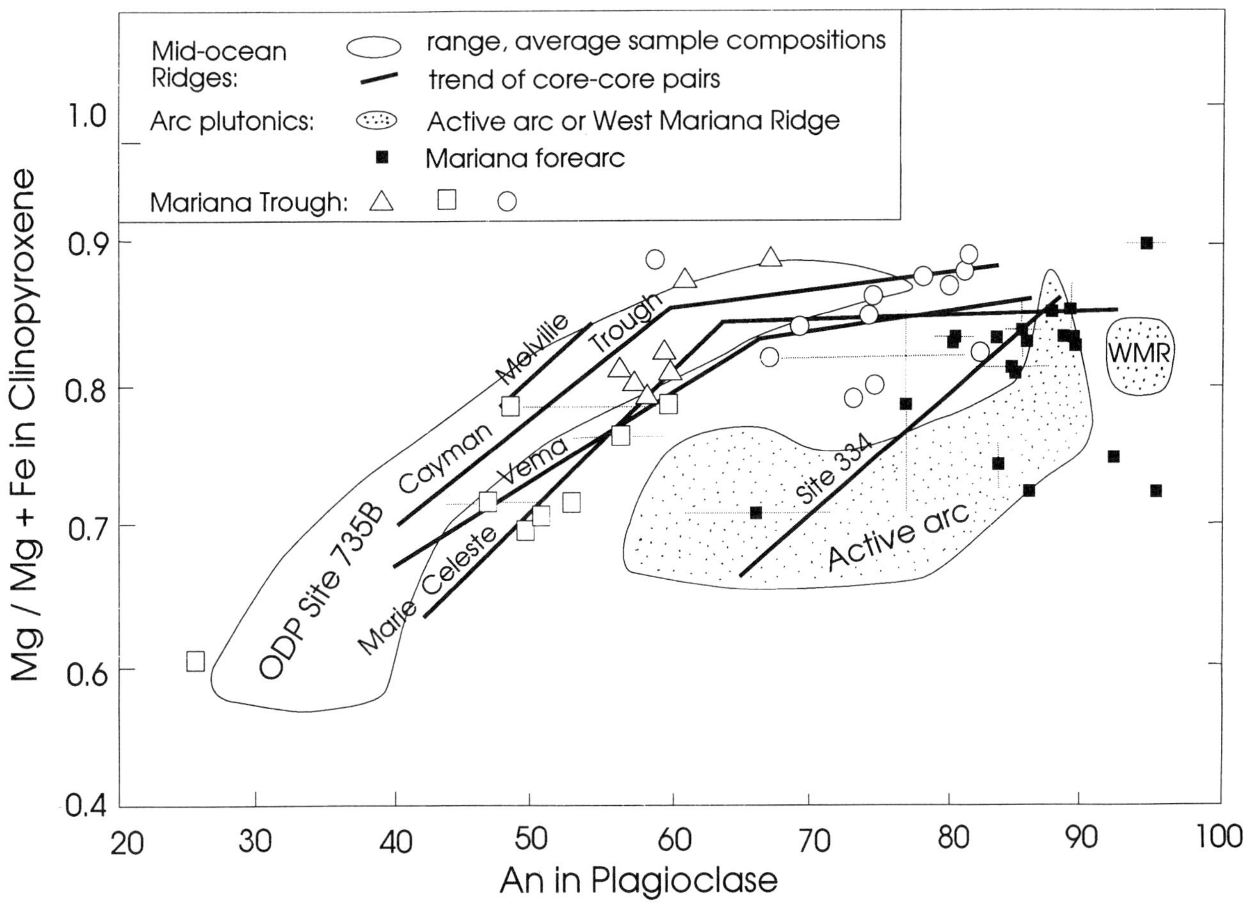

Fig. 9. Compositions of co-existing plagioclase and clinopyroxene in cumulate gabbros from the outer Mariana forearc compared to similar pairs in mid-ocean ridge gabbros [*Bloomer et al.*, 1989, 1991 and unpublished data], gabbroic clots in active arc lavas [*Nilsson*, 1987], gabbros from the West Mariana Ridge [a remnant arc; *Natland*, 1981], and gabbros dredged from the Mariana Trough backarc basin [Bloomer and Stern, unpublished data]. Note that the backarc gabbros are like gabbros from slow-spreading oceanic ridges, not like those from arcs.

breccias, welded tuffs, and lapilli tuffs and are date at 44 $\pm$ 2 Ma [*McDougall*, 1994]. The welded fragments, the absence of interbedded sediments and marine fauna, and the lack of density sorting within the pumice breccias indicate a subaerial origin for the rhyolitic complex. Sparsely distributed dredge hauls along the inner trench slope (Figure 2) have also recovered fragments of arc tholeiites and their derivatives above the Pacific plate, suggesting that the Tonga forearc, like that in the IBM region, is constructed primarily of arc volcanics.

The forearc volcanic rocks in Tonga, except for the poorly understood exposures in the north, are uniformly of tholeiitic derivation. The basalts and their fractionates are not as depleted as are similar rocks derived from a boninitic magma stem (Figure 11) and in most of the Tonga forearc there is no evidence of significant boninitic volcanism.

3.2.2. *Plutonic rocks*. There are no substantial collections of plutonic rocks from the outer Tonga forearc, except in the unusual crustal collection form the north end of the Trench. Fragments of very fresh dunite and harzburgite were dredged by *Fisher and Engel* [1969] just above the subducting Pacific plate at 20°20'S. The mineralogy and chemistry of those peridotites indicate that they are very depleted mantle residues, much like their counterparts in the IBM forearc [*Bloomer and Fisher*, 1987]. Small fragments of gabbro were recovered in the same dredge hauls, but they have no mineralogical features diagnostic of a particular geodynamic setting [*Bloomer and Fisher*, 1987].

3.3. *Evidence of older oceanic crust*

Surprisingly, there is little evidence of older mid-ocean ridge type oceanic crust in either the IBM or Tonga forearcs, given that both are thought to have developed in intraoceanic settings. What is now forearc crust must have replaced, or

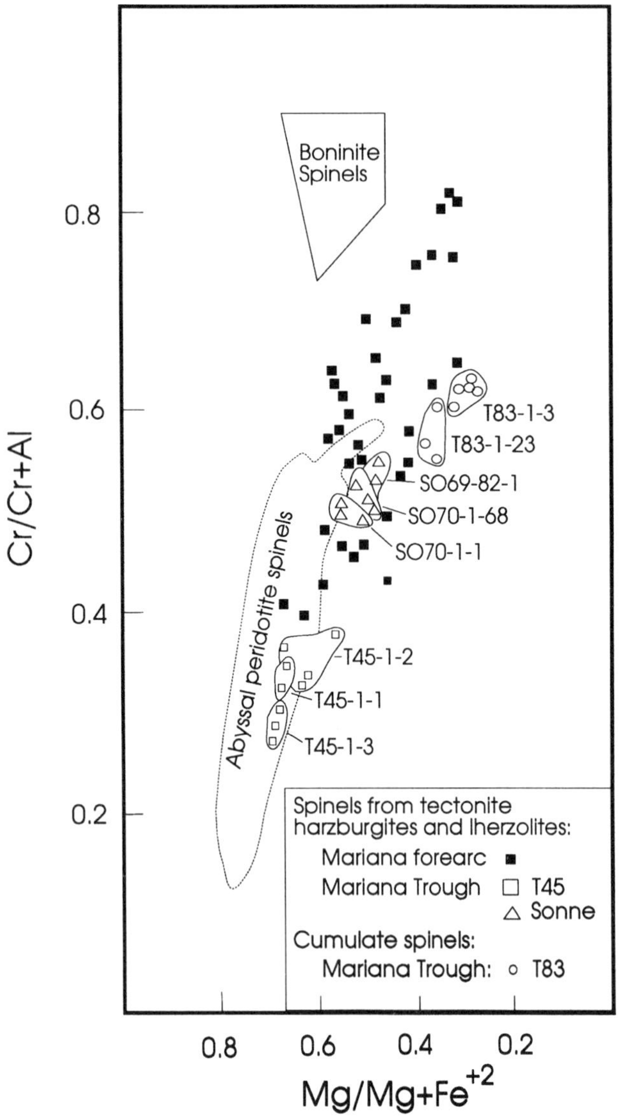

Fig. 10. Spinels in tectonite peridotites dredged from the outer Mariana forearc [*Bloomer and Hawkins*, 1983] compared to spinel compositions in boninites, spinels in abyssal peridotites from mid-ocean ridges [from *Dick and Bullen*, 1984], and to spinels in depleted peridotites from the Mariana Trough backarc basin [Bloomer, unpub. data]. Note that the backarc peridotites are similar to oceanic peridotites and are not as depleted as peridotites from the forearc.

been built on, preexisting oceanic crust, yet we are hard pressed to find evidence of that older crust.

In the IBM forearc, the most signficant samples of mid-ocean ridge type crust are found in dredge hauls along the inner trench slope [*Bloomer*, 1983; *Johnson et al.*, 1991] and in graben in the outer forearc [*Johnson and Fryer*, 1990]. The former occur with alkalic basalts and sedimentary rocks similar to those found on seamounts built on the subducting Mesozoic crust. The most reasonable interpretation of these exposures is that they are fault slices accreted to the lower slope from the downgoing plate. This is consistent with evidence cited by *Karig and Ranken* [1983] for episodic accretion along parts of the outer forearc and with the interpretations of seamount subduction and accretion of *Fryer et al.* [1985].

There are some clasts of mid-ocean ridge basaltic composition within serpentine diapirs drilled on leg 125 [Figure 12; *Johnson,* 1992]. The samples from the outer forearc graben are probably derived from serpentine melanges shed form the same diapirs. These clasts could represent fragments from older oceanic crust trapped within the forearc, but could also be pieces from accreted (or underplated) sections of oceanic crust or fragments from the downgoing plate. Blueschist-facies clasts [*Pearce et al.* 1990; *Maekawa et al.*, 1992] contained within the same serpentinite show that it is possible to transport material from near the plate boundary at depth to the surface in the serpentine diapirs. The lower-crustal clasts in the serpentinites and the serpentinites themselves are all of arc-like affinity; the absence of any oceanic lower-crustal components makes it more likely that the oceanic basalt clasts are derived in some manner from the subducting plate.

Certainly, the borehole and island sections of the forearc basement show no evidence of trapped ocean crust (Figure 5). It could be argued that the examined sections are biased towards large volcanic constructs, which might have been built on top of older ocean crust, and that to find trapped crust we should search between these larger edifices. However, all sections from the forearc—those that were originally submarine, those that were subaerial, those that were principally mafic and those that were principally silicic, as well as the bulk of the dredge hauls recovered between boreholes or islands—consist of arc volcanic rocks with no evidence of older oceanic basement. The most straighforward conclusion that can be drawn is that the IBM

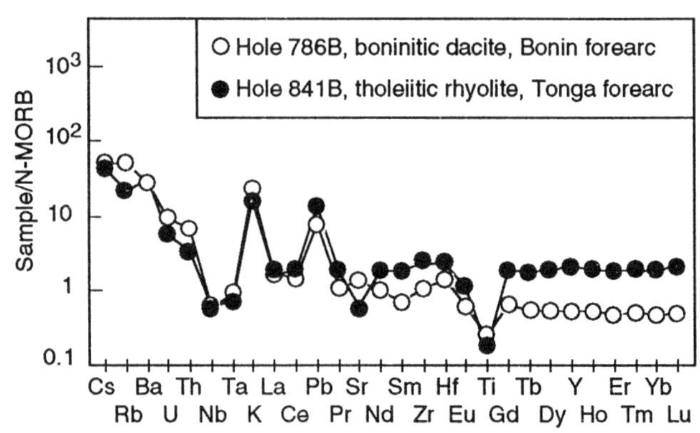

Fig. 11. MORB-normalized composition of a silicic volcanic from site 841 on the outer Tonga forearc [*Bloomer et al.*, 1994] compared to a silicic boninite series volcanic from the Bonin forearc [*Pearce et al.*, 1992b, *Murton et al.*, 1992].

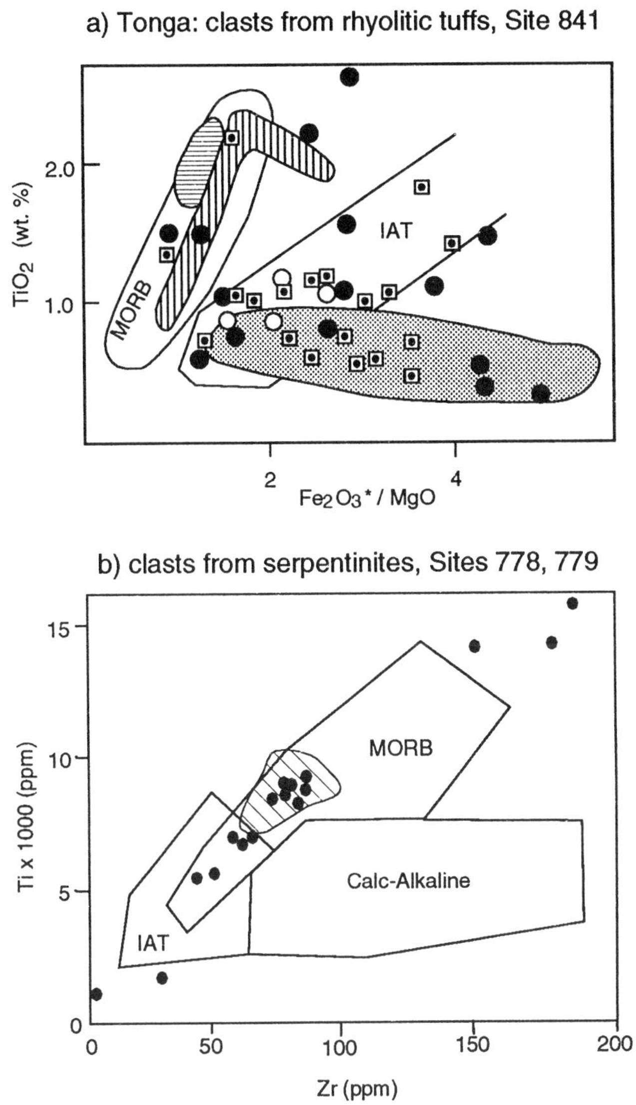

Fig. 12. Compositions of clasts found within forearc basement lavas in the Tonga forearc [*Bloomer et al.*, 1994] and in serpentinite diapirs in the Mariana forearc [*Johnson*, 1992]; lined field in bottom figure is for compositions of accreted MORB lavas dredged from the inner Mariana trench slope [*Johnson et al.*, 1991]. In top figure, shaded field is for lavas from'Eua [*Cunningham and Anscombe*, 1985] horizontally lined field is for MORB from the offshore trench slope and the diagonally ruled field is for E-MORB from the offshore slope [*Bloomer and Fisher*, 1987].

forearc is composed almost completely of arc tholeiitic and boninitic series lavas which displaced the older oceanic crust into which they were erupted.

Much the same appears to be true in the Tonga forearc, though the sampling density is much lower. There are no significant volumes of oceanic crustal compositions on 'Eua, at site 841, or dredged from the inner trench slope. However, within the rhyolitic basement at site 841 there are a few tiny volcanic clasts with mid-ocean ridge-like compositions (Figure 12). Their presence within this Eocene arc volcanic complex requires that there were some fragments of ocean crust present beneath the arc edifice as it grew. However, the volume of oceanic crust appears to have been very small, judging from the dominance of mafic to intermediate arc compositions within the clasts in the silicic basement (Figure 12).

4. DISTRIBUTION OF BASEMENT AGES

4.1. *Izu-Bonin-Mariana System*

The ages of the IBM forearc basement sites are shown in Figures 13 and 14 projected both along and across strike. Ranges for minimum basement ages from paleontologic data and reliable radiometric ages are both plotted; data sources are listed in the figure captions. There was clearly voluminous volcanic activity along the entire length of the IBM system in the mid- to late Eocene (Figure 13). Volcanism appears to have begun slightly earlier in the Bonin forearc than to the north or the south, but clearly developed rapidly along the entire 3000 km length of what would become the IBM forearc.

More striking, perhaps, is the distribution of activity *across* the forearc (Figure 14). Mid- to late Eocene arc volcanic activity occurred synchronously at both the inboard and the outboard portions of the present forearc. The apparent younging of volcanism across the forearc in Figure 14 is

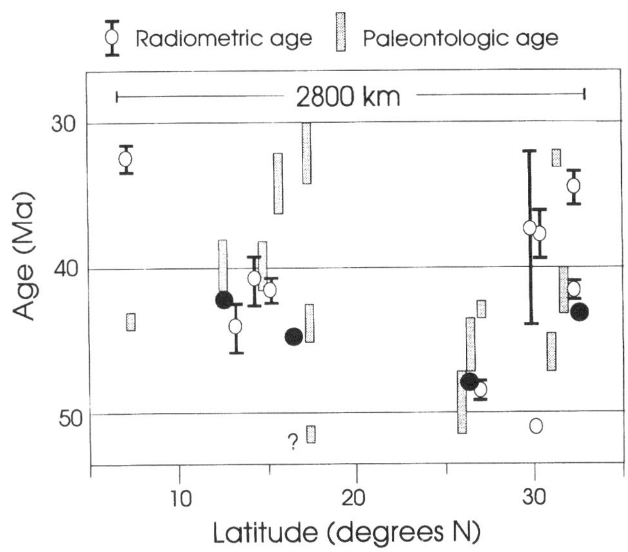

Fig. 13. Distribution of ages of the forearc basement sites shown in Figure 5 projected along the length of the IBM system. Solid dots indicate locations with a close coincidence between radiometric and paleontologic ages of the basement. Data sources as in Figures 4 and 5.

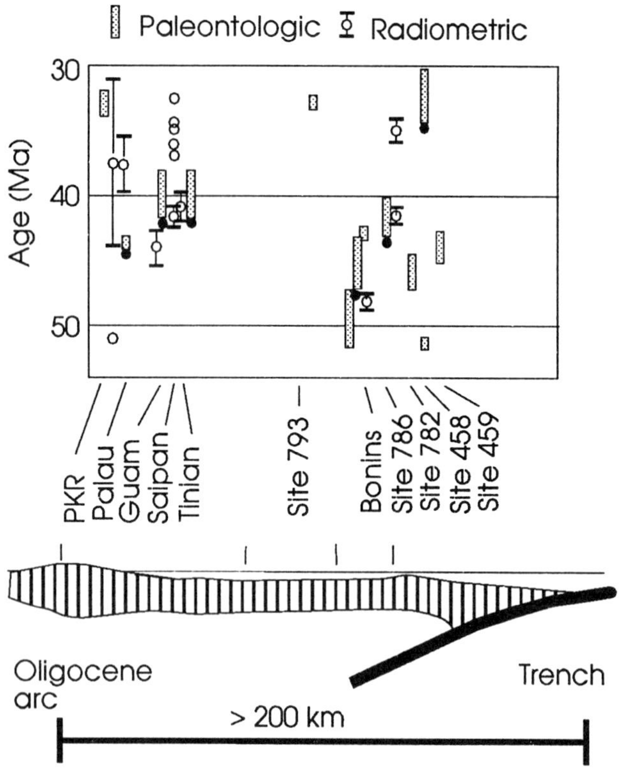

Fig. 14. Distribution of forearc basement ages projected on a single cross section of the forearc, illustrating the widespread distribution of arc volcanism in the mid- to late Eocene. Data sources as in Figures 4 and 5.

actually a latitudinal variation; the oldest pair of ages is in the north, the Bonin Islands on the outer forearc and the northern Palau-Kyushu Ridge on the inner forearc. From 45 Ma to about 35 Ma, arc crust of tholeiitic to boninitic composition was being constructed across a zone over 200 km wide. This kind of distributed arc volcanism has no analog in modern subduction zones.

Our favored interpretation of the biostratigraphic, radiometric and sedimentologic data is that voluminous volcanism ceased on the outer arc high by the end of the Eocene, but continued on the remnant arc/frontal arc high until 27-29 Ma (i.e., well into the rift phase prior to Shikoku and Parece Vela basin spreading). Determining the volcanic chronology is complicated by three factors: (1) redeposition of older fossil assemblages can increase apparent biostratigraphic ages, (2) K/Ar ages can be decreased by secondary alteration and/or argon loss, especially in K-poor, glassy rocks such as boninites [*Walker and McDougall,* 1982], and (3) intrusions much younger than the main phase of volcanism are common on the frontal arc high [in the middle Miocene Umatac and Fina Sisu formations on Guam and Saipan, respectively; *Meijer et al.*, 1983], in the forearc basin [such as the post-15 Ma basalt sill at site 792; *Taylor et al.,* 1990]; on the outer arc high [in the Quaternary at Mariana site 781; *Fryer et al.*, 1990], and along major forearc faults [*Fryer*, 1992].

We agree with *Dobson* [1986] that most of the young Bonin Island K/Ar dates, which range from 8 to 43 Ma [*Tsunakawa*, 1983], are minimum ages. The interpillow sediments have middle Eocene faunal assemblages and there is one K/Ar date of 48 Ma [*Dobson*, 1986]. Similarly, the 26-29 Ma K/Ar ages reported for the forearc basin basement at site 793 by *Taylor and Mitchell,* [1992], are too young for the 31 Ma magneto-biostratigraphic age. There is evidence, however, for early Oligocene outer arc high magmatism at sites 458 and 786. The nannofossil age of the overying sediments and a "b-quality" $^{40}Ar/^{39}Ar$ age for the lavas at site 458 are both 34 Ma [*Hussong et al.*, 1981, *Ozima et al.*, 1983], though reworked Eocene fauna within the Oligocene sediments on the basement high suggest that the volcanic basement could be Eocene in age. K/Ar dates on boninite and andesite intrusions within the basement at site 786 also cluster around 34 Ma [*Mitchell et al.*, in press), though these may be minimum ages. Nevertheless, the slow accumulation of dominantly pelagic Oligocene sediments at sites 458, 459, 782, and 786 (Figure 5) and of the Minamizaki limestone on the Bonin Islands [*Umino*, 1985], attests to the lack of major Oligocene volcanism along the outer arc high.

In contrast, volcanism continued into the late Oligocene along the remnant arc/frontal arc high. The late Eocene to early Oligocene Alutom Formation on Guam is composed of volcanic breccias, tuffaceous sandstones, and lava flows, with calc-alkaline as well as arc tholeiitic and boninitic chemistries [*Reagan and Meijer*, 1984]. K/Ar dates on the upper section of this formation range between 36 and 34 Ma, with sills intruded about 32 Ma [*Meijer et al.*, 1983]. Tholeiitic basalts, breccias and tuffs drilled at site 448 on the southern Palau-Kyushu Ridge (Figure 1) document vigorous volcanism there at 34-29 Ma [*Kroenke et al.*, 1980]. Maximum sedimentation at the West Philippine Basin site 447 (Figure 1) also occurs at 34-29 Ma, associated with volcaniclastic breccia, conglomerate, and tuff from the Palau-Kyushu Ridge volcaniclastic apron [*Kroenke et al.*, 1980]. Volcanic tuffs, lapilli tuffs, and sandstones drilled at site 296 on the northern Palau-Kyushu Ridge (Figure 1) indicate an overlapping, but slightly later volcanic maximum, from 33 to 27 Ma [*Karig et al.*, 1975]. Rapidly deposited volcaniclastic turbidites and debris flow deposits recovered at the leg 126 forearc sites 787, 792 and 793 [*Taylor et al.*, 1990] document intense Izu-Bonin volcanism 31-27 Ma. Site 792 was drilled on one of the frontal arc highs and proved to be an Oligocene volcano (Figure 1). The basement, K/Ar dated at 33 Ma [*Taylor and Mitchell*, 1992], is composed of massive flows, with intercalated hyaloclastite and breccia layers, that are plagioclase-rich two-pyroxene andesites and dacites with calc-alkaline affinities [*Taylor et al.*, 1990, *Taylor et al.*, 1992a]. Clearly, the frontal arc highs are the large arc volcanoes of the Oligocene volcanic front. By the early Oligocene (~33 Ma), the widespread Eocene volcanism that formed the IBM arc massif had evolved and focussed into an arc system similar to the one that exists now, and the locus of arc volcanism has remained fixed, relative to the trench, for most of the last 30 Ma (Figure 15).

4.2. *Tonga forearc*

The basement ages in the Tonga forearc, though limited in number, show the same kind of distribution as in the IBM forearc. Arc volcanic activity occurred nearly simultaneously at the inboard and the outboard ends of the forearc platform (Figure 16). The volcanism was of similar geochemistry (arc tholeiitic) and was succeeded by virtually identical sedimentary assemblages [*Cunningham and Anscombe*, 1985; *Hawkins et al.*, 1992].

4.3. *Temporal and Spatial Geochemical Trends*

The Eocene forearc tholeiitic and boninitic volcanics in the Tonga and IBM forearcs represent the most depleted compositions erupted in the arcs throughout their histories. Subsequent arc lavas in both arcs, from the Oligocene onwards, are more enriched than lavas erupted in the Eocene (Figure 17). This temporal trend is reflected in part in the across-arc geochemical trends identified by *Taylor et al.* [1992b] in the Izu-Bonin forearc (Figure 18). The lavas nearest the trench are the most depleted in Ba; those include the oldest, Eocene foreac lavas. Lavas erupted at increasing distances from the trench are increasingly enriched.

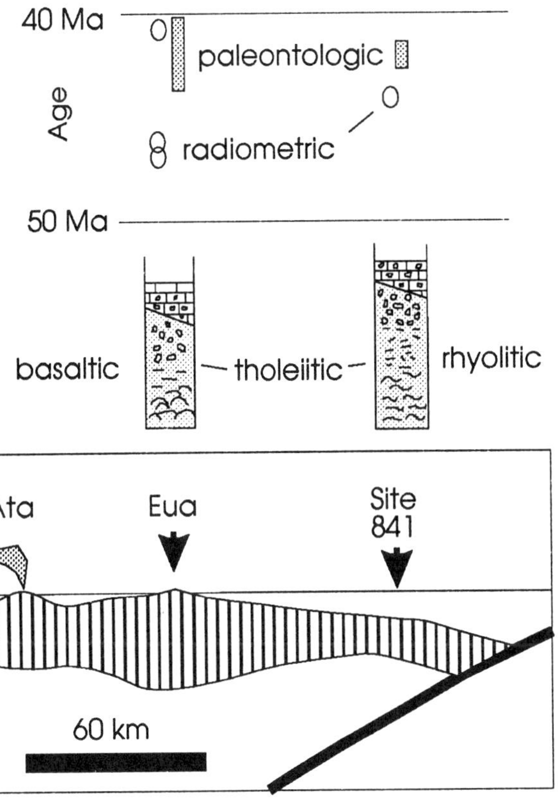

Fig. 16. Relative locations and compositions for the only two well-dated forearc samples from the Tonga forearc; data from *Bloomer et al.* [1994], *Hawkins and Falvey* [1985], *Duncan et al.* [1985], *Ewart and Bryan* [1972], and *McDougall* [1994].

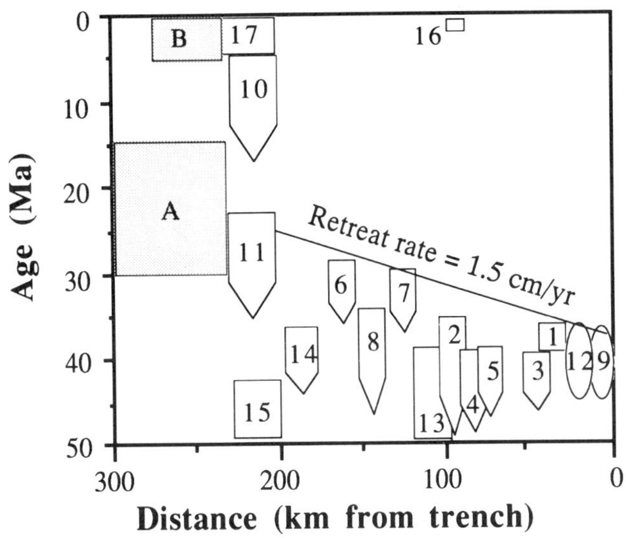

Fig. 15. A similar plot as in Figure 14 but for all arc units in the IBM system, illustrating the localization of arc activity at or close to the locus of the present arc since 30 Ma [from *Stern and Bloomer*, 1992].

These spatial and temporal trends probably result from increased enrichment of the sub-arc mantle with time, and the tapping of deeper, and hence more fertile lithospheric or asthenospheric mantle as the subduction zone evolves. The very depleted nature of the Eocene forearc volcanics is consistent with models for boninite genesis and arc initiation which require melting of depleted, shallow lithospheric mantle [*Meijer*, 1980; *Pearce et al.*, 1992a; *Stern and Bloomer*, 1992].

4.4. *Summary*

The mid- to late-Eocene arc volcanic platforms now preserved in the IBM and Tonga forearcs were constructed by a style of arc volcanism for which there are no modern analogs. In both cases, this unusual style of distributed volcanism was the first arc volcanism to occur in the system and marks the onset of subduction in the Western Pacific. This Eocene volcanism also produced the most depleted lavas that have appeared in either the Tonga or IBM arcs.

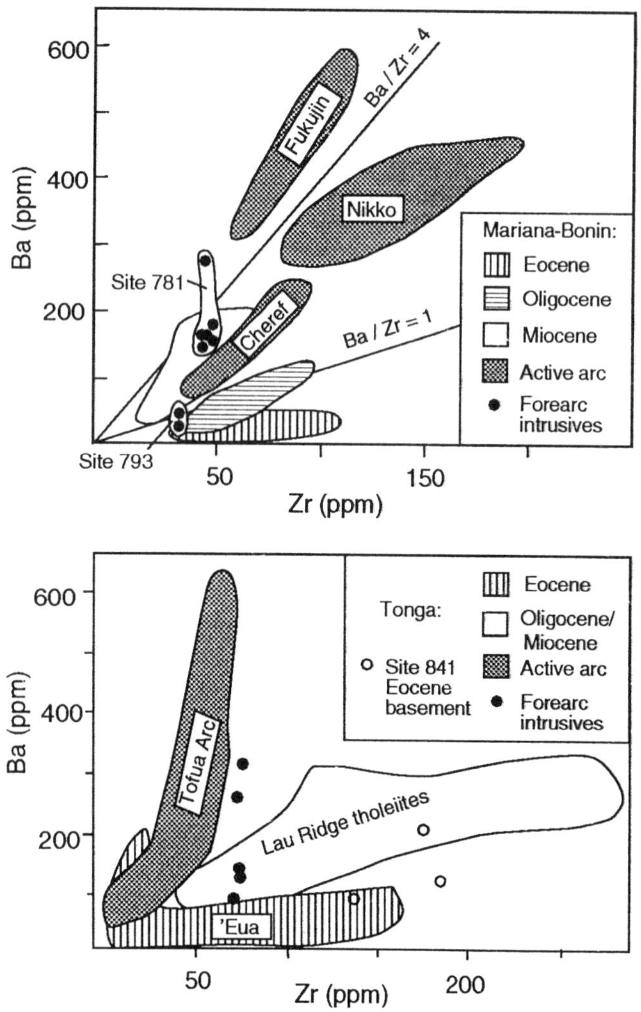

Fig. 17. Comparions of Ba-Zr systematics for Eocene (including 'Eua and forearc lavas), Miocene (including Lau Ridge lavas) and active arcs (including Tofua arc) in both the(a) Mariana and (b) Tonga systems. Data from *Bloomer et al.* [1989], *Cunningham and Anscombe*[1985], *Hawkins and Falvey* [1985], *Cole et al.* [1985], *Hawkins et al.* [1992], *Ewart and Hawkesworth* [1987] *Bloomer* [1987], *Wood et al.* [1980], *Meijer* [1980]. Fukujin, Nikko, and Cheref are active volcanoes in the Marianas arc. Unit 1 in Tonga is a series of dikes and sills that intrudes the Miocene forearc sediments.

5. MAGMA PRODUCTION RATES IN THE EOCENE

The distribution of forearc basement ages and compositions document a zone of arc volcanic production 200 to 300 km wide, and nearly 3000 km long, in the mid- to late Eocene. The peak production of this broad volcanic platform was between about 45 and 35 Ma. It is clear that this is a style of volcanism unlike what we see in modern arcs, and must have proceeded at unusually high eruption rates. However, to constrain magma production rates, we need some estimate of crustal thicknesses in the forearc. Most of the subsequent discussion follows *Stern and Bloomer* [1992].

There are a number of problems encountered in estimating the thickness of the Eocene arc crust. There are few crustal refraction measurements in the IBM forearc. The only such measurement that reached mantle velocities gave a crustal thickness of about 12 km in the outer forearc [*Murauchi et al.*, 1968]. Other measurements in the forearc yielded velocities of 7.2 km/sec at 7-9 km depth [*LaTraille and Hussong,* 1980; *Ambos*, 1984], but did not identify the base of the crust. *Ambos* [1984] identified a velocity of 8.2 km/sec at 19 km below the trench-break. The IBM forearc crust has undergone subduction-related hydration and has been pierced by a number of serpentinite diapirs on its eastern end. It is likely that these processes lowered crustal and upper mantle velocities and have made the igneous crust appear to be overly thick and to thicken to the east [*Murauchi et al.*, 1968; *LaTraille and Hussong*, 1980]. The western end of the IBM forearc has been thickened and overprinted by post-Eocene volcanic activity; other parts of the forearc were stretched during the Oligocene forearc rifting episode. There is, in short, no reliable measurement of the crustal thickness of the purely Eocene portion of the forearc crust. *Stern and Bloomer* [1992] used the crust of the Zambales ophiolite on the island of Luzon, Philippines, as an analogy to the IBM forearc crust. The Zambales ophiolite is interpreted to be an SSZ ophiolite [*Hawkins and Evans*, 1983; *Geary et al.*, 1989]. It formed in the late Eocene and is compositionally similar to the IBM forearc. Crustal thicknesses of the Zambales ophiolites are well-constrained between 5 and 8 km (*Hawkins and Evans*, 1983). If we assume a similar, 6 km thickness for the late Eocene IBM forearc crust, accept the evidence that there is little older oceanic crust trapped beneath the forearc, and allow a 10-15 Ma time span for volcanic activity in the forearc, we calculate magma production rates of 80 to 180 km^3/km-Ma (*Stern and Bloomer*, 1992). These rates are substantially higher than those estimated for mature arc eruption rates and mean arc growth rates (Figure 19) and are comparable to crustal production rates at slow spreading oceanic ridges (2-3 cm/yr half rates).

6. STRUCTURAL SETTINGS AND EVOLUTION OF THE FOREARCS

6.1. *Extension vs. Compression*

The volume of the earliest arc volcanism in the IBM and Tonga forearcs is clearly different than anything we see in modern volcanic arcs and must require a style of volcanism different than we think of as typical of arcs. The broadly distributed volcanism, high crustal production rates, apparent displacement of the older oceanic crust by Eocene arc crust,

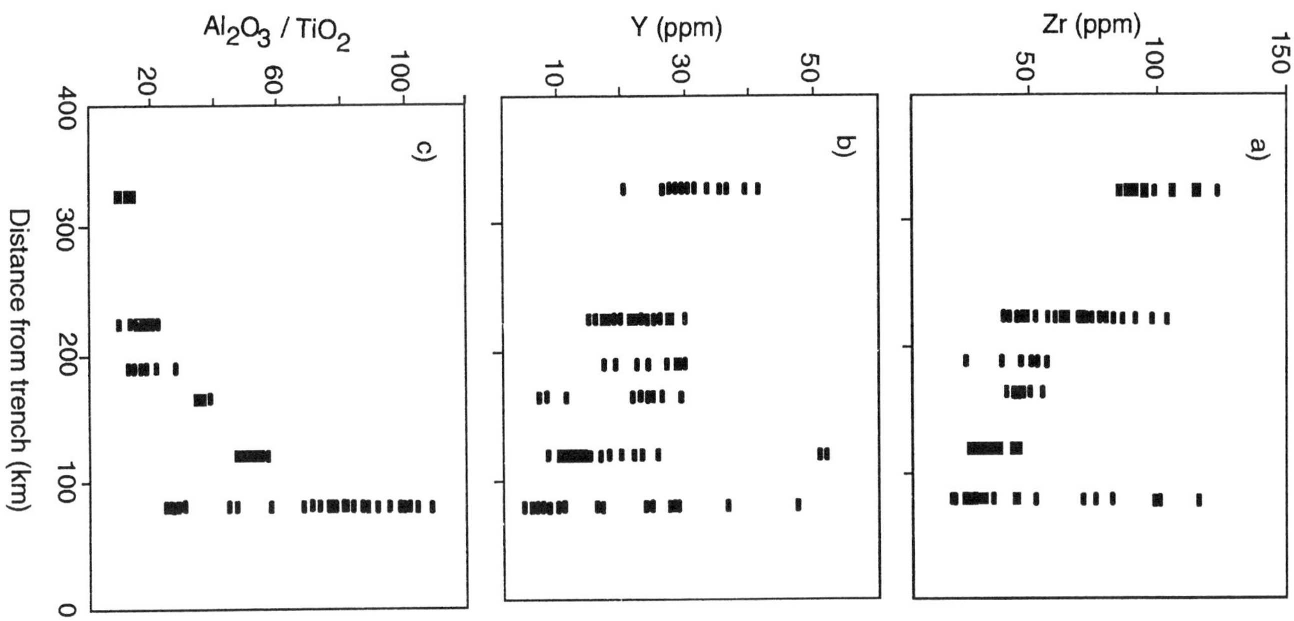

Fig. 18. Across-arc geochemical systematics in the Izu-Bonin forearc; lavas closest to the trench are the old, Eocene forearc lavas and are the most depleted in the region, from *Taylor et al.* [1992b].

and relatively thin (by analogy with the Zambales) crustal thicknesses, can all be most easily accommodated by concluding that the Eocene arc volcanism occurred in a strongly extensional environment, likely by a form of seafloor spreading. However, because of the limited exposures of the forearc basement, it is very difficult to find evidence to prove such a hypothesis. Most desirable, of course, would be documentation of sheeted dikes within the Eocene forearcs. There are dike swarms reported in the oldest basement of both 'Eua [*Ewart and Bryan,* 1972; *Cunningham and Anscombe*, 1985] and Chichi-jima [Arculus, 1992, pers. comm.; *Umino*, 1985; *Dobson*, 1986], but the exposures are too limited to deem them "sheeted".

There is evidence of normal faulting and extension throughout many of the forearc boreholes [e.g., hole 459B, *Natland and Mahoney*, 1981]. This evidence, of course, tells us only about deformation of the forearc after the Eocene, but it does give us a sense of the long term structural state of the forearc. One of the structurally best-documented boreholes is site 841 on the outer Tonga forearc. In cores at site 841 sedimentary bedding is often inclined in excess of 15°, even to within a meter of the mudline, and microfaults, with throws of <1 mm to several centimeters, are abundant throughout the sequence. The majority of these microfaults have normal displacements which, when their slip vectors can be determined with precision, are very close to pure dip slip. Normal microfaults have been identified as high in the succession as core 135-841A-2H, corresponding to a depth of only 9.3 m below seafloor (mbsf). Paleomagnetic data suggest that sediment from this depth has reversed polarity consistent with magnetization during the Jaramillo subchron, of early Pleistocene age [0.7-0.9 Ma; *Parson et al.* 1992]. In addition, at least five major fault zones cut the borehole (Figure 20) which, where offset directions can be determined, show normal senses of motion. The age of the first motion on these faults cannot be constrained. Normal faulting must have been taking place in the region at least until Early Pleistocene times, and is thought to be continuing at the present day on the evidence of seismic data [*von Huene and Scholl,* 1991; *Parson et al.*, 1992; *MacLeod*

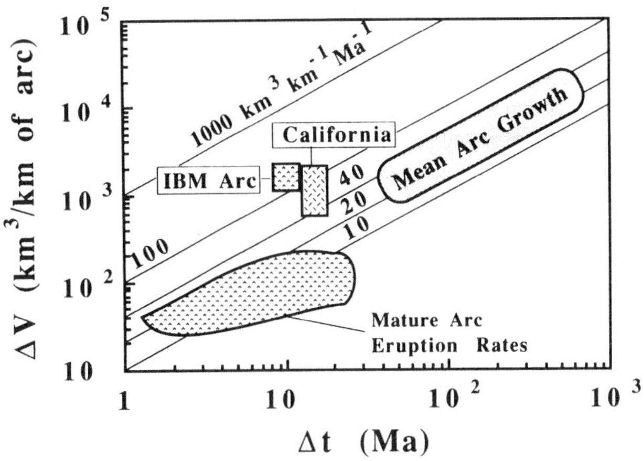

Fig. 19. Estimated magma production rates for the early IBM arc volcanism, from *Stern and Bloomer* [1992].

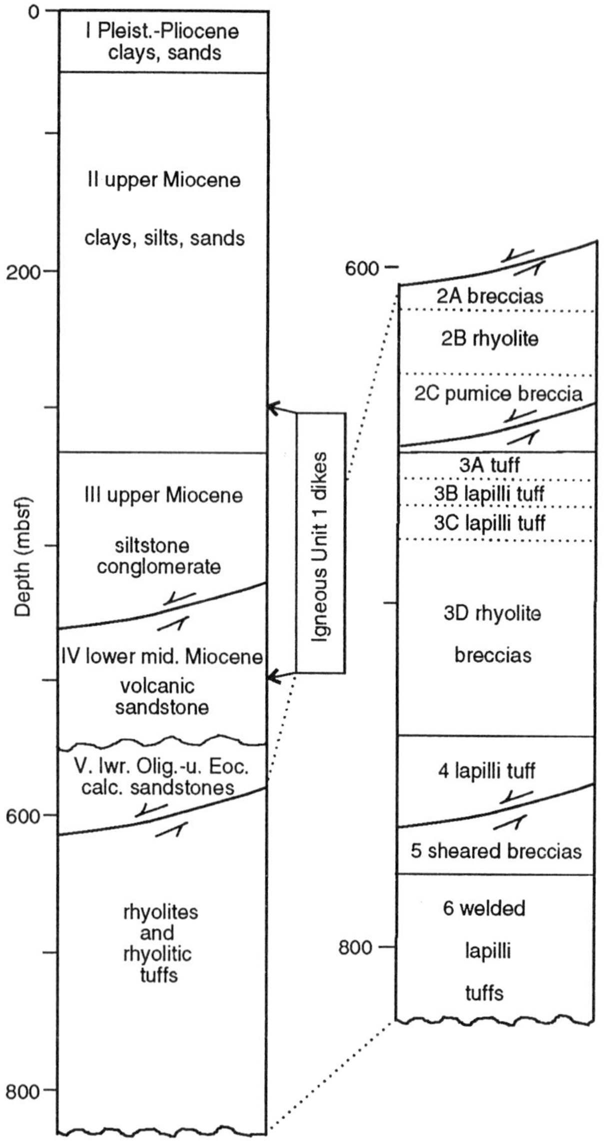

Fig. 20. Schematic section of core from site 841 in the Tonga forearc, illustrating the normal offsets in the section [after *Bloomer et al.*, 1994].

and Lothian, 1994], and GLORIA sidescan sonograph records [*Parson et al.*, 1990; *MacLeod and Lothian*, 1994]. In both pre- and post-collisional (relative to the Lousiville Ridge collision) areas along the inner trench slope the GLORIA sonographs have a typical even to finely-mottled texture, characterized by a high density of short (0.5-7 km), closely-spaced (<1.5 km separation), discontinuous high-backscattering linear features, interspersed with a smaller number of longer, high-backscattering lineaments. Comparison of the sonographs with the bathymetry data shows that many of the longer lineaments coincide with scarps of the order of 100 m in height, the vast majority of which step down eastwards towards the axis of the trench. Reference to multichannel seismic lines [USGS Line 82-12; *Scholl et al.*, 1985] shows that some of the lineaments coincide with the high angle discontinuities in the seismic records that are interpreted as recent faults [*Parson et al.*, 1992; and see above]. Together, these observations offer strong support that the larger lineaments on the inner slope indeed represent fault scarps.

Seismic reflection surveys of the Mariana forearc also show evidence of very recent extensional faulting [*Mrozowski and Hayes*, 1980] and structures in cores from sites 458 and 459 show tensional deformation on all scales [*Hussong et al.*, 1981]. There is also clear evidence that the Izu-Bonin forearc underwent substantial extension in the Oligocene [*Taylor*, 1992]. SeaMARC II studies of the northern Mariana forearc show wide regions of recent tensional deformation [*Wessel et al.*, 1994] and *Fryer* [in press] notes the extension of a forearc rift across the arc into the Northern Mariana Trough. Although we do not know the state-of-stress in the forearcs during the Eocene volcanic episode, it is clear that in their subsequent evolution they have been dominantly subjected to extension.

In situ stress determinations from ODP site 841 [*MacLeod and Pratt*, 1994] show that, across the Tonga platform and inner trench slope, the maximum horizontal compressive stress field is approximately parallel to the trend of the forearc (minimum horizontal compressive stress direction is parallel to the convergence direction), i.e., that the forearc is actively in extension, in accord with the seismic and GLORIA data, and suggests that there is little coupling and hence stress transmission across the main interplate subduction fault [*MacLeod and Pratt*, 1994]. This is, however, opposite to the relationship observed in the IBM [*Pezard et al.*, 1992], part of the Aleutian, and the Japanese forearcs [*Uyeda and Kanamori*, 1979], in which the maximum compressional stress direction is parallel to the convergence direction.

6.2. *Uplift/Subsidence Histories of the Forearc Basement*

Geological studies of the emergent parts of the IBM system in both the frontal arc, from Guam to Saipan, and the outer arc high, the Bonin Islands, show that these ridges formed volcanic highs in the middle and late Eocene. Some sections reached sea level, including volcanoes on and to the west of Guam and Saipan [*Cloud et al.*, 1956, *Tracey et al.*, 1964] and on Hahajima [*Hanzawa*, 1947], though Chichijima may not have been as shallow until the Oligocene [*Dobson*, 1986]. Large foraminifers dredged from the Izu- Bonin frontal arc high at 29°09'N indicate a shallow-water environment in the late Eocene [*Nishimura*, 1992].

Much the same is true in the Tonga forearc. The welded fragments, the absence of interbedded sediments and marine fauna, and the lack of density sorting within the pumice breccias of the basement at site 841 indicate a subaerial origin for the rhyolitic complex. The igneous basement, as is the basement at 'Eua, is overlain by calcareous sandstone bearing large foraminifers indicating a shallow-water environment [*Cunningham and Anscombe*, 1985; *Hawkins et al.*, 1992].

The subsequent uplift/subsidence history of the forearc crust is complicated. The Bonin Ridge became subdivided into a series of fault blocks that were variably tilted [*Keating and Helsley*, 1985] and differentially uplifted/downdropped, the frontal arc high by at least 1 km [*Ingle*, 1975; *Kaiho*, 1992] and the outer arc high by up to 3 km. The Bonin ridge and islands have been uplifted several hundred meters since the early Miocene [*Karig and Moore*, 1975; *Honza and Tamaki*, 1985], whereas the Mariana outer arc high at sites 458 and 459 has remained within 1 km of its present depth, just above and below the paleo-carbonate compensation depth (CCD) respectively [*Hussong* 1981].

However, in several sites where the earliest, post-volcanic motion of the basement can be documented, the Eocene basement, particularly on the outer edge of the forearc, underwent dramatic subsidence in the late Eocene and early Oligocene. Benthic foraminifers indicate that the basement at both sites 782 and 786 subsided nearly 2 km by the late Eocene, and 1.5 km since [Figure 21; *Kaiho*, 1992]. The basement at site 841 underwent nearly 5.4 km of subsidence since the Eocene; the dissolution of microfossils and carbonate phases, and sedimentary facies indicate that the section subsided into the lysocline by early Oligocene and certainly below the CCD by late Miocene [*Clift*, 1994]. In both forearcs, the amount of subsidence is in excess of that expected purely from thermal contraction; it could be consistent with, though does not necessarily require, subduction erosion or extensional thinning of the crust in the late Eocene.

In the Tonga forearc, there is evidence that subsidence of the trench-slope has continued. The attitude of sedimentary bedding shows continual trenchward tilting of this portion of forearc at a more-or-less constant rate since the Eocene [*MacLeod*, 1994] and requires slow removal of material from the underside of the forearc lithosphere [*Ballance et al.*, 1989].

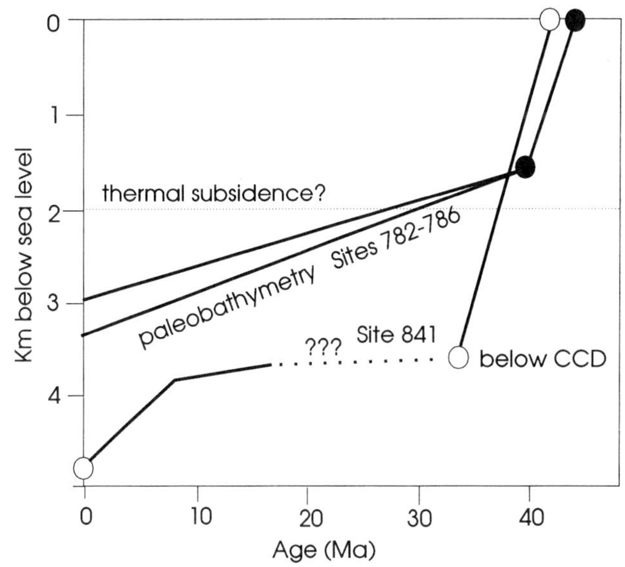

Fig. 21. Estimates for subsidence histories of two sites in the Bonin forearc [*Kaiho*, 1992] and site 841 in the Tonga forearc [*Hawkin et al.*, 1992].

6.3. *Tectonic Erosion or Accretion ?*

The dramatic subsidence of some of the forearc sites in the IBM and Tonga forearcs may be, in part, due to subduction erosion in the late Eocene or early Oligocene. A certain amount of such erosion is required to expose the arc basement which can be traced seismically into the inner trench slope and can be sampled right down to the contact with the Pacific plate. In even the most exotic thermal regime, volcanism adjacent to a nascent trench is highly unlikely; at least 10-20 km of erosion must have been required to remove the older oceanic crust between the paleo-trench and the nearest plausible site of early arc volcanism.

It is possible that subduction erosion has been an ongoing process and that the extent of the Eocene volcanic province was even greater than at present, if significant tectonic erosion of the forearc has occurred. The evidence for tectonic erosion or accretion is controversial however. *Hussong and Uyeda* [1981] argued for tectonic erosion because (1) island arc rocks form much of the trench inner wall, (2) the outer forearc is cut by high angle normal faults, and (3) sites 460 and 461 were interpreted to show massive subsidence of the central Mariana outer forearc. *Karig and Ranken* (1983) argued to the contrary that relative uplift and arcward rotation of strata on the inner trench slope further south is more compatible with intermittent Neogene accretion. The displacement of the shallow-water Paleogene section at sites 460 and 461 can be explained reasonably by slumping and canyon processes [*Karig and Ranken,* 1983]. The relative uplift of the lower slope strata also can be explained, however, by the formation of serpentinite ridges and volcanoes similar to those drilled on leg 125 [*Horine et al.*, 1990; *Fryer et al.*, 1990].

The known vertical motion of the outer arc high, summarised previously, does not require substantial tectonic accretion or erosion since the early Oligocene. There is seismic reflection evidence for minor accretion of sediments at the toe of the IBM trench [*Mrozowski et al.*, 1981; *Honza and Tamaki*, 1985; *Horine*, 1989], however this may be intermittent and short lived. The undisturbed deposition of hemipelagic silts and clays at or below the CCD since the middle Miocene onto the flank of the serpentinite seamount drilled at sites 783 and 784 [Figure 2, *Fryer et al.*, 1990] argues for the stability and lack of net Neogene tectonic erosion/accretion of the Izu-Bonin lower trench slope. The available data are compatible with little tectonic erosion or accretion of the IBM forearc since the mid-Oligocene.

Lower Cretaceous radiolarian cherts and mid-ocean ridge basalts recovered from the lower slopes of the Mariana Trench indicate that some sedimentary and igneous material from the subducting plate was accreted early into the Mariana forearc [*Azema and Blanchet*, 1981; *Johnson et al.*, 1991]. (Similar material found as clasts in the outer forearc serpentinite melanges may be older, trapped oceanic crust brought to the surface by diapirism). As yet there is no evidence for similar accretion in the region of initial subduction, along strike in the Izu-Bonins. The accreted material is now tectonically and/or magmatically surrounded by Eocene boninites, island arc tholeiites and their plutonic roots, which now extend to the trench inner wall. Given the anomalous character of the widespread Eocene volcanism, including the probability of very shallow (<10 km) boninite magmagenesis, the amount of tectonic erosion necessary to produce these near-trench exposures of arc rocks is unknown, but could be quite small, perhaps on the order of 10-20 km. The conclusion remains, therefore, that the middle and upper Eocene IBM volcanic province was at least 300 km wide, with contemporaneous volcanism extending from what is now the Palau-Kyushu remnant arc to the inner trench wall, and with subaerial volcanoes along some segments of the frontal arc and outer arc highs.

7. DISCUSSION

The detailed geological, geophysical, and drillship studies in the Izu-Bonin-Mariana forearc demonstrate that the forearc formed by the emplacement of voluminous middle and late Eocene boninites, island arc tholeiites, and their differentiates and plutonic roots. That volcanism formed an enormous province (20-300 km × 3000 km) of contemporaneous supra-subduction zone magmatism, which constructed crust at a rate like that typical of slow-spreading mid-ocean ridges. The rates of eruption, the displacement of older oceanic crust, and the available evidence on the tectonic character of the IBM forearc are most consistent with the hypothesis that crustal construction occurring in an extensional environment, perhaps by a form of seafloor spreading. This phase of forearc volcanism lasted 10-15 million years, after which the arc localized along a line 200-300 km behind the trench and built the centralized edifices diagnostic of a mature volcanic arc. The forearc crust has since been modified by faulting, subsidence, serpentine diapirism, and dike intrusion, obscuring many details of its Eocene history. Preliminary studies in the Tonga forearc support a very similar evolution for the earliest arc volcanism in the Tonga Trench.

We have, then, a fairly good picture of *what* was built at the beginning of subduction in the western Pacific. We also think that the IBM forearc was constructed in an extensional stress regime. However, we don't have a very clear idea about why subduction began. In fact, it is difficult to reconcile the onset of subduction, a process normally thought of as requiring convergence, with voluminous volcanism in an extensional setting.

7.1. *Subduction Initiation*

One model that reconciles the early extension and arc volcanism in the IBM forearc and the onset of subduction has been presented by *Stern and Bloomer* [1992]. That model assumes that the initial failure of the lithosphere is due to a density instability along a transform fault (Figure 22). The initial motion of the descending lithosphere is to sink vertically, producing a strong trench suction effect and creating tension within the upper plate and inducing decompression melting at depth. This phase of volcanism corresponds to the broad, Eocene zone of arc volcanism in the IBM system. After about 10-15 Ma, the slab begins to

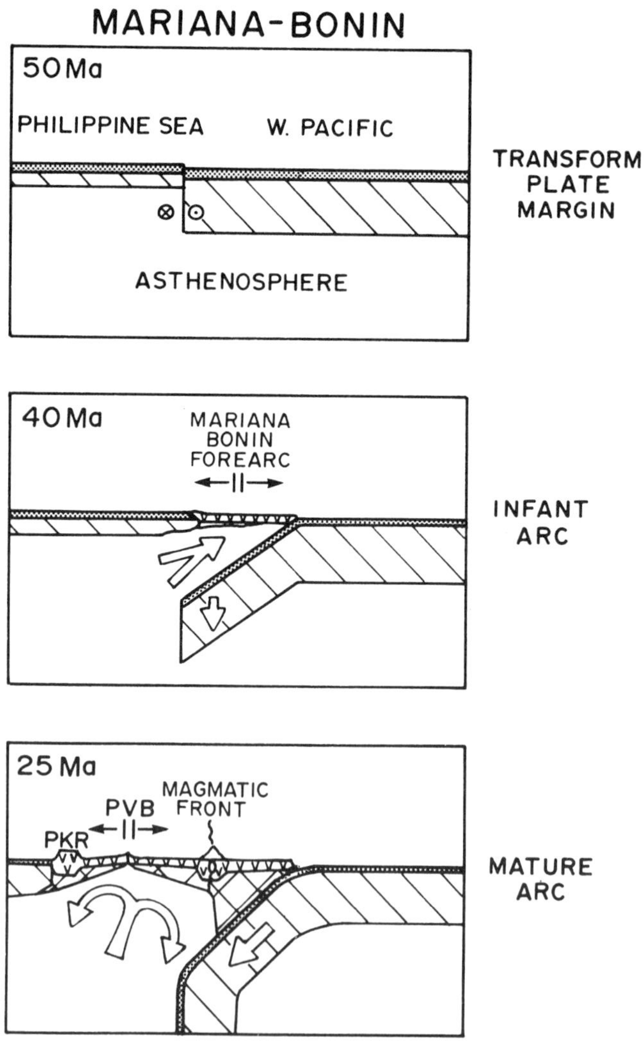

Fig. 22. A model for the evolution of early arc volcanism in the Mariana-Bonin forearc, after *Stern and Bloomer* [1992].

meet a substantial resistive force from the asthenosphere it is displacing, and develops a down-dip motion. This stabilizes the zone of melting and begins to build the line of centralized arc edifices which persist thorough the history of the subduction zone (Figure 22).

The model explains the origin of the broad, contemporaneous Eocene arc volcanism, but requires a pre-existing geometry of old lithosphere juxtaposed against young lithosphere which may not have existed in this part of the Pacific. *Uyeda and Ben Avraham*, [1972], and later *Hilde et al.* [1977], proposed such a geometry for the western Pacific, which assumed that the now roughly north-south trending IBM trenches originated as large north-south trending transform faults (Figure 23). However, there is a substantial volume of paleomagnetic data which indicate that the Philippine Sea Plate has rotated nearly 90° clockwise since the middle Eocene, which would require the original subduction boundary to run nearly east-west (Figure 24). In such an orientation, the Pacific Plate would be moving away from the subduction zone, not into it. To produce subduction beneath the Philippine Sea plate, *Seno and Maruyama* [1984] postulated the existence of an intervening New Guinea plate (Figure 23).

The initial boundary conditions in these two cases are very different—in one version [*Uyeda and Ben Avraham*, 1972; *Hilde et al.*, 1977], old, cold lithosphere subducts beneath young relatively hot lithosphere; in the Seno and Maruyama case, subduction begins with young, hot lithosphere subducting beneath young, hot lithosphere. It is clear that if we are to unravel the how and why of the unusual Eocene early arc volcanism, we must first constrain the initial plate geometries in the western Pacific. The Tonga-Kermedec region may provide a good place to pursue this problem. The forearc there has undergone less disruption and overprinting by post-Eocene processes than has the IBM forearc, and the little existing paleomagnetic data do not show such pronounced post-Eocene rotations.

7.2. *Variations in Early Arc Crust*

The model for subduction initiation proposed by *Stern and Bloomer* [1992], though it is yet to be fully verified, does highlight two important processes that must have contributed to the construction of the forearc crust we see in the IBM and Tonga forearcs. First, older oceanic crust must be stretched and thinned and, second, that thinned crust is partially or totally displaced and overbuilt by new arc crust (Figure 25). The type of crust that is constructed during early arc volcanism may vary substantially, depending upon the balance between the stretching and thinning of the pre-existing crust and the rate of magma generation in the mantle wedge. That balance will depend upon the thermal regimes in the upper and lower plates, and the rate of extension and strength of the upper plate. There may be cases, as in the IBM forearc, where extension is extreme, temperatures are very high at shallow depths in the mantle wedge, and the pre-existing oceanic crust is nearly totally

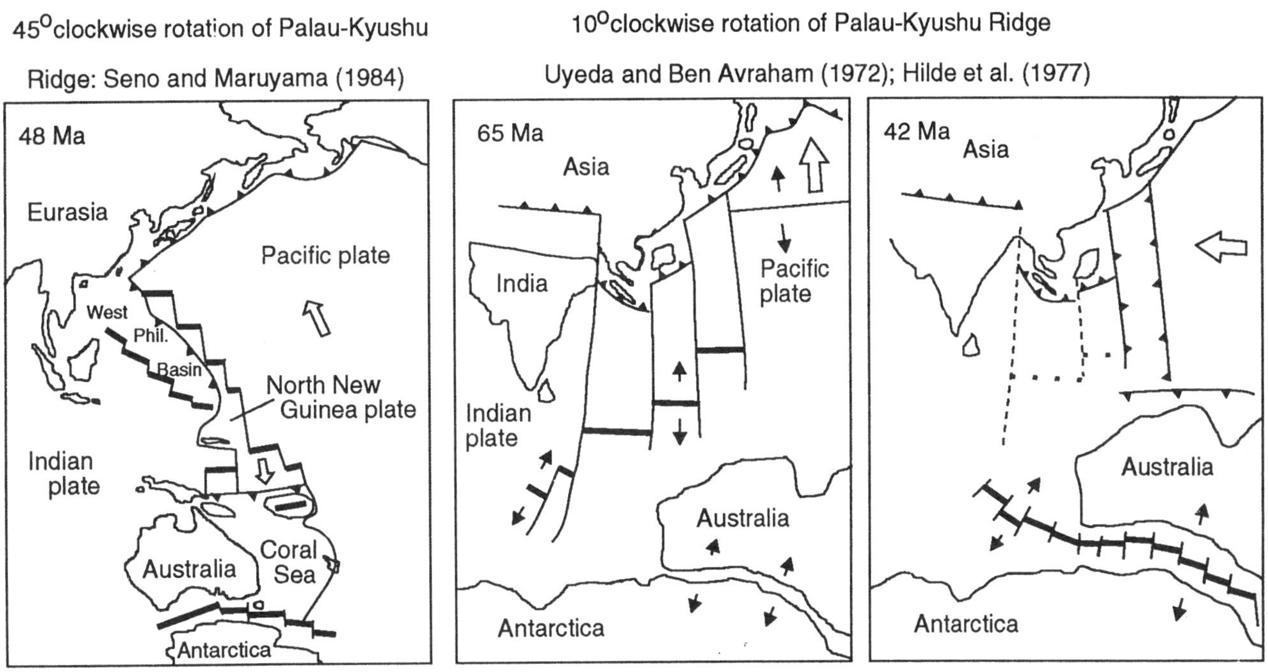

Fig. 23. Two models for plate geometries at or near the time of initiation of subduction in the western Pacific, after *Seno and Maruyama* [1984], *Uyeda and Ben Avraham* [1972], *Hilde et al.* [1977].

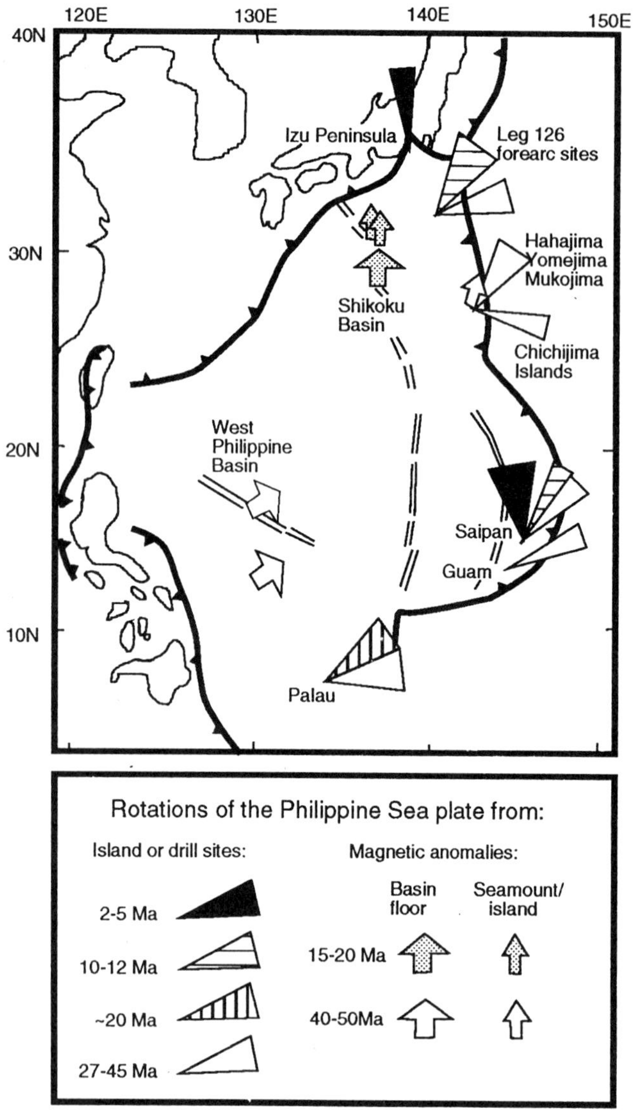

Fig. 24. Summary of paleomagnetic data around the Philippine Sea indicating substantial clockwise rotations since the Eocene, after *Koyama et al.* [1992].

replaced by very depleted boninitic arc crust. At the other extreme, with a much colder mantle wedge and less extension, no new arc crust may be formed at all; the forearc might eventually consist simply of stretched and faulted older oceanic crust. The Tonga forearc may represent an intermediate case—temperatures were high enough for substantial arc tholeiitic volcanism, but not high enough to create voluminous boninitic magmatism; neither was extension extreme enough to remove all vestiges of the older oceanic crust. The key aspect of the early arc evolution is that of extension in the upper plate, usually accompanied by depleted arc volcanism, across a broad (200-300 km) zone. The balance between volcanism and extension has the potential to create a variety of different crustal types.

7.3. *Timing of Pacific Plate Motions*

An interesting observation that derives from the drilling studies of western Pacific forearcs concerns the timing of subduction initiation and the change in motion of the Pacific plate, marked by the bend in the Hawaiian-Emperor seamount chain. It is clear that both in the IBM and the Tonga systems, there was substantial arc volcanism associated with subduction (in fact, there were emergent arc edifices) well before the Pacific plate changed direction (Figure 26). It is clear that the development of the western Pacific subduction zones preceded the major Eocene change in Pacific plate motion. This is consistent with analyses that show slab pull to be a major determinant of plate velocities [*Gordon et al.*, 1978; *Forsythe and Uyeda*, 1975].

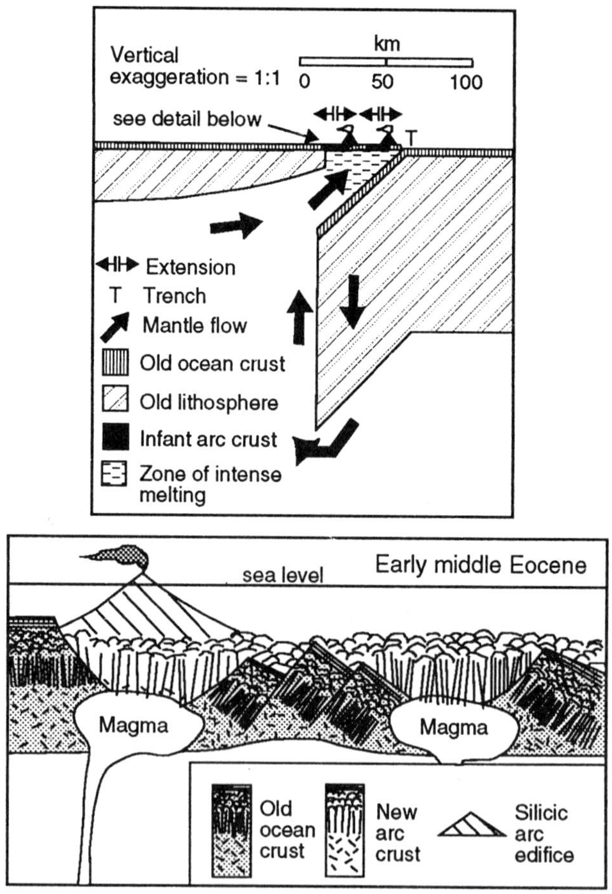

Fig. 25. A model for producing forearc extension and voluminous volcanism during the early stages of subduction and a schematic view of the kind of crust that might be produced by that process, from *Bloomer et al.* [1994].

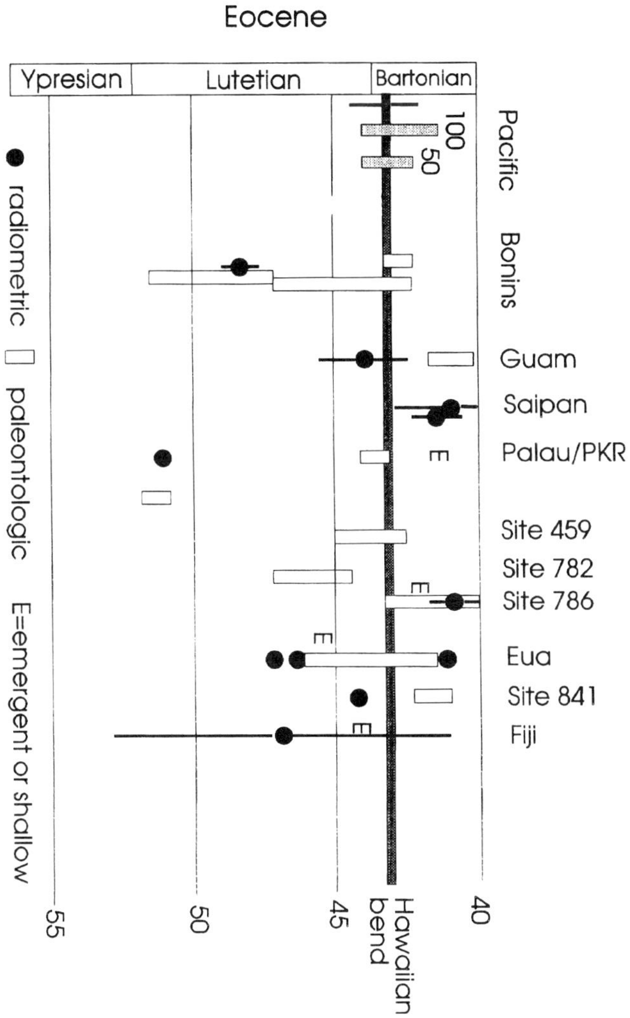

Fig. 26. Distribution of ages of arc volcanism around the western Pacific in relation to the age of the bend in the Hawaiian-Emperor Seamount chain [*Dalrymple and Clague*, 1976]; arc volcanism, and therefore subduction, clearly preceded the change in motion of the Pacific Plate. Age data are cited elsewhere in text.

7.4. *The Ophiolite Analogy*

The analogies between the forearc assemblages of the western Pacific and ophiolitic rock suites like those in Troodos, Oman, California, and Newfoundland have been pointed out by a number of workers [*Bloomer and Hawkins*, 1983; *Sharaskin et al.*, 1983; *Ishii*, 1985; *Dick and Bullen*, 1984; *Hall*, 1990; *Ballantrye and Hall*, 1990; *Shervais*, 1990; *Stern and Bloomer*, 1992]. The parallels between these geologic terranes have been drawn largely on the basis of rock types, and in response to the recognition that the petrology and geology of mid-ocean ridges do not correspond to what is found in these ophiolites. A number of studies have documented the "subduction-zone" trace element signature found in many ophiolitic lavas [e.g., *Pearce et al.*, 1984; *Rautenschlein et al.*, 1985]. The gabbroic rocks in many ophiolites show cumulate sequences unlike those in mid-ocean ridge gabbros, with early appearing pyroxenes and calcic plagioclase at a given clinopyroxene composition [Figure 27; *Thy*, 1987; *Thy et al.*, 1989; *Ernewein et al.*, 1988] whereas the ultramafic sections in the ophiolites are more depleted in both their modal clinopyroxene abundances and spinel compositions than is typical of peridotites dregded from mid-ocean ridges [Figure 27; *Dick and Bullen*, 1984].

The clear geochemical and mineralogic fingerprint of a subduction zone setting in so many ophiolitic lavas strongly suggests that they formed over a subducting slab. An examination of the geochemical and petrologic characteristics of volcanic and plutonic rocks from the three principal SSZ environments clearly show that the one most similar to the rock assemblages in ophiolites like Troodos and Oman occurs in the forearc (Table 2). What has been missing to bring this observation more widespread recognition was a way to reconcile those forearc rock compositions with the structural evidence for extension found in all of the ophiolites. *Natland and Tarney* [1981], *Pearce et al.* [1984], and *Robertson* [1990] had the prescience to hypothesize "pre-arc" spreading in a geometry much like that now documented by results from Ocean Drilling Program legs 125 and 126 in the IBM forearc. The forearc crust is constructed in an extensional setting, creating a broad platform similar in thickness and dimensions to the crust created at oceanic spreading centers. It is now clear that the non-accretionary forearcs of the western Pacific trenches are the best modern

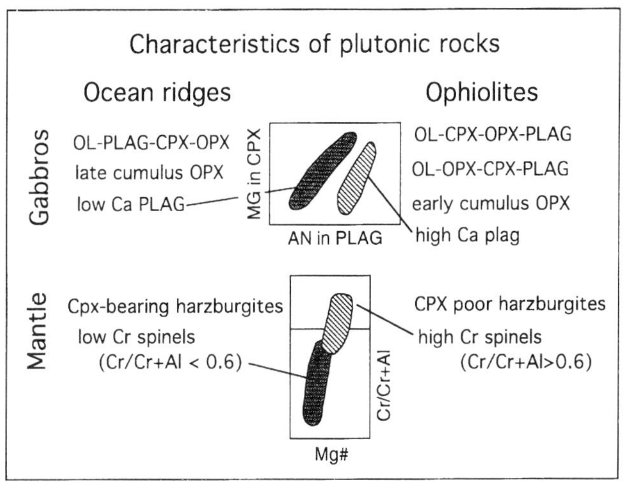

Fig. 27. A summary of some of the principal features of gabbroic and ultramafic (mantle) sections sampled from mid-ocean ridges [*Bloomer et al.*, 1989, 1991; *Dick and Bullen*, 1984] and from Troodos and Oman [*Dick and Bullen*, 1984; *Thy*, 1987; *Browning*, 1984].

TABLE 2. A Summary of the Geologic Features of Likely SSZ Ophiolite Analogs in Modern Subduction Zones like the IBM System.

Analog:	Characteristics:
Forearc:	low-Ti tholeiitic and boninitic volcanics with pronounced subduction-zone geochemical signature gabbroic cumulates with sequence of olivine+spinel-orthopyroxene-clinopyroxene-plagioclase calcic cumulate plagioclase with a given olivine or clinopyroxene composition magnesian cumulate orthopyroxene high Cr/Cr+Al spinels (> 0.60) in residual peridotites
Arc:	low-Ti tholeiitic to calcalkaline volcanics with pronounced subduction-zone geochemical signature gabbroic cumulates with sequence olivine-plagioclase-clinopyroxene-orthopyroxene or olivine-clinopyroxene-plagioclase-orthopyroxene-(minor amphibole) calcic cumulate plagioclase with a given olivine or clinopyroxene composition
Back-arc:	tholeiitic basalts with slight subduction zone signature gabbroic cumulates with sequence of olivine, plagioclase, clinopyroxene, orthopyroxene, low-magnesian cumulus orthopyroxene plagioclase with given clinopyroxene or olivine not unusually calcic--like ocean-ridge gabbros low Cr/Cr+Al spinels (< 0.60) in residual peridotites

analog we have for the origin of the large ophiolites like Troodos and Oman.

This is not the venue for a review and synthesis of the geology of those ophiolites. There are years of careful geologic studies in both, and the level of detailed observation available in them will never be reached in their Pacific analogs. The ophiolites at Troodos and Oman are also distinctly different. Troodos, for example, appears to have formed in a relatively magma-poor setting, with plastic deformation of gabbros [*George,* 1978; *Malpas and Robinson* , 1987], formation of grabens and detachments [viz. Solea graben, etc., and Kakopetria detachment: *Varga and Moores,* 1985; *Allerton and Vine,* 1987, 1991; *Varga* 1991; *Hurst et al.*, 1992] and intrusive boundaries between older, rotated, stretched crust cut by younger, unrotated crust [*Allerton and Vine,* 1990, 1992]. Troodos also shows clear evidence for multiple magma chambers from field evidence [*Benn and Laurent,* 1987] and from supposed intrusive boundaries in CY-4 borehole [*Malpas et al.*, 1989]. The Oman ophiolite has a complicated lava stratigraphy, with an apparently much higher rate of magmatism and multiple plutonic episodes [*Juteau et al.,* 1988; *Ernewein et al.,* 1988]; it is commonly taken to be a more magmatically robust setting than Troodos. These kinds of differences have commonly been taken to indicate formation at slow or fast-spreading ridges, but could also reflect variations in the rate and style of early arc volcanism.

It is important to note that there *are* significant differences between the Western Pacific forearc crust and the large Tethyan ophiolites. While the sedimentologic records in both the IBM and Tonga forearcs show pronounced subsidence, paleobathymetric evidence in Troodos indicates that the ophiolite was always at abyssal depths [*Robertson,* 1990; *Lippard et al.*, 1986]. Silicic volcanics are not as abundant in either Troodos or Oman as they appear to be in the Eocene forearcs [e.g., *Pearce et al.*, 1992a; *Bloomer et al.*, 1994]. The older crust in the Oman ophiolite appears to be quite homogeneous [e.g., *Nicolas et al.*, 1988; *Juteau et al.,* 1988] whereas the IBM crust seems to be laterally and vertically complex. There are clearly gaps in our understanding of the factors controlling crustal construction in ophiolites and in early arc crust which need to be explored if we are to resolve the debate about the origins of ophiolites.

It remains clear that existing chemical and petrologic data preclude drawing one-to-one analogies between modern ocean ridges and these ophiolites—the results we have reviewed from the western Pacific lead us to believe that the mafic magmatism which constructed the ophiolites occurred during early subduction. Nonetheless, even if the large, intact ophiolites do represent a variety of subduciton zone magmatism, they still provide important details about the construction of mafic crust created in an extensional stress regime, including crust formed at mid-ocean ridges. The interpretation of these ophiolites as early arc crust requires a re-examination of their tectonic significance, and alerts us to be careful when applying models derived from ophiolites to modern oceanic environments.

8. CONCLUSIONS AND SUMMARY

Marine geologic studies and Deep Sea Drilling Project and Ocean Drilling Program transects of the intraoceanic Izu-Bonin-Mariana forearcs have generated several important new ideas about the origin and evolution of these terranes. The first, and most striking, is the recognition that the initial phases of volcanism in these subduction zones developed nearly synchronously in the middle to late Eocene over a zone up to 300 km wide and thousands of kilometers long. This early, or "infant," arc volcanism was characterized by the eruption of very depleted boninitic and arc tholeiitic lava compositions and developed in an extensional setting (as evidenced by dikes on Chichi-jima in the Bonin Islands), possibly to the point of true seafloor spreading. This infant arc volcanism was built on, or replaced, the pre-existing oceanic crust and had production rates much higher than those of mature arcs--eruption rates on the order of those in slow-spreading ridges. This mode of arc volcanism is unlike that developed during the "normal" or mature phases of arc activity and is a plausible mechanism for developing supra-subduction zone ophiolites [what *Pearce et al.*, 1984, termed "pre-arc" ophiolites; *Robertson*, 1990; *Stern and Bloomer*, 1992]. In fact, the duration of volcanism, volcanic and plutonic rock compositions, and structural setting in the IBM forearc are virtually identical to those in the Troodos ophiolite.

The exposure of Eocene arc basement immediately adjacent to the axis of the Mariana-Bonin Trench requires some subduction erosion since the Eocene. However, if the Eocene basement did form by unusual, voluminous volcanism in an extensional environment, the required amount of erosion may only be 20 to 50 km; the vertical tectonics of the forearc indicate there has been little erosion of the forearc since the Oligocene. There is also clear evidence of minor, episodic, post-Eocene accretion of Mesozoic Pacific plate sediments and crustal fragments to the outer forearc.

There is a geochemical assymetry in the forearc basement, with more depleted lavas nearer the trench [*Taylor et al.*, 1992b]. This in part results from the age progression, with the oldest and most depleted lavas left in the outer forearc basement, and younger lavas from central edifices dominating the inboard side of the forearc. This assymmetry provides a potentially powerful tool for determining the polarity in supra-subduction zone ophiolites, if these do indeed represent ancient "infant arcs".

These observations in the Izu-Mariana-Bonin system have painted a far different picture of intraoceanic forearcs than existed nearly fifteen years ago before the results of DSDP leg 60 documented in-situ Eocene boninitic basement in the outer Mariana forearc [*Hussong and Uyeda, 1981*]. These forearcs are geologically complex, and are clearly direct products of subduction, not just trapped pieces of oceanic crust. They have geologic and geochemical similarities to supra-subduction zone ophiolites like those in Troodos, Oman, and California, and represent a previously unappreciated type of crustal construction. They are, in fact, the closest analog to supra-subduction zone ophiolites that has been found in *any* modern geologic environment. The conclusions drawn from the studies of the IBM system bear on an array of convergent margin problems—subduction initiation (subsidence vs. compression), the nature of early arc volcanism, the significance of ophiolites, the state of stress in the forearc, the long-term thermal evolution of forearcs, and mass balance (both magmatic, as reflected by the early arc production rates, and fluid, as reflected by the serpentinite diapirs). If these results apply only to the IBM system, they are still interesting. It is possible that they do—the system has a very complicated history and may have originated from an unusual initial plate geometry. However, if we can prove that these models apply to other subduction systems, we have discerned something fundamental about the structural and magmatic dynamics of subduction and may have solved the enigma of supra-subduction zone ophiolites

Acknowledgments. This work has been supported over several years by the National Science Foundation and the U.S. Science Advisory Committee. We would like to thank Tracy Vallier and James Natland for their detailed comments on the manuscript. The results reviewed here could never have been achieved without the dedication and skill of the scientists, technicians, and crews who have made the Ocean Drilling Program a model for cooperative international science. This is SOEST contribution no. 3661 and Hawaii Institute of Geophysics and Planetology contribution no. 784.

REFERENCES

Acland, A. S., S. V. Vysotsky, J. A. Pearce, and L. M. Parson, Geochemistry of a boninitic crustal section from the northern Tonga Trench, (abstract), at *Back-Arc development and crustal accretion in the western Pacific*, Cambridge, England, May 1990.

Allerton, S., and F. J. Vine, Spreading structure of the Troodos ophiolite, Cyprus: some palaeomagnetic constraints, *Geology, 15*, 593-597, 1987.

Allerton, S., and F. J. Vine, Palaeomagnetic and structural studies of the southeastern part of the Troodos complex, in *Ophiolites: Oceanic Crustal Analogues,* edited by J. Malpas, E. M. Moores, A. Panayiotou, and C. Xenophontos, pp. 99-111, Proc. Symp. Ophiolites and Oceanic Lithosphere, Troodos 87, Ministry Agric. Nat. Resources, Nicosia, Cyprus, 1990.

Allerton, S., and F. J. Vine, Spreading evolution of the Troodos ophiolite, Cyprus, *Geology, 19*, 637-640, 1991.

Allerton, S., and F. J. Vine, The structure of ridge-transform intersections: evidence from the Troodos ophiolite, Cyprus, in *Ophiolites and Their Modern Oceanic Analogues*, edited by L. M. Parson, B. J. Murton, and P. Browning, P., Geol. Soc., London, Spec. Publ. 60, 251-261, 1992.

Ambos, E. L., Applications of ocean bottom seismometer data to the study of forearc and transform fault systems, unpub. Ph.D. dissertation, University of Hawaii, Honolulu, Hawaii, 1984.

Arculus, R. J., J. A. Pearce, B. Murton, and S. van der Laan, Igneous stratigraphy and major element geochemistry of Holes 786A and 786B, in *Proc. ODP, Sci. Results*, *125*, P. Fryer, J. A. Pearce, L. B. Stokking, et al., pp. 143-170, Ocean Drilling Program, College Station, TX, 1992.

Azema, J., and R. Blanchet, The late Jurassic, early Cretaceous genus *Calpionella* in reworked pebbles from Deep Sea Drilling Project site 460, Mariana Transect, in *Init. Repts. DSDP*, 60, D. M. Hussong and S. Uyeda, et al., pp. 575-576, U.S. Govt. Printing Office, Washington, D.C., 1981.

Ballance, P. F., D. W. Scholl, T. L. Vallier, A. J. Stevensen, H. Ryan, and R. H. Herzer, Subduction of a late Cretaceous seamount of the Louisville Ridge at the Tonga Trench: a model of normal and accelerated tectonic erosion, *Tectonics, 8*, 953-962, 1989.

Ballantrye, P. D., and R. Hall, The petrology of the Halmahera ophiolite, Indonesia: an early Tertiary forearc, in *Ophiolites: Oceanic Crustal Analogues,* edited by J. Malpas, E. M. Moores, A. Panayiotou, and C. Xenophontos, pp. 461-476, Proc. Symp. Ophiolites and Oceanic Lithosphere, Troodos 87, Ministry Agric. Nat. Resources, Nicosia, Cyprus, 1990.

Benn, K., and R. Laurent, Intrusive suite documented in the Troodos ophiolite plutonic complex, Cyprus, *Geology 15,* 821-824, 1987.

Bloomer, S. H., Distribution and origin of igneous rocks from the landward slopes of the Mariana Trench: implications for its structure and evolution, *J. Geophys. Res., 88,* 7411-7428, 1983.

Bloomer, S. H., Geochemical characteristics of boninite- and tholeiite- series volcanic rocks from the Mariana forearc and the role of an incompatible element-enriched fluid in arc petrogenesis, *Geol Soc. Am., Spec. Pap., 215,* 151-164, 1987.

Bloomer, S. H., and R. L. Fisher, Petrology and geochemistry of igneous rocks from the Tonga Trench: implications for its structure, *J. Geology, 95,* 469-495, 1987.

Bloomer, S. H., and J. W. Hawkins, Gabbroic and ultramafic rocks from the Mariana trench: An island arc ophiolite, in *The Tectonic and Geologic Evolution of Southeast Asian Seas and Islands (Pt. II), Geophys. Monog. Ser.,* vol. 27, edited by D. E. Hayes, pp. 294-317, AGU, Washington, D.C., 1983.

Bloomer S., J. Natland, and R. L. Fisher, Mineral relationships in gabbroic rocks from fracture zones of the Indian Ocean Ridge: evidence for extensive fractionation, parental diversity and boundary layer recrystallization, in *Magmatism in the Ocean Basins,* edited by A .D. Saunders and M. J. Norry, *Geol. Soc., London, Spec. Publ., 42,* 107-124, 1989.

Bloomer S. H., P. S. Meyer, H. J. B. Dick, K. Ozawa, and J. H. Natland, Textural and mineralogic variations in gabbroic rocks from Hole 735B, in *Proc. ODP, Sci. Results, 118,* R. P. Von Herzen, P. T. Robinson, et al., pp. 21-41, Ocean Drilling Program, College Station, TX, 1991.

Bloomer, S. H., A. Ewart, J. M. Hergt, and W. B. Bryan, Geochemistry of igneous rocks from ODP site 841, Tonga forearc. in *Proc. ODP, Sci. Results 135,* J. Hawkins, L. Parson, J. Allan, et al., pp. 625-646, Ocean Drilling Program, College Station, TX, 1994.

Browning, P., Cryptic variation within the cumulate sequence of the Oman ophiolite: magma chamber depth and petrological implications, in *Ophiolites and Oceanic Lithosphere, Geol. Soc., London, Spec. Publ., 13,* 71-82, 1984.

Clift, P. D., Controls on the sedimentary and subsidence history of an active plate margin: an example from the Tonga Arc (Southwest Pacific), in *Proc. ODP, Sci. Result, 135,* J. Hawkins, L. Parson, J. Allan, et al., pp. 173-188, Ocean Drilling Program, College Station, TX, 1994.

Cloud, P. E., Jr., R. G. Schmidt, and H. W. Burke, Geology of Saipan, Mariana Islands, *U.S. Geol. Surv. Prof. Pap., 1350,* 5-54, 1956.

Cole, J. W., J. B. Gill, and D. Woodhall, Petrologic history of the Lau Ridge, Fiji, in *Geology and Offshore Resources of Pacific Island Arcs-Tonga Region,* Earth Science Series, vol. 2, edited by D. W. Scholl and T. L. Vallier, pp. 379-413, Circum-Pacific Council for Energy and Mineral Resources, Houston, TX, 1985.

Cole, J. W., I. J. Graham, and I. L. Gibson, Magmatic evolution of Late Cenozoic volcanic rocks of the Lau Ridge, Fiji, *Contrib. Mineral. Petrol., 104,* 540-554, 1990.

Cunningham, J. K., and K. J. Ansombe, Geology of 'Eua and other islands, Kingdom of Tonga, in *Geology and Offshore Resources of Pacific Island Arcs-Tonga Region,* Earth Science Series, vol. 2, edited by D. W. Scholl and T. L. Vallier, pp. 221-257, Circum-Pacific Council for Energy and Mineral Resources, Houston, TX, 1985.

Dalrymple, G. B., and D. A. Clague, Age of the Hawaiian-Emperor bend, *Earth Planet. Sci. Lett., 31,* 313-329, 1976.

Dick, H. J. B., and T. Bullen, Chromian spinel as a petrogenetic indicator in abyssal and alpine-type peridotites and spatially associated lavas, *Contrib. Mineral. Petrol., 86,* 54-76, 1984.

Dick, H. J. B., R. L. Fisher, and W. B. Bryan, Mineralogic variability of the uppermost mantle along mid-ocean ridges, *Earth Planet. Sci. Lett., 69,* 88-106, 1984.

Dietrich, V., R. Emmeran, R. Oberhansli, and H. Puchlet, Geochemistry of basaltic and gabbroic rocks from the West Mariana Basin and the Mariana Trench, *Earth Planet. Sci. Lett., 39,* 127-144, 1978.

Dobson, P. F., The petrogenesis of boninite: a field, petrologic and geochemical study of the volcanic rocks of Chichi-jima, Bonin Islands, Japan. [unpubl. Ph.D. dissert.] Stanford Univ., California, 1986.

Duncan, R. A., T. L. Vallier, and D. A. Falvey, Volcanic episodes at 'Eua, Tonga Islands, in *Geology and Offshore Resources of Pacific Island Arcs-Tonga Region,* Earth Science Series, vol. 2, edited by D. W. Scholl and T. L. Vallier, pp. 281-289, Circum-Pacific Council for Energy and Mineral Resources, Houston, TX, 1985.

Ellis, C. H., Calcareous nannofossil biostratigraphy--Deep Sea Drilling Project Leg 60, in *Init. Repts. DSDP, 60,* D. M. Hussong, S. Uyeda, et al., pp. 507-536, U.S. Govt. Printing Office, Washington, D.C., 1981.

Ernewein, M., C. Pflumio, and H. Whitechurch, The death of accretion zone as evidenced by the magmatic history of the Sumail ophiolite (Oman), *Tectonophysics, 151,* 247-274, 1988.

Ewart, A., and W. B. Bryan, Petrography and geochemistry of the igneous rocks from 'Eua, Tongan islands, *Geol. Soc. Am. Bull, 83,*3281-3298, 1972.

Ewart, A., and C. J. Hawkesworth, The Pleistocene-Recent Tonga-Kermedec arc lavas: interpretation of new isotopic and rare earth data in terms of a depleted mantle source model. *J. Petrol., 28,* 495-530, 1987.

Ewart, A., R. N. Brothers, and A. Mateen, An outline of the geology and geochemistry and the possible petrogenetic evolution of the volcanic rocks of the Tonga-Kermadec-New Zealand island arc, *J. Volcanol. Geothermal. Res., 2,* 205-250, 1977

Falloon, T. J., D. H. Green, and A. J. Crawford, Dredged igneous rocks from the northern termination of the Tofua magmatic arc, Tonga and adjacent Lau Basin, *Austral. J. Earth Sci., 34,* 487-506, 1987.

Fisher, R. L., and C. G. Engel, Ultramafic and basaltic rocks dredged from the nearshore flank of the Tonga Trench, *Geol. Soc. Am. Bull., 80,* 1373-1378, 1969.

Forsythe, D., and S. Uyeda, On the relative importance of the driving forces of plate motion, *Geophys. J. Roy. Astron. Soc., 43*, 163-200, 1975.

Fryer, P., A synthesis of Leg 125 drilling of sepentinite seamounts on the Mariana and Izu-Bonin forearcs, in *Proc. ODP, Sci. Results, 125*, P. Fryer, J. A. Pearce, L. B. Stokking et al., pp. 593-614, Ocean Drilling Program, College Station, TX, 1992.

Fryer, P., Geology of the Mariana Trough, in Backarc Basins: Tectonics and Magmatism, edited by B. Taylor, Plenum Press, New York, in press, 1994.

Fryer, P., and G. J. Fryer, Origins of nonvolcanic seamounts in a forearc environment. in *Seamounts, Islands, and Atolls, Geophysical Monograph Ser.*, vol. 43, edited by B. H. Keating, P. Fryer, R. Batiza, and G. W. Boehlert, pp. 61-72, AGU, Washington, D.C., 1985.

Fryer, P., E. L. Ambos, and D. M. Hussong, Origin and emplacement of Mariana forearc seamounts, *Geology, 13*, 774-777, 1985.

Fryer, P., J. A. Pearce, L. B. Stokking, et al., *Proc. ODP, Init. Repts., 125*, Ocean Drilling Program, College Station TX, 1990.

Geary, E. E., R. W. Kay, J. C. Reynolds, and S. M. Kay, Geochemistry of mafic rocks from the Coto block, Zambales ophiolite, Philippines: trace element evidence for two stages of crustal growth, *Tectonophysics, 168*, 43-63, 1989

George, R. P., Jr., Structural petrology of the Olympus Ultramafic Complex in the Troodos Ophiolite, Cyprus, *Geol. Soc. Am. Bull., 89*, 845-865, 1978.

Gnibidenko, H. S., G. I. Anosov, V. V. Argentov, and I. K. Pushchin, Tectonics of the Tonga-Kermedec Trench and Ozbourn Seamount junction area, *Tectonophysics, 112,* 357-383, 1985.

Gordon, R., A. Cox, and C. Harter, Absolute motion of an individual plate estimated from its ridge and trench boundaries, *Nature, 274*, 752-755, 1978

Greene, H. G., and F. L. Wong, Hydrocarbon resource studies of the southwestern Pacific, 1982, *U.S. Geol. Surv. Open File Rept. 82-293*, 24 pp., 1983.

Hall, R., Subduction-related ophiolite terrains: evidence from Southeast Asia, in *Ophiolites: Oceanic Crustal Analogues,* edited by J. Malpas, E. M. Moores, A. Panayiotou, and C. Xenophontos, pp. 449-460, Proc. Symp. Ophiolites and Oceanic Lithosphere, Troodos 87, Ministry Agric. Nat. Resources, Nicosia, Cyprus, 1990.

Hanzawa, S., Eocene foraminifera from Hahajima, *J. Paleontol., 21*, 254-259, 1947.

Hawkins, J. W., and C. A. Evans, Geology of the Zambales Range, Luzon, Philippine Islands: ophiolite derived from an island arc-backarc basin pair, in *The Tectonic and Geologic Evolution of Southeast Asian Seas and Islands (Pt. II), Geophys. Monog. Ser,* vol. 27, edited by D. E. Hayes, pp. 95-123, AGU, Washington, D.C., 1983

Hawkins, J. W., and D. A. Falvey, Petrology of andesitic dikes and flows from 'Eua, Tonga, in *Geology and Offshore Resources of Pacific Island Arcs-Tonga Region*, Earth Science Series, vol. 2, edited by D. W. Scholl and T. L. Vallier, pp. 269-279, Circum-Pacific Council for Energy and Mineral Resources, Houston, TX, 1985.

Hawkins, J. W., S. H. Bloomer, C. A. Evans, and J. T. Melchior, Evolution of intra-oceanic arc-trench systems, *Tectonophysics, 102*, 175-205, 1984.

Hawkins, J. W., L. M. Parson, J. F. Allan, J.F., et al., *Proc. ODP, Init. Repts, 135*, Ocean Drilling Program, College Station, TX, 1991.

Herzer, R. H., and N. Exon, Structure and basin analysis of the southern Tonga forearc, in *Geology and Offshore Resources of Pacific Island Arcs-Tonga Region*, Earth Science Series, vol. 2, edited by D. W. Scholl and T. L. Vallier, pp. 55-74, Circum-Pacific Council for Energy and Mineral Resources, Houston, TX, 1985.

Hilde, T. W. C., S. Uyeda, and L. Kroenke, Evolution of the Western Pacific and its margin, *Tectonophysics, 38*, 145-165, 1977.

Hodges, F. N., and J. J. Papike, DSDP Site 334; magmatic cumulates from oceanic layer 3, *J. Geophys. Res., 81*,4135-4151, 1976.

Honza, E., and K. Tamaki, The Bonin Arc, in *The Ocean Basins and Margins, (Vol. 7): The Pacific Ocean*, edited by A. E. M. Nairn and S. Uyeda, pp. 459-502, Plenum Press, New York, 1985.

Horine, R. L., Accretion and diapirism in the Izu-Bonin Arc [Ph.D. Dissert.]. Univ. Tulsa, Oklahoma, 1989.

Horine, R. L., G. F. Moore, and B. Taylor, Structure of the outer Izu-Bonin forearc from seismic-reflection profiling and gravity modeling, in *Proc. ODP, Init. Repts., 125,* P. Fryer, J. A. Pearce, L. B. Stokking, et al., pp. 81-94, Ocean Drilling Program, College Station, TX, 1990.

Hurst, S. D., K. L. Verosub, and E. M. Moores, Paleomagnetic constraints on the formation of the Solea graben, Troodos ophiolite, Cyprus, *Tectonophysics, 208*, 431-445, 1992.

Hussong, D. M., and P. Fryer, Structure and tectonics of the Mariana arc and fore-arc: drillsite selection surveys, in *Init. Repts. DSDP, 60*, D. M. Hussong, S. Uyeda, et al., pp. 33-44, U.S. Govt. Printing Office, Washington, D.C., 1981.

Hussong, D., and S. Uyeda, Tectonic processes and the history of the Mariana Arc: a synthesis of the results of Deep Sea Drilling Project Leg 60, in *Init. Repts. DSDP, 60*, D. M. Hussong, S. Uyeda, et al., pp. 909-929, U.S. Govt. Printing Office, Washington, D.C., 1981.

Hussong, D. M., S. Uyeda, et al., *Init. Repts. DSDP, 60*, U.S. Govt. Printing Office, Washington, D.C., 1981.

Ingle, J. C., Jr., Summary of late Paleogene-Neogene insular stratigraphy, paleobathymetry, and correlations, Philippine Sea and Sea of Japan region., in *Init. Repts. DSDP, 31*, D. E. Karig, J. C. Ingle, Jr., et al., pp. 857-879, U.S. Govt. Printing Office, Washington, D.C., 1975.

Ishii, T., Dredged samples from the Ogasawara fore-arc seamount or "Ogasawara Paleoland" - fore-arc ophiolite. in *Formation of Active Ocean Margins*, edited by N. Nasu, Tokyo (Terrapub), 307-342, 1985.

Johnson, L. E., Mafic clasts in serpentinite seamounts: petrology and geocemistry of a diverse crustal suite from the outer Mariana forearc, in *Proc. ODP, Sci. Results, 125*, P. Fryer, J. A. Pearce, L. B. Stokking et al., pp. 401-414, Ocean Drilling Program, College Station, TX, 1992.

Johnson, L. E., and P. Fryer, The first evidence of MORB-like lavas from the outer Mariana forearc: geochemistry, petrography and tectonic implications, *Earth Planet. Sci. Lett., 100*, 304-316, 1990.

Johnson, L. E., P. Fryer, B. Taylor, M. Silk, D. L. Jones, W. V. Sliter, T. Itaya, and T. Ishii, New evidence for crustal accretion in the outer Mariana forearc: Cretaceous radiolarian cherts and MORB-like lavas, *Geology, 19,* 811-814, 1991.

Juteau, T., M. Ernewein, I. Reuber, H. Whitechurch, and R. Dahl, Duality of magmatism in the plutonic sequence of the Sumail nappe, Oman, *Tectonophysics, 151,* 107-135, 1988.

Kaiho, K., Eocene to Quaternary benthic foraminifers and paleobathymetry of the Izu-Bonin Arc, Legs 125 and 126, in *Proc. ODP, Sci. Results, 126,* B. Taylor, K. Fujioka, et al., pp. 285-310, Ocean Drilling Program, College Station, TX, 1992.

Karig, D. E., Origin and development of marginal basins in the western Pacific, *J. Geophys. Res., 76,* 2542-2561, 1971.

Karig, D. E., and G. F. Moore, Tectonic complexities in the Bonin Arc system, *Tectonophysics, 27,* 97-118, 1975.

Karig, D. E., and B. Ranken, Marine geology of the forearc region, southern Mariana island arc, in *The Tectonic and Geologic Evolution of Southeast Asian Seas and Islands (Pt. II), Geophys. Monog. Ser.,* vol. 27, edited by D. E. Hayes, pp. 266-280, AGU, Washington, D.C., 1983.

Karig, D. E., J. C. Ingle, Jr., et al., in *Init. Repts. DSDP, 31,* U.S. Govt. Printing Office, Washington, D.C., 1975.

Keating, B. H., and C. E. Helsley, Implications of island arc rotations to the studies of marginal terranes, *J. Geodynamics, 2,* 159-181, 1985.

Klein, G. deVries, K. Kobayashi, et al., *Init. Repts. DSDP, 58,* U.S. Govt. Printing Office, Washington, D.C., 1980.

Kling, S. A., Radiolarians from the Mariana Trough and Trench region: Deep Sea Drilling Project Leg 60, *Init. Repts. DSDP, 60,* D. M. Hussong and S. Uyeda, pp. 537-557, U.S. Govt. Printing Office, Washington, D.C., 1981.

Koyama, M., S. M. Cisowski, and P. Pezard, Paleomagnetic evidence for northward drift and clockwise rotation of the Izu-Bonin forearc since the early Oligocene, in *Proc. ODP, Sci. Results, 126,* B. Taylor, K. Fujioka, et al., pp. 371-382, Ocean Drilling Program, College Station, TX, 1992.

Kroenke, L., R. Scott, et al., *Init. Repts. DSDP, 59,* U.S. Govt. Printing Office, Washington, D.C., 1980.

Lagabrielle, Y., J.-P. Sizun, and R. J. Arculus, The constructional and deformational history of the igneous basement penetrated at site 786, in *Proc. ODP, Sci. Results, 125,* P. Fryer, J. A. Pearce, L. B. Stokking, et al., pp. 263-276, Ocean Drilling Program, College Station, TX, 1992.

LaTraille, S. L., and D. M. Hussong, Crustal structure across the Mariana Island arc, in *The Tectonic and Geologic Evolution of Southeast Asian Seas and Islands, Geophys. Monogr. Ser.,* vol. 23, edited by D. E. Hayes, pp. 209-222, AGU, Washington, D.C., 1980.

Lippard, S. J., A. W. Shelton, and I. G. Gass, The ophiolite of northern Oman, *Geol. Soc., London, Mem., 11,* 178 pp., 1986.

Lonsdale, P., A multibeam reconnaissance of the Tonga Trench axis and its intersection with the Louisville Ridge Guyot chain, *Mar. Geophys. Res., 8,* 295-327, 1986.

McDougall, I., Dating of rhyolitic glass in the Tonga forearc (Hole 841B), in *Proc. ODP, Sci. Results, 135,* J. W. Hawkins, L. M. Parson, J. F. Allan, et al., pp. 923-924, Ocean Drilling Program, College Station, TX, 1994.

MacLeod, C. J., Structure of the outer Tonga forearc at site 841, in *Proc. ODP, Sci. Results, 135,* J. W. Hawkins, L. M. Parson, J. F. Allan, et al., pp. 313-329, Ocean Drilling Program, College Station, TX, 1994.

MacLeod, C. J., and A. M. Lothian, Comparison of GLORIA sidescan sonar and core-derived structural data from site 841, Tonga Trench, in *Proc. ODP, Sci. Results, 135,* J. W. Hawkins, L. M. Parson, J. F. Allan, et al., pp. 373-382, Ocean Drilling Program, College Station, TX, 1994.

MacLeod, C. J., and C. E. Pratt, In situ stresses in the Lau Basin and Tonga forearc, Southwest Pacific, in *Proc. ODP, Sci. Results, 135,* J. W. Hawkins, L. M. Parson, J. F. Allan, et al., pp. 287-299, Ocean Drilling Program, College Station, TX, 1994.

Maekawa, H., M. Shozui, T. Ishii, K. L. Saboda, and Y. Ogawa, Metamorphic rocks from the serpentinite seamounts in the Mariana and Izu-Ogasawars forearcs, in *Proc. ODP, Sci. Results, 125,* P. Fryer, J. A. Pearce, L. B. Stokking, et al., pp. 415-430, Ocean Drilling Program, College Station, TX, 1992.

Malpas, J. and Robinson, P.T., Chromite Mineralisation in the Troodos Ophiolite, Cyprus, in *Evolution of Chromium Ore Fields,* edited by C. W. Stowe, pp. 220-237, Van Nostrand Reinhold, New York, 1987.

Malpas, J., P. T. Robinson, and M. Salisbury, Geology and geophysics of borehole CY-4 of the Cyprus Crustal Study Project: summary, in *Cyprus Crustal Study Project: Initial Report, Hole CY-4,* edited by I. L. Gibson, J. Malpas, P. T. Robinson, and C. Xenophontos, pp. 381-393, *Geol. Surv. Can. Pap. 88-9,* 1989.

Martini, E., M. E. Heiman, and F. Theyer, Synthesis of Oligocene to Quaternary biostratigraphy of the Philippine Sea, Deep Sea Drilling Project Leg 59, in *Init. Repts. DSDP, 59,* L. Kroenke, R. Scott, et al., pp. 587-593, U.S. Govt. Printing Office, Washington, D.C., 1980.

Meijer, A., Primitive island arc volcanism and a boninite series: examples from western Pacific island arcs, in *The Tectonic and Geologic Evolution of Southeast Asian Seas and Islands, Geophys. Monogr. Ser.,* vol. 23, edited by D. E. Hayes, pp. 271-282, AGU, Washington, D.C., 1980.

Meijer, A., The origin of low-K rhyolites from the Mariana frontal arc, *Contrib. Mineral. Petrol., 83,* 45-51, 1983.

Meijer, A., M. Reagan, H. Ellis, M. Shafiqullah, J. Sutter, P. Damon, and J. Kling, Chronology of volcanic events in the eastern Philippine Sea, in *The Tectonic and Geologic Evolution of Southeast Asian Seas and Islands (Pt. II). Geophys. Monogr. Ser.,* vol. 27, edited by D. E. Hayes, pp. 349-359, AGU, Washington, D.C., 1983.

Mrozowski, C. L., and D. E. Hayes, A seismic reflection study of faulting in the Mariana fore arc, in *The Tectonic and Geologic Evolution of Southeast Asian Seas and Islands, Geophys. Monogr. Ser.,* vol. 23, edited by D. E. Hayes, pp. 223-234, AGU, Washington, D.C., 1980.

Mrozowski, C. L., D. E. Hayes, and B. Taylor, Multichannel seismic reflection surveys of Leg 60 sites, Deep Sea Drilling Project, in *Init. Repts. DSDP, 60,* D. M. Hussong, S. Uyeda, et al., pp. 57-69, U.S. Govt. Printing Office, Washington, D.C., 1981.

Murauchi, S., N. Den, S. Asano, et al., Crystal structure of the Philippine Sea, *J. Geophys. Res., 73,* 3143-3171, 1968.

Murton, B. J., D. W. Peete, R. J. Arculus, J. A. Pearce, and S. van der Laan, Trace-element geochemistry of volcanic rocks from site 786: the Izu-Bonin forearc, in *Proc. ODP, Sci. Results, 125,* P. Fryer, J. A. Pearce, L. B. Stokking, et al., pp. 211-236, Ocean Drilling Program, College Station, TX, 1992.

Natland, J. H., Petrography and mineral compositins of gabbros recovered in DSDP Hole 453 on the western side of the Mariana Trough, in *Init. Repts. DSDP, 60,* D. M. Hussong, S. Uyeda, et al., pp. 579-599, U.S. Govt. Printing Office, Washington, D.C., 1981.

Natland, J. H., and J. J. Mahoney, Alteration of igneous rocks at Deep Sea Drilling Project sites 458 and 459, Mariana fore-arc region: relationship to basement structure, in *Init. Repts. DSDP, 60,* D. M. Hussong, S. Uyeda, et al., pp. 769-788, U.S. Govt. Printing Office, Washington, D.C., 1981.

Natland, J. H., and J. Tarney, Petrologic evolution of the Mariana arc and back-arc system-a synthesis of drilling results in the southern Philippine Sea, in *Init. Repts. DSDP, 60,* D. M. Hussong, S. Uyeda, et al., pp. 681-708, U.S. Govt. Printing Office, Washington, D.C., 1981.

Nicolas, A., F. Boudier, and G. Ceuleneer, Mantle flow patterns and magma chambers at ocean ridges: evidence from Oman ophiolite, *Mar. Geophys. Res., 9,* 293-310, 1988.

Nilsson, K., Plutonic and volcanic inclusions in rocks from the northern Mariana island arc: crustal assimilation in an intraoceanic arc, unpubl. M.A. thesis, Duke University, Durham, NC, 110 pp., 1987

Nishimura, A., Carbonate bioclasts of shallow-water origin at site 793, in *Proc. ODP, Sci. Results, 126,* B. Taylor, K. Fujioka, et al., pp. 231-236, Ocean Drilling Program, College Station, TX, 1992.

Ozima, M., I. Kaneoka, and H. Ujie, ^{40}Ar-^{39}Ar age of rocks, and development mode of the Philippine Sea, *Nature, 267,* 816-818, 1977.

Ozima, M., I. Kaneoka, K. Saito, M. Honda, M. Yanagisawa, and Y. Takigami, Summary of geochronological studies of submarine rocks from the western Pacific Ocean, in *Geodynamics of the Western Pacific-Indonesian Region, Geodynam. Ser.,* vol. 11, edited by T. W. C. Hilde, and S. Uyeda, pp. 137-142, AGU, Washington, D.C., 1983.

Packham, G. H., and D. A. Falvey, An hypothesis for the formation of marginal seas in the western Pacific, *Tectonophysics, 11,* 79-109, 1971.

Parson, L. M., J. A. Pearce, B. J. Murton, R. A. Hodkinson, S. Bloomer, M. Ernewein, Q. J. Huggett, S. Miller, L. Johnson, P. Rodda, and S. Helu, Role of ridge jumps and ridge propagation in the tectonic evolution of the Lau back-arc basin, southwest Pacific, *Geology, 18,* 470-473, 1990.

Parson, L., J. W. Hawkins, J F. Allan, et al., in Proc. *ODP, Init. Repts., 135,* Ocean Drilling Program, College Station, TX, 1992.

Pearce, J. A., S. J. Lippard, and S. Roberts, Characteristics and tectonic significance of supra-subduction zone ophiolites, *Geol. Soc., London, Spec. Publ., 16,* 77-96, 1984.

Pearce, J. A., M. F. Thirlwall, G. Ingram, B. J. Murton, R. J. Arculus, and S. van der Laan, Isotopic evidence for the origin of boninites and related rocks drilled in the Izu-Bonin (Ogasawara) forearc, Leg 125, in *Proc. ODP, Sci. Results, 125,* P. Fryer, J. A. Pearce, L. B. Stokking, et al., pp. 237-263, Ocean Drilling Program, College Station, TX, 1992a.

Pearce, J. A., S. R. van der Laan, R. J. Arculus, B. J. Murton, T. Ishii, D. W. Peate, and I. J. Parkinson, Boninite and harzburgite from Leg 125 (Bonin-Mariana forearc): a case study of magma genesis during the initial stages of subduction, in Proc. *ODP, Sci. Results, 125,* P. Fryer, J. A. Pearce, L. B. Stokking, et al., pp. 623-662, Ocean Drilling Program, College Station, TX, 1992b.

Pezard, P. A., R. N. Hiscott, M. A. Lovell, A. Colella, and A. Malinverno, Evolution of the Izu-Bonin intraoceanic forearc basin, western Pacific, from cores and FMS images, in *Geological Applications of Wireline Logs II,* edited by A. Hurst, C. M. Griffiths, and P. F. Worthington, pp. 43-69, *Geol. Soc., London, Spec. Publ., 65,* 1992.

Rautenschlein, M., G. A. Jenner, J. Hertogen, A. Hoffman, R. Kerrich, H. -U. Schmincke, and W. M. White, Isotopic and trace element composition of volcanic glasses from the Akaki Canyon, Cyprus: implications for the origin of the Troodos ophiolite, *Earth Planet. Sci. Lett., 75,* 369-383, 1985.

Reagan, M. K., and A. Meijer, Geology and geochemistry of early arc-volcanic rocks from Guam, *Geol. Soc. Am. Bull., 95,* 701-713, 1984.

Robertson, A. H. F., Tectonic evolution of Cyprus, in *Ophiolites: Oceanic Crustal Analogues,* edited by J. Malpas, E. M. Moores, A. Panayiotou, and C. Xenophontos, pp. 235-249, Proc. Symp. Ophiolites and Oceanic Lithosphere, Troodos 87, Ministry Agric. Nat. Resources, Nicosia, Cyprus, 1990.

Schmidt, R. G., Petrology of the volcanic rocks, in *Geology of Saipan, Mariana Islands, Pt. 2, Chapt. B,* edited by P. E. Cloud, et al., U.S. Geol. Surv. Prof. Pap. 280, 127-172, 1957.

Scholl, D. W., T. L. Vallier, and G. H. Packham, Framework geology and resources potential of southern Tonga platform and adjacent terranes-a synthesis, in *Geology and Offshore Resources of Pacific Island Arcs-Tonga Region,* Earth Science Series, vol. 2, edited by D. W. Scholl and T. L. Vallier, pp. 457-488, Circum-Pacific Council for Energy and Mineral Resources, Houston, TX, 1985.

Seno, T., and S. Maruyama, Paleogeographic reconstruction and origin of the Philippine Sea, *Tectonophysics, 102,* 53-84, 1984.

Serri, G., The petrochemistry of ophiolitic gabbroic complexes: a key for the classification of ophiolites into low-Ti and high-Ti types, *Earth Planet. Sci. Lett., 52,* 203-212, 1981.

Sharaskin, A. Ya, I. K. Pustchin, S. K. Zlobin, and G. M. Kolesov, Two ophiolitic sequences from the basement of the northern Tonga Arc, *Ofioliti, 8,* 411-430, 1983.

Shervais, J. W., Island arc and ocean crust ophiolites: contrasts in the petrology, geochemistry and tectonic style of ophiolite assemblages in the California Coast Ranges, in *Ophiolites: Oceanic Crustal Analogues,* edited by J. Malpas, E. M. Moores, A. Panayiotou, and C. Xenophontos, pp. 507-520, Proc. Symp. Ophiolites and Oceanic Lithosphere, Troodos 87, Ministry Agric. Nat. Resources, Nicosia, Cyprus, 1990.

Shiraki, K., N. Kuroda, H. Urano, and S. Maruyama, Clinoenstatite in boninites from the Bonin Islands, Japan, *Nature, 285,* 31-32, 1980.

Shiki, T. (Ed.), *Geology of the Northern Philippine Sea,* Tokai University Press, Tokyo, 1985.

Stern, R. J., and S. H. Bloomer, Subduction zone infancy: examples from the Eocene Izu-Bonin-Mariana and Jurassic California Arcs, *Geol. Soc. Am. Bull., 104,* 1621-1636, 1992.

Stern, R. J., J. Morris, S. H. Bloomer, and J. W. Hawkins, The source of the subduction component in convergent margin magmas: trace element and radiogenic isotope evidence from Eocene boninites, Mariana forearc, *Geochim. Cosmochim. Acta, 55,* 1467-1481, 1991.

Taylor, B., Rifting and the volcanic-tectonic evolution of the Izu-Bonin-Mariana arc, in *Proc. ODP, Sci. Results, 126,* B. Taylor, K. Fujioka, et al., pp. 627-653, Ocean Drilling Program, College Station, TX, 1992.

Taylor, B., K. Fujioka, et al., *Proc. ODP Init. Repts., 126,* Ocean Drilling Program, College Station, TX, 1990.

Taylor, R. N., and J. G. Mitchell, K-Ar results from whole-rock and mineral separates of the Izu-Bonin forearc basement, Leg 126, in Proc. *ODP, Sci. Results, 126,* B. Taylor, K. Fujioka, et al., pp. 677-682, Ocean Drilling Program, College Station, TX, 1992.

Taylor, R. N., B. J. Murton, and R. W. Nesbitt, Chemical transects across intra-oceanic arcs: implications for the tectonic setting of ophiolites, *Geol. Soc., London, Spec. Publ. 60,* 117-132, 1992b.

Taylor, R. N., H. Lapierre, P. Vidal, R. W. Nesbitt, and I. W. Croudace, Igneous geochemistry and petrogenesis of the Izu-Bonin forearc basin, in *Proc. ODP, Sci. Results, 126,* B. Taylor, K. Fujioka, et al., pp. 405-430, Ocean Drilling Program, College Station, TX, 1992a.

Thy, P., Magmas and magma chamber evolution, Troodos ophiolite, Cyprus, *Geology, 16,* 448-451, 1987.

Thy, P., P. Schiffman, and E. M. Moores, Igneous mineral stratigraphy and chemistry of the Cyprus Crustal Study Project drill core in the plutonic sequence of the Troodos ophiolite, in *Cyprus Crustal Study Project: Initial Report, Hole CY-4,* edited by I. L., Gibson, J. Malpas, P. T. Robinson, and C. Xenophontos, *Geol. Surv. Can. Pap., 88-9,* 147-185, 1989.

Tracey, J. L., S. O. Schlanger, J. T. Stark, and D. B. Doan, General Geology of Guam, *U.S. Geol. Surv. Prof. Pap., 403-A,* 1964.

Tsunakawa, H., K-Ar dating on volcanic rocks in the Bonin Islands and its tectonic implications, *Tectonophysics, 95,* 221-232, 1983.

Umino, S., Volcanic geology of Chichijima, the Bonin Islands (Ogasawara Islands), *J. Geol. Soc. Japan, 91,* 505-523, 1985.

Uyeda, S., and Z. Ben Avraham, Origin and development of the Philippine Sea, *Nature, 240,* 176-178, 1972.

Uyeda, S., and H. Kanamori, Back-arc opening and the mode of subduction, *J. Geophys. Res., 84,* 1049-1061, 1979.

Vallier, T. L., R. M. O'Conner, D. W. Scholl, A. J. Stevenson, and P. J. Quintero, Petrology of rocks dredged from the landward slope of the Tonga Trench: implications for middle Miocene volcanism and subsidence of the Tonga Ridge, in *Geology and Offshore Resources of Pacific Island Arcs-Tonga Region,* Earth Science Series, vol. 2, edited by D. W. Scholl and T. L. Vallier, pp. 109-120, Circum-Pacific Council for Energy and Mineral Resources, Houston, TX, 1985.

Varga, R. J., Modes of extension at oceanic spreading centers: evidence from the Solea graben, Troodos ophiolite, Cyprus, *J. Struct. Geol., 13,* 517-537, 1991.

Varga, R. J., and E. M. Moores, Spreading structure of the Troodos ophiolite, Cyprus, *Geology, 13,* 846-850, 1985.

von Huene, R., and D. W. Scholl, Observations at convergent margins concerning sediment subduction, subduction erosion, and the growth of continental crust, *Rev. Geophys., 29,* 279-316, 1991.

Walker, D. A., and I. McDougall, $^{40}Ar/^{39}Ar$ and K-Ar dating of altered glassy volcanic rocks: the Dabi volcanics, P.N.G., *Geochim. Cosmochim. Acta, 46,* 2181-2190, 1982.

Wessel, J. K., P. Fryer, P. Wessel, and B. Taylor, Extension in the northern Mariana forearc, *J. Geophys. Res., 99,* 15,181-15,203, 1994.

Wood, D. A., D. P. Mattey, J. L. Joron, N. G. Marsh, J. Tarney, and M. Treuil, A geochemical study of 17 selected samples from basement cores recovered at sites 447, 448, 449, 450, and 451, Deep Sea Drilling Project Leg 59, *Init. Repts. DSDP, 59,* L. Kroenke, R. Scott, et al., pp. 743-752, U.S. Govt. Printing Office, Washington, D.C., 1980.

S. H. Bloomer, Department of Earth Sciences, Boston University, 675 Commonwealth Ave., Boston, MA 02215. Now at Department of Geosciences, Oregon State University, Corvallis, OR 97331

J. W. Hawkins, Geological Research Division, Scripps Institution of Oceanography, University of California, San Diego, La Jolla, CA 92092

L. Johnson, Marine Physics Branch, Naval Research Laboratory Code 7420, Washington, DC 20375-5000

C. J. MacLeod, Institute of Oceanographic Sciences, Brook Road, Wormley, Godalming, Surrey GU8 5UB U.K. Now at Department of Geology, University of Wales, Cardiff, P.O. Box 914, Cathays Park, Cardiff CF1 3YE, UK

R. J. Stern, Center for Lithospheric Studies, University of Texas at Dallas, Richardson, TX, 75083-0688

B. Taylor and P. Fryer, School of Ocean and Earth Science and Technology, University of Hawaii, 2525 Correa Road, Honolulu, HI 96822

Intrusive Volcanic Rocks in Western Pacific Forearcs

Rex N. Taylor [1], Michael S. Marlow [2], Lynn E. Johnson,
Brian Taylor[4], Sherman H. Bloomer [5], and John G. Mitchell[6]

Neogene intrusives were drilled in three Ocean Drilling Program (ODP) holes located in separate segments of the western Pacific intra-oceanic arcs. This paper summarizes the geochemical and physical data from each of the intrusions. New data including K-Ar ages and trace element analyses from the Izu-Bonin sill are presented. Major and trace element data from each intrusion are compared with analyses of corresponding regional volcanics to establish the most likely magmatic source. The intrusives petrographically and geochemically resemble the contemporaneous arc lavas which erupted along the frontal arc 50-100 km to the west. Furthermore, the sills most closely resemble the subjacent volcanoes along the Quaternary arc. They do not resemble backarc basalts nor the spectrum of lava types that erupted in the forearc. The sills were most likely to have been fed from dikes which propagated from the arc through basement rocks at deeper crustal levels.

1. INTRODUCTION

In the period 1989-1991, the Ocean Drilling Program (ODP) investigated the arc, trench and marginal basin systems of the Western Pacific. The objectives of ODP in this region included an examination of the nature and petrogenesis of magmatism in the arc-trench terrain (the forearc). An important discovery during the drilling was that basaltic sills of Neogene age had intruded the forearc sediments. These intrusives were found on ODP Legs, 125, 126 and 135, which investigated the Mariana, Izu-Bonin and Tonga systems respectively. Although not previously reported, these intrusives appear to be a common feature of the western Pacific intra-oceanic forearcs.

[1] Department of Geology, The University of Southampton, Southampton, United Kingdom; also at: Royal Holloway, University of London, Egham, Surrey, TW20 0EX, UK
[2] U.S. Geological Survey, Menlo Park, CA
[3] Naval Research Laboratory, Washington, DC
[4] School of Ocean and Earth Science and Technology, University of Hawaii, Honolulu, HI
[5] Department of Geology, Boston University, Boston, MA
[6]Department of Physics, The University, Newcastle upon Tyne, NE1 7RU, UK

Active Margins and Marginal Basins of the Western Pacific
Geophysical Monograph 88

The discovery of these forearc igneous bodies raises some important questions, namely: do they represent magmatism derived from melting of the forearc mantle wedge? If so, what are the implications for our perception of the thermal conditions above subducting plates? If the intrusives are not derived from the forearc mantle, where did the melts originate and how did they travel to their forearc location? In an attempt to answer these questions, this study brings together and analyses the relevant data from the ODP and other western Pacific investigations.

2. FOREARC MAGMATISM—THE EVIDENCE

Arc magmatism closest to the trench is commonly manifested as a line of volcanic constructs, some 150-250 km from the trench. The forearc has normally been considered a volcanically inactive terrain, dominated by sedimentation derived from the arc (Hussong and Uyeda, 1981). However, studies of ophiolite complexes, ODP exploration in the Western Pacific and experimental work all suggest that forearcs have been the focus of magmatism during their development. ODP Leg 126 demonstrated that the Izu-Bonin forearc basin has Oligocene volcanics at the base of its sedimentary sequence (Taylor, Fujioka, et al., 1990). These extrusives were coeval with, or erupted slightly later than, the Oligocene volcanic front, which lay to the west of the forearc basin. A similar volcano-tectonic arrangement is envisaged for the Eocene volcanics found along the present Izu-Bonin-Mariana forearc (e.g., Chichijima, site 786, site 459). These volcanics were coeval with the Palau-Kyushu Ridge system, which is a sub-parallel line of bathymetric highs, now displaced to the west by two successive backarc rifting events. These spatial observations regarding the Oligocene and Eocene forearc confirm the idea that volcanism has occurred and can occur on the trench side of an active arc.

From the volcano-tectonic relationships now recognized in the western Pacific arc-trench systems, the early stages of supra-subduction zone magmatism have been interpreted as being marked by a period of widespread volcanism (Taylor et al., 1992b). This volcanism apparently spanned some 200 km, the distance across the present forearc massif. Furthermore, volcanism occurred much closer to the trench than it does at the currently active Izu-Bonin-Mariana arcs. It has been proposed that such a volcanic province could have developed as a result of intra-oceanic extension above a subsiding, transform-bounded section of ocean crust (Bloomer and Hawkins, 1983; Stern and Bloomer, 1992). In this model the entire crustal fabric of the supra-subduction zone region would be generated by a style of volcanic construction similar to a mid-ocean spreading centre (Natland and Tarney, 1981). Another model for widespread across-arc magmatism involves the initiation of subduction at an intra-oceanic transform close to an active spreading centre (e.g., Meijer, 1980; Pearce et al., 1992). This model invokes the subduction of young hot crust beneath similarly young crust. Both models to explain the nature and dispersion of early western Pacific arc magmatism require anomalous thermal conditions in the proto-mantle wedge. The rationale behind this requirement is that a steeper geothermal gradient in the initial mantle wedge would result in peridotite melting at shallow depths (<90 km). This in turn would lead to volcanism closer to the trench than the current arc. At the inception of subduction the higher temperatures in the mantle wedge could have been provided by proximity of the active spreading Philippine Sea plate. This plate was generated by spreading between 60 and 35Ma (Hilde and Lee, 1984). Therefore, the mantle upon which the Eocene arc was constructed would have had a geotherm closer to a mid-ocean ridge than an older, cooler oceanic plate.

Although evidence suggests that forearc volcanism was prevalent during the Eocene and Oligocene, drilling has not revealed any post-Oligocene *extrusive* magmatism. This could indicate that with time the mantle wedge cooled and reached a steady thermal state. At this point, probably in the late Oligocene, a true volcanic front developed as the focus of arc magmatism and the thermal structure of the Neogene arc would have reached conditions similar to those of the present. Thus, the thermal conditions in the forearc from the Miocene onwards had probably become incapable of forming melt from the mantle wedge directly underlying the forearc. If these assumptions regarding the thermal conditions in the forearc mantle wedge are correct, then some mechanism to allow the intrusion of Neogene magmas into the forearc must be contemplated. As the closest source of basaltic melts at that time would appear to have been the active arc, magma would have had to travel over 100 km to reach the outermost locations in the forearc.

3. FOREARC INTRUSIONS DRILLED BY THE ODP

3.1. *Size, Position, and Age*

Details concerning the location, size, distribution and age of the forearc intrusions recovered by the ODP are given in Figure 1 and Table 1. Holes 781A (Mariana) and 793B (Izu-Bonin) each drilled through single sills, which are 23 m and 6 m thick respectively. Hole 841B (Tonga) drilled though a series of nine sills, ranging in thickness from 7 cm to 18 m over a vertical distance of 170 m. These sills are all over 75 km from the active volcanic front, and their ages can be constrained as late Miocene or younger (Table 1).

K-Ar analysis of the Mariana sill in Hole 781A gave an age of 1.7 Ma (Marlow et al., 1992). New K-Ar data for the sill cored in Hole 793B are presented in Table 2. Two whole rock samples were chosen, one from the crystal-rich base (sample 011:- Leg 126,1R-3,77-82cm) and another from the lower center of the sill (sample 008:- Leg 126, 1R-2, 91-95cm), because they are from different sections of the sill and have differing K_2O contents. Conventional K-Ar ages for both samples are older than the sediments that they intruded (cf. Tables 1 and 2). In light of these anomalous ages, it is likely that the samples have trapped "excess" ^{40}Ar, either as a result of marginal chilling or because the ^{40}Ar was retained as inclusions in phenocrysts within the sill. The problem of excess Ar can be avoided by using the isochron method of dating developed by Roddick (1978). An isochron plot for the sill data is shown in Figure 2. The 011 sample analyses cluster as a tight group, while the 008 analyses lie along a mixing line between the present-day "air" $^{40}Ar/^{36}Ar$ value of 295.5 and an inherent value for the uncontaminated sample. The line drawn is a best fit through the 008 analyses and the "air" value. This spread of 008 data is likely to be a consequence of varying amounts of modern atmospheric argon which have contributed to the analysis. The most feasible isochron that is consistent with the data set can be drawn through the least atmospheric-argon

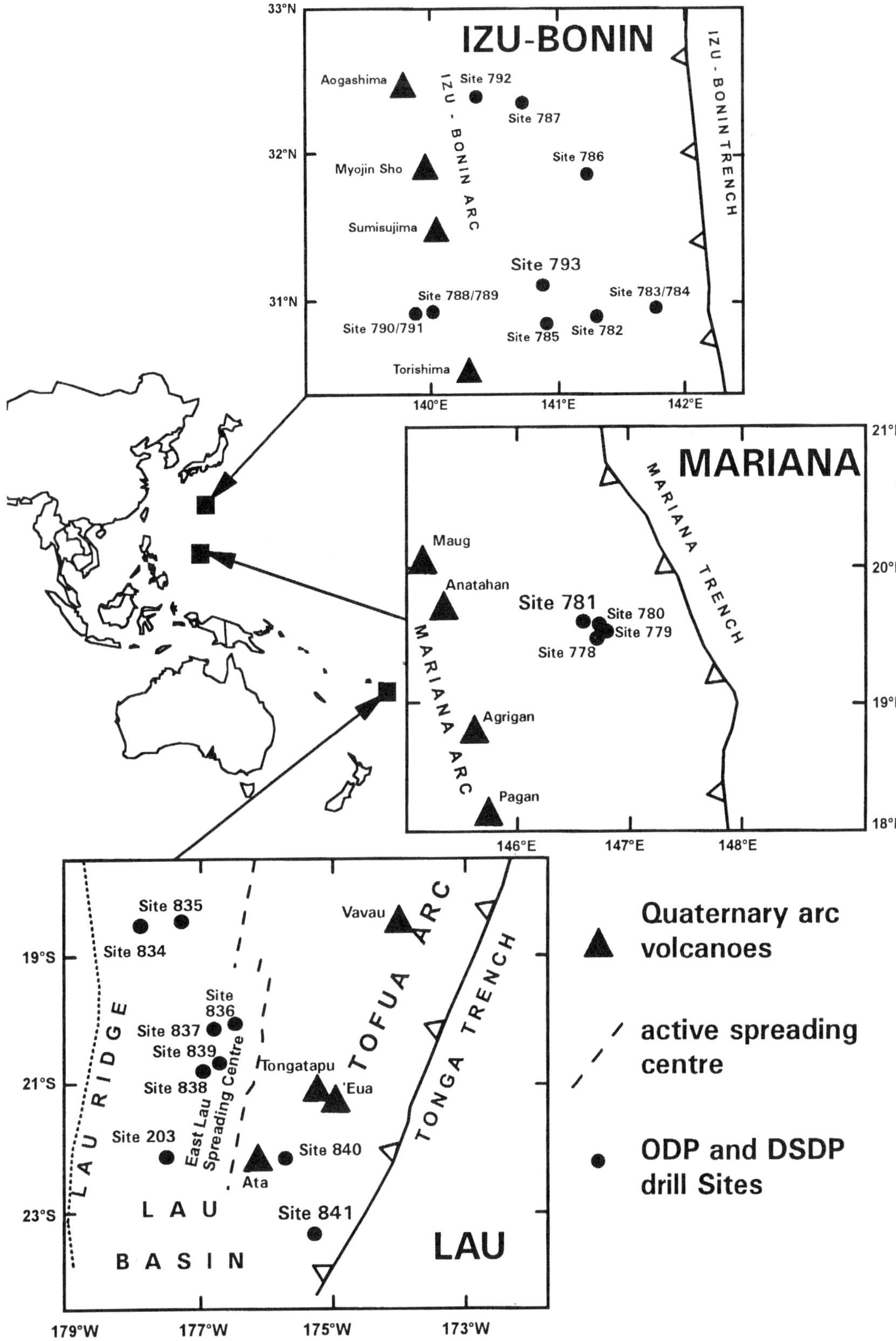

Fig. 1. Location map of the western Pacific showing the location of ODP drill sites in the Izu-Bonin, Mariana, and Lau regions.

TABLE 1. Details of Forearc Intrusions on Legs 125, 126 and 135.

Region	Mariana	Izu-Bonin	Tonga
ODP Leg	125	126	135
Hole	781A	793B	841B
Latitude	19°37.91' N	31°06.33' N	23°20.74' S
Longitude	146°32.56' E	140°53.27' E	175°17.87' W
Core intervals	8R-CC to 10R-3	1R-1 to 1R-3	18R-1 to 36R-1
Depth range (mbsf)	67 - ª90m	586.5 - 592m	325 - 498m
Number of units	1?	1	9
Unit thickness	ª20m	ª5m	7cm - 18m
Sediment age	late Pliocene (2.5Ma)	mid Miocene (13-15Ma)	upper Miocene (6-12Ma)
K-Ar age	1.68 ± 0.37 Ma	>6.3 Ma	-
Distance from arc at intrusion	105km	75km	90 km

contaminated analysis of 008 (highest $^{40}Ar/^{36}Ar$ value) and the average 011 value. This reference isochron has an $^{40}Ar/^{36}Ar$ intercept of 304.5 and has a slope equivalent to 6.3 Ma. A reference isochron with a slope equivalent to 15 Ma is also shown. Although the data are ambiguous, one possible interpretation of the isochron is that this sill has a minimum age of 6.3 Ma. No radiometric age determinations of the sills at site 841 have been undertaken.

The thickness of individual intrusions is variable, from <10 cm in the case of one unit in Hole 841B, to an estimated 13-25 m for the Mariana sill (Hole 781A). Some idea of the lateral extent of these intrusions is indicated by multichannel seismic reflection profiles taken across the forearc. A feature interpreted as an intrusion can be seen in the center of the profile shown in Figure 3, although this feature was never drilled. This strong reflector extends for over 20 km across the Izu-Bonin forearc basin. The ≈5 m thick sill drilled at site 793 (some 49km north of the Figure 3 profile) was not recognized on a multichannel seismic reflection profile evidently because it is too thin. Single channel seismic reflection data across site 781 show a strong reflector at a time and depth interval coincident with the drilled intrusion (which is around 20 m thick). This reflector can be recognized for 5-10 km along the center of the Mariana forearc high.

3.2. PETROLOGY AND GEOCHEMISTRY

Specimens from each of the three ODP sites have similar mineralogical characteristics. All are plagioclase and clinopyroxene ± olivine ± orthopyroxene phyric basalts and basaltic andesites. Each intrusion has been affected by low temperature alteration to some extent. Groundmass glass has been partially altered to clay minerals and, in the case of site 793, olivine and orthopyroxene have been pseudomorphed by saponite. Plagioclase phenocryst compositions from the 781 and 793 intrusions range between An_{41}-An_{97} and An_{60}-An_{92} respectively (Marlow et al., 1992; Lapierre et al., 1992). Clinopyroxene phenocrysts from these two sites have been confirmed as augite, all with < Fs_{37}. Fresh olivine analyzed from site 781 is $Fo_{69\text{-}71}$.

The whole-rock geochemical data from the intrusions are summarized in Table 3. All samples are low-K basalts and basaltic andesites. The sill from Hole 793B is more primitive (MgO 8.5-9.5 wt%) than the Mariana and Tonga

TABLE 2. K-Ar Dating Results of Whole Rocks from 793B Intrusion, Izu-Bonin Forearc Basin.

Sample number	K_2O wt%	Radiogenic ^{40}Ar (mm^3g^{-1})	% atmospheric contamination	Age Ma±1	$^{40}Ar_{TOTAL}$ ($10^{-3}mm^3g^{-1}$)	$^{36}Ar_{TOTAL}$ ($10^{-5}mm^3g^{-1}$)	$^{40}Ar_{TOTAL}/^{36}Ar_{TOTAL}$	$K_2O/^{36}Ar_{TOTAL}$
126 793B	0.25±0.01	(2.68±0.27)10^{-4}	96.3	32.9±3.6	7.24	2.36	306.8	1.06
1-R3 77-82	average of	(2.48±0.25)10^{-4}	96.5	30.5±3.3	7.09	2.31	306.5	1.08
(011)	four	(2.71±0.28)10^{-4}	93.3	33.3±3.7	7.32	2.39	306.8	1.05
126-793B	0.31±0.01	(1.75±0.20)10^{-4}	95.6	17.4±2.1	3.98	1.29	309.4	2.41
1R2 61-65	average of	(1.93±0.21)10^{-4}	96.4	19.2±2.2	5.36	1.75	306.6	1.77
(008)	four	(1.76±0.20)10^{-4}	95.9	17.5±2.1	4.29	1.39	308.1	2.23

Details of the analytical methods are given in Taylor and Mitchell, 1992.

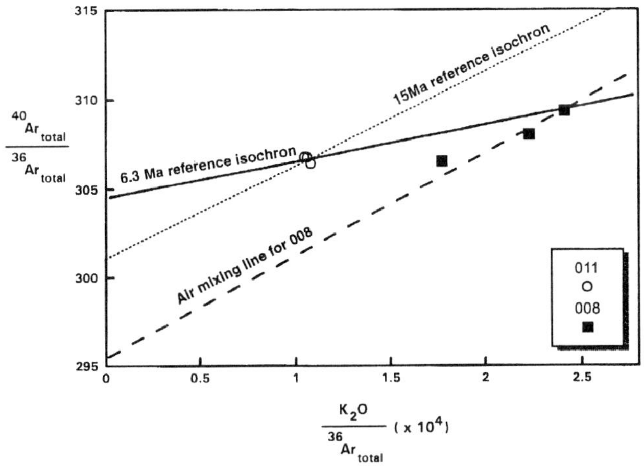

Fig. 2. $^{40}Ar_{total}/^{36}Ar_{total}$ versus $K_2O/^{36}Ar_{total}$ isochron plot for two samples from site 793 forearc intrusion. Reference isochrons passing through the average 011 sample with slopes equivalent to 6.3 Ma and 15 Ma are shown. A best-fit line through modern "air" value and the three 008 analyses is also shown. Further details are given in the text and in Table 2.

intrusions (MgO 4.4-5.0 wt%). Each intrusion has a significant enrichment of Ba, Rb, Sr, and K relative to the rare earth elements (REE) and high field strength elements (HFSE) as shown in the multi-element plot Figure 4. In addition Nb, and to a slight extent Zr, produce negative anomalies relative to the REE in Figure 4. These trace element characteristics are typical of mantle-derived melts in volcanic arcs (e.g. Pearce, 1982). Elements such as Ba and K are known to be mobile during low-temperature alteration to which each sill has been subjected (e.g., Taylor et al., 1992a). This mobility is suggested in the Ba and K data from the sills, as these elements have the widest variation in concentrations relative to the mean values (Table 3). Bloomer et al. (1994) have noted that some samples from the sill at site 841 have anomalously high Ba concentrations, which correspond to the occurrence of extensive secondary veining. If such samples are removed from the average calculations then concentrations are likely to be relatively close to magmatic values. In this sense the values shown in Figure 4 can be considered as a best estimate of the original liquid compositions.

4. GEOCHEMICAL ORIGINS OF THE INTRUSIONS

Three possible sources are available in the supra-subduction zone that could have supplied the forearc intrusions. Firstly, magmatism related to melting of the forearc mantle. The examples of such magmatism include the Eocene volcanics of the Izu-Bonin and Mariana forearc highs (e.g., Chichijima, site 786 and Guam, and the syn-rift volcanism of the Oligocene forearc basin (sites 793 and 458). Secondly, partial melts from a similar mantle source to the frontal arc and cross-backarc volcanic chains (e.g., the Izu arc and Zenisu ridge respectively). These arc melts may be dominantly focused at the volcanic front, but the melt supply can also encroach into the forearc and backarc regions (e.g., Stern et al., 1993). Finally, the intrusives may have been supplied as a result of extension and melting during the formation of backarc basins or backarc rifts. As a result of the ODP investigations along the Pacific margin, data from volcanics representative of all three potential magma sources are available for comparison.

4.1. Major Element Constraints

Major element characteristics are likely to be distinct between the arc and forearc sources because of differing depths of melt generation and potentially differing source compositions. This stems from the plate geometry of subduction zones inasmuch as the mantle wedge becomes thinner closer to the trench. If the subduction angle and trench distance have remained constant since the forearc intrusions were emplaced (≤6.3 Ma), then the vertical depth to the slab surface from the sills would have been 25-70 km. This contrasts with the typical mantle wedge thickness of >120 km beneath the present frontal arc.

If melts do form by shallow melting of supra-subduction zone peridotite, then they should have appropriate compositions. Melts formed in such an environment should have characteristics such as high volatile content and silica saturation at relatively high MgO contents. It is also postulated that shallow (<50 km) intra-oceanic mantle should have depleted concentrations of incompatible elements compared to a MORB-source mantle. This depletion is based on the notion that the oceanic lithosphere (upon which the arc-trench massif is sited) developed its chemical characteristics as a result of progressive melting of upwelling mantle at an oceanic spreading centre. As a consequence of such a melting regime, the uppermost levels of the mantle should have experienced the greatest melt extraction, and be the most depleted in incompatible elements (e.g., Kostopoulos and James, 1992; Taylor et al., 1992b). A remelting of this depletion-stratified mantle in an arc-trench system would give rise to arc and forearc melts with ratios such as Al_2O_3/TiO_2 significantly greater than MORB.

These features (high volatile and MgO contents, and high Al_2O_3/TiO_2) are typical of the boninite series volcanics. These lavas have been demonstrated to be experimentally and spatially of a shallow, forearc origin

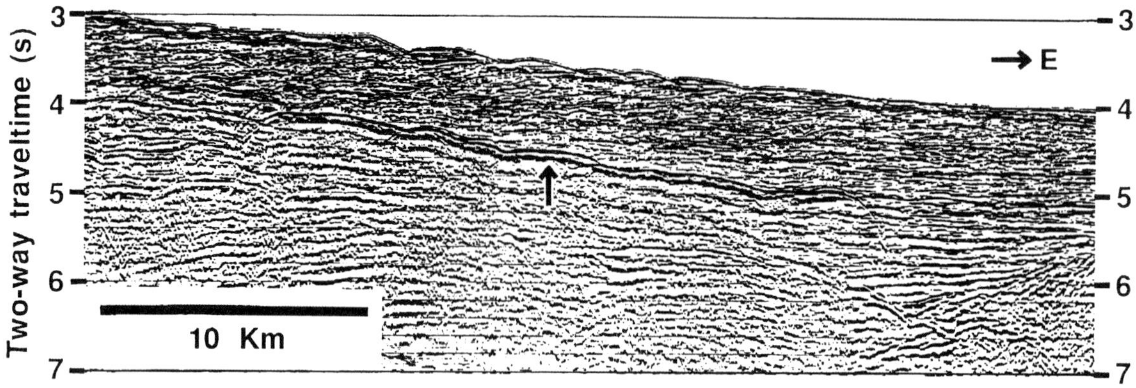

Fig. 3. Stacked multichannel seismic section across the center of the Izu-Bonin forearc basin from 30°44'N, 140°22'E to 30°39'N, 141°44'E (courtesy of Japan National Oil Co.) showing a strong stratiform reflector (above the vertical arrow) that is 21km long and is interpreted as an igneous intrusion. The complete seismic profile is published in Taylor (1992).

(van der Laan et al., 1989; Taylor et al., 1992b). Boninites and boninite-like rocks are well represented in the western Pacific forearcs (Meijer, 1980; Hickey and Frey, 1982; Umino, 1985). However, there are however a number of occurrences of more typical arc-tholeiite and calc-alkaline volcanic rocks associated with forearc boninitic magmatism (e.g., site 459, Hussong, Uyeda et al., 1981; the Mikazukiyama formation lavas on Chichijima, Taylor et al., 1994; site 786, Pearce et al., 1992, Murton et al., 1992). These associated tholeiitic and calc-alkaline lavas are also likely to have been derived from a shallow mantle source, because of the wedge-slab geometry. So how do the major element compositions of these lavas compare with those of the Neogene forearc intrusives?

Figure 5 compares $Na_{6.0}$ versus $Ca_{6.0}$ variation for various volcanics in the Izu-Bonin arc, the Izu-Bonin forearc, and the sill samples from Hole 793B. The sill from Hole 793B is used as an example as it is the most primitive (high MgO) and has the most geochemical data from the local region available. This plot utilizes the technique described by Plank and Langmuir (1988) of normalizing certain elements (in this case Na_2O and CaO) to a MgO content of 6 wt% ($Na_{6.0}$ and $Ca_{6.0}$). This procedure removes most effects of crystal fractionation within any suite, and should allow comparison of major chemical components with respect to mantle processes. Figure 5 shows that

1. The intrusives lie within the field of the Quaternary Izu-Bonin arc and close to the Oligocene-Recent arc.

2. The forearc tholeiites, boninites, and calc-alkaline lavas have higher $Na_{6.0}$ and lower $Ca_{6.0}$ than the arc lavas.

The above major-element data indicate that the forearc magmas (be they tholeiitic, calcalkaline, or boninitic) have a higher MgO content at similar CaO or Na_2O content than the main frontal Izu-Bonin arc. Relatively high MgO contents appear to be a feature of all the forearc magmas studied in this region (Taylor and Nesbitt, 1994). This is most likely to be the result of melting a shallow, depleted forearc mantle with a steeper Eocene-Oligocene thermal gradient. As the forearc sills are geochemically distinct (in terms of MgO-CaO-Na_2O systematics) from the forearc volcanic rocks, we suggest that the sills are not the product of melting in the forearc mantle wedge. By the same token, the geochemical similarity of the sills to the Recent frontal arc lavas demonstrates their close magmatic association.

4.2. Trace Element Constraints

A further test of the potential mantle sources for the sills is to compare element concentrations or ratios which reflect variations in source components. Useful parameters include Ba/Zr, Zr/Y, and La/Sm because they show large variations among the potential magma sources. A plot of Ba/Zr-Zr/Y distinguishes between backarc, arc, and forearc lavas. Backarc lavas have low Ba/Zr and have Zr/Y similar to mid-ocean ridge basalts (MORB); forearc volcanic rocks have high Zr/Y and moderate Zr/Y compared to the other lava types, whereas arc volcanic rocks have higher Ba/Zr and lower Zr/Y than the other two lava types.

4.2.1. *The Izu-Bonin system and Hole 793B.* Trace element ratios for samples in the Hole 793B forearc intrusion are compared with ratios from the arc and back-arc of the Izu-Bonin region in Figure 6. The backarc samples are taken from the Sumisu and adjacent rifts and have low Ba/Zr (<0.7) and Zr/Y in the range 2.5-5, which are typical mid-ocean ridge basalt (MORB) characteristics. Lavas from the Quaternary, Miocene, and Oligocene, Izu-Bonin arcs

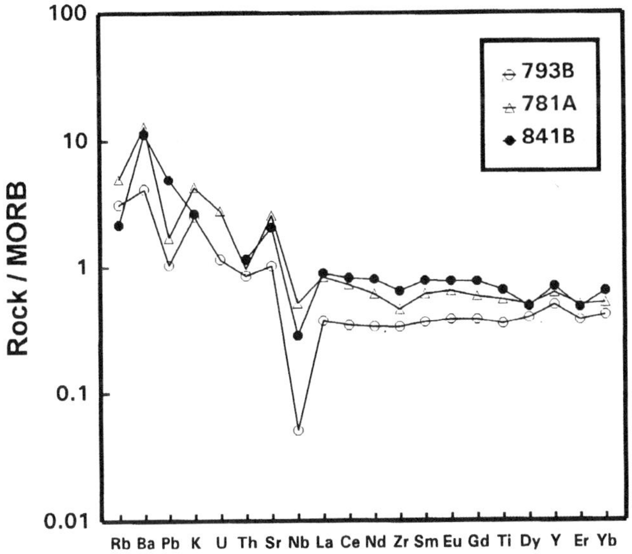

Fig. 4. MORB-normalized multi-element plot of representative analyses from intrusions from ODP Holes 793B, 781A and 841B. MORB values from Hofmann (1988). Element ordering used is that of McCulloch and Gamble (1991).

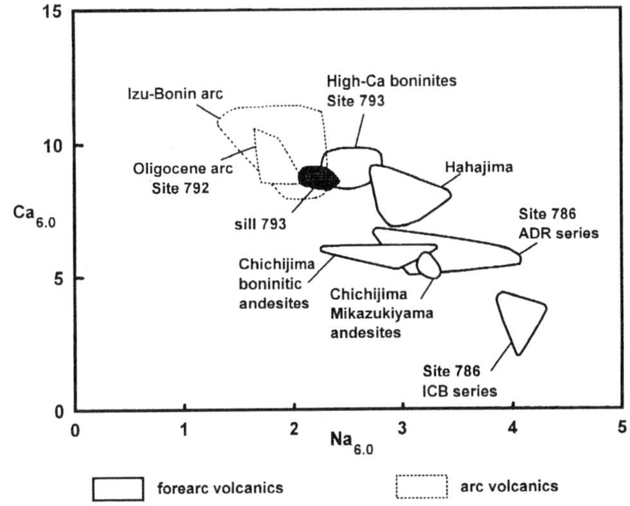

Fig. 5. $Ca_{6.0}$ versus $Na_{6.0}$ for arc and forearc volcanics within the Izu-Bonin region. Shaded field encloses data from the Hole 793B intrusion. Parameters calculated using the method of Plank and Langmuir (1988).

have elevated Ba/Zr (1-5) and low Zr/Y (<2.3). Pearce (1982) considers that these ratios reflect an input of subduction zone fluids (high Ba) into a depleted (low Zr/Y) mantle wedge. The Eocene-Oligocene forearc volcanics have relatively high Ba/Zr and higher Zr/Y, indicating subduction fluid-influence (enriched Ba) on a less depleted or perhaps re-enriched mantle wedge.

In terms of Ba/Zr and Zr/Y ratios (Figure 6), the intrusion of Hole 793B most closely resembles lavas of the Miocene-Quaternary arc. However, this arc displays remarkable along-strike variations in Ba/Zr (Taylor and Nesbitt, unpublished information). Along the Izu-Bonin-Volcano arc Ba/Zr decreases southward from Oshima to a minimum at Torishima, before increasing again through Nishinoshima. If the position of Hole 793B is projected west perpendicular to the arc, the projection line intersects the arc between the islands of Torishima and Sumisujima. The field for these adjacent volcanoes is shown on Figure 6. The majority of the sill sample measurements coincide with this field.

The sill of Hole 793B may have assimilated host sediments during its propagation. Indeed, the recovered upper contact shows a mingling of the diabase magma with host calcareous claystone. If assimilation of sediment did occur during sill emplacement then the igneous trace element ratios may be perturbed. However, the dominant lithologies within the forearc basin are volcanogenic turbidites, derived from the arc ridge to the west (Nishimura et al., 1991). As a consequence, these turbidites have Ba, Zr, and Y concentrations comparable to the sill itself (Hiscott and Gill, 1992). Any assimilation of such sediments would result in very little change in ratios such as Ba/Zr or Zr/Y.

In general, the sill found in Hole 793B has similar geochemical characteristics to the active frontal arc (in particular to the adjacent volcanoes) and is quite distinct from the forearc and backarc volcanic rocks. This suggests an origin from a mantle source similar to that which supplied the frontal arc.

4.2.2. Mariana (site 781) and Tonga (site 841). In terms of structure, composition and evolution the Mariana region shows strong similarities to the Izu-Bonin system. For this reason, the sill at site 781 is compared with the local volcanic suites using Ba/Zr - Zr/Y systematics (Figure 7). In general, the distribution of data fields in Figure 7 is similar to the Izu-Bonin system. For example, the Oligocene Palau-Kyushu arc (site 448) is characterized by low Ba/Zr relative to the Quaternary Mariana arc, which is the same as the age-geochemistry relationship observed for Izu-Bonin system in Figure 6. The data from the sill at site 781 plot closest to the Mariana Quaternary arc field, and are most similar to the four adjacent Quaternary arc volcanoes Maug, Ascuncion, Agrigan, and Pagan.

TABLE 3. Representative major and trace element and Sr, Nd, Pb isotopic ratios for the 793, 781 and 841 intrusions. Average values for each site are provided.

Hole	793B	793B	793B	793B	793B	781A	781A	781A	781A
core	1R-1	1R-1	1R-2	1R-3	average	9R-1	10R-2	10R-3	average
interval	35-39	132-1	61-65	77-82	of 11	69-71, 72-77	68-72	65-67	of 11
sample #	001	004	007	011					
SiO_2	52.84	52.07	53.10	52.58	52.68	49.56	49.53	50.35	49.99
TiO_2	0.58	0.53	0.58	0.56	0.56	0.88	0.85	0.90	0.88
Al_2O_3	14.85	14.46	14.68	14.63	14.64	17.69	17.57	17.28	17.60
Fe_2O_3	10.61	10.03	9.80	9.65	10.06	12.43	12.02	12.73	12.26
MnO	0.16	0.19	0.16	0.17	0.17	0.23	0.22	0.24	0.23
MgO	8.56	9.41	8.99	9.26	9.04	4.40	4.45	4.45	4.43
CaO	10.19	11.01	10.49	10.76	10.62	11.32	11.08	10.89	11.11
Na_2O	1.94	1.64	1.88	1.96	1.84	2.02	2.31	2.19	2.17
K_2O	0.26	0.26	0.32	0.28	0.26	0.42	0.39	0.50	0.45
P_2O_5	0.065	0.057	0.068	0.064	0.064	0.06	0.06	0.06	0.07
Total	100.05	99.66	100.07	99.91	99.94	99.01	98.48	99.59	99.17
LOI	0.48	0.55	0.30	0.68	0.41	1.28	0.75	0.79	0.78
Nb	0.13		0.21	0.20	0.18	1	1	1	1
Zr	34.4	32.6	36.4	34.2	34.7	46	46	49	48
Y	17.9	17.4	18.1	17.7	17.75	22	22	23	22
Rb	2.8	4.1	4.3	4.8	3.9	6	2	7	6.3
Sr	116	113	116	114	116	294	294	293	291
Ba	37	45	66	45	57	142	280	177	181
V	288	266	277	290	275	428	408	370	390
Sc									
Ni	76	104	100	106	96	9	11	10	10
Cr	167	260	255	275	233	11	9	11	15
Zn	69	70	73	72	70	92	77	82	85
Cu	58	112	99	88	93	184	141	196	160
Hf						1.37			1.37
Ta						0.12			0.12
Pb	0.54		0.77	0.21	0.51	2.52			2.52
Th	0.14		0.18	0.16	0.16	0.18			0.18
U	0.06		0.11	0.08	0.08	0.20			0.20
La	1.26		1.64	1.45	1.45	2.89	3.37	3.17	3.26
Ce	3.85		4.47	4.06	4.12	8.54	9.11	9.46	8.69
Nd	3.51		4.04	3.70	3.75	6.19	7.60	7.44	6.85
Sm	1.28		1.43	1.36	1.36	2.09	2.51	2.42	2.30
Eu	0.47		0.53	0.52	0.51	0.81	0.92	0.92	0.86
Gd	1.85		1.99	1.93	1.92	2.92	2.99	3.18	2.96
Dy	2.36		2.56	2.49	2.47	2.80	3.58	3.48	3.18
Er	1.48		1.62	1.58	1.56	1.82	2.56	2.23	2.08
Yb	1.51		1.72	1.63	1.62	1.80	2.26	2.20	2.01
Lu	0.23		0.30	1.25	0.26	0.30	0.36	0.34	0.33
$^{87}Sr/^{86}Sr$			0.70381			0.703487			
$^{143}Nd/^{144}Nd$			0.513086			0.513048			
$^{206}Pb/^{204}Pb$			18.330			18.740			
$^{207}Pb/^{204}Pb$			15.518			15.569			
$^{208}Pb/^{204}Pb$			38.217			38.386			

Data sources: 793, Taylor et al. (1992a); 781, Marlow et al. (1992); 841, Bloomer et al. (1994). Site 793 rare-earth element, Pb, Th, U and Nb concentrations were measured by ICP-MS at the Department of Geology, University of Southampton. Raw ICP-MS counts averaged from four runs per sample were drift corrected and initially calibrated using multi-element internal standards. Data was then re-calibrated using in-run international standards BCR-1, BHVO-1, BIR-1, BEN-1, and JB-2. Reproducibility is better than 5% for all ICP-MS elements except U at 8%. Accuracy is better than 10%.

TABLE 3. continued

Hole core interval sample #	841B 18R-2 28-32	841B 23R-4 17-21, 8-12	841B 25R-4 21-22, 38-46	841B 25R-4 42-46	841B average of 11
SiO_2	54.13	51.63	49.66	50.25	51.52
TiO_2	1.01	1.02	1.04	1.21	1.05
Al_2O_3	16.04	16.26	16.80	17.36	16.33
Fe_2O_3	12.53	12.51	12.51	13.43	12.72
MnO	0.22	0.26	0.23	0.26	0.24
MgO	4.55	4.65	4.64	4.98	4.67
CaO	9.52	9.79	9.54	9.69	9.59
Na_2O	2.78	3.04	3.60	3.26	3.09
K_2O	0.16	0.18	0.17	0.17	0.28
P_2O_5	0.15	0.17	0.18	0.15	0.16
Total	101.09	99.48	98.37	100.76	99.65
LOI	1.29	0.51			
Nb	1	1	0.67	0.77	1
Zr	71	68.5	66	69	67
Y	23	24.5	26	23.5	25
Rb	2	2.5		2	2.74
Sr	216	239	250	225	234
Ba	90	168	124	145	157
V	345	367	380	416	376
Sc		20	40	43	40
Ni	20	21	22	18	20
Cr	14	22	28	26	20
Zn	123	161	128	186	144
Cu	122	114	133	115	124
Hf			1.68	2.9	
Ta			0.12	0.06	
Pb			1.67	4.58	2.41
Th			0.25	0.15	
U			0.18	0.18	
La			3.24	4.17	3.49
Ce			9.08	11.70	9.81
Nd			8.42	10.80	8.93
Sm			2.84	3.46	2.93
Eu			0.97	1.16	1.03
Gd			3.95	4.08	3.90
Dy			3.98	4.36	3.07
Er			2.69	2.63	1.98
Yb			2.48	2.52	2.50
Lu			0.48	0.41	0.45
$^{87}Sr/^{86}Sr$			0.703466		
$^{143}Nd/^{144}Nd$			0.513085		
$^{206}Pb/^{204}Pb$			18.709		
$^{207}Pb/^{204}Pb$			15.547		
$^{208}Pb/^{204}Pb$			38.310		

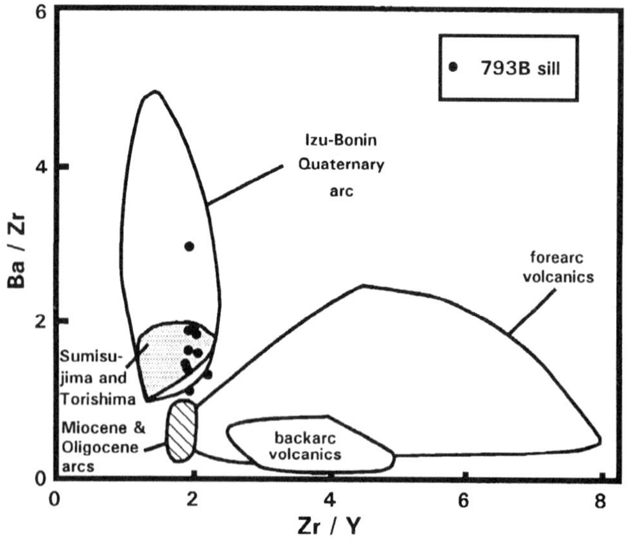

Fig. 6. Ba/Zr versus Zr/Y. Data from 793B sill (Taylor et al., 1992a) are plotted with fields for backarc, arc and forearc Izu-Bonin volcanics. Data for Izu-Bonin Quaternary arc, Taylor and Nesbitt (unpublished); Miocene and Oligocene arcs, Taylor et al. (1992a); backarc volcanics, Fryer et al. (1990), Ikeda and Yuasa (1989); forearc volcanics, Taylor et al. (1994), Murton et al. (1992), Pearce et al. (1992). Data filtered to remove samples with visible alteration and extreme Ba concentrations.

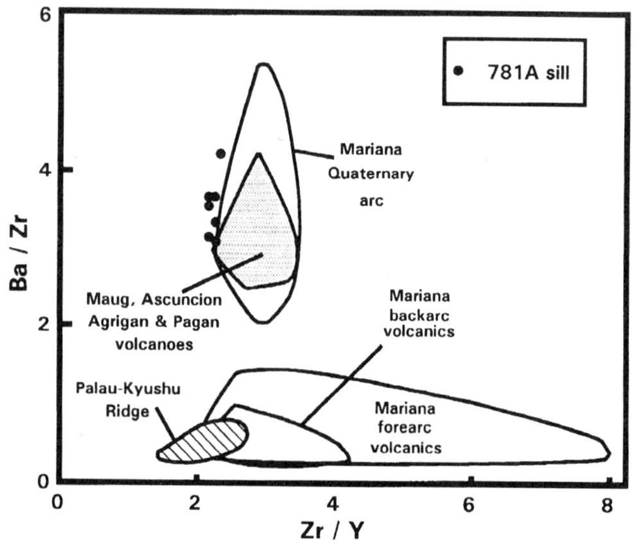

Fig. 7. Ba/Zr versus Zr/Y. Data from 781A sill (Marlow et al., 1992) are plotted with fields for backarc, arc and forearc Mariana volcanics. Data for Mariana Quaternary arc, Woodhead (1988,1989); Palau-Kyushu Ridge, Wood et al. (1980); backarc volcanics, Mattey et al. (1981), Bougault et al. (1981), Volpe et al. (1987); forearc volcanics, Hickey-Vargas and Reagan (1987), Bloomer and Hawkins (1987), Hussong, Uyeda et al. (1981) and references therein. Data filtered to remove samples with extreme Ba concentrations.

The structure of the Tonga arc-forearc region is similar to the Izu-Bonin and Mariana systems. Site 841 is located on the trench slope to the east of a forearc basin and forearc high (the Tonga Platform). Geochemical comparisons between the site 841 intrusive rocks and regional volcanics were made by Bloomer et al. (1994). They demonstrate that the basaltic material has HFSE and LFSE concentrations and ratios most similar to the Lau Ridge remnant arc which now lies to the west of the active Tofua arc. Furthermore, the site 841 material is most similar to the Korobasaga Group, a lava series that developed late in the evolution of the Lau Ridge (Cole et al., 1985).

5. CONDITIONS AND METHOD OF EMPLACEMENT

The geochemical evidence from the forearc intrusions indicates that their magmatic source was located below the contemporaneous frontal volcanic arc. If this is the case then the magmas would have travelled through the forearc crust for distances up to 100 km. Although such distances seem substantial, the geological record contains many instances of sills or dikes extending for distances similar to, or greater than, the western Pacific arc-trench gap. Notable examples are the Paleocene dike swarms of the British Tertiary volcanic province and the late Carboniferous tholeiite sills of northern Britain (the Midland Valley and Whin sills), which extend for several hundred kilometers through crystalline basement and shallower sediments (Francis, 1982). However, it is worth investigating the potential for extensive forearc intrusions in the light of the thickness of the sills and the conditions specific to the arc-trench region.

If magma was transported out from the arc entirely along a sill, then its journey would predominantly be through low-density sediments at sub-bottom depths <500 m. Typical width/length ratios for shallow intrusions emplaced by hydraulic fracture in consolidated sediments are 10^{-2} to 10^{-3} (Fedotov, 1978). Using the lower ratio the thickest sills encountered in the forearc should only have traveled up to 25 km. This is not likely since the host sediments were wet, which would have resulted in high cooling rates and shorter propagation lengths (Bloomer et al., 1994). Such distances are consistent with reflectors interpreted as sills on multi-channel seismic reflection profiles, which can be traced for less than 25 km (e.g., Figure 3). An alternative mode of emplacement involves the passage of magma through dike injection from deeper crustal levels and would presumably allow the sill to pass through higher density, drier lithologies. Basaltic dikes in deeper crust typically have length/width ratios in the range 10^{-3} to 10^{-4} (Fedotov,

1978). The lower number would permit intrusions to reach the outermost sections of the forearc.

Lister and Kerr (1990) have calculated theoretical solutions for the scale of igneous intrusions by assuming that the effect of temperature on melt rheology and the effects of melting and solidification at dike walls are minimal. Their results predict that a dike carrying a flux of 100 m^3s^{-1} would extend laterally for 18 km with a width of 18 cm. In the case of a larger magma flux (~$10^6m^3s^{-1}$) the dike extends to 46 km long and 2.8 m wide. These calculations suggest dikes that are narrower, and of a greater horizontal extent, than is likely for magma transport through cold lithosphere. On this basis, Lister and Kerr (1990) suggest that thermal erosion at the center of the dike and solidification at its edges are likely to limit the lateral extent of flow, resulting in a concentration of the flow into a wider dike capable of melt transport without further solidification. The widths of the forearc intrusions are predominantly in the range 5-20 m. In light of the theoretical dimensions above, it is plausible that with sufficiently large magma flux that dikes or sills could propagate large distances into the forearc from the arc. Some of the thin intrusions found in Hole 841 (<50 cm) are clearly not thermally capable of crossing the forearc. These intrusions must therefore represent a "fingering" of a wider intrusion or intrusions, as documented by Grapes et al., (1972) and Pollard et al. (1975). This is consistent with the geochemical evidence from the intrusions in Hole 841, which indicates a remarkably similar geochemistry for each of the recovered units (Bloomer et al., 1994).

6. SUMMARY

Neogene intrusions appear to be an integral feature of western Pacific forearcs. Geochemically the intrusions show strong affinities to the Quaternary arc. Furthermore each intrusion is most similar to the adjacent arc volcanoes. This points to a magmatic source beneath the frontal arc and not within the forearc mantle wedge. To reach the distal locations in the forearc, magma must have travelled up to 100 km from its source. The most likely mode of intrusion to achieve such distances are via dikes passing through basal forearc sediments and basement rocks with higher densities and lower water contents than the upper sedimentary sequences of the forearc. These dikes would then feed sills or fingering sill complexes at higher crustal levels.

Acknowledgments. John Mitchell is thanked for K-Ar analysis and advice. Bob Nesbitt is thanked for the use of the ICPMS facility at the Department of Geology, University of Southampton. We are indebted to Bramley Murton and Richard Jolly for helpful discussion and to Jon Woodhead and Jim Natland for constructive comments. R.N.T. was supported by NERC grant GST/02/558.

REFERENCES

Bloomer, S. H. and J. W. Hawkins, Gabbroic and ultramafic rocks from the Mariana Trench: and island arc ophiolite, in *The Tectonic and Geologic Evolution of Southeast Asian Seas and Islands* (pt. 2) *Geophys. Monogr. Ser.*, vol. 27, edited by D. E. Hayes, pp. 294-317, AGU, Washington, D.C., 1983.

Bloomer, S. H. and J. W. Hawkins, Petrology and geochemistry of boninite series volcanic rocks from the Mariana trench. *Contrib. Mineral. Petrol. 97*, 361-377. 1987.

Bloomer, S. H., A. Ewart, J. Hergt, and W. B. Bryan, Geochemistry and origin of igneous rocks from the outer Tonga forearc, Site 841, in *Proc. ODP, Sci Results, 135,* J. Hawkins, L. Parson, J. Allan, et al., pp. 625-646, Ocean Drilling Program College Station, TX, 1994.

Bougault, H., R. C. Maury, M. El Azzouzi, J.-L. Joron, and M. Treuil, Tholeiites, basaltic andesites, and andesites from Leg 60 Sites: geochemistry, mineralogy, and low partition coefficient elements, in *Init. Repts. DSDP, 60,* D. M. Hussong,, S. Uyeda, et al., pp. 657-677, U.S. Govt. Printing Office, Washington, D.C., 1981.

Cole, J. W., J. B. Gill, and D. Woodhall, Petrologic history of the Lau Ridge, Fiji, in *Geology and offshore resources of Pacific island arcs-Tonga region,* edited by D. W. Scholl and T. L. Vallier, pp. 379-414, Circum-Pacific Council for Energy and Mineral Resources Earth Science Series, vol. 2, Houston, TX, 1985.

Fedotov, S. A., Ascent of basic magmas in the crust and the mechanism of basaltic fissure eruptions, *Int. Geol. Rev.*, 20, 33-47, 1978.

Francis, E. H., Magma and sediment-I, Emplacement mechanism of late Carboniferous tholeiite sills in northern Britain, *J. geol. Soc. London, 139*, 1-20, 1982.

Fryer, P., B. Taylor, C. H. Langmuir, and A. G. Hochstaedter, Petrology and geochemistry of lavas from the Sumisu and Torishima backarc rifts, *Earth. Planet. Sci. Lett., 100*, 161-178, 1990

Grapes, R. H., D. L. Reid, and J. G. McPherson, Shallow dolerite intrusion and phreatic eruption in the Allan Hills region, Antarctica, *N.Z. J. Geol. Geophys.*, 17, 563-577, 1972.

Hickey, R., and F. A. Frey, Geochemical characteristics of boninite series volcanics: implications for their source: *Geochim. Cosmochim. Acta*, 46, 2099-2115, 1982.

Hickey-Vargas, R. L., and M. Reagan, Temporal variation of isotope and rare earth element abundances from Guam: implications for the evolution of the Mariana arc. *Contrib. Mineral. Petrol., 97*, 497-508, 1987.

Hilde, T. W. C. and C-S. Lee, Origin and evolution of the West Philippine Basin: A new interpretation. *Tectonophysics, 102*, 85-104, 1984.

Hiscott, R. N., and J. B. Gill, Major and trace element geochemistry of Oligocene to Quaternary volcaniclastic sands and sandstones from the Izu-Bonin Arc. *Proc. ODP, Sci. Results,* 126, B. Taylor, K. Fujioka, et al., Ocean Drilling Program, College Station, TX, pp.467-485, 1992.

Hofmann, A. W., Chemical differentiation of the Earth: the relationship between mantle, continental crust, and oceanic crust. *Earth. Planet. Sci. Lett. 90,* 297-314, 1988.

Hussong, D. M., S. Uyeda, et al., 1981. *Init. Rept. DSDP,* 60, U.S. Govt. Printing Office, Washington, D.C., 1981.

Ikeda, Y., and M. Yuasa, Volcanism in nascent back-arc basins behind the Shichito Ridge and adjacent areas in the Izu-Ogasawara arc, northwest Pacific: evidence for mixing between E-type MORB and island arc magmas at the initiation of back-arc rifting. *Contrib. Mineral. Petrol. 101,* 377-393, 1989.

Kostopoulos, D. K., and S. D. James, Parameterization of the melting regime of the shallow upper mantle and the effects of variable lithospheric stretching on mantle modal stratification and trace element concentrations in magmas. *J. Petrol,* 33, pt. 3, 665-691, 1992.

Lapierre, H., R. N. Taylor, O. Rouer, and H. Chaisemartin, Mineral chemistry of fore-arc volcanic rocks from the Izu-Bonin Arc, Holes 792E and 793B, *Proc. ODP, Sci. Results, 126,* B. Taylor, K. Fujioka, et al., pp. 431-447, Ocean Drilling Program, College Station, TX, 1992.

Lister, J. R. and R. C. Kerr, Fluid-mechanical models of dike propagation and magma transport. *Mafic Dykes and Emplacement Mechanisms*, edited by: A. J. Parker et al., pp. 69-80, Balkema, Rotterdam, 1990.

Marlow, M. S., L. E. Johnson, J. A. Pearce, P. B. Fryer, L. G. Pickthorn, and B. J. Murton, Pleistocene volcanic rocks in the Mariana forearc revealed by drilling at Site 781, *Proc. ODP, Sci. Results,* 125, P. B. Fryer, J. A. Pearce, L. B. Stokking, et al., pp. 293-310, Ocean Drilling Program, College Station, TX, 1992.

Mattey, D. P., N. G. Marsh, and J. Tarney, The geochemistry, mineralogy and petrology of basalts from the west Philippine and Parece Vela basins and from the Palau-Kyushu and west Mariana ridges, Deep Sea Drilling Project Leg 59, *Init. Rept. DSDP, 59,* L. Kroenke, R. Scott, et al., pp. 753-800, U.S. Govt. Printing Office, Washington, D.C. 1981.

McCulloch, M. T., and J. A. Gamble, Geochemical and geodynamical constraints on subduction zone magmatism, *Earth. Planet. Sci. Lett., 102,* 321-355, 1991.

Meijer, A., Primitive arc volcanism and a boninite series: Examples from Western Pacific island arcs, In *The Tectonic and Geological Evolution of Southeast Asian Seas and Islands, Part 1, Geophys. Monogr. Ser.*, vol. 23, edited by D. E. Hayes, pp. 271-282, 1980.

Murton, B. J., D. W. Peete, R. J. Arculus, J. A. Pearce, and S. Van der Lann, Trace element geochemistry of volcanic rocks from Site 786: the Izu-Bonin forearc, Leg 125, in *Proc. ODP, Sci. Results,* 125, P. B. Fryer, J. A. Pearce, L. B. Stokking, et al., pp. 211-235, Ocean Drilling Program, College Station, TX, 1992.

Natland, J. H., and J. Tarney, Petrologic evolution of the Mariana arc and back-arc system—A synthesis of drilling results in the south Philippine Sea, in *Initial. Repts. DSDP,* 60, D. M. Hussong and S. Uyeda, et al., pp. 877-908, U.S. Government Printing Office, Washington, D.C., 1981.

Nishimura, A., K. M. Marsaglia, K.S. Rodolfo, A. Colella, R. N. Hiscott, K. Tazaki, J. B. Gill, T. Janecek, J. Firth, M. Isiminger-Kelso, Y. Herman, R. N. Taylor, B. Taylor, K. Fujioka, and Leg 126 Scientific Party, Pliocene-Quaternary submarine pumice deposits in the Sumisu Rift area, Izu-Bonin Arc. In: *Sedimentation in Volcanic Settings, SEPM Special Publication* No. 45. 201-208, 1991.

Pearce, J. A., Trace element characteristics of lavas from destructive plate boundaries, in *Andesites, Orogenic Andesites and Related Rocks,* edited by R. S. Thorpe, pp. 525-548, John Wiley and sons, Chichester, 1982.

Pearce, J. A., S. R. Van der Laan, R. J. Arculus, B. J. Murton, T. Ishii, D. W. Peate, and I. J. Parkinson, Boninite and harzburgite from ODP Leg 125 (Bonin-Mariana forearc): A case study of magma genesis during the initial stages of subduction, in *Proc. ODP, Sci. Results,* 125, P. B. Fryer, J. A. Pearce, L. B. Stokking, et al. pp. 623-659, Ocean Drilling Program, College Station, TX, 1992.

Plank, T., and C. H. Langmuir, An evaluation of the global variations in the major element chemistry of arc basalts, *Earth Planet. Sci. Lett., 90,* 349-370, 1988.

Pollard, D. D, O. H. Muller, and D. R. Dockstader, The form and growth of fingered sheet intrusions, *Bull. Geol. Soc. Am.,* 86, 351-363, 1975.

Roddick, J. C, The application of isochron diagrams in ^{40}Ar-^{39}Ar dating: A discussion, *Earth Planet. Sci. Lett. 41,* 233-244, 1978.

Stern, R. J. and S. H. Bloomer, Subduction zone infancy: Examples from the Eocene Izu-Bonin-Mariana and Jurassic California arcs, *Geol. Soc. Am. Bull., 104,* 1621-1636, 1992.

Stern, R. J., M. C. Jackson, P. B. Fryer, and E. Ito, O, Sr, Nd and Pb isotopic composition of the Kasuga cross-chain in the Mariana Arc: A new perspective on the K-*h* relationship, *Earth Planet. Sci. Lett., 119,* 459-475, 1993.

Taylor, B., K. Fujioka, et al., *Proc. ODP, Init. Repts.*, 126; Ocean Drilling Program, College Station, TX, 1990.

Taylor, B., Rifting and the volcanic-tectonic evolution of the Izu-Bonin-Mariana arc, in *Proc. ODP, Sci. Results,* 126, B. Taylor, K. Fujioka, et al., pp. 627-651, Ocean Drilling Program, College Station, TX, 1992.

Taylor, R. N., H. Lapierre, P. Vidal, R. W. Nesbitt, and I. W. Croudace, Igneous geochemistry and petrogenesis of the Izu-Bonin fore-arc basin, in *Proc. ODP, Sci. Results,* 126, B. Taylor, K. Fujioka, et al., pp. 405-430, Ocean Drilling Program, College Station, TX, 1992a.

Taylor, R. N., B. J. Murton, and R. W. Nesbitt, Chemical transects across intra-oceanic arcs: Implications for the tectonic setting of ophiolites, In *Ophiolites and Their Modern Oceanic Analogues*, edited by L. M. Parson, B. J. Murton, and P. J. Browning, *Geological Society Special Publication* No. 60, 117-132, 1992b.

Taylor, R. N., and J. Mitchell, K-Ar dating results from whole rock and mineral separates of Izu-Bonin fore-arc basement, Leg 126. in *Proc. ODP, Sci. Results,* 126, B. Taylor, K. Fujioka, et al., pp. 677-680, Ocean Drilling Program, College Station, TX, 1992.

Taylor, R. N., R. W. Nesbitt, P. Vidal, R. Harmon, B. Auvray, and I. W. Croudace, Mineralogy, chemistry and genesis of the boninite series volcanics, Bonin Islands, Japan, *J. Petrol, 35,* 577-617, 1994.

Taylor, R. N., and R. W. Nesbitt, Arc volcanism in an extensional regime at the initiation of subduction: a geochemical study of Hahajima, Bonin Islands, Japan, in *Volcanism Associated with Extension at Consuming Plate Margins*, edited by J. Smellie, *Geological Society Special Publication*, in press, 1994.

Umino, S., Volcanic geology of Chichihima, the Bonin Islands (Ogasawara Islands), *J. Geol. Soc. Jpn*, 91, 505-523, 1985.

Van der Laan, S. R., M. F. Flower, and F. K. Van Groosa, Experimental evidence for the origin of boninites: near-liquidus phase relations to 7.5 Kbar, in *Boninites,* edited by A. J. Crawford, pp. 112-147, Unwin Hyman, 1989.

Volpe, A. M., J. D. MacDougall, and J. W. Hawkins, Mariana Trough basalts (MTB): trace element and Sr-Nd isotopic evidence for mixing between MORB-like and arc-like melts,. *Earth. Planet. Sci. Lett., 82*, 241-254, 1987.

Wood, D. A., J.-L. Joron, N. G. Marsh, J. Tarney, and M. Treuil, Major- and trace-element variations in basalts from the north Philippine Sea drilled during Deep Sea Drilling Project Leg 58: a comparative study of back-arc basin basalts with lava series from Japan and mid-ocean ridges, in *Init. Repts. DSDP,* 58, G. DeVries Klein, K. Kobayashi, et al., pp. 873-894, U.S. Govt. Printing Office, Washington, D.C., 1980.

Woodhead, J. D., The origin of geochemical variations in Mariana lavas: a general model for petrogenesis in intra-oceanic island arcs? *J. Petrol., 29,* 805-830, 1988.

Woodhead, J. D., Geochemistry of the Mariana arc (western Pacific): Source composition and processes, Chem. Geol., 76, 1-24, 1989.

S. H. Bloomer, Department of Geology, Boston University, Boston, MA 02215

L. E. Johnson, Naval Research Laboratory, Code 5110, Washington, DC 20375-5000

M. S. Marlow, U.S. Geological Survey, 345 Middlefield Road, Menlo Park, CA 94025

J. H. Mitchell, Department of Physics, The University, Newcastle upon Tyne, UK

B. Taylor, School of Ocean and Earth Science and Technology, University of Hawaii, Honolulu, HI 96822

R. N. Taylor, Department of Geology, The University of Southampton, Southampton, SO9 5NH, U.K.

Geochemical Evolution of Arc Systems in the Western Pacific: the Ash and Turbidite Record Recovered by Drilling

R. J. Arculus[1], J. B. Gill[2], H. Cambray[3], W. Chen[4], R. J. Stern[5]

The Ocean Drilling Program and its predecessor, the Deep Sea Drilling Project, have successfully recovered extensive cores of volcanic ashes and turbidites overlying basement in a number of the forearc-arc-backarc systems and more distal plinian deposits in the western Pacific. These materials are particularly useful for reconstructions of spatial and temporal aspects of the chemical evolution of island arcs. Notable results from studies of volcaniclastic materials derived from the Vanuatu, Sulu-Celebes, Izu-Bonin, Japan and Kurile-Kamchatka systems are (1) absence of a consistent and monotonic increase in alkalinity from older to younger deposits; (2) a change in comparative fertility of mantle wedge sources in the case of the Izu-Bonin arc from relatively fertile to refractory subsequent to backarc (Shikoku Basin) formation; (3) the absence of such an effect in the Marianas where more extensive backarc basin spreading has occurred; (4) the appearance of alkaline compositions in the Marianas either related to arc rennaissance subsequent to initial stages of backarc rifting, or through the tapping of a long-lived wedge heterogeneity centered at present in the vicinity of Iwo Jima near the junction of the Izu-Bonin and Mariana systems; (5) the absence of any evidence of crustal growth in these arc systems on the basis of consistency of major element characteristics (e.g., CaO concentration at specific MgO) believed to be a function of crustal thickness; (6) apparent peaks in explosive volcanism coincident with initial stages of backarc spreading in the Japan Sea, Parece Vela and Shikoku Basins, followed by quiescence, or even termination of arc activity in the Izu-Bonin system during full establishment of spreading in the Shikoku Basin; (7) indications of significant compositional controls on the nature of arc volcanism exerted by a lithospheric filter effect, possibly most pronounced in the case of the Marianas (comparatively fertile filter) compared with the Izu-Bonin arc (comparatively refractory lithospheric filter); (8) two marked periods of enhanced explosivity in circum-Pacific arcs at 18-13 Ma and at 5 Ma to present.

[1]Department of Geology, Australian National University, Canberra, Australia
[2]Earth Sciences Department, University of California, Santa Cruz, California
[3]Unité de Genié Océanique, Institut Méditerranean de Technologie, Marseille, France
[4]Department of Geology and Geophysics, University of New England, Armidale, Australia
[5]Center for Lithospheric Studies, University of Texas at Dallas, Richardson, Texas

Active Margins and Marginal Basins of the Western Pacific
Geophysical Monograph 88

1. INTRODUCTION

1.1. *Scope of this Paper*

Understanding the geochemical evolution of arc systems has been a specific goal of geochemists and petrologists interested in the processes of supra-subduction zone (SSZ) magmatism. These systems have been regarded as the locus of initial growth of the continents [*Taylor*, 1967], so that identification of long-term changes within a specific arc has been viewed as crucial for understanding the process of formation and modification of the continental crust. Following the documentation of spatial variety of characteristics within island arcs [*Kuno*, 1959; *Sugimura*,

1960], geochemical evolution has been claimed for example, from early low-alkali through intermediate- to highly-alkaline compositions with "maturing" of the arc in the sense of changes in crustal character from intra-oceanic to continental [*Baker*, 1968; *Gill*, 1970]. Within any individual arc, however, onland exposures are typically incomplete or poor, rendering decipherment of the temporal record an awkward task and restricting our view to a narrow period of development. Experience has shown that lengthy records of arc activity, either as ash or volcaniclastic-rich turbidite layers, can be obtained by drilling of deposits in the marine realm, optimally sited to avoid disturbance, reworking, and with reasonable penetration thicknesses. Disadvantages include the difficulty of precisely identifying the sources of these materials, technical drilling problems with recovery of coarse clastic sequences, the possibility of incomplete records through erosion and non-deposition, and the problems of diagenetic change as a result of exposure to sea water.

The Ocean Drilling Program (ODP) successfully recovered extensive cores of volcaniclastic materials overlying basement in a number of the SSZ systems of the western Pacific (Figure 1) during late 1988 through late 1992. Allied with the results of analytical and interpretative efforts targeted on cores obtained during the predecessor efforts of the Deep Sea Drilling Project (DSDP) in some of these systems, we now have an improved appreciation for the factors that appear to be critical in controlling aspects of the geochemistry of the arcs of the region.

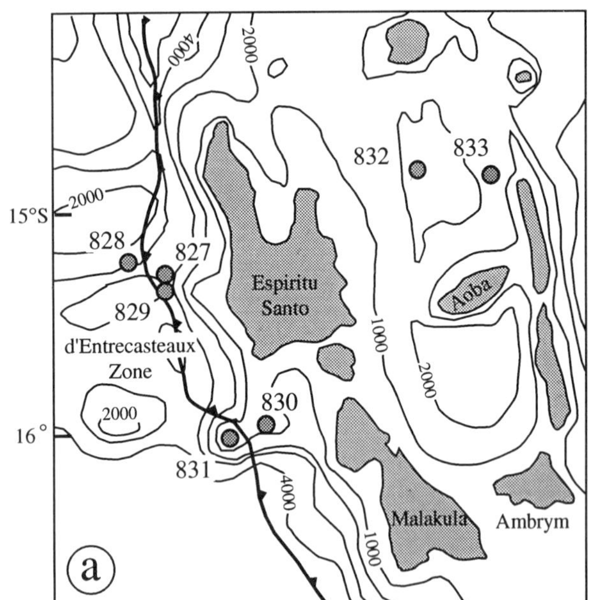

Fig. 1a. Ocean Drilling Program leg 134 (Vanuatu transect) site (827-833) location map. Depths in meters; location of trench in this and subsequent figures indicated as a heavy continuous line with triangle decoration.

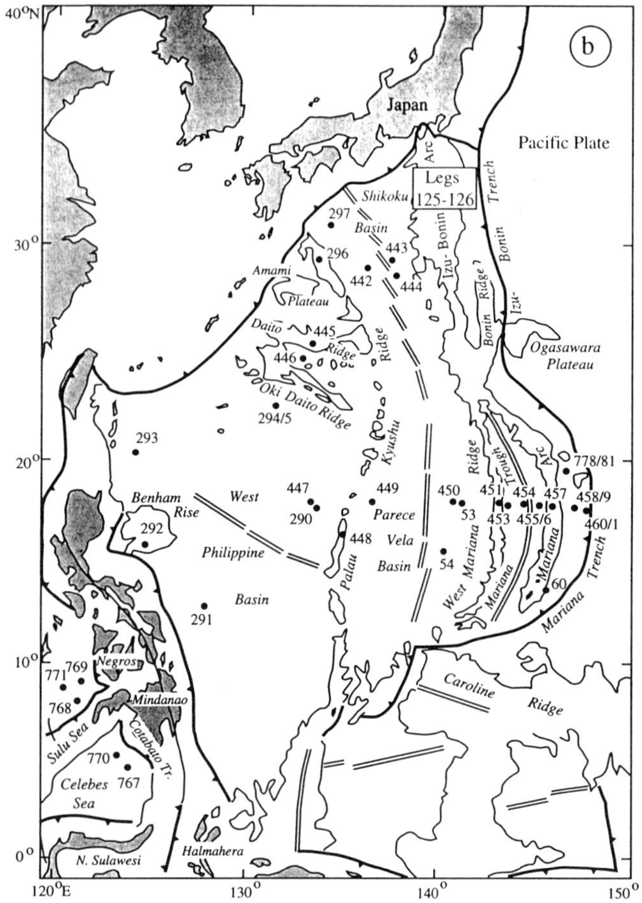

Fig. 1b. Location map of Deep Sea Drilling Project (legs 6, 31, 58, 59, and 60) and Ocean Drilling Program (legs 124, 125, and 126) sites in the Izu-Bonin, Mariana, and Philippine regions. A more detailed map of the ODP legs 125 and 126 sites in the central Izu-Bonin system is displayed in Figure 1c. Sites are indicated as filled dots, spreading (backarc) ridges, both active and inactive, are shown as double lines. Basins and ridges are outlined by the 4-km bathymetric contour, except for the Izu-Bonin, West Mariana and Mariana arcs, which are outlined by the 3-km contour.

Some of the questions and issues that have been addressed by these efforts include a number that have been on the research agenda for students of SSZ magmatism for a long time:

(1) Are there any temporal changes in the geochemistry of arc magmas;

(2) If so, are these regular, erratic, or blends of these end-member behaviors;

(3) Can we detect crustal growth using geochemical parameters believed to correlate at least globally with crustal thicknesses;

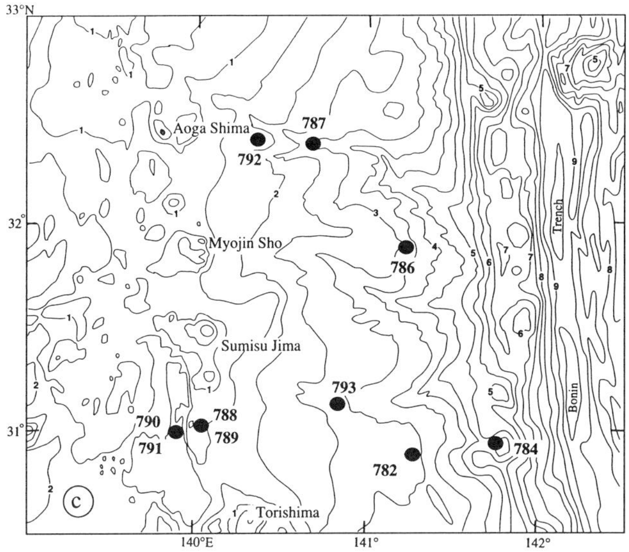

Fig. 1c. Location map of Ocean Drilling Program legs 125 and 126 sites in the Izu-Bonin system. Depths in kilometers.

(4) Can changes in the style of eruptive activity and/or geochemistry be attributed to distinct tectonic events;

(5) Does arc volcanism change character or cease for any part of the rift-to-spread-to-cessation stages of backarc basin evolution;

(6) Does the concurrent or former existence of adjacent backarc magmatism affect the nature of arc magma sources, possibly through advection of melt-depleted mantle into the path of slab-derived fluids;

(7) Is there inter-arc synchroneity of major periods of explosive activity;

(8) Do episodes of major explosive activity in arcs correlate with climate change?

In this paper, we summarize the contributions relating to these issues published in the Scientific Results volumes of the ODP in a geographic distribution from south-to-north; interweave and review subsequent progress achieved with specific sample sets both from the ODP and DSDP legs, drawing on published and unpublished data; discuss some of the problems identified by these efforts; and indicate directions for further progress. We consider each of the questions (1 to 8) in the discussion. Details of the volcaniclastic record of arc evolution in the Lau-Tonga region (ODP leg 135) are not pursued here as this topic is addressed by *Clift et al.* [this volume].

1.2. *Ash Deposition and Lithologies*

We attempt an integration of studies of ashes and turbidites rich in volcaniclastic materials. *Cambray et al.* [1993] and *Gill et al.* [1994] have described the advantages, difficulties and potential ambiguities of working with these different sample types. In the case of ashes, *Cambray et al.* [1993] demonstrate that although the violence of eruption and altitude attained by the eruption column, distance from vent, prevailing wind directions at different altitudes and strengths (glacial -stronger vs. interglacial- weaker) can be significant factors controlling the distribution of deposits, a remarkable similarity of ash layer frequency east (DSDP legs 57 and 87A) and west (ODP legs 127 and 128) of Honshu, Japan, exists. We recognize that latitudinal migration of major jet streams (influenced by global climate) can be significant in controlling transportation of ashes injected into the stratosphere (*J. Natland*, personal communication, 1994), but at a time scale of several 100 k.y., the influence of wind direction on distribution of tephra east and west of the Japanese archipelago does not negate the usefulness of thickness/frequency studies of ash layers confined to one flank of an arc system. We do nevertheless face the major problem of relating true volumes of eruptive materials with the potentially patchy record of distal tephra deposits [*Karig*, 1983].

Concentrations of ash and distinct ash layers can persist despite slumping and/or erosion by bottom currents, and bioturbation. Our collective experience is that fresh volcanic

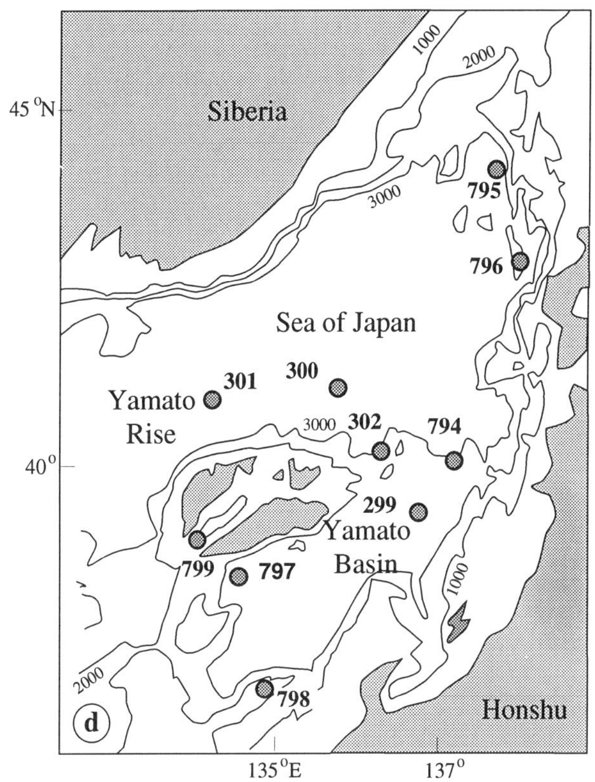

Fig. 1d. Location map of Deep Sea Drilling Project (leg 31) and Ocean Drilling Program (legs 127 and 128) sites in the Japan Sea. Bathymetry in meters. The Yamato Rise is shaded as well as the emergent land masses.

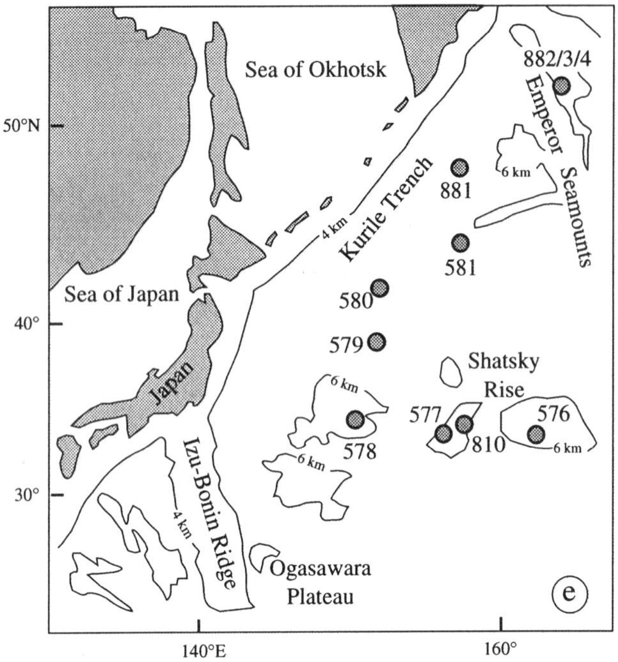

Fig. 1e. Location map of selected Deep Sea Drilling Project (leg 86) and Ocean Drilling Program (legs 132 and 145) sites in the northwestern Pacific. Depths in kilometers.

glass can persist even in layers up to 35 m.y. old, and sometimes even older. Both discrete homogeneous and heterogeneous ash layers are reported from the Lau-Tonga, Vanuatu, Sulu-Celebes Seas, Izu-Bonin-Mariana forearc, Japan Sea, and northern Pacific. Single eruptive episodes were probably responsible for the homogeneous layers (type I of *Huang*, 1980) whereas multiple eruptions from different sources or mixed magmas are likely the origin of heterogeneous layers (type II of *Huang*, 1980). Shards, tubular-elongate micropumice, bubble wall, and U- to Y-shaped fragments are common and probably associated with Plinian eruptions. Apparently pristine igneous trends are apparent from analysis of many of these materials, and magmatic Sr isotopic ratios can be recovered with comparatively mild leaching treatment of glass shards.

Arc-derived, medium- to coarse-grained sands and sandstones dominantly of vitric and lithic fragments are the primary turbidite lithologies studied by *Gill et al.* [1994] from the Izu-Bonin arc. Samples range from largely undiluted juvenile debris probably from single eruptions, to mixtures from several eruptions and sources. In fact, sedimentary mixing may be an advantage in the sense of representing composites of igneous processes over some time period in the arc. Despite intense diagenesis in some sedimentary sections of the Izu-Bonin forearc, it proved possible for *Gill et al.* [1994] to identify the original igneous characteristics of the turbidite deposits. Over periods of time when turbidite and ash records coincide in the Izu-Bonin arc (0 to ~15 Ma), a close correspondence in geochemistry is apparent (see below). This inspires confidence in the sole use of turbidites for earlier periods of activity in this arc where the ash record, in the forearc at least, is absent.

1.3. *Types of Data Evaluated*

Electron microprobe (major elements) and proton-induced X-ray emission (trace elements) analysis of individual glass shards, bulk analysis by atomic absorption (majors), X-ray fluorescence and inductively coupled plasma source mass spectrometry (various traces), and thermal ionization mass spectrometry for Sr, Pb, and Nd isotopes have been variously employed by the studies cited here. Of particular analytical concern is the apparent migration of Na in water-rich glasses, especially under conditions of relatively high beam currents employed during wavelength-dispersive electron microprobe analysis. Comparative tests of energy dispersive *vs.* wavelength dispersive analysis has shown that care is required for acquisition of meaningful Na concentrations even with defocused and rastered beams.

2. SUMMARY OF PUBLISHED STUDIES

2.1. *ODP Leg 134—Vanuatu Transect*

The effects of ridge collisions and processes of intra arc rifting were primary aims of the leg 134 transect across the Vanuatu arc [*Collot et al.*, 1992]. Volcanic ash layers (1-3 cm thick) are abundant in the north Aoba basin sites (832, 833) but are rare, possibly swept clear by turbidite action, in the forearc; sparse at sites 828 (north d'Entrecasteaux ridge) and 831 (south d'Entrecasteaux chain); and absent on the forearc slope west of Espiritu Santo (sites 827 and 829) (Figure 1a). All of the ashes recovered on leg 134 [*Baker et al.*, 1994] are believed to have originated from the central chain volcanoes and include high-K and low-K series previously recognized from land-based studies [*Mallick*, 1973; *Gorton*, 1977; *Crawford et al.*, 1988].

With the exception of a calcalkaline medium-K rhyodacite ash from site 830 that was probably derived from major caldera-forming eruptions in the vicinity of Epia-Epi-Tongoa [*Crawford et al.*, 1988], all other samples are tholeiitic on the basis of *Miyashiro's* [1974] FeO*/MgO vs. SiO_2 discriminant. All samples form a tight trend of wt% CaO vs. MgO (Figure 2) and define a crustal thickness for the Vanuatu arc of ~14 km according to the *Plank and Langmuir* [1988] correlation. Crystalline phases are primarily plagioclase, clinopyroxene, olivine and titanomagnetite with minor orthopyroxene, apatite, and quartz.

2.2. *ODP Leg 124—Celebes and Sulu Seas*

A number of ash layers together with pyroclastic deposits were recovered by drilling in the Celebes and Sulu Seas

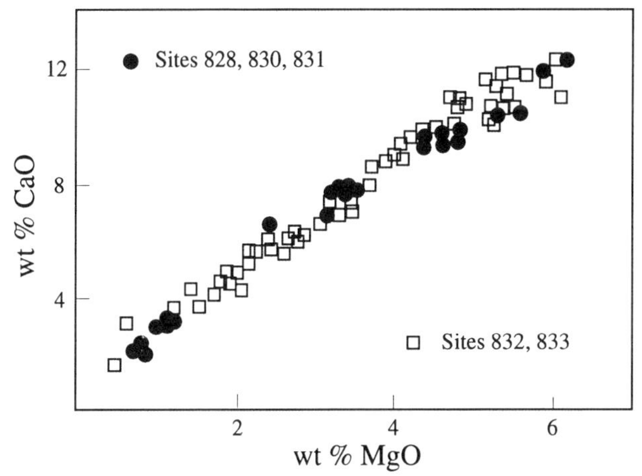

Fig. 2. Plot of wt% CaO vs. MgO for ashes from the ODP Leg 134 Vanuatu sites.

(Figure 1b). These marginal basins are surrounded on three quarters of the compass by a festoon of island arcs—clockwise from the Sulu Ridge, Negros (proximal) and Philippine (distal) to the north of these basins, Mindanao, Cotabato, Sangihe (proximal) and Halmahera (distal) to the east, and the North Sulawesi arc to the south. The Celebes Sea is mid-Eocene in age, but of uncertain environment of formation. A fragment of major ocean floor (e.g., Indian Ocean or Molucca Sea) or backarc basin are possibilities [*Rangin and Silver*, 1991]. The Sulu Sea formed in the early Miocene, most likely as a backarc basin south of the Cagayan ridge.

Identification of specific ash sources in these basins is not straightforward given the number of neighboring arcs, the tectonic complexity of the region including polarity flips that have occurred in the Tertiary, and the waxing and waning of various arcs. Resolution of these problems has been aided by a program of isotopic dating of volcanic arc sequences on-land [*Bellon and Rangin*, 1991], which recognizes two main arc types: a group related to the closing of the Celebes and Sulu Basins and another related to the Philippine Sea plate. Tephra correlations by *Pubellier et al.* [1991] and in particular a comparison of DSDP site 292 in the West Philippine Basin with ODP site 767 in the Celebes Basin are of particular significance with respect to understanding the synchroneity or otherwise of concurrent arc vs. backarc activity (Figure 3). These authors make the point that the major pulse of explosive activity recorded by the ash layers in the time interval 23 to 30 Ma at site 292, which is only weakly present at site 767, was most probably derived from the eastern margin of the Philippine Sea plate, i.e., West Mariana Ridge activity concurrent with backarc spreading in the Parece Vela Basin, rather than magmatism associated with subduction of the Indo-Australian beneath the western margin of the Philippine Sea plate.

A variety of magmatic lineages is represented within ashes that occur in both basins [*Pouclet et al.*, 1991]. Integration of mineralogical and geochemical data shows that the series and fractionating phases (in addition to titanomagnetite) listed below are involved in the generation of dacitic from andesitic parent magmas: (1) low-Al, medium-K andesitic series involving magnesian hornblende > clinopyroxene + plagioclase; (2) high-Al, medium-K basalt series involving plagioclase + pargasitic hornblende or biotite > clinopyroxene; and (3) high-Al, high-K andesitic series involving clinopyroxene + plagioclase > pargasitic hornblende.

Further fractionation of Fe-Ti oxides, hornblende, biotite, plagioclase and alkali feldspars from dacitic parents resulted in the formation of rhyolitic magmas. The high-K (shoshonitic) character of a number of ash layers derived from the Manila, Sulu, Negros, and central Mindanao arcs and the relative abundance of amphibole are noteworthy in comparison with the other SSZ magmatic systems examined by the ODP in the western Pacific. In addition, fresh but highly oxidized andesitic magmas (oxygen fugacity (fO_2) equivalent to the synthetic fayalite-magnetite-quartz (FMQ) buffer + 4 $\log_{10}$ units) were recovered from site 767 in the Celebes Sea. *Pouclet et al.* [1991] did not recognize any

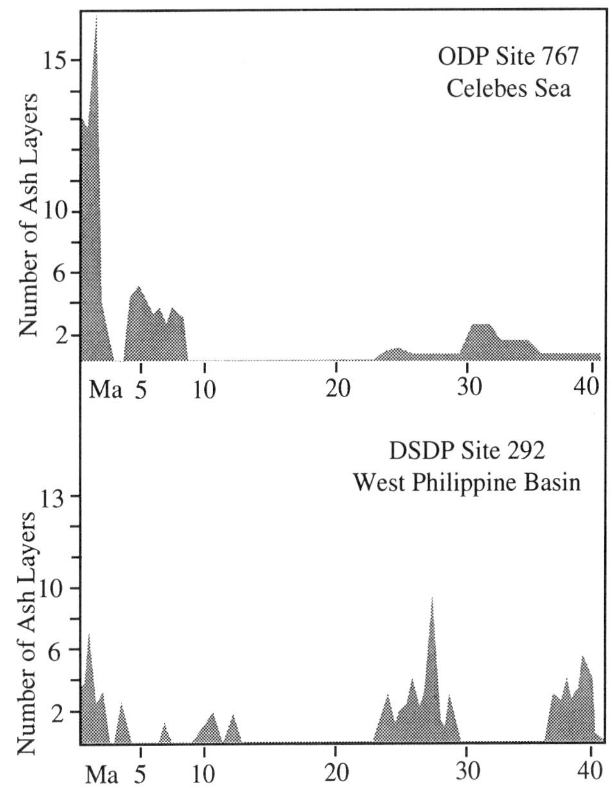

Fig. 3. Comparison of numbers of ash layers grouped by 0.5 Ma intervals for DSDP site 292 (Benham Rise, West Philippine Sea) and the ODP site 767 (Celebes Sea).

consistent temporal sequence of geochemical evolution in the arcs of the region.

2.3. *ODP Legs 125 and 126—Mariana Forearc and Izu-Bonin Backarc to Forearc Transect*

Following a major marine geophysical, dredging, submersible and drilling effort including DSDP legs 6, 31, 58, 59, 60 [*Fischer et al.*, 1971; *Karig et al.*, 1977; *Klein et al.*, 1979; *Kroenke et al.*, 1980; *Packham and Williams*, 1981; *Hussong and Uyeda*, 1981], and ODP legs 125 and 126 [*Fryer et al.*, 1992; *Taylor et al.*, 1992], we have a good understanding of many aspects of the evolution of the Philippine Sea plate (See *Taylor*, 1992 for review and this volume). Biostratigraphic and radiometric dating indicate the inception of arc magmatism in the Izu-Bonin system at about 48-49 Ma, preceding the change in Pacific plate motion at about 43.5 Ma. An Ar-Ar date of 45.8 Ma (*M. Cosca*, personal communication, 1993) acquired for a sample from the boninite basement drilled at site 786 in the forearc basement high (Figures 1b,c) confirms the general timing of this arc initiation, but no ages older than about 45 Ma have been determined for the Marianas. A major point made by *Taylor* [1992] is that the juvenile arc terrane developed at the margin of the West Philippine Basin during the period ~48-38 Ma was remarkable in terms of its across-strike active width of >300 km, dominated by boninite and island arc tholeiite series rocks, similar in many respects to many ophiolites. The Palau-Kyushu Ridge, the forearc and possibly much of the thick (~20 km) arc crust [*Shinohara et al.*, 1992; *Takahashi et al.*, 1993] of the Izu-Bonin-Mariana system was probably formed at this time.

In the late Eocene to early Oligocene, volcanism must have contracted to a narrower zone of eruptive activity. Rifting affected this major arc structure before 31 Ma, producing a forearc basin in the Izu-Bonin system that is locally floored by pillow lavas emplaced during maximum stretching. Spreading leading to formation of oceanic crust began at ~30 Ma behind the Mariana arc leading to the formation of the Parece Vela Basin, and ~25-22 Ma behind the Bonin arc leading to the formation of the Shikoku Basin (Figure 1). Backarc spreading had ceased in both basins at 15 Ma. Rifting followed by spreading may have reappeared in the Marianas as early as ~8 Ma, continuing to the present and leading to the formation of the Mariana Trough and isolation of the West Mariana Ridge as a remnant arc. Rifting also recommenced in the central Izu-Bonin arc at about 2 Ma. The total width of backarc basin crust is of the order of 1000 km in the case of the Marianas and 600 km behind the Izu-Bonin arc.

Understanding the nature of the Mariana and Izu-Bonin forearc basement was a primary and in part, successfully achieved aim of leg 125 [*Fryer et al.*, 1992; this volume] but in addition, numerous ash layers (Figure 4) were recovered from sites 782, 784 and 786 in the Izu-Bonin forearc (Figure 1c). These sites are presently located to the east of Sumisujima to Torishima volcanic centers in the active arc. The majority of the ashes recovered are Miocene to Recent in age with a single exception of a middle Eocene layer at site 786.

In a preliminary study of the ash sequences, *Arculus and Bloomfield* [1992] made the following observations:
(1) The compositional range is from basalt to rhyolite (Figure 5a);
(2) The majority are low-K tholeiitic with a minority of high-K character (Figures 5a and 5b);
(3) Crystalline phases associated with vitric shards are predominantly plagioclase, ortho- and clinopyroxene, titanomagnetite and olivine; no amphibole has been found;
(4) Within the age range represented by these sequences, there is neither evidence for a break in explosive activity of the arc nor evidence of consistent secular variation in composition (Figure 5c);
(5) Projections of the glass compositions cast in the form of molecular proportions coincide closely with the experimentally determined, atmospheric pressure cotectics for melt-olivine-pyroxene-plagioclase equilibria (Figures 5d and 5e);
(6) Distinct differences in petrology and geochemistry of the ash sequences compared with the volcanic and hypabyssal rocks forming the Eocene-Oligocene forearc basement [*Arculus et al.*, 1992a; *Murton et al.*, 1992] implicate the involvement of different mantle sources and indicate that across-arc strike variability in mantle wedge composition exists.

A temporally more extensive (0.1 to 31 Ma) record of activity in the Izu-Bonin arc was recovered during ODP leg 126 from the backarc rift (sites 790, 791), within the modern volcanic arc (site 788), and in the forearc from sites 787, 792, and 793 (Figure 1) in the form of volcaniclastic turbidites [*Hiscott and Gill*, 1992; *Gill et al.*, 1994]. Judging from the paucity of ashes and volcaniclastic turbidites, it appears from the forearc record that Izu-Bonin arc volcanism was minimal between 27 and 17 Ma and totally ceased between 23 and 20 Ma according to *Taylor* [1992]. The absence of ashes alone could not be regarded as conclusive evidence given the overlap of this period with possible subsidence of the arc concurrent with Shikoku Basin backarc spreading. But turbidites may still be shed from a submerged edifice, and their absence is significant. Although the turbidite samples analyzed are almost entirely volcanogenic in origin, the components range from products of single eruptions to mixtures of several.

Bulk compositions of the Oligocene turbidites range from basalt to dacite, many appear to be calcalkaline in character (Figure 6), and rare earth element (REE) patterns range from flat to slightly light REE-enriched (Figure 7). Isotopically, the Oligocene turbidites are very similar to the igneous forearc basement of the same age at site 792: $^{143}Nd/^{144}Nd$ = 0.51300 to 0.513115 (ε_{Nd} ~+7.4 to 9.6), ^{206}Pb = 18.14 to 18.22, and $\delta^{207}Pb$, ^{208}Pb are +2 and +40 respectively [*Gill et al.*, 1994] where the δ notation refers to deviation from

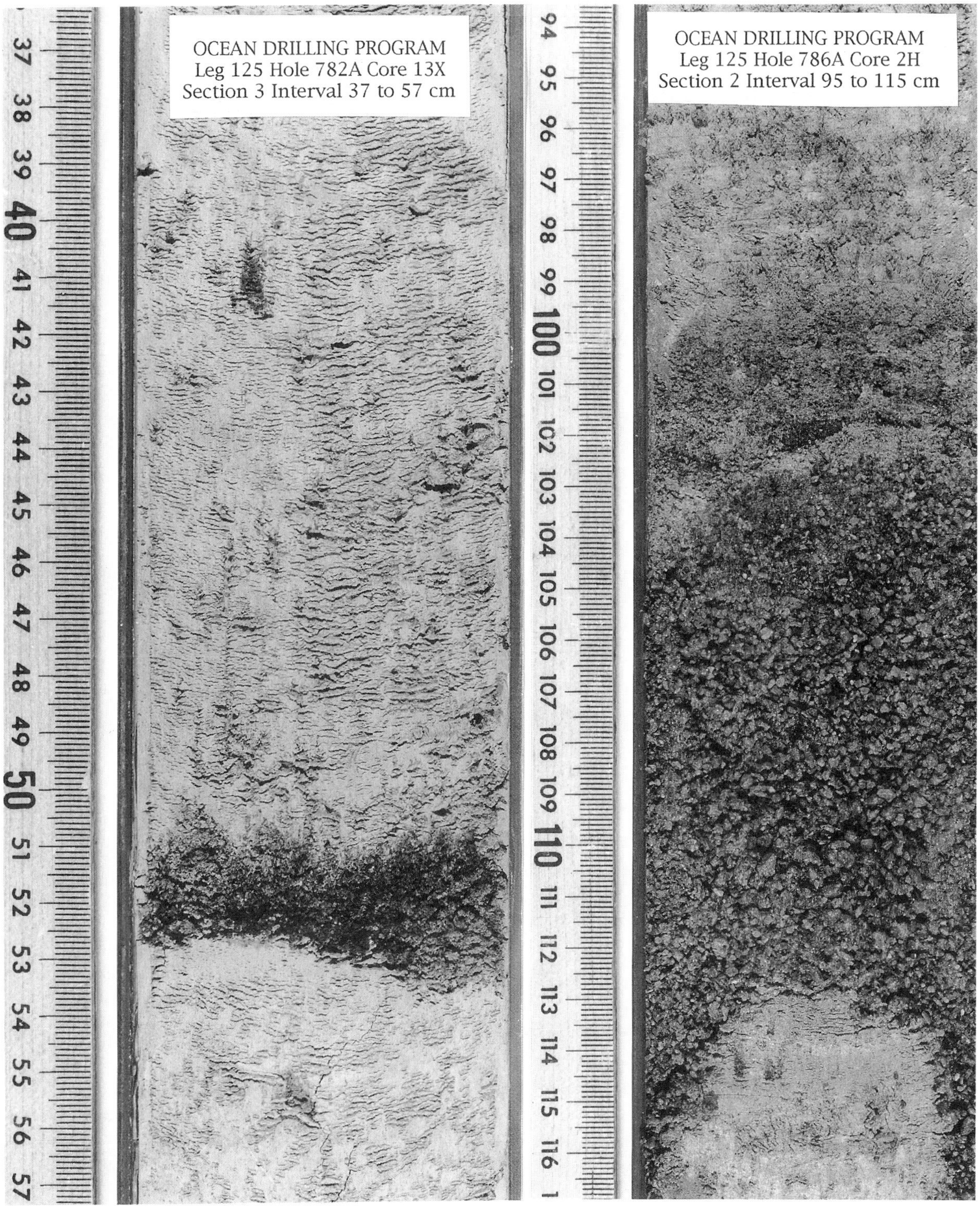

Fig. 4. Photographs of representative ash layers from leg 125 in the Izu-Bonin forearc. The sample from 782A, core 13X, Section 3, 51-53 cm is basaltic andesite in composition, and the sample from 786A, 2H, 2, 95-115 cm is rhyolitic.

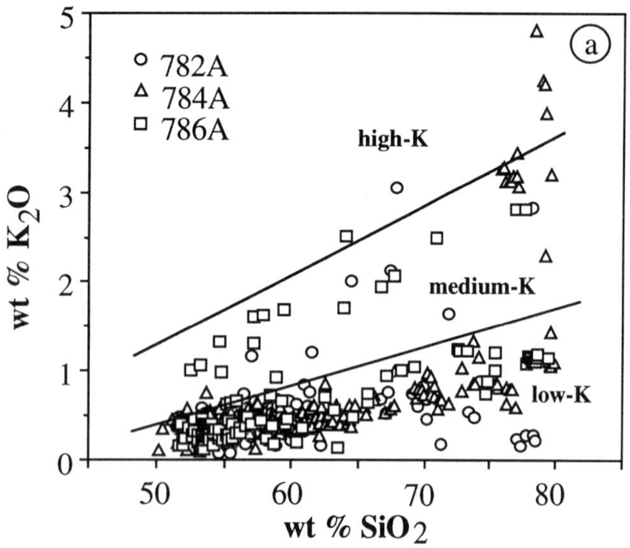

Fig. 5a. Plot of wt% K_2O vs. SiO_2 of individual glass shards from ODP leg 125 holes 782A, 784A and 786A, showing predominance of low-K compositions (*Gill*, 1981, discriminant boundaries) and a range from basalt to rhyolite.

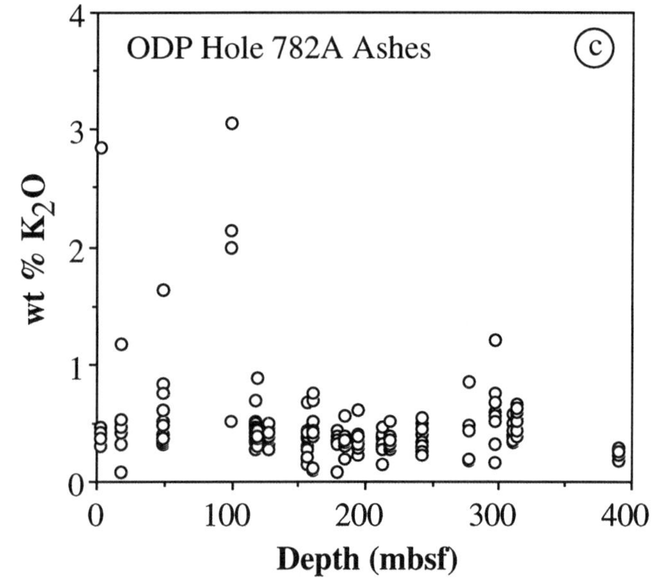

Fig. 5c. Plot of wt% K_2O vs. depth (meters below sea floor-mbsf) for individual glass shards from ODP leg 125 Site 782.

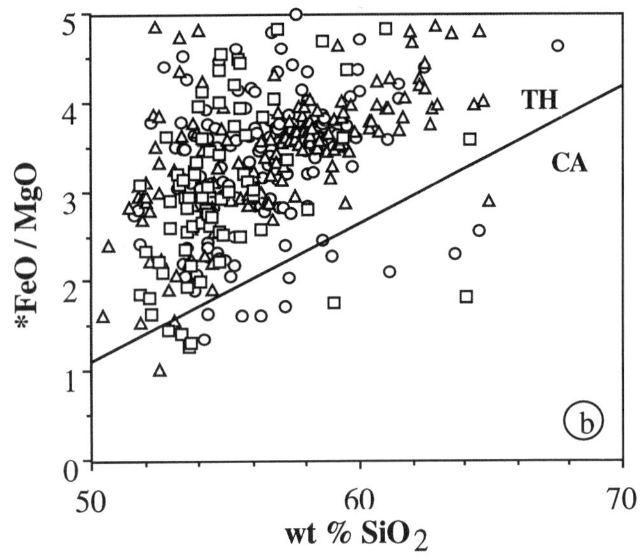

Fig. 5b. Plot of FeO*/MgO vs. SiO_{22} for individual glass shards from ODP leg 125 holes 782A, 784A and 786A (symbols as in Figure 5a) showing predominance of tholeiitic (TH) (*Miyashiro's* 1974 discriminant) compositions and a minority of calcalkaline (CA) samples.

the Northern Hemisphere Pb isotopic Reference Line (NHRL) of *Hart* [1984].

Resumption of arc activity after 15 Ma was characterized by different geochemistry. For example, these younger turbidites are generally tholeiitic, more felsic, and depleted in Zr and light REE (Figure 7). Isotopically, the < 15 Ma turbidites are similar to the older group with ε_{Nd} +6 to 9.4, δ^{207}Pb, ^{208}Pb of + 3.2 to 6.2 and +34 to 56 respectively, but with higher ^{206}Pb of 18.33 to 18.43. There is a good correspondence between the chemistry of the ashes (Figure 8) and turbidites <15 Ma, but we note that these younger ashes are predominantly tholeiitic unlike the more even distribution of turbidites between calcalkalic and tholeiitic (Figures 5b, 6, 7, and 8).

Subsequent to the initial study of leg 125 ashes by *Arculus and Bloomfield* [1992], several more detailed analytical efforts have been undertaken by different research groups on the leg 125 ashes as well as material from previous DSDP legs in the Izu-Bonin-Mariana region [*Arculus et al.*, 1992; *Chen and Arculus*, 1993; *Lee and Stern*, 1993; *Straub*, 1993]. Some important comparative features have emerged from these studies:

(1) The ash record from DSDP sites 458 and 459 (east of the currently active volcanoes of Pagan, Alamagan, and Guguan) is more extensive temporally in the Marianas than is the case for the Izu-Bonin system, ranging from the present to ~35 Ma, but with sedimentation breaks particularly at 32-30 Ma and 6-3 Ma [*Hussong and Uyeda*, 1981];

(2) Although the majority of the Mariana ashes are of low-K tholeiitic character, a significant proportion are of medium-K and a greater proportion than the Izu-Bonin system are high-K (Figures 9a and 9b);

(3) no evidence for consistent secular trends in ash composition can be discerned in the Marianas-a high- K pulse appears at ~8-11 Ma (~50-80 mbsf in DSDP Hole 459B) (Figure 9c);

(4) the Mariana ashes are predominantly derived from intrinsically more fertile mantle sources than those of the

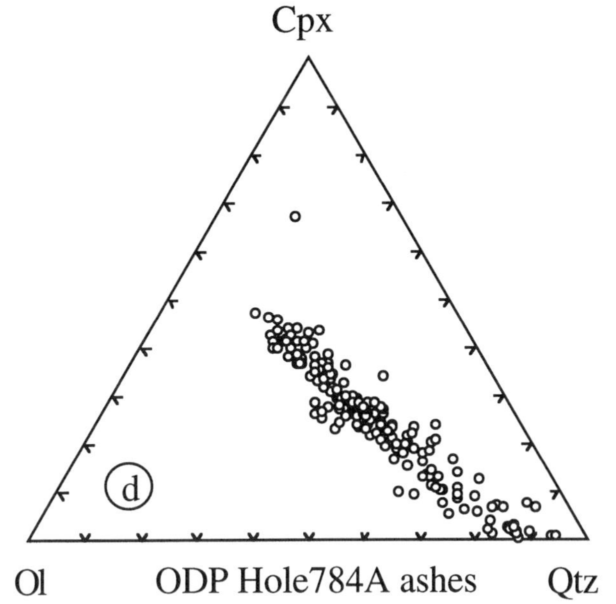

Fig. 5d. Oxygen-weighted molar projections (from plagioclase) of ash compositions from ODP leg 125 site 784, in the system olivine (Ol)-clinopyroxene (Cpx)-SiO_2(Qtz) with the experimentally determined atmospheric pressure cotectics after *Grove et al.* (1982) shown in Figure 5e.

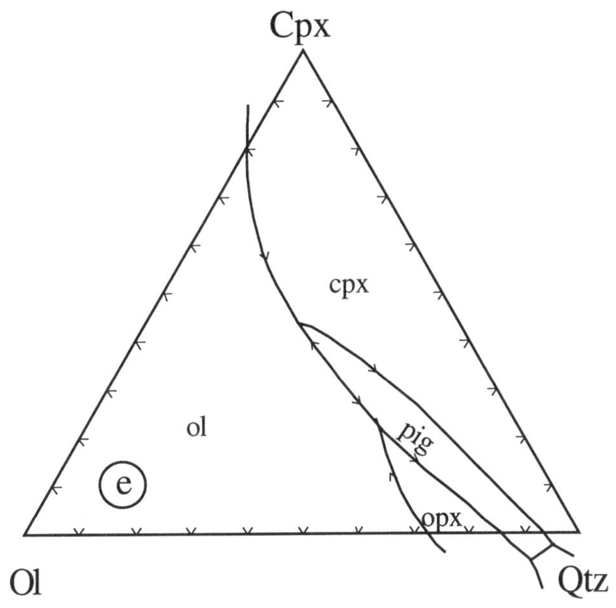

Fig. 5e. Oxygen-weighted molar projections (from plagioclase) of the experimentally-determined atmospheric pressure cotectics and primary liquidus phase fields in the system olivine (Ol) - clinopyroxene (Cpx)-SiO_2(Qtz), after *Grove et al.* (1982).

Izu-Bonin, as documented by lower CaO/K_2O at a given MgO content (Figures 9d and 9e). The <15 Ma Izu-Bonin ashes also contain consistently lower La/Sm than the Marianas (Figure 8);

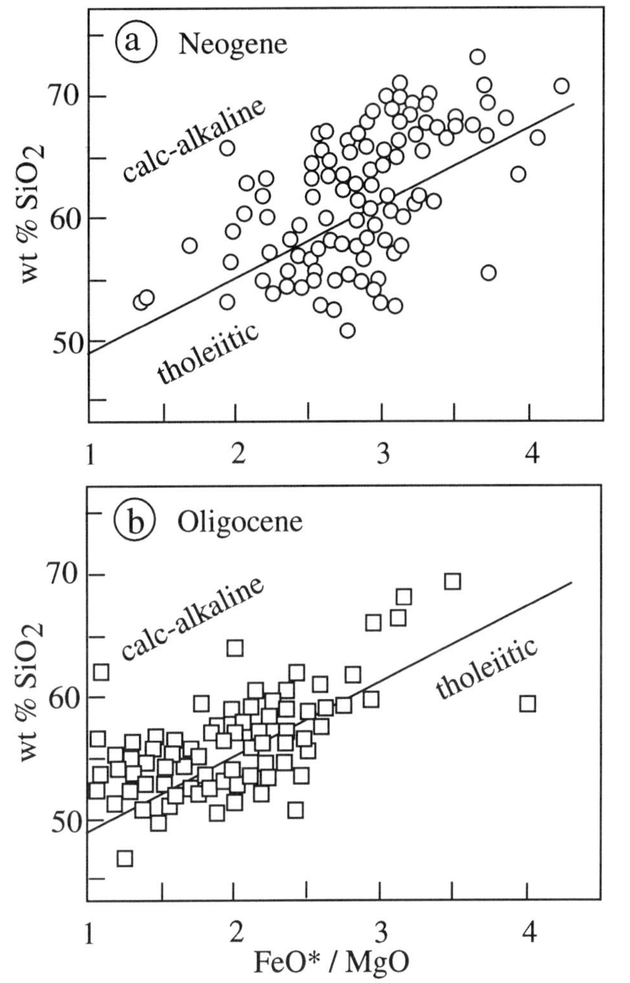

Fig. 6. Tholeiitic vs. calcalkaline comparison of Neogene (a) and Paleogene (b) turbidites recovered by ODP leg 126 from the Izu-Bonin arc using *Miyashiro's* (1974) FeO*/MgO vs. wt % SiO_2 discriminant.

(5) For both arcs, the trends of CaO vs. MgO variation of the modern arcs are essentially coincident with the whole of the ash record, and indicate comparable and unchanging crustal thicknesses of ~12 km according to the global averaging approach of *Plank and Langmuir* [1988] (Figure 10);

(6) Crystalline phases associated with Mariana vitric shards are predominantly plagioclase, ortho- and clinopyroxene, titanomagnetite and olivine; no amphibole or biotite has been found despite their presence in the modern arc especially in the northern part and the region of Iwo Jima [*Lin et al.* 1990]. No differences in pyroxene (Figure 11a), plagioclase or oxide (Figure 11b) compositions between the two arcs can be detected. Sparse occurrences of coexisting ilmenite-magnetite pairs define fO_2's = ~FMQ + 2 log_{10} units;

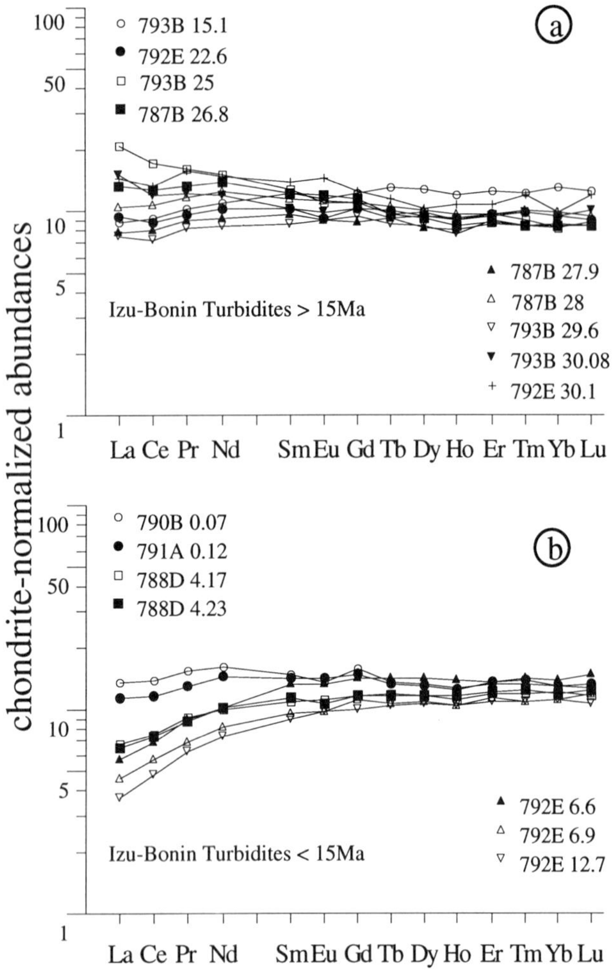

Fig. 7. Representative chondrite-normalized rare earth element abundances of >15 Ma (a) and <15 Ma (b) turbidites from Izu-Bonin recovered by ODP leg 126 in the forearc sites. Notation by the specific sample symbols is hole number followed by age in Ma.

(7) projections of the Mariana glass compositions cast in the form of molecular proportions again coincide closely with the experimentally-determined, atmospheric pressure cotectics for melt-olivine-pyroxene-plagioclase equilibria [see also *Baker*, 1987];

(8) in general, the Izu-Bonin ashes are characterized by higher $^{143}Nd/^{144}Nd$ (ε_{Nd} +8.5 to +9.5) than the Marianas (ε_{Nd} +7 to +8.5) (Figure 12) but the range of $^{87}Sr/^{86}Sr$ overlaps. A distinctive group of low $^{143}Nd/^{144}Nd$ (ε_{Nd} +5.5) and high-K ashes, similar to the modern Iwo Jima compositions, appears at ~8 Ma);

(9) *Warner et al.* (1987) report a brief period of boninitic ash deposition at site 450 in the Parece Vela Basin at 15-14 Ma, supposedly derived from the West Mariana Ridge; this episode is not recorded in the forearc sites (458 and 459) and we have not been able to identify this pulse in our studies of site 450.

In addition to the occurrences of ash layers at DSDP site 292, interpreted by *Pubellier et al.* [1992] to be derived from activity of the West Mariana Ridge between 30 and 23 Ma, important explosive episodes are also recorded from site 296 on the northernmost Palau-Kyushu Ridge and sites 445-446 (Daito Ridge and Basin) (Figure 1b). Ninety ash layers are reported by *Cambray* [1991] in nannofossil-carbonate-rich pelagic sediments at site 296. A major eruptive episode occurred between 30 and 23 Ma (less strongly developed at sites 445-446), and is regarded by *Cambray* [1991] as derived from the Palau-Kyushu Ridge prior to true spreading in the Shikoku Basin. Following a lull from 23 to 17 Ma with one pulse at ~21 Ma, a major episode is recorded at site 296

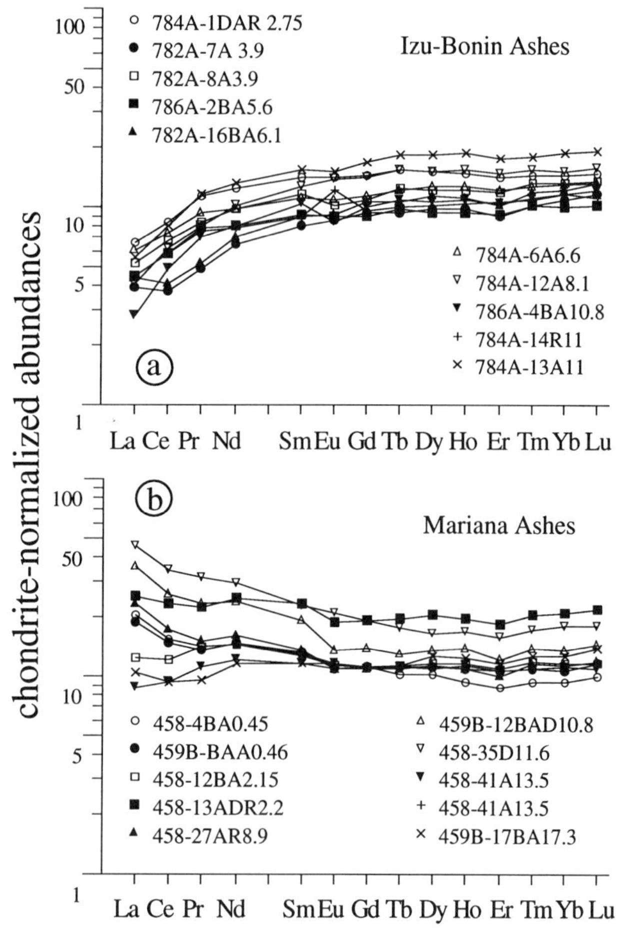

Fig. 8. Representative chondrite-normalized rare earth element abundances of ashes from the Izu-Bonin (a) forearc sites (ODP leg 125 - 782, 784 and 786) compared with the Mariana (b) forearc sites (DSDP leg 60 - 458 and 459). Notation by the specific sample symbols is hole number followed by layer number, compositional type (B-basalt, BA-basaltic andesite, A-andesite, D-dacite, R-rhyolite) and age in Ma.

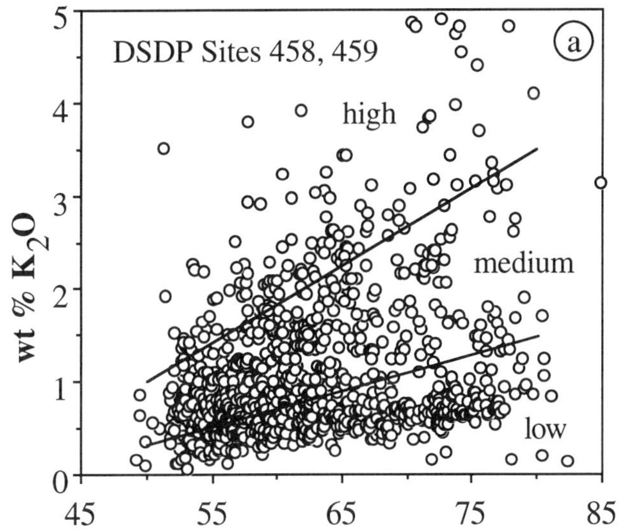

Fig. 9a. Plot of wt% K_2O vs. SiO_2 of individual glass shards from DSDP leg 60 sites 458 and 459, showing predominance of low-K compositions (*Gill,* 1981, discriminant boundaries), and a range from basalt to rhyolite.

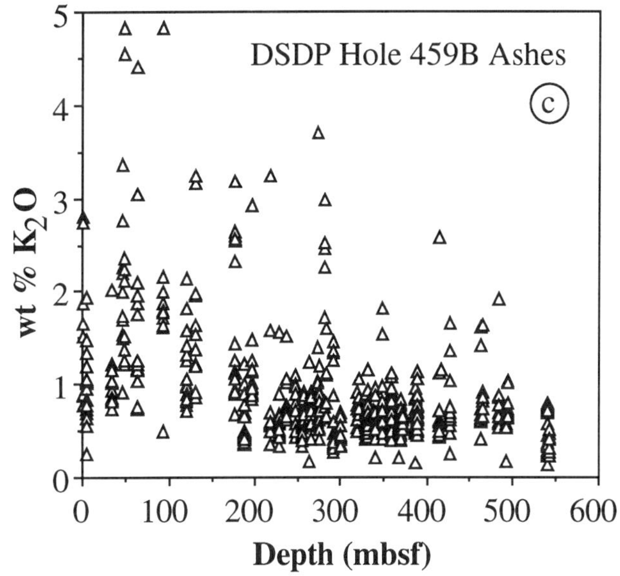

Fig. 9c. Plot of wt% K_2O vs. depth (meters below sea floor - mbsf) for ash compositions from DSDP leg 60 site 459.

from 17-14 Ma, which is even more strongly developed in the Shikoku Basin sites (442-443-444) arguing in favor of a derivation from the east or Izu-Bonin arc. Another major episode sourced from the Izu-Bonin arc is recorded in these Shikoku Basin sites between 12-10 Ma. From the Pliocene to the present, it is likely that the Ryukyu arc also became a prominent ash source. Note that the record from the backarc is somewhat at odds with the Izu-Bonin forearc as far as the period between 30 and 23 Ma is concerned. Prominent explosive eruptions are recorded in the backarc whereas *Taylor* [1992] regards this as a period of arc quiescence on the basis of the low sedimentation rate and paucity of preserved materials in the forearc.

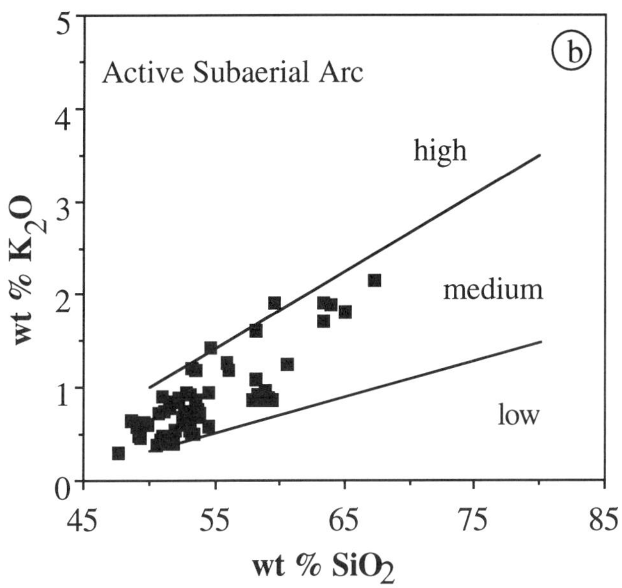

Fig. 9b. Plot of wt% K_2O vs. SiO_2 showing predominance of medium-K compositions in the active Mariana subaerial arc.

2.4. *ODP Legs 127 and 128—Japan Sea*

Understanding the mechanisms and consequences of the opening of the backarc basin were the primary aims of the twin leg campaign in the Japan Sea (this volume), but a diverse population of ash layers [*Pisciotto et al.*, 1992] was also recovered in sediments Miocene and younger at northern 794-795, and southern sites 797- 799 (Figure 1d). For the northern sites, alternations of eruptive episodes (12-10 Ma, and 2-0 Ma) and hiatuses occur. Similarly, *Cambray* [1991] identifies two major peaks of activity at the southern sites in the lower Miocene (19-14 Ma) and Pliocene-Recent (3-0 Ma).

Three major types of source contribute to the ash populations in the Sea of Japan: (i) low- to high-K dacite-rhyolite types derived from the Tohoku arc to the east; (ii) a Quaternary group of alkaline (acmite- and sometimes nepheline-normative) intermediate-SiO_2 types, probably derived at least in part from volcanoes in the western Japan Sea near Korea (e.g., Ulreung) and Chang Bai Shan on the Korea-China border; (iii) medium- to high-K shoshonitic -to-rhyolitic types derived from Kyushu, including some correlative with the major Quaternary deposits (Kikai-Akahoya, Aira-Tn and Aso-4) [*Fujioka*, 1985; *Machida and Arai*, 1983].

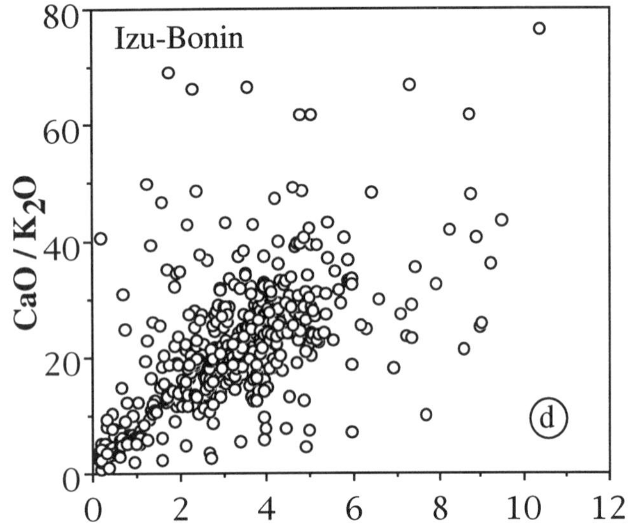

Fig 9d. Plot of CaO/K_2O vs. wt% MgO for individual shards recovered by ODP leg 125 at sites 782, 784 and 786 in the Izu-Bonin forearc; at any specific MgO content, increasing CaO/K_2O implies derivation from more refractory sources.

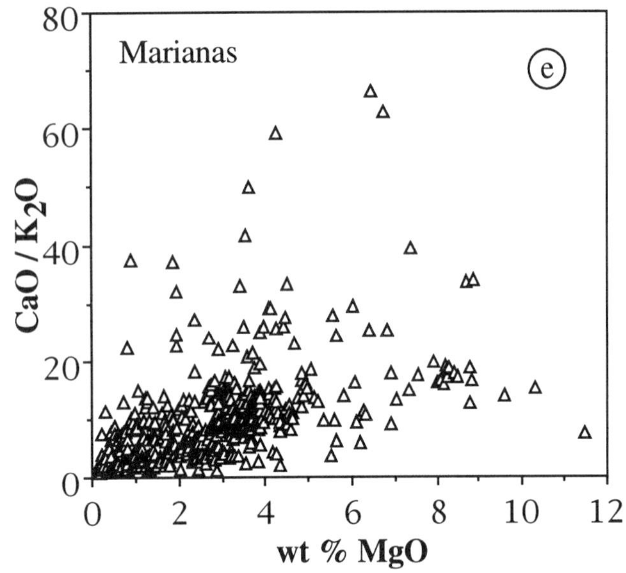

Fig. 9e. Plot of CaO/K_2O vs. wt% MgO for individual shards recovered by DSDP leg 60 at sites 458 and 459 in the Mariana forearc.

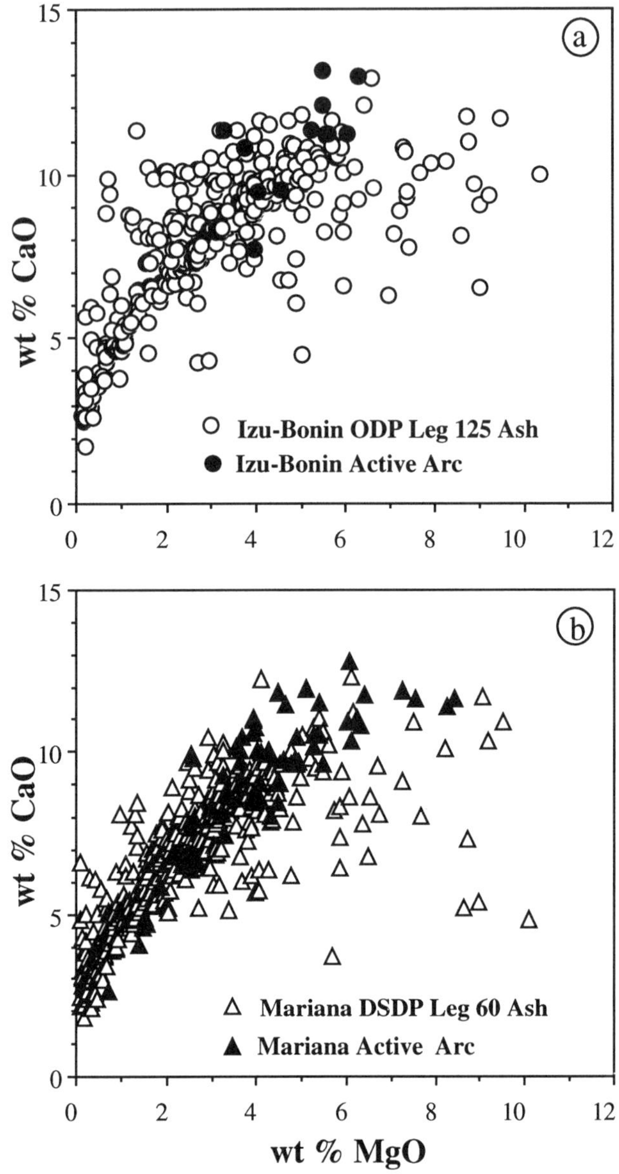

Fig. 10. Comparison of Izu-Bonin and Mariana ashes with active subaerial arc compositions in terms of wt% CaO vs. MgO. Note the coincidence of the majority of the ash compositions with the modern arcs and the overall similarity of the two arcs with respect to putative crustal thickness measures (specific wt% CaO at 6 wt% MgO).

Important comparative analyses of the frequency and composition of ashes deposited to the east of the Tohoku (northern Honshu) arc in the forearc (DSDP legs 56, 57 and 87A) and on the incoming Pacific plate (DSDP leg 86) have also been completed by *Cambray* [1991]. Important features are these:

(1) Major explosive pulses in the intervals 17-14 Ma and 5 Ma to the present coincide for the most part with the explosive record in the Japan Sea;

(2) The compositional range is restricted to dacite-rhyolite, with a range from low- through medium- to high-K ($0.5 < wt\%\ K_2O < 4$) character, both tholeiitic and calcalkaline, matching the present spread of compositions across the strike of the Tohoku arc;

(3) There is no consistent secular trend in composition.

Ohki et al. [1993] have demonstrated that the volcanic

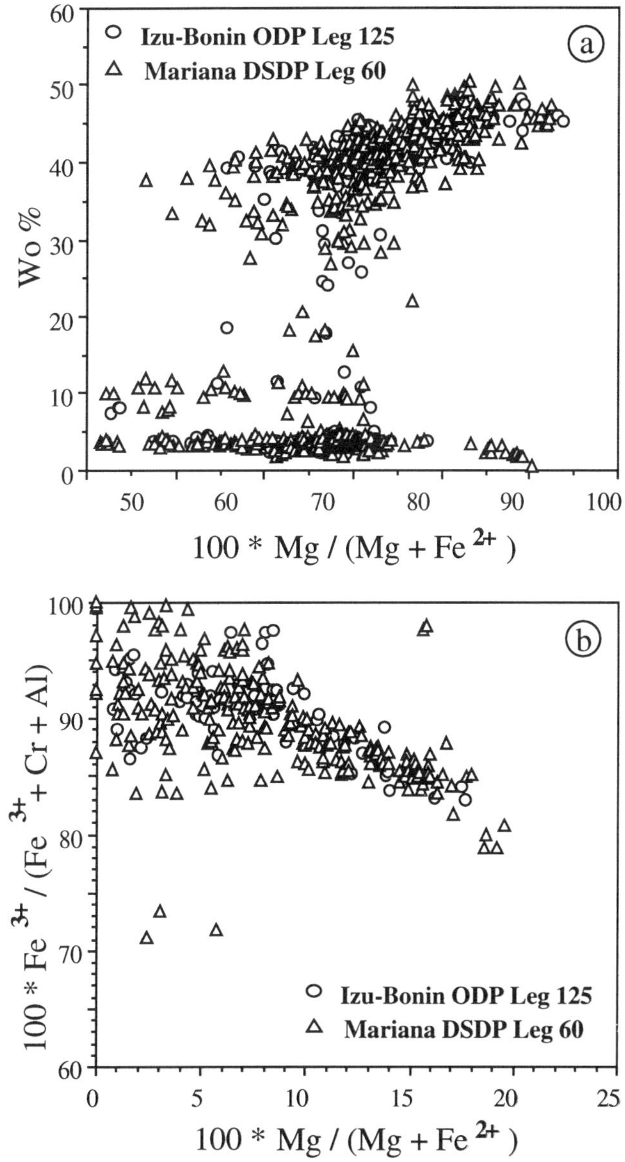

Fig. 11. Comparison of compositions of crystalline components in ashes from the Izu-Bonin (ODP leg 125) and Mariana (DSDP leg 60) systems: (a) % wollastonite (Wo) vs. Mg/(Mg + Fe^{2+}) of pyroxenes; note the cluster of Mariana compositions at high Mg contents are for bronzites probably derived by fragmentation of the basement bronzite andesite at the base of hole 458; (b) titanomagnetite compositions expressed as a function of proportions of trivalent cations and Mg/(Mg + Fe^{2+}).

front in the Tohoku arc has migrated spatially and temporally, complicating interpretations of geochemical evolution of the arc system. From 16-13 Ma, the volcanic front lay 30 to 50 km east of the Quaternary front and may have migrated westwards as a result of a shallowing of the angle of subduction of the Pacific plate.

Natland [1993] reported a sparse (12 ash layers and one rafted pumice bed) ash record in sediments <4.5 Ma in age, cored on the Shatsky Rise at site 810 (ODP leg 132), about 1600 km east of Japan (Figure 1e). The majority of the ashes are again dacite-rhyolite with a minority of andesitic composition, correlatable in part with layers cored on leg 86, and derived from the Tohoku-Kurile arc systems. Note that ash layers are restricted to sediments younger than 5 Ma at DSDP leg 86, sites 578 to 580, about 1900 km east of Japan.

2.5. *ODP Leg 145—Northern Pacific Transect*

The record of Tertiary-Quaternary geochemical, biological and sedimentologic changes in the northern Pacific together with the history of glaciation in the land masses surrounding the ocean were primary drilling aims of leg 145 [*Rea et al.*, 1993]. Sites chosen in the northwestern Pacific complemented a south-north transect initiated by DSDP leg 86 (Figure 1e). Site 881 and others (sites 882, 883, 884) located on the Detroit Seamount of the Emperor chain are presently about 600 kilometers to the east of the Kurile-Kamchatka arc system. Bold deployment of the advanced piston coring technique recovered temporally extensive siliceous oozes containing numerous ash layers derived from explosive activity in this arc. Several of these layers exceed a meter in thickness and if solely airfall, must represent colossal eruptive episodes. A particularly striking feature of the ash occurrence is the marked coincidence of a major increase in number of layers at 2.7 Ma, correlative with an increase in the amount of ice-rafted debris in the cores and

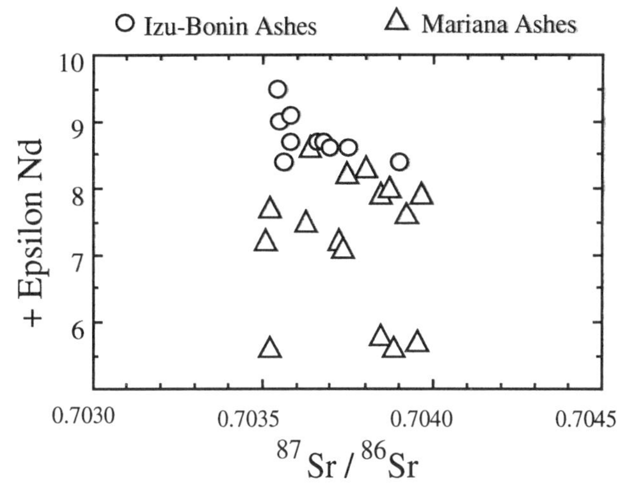

Fig. 12. Comparison of $_{Nd}$ (deviation of sample $^{143}Nd/^{144}Nd$ in parts in 10,000 from a chondritic value at present of 0.512638) vs. $^{87}Sr/^{86}Sr$ for representative Izu-Bonin (ODP leg 125) and Mariana (DSDP leg 60) ashes. Note the generally more radiogenic (Nd) values of the Izu-Bonin ashes, and the distinctive group of low $_{Nd}$ (~+ 5.5) shoshonitic ashes from the Marianas.

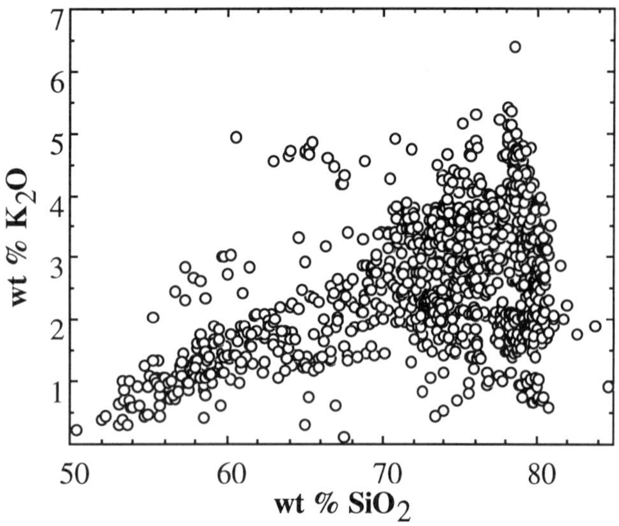

Fig. 13. Plot of wt% K_2O vs. SiO_2 for ashes from ODP leg 145 sites 881-884, showing extreme compositional variations from low- through medium- to high-K, and a range from basaltic andesite to rhyolite.

the onset of northern hemisphere glaciation [*Rea et al.*, 1993]. Geochemical analyses for a number of ash layers derived from explosive activity in the Aleutians that were recovered at site 887, located in the north eastern Pacific on the Patton-Murray Seamount chain, are not yet complete.

Cao et al. [1993] have described preliminary results of a geochemical and petrologic study of the ashes from the north western sites. Ash layers are sparsely distributed through the Miocene, but become more prominent in the late Miocene (at ~6.5 Ma). Major pulses are present at ~2.5-2.4, ~1.8-1.5, ~1.3-1.0, ~0.9-0.7, and ~0.3-0.4 Ma. Bulk compositions range from intermediate to high-SiO_2 in composition, but most are dacite-rhyolite with a considerable range of K content (low-, medium- and high-K) (Figure 13). The great majority of these ashes are tholeiitic on the basis of *Miyashiro's* [1974] criterion. We have not identified any regularity in temporal variations in composition. At least 12 compositional groups have been recognized on the basis of distinctive characteristics of major and trace element (including the REE) geochemistry (Figure 14). Isotopically, the ashes are similar in terms of $^{143}Nd/^{144}Nd$ and $^{87}Sr/^{86}Sr$ to the range known in the currently active Kurile-Kamchatka arc. The most abundant crystalline phase in the ashes is plagioclase but alkali feldspar is surprisingly rare. Pyroxene, Fe-Ti oxides, and amphibole are minor crystalline components.

We believe eventual integration of the ash record with known periods of activity of the different volcanic chains in Kamchatka (Sredinny range, central Kamchatka depression, eastern Kamchatka) will lead to an improvement in our understanding of the complex volcanic development in this arc. For example, it is known that the relatively alkaline Sredinny range was particularly active prior to the Quaternary, but only one major center (Ichinsky) is now active [*Fedotov and Masurenkov*, 1991]. Conversely, the central Kamchatka depression now contains a cluster of highly active volcanoes including Klyuchevskoy, the most active subaerial arc volcano in the world at present. The record of alkalinity *vs.*. time in the ash record accordingly is not a simple measure of the geochemical flux from a single geographic/tectonic location.

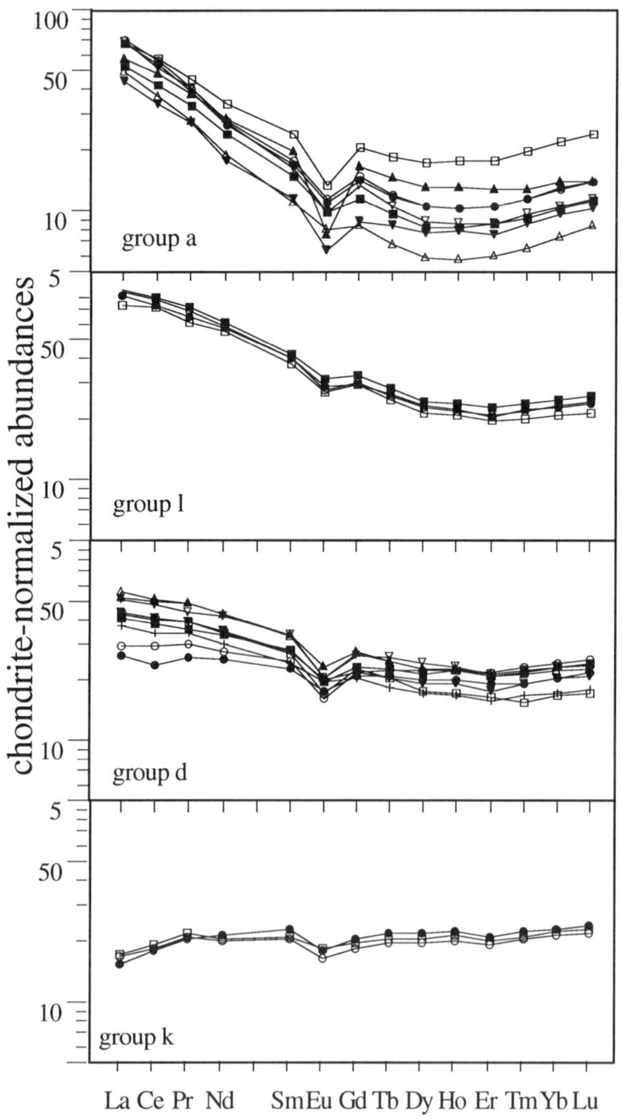

Fig. 14. Representative chondrite-normalized rare earth element abundances for four (groups a, l, d and k of *Cao et al.*, 1993) of the 12 compositional groups of ashes recovered by ODP leg 145 at sites 881-884, derived from the Kurile-Kamchatka arc system.

3. DISCUSSION

Emergent from the details of individual studies of the ash-turbidite record of arcs are points of general discussion relating to the questions and issues addressed in the introduction. It will be obvious from the varied level of geochemical and petrologic detail presented for individual legs that we lack consistent coverage of all of the parameters that are desirable. We also note that the drilling strategy is generally not optimized for physical volcanological studies so that sources of eruption cannot be identified precisely. Furthermore, the percentage of core recovery varies considerably but has generally improved during the ODP legs. Taking the issues in order:

3.1. *Are there any temporal changes in the geochemistry of arc magmas ?*

Recognizing the difficulty inherent in all of the ash-turbidite studies of detecting the exact location of eruptive sources, we clearly have evidence of temporal change for a number of arcs. However, a monotonic temporal increase in alkalinity from past towards present has not been found in any of these western Pacific arcs. Rather, magmas of varying alkalinity appear in the drilled record at different times in a number of these systems. In the case of the Sulu-Celebes Seas, Japan, and Kamchatka, interpretation of the changes is not simple, because a number of different volcanic loci have supplied material to the drilled locations. Vanuatu, furthermore, has experienced a flip in the polarity of subduction, albeit that the drilling has not recovered older ash deposits dating from the era of westward subduction of the Pacific plate. The least complex situations would appear to be the Izu-Bonin-Mariana and Lau-Tonga systems, but the apparent geographic simplicity belies a complex tectonic history.

It appears from the combined turbidite-ash record for the Izu-Bonin arc that a distinct change occurred in the relative fertility of the mantle sources tapped in the genesis of the magmas following the formation of the Shikoku Basin at 25-15 Ma. *Gill et al.* [1994] argue a straightforward model for arc magma generation during the period 15 Ma to present: (i) Melt was extracted from an upper mantle asthenospheric source to supply the Shikoku Basin; (ii) The residual (and relatively refractory) mantle is advected towards the arc by the induced flow of the subducted Pacific lithosphere; (iii) Ingress of subducted slab-derived fluid triggers melting of this refractory mantle, generating arc magmas. In general, the hypothesis that the relative refractoriness of the mantle sources of a number of arcs is controlled by prior backarc melt extraction has been advanced by a number of workers [e.g., *Ewart and Hawkesworth*, 1987]. Other aspects of this specific model are discussed below, but it is clear in a general sense that no overall increase in alkalinity during "maturing" of the Izu-Bonin arc can be detected—in fact, the opposite seems to be the case.

With a more complete temporal ash record from the Marianas, it appears that the products of the currently active arc are more alkaline than the majority of the explosive products through time (see Figures 9a and 9b). There are of course a number of complexities clouding a straightforward interpretation of this observation, not the least that considerable along-strike variation in alkalinity occurs in the active Izu-Bonin-Mariana system [e.g., *Lin et al.* 1990; *Yuasa*, 1992; *Yuasa and Nohara*, 1992]. Quite markedly alkaline lavas are being erupted in the zone near the junction of the two arcs in the vicinity of Iwo Jima. However, even the magmas of the active volcanoes of the central portion of the Mariana chain, adjacent to the maximum total backarc spreading of the Parece Vela and Mariana Trough, are derived from more fertile sources than those in the central Izu-Bonin arc.

3.2. *If so, are these regular, erratic, or blends of these end-member behaviors ?*

The previous melting/metasomatic history of the mantle, percentages of melting (possibly controlled by length of melting column or, to a first approximation, the distance from the top of the subducted lithosphere to the base of the arc lithosphere, see *Plank and Langmuir*, 1988), nature of crystalline phase-fluid-melt equilibria during melting, blend of subduction zone inputs, depth of arrest/ fractionation of magmas within the lithosphere, and tectonic/ structural effects in the arc lithosphere may all be important factors in controlling the alkalinity and other geochemical characteristics of arc magmas. Conditions promoting the longevity of a specific arc may not be simply related to those conducive to increasing alkalinity with time. Any degree of perceived erratic character of temporal geochemical changes is likely to be symptomatic of failures on the part of the observer to appreciate the interrelationships of some of these factors. Nevertheless, we are still some distance from understanding the interplay of these specific factors (and others) for any arc system.

3.3. *Can we detect crustal growth using geochemical parameters believed to correlate at least globally with crustal thicknesses in arcs?*

Correlations of geochemical parameters of mid-ocean ridge basalts (MORB) with physical factors such as ridge crest depths and inferred percentages of melting in the mantle have proved to be successful on a global scale [*Klein and Langmuir*, 1987] even if problematic in detail. In a survey of the geochemistry of the world's arc systems, *Plank and Langmuir* [1988] asserted a remarkable degree of overlap exists between MORB and arc basalts [see also *Perfit et al.*, 1980], and suggested degrees of melting are controlled at least in part by arc crustal (or lithospheric) thicknesses. Given that island arcs are believed to be the locus of continental crustal growth, we might anticipate some indication of crustal thickening as a function of arc longevity (e.g., decreasing wt% CaO and increasing Na_2O at specific MgO). Concentrating on the Izu-Bonin-Mariana system, we find in fact no evidence for temporal changes of this nature (Figure 10). In detail, in the case of the Izu-Bonin system, the inferred crustal thickness from the global correlation of *Plank and Langmuir* [1988] is ~12 km whereas measured thicknesses are ~20 km [*Shinohara et al.*, 1992; *Takahashi et al.*, 1993]. In fact, our knowledge of crustal thicknesses in intraoceanic arcs where we might expect the bulk of the crust to have been constructed solely by arc magmatism, is generally poor, and the recent Izu-Bonin study is a notable contribution to our understanding of crustal structure in arcs. We clearly need more data of these types.

Given the evidence for extension in the Izu-Bonin system from back- to forearc [*Taylor*, 1992], we might argue that crustal thinning concurrent with crustal growth has been important. However, the absence of any crustal thinning in the region of the forearc rift (Bonin Trough) and nascent backarc rifts [*Shinohara et al.*, 1992] is surprising in this context.

It is possible that the major period of crustal growth occurred in the initial stages of the Izu-Bonin-Mariana arc in mid-Eocene-early Oligocene, and that the extent of crustal growth subsequently has been minimal. However, determinations of crustal thicknesses are needed for the remnant arcs such as the Palau-Kyushu and West Mariana Ridges before we can ascertain whether or not the bulk of crustal growth occurred early in the history of the Izu-Bonin-Mariana system, or whether a more linear growth with time has occurred.

3.4. *Can changes in the style of eruptive activity and/or geochemistry be attributed to distinct tectonic events, and does arc volcanism change character or cease for any part of the rift-to-spread-to-cessation stages of backarc basin evolution ?*

Whether or not backarc basin activity signals a cessation of magmatism at the locus of the arc has been of interest since the nature of backarc basins was recognized [*Karig*, 1971, 1974], and the results of DSDP drilling on the Philippine plate became known [*Scott and Kroenke*, 1981; *Hussong and Uyeda*, 1981; *Karig*, 1983]. The channeling or otherwise of magmas from source regions in the upper mantle to the surface is currently believed to reflect in part the nature of the solid phase flow in the mantle [e.g., *Spiegelman and Mackenzie*, 1987]. We might anticipate that lithosphere spreading adjacent to an arc would modify the melt-residue flow regime in the mantle beneath the arc. But obtaining sufficient direct evidence of synchroneity or otherwise of arc vs. backarc magmatism has been difficult. In part, this may reflect the fact that during stretching and sundering of the arc, subaerial explosive activity vanishes because of deep submergence of the vents. In this sense, integration of the turbidite record is essential because juvenile volcanism may still be manifest by this style of sedimentation on the flanks of the arc [*Hiscott and Gill*, 1992].

Based on an analysis of the occurrence of ash layers at the DSDP sites on the western margin of the Philippine Sea plate (292 - Benham Rise; 296 - northern Palau-Kyushu Ridge - Figure 1b), *Cambray* [1991] argues that the peak of volcanism coincides with the first true spreading stages of the Parece Vela (~28 Ma) and Shikoku Basins (~25 Ma) respectively. As spreading becomes established, subaerial explosive activity diminishes. A renewed explosive pulse becomes apparent at site 296 at 15 Ma coincident with the termination of Shikoku Basin spreading. *Taylor* [1992] and *Hiscott and Gill* [1992] have also argued for a complete cessation of Izu-Bonin arc activity in the interval 23-20 Ma, based on the turbidite/sedimentary history of the forearc.

Integration of the DSDP and ODP explosive record to the east and west of northern Honshu [*Cambray*, 1991] indicates a similar general sequence of events accompanying the opening of the Sea of Japan. A major explosive episode (including the "Green Tuff") concurrent with the first stages of spreading in the Sea of Japan followed by apparent quiescence during the final stages of opening and uplift of the arc. On the other hand, it appears that arc volcanism in the Marianas has generally not been interrupted since the establishment of the Mariana Trough ≥5 Ma ago [see also *Karig*, 1983]. However, the situation may be more intricate in detail. For example, brief cessation of arc magmatism following the initiation of rifting and prompt re-establishment of shoshonitic magmatism in the arc has been argued for the Marianas [*Stern et al.*, 1988; *Lin et al*, 1990]. A swift temporal change ensues from shoshonitic to tholeiitic.

We find that potassic ashes with the Sr-Nd isotopic signature of Iwo-Jima character do indeed appear in the Mariana DSDP forearc sites at ~8 Ma. However, the source of these ashes has not been precisely located, and an alternative explanation for the generality of shoshonitic magmatism in the Izu-Bonin-Mariana system is that it has always been located at the junction of the two arcs (currently Iwo-Jima), possibly as the result of a long-lived mantle heterogeneity [*Arculus et al.*, 1992b]. This implies that a wave of shoshonitic magmatism should not have swept along the strike-length of the Marianas as arc resuscitation followed the propagating rifting.

3.5. *Does the concurrent or former existence of adjacent backarc magmatism affect the nature of arc magma sources, possibly through advection of melt-depleted mantle into the path of slab-derived fluids ?*

Pearce and Parkinson [1993] have recently reviewed the systematics of trace element distributions during mantle melting, especially those that are incompatible to varying degrees in residual phases but not hydrophilic. They further compare the predicted melt abundances with different initial abundances in the mantle prior to melting (modelling for example fertile MORB mantle, depleted MORB mantle and enriched mantle) with observed trace element systematics in various arc systems world-wide. From this comparative exercise, it is clear that arcs with active backarc basins (Mariana, New Britain, Tonga, Kermadec, Scotia) have the most depleted mantle sources, and this can best be attributed to prior loss of melt in the adjacent backarc basin [*Ewart and Hawkesworth*, 1987; *McCulloch and Gamble*, 1991]. The cases of Vanuatu, Ryukyu and Andaman appear to be exceptions, but tectonic complications, youthfulness and lack of current arc activity given the polarity of subduction (respectively) may be complicating factors Comparing the geochemistry of the Izu-Bonin arc volcanism pre- and post-Shikoku Basin spreading, *Gill et al.* [1994] conclude that extraction of backarc basin melt has rendered the potential mantle wedge source regions of the arc more refractory (see above).

However, there may be some additional complexities involved in this relatively straightforward accounting for the degree of fertility of arc vs. backarc sources. For example, one of the most striking features of the ~30 Ma ash record from the Mariana DSDP forearc sites (458 and 459) is the general consistency of compositions through time, with the exception of the pulses of K-rich (shoshonitic) activity. Despite the episodic but prolonged nature of backarc (Parece Vela and Mariana Trough) activity, it is surprising that we observe no tendency towards more refractory sources in the arc consequent to episodes of backarc activity.

It is possible that the nature and durability of the arc lithosphere and the locus of the active arc with respect to the lithosphere boundaries are also important in this respect. Despite the general isotopic (Sr-Nd-Pb) similarity of the West Philippine, Shikoku and Parece Vela Basins [*Hickey-Vargas*, 1991; *Hickey-Vargas et al.*, this volume], it is clear that the mantle sources of the West Philippine lavas were more depleted than those supplying the latter two backarc basins (Figure 15). Comparison of the Zr-Y geochemistry of the Izu-Bonin and Mariana ashes with the basin lavas shows a marked coincidence of the <15 Ma Izu-Bonin ashes and West Philippine Sea sources, and an overlap of Mariana ash chemistry with the Parece Vela and Mariana Trough lavas. In this context, it is interesting to note that the Eocene-Oligocene Palau-Kyushu Ridge arc magmas (DSDP site 448 basement) are also similar to the West Philippine Sea rather than the successor backarcs and currently active Mariana arc.

The tectonic locus of the active arc could be significant. The currently active Mariana arc is in fact located for the most part on the easternmost part of the Mariana Trough backarc crust [*Hussong and Uyeda*, 1981]—by itself, an indication perhaps that arc volcanism at least ceased for an interval during the initial stages of formation of the Mariana Trough. In contrast, the Izu-Bonin arc is located well to the east of the boundary of the Eocene-Oligocene arc crust with the eastern boundary of the Shikoku Basin [*Honza and Tamaki*, 1985; *Shinohara et al.*, 1992). In view of the persistently more refractory character of the <15 Ma ashes and active arc magmas in the Izu-Bonin compared with the Marianas, we suggest that the subarc lithosphere may be acting as a refractory filter to the passage of arc magmas in the general manner advocated by *Kelemen* [1986]. Note however, that the geochemistry and petrology of this Izu-Bonin lithosphere is also probably heterogeneous across-strike with strongly refractory rocks in the forearc (boninite-harzburgite association) and a more fertile suite of rocks towards the backarc region [*Arculus and Bloomfield*, 1992; *Pearce et al.*, 1992].

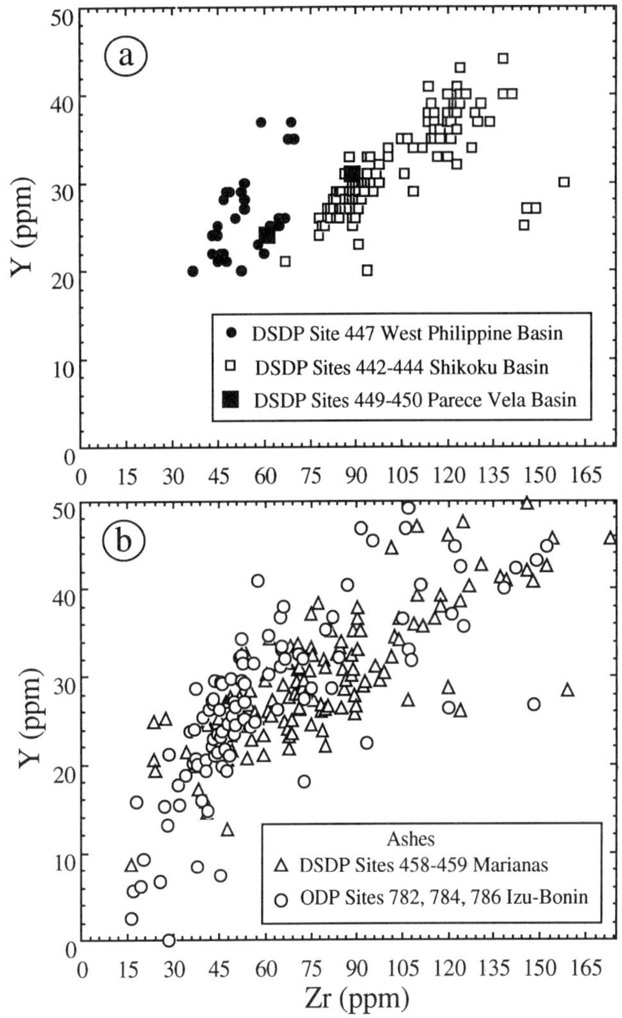

Fig. 15. Plot of Y vs. Zr of backarc and West Philippine Basin basement samples (analyses from DSDP Scientific Results volumes); (a) compared with proton-probe and ICP-MS data for ashes; (b) from the Izu-Bonin (ODP leg 125) and Mariana (DSDP leg 60) forearc sites. Higher Zr/Y implies increasing fertility of the mantle sources involved.

3.6. *Is there inter-arc synchroneity of major periods of explosive activity ?*

Whether some level of global synchroneity of explosive volcanism exists and whether such pulses could be climatically significant [*Kennett and Thunell*, 1975; *Ninkovich and Donn*, 1976; *Kennett et al.*, 1977] have been controversial issues. Excluding localized effects such as the subduction of adjacent backarc basins in the Philippine region for example, *Cambray* [1991] advocates synchroneity of explosive activity on the basis of a comprehensive analysis of the principal arcs bordering the Pacific plate. His analysis involved systematic frequency/ thickness ash layer determinations, and discrimination between adjacent arc and remote inputs. Two marked periods of concentrated explosive activity in arcs emerge:

1. lower to middle Miocene (18-13 Ma);
2. Pliocene-Quaternary (5 Ma-present).

But what the *global* trigger for these enhanced periods of explosive activity might have been is not clear.

3.7. *Do episodes of major explosive activity in arcs correlate with climate change ?*

It is of course clear from the recent direct experiences with sulfur-laden injections into the stratosphere of the products of El Chichon and Pinatubo during the past decade, that climate can be affected by violently explosive arc eruptions. A marked coincidence has been recognized between northern hemisphere major cooling, ice rafted debris distributed through the northern Pacific, and the dramatic increase of Kamchatkan volcanism [*Rea et al.*, 1993]. We note, however, that we are not yet in a position to document quantitatively the volumes (and eruption column elevations) of the major Pliocene eruptive episodes in the Kurile-Kamchatka system. A physical, volcanologically based drilling strategy is required to obtain these data prior to attempts at further quantification of stratospheric ash inputs and golobal climate change.

4. CONCLUDING REMARKS

Advent of microanalytical trace element techniques such as the proton and ion probes, and lately the laser ablation systems attached to inductively coupled plasma source mass spectrometers, are revolutionizing our capabilities of handling extremely fine-grained volcanic samples such as distal ashes. It is clear that a broad-based analytical approach (major-, trace-element, isotopic, mineralogical, phase equilibria) coupled with thorough understanding of the terrestrial and marine geology of the island arcs in the western Pacific will repay large dividends in terms of our knowledge of the temporal evolution of these systems. We have attempted in this review to provide a summary of the "state of play" in this endeavor, providing references to key literature and ongoing work, fully recognizing that research proceeds apace not only on the more recently acquired ODP samples but also the valuable resource represented by older DSDP materials.

Progress can clearly be made with a number of the sample sets from the various arc systems by a systematic approach to the generation of comprehensive and comparable data. As one of many possible examples, the interplay between sources in the mantle wedge (various MORB and ocean island reservoirs) coupled with subducted slab input described by *Hickey et al.* [this volume] as important in active arc petrogenesis, requires Pb isotopic analysis for resolution. Only in the case of the turbidite record from the Izu-Bonin system have such data been acquired [*Gill et al.*, 1994]. A specific physical volcanologic drilling strategy might also return dividends in the sense of sufficiently precise quantification of volumes of some of the major eruptions and their sources. Documentation furthermore, of the geochemical consequences of arc polarity reversal in the prime example of the Solomons-Vanuatu region, is also eminently approachable with a strategy of ash recovery. Sites displaced from maximum sedimentation in proximal arc locations may be necessary.

Acknowledgments. Our research in the area of ashes and turbidites has been variously supported by Australian Research Council and National Science Foundation Grants. Weibang (*aka* Eric) Chen gratefully acknowledges the receipt of an Overseas Postgraduate and University of New England scholarship. We enjoyed the vigorous debate of the Asilomar review meeting early in 1993 that forced us to view the problems holistically. Brian Taylor is especially thanked for the energy with which he has marshalled the western Pacific arc community, and his patience through another year. And finally, we appreciate Jim Natland's insights and critical review completed during a period of acute professional and personal stress.

REFERENCES

Arculus, R. J., and A. L. Bloomfield, Major element geochemistry of ashes from Sites 782, 784, and 786 in the Bonin forearc, in *Proc. ODP, Sci. Results*, *125*, P. Fryer, J. A. Pearce, L. B. Stokking, et al., pp. 277-292, Ocean Drilling Program, College Station, TX, 1992.

Arculus, R. J., J. A. Pearce, B. J. Murton, and S. R. van der Laan, Igneous stratigraphy and major-element geochemistry of Holes 786A and 786B, in *Proc. ODP, Sci. Results*, *125*, P. Fryer, J. A. Pearce, L. B. Stokking, et al., pp. 143-169, Ocean Drilling Program, College Station, TX, 1992a.

Arculus, R. J., C. Weibang, and W. L. Griffin, Bonin and Mariana island arc ash geochemistry : evidence for long-term mantle source distinctiveness, *EOS*, *73*, 646, 1992b.

Baker, D. R., Depths and water contents of magma chambers in the Aleutian and Mariana island arcs, *Geology*, *15*, 496-499, 1987.

Baker, P. E., Comparative volcanology and petrology of the

Atlantic island-arcs, *Bull. Volcanol.*, *32*, 189-206, 1968.

Baker, P. E., M. Coltorti, L. Briqueu, T. Hasenaka, and E. Condliffe, Volcanic ash layers from Sites 828, 830, 831, 832 and 833, Leg 134, New Hebrides island arc, in *Proc. ODP Sci. Results*, *13,* H. G. Greene, J-Y. Collot, L. Stokking, L., et al., Ocean Drilling Program, College Station, TX, in press, 1994.

Bellon, H., and C. Rangin, Geochemistry and isotopic dating of Cenozoic volcanic arc sequences around the Celebes and Sulu Seas, in *Proc ODP, Sci. Results*, *124*, C. Rangin, E. Silver, M. T. von Breymann, M. T., et al., pp. 321-338, Ocean Drilling Program, College Station, TX, 1991.

Cambray, H., Etude de téphra des sédiments marins: comparaison entre les épisodes d'activité volcanique et l'évolution tectonique, Unpub. PhD thesis, Univ. Pierre et Marie Curie, Paris VI, 347 pp, 1991.

Cambray, H., J-P. Cadet, and A. Pouclet, Ash layers in deep sea sediments as tracers of volcanic activity : Japan and central America as case studies, *The Island Arc*, 2, 72-86, 1993.

Cao, L. Q., R. J. Arculus, B. McKelvey, and Leg 145 Shipboard Scientific Party, Kamchatka and Kurile volcanism: the explosive record recovered during ODP Leg 145, *EOS*, *74,* 357, 1993

Chen, W., and R. J. Arculus, Ash geochemistry of Bonin and Mariana island arcs, *1993 IAVCEI General Assembly*, Abstr. Vol.,19, 1993

Collot, J-Y., H. G. Greene, L. B. Stokking, et al., *Proc. ODP Init. Repts.*, *134,* Ocean Drilling Program, College Station, TX, 1992.

Crawford, A.J., H. G. Greene, and N. J. Exon, Geology, petrology and geochemistry of submarine volcanoes around Epi island, New Hebrides island arc, in *Geology and Offshore Resources of Pacific Island Arcs - Vanuatu Region, Earth Science Series, 8*, edited by H. G. Greene and F. L. Wong, pp. 301-327, Circum-Pacific Council for Energy and Mineral Resources, Houston, TX, 1988.

Ewart, A., and C. J. Hawkesworth, The Pleistocene-Recent Tonga-Kermadec lavas: interpretation of new isotopic and rare earth data in terms of a depleted mantle source model, *J. Petrol.y*, *28*, 495-530, 1987

Fedotov, S. A., and Yu. P. Masurenkov, *Active Volcanoes of Kamchatka*, Nauka, Moscow, 432 pp., 1991.

Fischer, A. G., B. C. Heezen, et al., *Init Repts. DSDP*, *6*: Washington (U.S. Govt. Printing Office), 1971.

Fryer, P., J. A. Pearce, L. B. Stokking, et al., *Proc. ODP, Sci. Results*, *125*, Ocean Drilling Program, College Station, TX, 1992.

Fujioka, K., Synthesis of Neogene explosive volcanism of the Tohoku arc, deduced from marine tephra drilled around the Japan trench region, DSDP, Legs 56, 57 and 87A, in *Init. Repts. DSDP*, *87*, H. Kagami, D. E. Karig, W. T. Coulburn, et al., pp. 703-721, U.S. Govt. Printing Office, Washington, D.C., 1985.

Gill, J. B., Geochemistry of Viti Levu, Fiji, and its evolution as an island arc, *Contrib. Mineral. Petrol.*, *44*, 179-203, 1970.

Gill, J. B., *Orogenic Andesites and Plate Tectonics*, Springer-Verlag, New York, 390 pp., 1981.

Gill, J. B., R. N. Hiscott, and Ph. Vidal, Turbidite geochemistry and evolution of the Izu-Bonin arc and continents, *Lithos*, in press, 1994.

Gorton, M. P., The geochemistry and origin of Quaternary volcanism in the New Hebrides, *Geochim. Cosmochim. Acta*, *41*, 1257-1270, 1977.

Grove, T. L., D. C. Gerlach, and T. W. Sando, Origin of calc-alkaline series lavas at Medicine Lake Volcano by fractionation, assimilation and mixing, *Contrib. Mineral. Petrol.*, *80*, 160-182, 1982.

Hart, S. R., A large-scale anomaly in the southern hemisphere mantle, *Nature*, *309*, 753-757, 1984.

Hickey-Vargas, R. L., Isotopic characteristics of submarine lavas from the Philippine Sea: implications for the origin of arc and basin magmas of the Philippine tectonic plate, *Earth Planet. Sci. Lett.*, *107*, 290-304, 1991.

Hiscott, R. N., and J. B. Gill, Major- and trace- element geochemistry of Oligocene to Quaternary volcaniclastic sands and sandstones from the Izu-Bonin arc, in *Proc. ODP, Sci. Results*, *12,* B. Taylor, K. Fujioka, et al., pp. 467-486, Ocean Drilling Program, College Station, TX, 1992.

Honza, E., and K. Tamaki, The Bonin arc, in *The Ocean Basins and Margin : The Pacific Ocean, 7A*, edited by A. E. M. Nair, F. G. Stehli, and S. Uyeda, pp. 459-502, Plenum Press, New York), 1985.

Huang, T. C., A volcanic sedimentation model: implication of processes and responses of deep sea ashes, *Mar. Geol.*, *38*, 103-122, 1980.

Hussong, D. M., and S. Uyeda, Tectonic processes and the history of the Mariana arc: a synthesis of the results of Deep Sea Drilling Project Leg 60, in *Init Repts. DSDP*, *60,* D. M. Hussong, S. Uyeda, S., et al., pp. 909-929, U. S. Govt. Printing Office, Washington, D.C., 1981.

Karig, D. E., Origin and development of marginal basins in the western Pacific, *J. Geophys. Res.*, *76*, 2542-2561, 1971.

Karig, D. E., Evolution of arc systems in the western Pacific, *Ann. Rev. Earth Planet. Sci.*, *2*, 51-75, 1974 .

Karig, D. E., Temporal relationships between back arc basin formation and arc volcanism with special reference to the Philippine Sea, in *The Tectonic and Geologic Evolution of southeast Asian Seas and islands: Part 2*, edited by D. E. Hayes, *Geophys. Monograph 27*, pp. 318-325, Am. Geophys. Un., Washington, D.C., 1983.

Karig, D. E., Ingle, J. C., Jr., et al., *Init. Repts. DSDP*, *31*, U.S. Govt. Printing Office, Washington, D.C., 1975.

Kelemen, P. B., Assimilation of ultramafic rock in subduction-related magmatic arcs, *J. Geol.*, *94*, 829-843, 1986.

Kennett, J. P., and R. C. Thunell, On explosive Cenozoic volcanism and climatic implications, *Science*, *193*, 1231-1234, 1975.

Kennett, J. P., A. R. McBirney, and R. C. Thunell, Episodes of Cenozoic volcanism in the circum-Pacific, *J. Volcanol. Geotherm. Res.*, *2*, 145-163, 1977.

Klein, E. M., and C. H. Langmuir, Global correlations of ocean ridge basalt chemistry with axial depth and crustal thickness, *J. Geophys. Res.*, *92*, 8089-8115, 1987.

Klein, G. D., K. Kobayashi, et al., *Init. Repts. DSDP*, *58,* U.S. Govt. Printing Office, Washington, D.C., 1979.

Kroenke, L. W., R. B. Scott, et al., *Init. Repts. DSDP*, *59,* U.S. Govt. Printing Office, Washington, D.C., 1980.

Kuno, H., Origin of Cenozoic petrographic provinces of Japan and surrounding areas, *Bull. Volcanol.*, *20*, 37-76, 1959.

Lee, J. M., and R. J. Stern, Tracking the evolution of the Mariana arc using major element compositions of tephra glasses, *EOS, Trans. AGU*, *74*, 681, 1993

Lin, P. N., R. J. Stern, J. Morris, and S. H. Bloomer, Nd- and Sr-isotopic compositions of lavas from the northern Mariana and southern Volcano arcs: implications for the origin of island arc melts, *Contrib. Mineral. Petrol., 105*, 381-392, 1990

McCulloch, M. T., and J. A. Gamble, Geochemical and geodynamical constraints on subduction zone magmatism, *Earth Planet. Sci. Lett., 102*, 358-374, 1991.

Machida, H., and F. Arai, Extensive ash falls in and around the Sea of Japan from large Quaternary eruptions, *J. Volcanol. Geotherm. Res., 18*, 151-164, 1983.

Mallick, D. I. J., Some petrological and structural variations in the New Hebrides, in *The Western Pacific: Island Arcs, Marginal Seas, Geochemistry*, edited by P. J. Coleman, pp. 193-211, University of Western Australia Press (Perth), 1973.

Miyashiro, A., Volcanic rock series in island arcs and active continental margins, *Am. J. Sci., 274*, 321-355, 1974.

Murton, B. J., D. W. Peate, R. J. Arculus, J. A. Pearce, and S. R. van der Laan, Trace element geochemistry of volcanic rocks from Site 786: the Izu-Bonin forearc, in *Proc. ODP, Sci. Results, 125*, P. Fryer, J. A. Pearce, L. B. Stokking, et al., pp. 211-235, Ocean Drilling Program, College Station, TX, 1992.

Natland, J. H., Volcanic ash and pumice at Shatsky Rise : sources, mechanisms of transport, and bearing on atmosphere circulation, in *Proc. ODP Sci. Results, 132,* J. H. Natland, M. A. Storms, M. A., et al., pp. 57-66, Ocean Drillng Program, College Station, TX, 1993 .

Ninkovich, D., and W. L. Donn, Explosive Cenozoic volcanism and climatic implications, *Science, 194*, 899-906, 1976.

Ohki, J., N. Watanabe, K. Shuto, and T. Itaya, Shifting of the volcanic fronts during early to late Miocene in the northeast Japan arc, *The Island Arc, 2*, 87-93, 1993.

Packham, G. H., and K. L. Williams, Volcanic glasses from sediments from Sites 453 and 454 in the Mariana Trough, in *Init Repts. DSDP, 60,* D. M. Hussong, S. Uyeda, S., et al., pp. 483-495, U. S. Govt. Printing Office, Washington, D.C., 1981.

Pearce, J. A., S. R. van der Laan, R. J. Arculus, B. J. Murton, T. Ishii, D. W. Peate, and I. J. Parkinson, Boninite and harzburgite from Leg 125 (Bonin-Mariana forearc) : a case study of magma genesis during the initial stages of subduction, in *Proc. ODP, Sci. Results, 125*, P. Fryer, J. A. Pearce, L. B. Stokking, et al., pp. 623-659, Ocean Drilling Program, College Station, TX, 1992.

Pearce, J. A., and I. J. Parkinson, Trace element models for mantle melting : application to volcanic arc petrogenesis, in *Magmatic Processes and Plate Tectonics*, edited by H. M. Pritchard, T. Alabaster, N. B. W. Harris, C. R. Neary, *Geol. Soc. Spec. Publ., 76*, 373-403, 1993.

Perfit, M. R., D. A. Gust, A. E. Bence, R. J. Arculus, and S. R. Taylor, Chemical characteristics of island-arc basalts: implications for mantle sources, *Chem. Geol., 30*, 227-256, 1980.

Pisciotto, K. A., J. C. Ingle, Jr., M. T. von Breymann, J. Barron, et al., *Proc. ODP, Sci. Results, 127/128*, Ocean Drilling Program, College Station, TX, 1992.

Plank, T., and C. H. Langmuir, An evaluation of the global variations in the major element chemistry of arc basalts, *Earth Planet. Sci. Lett., 90*, 349-370, 1988.

Pouclet, A., M. Pubellier, and P. Spadea, Volcanic ash from Celebes and Sulu Sea basins off the Philippines (Leg 124) : petrography and geochemistry, in *Proc. ODP, Sci. Results, 124*, C. Rangin, E. Silver, M. T. Von Breymann, et al., pp. 467-487, Ocean Drilling Program, College Station, TX, 1991.

Pubellier, M., P. Spadea, A. Pouclet, R. Solidum, A.. Desprairie, and H. Cambray, Correlations of tephra in Celebes and Sulu Basins, constraints on geodynamics, in *Proc. ODP, Sci. Results, 124*, C. Rangin, E. Silver, M. T. Von Breymann, et al., pp. 459-465, Ocean Drilling Program, College Station, TX, 1991.

Rangin, C., and E. A. Silver, Neogene tectonic evolution of the Celebes -Sulu Basins : new insights from Leg 124 drilling, in *Proc. ODP, Sci. Results, 124*, C. Rangin, E. Silver, M. T. Von Breymann, et al., pp. 51-63, Ocean Drilling Program, College Station, TX, 1991.

Rea, D. K., I. A. Basov, T. R. Janacek, et al., *Proc. ODP, Init. Repts., 145,* Ocean Drilling Program, College Station, TX, 1993.

Scott, R., and L. Kroenke, Evolution of back arc spreading and arc volcanism in the Philipppine Sea: interpretation of Leg 59 DSDP results, in *The Tectonic and Geologic Evolution of Southeast Asian Seas and Islands: Part 1, Geophys. Monograph 23,* edited by D. E. Hayes, pp. 283-291, Am. Geophys. Un., Washington, D.C., 1981.

Shinohara, M., K. Suyehiro, H. Tokuyama, A. Taira, Y. Kaiho, T. Kanazawa, N. Takahashi, N. Hirata, and R. Hino, High resolution seismic crustal structure on northern Izu-Bonin arc trench system, *EOS, Trans. AGU, 73*, 489, 1992.

Spiegelman, M., and D. McKenzie, Simple 2-D models for melt extraction at mid-ocean ridges and island arcs, *Earth Planet. Sci. Lett., 83*, 137-152, 1987.

Stern, R. J., S. H. Bloomer, P. N. Lin, E. Ito, and J. Morris, Shoshonitic magmas in nascent arcs: new evidence from submarine volcanoes in the northern Marianas, *Geology, 16*, 426-430, 1988.

Straub, S. M, Volcanic glasses in tephra fallout layers from the Mariana Trough: a mirror of Mariana arc magma sources ? *1993 IAVCEI General Assembly*, Abstr. Vol.,109, 1993.

Sugimura, A., Zonal arrangement of some geophysical and petrological features in Japan and its environs, *J. Fac. Sci. Univ. Tokyo, 12*, 133-153, 1960.

Takahashi, N., M. Shinohara, K. Suyehiro, A. Taira, R. Hino, N. Hirata, and T. Kanazawa, Seismic structure of the Izu-Ogasawara arc using airgun and explosion-OBS data, *EOS, Trans. AGU, 74*, 415, 1993.

Taylor, B., Rifting and the volcanic-tectonic evolution of the Izu-Bonin-Mariana arc, in *Proc. ODP, Sci. Results, 12,* B. Taylor, K. Fujioka, et al., pp. 627-651, Ocean Drilling Program, College Station, TX, 1992.

Taylor, S. R., The origin and growth of continents, *Tectonophysics, 4,* 17-34, 1967.

Warner, R. J., M. F. J. Flower, and K. S. Rodolfo, Geochemical evidence for sundering of the west Mariana arc in Miocene ash from the Parece Vela Basin, *J. Volcanol. Geotherm. Res., 33*, 263-281, 1987.

Yuasa, M., Origin of along-arc geologic variations on the volcanic front of the Izu-Ogasawara (Bonin) arc, *Bull. Geol. Survey Japan, 43*, 457-466, 1992.

Yuasa, M., and M. Nohara, Petrographic and geochemical along-arc variations of volcanic rocks on the volcanic front of the Izu-Ogasawara (Bonin) arc, *Bull. Geol. Survey Japan, 43*, 421-456, 1992.

R. J. Arculus, Department of Geology, Australian National University, Canberra, A.C.T., 0200, Australia.

J. B. Gill, Earth Sciences Department, University of California, Santa Cruz, CA 95064, U.S.A.

H. Cambray, Unité de Genié Océanique, Institut Méditerranean de Technologie, 13451 Marseille cedex, France.

W. Chen, Department of Geology and Geophysics, University of New England, Armidale, N.S.W., 2351, Australia.

R. J. Stern, Center for Lithospheric Studies, University of Texas at Dallas, Richardson, TX 75083-0688, U.S.A.

Volcaniclastic Sedimentation and Volcanism During the Rifting of Western Pacific Backarc Basins

Peter D. Clift

Ocean Drilling Program, 1000 Discovery Drive,
Texas A&M University, College Station, Texas 77845, USA

The episodic rifting of both continental and oceanic volcanic arc complexes to form marginal basins has characterized the tectonic evolution of the western Pacific since the Eocene. The sedimentary sequences that have accumulated within the backarc and forearc basins provide a record of the nature, volume, and chemistry of volcanism in different parts of the arc system during the pre-rift, syn-rift, and spreading stages of basin opening. The location of rifting relative to the active volcanic front is a significant factor in governing the nature of basin opening, which is therefore not uniform between different arcs. High silica volcanism in the Japan, Izu-Bonin, Mariana and Tonga arcs reached a peak of intensity during earliest rifting and was then followed by a brief (approximately 1 m.y.) hiatus during creation of the earliest backarc crust in the Sumisu Rift, Sulu Sea, Lau Basin, and South Fiji Basin. Elsewhere volcanism in the Japan and Ryuku arcs continued through the initial breakup period. Eruption of MORB-like basalts marked the early rifting stage in the Sumisu Rift, Japan Basin, Sulu Sea, and Lau Basin, while the adjacent arc produced voluminous acidic and andesitic volcanics. However, after rifting of the Parece Vela, Shikoku, and South Fiji Basins, arc volcanism declined markedly. In the Lau Basin the early rift stage was prolonged and volcanism on the remnant arc was modest. Volcanoes within the extending arc crust were the focus of explosive volcanism. A new arc, the Tofua arc, formed as propagating spreading centers developed during the transition to the spreading stage. However, in most cases the rift stage did not involve voluminous volcanism in the basin itself. In the Ryuku arc volcanism continued through the rifting of the Okinawa Trough, but retreated as propagating spreading ridges developed. Trace and rare earth element data from the Tonga and Izu-Bonin arcs show trends to increasing depletion of incompatible elements prior to rifting of the Lau and Shikoku Basins and a subsequent enrichment with time after activity resumed on the arc. These trends are principally the result of arc lithospheric thickness controlling the degree of partial melting. Volcanism in the basin is generally more enriched than that on the volcanic front, possibly a result of melt extraction along backarc spreading ridges. Pb, Sr, Nd isotopes from these basins also show the strong heterogeneity between early backarc basin and arc mantle sources, as well as temporal changes in the mantle source under the arc caused by roll back of the subducting plate.

INTRODUCTION

Continuous subduction and related arc magmatism have been active in the western Pacific since the middle Eocene [45-50 Ma; Karig, 1975] and represent the classic examples of these geotectonic phenomena in the modern oceans. Episodes of rifting, which result in the formation of marginal or backarc basins, punctuate long periods of arc volcanism in both oceanic and continental arcs. The aim of this paper is to document how the nature of arc volcanism and volcaniclastic sedimentation has responded to this process and to offer some explanations as to what may be controlling those characteristics common to all backarc rifting events. Figure 1 highlights the volcanic arcs that form the basis of this study, some of which have been the subject of recent investigation by the Ocean Drilling Program during legs 124-128, 134 and 135. Also considered are data collected during earlier campaigns by the Deep Sea Drilling Project. Figure 2 shows a schematic cross section of an oceanic island arc and its various component pieces.

Active Margins and Marginal Basins of the Western Pacific
Geophysical Monograph 88

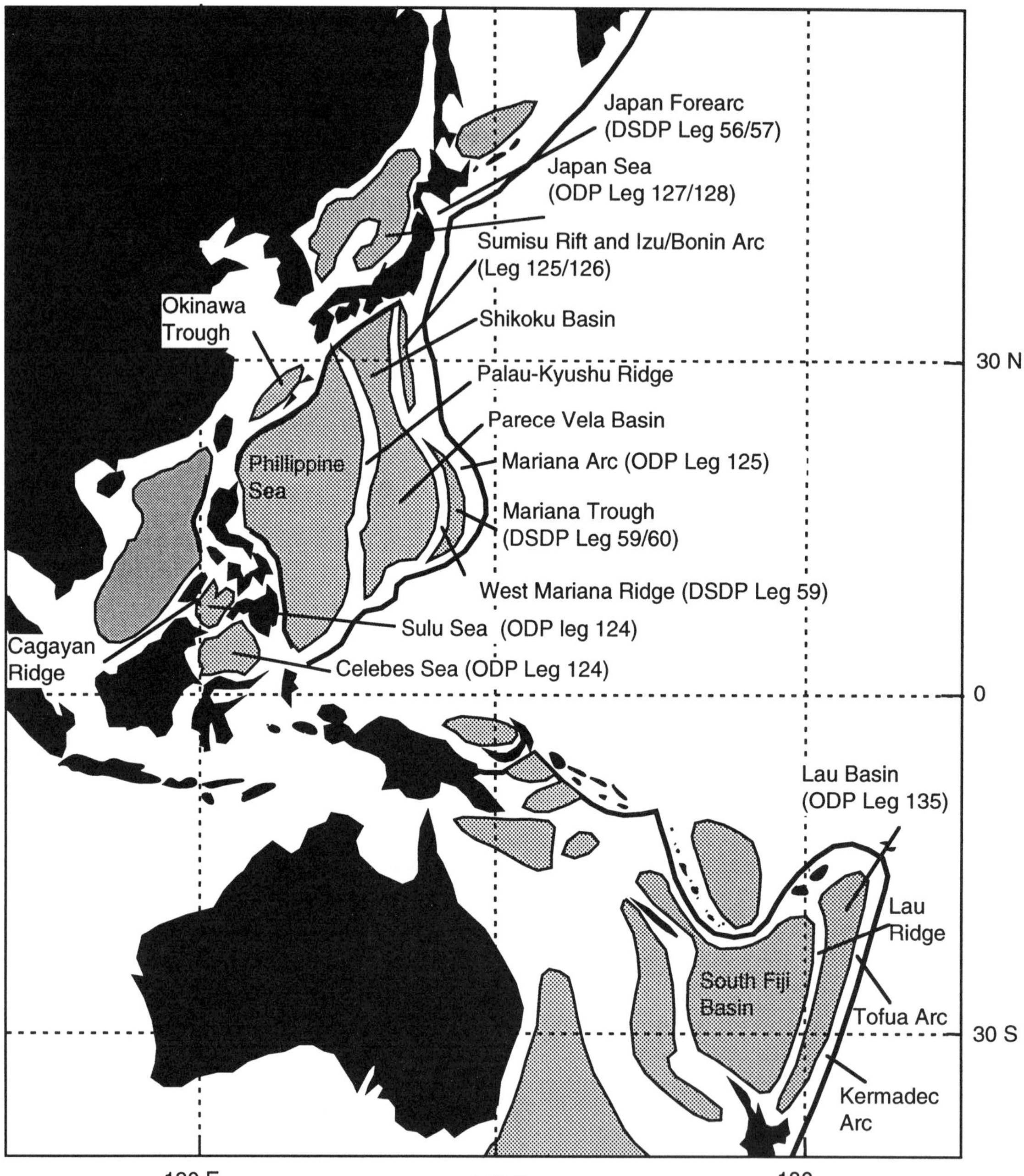

Fig. 1. Simplified geotectonic map of the western Pacific region showing active modern subduction zones. Volcanic arcs targeted by ODP and discussed in the text are Japan, Izu-Bonin, Marianas, Cagayan Ridge, and Tonga. The Kermadec arc is discussed but has not been drilled.

The formation of backarc basins is a two-part process: an initial stage of rifting, followed by a later stage of spreading. Basin formation involves the splitting of an original arc edifice into two blocks. The remnant arc (Figure 2) is separated from another ridge on the trench side of the basin. Breakup may leave the line of active arc volcanoes, the volcanic front, on either side of the rift. Regional extension similar in style to a continental "basin and range" style tectonism characterizes the arc breakup, defined as the start of rifting. It typically produces a block

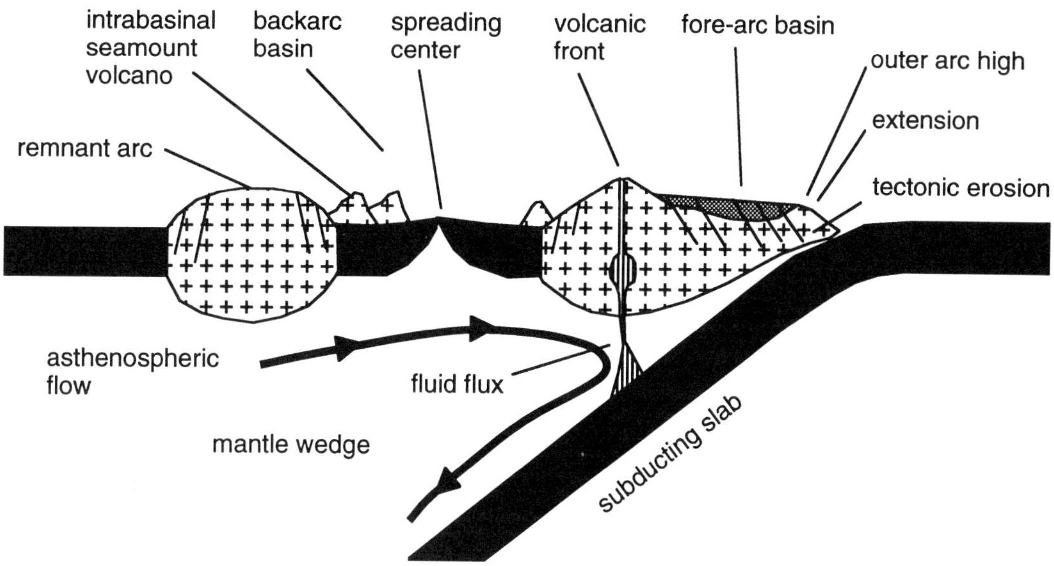

Fig. 2. Simplified tectonic diagram of an ideal arc system in the western Pacific showing the essential components discussed in the text.

and basin topography of arc crust, separated by graben floored with basaltic material. No identifiable magnetic anomaly pattern can be observed. Dates of rifting cited in this paper refer to the age of the oldest known crust generated in the backarc basin. The duration of any such rifting is quite variable, and may rapidly progress to the spreading stage, which may be said to have started when extension becomes focused along a spreading center, producing basaltic crust with a regular magnetic anomaly pattern, similar to a normal major ocean basin.

This paper principally draws on the marine volcaniclastic record recovered by the ODP/DSDP drilling in both the forearc and backarc regions of western Pacific arcs in an attempt to reconstruct the volcanic history of each area (Figure 3). Recent work in the forearc regions of the Tonga [Clift and Dixon, 1994], and Izu-Bonin arcs [Hiscott and Gill, 1992; Arculus and Bloomfield, 1992; Fujioka and Saito, 1992] indicates that the sedimentary sequences that have accumulated in these places can be used to reconstruct the nature of the island arc volcanism since the initiation of subduction (Table 1; Figure 4). The forearc basins are often far removed from continental or other clastic sediment sources, and it is thus not surprising that they should be filled with the erosive products of the adjacent arc. What is important is that the sediments deposited on the forearc at any one time tend to reflect the nature of the arc volcanism at that time. This feature of forearc basins reflects the fact that the basins of the western Pacific show steady subsidence throughout the period of subduction, possibly due to tectonic erosion of the underside of the plate. This contrasts with the behavior of the outer arc high. Drill sites and seismic data from these regions clearly show that the high is not subject to the same degree of subsidence and typically accumulates a more reduced sedimentary cover often containing a number of hiatuses. Figure 5 shows a comparison between sedimentary logs from the Izu-Bonin forearc basin at site 792 and the outer arc high at site 782. The Miocene to Eocene section at site 782 is reduced compared to that of site 792.

It is noteworthy that volumetrically the contribution of primary airfall ash layers to forearc basin sequences is very small and the majority of the volcaniclastic sediment comprises turbidite and mass flow sands and gravels. Despite this mode of sedimentation, sedimentary reworking of older parts of the volcaniclastic stratigraphy is negligible. Study of the marine ash record provides a better regional picture of the arc volcanism than direct examination of the volcanic edifices themselves, as sediment at any one place may be derived from a number of volcanoes along the length of the arc. Although direct mass wasting of the closest arc volcano might be expected to form the single greatest source of sediment at any one point in the forearc basin, ash deposits may spread debris up to 2000 km downwind of the point of eruption [e.g., Shane and Froggatt, 1991]. The result is that the volcaniclastic sediment at any one point may represent the products of volcanism from a number of arc centers. However, such

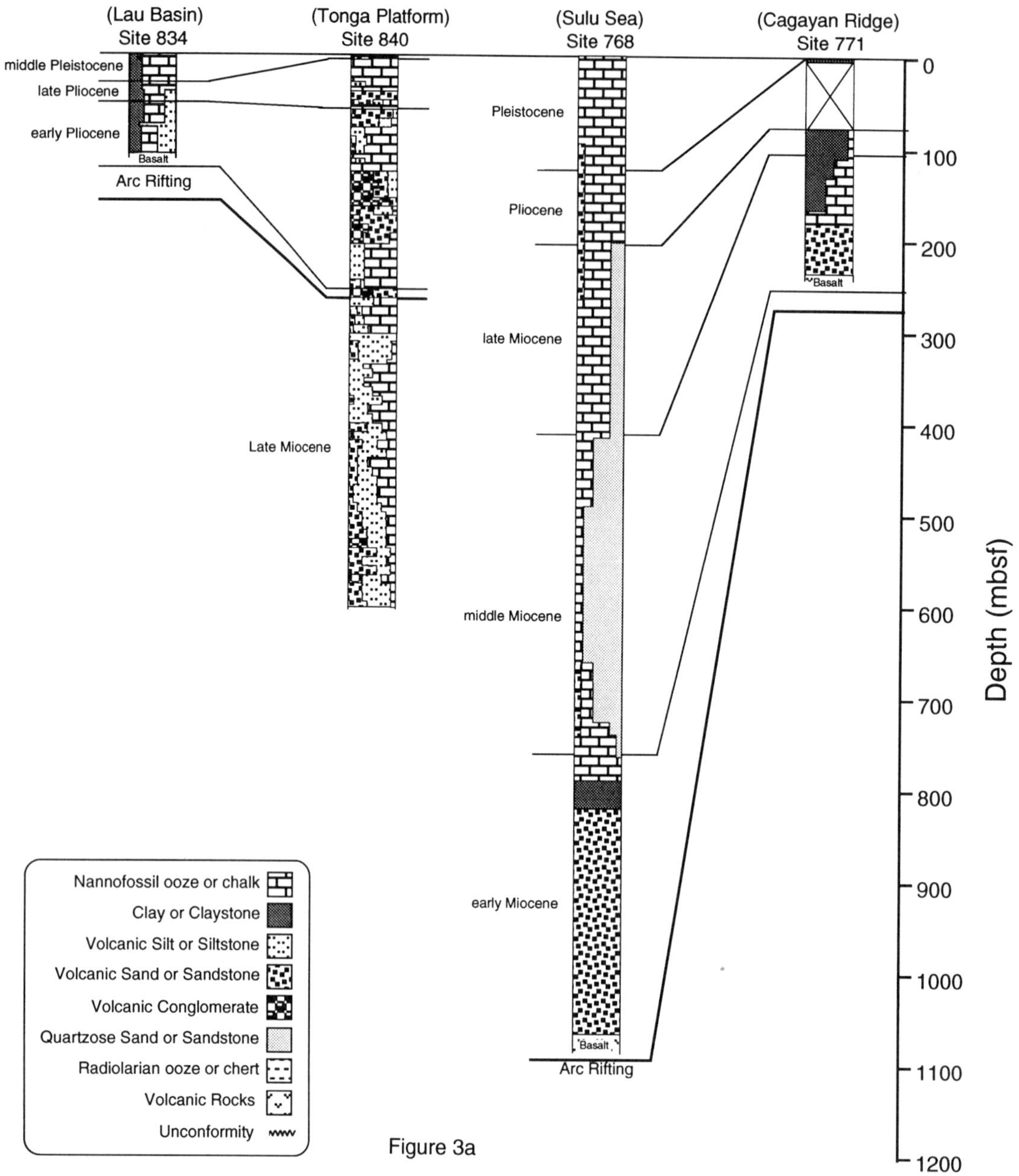

Fig. 3. Representative sedimentary logs from the backarc basin and forearc of each of the arc systems drilled during the ODP/DSDP campaign. Thick dark lines shows the age of arc breakup during rifting and backarc formation in each system. Logs are not to scale and are derived from Parson et al. [1992], Rangin et al. [1990], Hussong et al. [1982], Taylor et al. [1990], Fryer et al. [1990], Tamaki et al. [1990] and Shipboard Scientific Party [1980].

sampling naturally favors the more explosive, high-silica volcanism, which may result in a biased record. Another complexity with the sedimentary record is that subaerial volcanic eruptions are more likely to spread ash over wide areas than submarine eruptions. Submarine activity may result in apparently contradictory ash records being deposited in different parts of a given arc system. Subaerial, high-silica volcanism is represented throughout the arc system, while submarine and/or low silica activity will tend only be represented close to the site of eruption.

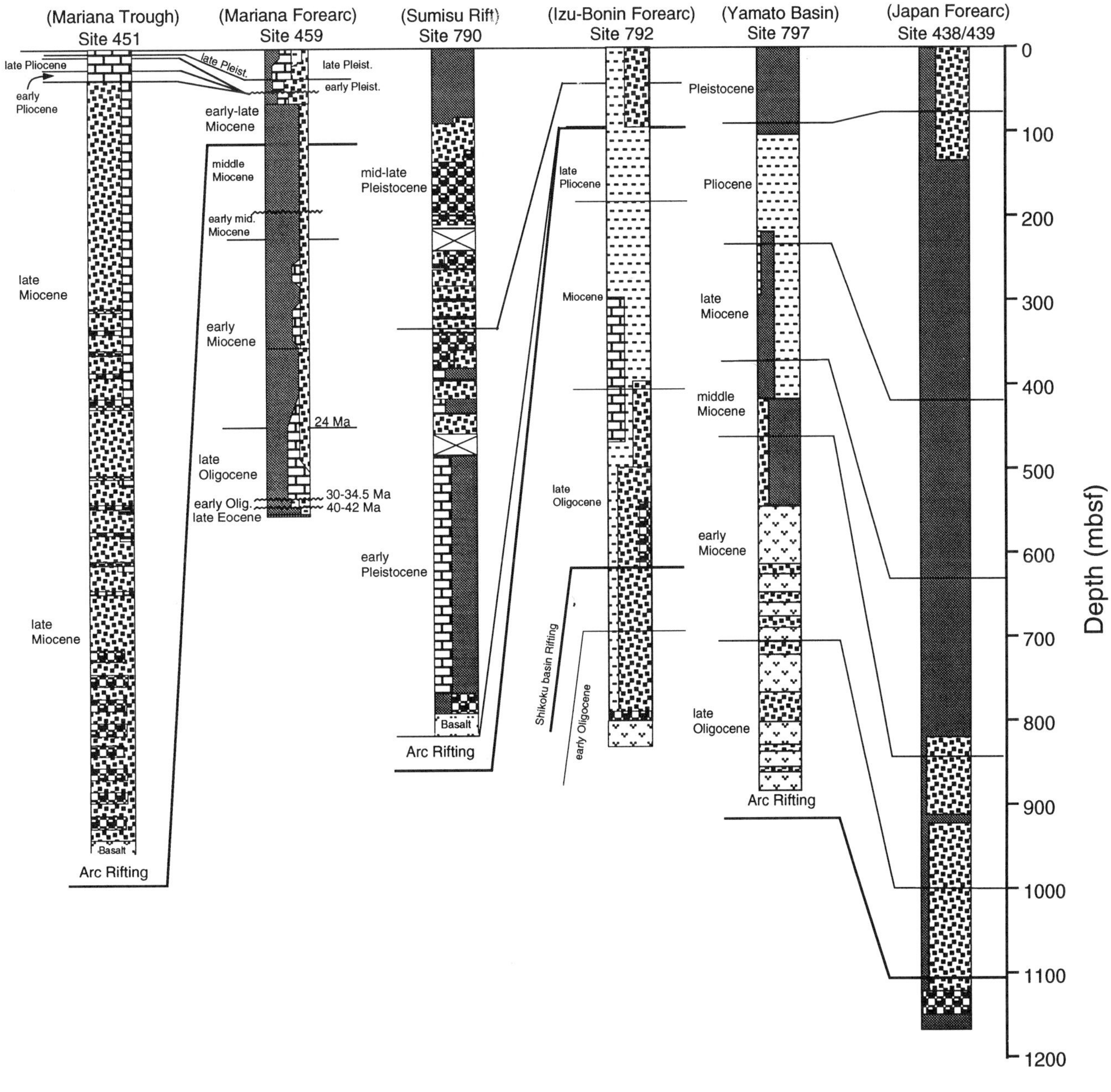

Figure 3b

Although the forearc sequences allow the activity along the volcanic front to be reconstructed, cores from the backarc basin of each system have also been employed in tracing the nature of the early rift-related volcanism. Although clearly these sections cannot contribute to the reconstruction of the pre-rift activity, the oldest sediments in these basins reflect the volcanism during the early rifting period, which generally is located on the adjacent volcanic arc (e.g., Izu-Bonin, Marianas, Bismarck, New Hebrides). Volcanism and associated sedimentation are later focused in the rift axis, usually along a spreading ridge, although central volcanic edifices are also known from the early Sumisu Rift and Lau Basin prior to the onset of ridge propagation. Most volcaniclastic sediment in a backarc basin is interpreted as being derived directly from active intrabasinal volcanic centers, although when the basin is still narrow, with little volcanic activity, e.g., modern Sumisu Rift, the arc can be demonstrated to supply the

TABLE 1. Summary of Volcanic History in The Western Pacific Arc Systems Drilled by DSDP/ODP.

Arc Backarc Basin	Japan Japan Sea N	Japan Japan Sea S	Ryuku Okinawa Trough	Bonin Shikoku Basin	Bonin Sumisu Rift	Mariana Parece Vela Basin	Mariana Mariana Trough	Cagayan Sulu Sea	Tonga Lau Basin
Arc basement	continental	continental	continental	oceanic	oceanic	oceanic	oceanic	continental	oceanic
Position of rifting relative to arc	behind	behind	behind	central	behind and trenchward	central	central	behind	trenchward
Volcanic gap	no	yes	arc becomes extinct after rifting	yes	no, decrease only	yes	no, decrease only	no	yes
Duration of gap	4.0 m.y.	8.0 m.y.		3.0 m.y.		2.0 m.y.			2.0 m.y.
Arc subsidence after rifting ?	yes	yes	yes, after mid-late Miocene uplift	yes	basinal subsidence, flank uplift	yes	yes	yes	yes
Prerift volcanism									
Major elements	calcalkaline dacitic	dacitic	calcalkaline dacitic	low-K tholeiite	low-K tholeiite	calcalkaline basalt	calcalkaline basalt	calcalkaline	tholeiitic andesite LILE depleted
Trace elements				LREE, HFSE depleted		LREE, HFSE depleted		LREE, HFSE depleted	
Post-rift volcanism									
Position	volcanic front	volcanic front	volcanic front become extinct	volcanic front	volcanic front	volcanic front	volcanic front	volcanic front	remnant arc
Major elements	tholeiitic bimodal	tholeiitic bimodal		tholeiitic basalt	low-K, then high-K tholeiite	tholeiitic basalt	calcalkaline basalt	silica-rich calcalkaline	tholeiitic becoming alkaline
Trace elements				LREE, HFSE enrichment with time					volcanic front tholeiitic basalt
First backarc volcanism									
Major elements	tholeiites	tholeiites	tholeiitic basalt	tholeiitic basalt	tholeiitic basalt	tholeiitic basalt	tholeiitic basalt	tholeiitic basalt	bimodal tholeiitic
Trace elements	LREE depleted MORB, becomes arc-like upsection	LREE depleted MORB, becomes arc-like upsection	transitional MORB to IAT	transitional MORB to IAT	transitional MORB to IAT	transitional MORB to IAT	transitional MORB to IAT	transitional MORB to IAT	transitional MORB to IAT

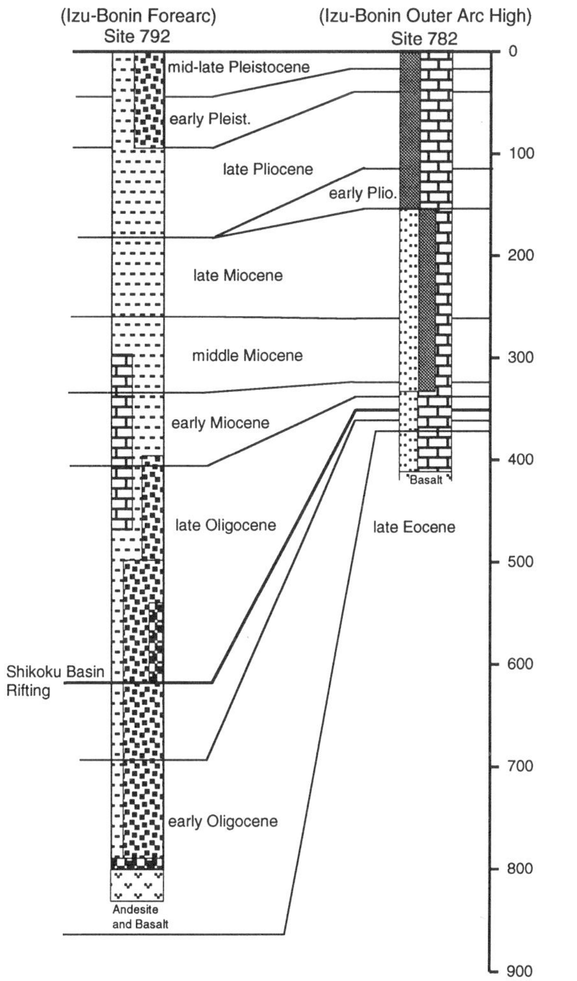

Fig. 4. Summary of arc volcanism in the major arc systems of the western Pacific sampled by ODP/DSDP drilling. Note the common minimum of volcanic intensity immediately prior to arc breakup and the associated peak in high silica volcanism in the early rift stages.

bulk of the sediment. Therefore in the oldest parts of more mature backarc basins the sediment is also interpreted to have been derived from the adjacent volcanic arc. Mature backarc basins often have smooth basin floors, due to the infilling effect of sedimentation. These basins can thus derive clastic and volcaniclastic sediments over long distances. However, there is evidence that in some newly rifted basins the active tectonism produces a rough topography, which inhibits sediment transport across basin and thus results in ponding of sediments close to their sources [e.g., Lau Basin; Clift et al., 1994]. In this case the volcaniclastic sedimentation in any one sub-basin within the overall backarc basin principally records the local volcanism, accumulating either hyaloclastites from basaltic volcanism on the spreading ridge or ashes reworked from local eruptions from submarine volcanoes.

TECTONICS OF ARC BREAKUP

Backarc Rifting

Recent syntheses of the results of ODP and other oceanographic surveys in the western Pacific are providing an increasingly clear picture of the nature of arc rifting and marginal basin formation. A modest uplift of the arc edifice usually precedes rifting [e.g., Tappin et al., 1991], followed by breakup and extension of the original structure. Extensional deformation more typical of continental rift settings characterizes the early stage of basin opening, rather than the focused extension along a rift valley, seen along mid ocean and backarc spreading ridges [e.g., Jolivet and Tamaki, 1992; Taylor, 1992; Parson et al., 1992]. This early phase of extension may last up to several million years before a true backarc spreading center forms. The crust of the backarc basin is thus of two markedly different varieties and its generation is accompanied by two different styles of volcanism, discussed below.

Evidence from the Japan and Tonga backarc regions indicates that in these cases the changeover in the style of extension is achieved following the nucleation of a spreading center along a transform fault [Jolivet and Tamaki, 1992; Parson and Hawkins, 1994]. The spreading center then propagates into the rifted arc crust of the early backarc basin, forming a wedge of new "spreading center-type" crust as it proceeds.

The Location of Rifting and Volcanism

Exactly where rifting occurs within an active arc system has important implications for the nature of the syn-rift and immediate post-rift volcanism and thus on the nature of the volcaniclastic record. The sedimentary record deposited in the forearc basin will be critically dependent on whether rifting involves separation of the forearc basin and the major arc volcanic centers. In general, the central part of the arc edifice, on one side or other of the arc volcanoes, is the location of rifting. This reflects the mechanical weakness of the thick arc lithosphere. Deep levels of island arcs accreted into mountain belts, such as the Tonsina arc of Alaska, show a high proportion of mafic and ultramafic rocks within the lower arc crust and lithospheric mantle [DeBari and Coleman, 1989]. However chemical similarities between arc and the andesitic continental crust [Kelemen et al., 1993] suggest that the exposed mafic and ultramafic assemblages of this example may not be fully

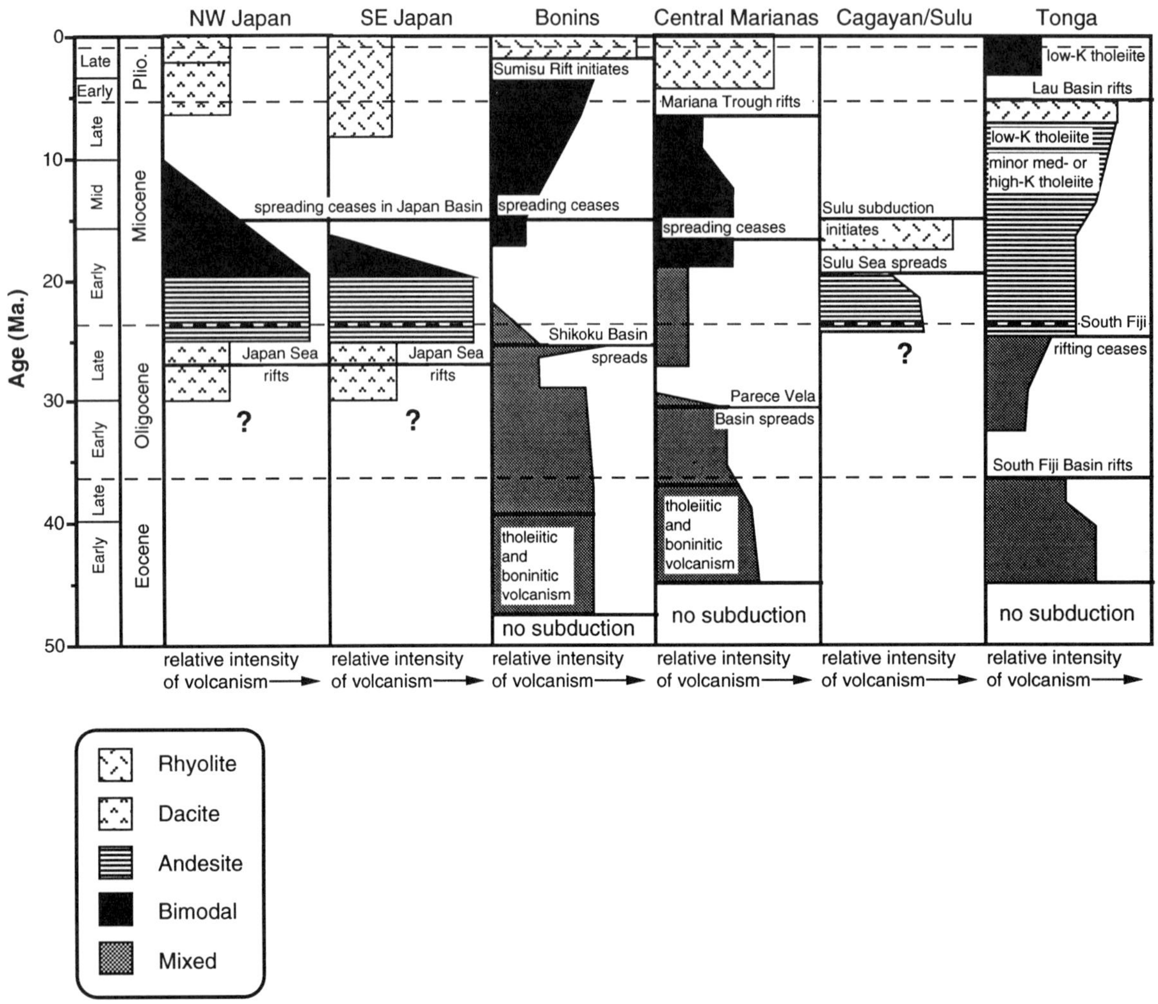

Fig. 5. Sedimentary logs from the Izu-Bonin forearc basin (site 792) and outer arc high (site 782). Note their differing stratigraphies and in particular the finer grained nature of the sediments and reduced stratigraphic thicknesses on the outer arc high.

representative, presumably a result of tectonic dissection. Mechanically weaker andesitic or acidic crustal rocks would also provide a weak point at which to initiate rifting. The thermal anomaly associated with the volcanic front accentuates the mechanical difference and makes this the weakest section of the arc.

Rifting can take place trenchward of the volcanic front (e.g., Tonga; Figure 6), behind the volcanic front (e.g., Izu-Bonin, Japan, Cagayan). Modest volcanism within the basin and more vigorous activity on an adjacent arc or remnant arc mark the early rift period. This situation is unstable, but may change or stabilize with time depending on the relative positions of the remnant arc, basin and trench. When rifting occurs trenchward of the original volcanic front, volcanism may continue on the remnant arc for some time after rifting. However, opening of the basin will necessarily remove the remnant arc from the trench and lead to its eventual extinction. In the case of the Lau Ridge (Figure 1), the remnant Miocene Tonga arc, activity continues for more than 4.0 m.y. after the initial rifting of the Lau Basin [Cole et al., 1985]. At the same time, the Tofua arc, a Pliocene-Recent volcanic arc, was constructed on the trench side of the basin after 3.0 Ma [Tappin et al., 1994; Clift and Dixon, 1994]. Some workers [e.g., Parson and Hawkins, 1994] have suggested that the southward propagation of the eastern Lau spreading center from the Peggy Ridge transform fault into the basin parallels, and is linked to, the development of the new Tofua arc. Thus, 2

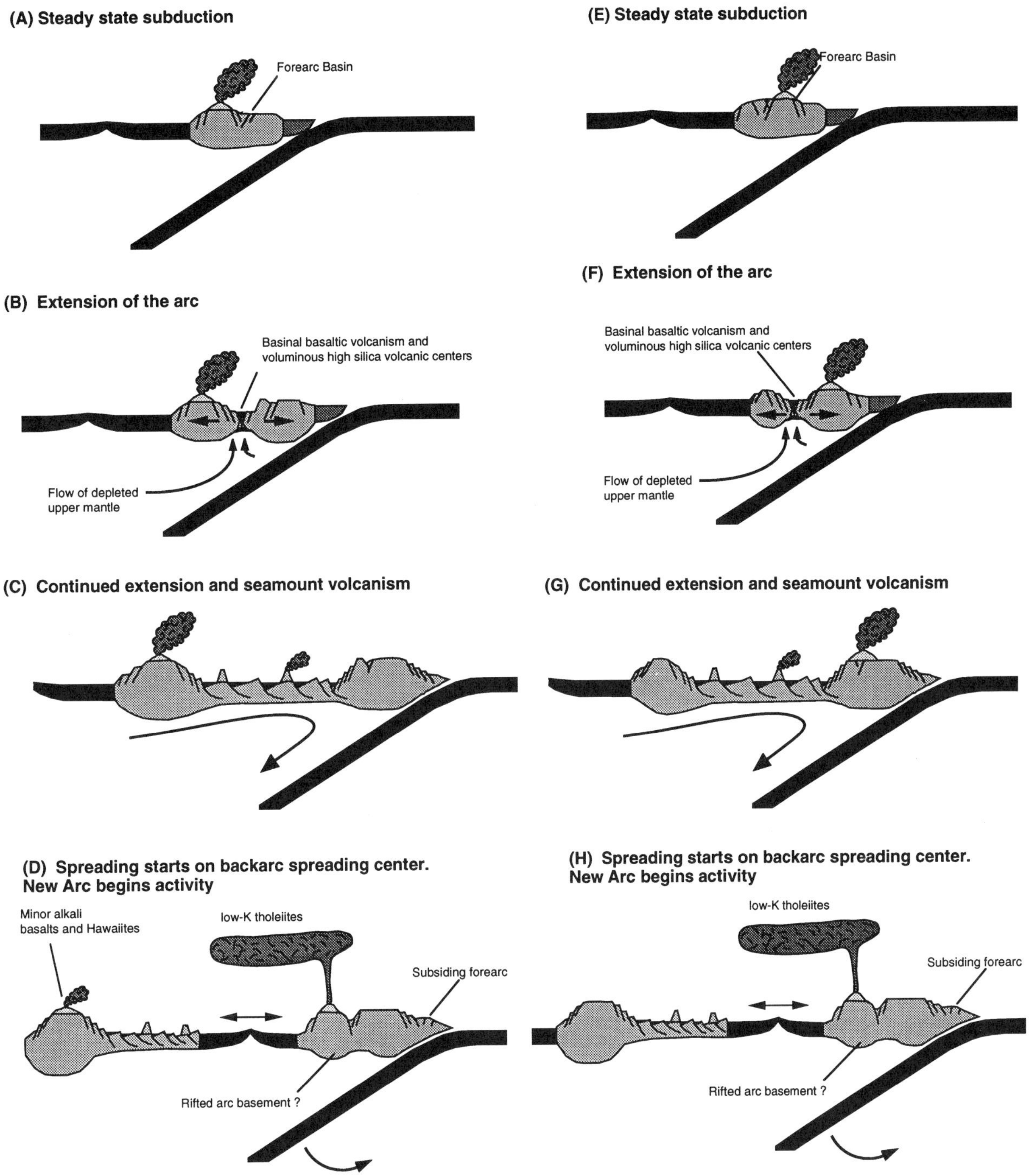

Fig. 6. Cartoon demonstrating the location and nature of volcanism in a rifting oceanic arc environment, when rifting occurs trenchward of (A-D) or behind (E-H) the arc volcanic front.

m.y. after the start of rifting in Tonga there was a short time span when there was active volcanism within the basin and on both the new Tofua arc and remnant Lau Ridge. Whether such widespread volcanism is common is not clear, as the nearby presence of the Samoan hotspot affects the northern, best known, part of the Lau Ridge during eruption of the Mago Volcanic Group. The thermal input from the hotspot may be responsible for prolonging activity into the late Pleistocene [Gill, 1976].

In the case of rifting behind the volcanic front, such as in the Japan [Tamaki et al., 1990] and Sulu Seas [Rangin and Silver, 1992], as well as the Sumisu Rift of the Izu-Bonin arc [Taylor et al., 1990], volcanism on the original arc edifice shows a hiatus or a minimum during the early rifting period. In the Izu-Bonin forearc basin, at site 792, a hiatus in sedimentation from 2.25-1.0 Ma immediately postdates the start of the Sumisu Rift and probably reflects cessation of arc volcanism [Taylor et al., 1990]. However, the length of the hiatus is quite short and volcanism resumes shortly after breakup started. The modern Izu-Bonin arc represents an example of an active arc next to a young backarc basin. The Havre Trough behind the Kermadec arc, dating from 5.0 Ma, is similar in showing significant modern activity along the arc volcanic front [Wright, 1993] but, like the Sumisu Rift, little or no activity in the basin. In contrast, the northern Japan arc does not appear to show a decrease in volcanism, and in the Japan forearc volcaniclastic sedimentation continues throughout the rift period at site 438 and 439.

The Havre Trough and Sumisu Rift provide good examples of backarc basins with little volcanism within the rift. However, if extension in the basin continues for a significant period prior to the propagation of oceanic spreading centers into the area then volcanism occurs within the basin (e.g., Lau Basin). Intra-basinal volcanism during the rifting stage of basin opening is in two forms. Basaltic volcanism floors the bottoms of newly extending grabens and half grabens, with the horsts comprising pre-existing arc crust. More silicic volcanism occurs at volcanic centers on the flanks of individual grabens within the basin (Figure 7). These form intrabasinal volcanic seamounts, similar to submarine arc volcanoes except for their position within the backarc basin. Volcanism of this type is active in the rifting stages of both oceanic and continental margin backarc basin, and may start earlier when rifting starts closer to the volcanic front. If, in the Marianas, rifting truly dismembers the arc volcanic front during rifting of the Mariana Trough as proposed by Hussong and Uyeda [1982], then this could be the only type of volcanism active prior to foundation of a new arc on the trench side of the basin.

Sea Beam and GLORIA data from the Lau Basin [Parson et al., 1992] indicate that early syn-rift intrabasinal volcanoes are founded on rifted blocks of the Miocene Tonga arc (Lau Ridge) and are the principal source of volcaniclastic debris into the adjoining grabens [Figure 7: Clift et al., in press; Bednarz and Schmincke, 1994], rather than the remnant arc, which declined in activity with time. While the eruption of the basin floor basalts does not last long and migrates towards the trench with time, eruption on these central volcanoes continues longer, lasting up to 2-3 m.y. As a result, long-lived magma chambers are able to develop and high-silica liquids evolve, giving the eruptive products a dominantly dacitic character. The early history of the Mariana Trough is similar, but without adjacent activity on the remnant arc [Packham and Williams, 1982]. In the Sumisu Rift volcanism is not centered on the rift axis but occurs on cross-rift transfer zones. At one place (28.3° N) a large shield volcano appears to have infilled a rift basin. Elsewhere volcanism is present along faults, creating predominantly basaltic edifices up to 700 m high [Taylor, 1992]. Locally, en echelon ridges, up to 600 m high and 25 km long, lie in the inner rift basins, although these features are less well developed than those in backarc basins with more extension.

In the Lau Basin the advance of propagating spreading centers through the extended arc crust caused a radical change in the intra-basinal volcanism. A resurgence of arc volcanism on the trench side of the basin shadowed the propagating rift tip. Concentration of the extension at the spreading ridge caused intrabasinal volcanism to become focused there. Volcanism in this spreading stage is in the form of non-explosive pillow and sheet flow basaltic eruptions. Basaltic hyaloclastites are the typical volcaniclastic sediments immediately overlying the basement during this phase of basin opening. The new volcanic arc is the dominant source for small volumes of distal airfall ashes, of various compositions, that accumulate higher in the section. Other examples show almost the opposite trend. In the Okinawa Trough the northward propagation of en echelon spreading ridges causes a retreat of activity on the adjacent Ryukyu arc [Sibuet et al., 1992]

Volcanic Productivity during Rifting

The volume of material being erupted along the volcanic front during the early stages of arc rifting differs from that in the steady state subduction situation. In many cases a steady increase in the volume of volcanism along the volcanic front marks the period immediately prior to rifting. Thick piles of volcaniclastic conglomerates and mass flows in the forearc and new backarc rift evidence a climax of

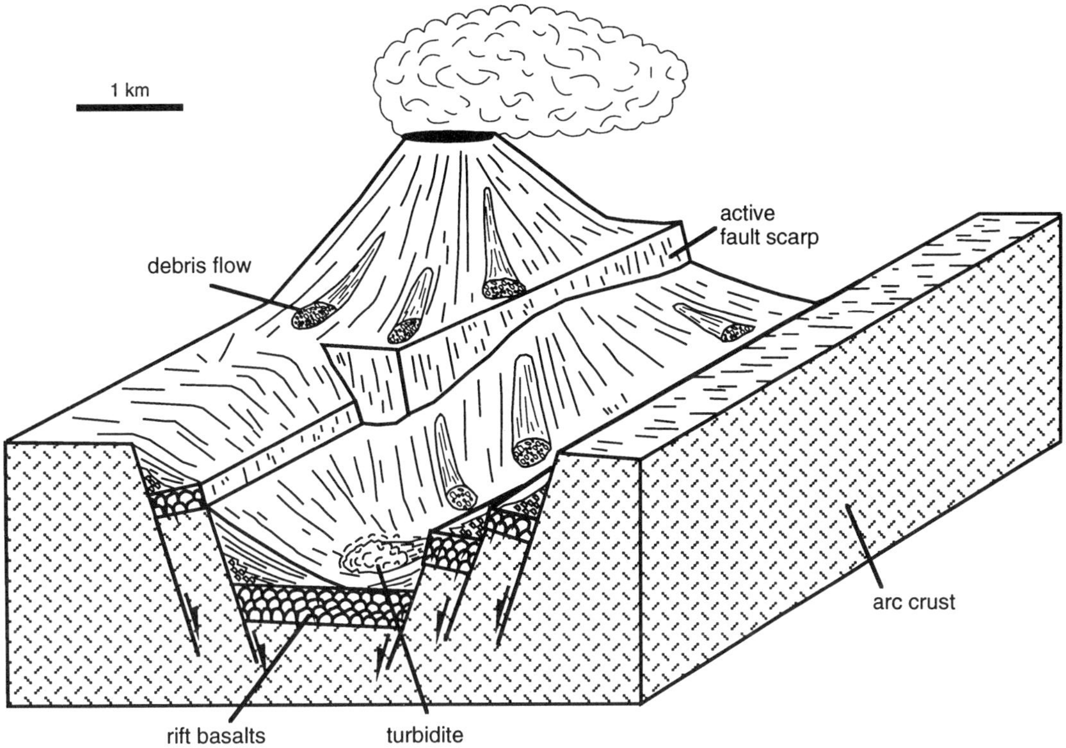

Fig. 7. Schematic representation of a volcanic sub-basin and adjacent submarine seamount volcano. Lau Basin crust produced prior to spreading center propagation appears to be composed of a series of such sub-basins and associated volcanic centers. The sub-basin is floored by MORB/IAT transitional basalts. More evolved lavas were erupted from the volcano sited on extended arc crust on the sub-basin margins, forming locally large volumes of volcaniclastic sediment (from Clift et al., in press).

activity in the early syn-rift period [Gill et al., 1992; Fujioka et al., 1992a; Tappin et al., 1994]. Figure 8 shows the changing rates of sedimentation in the Mariana, Izu-Bonin, Cagayan, and Tonga forearc basins around the time of arc breakup. The Izu-Bonin and Tonga examples (sites 792 and 840) both show high rates of sedimentation prior to breakup, followed by reduced sedimentation. In contrast the Mariana arc appears to peak in volcaniclastic sedimentation towards the end of rifting of the Parece Vela Basin and Mariana Trough. Sedimentation prior to rifting of the Sulu Sea is unconstrained, but sedimentation at site 771 is rapid and coarse grained during the rifting of the Sulu Sea. The "Green Tuff" of the Japan arc represents the classic example of this maximum of high silica volcanism in the continental arc setting [Sugimura et al., 1963]. Following this time, productivity along the Japan arc markedly decreases, involving a complete hiatus of activity in the northern part of the arc [Fujioka, 1986]. The Izu-Bonin arc displayed voluminous volcanism early during the rifting period, followed by a decrease in volume as spreading initiated. The northward propagation of the spreading center in the Okinawa Trough is an even more extreme example, which resulted in extinction of the volcanic arc [Sibuet et al., 1992]. Voluminous productivity on the Ryukyu arc continues farther north where the basin is in the rift stage.

The volume of explosive volcanic products in the basin diminishes when backarc spreading centers develop. Once spreading has started, magmatic accretion along the ridge crest takes up all the extension in the basin, compared to the pre-spreading mixture of crustal thinning and magmatic extension. The total amount of magmatism must increase as spreading begins. In some cases, the parallel development of an arc on the trench side of the basin accentuates the total volume of magmatism, e.g., Tofua arc [Parson and Hawkins, 1994], and Izu-Bonin and Mariana arcs [Taylor, 1992]. In other examples of spreading backarc basins the arc becomes extinct or less active [e.g., Okinawa Trough; Sibuet et al., 1992]. However, such a gap may be short-lived and represent the time necessary for a new volcanic front to become established. The forearc sedimentary record thus often shows a syn-rift maximum in coarse clastic material (e.g., sites 459, 792, 840; Figure 3),

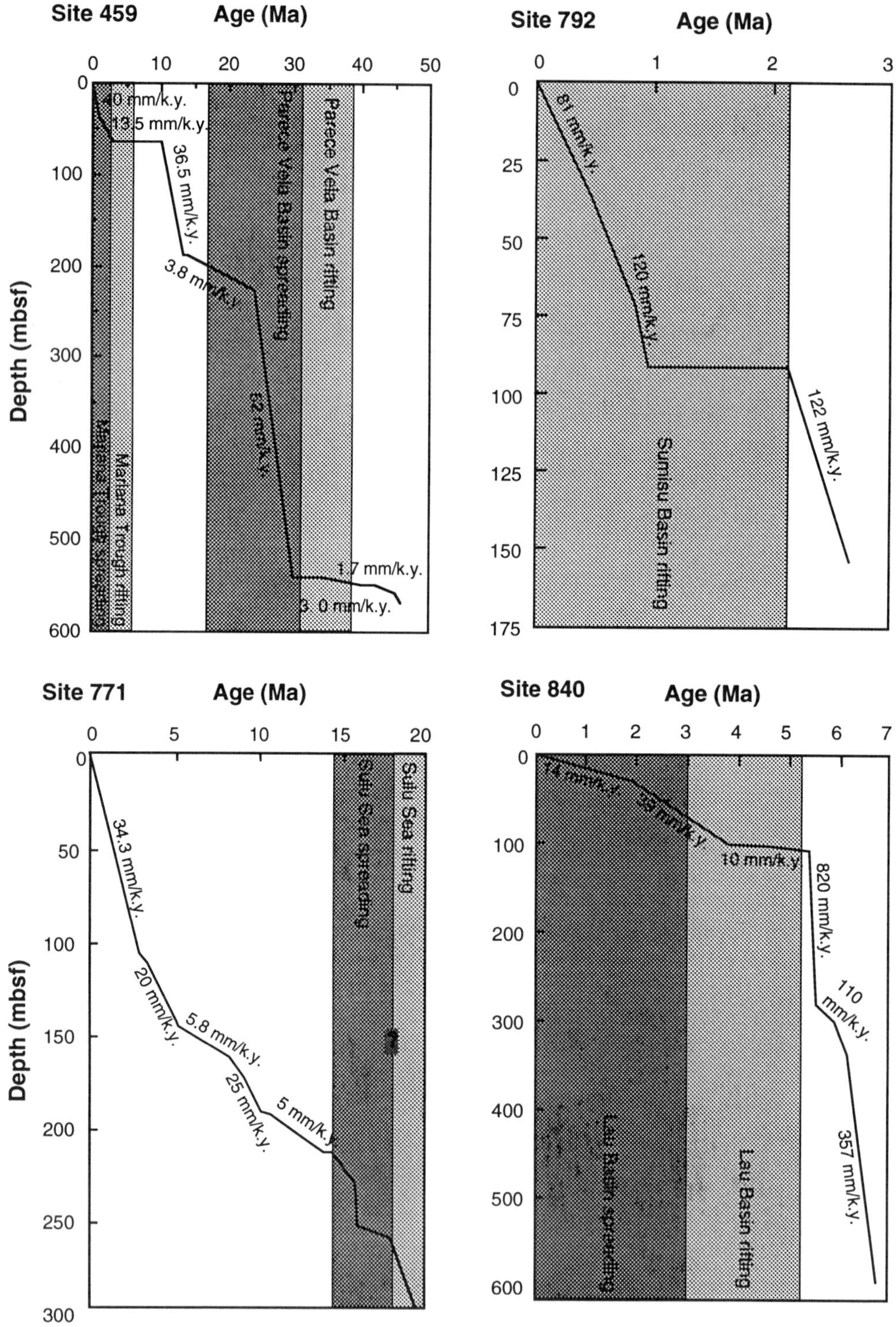

Fig. 8. Diagrams showing the maxima in volcaniclastic sedimentation in forearc basins associated with arc rifting. Peaks in volcaniclastic sedimentation are inferred to result from an increase in volcanic productivity during this period. Data for site 792 from Taylor et al. [1990], for site 459 from Hussong et al. [1982], for site 771 from Rangin et al. [1990], and for site 840 from Parson et al. [1992]. Age of rifting is taken as the age of the oldest crust within the basin.

followed by a hiatus or volcanic minimum, and eventually returning to a steady state level. This record is unrepresentative of the arc system as a whole, as it ignores activity within the basin (e.g., sites 451, 790, 797, 834; Figure 3). Overall volcanism is generally at a maximum throughout the rifting period, so that the term "volcanic hiatus" is only appropriate for the activity along the volcanic front.

SEDIMENTATION IN ARC ENVIRONMENTS

The differing modes of arc break-up described above result in different styles of volcanism, and thus different ash records developing over short distances. Volcanism within the basin itself will tend to be recorded close to the site of eruption [e.g., Hussong et al., 1982; Clift et al., in press]. Submarine eruptions within a backarc basin are often incapable of depositing material up and out of the basin and on to the forearc area. Thus a lack of volcaniclastic debris within the forearc basin, or on the remnant arc, may not necessarily indicate a gap in volcanism, but merely submarine volcanism focused within the new basin. As discussed above, the presence of a volcaniclastic hiatus in some parts of the arc system is a common characteristic of volcanism during arc rifting [e.g., Scott et al., 1980, Tappin et al., 1994]. The sedimentation in forearc basins will reflect principally the nature of the adjacent volcanic arc. Subsidence of the basin provides the space for sediment derived straight from the arc to accumulate. Seismic surveys [e.g., Austin et al., 1989, Taylor et al., 1992] do not indicate major erosional features in the stratigraphy of the central forearc basins in Tonga or the Izu-Bonin arc. Similarly, chemical data from sediments in the Tonga, Izu-Bonin, and Mariana forearc basins, discussed below, show trends that would not be visible if major reworking was taking place. Unfortunately sediments in the backarc basin are usually less easy to interpret, because intrabasinal volcanoes, the remnant arc, or the active arc volcanic front may all contribute to the sediments accumulating at any one place within the basin. This is especially true when the basin is mature and sedimentation has infilled any topography that caused localized sediment ponding in the early stages of rifting.

Another cause of apparent volcanic hiatuses in forearc basins is the separation of active arc volcanoes from the forearc during the rifting process. This results in a major reduction in the volume of volcaniclastic sedimentation on to the forearc. Turbidite and mass-flow deposition on submarine fans typifies sedimentation in forearc basins [e.g., Bøe, 1994], so that the separation of arc from forearc will create a barrier to sedimentation in that area. Primary airfall deposition will be the sole form of volcaniclastic sedimentation. Should separation also result in subsidence of the remnant arc volcanoes into the submarine environment, as it often does, then the forearc may record an almost total absence of volcanic input, while the volcanism on the old volcanic front may continue at high volume. In contrast, where rifting is asymmetric and behind the volcanic front (Figure 6), the continuation of activity on the original arc means that some volcanic sediment will accumulate on the forearc during the early rift period. In this case reduced sedimentation is an accurate mirror of decreasing volcanism on the volcanic front, but does not reflect any activity within the basin. However, if rifting separates the forearc region from the arc (Figure 6) then the sedimentary succession in the forearc provides little information on the nature of the volcanism in the system as a whole prior to re-foundation of a new arc edifice.

Within the basin the rough topography and submarine nature of the eruptions tend to result in the ponding of volcaniclastic debris close to the source, rather than permitting its distribution throughout the basin or onto the forearc. Thick piles of dacite-dominated volcaniclastic mass flow sands and gravels typify the earliest sediments in many backarc basins (Figure 3) and point to voluminous high silica volcanism on the volcanic front during the early rift stage and subsequently within the basin [Gill et al., 1992; Nishimura et al., 1991].

CHEMISTRY OF ARC VOLCANISM

The Volcanic Front

Major element variations. The chemistry of lavas being erupted along the volcanic front of some western Pacific arc systems shows a number of important trends during the rifting period. In general the major element evolution of the arcs reflects the tectonic division between oceanic and continental margin arcs. The best known oceanic arcs, the Izu-Bonin, Mariana, and Tonga arcs, do not show large changes in their major element chemistry during rifting. Low-K tholeiite dominates the Izu-Bonin and Tonga arcs [Figure 9] both before, during, and after rifting of the Sumisu Rift and Lau Basin, as recorded in the volcaniclastic sediments in their forearc basin [Clift and Dixon, 1994; Bøe, 1994; Fujioka et al., 1992a, b; Arculus and Bloomfield, 1992]. Arculus and Bloomfield [1992] were unable to find any link to rifting of the alkali content of sediment samples from sites 782, 784, and 786 on the Bonin forearc. Figure 10 shows that glasses at site 784 do not show a strong correlation between age and potassium

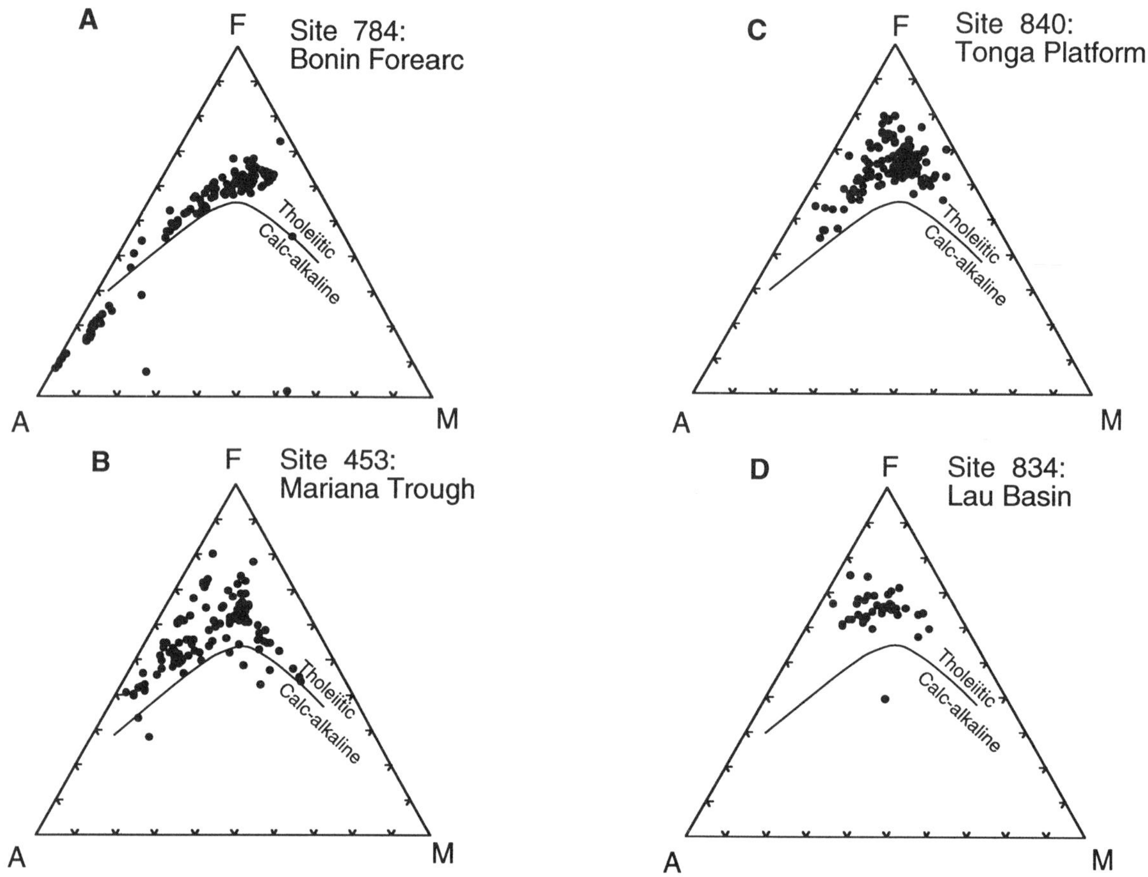

Fig. 9. Alkali/iron/magnesium variation diagrams for ashes erupted during the early arc rifting stage and recovered from (A) Bonin-Izu forearc, site 784; data from Arculus and Bloomfield [1992]; (B) Mariana Trough, site 453; data from Packham and Williams [1982]; (C) the Tonga Platform, site 840; data from Clift and Dixon [1994]; (D) Lau Basin, site 834; data from Clift and Dixon [1994]. Tholeiite and Calc-alkaline fields are defined by Irvine and Baragar [1971].

content, although low-K volcanism characterizes the immediate pre-rift period. Episodes of higher alkali content do exist at various levels of the forearc stratigraphy, but these do not correlate easily with the arc tectonism. The Mariana arc is presently erupting calc-alkaline compositions and the West Mariana Ridge was also doing so prior to rifting of the Mariana Trough [site 451; Mattey et al., 1981; Natland and Tarney, 1982], suggesting no major change during this time. Earlier rifting on this arc, during the formation of the Parece Vela Basin, was characterized by tholeiitic volcanism on the Palau-Kyushu Ridge prior to breakup [site 448; Kroenke et al., 1980] and in the post-rift period on the West Mariana Ridge arc, which became calc-alkaline as volcanism proceeded. Arc volcanism changes at the point of breakup at the northernmost termination of the propagating Mariana Trough rift, adjacent to the Volcano Islands. Anomalous trachy-andesitic compositions on the volcanic island of Iwo Jima characterize the arc here. Stern et al. [1984] attributed this anomaly to the onset of rifting in this part of the arc.

Although volcanism on the Tonga arc was tholeiitic throughout the rifting of the Lau Basin [Cawood, 1991], analyses of glass grains from Tonga arc site 840 identified a progressive change over 2 m.y. from the pre-rift situation with a minority of medium and high-K glasses (Figure 10) and a majority of low-K glass, to an exclusively low-K volcanism by the time of breakup. Fujioka et al. [1992a] used electron probe analyses of glasses from sites 792 and 793 on the Bonin forearc to show that 2 m.y. after rifting began medium and high-K glasses were being deposited on the forearc. Fujioka et al. [1992a] interpreted this trend to be the result of arc lithospheric thickening during this time. Plank and Langmuir [1987, 1988] demonstrated that the Na

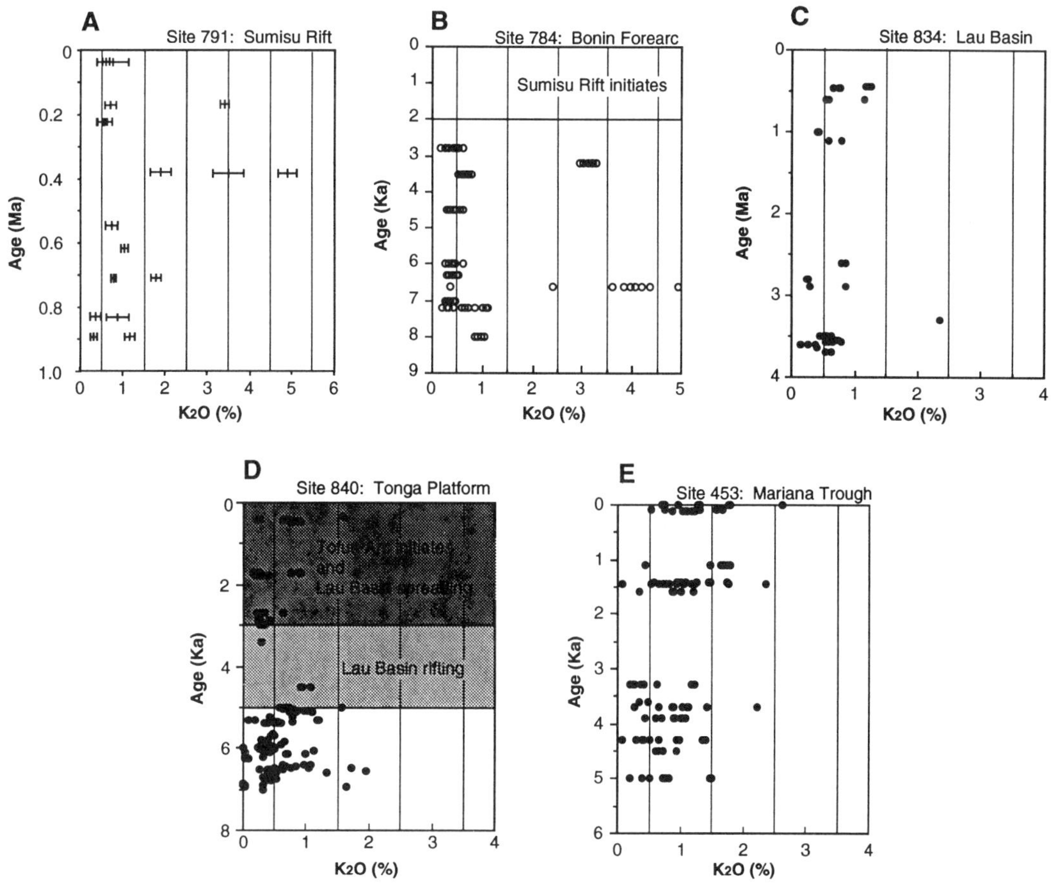

Fig. 10. Diagram showing the changes in potassium content of volcaniclastic glasses erupted during the early rifting process (A) Sumisu Rift, site 791; data from Rudolfo et al. [1992]; (B) Bonin forearc, site 784; data from Arculus and Bloomfield [1992]; (C) Lau Basin, site 834; data from Clift and Dixon [1994]; (D) Tonga Platform, site 840; data from Clift and Dixon [1994]; (E) Mariana Trough, site 453; data from Packham and Williams [1982]. Solid dots represent individual electron probe chemical analyses. Horizontal bars represent the range of potassium values from any one sample. Age of rifting is taken as the age of the oldest crust within the basin.

and Ca contents of basalt arc lavas (normalized to MgO = 6.0%) were related to the crustal thickness, as measured by the seismically imaged depth to the Moho. Looking at the high Mg glasses analyzed from site 840 [Clift and Dixon, 1994] shows that the pre-breakup maximum value of $Na_{6.0}$ at 2.2 % falls to 1.2 % at the time of first backarc crust formation. Comparing this trend with worldwide arc data [Plank and Langmuir, 1988] suggests that the crust was probably approximately 25 km thick prior to rifting and thinned to less than 12 km by 5 Ma. While this seems a reasonable estimate there is presently no independent method of checking this degree of crustal extension. However, if this is correct then such a stretching factor would also have had a large effect on the volume of melt produced along the volcanic front. The effects of variations in the degree of partial melting have not been calculated for arc environments; however, reference to the melt thicknesses calculated for a variety of stretching factors in rift settings [McKenzie and Bickle, 1988] suggests that when $ß = 2$ at elevated mantle temperatures (1450-1500°C) the melt thickness generated will increase by a factor of approximately five. Although there is no suggestion that mantle temperatures are elevated below arcs, the presence of fluids has the effect of lowering the solidus and producing a similar effect. Such a large change would explain why such profuse volcanism is known throughout the arc system during its rifting. It should however be noted that arc settings are probably more complicated than the simple rift model, because of complicated flow in the mantle wedge and the presence of fluids, which have yet to be quantified. In addition, some caution must be exercised with regard to the potassium contents, a result of the easy remobilization of the alkali elements both during diagenesis and during probe analysis [Neilson and Sigurdsson, 1981].

Dacitic and rhyolitic compositions are typical of the increased volcanism during the early rift period. Many western Pacific arcs, both oceanic and continental, show this phenomenon. In the Tonga arc early rifting is coincident with the deposition of thick, mass-flow conglomerates, dominated by dacitic material at site 840 [Tappin et al, 1994; Clift and Dixon 1994]. In this case biostratigraphic dating shows that the volcanism immediately predates the generation of the oldest backarc basin basalts at site 834. In a similar fashion, farther north in the Izu-Bonin arc, initial rifting of the Sumisu Rift has caused a change from a mixed bimodal tholeiite suite to abundant rhyolitic activity. Thick pumiceous gravels were deposited as a result on the forearc at sites 787 and 793 and within the Sumisu Rift itself at sites 790 and 791 [Taylor et al., 1990]. Farther north again, an increase in volcanism of andesitic and rhyolitic composition accompanies rifting of the Japan arc from mainland Asia [Cadet and Fujioka, 1980]. The famous "Green Tuff" [Sugimura et al., 1963], which represents the zenith of this phase of activity, outcrops extensively on-land in Honshu and around the rim of the Japan Sea [e.g., Lee and Pouclet 1988]. The tuff is also found as an altered volcanic tuff on the Yamato Bank [site 799; Ingle et al., 1990] and as tuffaceous sediments and associated dykes at site 794 in the Yamato Basin [Tamaki et al., 1990].

The total distribution of silica values does not appears to change in a consistent fashion with time during arc rifting. Probe data from the Izu-Bonin forearc [Figure 11: Arculus and Bloomfield, 1992; Fujioka et al., 1992a] show a complete range of silica contents. However, whole rock analyses of breccias and lavas from sites 788 and 793 next to the Izu-Bonin arc, as well as sites 790 and 791 in the Sumisu Rift, indicate that the arc has erupted more siliceous lavas since about 1.5 m.y. prior to Sumisu basin rifting. Gill et al. [1992] suggest this reflects the new extensional nature of the arc that would permit fractionated magma more easily to rise to the surface. This is similar to the pre- and early syn-rift volcanism recorded at site 840 on the Tonga forearc. Electron probe analyses [Clift and Dixon, 1994] indicate no change in the total spread of silica values prior to rifting, but Bøe [1994] used whole rock analyses to show the overall siliceous character of this phase of volcanism. At site 840 the first ashes deposited some 2.0 m.y. after backarc basin rifting show a change to a bimodal silica population; the significance of this distribution to the rifting process is presently unclear.

Trace element variations. Trace and rare earth element fluctuations in arc volcanism are less well documented in the western Pacific but are of significance where they do occur. In particular, the association of some trace element groups with certain mantle processes provides important information on changes in the arc petrogenesis during rifting. Levels of water-mobile large ion lithophile elements (LILEs) are susceptible to fluctuations in the degree of fluid flux from the subducting oceanic lithosphere. Aqueous fluids released from the slab readily transport LILEs in the subducting slab and overlying mantle wedge and concentrate them in the arc lavas. Thus high values of LILEs in arc lavas may in part indicate significant fluid flux from the subducting plate. In comparison the high field strength elements (HFSEs) are water immobile, so that melting processes in the mantle wedge are the dominant influence on their concentrations. The rare earth elements (REEs) are more difficult to interpret, as they are principally derived by melting of the mantle wedge. However, aqueous fluids derived from the slab may enrich the light REEs (LREEs) content [e.g., Hole et al., 1984]. Trace and rare earth element analyses of volcaniclastic sediments have been made from the Bonin arc [sites 787, 788, and 790-793; Hiscott and Gill, 1992] and Tonga arc [site 840; Clift and Dixon, 1994] samples. The Bonin arc whole rock sediment analyses were made using atomic absorption for major elements, XRF for trace elements, and ICP-MS for rare earth elements. In contrast, the Tonga arc data are derived from probe analyses of individual basaltic glass grains. An electron microprobe was used to determine major elements and a secondary ion mass spectrometer (ion microprobe) to determine trace and rare earth elements. Implicit to both these methods was the assumption that the sediments deposited in the forearc basin reflected the nature of the volcanic arc at the time of deposition, even when these sediments are not primary airfall ash layers. The identification of consistent chemical trends upsection supports this idea and argues against reworking being a major factor in the deposition of these sediments. Exactly how comparable the data are is not clear. The basaltic glass fragments from site 840 represent the most primitive liquids erupted at any one time from the arc, which are thus the easiest to relate to their source mantle and thus to deduce their petrogenesis. However, volumetrically they represent only a small fraction of the volcaniclastic sediment and thus of the arc volcanism. On the other hand, the whole rock analyses of the turbidites from Izu-Bonin and Sumisu sites 787, 788, and 790-793 will represent more fully the products of the arc volcanism but will be more difficult to interpret because of the high proportion of evolved lavas, which will have elemental characteristics less easy to relate to petrogenesis because of the effects of fractional crystallization. In both cases the ICP-MS and ion microprobe provide high levels of analytical accuracy. In

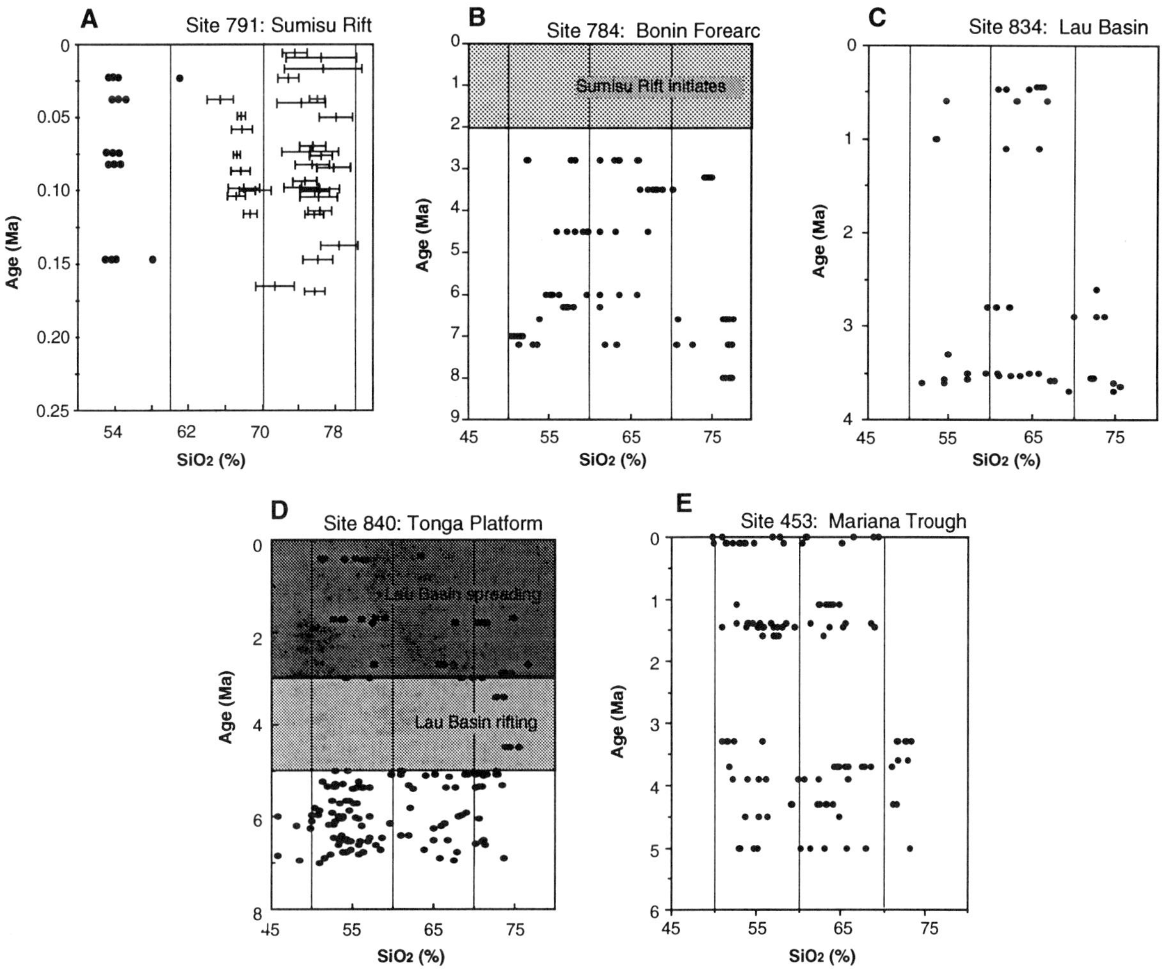

Fig. 11. Diagram showing the changes in silica content of volcaniclastic glasses erupted during the early rifting process (A) Sumisu Rift, site 791; data from Rudolfo et al. [1992]; (B) Bonin forearc, site 784; data from Arculus and Bloomfield [1992]; (C) Lau Basin, site 834; data from Clift and Dixon [1994]; (D) Tonga Platform, site 840; data from Clift and Dixon [1994]; (E) Mariana Trough, site 453; data from Packham and Williams [1982]. Solid dots represent individual electron probe chemical analyses. Horizontal bars represent the range of silica values from any one sample. Age of rifting is taken as the age of the oldest crust within the basin.

the case of the ion microprobe analyses, counting statistics indicate that for certain geochemically significant elements, such as Nb, errors will be < 5% at concentrations of 1 ppm in basaltic glass. Errors of approximately 1% for typical Zr concentrations mean that errors in plotted elemental ratios are low, even when the absolute concentrations are very low. Table 2 shows typical counting statistical errors from glass grains from site 840. Clift and Dixon [1994] demonstrate that the ion microprobe method is accurate to 5% error down to Nb = 0.4 ppm and possibly as low as 0.1 ppm.

The elemental ratios Nb/Zr, La/Sm, and Ba/Sr were chosen to model the depletion in the HFSEs, REEs, and LILEs respectively. In each case these ratios plot an incompatible element against a more compatible element of the same group. A low ratio value thus indicates relative depletion. These pairs of elements are suitable for consideration because of the large differences in their compatibility in mantle phases, and the relative confidence that both the whole rock and microprobe techniques produce reliable values at the concentration levels present. The Tonga forearc probe analyses indicate similar trends to increasing chemical depletion with time within the LILEs, REEs and HFSEs during the pre-rift period, 7.0-5.0 Ma (Figures 12A, B, C). Between 5.0 and 3.0 Ma few data are available, because of a volcaniclastic hiatus on the forearc.

TABLE 2. Statistical Errors in Ion Microprobe Count Rates for Samples from Site 840

Element	Typical % error in count rates	Concentration (ppm)
Rb	2.4	7.040
Sr	0.4	214.0
Y	1.0	14.40
Zr	0.9	60.00
Nb	4.7	0.576
Ba	1.0	188.0
La	3.3	3.920
Ce	2.3	9.780
Nd	7.5	6.100
Sm	12.1	1.670
Eu	8.6	0.112
Gd	9.3	2.900
Hf	8.5	1.990
Tb	9.3	0.406
Dy	10.0	1.800
Ho	9.1	0.522
Er	11.5	1.550
Tm	13.7	0.245
Yb	11.5	1.560
Lu	15.8	0.181
Pb	2.6	1.160
Th	26.7	0.174
U	24.3	0.226

Count rates for ion microprobe analyses of 135-840A-1H-3,45cm; Hf from 135-840B-35X-3,22cm

However, after initiation of the Tofua arc at 3.0 Ma this trend reversed, with the youngest (450 ka) samples showing the highest values of each ratio. Importantly, the degree to which each ratio "recovers" after the syn-rift minimum differs, with the Nb/Zr ratio showing a return to the pre-rift values, while Ba/Sr increases to far exceed this level. Nevertheless the broadly similar pattern, especially in the pre- and syn-rift period, suggests that a single, common process controls the degree of depletion in each elemental group. Since fluid flux from the slab cannot affect the HFSEs, the parallel depletion in each chemical group rules out the slab as the dominant control on magmatic chemistry. Instead, melting processes in the mantle wedge must dominate the initial stage of rifting. This situation changes in the post-rift period when greater degrees of enrichment in the LILEs and REEs, compared to the immobile HFSEs, suggest that increasing slab fluid flux marked this period. The fact that the LILEs increase in unison with the LREEs argues for LREEs being susceptible to enrichment by slab flux, although not as strongly as the LILEs. Clift and Dixon [1994] showed that throughout the arc and backarc regions the level of mobile LILEs compared to immobile HFSEs underwent a marked increase with time after arc breakup up to the present day. This was considered to reflect an increasing input of fluid from the slab with time.

During the initial breakup period, fluctuations in the degree of partial melting and/or the chemical enrichment of the mantle wedge governed the degree of enrichment in incompatible elements within the arc magmas. Multiple melt extractions from a limited volume of mantle, possibly caused by stagnation of asthenospheric flow in the mantle wedge, might be responsible for the progressive depletion seen. Subsequent re-establishment of asthenospheric circulation after breakup could then bring in new source mantle and cause the eruption of less depleted chemistries. Alternatively Clift and Dixon [1994] suggested that tectonic extension of the arc prior to breakup would cause a reduction in the thickness of the arc lithosphere and thus an increase in the height of the melting column under the arc. This would in turn have the effect of increasing the total degree of partial melting. A re-thickening of the lithosphere, possibly by magmatic underplating by the new Tofua arc caused a reversal in the trend after breakup. If this is the case then the chemical data would indicate a progressive thinning of the Tonga arc lithosphere from 7.0 Ma up to the point of arc breakup at 5.0 Ma, i.e., 2.0 m.y. pre-breakup and a rethickening of the lithosphere since 3.0 Ma.

Strangely, whole rock analyses of lavas from the modern Tofua arc show a recent halt in the post-rift trend toward enrichment in Nb/Zr at site 840. This trend is even more strongly developed for La/Sm or Ba/Sr, which are typically more depleted than the youngest glasses found at site 840 (Figure 12). Interestingly, the nearest arc volcano to site 840, Ata, is less depleted than more northerly members of the Tofua arc [Vallier et al., 1985; Ewart and Hawkesworth, 1987]. This suggests a progressive trend to depletion migrating south along the arc that is counteracting the enriching effect of lithospheric thickening. Ewart and Hawkesworth [1987] proposed that the northern Lau Basin and sub-Tofua arc mantle is more depleted in incompatible elements than that of the southern parts of the arc. Higher degrees of melt extraction along backarc spreading centers in the faster spreading northern part of the Lau Basin may be the cause of the depletion and generally explain why backarc lavas are generally less depleted than their arc counterparts [e.g., Woodhead et al., 1993]. Such a model would be in accord with isotopic data from Hergt and Hawkesworth [1994], discussed below, showing the southward propagation of a new mantle domain under the arc system in conjunction with spreading center propagation. In this

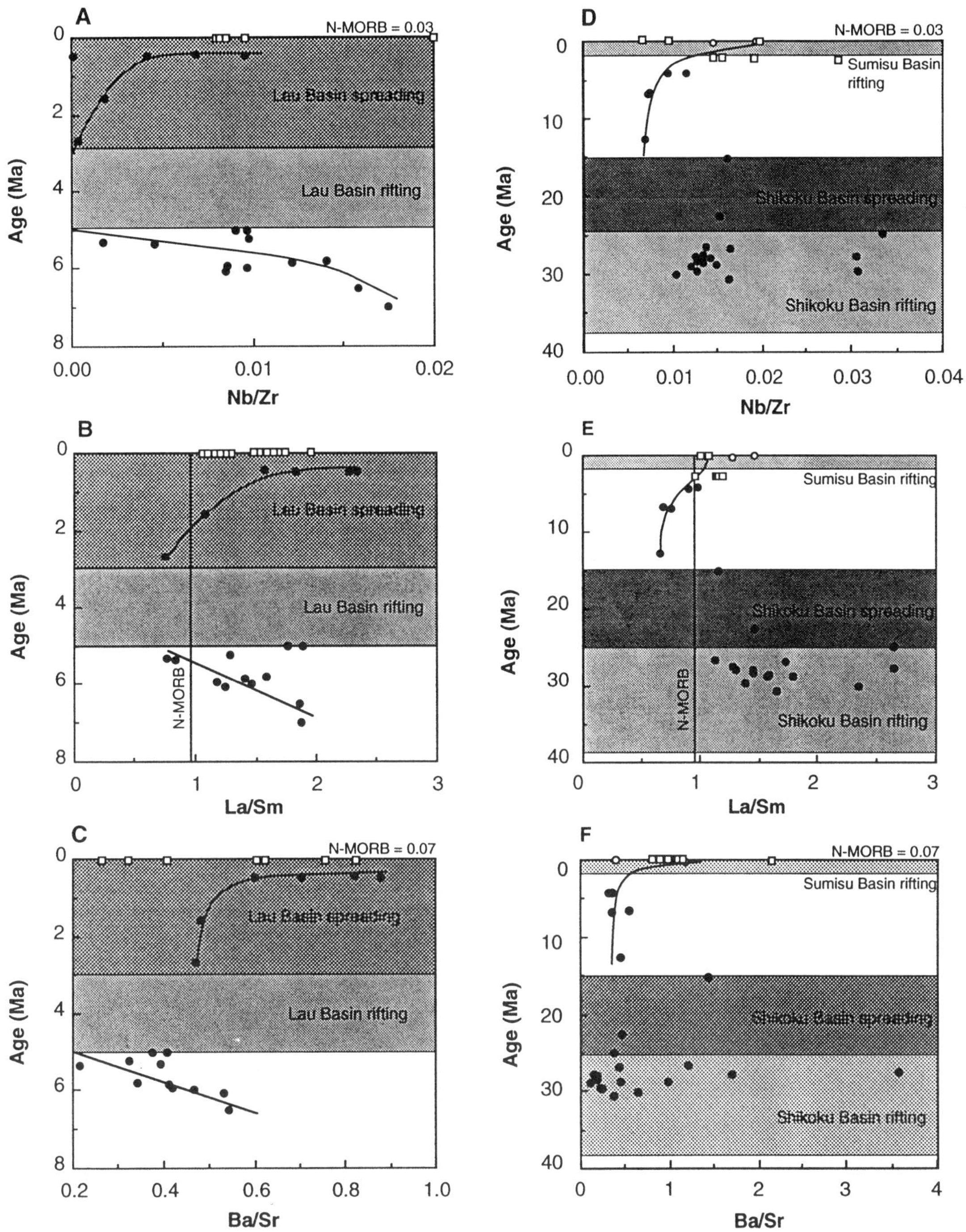

Fig. 12. Diagrams showing the similar temporal variation in (A) Nb/Zr; (B) La/Sm; and (C) Ba/Sr in individual basaltic glass grains from sediments at site 840 on the Tonga forearc measured by ion microprobe [black circles; Clift and Dixon, 1994] and modern Tofua arc lavas measured by ICP-MS [white squares; Ewart et al., 1994]. Ba/Sr ranges up to 2.586 on the Metis Shoal of the Tofua arc, and Nb/Zr ranges up to 0.02167 on Hunga Ha'apai in the southern Tofua arc. Note trend to increasing depletion prior to arc breakup at 5.0 Ma. (D) Nb/Zr; (E) La/Sm; and (F) Ba/Sr for whole rock analyses of turbidite sands from sites 787, 788, and 790-793 from the Izu-Bonin forearc (black circles) and Sumisu Rift (white circles) [Hiscott and Gill, 1992] and Pliocene to Pleistocene arc lavas [white squares; Gill et al., 1992], all measured by ICP-MS. Note general trend to increasing enrichment since the end of Shikoku basin spreading. Age of rifting is taken as the age of the oldest crust within the basin. Elemental ratios for N-MORB are from Sun and McDonough [1989].

model Ata Island and site 840 lie near the modern day junction between the two mantle domains.

Sediments recovered from sites 790, 791, 792, and 793 on the Izu-Bonin forearc span two rifting events, which must be distinguished if a comparison with the Lau rifting is to be attempted. Following the start of the Oligocene rifting of the Shikoku Basin, the range of Nb/Zr and especially La/Sm (modeling depletion in HFSEs and REEs respectively) falls dramatically to lower values, i.e., the most depleted volcaniclastic sediments were deposited after rifting initiated [Gill et al., in press]. This may reflect greater degrees of depletion of the source mantle and/or increased degrees of partial melting due to thinning of the arc lithosphere. Since 13 Ma the trend has reversed toward increasing Nb/Zr and La/Sm, which may reflect a thickening of the arc lithosphere by magmatic underplating since cessation of Shikoku Basin spreading at 17 Ma. Noticeably the present levels of enrichment are still below those which the arc was erupting prior to Shikoku Basin rifting. The lack of a corresponding trend in Ba/Sr may reflect greater mobility of these elements since deposition as a result of alteration, although when modern arc lavas are also considered the trend to enrichment in the LILEs can be seen (Figure 12). It is noteworthy that no relationship between chemistry and rifting of the Sumisu Rift has yet emerged. Any thinning of the arc during Sumisu rifting does not appear to affect the chemistry. Quite how much thinning has occurred is questionable, as no shallowing of the base of the lithosphere has been imaged by deep seismic surveys (ocean bottom seismometers and expanding spread seismic profiles) under the active Aoga rift, north of the main Sumisu Rift [A. Klaus, pers. comm., 1993].

Isotopic variations at the volcanic front. Consideration of isotopic data demonstrates changes in the nature of the Tonga/Tofua arc mantle wedge during arc breakup. New $^{208}Pb/^{204}Pb$ and $^{207}Pb/^{204}Pb$ data from site 840 [Clift, unpublished data] indicate that volcanism prior to rifting involved Pb contributions from Pacific pelagic sediments and an upper mantle source, falling broadly within a Pacific asthenospheric realm. Following re-establishment of arc volcanism along the Tofua arc at 3.0 Ma, 2.0 m.y. after the generation of the first backarc crust, the mantle component has changed (Figure 13). While still being within the Pacific realm defined by Hergt and Hawkesworth [1994], the new $^{208}Pb/^{204}Pb$ and $^{207}Pb/^{204}Pb$ values for post-rift volcanics lie on a trend between trench sediments and a distinct, but still Pacific, asthenospheric source. This implies that new asthenosphere had been drawn in under the island arc during the initial rift period. Hergt and Hawkesworth [1994] further demonstrated that the propagation of oceanic spreading centers in the backarc basin heralded the change from a mantle source of Pacific to one of Indian characteristics. The $^{208}Pb/^{204}Pb$ and $^{207}Pb/^{204}Pb$ values found along the modern eastern Lau spreading center and on the Tofua arc north of Ata shows this change in mantle character. The values in these places are distinctively Indian Ocean in character compared to those on the Kermadec arc and the Havre Trough [Ewart and Hawkesworth, 1987]. Figure 13 also shows changes in the nature of the sub-arc mantle during Sumisu Basin rifting [data from Gill et al., in press].

In summary, activity along the volcanic front shows a predominance of high-productivity, high-silica lavas in the early rift phase. Silica contents usually span the complete range of compositions, although the Tofua arc seems to develop a bimodal distribution following breakup. Eruption of trachy-andesite at the volcanic front occurs at the point of initial rifting. A reduction in the degree of alkali content has been measured in continental and some oceanic island arcs prior to arc breakup. Arc lavas retain a strong LILE enrichment and Nb depletion during the rifting period. However, the Tonga and Izu-Bonin arcs show evidence of increasing depletion in incompatible trace and REEs during rifting of the Lau and Shikoku Basins, followed by a reversal to more enriched chemistries. This may relate to tectonic thinning of the arc lithosphere prior to breakup and subsequent underplating by arc activity on the trench side of the basin. Following breakup lavas in the backarc basin are typically more enriched than those of the arc volcanic front, reflecting large scale melting of the asthenosphere under backarc spreading centers depleting the mantle from which the arc then derives its magma. Limited isotopic data suggest a change in the nature of the mantle wedge during the rifting process. Influence from the subducting slab would seem to be minor in the early rifting stage in the Tonga arc, but may increase with time.

The Backarc Basin

Volcanism within the backarc basin during the initial opening period usually occurs in two different forms. The first stages of extension are generally amagmatic; however, as extension continues, pillow and sheet-flow basaltic volcanism floors individual grabens, while more evolved lavas erupt on volcanoes to form seamounts on the graben flanks. In terms of their major elements the first pillow basalts are typically low-K tholeiites and show little variation as the basin opens. The ashes that accumulate next to the central volcanoes usually overlie the basalts and are also uniformly of low-K tholeiitic composition (Figures 9, 10). In this respect the first intra-basinal ashes from the

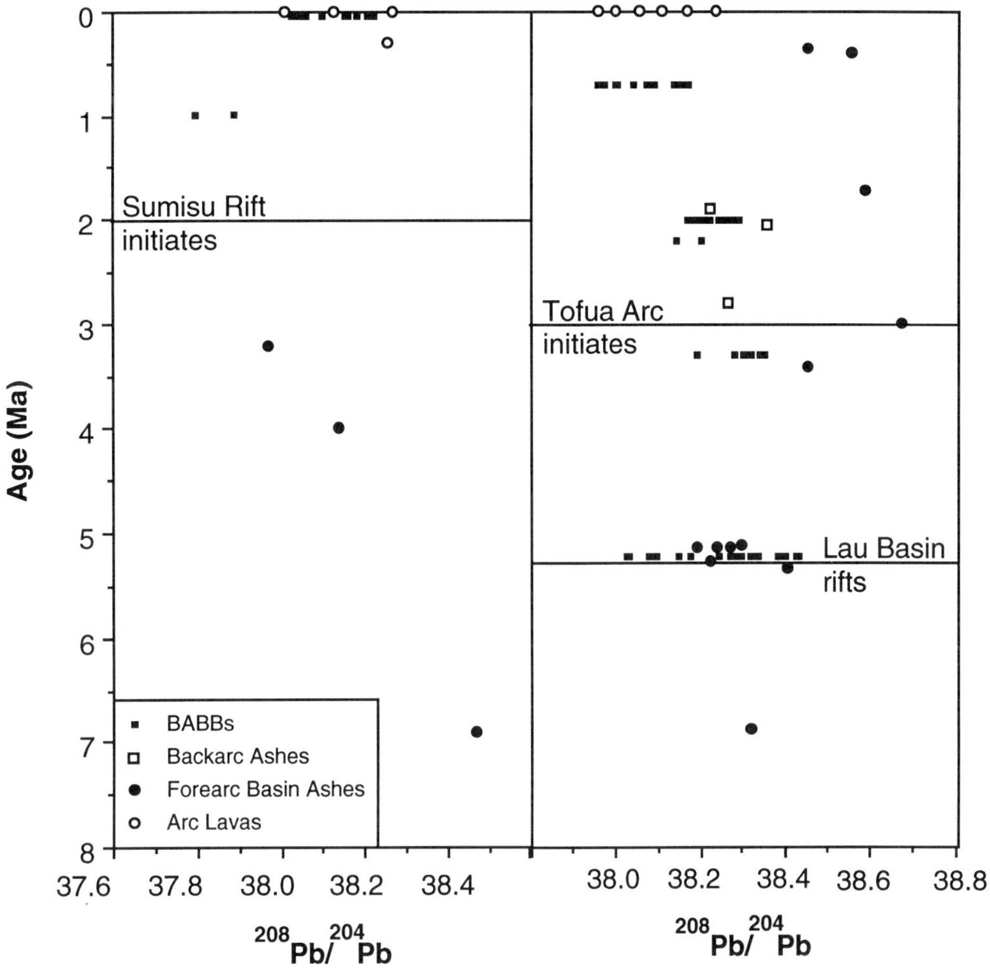

Fig. 13. Diagram showing the change in $^{208}Pb/^{204}Pb$ in ashes and lavas in the forearc basin and in the backarc rifts during rifting of the Sumisu Rift and Lau Basin. Data from Gill et al. [in press], Clift [unpublished data] and Hergt and Hawkesworth [1994]. BABBs = backarc basin basalts.

Izu-Bonin and Tonga arcs are identical to those erupted along the arc volcanic front prior to rifting. However, in the Marianas the tholeiitic ashes recorded in the basin [Packham and Williams, 1982] mark a change from the pre-rift calc-alkaline volcanism of the West Mariana Ridge [e.g., Mattey et al., 1981]. In terms of their silica contents, microprobe analyses indicate a complete spectrum of compositions for ashes in the Mariana Trough [Figure 11; Packham and Williams, 1982] and the earliest Lau Basin [Bednarz and Schmincke, 1994; Clift and Dixon, 1994]. In contrast, the Sumisu Rift has an early history characterized by a bimodal silica distribution of basalts/basaltic andesites and rhyolites/dacites [Rudolfo et al., 1992; Gill et al., 1992; Nishimura et al., 1991]. The situation in the Lau Basin is complicated by a changeover from a broad range of silica distribution to a bimodal one 2.0 m.y. after arc breakup [sampled at sites 837-839; Clift et al., 1994; Bednarz and Schmincke, 1994]. The geological significance of this change is unknown, but it serves to chemically distinguish between old and newer sub-basins and indicates that the sediment accumulating in each is derived locally from nearby submarine intra-basinal volcanoes [Clift et al., 1994]. As opening continues and extension is taken up along propagating spreading centers, the amount of ash accumulating in the backarc basin reduces markedly due to extinction of volcanism at central volcanic edifices. The only significant volcaniclastic material is hyaloclastite breccias, found immediately above the basement and of identical low-K tholeiitic basalt composition. Any arc volcanism that accompanies backarc spreading is then responsible for the input of thinner airfall ashes and pumice into the otherwise pelagic sediments of the backarc basin.

Fluctuations in trace element contents are of great petrogenetic significance. Since lavas similar to N-MORB are erupted along backarc spreading centers in the central Lau Basin and Mariana Trough [Hawkins and Melchior, 1985], it was predicted that the first backarc basin basalts (BABBs) would show the first stage of transition from island arc to MORB tholeiite [Tarney et al., 1981]. The eruption of arc-like tholeiitic andesites along the Valu Fa Ridge in the southern Lau Basin, where the spreading center is younger [Jenner et al., 1987; Vallier et al., 1991], served to support this theory. However, drilling of the very oldest backarc crust, predating the initiation of spreading centers, in a range of basins shows that in fact this simplistic model is not correct. Instead, Figure 14 shows multi-element spider diagrams from the oldest rift lavas sampled in each basin, with only a weak arc signature observed in each case. The typical Nb depletion relative to other incompatible trace elements is much reduced compared to normal arc tholeiites, or even to normal BABBs erupted later in the opening. Nevertheless there is some variability in the form of the multi-element diagrams between individual basins. While the Lau and Sumisu BABBs show relatively flat trends, and are thus more MORB-like, the oldest Japan Sea and Sulu Sea basalts show greater enrichment of the most incompatible elements, as well as the LILEs and LREEs. Relative enrichment in these elements may be due either to the asthenospheric mantle being relatively enriched compared to normal subarc mantle or to the addition of a more enriched component from melting of the continental lithosphere. Alternatively, the degree of partial melting in the Japan and Sulu Seas may have been less than that in the Lau and Sumisu cases. It is most likely that enriched melts frozen into the base of the lithosphere during its life span are prone to being remobilized by adiabatic melting during stretching and uplift [Fitton et al., 1991]. This is typical of continental rift basins and probably occurs in the rifting arc environment, thus enhancing the flat MORB-type trend of the oceanic arc BABBs. Another factor may be the greater thickness of the continental lithosphere compared to the oceanic arc lithosphere, resulting in a shorter melting column and lower degrees of partial melting.

Relative enrichment in incompatible elements in the earliest BABBs may indicate the presence of a fresh, undepleted mantle source under the arc at the point of breakup. It has been suggested that the MORB-like nature of the first BABBs relates to diapirism in the asthenosphere under the rifting arc, with normal MORB-type, upper mantle upwelling under the arc as breakup occurs [Crawford, 1989]. The results from the ODP drilling broadly support such a model and is supported by Pb isotopic evidence from BABBs in the Lau Basin, which indicate a change in mantle sources under the new basin [Hergt and Hawkesworth, 1994] during rifting (Figure 13). The $^{208}Pb/^{204}Pb$ and $^{207}Pb/^{204}Pb$ data indicate that rifting and the corresponding roll-back of the subducting Pacific lithosphere caused fresh Indian Ocean mantle to be drawn under the arc and backarc. The appearance of this new mantle source is further associated with the propagation of spreading centers into the region of rifted arc crust. Similarly, Pb isotopic data from fresh basalts collected by ALVIN dives [Hochstaedter et al., 1990a] in the Sumisu Rift show e_{Nd}, $^{208}Pb/^{204}Pb$ and $^{207}Pb/^{204}Pb$ characteristics suggestive of a mantle source under the rift distinctive from the arc volcanoes sited close by. However, in contrast to the Lau Basin, the Pb data imply an enriched mantle source, rather than the standard Pacific N-MORB source. HREE concentrations in the Sumisu Rift basalts are also incompatible with low degrees of partial melting of a normal arc mantle source.

It is noteworthy that few lavas of boninitic affinity are found in any of the early backarc sites. Boninites interbedded with island arc tholeiites in DSDP sites 458 and 459 in the Mariana forearc and onshore on Guam were interpreted by Hickey and Frey [1982] to be the result of shallow melting of a refractory mantle around an ascending mantle diapir, prior to rifting of backarc basin. Falloon et al. [1987] also explained the presence of boninites in the northern Tonga forearc to be the result of backarc ridge propagation into the forearc. Rising mantle diapirs under these spreading centers would necessarily cause melting of the refractory forearc mantle. Similar rocks have also been dredged in the northern Lau Basin, again a result of the action of propagating spreading centers causing the remelting of refractory mantle of the pre-existing backarc lithosphere [Falloon et al, 1992]. Boninites are not known from newly rifted backarc basins themselves. Instead, it appears that the earliest BABBs are of MORB-like composition, the direct product of melting from the upwelling diapir, and are rapidly followed by eruption of more arc-like compositions. Finally, as the backarc basin becomes wide and mature the basalts become more MORB-like again, as their source is removed from close proximity to the subduction zone.

Trace and REE data from volcanoes in rifting backarc basins are more scarce and are limited to the Lau Basin and Sumisu Rift [e.g., Figure 14: Clift and Dixon, 1994; Hochstaedter et al., 1990b; Fryer et al., 1990; Ikeda and Yuasa, 1989]. Sumisu Rift volcanoes have been sampled by *Alvin* dives and dredging and can be seen to have less subduction input (i.e., LILEs and LREEs) that the adjacent Izu-Bonin arc lavas, while still showing the distinctive Nb depletion and some LILE enrichment (Figure 15). Trace and

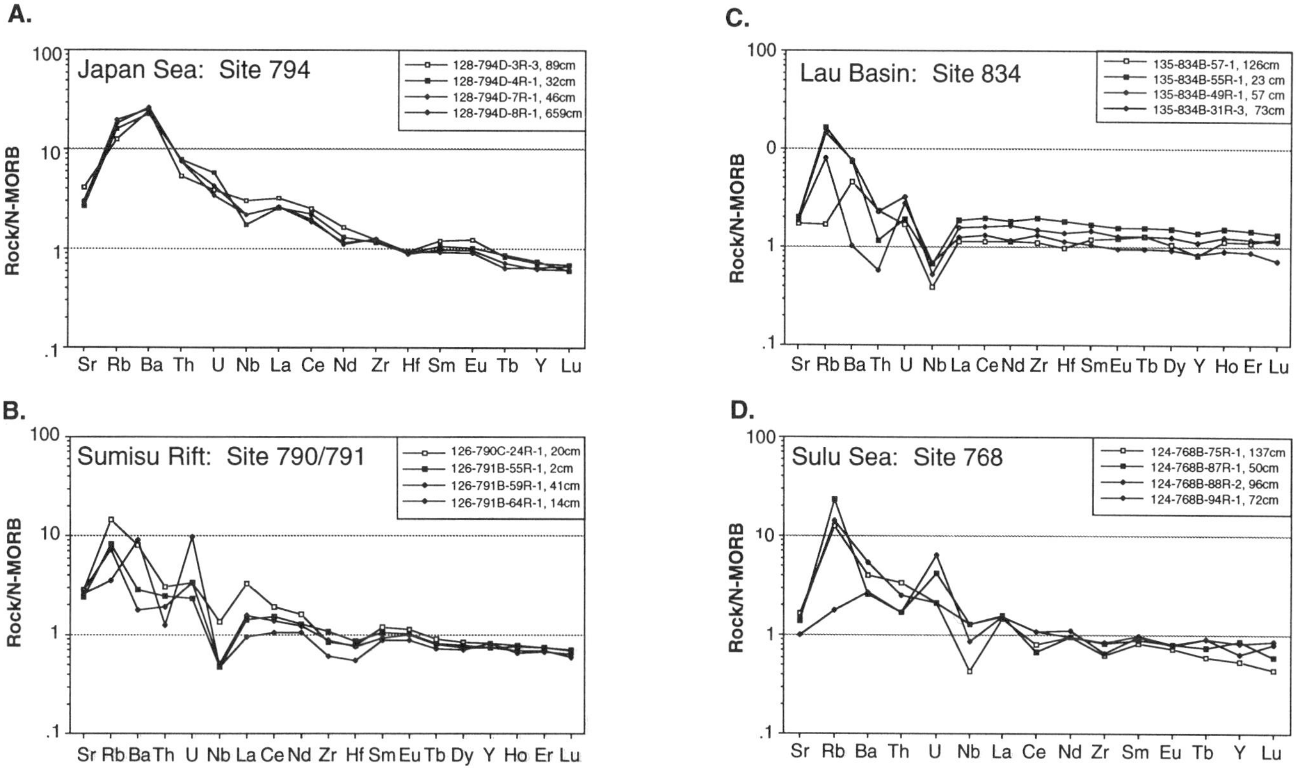

Fig. 14. Spider diagrams showing the chemical diversity of backarc basin basalts erupted in the first stages of arc rifting (A) Japan Sea, data from Pouclet and Bellon [1992], (B) Sumisu Rift, data from Gill et al. [1992], (C) Lau Basin, data from Ewart et al. [1994], and (D) Sulu Sea, data from Smith et al. [1991].

isotopic data from "Shadow Mountain" within the Sumisu Rift indicate a mixed N-MORB source and an arc mantle source, showing that these volcanoes are chemically similar to basalts erupted in more mature backarc basins. Data from the Lau Basin were mostly derived by ion microprobe analysis of individual basaltic glass shards within reworked turbidite sands (Figure 15). These also show Nb depletion and LILE enrichment, but do not show the regular temporal changes in enrichment recorded on the forearc [site 840; Clift and Dixon, 1994]. Calculated elemental ratios from across the Lau Basin display marked variability with time, with both incompatible depleted and enriched glasses being erupted and deposited over short time spans. Plotting Ba/Nb (LILE/HFSE) against La/Nb (REE/HFSE) allows discrimination to be made between the BABBs and the ashes from all the drilled sites (Figure 16). These ratios produce enough scatter to allow resolution between individual sites. This scatter reflects varying degrees of partial melting and slab influence. Interestingly, it can be seen that there is significant overlap between much of the Lau Basin ash and the pre-rift volcanic arc, as represented by site 840. This indicates that seamount volcanic edifices in the early Lau Basin are chemically indistinguishable from the arc in terms of their trace element, as well as major element compositions.

Since ashes in the early Lau Basin are interpreted as being derived into individual sub-basins from nearby central volcanoes, it might be expected that these ashes would have similar trace element characteristics to the basalts that form the basement of the sub-basin. If this were not the case it would require mantle heterogeneity on the small scale between basin and adjacent central volcano, a matter of only a few kilometers. In Figure 16 it may be seen that the BABBs and ashes often do show the same chemical characteristics. The most noticeable exception is site 834, the earliest backarc site, which has strongly arc-like ashes within its sedimentary cover, but a basement displaced towards a more MORB-like composition (bottom left corner of Figure 16). The MORB-like geochemistry of the earliest BABBs is not seen within the ash sequence.

Slab Contribution to Magma Genesis

Estimates of the contribution of the subducting slab to magma genesis during arc rifting can be made from the

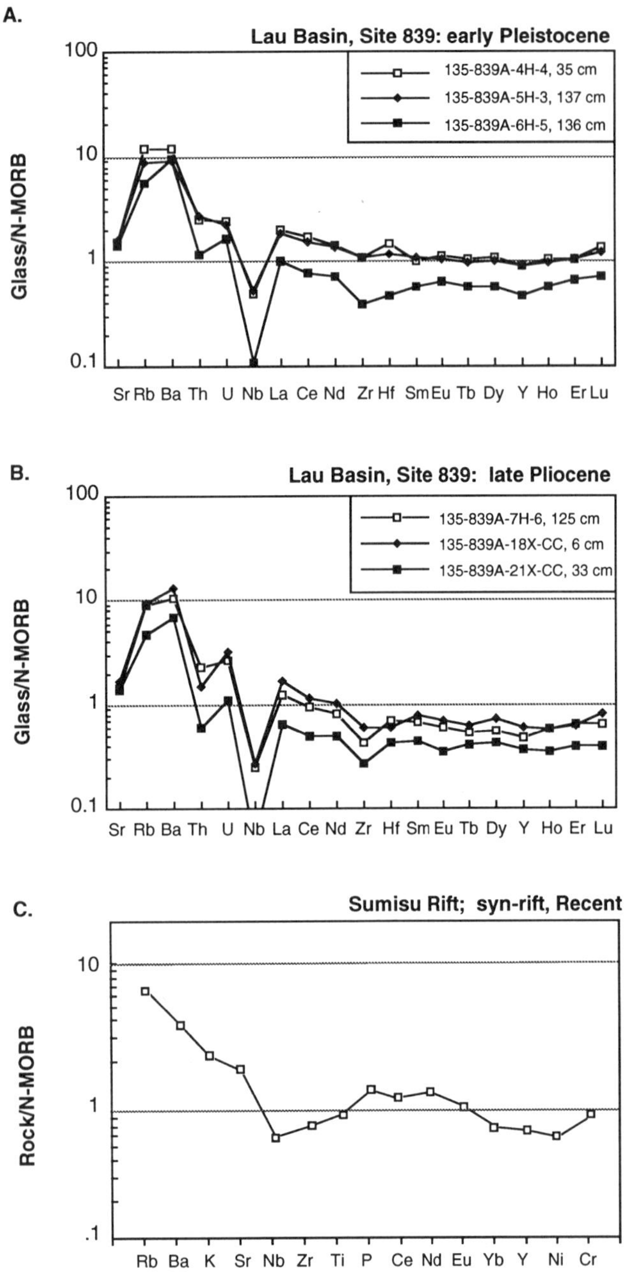

Fig. 15. Spidergrams of lavas from intrabasinal seamount volcanoes within the Lau Basin and Sumisu rift. Data from Lau Basin are from probe analysis of ashes from sites 839, showing the strongly arc character of the backarc ashes derived from seamount volcanoes within the early basin [Clift and Dixon, 1994]. Sumisu Rift analysis is a whole rock ICP-MS analysis of a dredged basalt [Hochstaedter et al., 1990a].

REE, trace element, and isotope characteristics of the volcanics within the new backarc rift. As noted above, the most important control on magma genesis during arc breakup along the Tongan volcanic front appeared to be the degree of melting in the mantle wedge, governed by the tectonic thinning of the arc lithosphere. Similarly, Pb isotopic and e_{Nd} data for Sumisu Rift basalts rule out subducted sediment sources as being important in their petrogenesis, and suggest that slab flux was small in the early stages of rifting of the Izu-Bonin arc. Changes in the slab flux with time may be traced by consideration of the concentration of a water-mobile LILE, such as Ba, relative to an immobile HFSE, such as Zr. Figure 17 shows the changes in Ba/Zr through time in the Lau Basin (sites 834-839), as well as along the volcanic front, represented by site 840. Prior to rifting, Ba/Zr had low, constant values. However, since that time Ba/Zr has risen steadily to a maximum value in both arc and backarc regions. This effect can also be seen on Figures 12B and C where the post-rift increase in enrichment of the REEs and LILEs is seen to outstrip that of the HFSEs. This extra enrichment can therefore not be attributed to processes in the mantle wedge, but must represent an additional slab flux component. The implication of the increasing Ba/Zr trend in the Tonga arc/backarc system is that fluid flux from the subducting slab to the arc and backarc has steadily increased from minimal values at the point of arc rifting to reach a maximum in the recent.

The Remnant arc

The nature of activity on a remnant arc is strongly dependent on the location of arc splitting. When rifting occurs trenchward of the volcanic front, activity on the remnant arc reduces in volume and also changes chemically with time as the basin widens. The best example of this in the western Pacific is the Lau Ridge, the remnant arc created by the opening of the Lau Basin [Karig, 1972]. In this case volcanism prior to rifting was characterized by dacitic, low-K, tholeiitic volcanism, exposed on land in the Lau Islands [Gill, 1976] and preserved in the volcaniclastic sediments of the Tonga forearc [site 840; Bøe, 1994; Clift and Dixon, 1994]. Following the 5.2 Ma rifting of the Tonga arc, volcanism became more basic, with basalt and andesite dominating the eruptives of the Korobasaga Volcanic Group of Fiji [Gill, 1976; Cole et al., 1985], which erupted from 4.4 to 2.4 Ma. This group displays the typical chemistry of an island arc tholeiite suite, and was followed by the volumetrically small Mago Volcanic Group, which is chemically distinctive in having an isotopically enriched character, more akin to hotspot lavas, and none of the Ta or

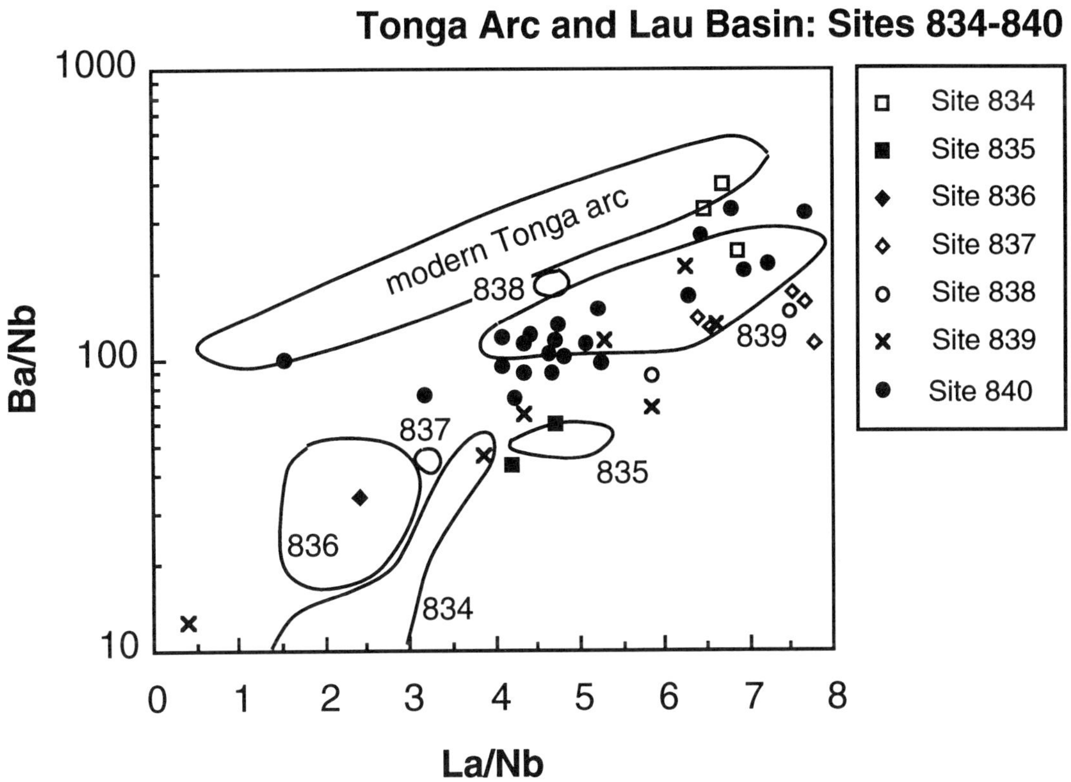

Fig. 16. Ba/Nb versus La/Nb for Lau Basin and Tonga arc (site 834-840) ashes. Note the dominantly arc-like character of the basaltic ashes from the backarc as well as from the Tonga Platform. Fields represent BABB composition. MORB composition lies towards the bottom left-hand corner of the diagram. Data points from Clift and Dixon [1994], fields from Ewart et al. [1994].

Zr depletion seen previously [Cole et al., 1990]. The Mago Volcanic Group shows a shift first to calc-alkaline and sub-alkaline andesites, before culminating in late Pleistocene alkali basalts and hawaiites. Most of the remnant arc products are generally deposited only in those parts of the Lau Basin adjacent to the ridge. Indeed, the volcaniclastic hiatus on the Tonga forearc between 5.0 and 3.0 Ma indicates that soon after rifting began the products of volcanism were only locally dispersed, possibly caused by subsidence of the Lau Ridge below sea level at that time. In addition, explosive submarine volcanism within the rifting Lau Basin tends to dominate the volcaniclastic record there and so masks any contribution from the Lau Ridge. Furthermore, in the early stages of rifting, the volcanic products of the Lau Ridge are impossible to distinguish chemically from other volcanic rocks, as they, and the products of intrabasinal volcanoes within the Lau Basin, are of strongly arc-like character.

The marine volcaniclastic record in the Lau Basin adjacent to the Lau Ridge (site 834)shows a compositional divergence from other material in the Lau Basin (sites 835-839) or Tonga forearc (sites 840-841) during the Pleistocene and is presumed to reflect changes in the chemistry of volcanism on the Lau Ridge. A small number of high-K andesite glass grains are found in volcaniclastic turbidites [Bednarz and Schmincke, 1994; Clift and Dixon, 1994]. Most importantly, ion microprobe analyses for trace and REEs in Pleistocene basaltic grains at site 834 show a major departure from the typical island arc tholeiite trend seen elsewhere in the basin. Figure 18 shows the contrast between the flat trend of the N-MORB normalized incompatible elements of the immediate post-rift volcanics compared to the depletion in the same elements in the glasses erupted immediately prior to remnant arc extinction [Clift and Dixon, 1994]. The significance of this trend is unclear, but may indicate successive melt extractions from the same volume of source mantle towards the end of remnant arc activity.

While activity on the Lau Ridge represents one of the best documented and longest lived examples of remnant arc volcanism in the western Pacific, it is noteworthy that most of the data come from onland exposures in the Lau

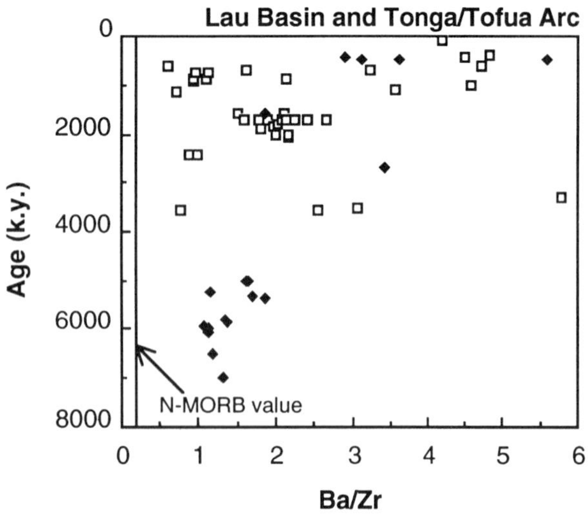

Fig. 17. Ba/Zr in Lau Basin and Tonga Platform (sites 834-840) ashes through time, showing the trend to increasing values and, by implication, increasing slab fluid flux since arc breakup. Squares indicate analyses from the backarc basin, while solid diamonds represent the volcanic arc measured at site 840. Data from Clift and Dixon [1994].

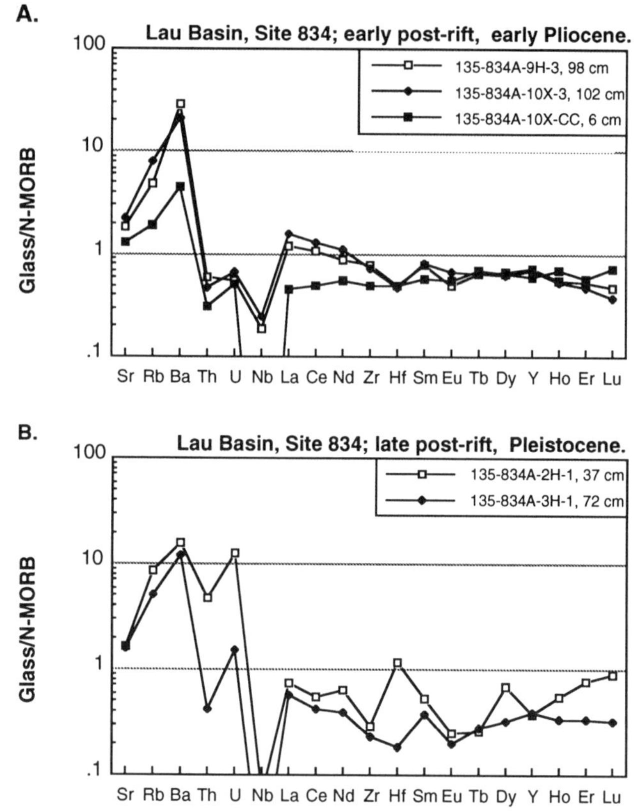

Fig. 18. N-MORB normalized spidergrams for basaltic glass grains in volcaniclastic sediments at site 834 in the Lau Basin. site 834 lies next to the remnant arc of the Lau Ridge and records changes in the nature of its volcanism. The post-rift period is characterized by a shift from flat incompatible trends, typical of IATs to an incompatible depleted pattern immediately prior to its extinction. Data from Clift and Dixon [1994].

and Fiji islands at the north end of the edifice. Since rifting of the Lau Basin, a tear in the Pacific lithosphere at the north end of the Tofua arc has allowed enriched upper mantle of the Samoan plume to flow under the northern Lau Basin [Volpe et al., 1988]. This extra effect, unknown elsewhere in the Western Pacific, may well be responsible for the longevity and alkaline character of the post-rift volcanism on the northern Lau Ridge. Some caution must be applied in making comparison between the Lau Ridge and other remnant arcs.

Volcanism on other remnant arcs is known in the Mariana system, where DSDP Leg 59 drilled holes on the West Mariana Ridge (site 451) and Palau-Kyushu Ridge (site 448). These form the remnant arcs to the Mariana Trough and Parece Vela Basins respectively. A series of calc-alkaline basaltic breccias recovered from the West Mariana Ridge indicates activity on the remnant arc up to the point of arc rifting. Dating is not well constrained at this site, but what is available suggests that activity ceased at, or shortly after, arc breakup. Furthermore, the similarity of the West Mariana Ridge basalts to calc-alkali basalts on the modern Mariana arc [Mattey et al., 1981] does not indicate a long post-rift volcanic history of evolved, alkaline character, such as found on the Lau Ridge. Similarly, depleted arc tholeiites recovered at site 448 on the Palau-Kyushu Ridge are covered only by pelagic oozes and distal airfall ashes, thus suggesting that arc volcanism there was rapidly curtailed during rifting of the Parece Vela Basin. No transitional volcanic chemistries are found at either site, reflecting the fact that arc rifting in each case split the volcanic front, leaving the opening basin as the only remaining locus of volcanism.

No volcanism is found on remnant arc ridges, where rifting occurs behind the arc volcanic front. In this case the remnant ridge is merely a rifted fragment of the old arc edifice. All volcanism is concentrated in the basin or on the trench side of the basin.

CONCLUSIONS

The rocks of the western Pacific arcs and backarc regions discussed above and sampled during ODP/DSDP drilling illustrate how the nature of arc volcanism changes during rifting events. The location of rifting has a profound effect on the distribution of volcaniclastic sediment throughout the arc system. Extension of a volcanic arc results in increasing partial melting and voluminous volcanism along the volcanic front. A volcaniclastic minimum or hiatus in the forearc basin early in the rift process may be often recognized. This can be due to rifting dissecting the original volcanic front (e.g., Mariana Trough rifting) or separating it from the forearc area (e.g., Lau Basin rifting). In cases when the rifting is behind the arc (e.g., Sumisu Rift, Parece Vela Basin, and Shikoku Basin rifting) a hiatus truly reflects a brief cessation of arc volcanism. After an early amagmatic phase, intrabasinal volcanism began in the form of tholeiitic pillow and sheet-flow basalt extrusions within grabens of the original arc edifice. These lavas are chemically more enriched in incompatible elements compared to the arc volcanic front. Voluminous dacitic and rhyolitic lavas were erupted from submarine volcanic centers on the flanks of individual sub-basins. Volcaniclastic debris from submarine eruptions ponded as turbidite and mass-flow sands and gravels in the small basins that surround each volcanic center. Continued extension in the Lau Basin resulted in a trenchward migration of the intra-basinal volcanoes, effectively as a submarine arc. Such activity continued until the propagation of a backarc spreading center into the region, after nucleation at a transform fault. In the Lau Basin the rifting phase was long, lasting for 2 m.y. before any spreading began. The proportion of lava erupted in the basin compared to the volcanic front is variable. In the Lau Basin intrabasinal volcanism is dominant, while in the Sumisu Rift, Mariana Trough, and Japan sea the arc is vigorous and intrabasinal volcanism is modest and restricted to across arc chains. In the southern Izu-Bonin arc ridge propagation occurred earlier and may have caused a cessation of activity along the arc volcanic front. Ridge propagation in the Okinawa Trough has caused the entire abandonment of activity on the Ryukyu arc and the concentration of all volcanism in the basin, while in the Lau Basin propagation is related to the resurgence of subaerial arc volcanism on the trench side of the basin. Commonly, the new arc was founded on the rifted remnants of the earlier edifice. At this spreading stage of basin opening airfall ashes derived from the new arc edifice were deposited throughout the backarc basin, usually against a background of pelagic sediment. Although uncommon, evolved alkaline volcanism is known to have continued on the remnant Lau Ridge arc for 4 m.y. after rifting and even overlapped with volcanism on the new arc front.

Most oceanic arcs show little change in their major element chemistry during rifting, remaining tholeiitic or calc-alkaline. There is a suggestion that prior to breakup in the Tonga, Japan and Izu-Bonin arcs there was a reduction in the frequency of medium and high-K lavas. The Izu-Bonin arc shows a return to medium and high-K lavas in the post-rift period. These changes may relate to degrees of partial melting governed by thinning and thickening of the arc lithosphere. Voluminous dacitic and rhyolitic volcanism in particular is a feature of the early pre- and syn-rift period, both along the volcanic front and later on central volcanoes within the rifting basin (e.g., Tonga, Izu-Bonin, Mariana, Cagayan and Japan arcs). Trachy-andesitic volcanism is also recorded at the point of arc breakup at the northern termination of the Mariana Trough. Volcanism within the new basins is almost always low-K tholeiitic. The first BABBs often show only a modest arc-related signature in terms of their trace elements, which increases with time during the first 5 m.y. of extension. The degree of Nb depletion and LILE enrichment show less subduction influence in the first BABBs than in the pre- and syn-rift volcanics rocks along the volcanic front. The first BABBs may show flat or LREE-enriched patterns in spider diagrams indicative of their derivation from a body of less depleted upper mantle, such as a diapir. In the Sumisu Rift an enriched asthenospheric source for the first BABBs is suggested by Pb and Nd isotopic data.

REE data from the Tonga and Izu-Bonin forearcs show a trend to increasing LREE depletion prior to rifting. This trend is accompanied by depletion in the HFSEs, and possibly the LILEs, over the same time span. It is thought that a common process of arc lithospheric thinning and the resultant increasing height of the melting column is responsible for these changes and would also explain the increased volume of volcanism at the point of arc breakup. The degree of slab input into the arc volcanics does not seem to be significant during the early rifting, but is seen to increase in importance following initial breakup in the Tonga/Lau system. Magmatic underplating of the new Tofua and Izu-Bonin arcs may cause the thickness of the lithosphere to increase, and so decrease the height of the melting column. This would explain the trend to increasing LREE, HFSE and LILE enrichment during the post-breakup period. In the Izu-Bonin forearc, sediments do not show any response to the rifting of the Sumisu Rift, which may be too young to have caused strong lithospheric thinning. Pb, Sr, and Nd isotopic data from both Tonga and Izu-Bonin arcs indicate that arc rifting draws in asthenosphere under the opening basin, sometimes resulting in a permanent change in the mantle reservoir from which arc and backarc melts were extracted. It is noteworthy that lithospheric extension appears to control the volcanism in the early stages of breakup; its effect, however, is limited, as the extrapolation of this model would predict the backarc basin as the focus of the vast majority of volcanism, rather than the non-extensional arc. That the volcanic front often continues to be active in the rift and spreading stage indicates that other factors, such as fluid flux from the subducting slab, may be at least as important.

Acknowledgments. PC would like to thank NERC and Lin Kay for their support of his involvement in the ODP, and particularly for the support to attend the WestPac synthesis meeting in Monterey, CA, in January 1993. PC would also like to thank the Royal Society of Edinburgh for the fellowship during which much of this work was done, and Science Operations at ODP for the time and facilities to complete this undertaking. John E. Dixon, Adam Klaus, Kantaro Fujioka, Tracy Vallier and Brian Taylor are thanked for their stimulating discussion and critical appraisal of many of the ideas presented in this paper. Rick Hiscott and Alfred Hochstaedter are thanked for their constructive reviews, which helped to greatly improve the paper.

REFERENCES

Arculus, R. J., and A. L. Bloomfield, Major-element geochemistry of ashes from sites 782, 784, and 786 in the Bonin forearc, in *Proc. ODP, Sci. Results, 125,* P. Fryer, J. A. Pearce, L. B. Stokking, et al., pp. 277-292, Ocean Drilling Program, College Station, TX, 1992.

Austin, J., F. W. Taylor, and C. D. Cagle, Seismic stratigraphy of the central Tonga Ridge. *Mar.Petrol. Geol., 6,* 71-92, 1989.

Bednarz, U., and H.-U. Schmincke, Composition and origin of volcaniclastic sediments in the Lau Basin (southwest Pacific), Leg 135 (site 834-839) in *Proc. ODP, Sci. Results, 135,* J. W. Hawkins, L. M. Parson, J. F. Allan, et al., pp. 51-74, Ocean Drilling Program, College Station, TX, 1994.

Bøe, R. Nature and record of late Miocene mass-flow deposits of the Lau-Tonga forearc basin, Tonga Platform (Hole 840B) in *Proc. ODP, Sci. Results, 135,* J. W. Hawkins, L. M. Parson, J. F. Allan, et al., pp. 87-100, Ocean Drilling Program, College Station, TX, 1994.

Cadet, J. P., and K. Fujioka, Neogene volcanic ashes and explosive volcanism: Japan Trench transect, Leg 57, in Scientific Party, *Init. Repts. DSDP, 56, 57,* 1027-1041, U.S. Govt. Printing Office, Washington, D.C., 1980.

Cawood, P. A., Nature and record of igneous activity in the Tonga arc, SW Pacific, deduced from the phase chemistry of derived detrital grains, in *Developments in Sedimentary Provenance Studies,* edited by A. C. Morton, S. P. Todd, and P. D. W. Haughton, Geol. Soc. London, Spec. Publ., *57,* 305-323, 1991.

Clift, P. D., and J. E. Dixon, Variations in arc volcanism and sedimentation related to rifting of the Lau Basin, SW Pacific, in *Proc. ODP, Sci. Results, 135,* J. W. Hawkins, L. M. Parson, J. F. Allan, et al., pp. 23-49, Ocean Drilling Program, College Station, TX, 1994.

Clift, P. D., and ODP Leg 135 Scientific Party, Volcanism and sedimentation in a rifting island arc terrain; an example from Tonga, SW Pacific, in *Volcanism Associated with Extension at Consuming Plate Margins,* edited by J. L. Smellie, Geol. Soc. London Spec. Publ., in press.

Cole, J. W., J. B. Gill, and D. Woodhall, Petrological history of the Lau Ridge, Fiji, in *Geology and Offshor eResources of the Pacific Island Arcs - Tonga Region,* edited by D. W. Scholl and T. L. Vallier, pp. 379-391, Circum-Pacific Council for Energy and Mineral Resources, Earth Science Series, 2, Houston, TX, 1985.

Cole, J. W., I. J. Graham, and I. L. Gibson, Magmatic evolution of Late Cenozoic volcanic rocks of the Lau Ridge, Fiji, *Contrib. Mineral. Petrol., 104,* 540-554, 1990.

Crawford, A. J., *Boninites,* Unwin Hyman, London, 1989.

DeBari, S., and R. G. Coleman, Examination of the deep levels of an island arc: evidence from the Tonsina ultramafic-mafic assemblage, Tonsina, Alaska, *J. Geophys. Res. 94, B4,* 4373-4391, 1989.

Ewart, A., W. B. Bryan, B. W. Chappell, and R. L. Rudnick, Regional geochemistry of the Lau-Tonga arc and Backarc Systems, in *Proc. ODP, Sci. Results, 135,* J. W. Hawkins, L. M. Parson, J. F. Allan, et al., pp. 385-425, Ocean Drilling Program, College Station, TX, 1994.

Ewart, A., and C. J. Hawkesworth, The Pleistocene-Recent Tonga-Kermadec arc lavas: Interpretation of new isotopic and rare earth data in terms of a depleted source mantle, *J. Petrol., 28,* 3, 495-530, 1987.

Falloon, T. J., D. H. Green, and A. J. Crawford, Dredged igneous rocks from the northern termination of the Tofua magmatic arc, Tonga and adjacent Lau Basin, *Austr. J. Earth Sci. 34,* 487-506, 1987.

Falloon, T. J., A. Malahoff, L. P. Zonenshain, and Y. Bogdanov, Petrology and Geochemistry of Back-arc Basin Basalts from Lau Basin Spreading Ridges at 15°, 18°, and 19°S, *Mineral. Petrol. 47,* 1, 1-36, 1992.

Fitton, J. G., D. James, and W. P. Leeman, Basic magmatism associated with late Cenozoic extension in the Western United States: compositional variations in space and time, *J. Geophys. Res., 96,* 13,963-13,711, 1991.

Fryer, P., J. A. Pearce, L. B. Stokking, et al., *Proc. ODP, Init. Repts., 125,* Ocean Drilling Program, College Station, TX, 1990.

Fryer, P., B. Taylor, C. H. Langmuir, and A. G. Hochstaedter, Petrology and geochemistry of lavas from the Sumisu and Torishima backarc rifts, *Earth Planet. Sci. Lett., 100,* 161-178, 1990.

Fujioka, K., Synthesis of Neogene explosive volcanism of the Tohoku arc, deduced from the marine tephra drilled around the Japan Trench region, Deep Sea Drilling Project legs 56, 57 and 87B, in *Init. Repts. DSDP, 87,* H. Kagami, D. E. Karig, W. T. Coulbourn, et al., pp. 703-723, U.S. Govt. Printing, Office, Washington, D.C., 1986.

Fujioka, K., Y. Matsuo, A. Nishimura, M. Koyama, and K. S. Rudolfo, Tephras of the Izu-Bonin forearc (sites 787, 792, and 793), in *Proc. ODP, Sci. Results, 126,* B. Taylor, K. Fujioka, et al., pp. 47-74, Ocean Drilling Program, College Station, TX, 1992a.

Fujioka, K., A. Nishimura, Y. Matsuo, and K. S. Rudolfo, Correlation of Quaternary tephras throughout the Izu-Bonin areas, in *Proc. ODP, Sci. Results, 126,* B. Taylor, K. Fujioka, et al., pp. 23-45, Ocean Drilling Program, College Station, TX, 1992b.

Fujioka, K., and S. Saito, Composition of heavy minerals from sands and sandstones of the Izu-Bonin arc, Leg 126, in *Proc.*

ODP, Sci. Results, 126, B. Taylor, K. Fujioka, et al., pp. 155-169, Ocean Drilling Program, College Station, TX, 1992.

Gill, J. B., Composition and age of Lau Basin and Ridge volcanic rocks: Implications for evolution of an inter-arc basin and remnant arc, *Geol. Soc. Am. Bull., 87*, 1384-1395, 1976.

Gill, J. B., C. Seales, P. Thompson, A. G. Hochstaedter, and C. Dunlap, C., Petrology and Geochemistry of Pliocene-Pleistocene volcanic rocks from the Izu arc, Leg 126, in *Proc. ODP, Sci. Results, 126*, B. Taylor, K. Fujioka, et al., pp. 383-404, Ocean Drilling Program, College Station, TX, 1992.

Gill, J. B., R. N. Hiscott, and Ph. Vidal, Turbidite geochemistry and evolution of the Izu-Bonin arc and Continents. *Lithos*, in press.

Hawkins, J. W., and J. T. Melchior, Petrology of Mariana Trough and Lau Basin basalts, J. *Geophys. Res., 90*, 11,431-11,468, 1985.

Hergt, J. M., and C. J. Hawkesworth, The Pb, Sr, and Nd isotopic evolution of the Lau Basin: Implications for mantle dynamics during back-arc opening, in *Proc. ODP, Sci. Results, 135*, J. W. Hawkins, L. M. Parson, J. F. Allan, et al., pp. 505-517, Ocean Drilling Program, College Station, TX, 1994.

Hickey, R. L., and F. A. Frey, Rare-earth element geochemistry of Mariana fore-arc volcanics: Deep Sea Drilling Project site 458 and Hole 459B, in *Init. Repts. DSDP, 60*, D. M. Hussong, S. Uyeda, et al., pp. 735-742, U.S. Govt. Printing Office, Washington, D.C., 1982.

Hiscott, R. N. and J. B. Gill, Major and trace element geochemistry of Oligocene to Quaternary volcaniclastic sands and sandstones from the Izu-Bonin arc, in *Proc. ODP, Sci. Results, 126*, B. Taylor, K. Fujioka, et al., pp. 467-486, Ocean Drilling Program, College Station, TX, 1992.

Hochstaedter, A. G., J. B. Gill, and J. D. Morris, Volcanism in the Sumisu Rift, II. Subduction and non-subduction related components, *Earth Planet. Sci. Lett., 100*, 195-209, 1990a.

Hochstaedter, A. G., J. B. Gill, M. Kusakabe, S. Newman, M. Pringle, B. Taylor, and P. Fryer, Volcanism in the Sumisu Rift, I. Major element, volatile and stable isotope geochemistry, *Earth Planet. Sci. Lett., 100*, 179-194, 1990b.

Hole, M. J., A. D. Saunders, G. F. Marriner, and J. Tarney, Subduction of pelagic sediment: implications for the origin of Ce-anomalous basalts from the Mariana Islands, *J. Geol. Soc. London, 141*, 453-472, 1984.

Hussong, D. M., and S. Uyeda, Tectonic processes and the history of the Mariana arc: A synthesis of the results of Deep Sea Drilling Project Leg 60, in *Init. Repts. DSDP, 60*, D. M. Hussong, S. Uyeda, et al., pp. 909-929, U.S. Govt. Printing Office, Washington, D.C., 1982.

Hussong, D. M., S. Uyeda, et al., *Init. Repts. DSDP, 60*, U.S. Govt. Printing Office, Washington, D.C., 1982.

Ikeda, Y., and M. Yuasa, Volcanism in nascent back-arc basins behind the Shichito Ridge and adjacent areas in the Izu-Ogasawara arc, northwest Pacific: evidence for mixing between E-type MORB and island arc magmas at the initiation of back-arc rifting, *Contrib. Mineral. Petrol. 101*, 377-393, 1989.

Ingle, J. C., K. Suyehiro, M. T. von Breymann, et al., *Proc. ODP, Init. Repts., 128,* Ocean Drilling Program, College Station , TX, 1990.

Irvine, T. N., and W. R. Baragar, A guide to the chemical classification of common igneous rocks, *Can. J. Earth Sci., 8*, 523-548, 1971.

Jenner, G. A., P. A. Cawood, M. Rautenschlein, and W. M. White, Composition of back-arc basin volcanics, Valu Fa Ridge, Lau Basin: Evidence for a slab-derived component in their mantle source, *J. Volcanol. Geotherm. Res.*, 32, 209-222, 1987.

Jolivet, L., and K. Tamaki, Neogene kinematics in the Japan Sea region an volcanic activity of the northeast Japan arc., in *Proc. ODP, Sci. Results, 127/128, Pt. 2*, K. Tamaki, K. Suyehiro, J. Allan, M. McWilliams, et al., pp. 1311-1329, Ocean Drilling Program, College Station, TX, 1992.

Karig, D. E., Remnant arcs, *Bull. Geol. Soc. Am.*, 83, 1057-1068, 1972

Karig, D.E., Basin genesis in the Philippine Sea, in *Init. Repts. DSDP, 31*, J. C. Ingle, D. E. Karig, et al., pp. 857-879, U.S. Govt. Printing Office, Washington, D.C., 1975.

Kelemen, P. B., N. Shimizu, and T. Dunn, Relative depletion of niobium in some arc magmas and the continental crust: partitioning of K, Nb, La and Ce during melt/rock reaction in the upper mantle, *Earth Planet. Sci. Lett., 120*, 111-134, 1993.

Lee, J.-S. and A. Pouclet, Le volcanisme Neogene de Pohang (SE Coree), nouvelles constraintes geochronologiques pour l'ouverture de la Mer du Japon, *C.R. Acad. Sci. Paris, ser. II 307*, 1405-1411, 1988.

Mattey, D. P., N. G. Marsh, and J. Tarney, The geochemistry, mineralogy, and petrology of basalts from the West Philippine and Parece Vela basins and from the Palau-Kyushu and West Mariana Ridges, Deep Sea Drilling Project Leg 59, in *Init. Repts. DSDP, 59*, L. W. Kroenke, R. B. Scott, et al., pp. 753-800, U.S. Govt. Printing Office, Washington, D.C., 1981.

McKenzie, D. P., and M. J. Bickle, The volume and composition of melt generated by extension of the lithosphere, J. *Petrol., 29*, 625-679, 1988

Natland, J. H., and J. Tarney, Petrological evolution of the Mariana arc and back-arc system: A synthesis of drilling results in the South Philippine Sea, in *Init. Repts. DSDP, 60*, D. M. Hussong, S. Uyeda, et al., pp. 877-908, U.S. Govt. Printing Office, Washington, D.C., 1982.

Neilson, C. H., and H. Sigurdsson, Quantitative methods for electron microprobe analysis of sodium in natural and synthetic glasses, *Am. Mineral., 66*, 547-552, 1981.

Nishimura, A., K. M. Marsaglia, K. S. Rudolfo, A, Colella, R. N. Hiscott, K. Tazaki, J. B. Gill, T. Janecek, J. Firth, M. Isiminger-Kelso, Y. Herman, R. N. Taylor, B. Taylor, K. Fujioka, and Leg 126 Scientific Party, Pliocene-Quaternary Submarine Pumice Deposits in the Sumisu Rift Area, Izu-Bonin arc, in *Sedimentation in Volcanic Settings,* SEPM Spec. Publ., *45*, 201-208, 1991.

Packham, G. and R. W. Williams, Volcanic glasses from sediments from sites 453 and 454 in the Mariana Trough, in *Init. Repts. DSDP, 60*, D. M. Hussong, S. Uyeda, et al., pp. 483-495, U.S. Govt. Printing Office, Washington, D.C., 1982.

Parson, L. M., J. W. Hawkins, J. F. Allan, et al., *Proc. ODP, Init. Repts, 135,* Ocean Drilling Program, College Station, TX, 1992.

Parson, L. M., J. W. Hawkins, and P. M Hunter, Morphotectonics of the Lau Basin Seafloor - Implications for the opening history of backarc basins, in *Proc. ODP, Init. Repts, 135*, L. M. Parsons, J. W. Hawkins, J. F. Allan, et al , pp. 81-84, Ocean Drilling Program, College Station, TX, 1992.

Parson, L. M., and J. W. Hawkins, Two stage ridge propagation and the geological history of the Lau backarc basin, in *Proc. ODP, Sci. Results, 135*, J. W. Hawkins, L. M. Parson, J. F. Allan, et al., pp. 819-828, Ocean Drilling Program, College Station, TX, 1994.

Plank, T., and C. Langmuir, What controls global variations in the chemistry of arc volcanics ? *EOS Trans. AGU, 68*, 44, 1987.

Plank, T., and C. H. Langmuir, An evaluation of the global variations in the major element chemistry of arc basalts, *Earth Planet. Sci. Lett. 90*, 349-370, 1988.

Pouclet, A., and H. Bellon, Geochemistry and isotopic composition of volcanic rocks from the Yamato Basin: Hole 794D, Sea of Japan, in *Proc ODP, Sci. Results, 127*, K. Tamaki, K. A. Pisciotto, J. F. Allan, et al., pp. 770-790, Ocean Drilling Program College Station, TX, 1992.

Rangin, C., E. A. Silver, M. T. von Breymann, et al., *Proc. ODP, Init. Repts, 124,* Ocean Drilling Program, College Station, TX, 1990.

Rangin, C., and E. A. Silver, Neogene tectonic evolution of the Celebes-Sulu Basins: New Insights from Leg 124 Drilling, in *Proc. ODP, Sci. Results, 124*, E. A. Silver, C. Rangin, M. T. von Breymann, et al., pp. 51-64. Ocean Drilling Program, College Station, TX, 1992.

Rudolfo, K. S., R. U. Solidum, A. Nishimura, Y. Matsuo, and K. Fujioka, Major element oxide stratigraphy of glass shards in volcanic ash layers of the Izu-Bonin arc-backarc sites (sites 788/789 and 790/791), in *Proc. ODP, Sci. Results, 126,* B. Taylor, K. Fujioka, et al., pp. 505-517, Ocean Drilling Program, College Station, TX, 1992.

Scott, R. B., L. W. Kroenke, G. Zakariadze, and A. Sharaskin, Evolution of the South Philippine Sea: Deep Sea Drilling Project Leg 59 results, in *Init. Repts. DSDP, 59*, L. Kroenke, R. B. Scott, et al., pp. 803-817, U.S. Govt. Printing Office, Washington D.C., 1980.

Shane, P. A. R., and P. C Froggatt, Glass chemistry, paleomagnetism and correlation of middle Pleistocene tuffs in southern North Island, New Zealand and western Pacific, *N.Z. J. Geol. Geophys., 34*, 203-211, 1991.

Shipboard Scientific Party, sites 438 and 439, Japan deep sea terrace, in *Init. Repts. DSDP, 56, 57,* Scientific Party, pp. 23-192, U.S. Govt. Printing Office, Washington, D.C., 1980.

Sibuet, J.-C., J. Letouzey, F. Barbier, J. Charvet, J.-P. Foucher, T. W. C. Hilde, M. Kimura, C. Ling-Yun, B. Marsset, C. Muller, and J.-F. Stephan, Back arc Extension in the Okinawa Trough, *J. Geophys. Res., 92*,B13,14,041-14,063, 1992.

Smith, T. E., C. H. Huang, and F. G. Sajona, Geochemistry and Petrogenesis of the volcanic rocks from Holes 768 and 769, Sulu Sea, in *Proc. ODP, Sci. Results, 124*, E. A. Silver, C. Rangin, M. T. von Breymann, et al., pp. 297-320. Ocean Drilling Program, College Station, TX, 1991.

Stern, R. J., N. C. Smoot, and M. Rubin, Unzipping of the Volcano arc, Japan, *Tectonophysics, 102*, 153-174, 1984.

Sugimura, A., T. Matsuda, K. Chinzei, and K. Nakamura, Quantitative distribution of late Cenozoic volcanic material in Japan, *Bull. Volcanol., 26*, 125-140, 1963.

Sun, S., and W. F. McDonough, Chemical and isotopic systematics of oceanic basalts: implications for mantle composition and processes, in *Magmatism in the Ocean Basins*, edited by A. D. Saunders, and M. J. Norry, Geol. Soc. Spec. Publ., 42, 313-346, 1989.

Tamaki, K., K. A. Pisciotto, J. F. Allan, et al., *Proc ODP, Init. Results, 127,* Ocean Drilling Program, College Station, TX, 1990.

Tappin, D. R., R. H. Herzer, and A. J. Stevenson, Structure and history of an oceanic forearc - the Tonga Ridge 22° to 26° south, in *Tonga arc,* edited by A. Stevenson and P. Ballance, Springer Verlag, Berlin, 1991.

Tappin, D. R., T. Bruns, and E. L. Geist, Rifting of the Tonga/Lau ridge and formation of the Lau backarc basin: evidence from site 840 on the Tonga Ridge, in *Proc. ODP, Sci. Results. 135,* J. W. Hawkins, L. M. Parson, J. F. Allan, et al., pp. 367-371, Ocean Drilling Program, College Station, TX, 1994.

Tarney, J., A. D. Saunders, D. P. Mattey, D. A. Wood, and N. G. Marsh, Geochemical aspects of back-arc spreading in the Scotia Sea and western Pacific, *Phil. Trans. R. Soc. London A, 300*, 263-285, 1981

Taylor, B., Rifting and the Volcanic-Tectonic Evolution of the Izu-Bonin-Mariana arc, in *Proc. ODP, Sci. Results, 126*, B. Taylor, K. Fujioka, et al., pp. 627-651, Ocean Drilling Program, College Station, TX, 1992.

Taylor, B., K. Fujioka, et al., *Proc ODP, Init. Results, 126,* Ocean Drilling Program, College Station, TX, 1990.

Taylor, B., G. Moore, A. Klaus, M. Systrom, P. Cooper, and M. MacKay, Multichannel seismic survey of the central Izu-Bonin arc, in *Proc. ODP, Sci. Results, 126*, B. Taylor, K. Fujioka, et al., pp. 51-60, Ocean Drilling Program, College Station, TX, 1992.

Vallier, T. L., A. J. Stevenson, and D. W. Scholl, Petrology of igneous rocks from Ata Island, Kingdom of Tonga, in *Geology and Offshore Resources of Pacific Island Arcs - Tonga region*, edited by D. W. Scholl and T. L. Vallier, Circum-Pacific Council for Energy and Mineral Resources, Earth Science Series, 2, 301-316, 1985.

Vallier, T. L., G. A. Jenner, F. A. Frey, J. B. Gill, A. S. Davis, A. M. Volpe, J. W. Hawkins, J. D. Morris, P. A. Cawood, J. L. Morton, D. W. Scholl, M. Rautenschlein, W. M. White, R. W. Williams, A. J. Stevenson, and L. D. White, Subalkaline andesite from Valu Fa Ridge, a back-arc spreading center in southern Lau basin: petrogenesis, comparative chemistry and tectonic implications, *Chem Geol., 91*,227-256, 1991.

Volpe, A. M., D. Macdougall, and J. W. Hawkins, Lau Basin Basalts (LBB): trace element and Sr-Nd isotopic evidence for heterogeneity in the back-arc basin mantle, *Earth Planet. Sci. Lett. 90*, 174-186, 1988.

Woodhead, J., S. Eggins, and J. Gamble, High field strength and transition element systematics in island arc and back-arc basin basalts: evidence for multi-phase melt extraction and a depleted mantle wedge, *Earth Planet. Sci. Lett., 114*, 491-504, 1993

Wright, I. C., Pre-spread rifting and heterogeneous volcanism in the southern Havre Trough back-arc basin, *Mar. Geol., 113*,179-200, 1993.

Peter D. Clift, Ocean Drilling Program, 1000 Discovery Drive, Texas A&M University, College Station, Texas 77845, USA

Volcanic Activity Recorded in Deep-Sea Sediments and the Geodynamic Evolution of Western Pacific Island Arcs

Hervé Cambray[1], Manuel Pubellier[2], Laurent Jolivet[3], and André Pouclet[4]

A compilation of volcanic ashes interbedded in deep-sea sediments was carried out from DSDP-IPOD and ODP data collected along the western Pacific margin. Using a tephrochronological method, we attempted to reconstruct the Cenozoic and Quaternary volcanic activity of major western Pacific arcs. For every arc, established volcanic episodes and volcanic-tectonic evolution recorded on land were compared. This study reveals close connections between tectonic events and volcanic activity of arcs, as well a temporal relationship between the opening of marginal basins and arc volcanism. In the Tohoku (NE Japan) and Bonin arcs (SE Japan), arc volcanic activity clearly vanishes during backarc spreading. In contrast, intense volcanism occurs during both arc rifting and intervals of no spreading. Detailed comparisons show that the maximum volcanic output is closely connected with the stress field evolution recorded on land. The case of Seinan arc (SW Japan) shows a good fit between volcanic episodes and periods of release of the compressional stress field after major orogenic events. Furthermore, in the marine sediments off Japan, a systematic late Miocene volcanic hiatus interpreted as a quiescence of volcanic activity corresponds to a changing stress field on the Tohoku and Bonin arcs. These correlations between volcanic episodicity and tectonic evolution of island arcs allow us to discuss the influence of subduction process on arc volcanism. In the Philippines, the volcanic signal in marine sediments is compromised by rapid alteration and diagenesis of ashes. Nonetheless, only the main events of arc volcanic activity are preserved. A comparison with on-land volcanism shows that this filtered volcanic signal in different places corresponds to incipient subduction (transition from passive to active margins) or to the final stages of basin closure.

1. INTRODUCTION

Compared to the volcanic episodicity of arcs [*Cambray and Cadet*, 1994], rift accretion and subduction can be considered as steady and regular processes through geologic time, because of the continuous formation and destruction of oceanic crust. Linked to subduction, arc magmatism originates directly from partial melting induced by the dehydration of the downgoing slab and, therefore, a causal link exists between subduction and arc volcanism. A paradox of this causal link is the discrepancy between the steady subduction and the episodic arc volcanism. The purpose of this paper is to discuss the paradox by stressing the factors that control the major episodes of volcanism, arc by arc in the western Pacific region.

The DSDP-IPOD and ODP databases are an inexhaustible source of data for those who are interested in comparing phenomena at a large scale. The volcanic products in the deep-sea sedimentary record resulted in numerous publications related to specific arcs [*Huang et al.*, 1975; *Cadet and Fujioka*, 1980; *Donn and Ninkovich*, 1980; *Fujioka et al.*, 1989; *Pouclet et al.*, 1990; *Cadet et al.*, 1990; *Pubellier et al.*, 1991; *Jolivet and Tamaki*, 1992]. There were several syntheses of DSDP data [*Shaw et al.*, 1974; *Kennett et al.*, 1977; *Vogt*, 1979], but none incorporating results of recent ODP drilling. In order to process recent data, a method to quantify the volcanic signal recorded from the occurrence of ash layers or tephra in deep-sea sediments was elaborated [*Cambray et al.*, 1993]. This method was used to test the hypothesis of a volcanic

[1]Laboratoire de Mesures en Forage, Institut Méditerranéen de technologie, Marseille, France.
[2]Département de Géotectonique, Université P. et M. Curie, Paris, France.
[3]Département de Géologie, Ecole Normale Supérieure, Paris, France.
[4]Laboratoire de Géotectonique et de Pétrologie, Université d'Orléans, Orléans, France.

Active Margins and Marginal Basins of the Western Pacific
Geophysical Monograph 88

TABLE 1. Main characteristics of sites drilled in the NW Pacific basin

Leg-Site number	Water depth (m)	Recovery (%)	Thickness of sediments (m)	Oldest sediments	Total number of ash layers
86-578	6010	98	177	Mioc >	46
86-579	5737	87	150	Plioc <	54
86-580	5375	91	155	Plioc >	87

episodicity of circum-Pacific arcs [*Cambray and Cadet*, 1994].

Following this global study, we tackle in this paper the regional approach by focussing on the western Pacific island arcs. Acquisition of the data implies a preliminary processing of the sedimentary records from the recovered cores. Origins of volcanic products of each site are determined using our petrological and geochemical analyses. For the older period, we traced the paleo-location of sites relative to the active volcanic arc using a model of geodynamic evolution. This model is not supposed to be universal, but corresponds to current assumptions [*Jolivet et al.*, 1989; *Rangin et al.*, 1990; *Daly et al.*, 1991]. Then, we examine the meaning of the volcanic episodes recorded in the marine sediments. Comparisons of the volcanic signal recorded by the drilling data with the volcanic and tectonic evolution of arcs allow us to discuss arc volcanism related to backarc opening. For the Tohoku, Seinan-Ryukyu and Bonin arcs, results suggest that episodicity of arc volcanism is mainly controlled by the tectonic evolution of arc substratum. On the other hand, in the Philippine region, volcanic pulses are connected to incipient subduction or to the final stage of basin closure.

2. VOLCANIC ACTIVITY AND EVOLUTION OF WESTERN PACIFIC ISLAND ARCS

2.1. *Record of the Volcanic Activity from Drilling Data*

In order to reconstruct the volcanic activity of arcs, sites in the western Pacific (Figure 1) were selected on the following criteria: continuous sedimentary record through long time intervals and accurate biostratigraphic dating. As a consequence, 22 sites were selected; off north-east Japan, legs 57 [*Scientific Party*, 1980], 86 [*Heath et al.*, 1985] and 87A [*Kagami et al.*, 1985]; in the Japan Sea, legs 127 [*Tamaki et al.*, 1990] and 128 [*Ingle et al.*, 1990]; in the Shikoku Basin, legs 31 [*Karig et al.*, 1975] and 58 [*Klein et al.*, 1980]; off the Philippine archipelago, legs 31 and 124 [*Rangin et al.*, 1990]. The core data were processed to get for each site the frequency of volcanic components versus time. Petrographical and geochemical analyses were used to trace the potential sources of volcanic products for each site, in order to reconstruct the frequency of volcanic activity for each volcanic arc.

2.1.1. *Leg 86, northwest Pacific basin.* Sites drilled during leg 86 (Table 1) are located along a north-south transect about 1000 km east of the Japan Trench (Figure 1). From piston cores, *Horn et al.* [1969] demonstrated that the volcanic products issued from the northwest Pacific arcs spread beyond longitude 160°E. As a consequence, the volcanogenic (Figure 2) sediments of leg 86 may have mainly derived from the Tohoku arc. Chemical analyses carried out on ash layers show differentiated high potassic rhyolites (Figure 3), which can be correlated to the calc-alkaline volcanism located in the inner part of the Tohoku arc. At all three sites, there was a major peak in ash deposition during Quaternary times; this is linked to the recent activity of the Tohoku arc. Episodes at each site can be correlated to the others despite the large distance between sites. We infer from this observation that variations of recent arc volcanism affected at least a large part of the Tohoku arc.

2.1.2. *Legs 56, 57 and 87A, Tohoku Pacific margin.* The Tohoku margin is located between the active volcanic arc and the Japan trench. Sites drilled during legs 56, 57, and 87A (Table 2) are located on the edge of the Tohoku margin (Figure 1). Tephrochronological data processed from these sites provide a nearly complete record of volcanic activity of the Tohoku arc. This is supported by comparison of geochemical analyses (Figure 3) which are characterized by rhyodacitic glasses ($69.0<SiO_2<76.0\%$) with a dispersed alkaline content ($1.8<Na_2O<3.6\%$ and $0.5<K_2O<4.0\%$). This dispersion in alkalis can be linked with the spatial evolution described by *Aramaki and Ui* [1982] along the width of the Tohoku arc. Arc tholeiites are observed next to the volcanic front, whereas, in the inner arc, lavas are more

TABLE 2. Main characteristics of sites drilled on the Pacific margin of the Tohoku arc

Leg-Site number	Water depth (m)	Recovery (%)	Thickness of sediments (m)	Oldest sediments	Total number of ash layers
57-438	1500	55	1050	Mioc <	129
57-439	1650	48	1160	Cretaceous	29
57-440	4507	51	820	Mioc >	111
87A-584	4080	51	950	Mioc <	58

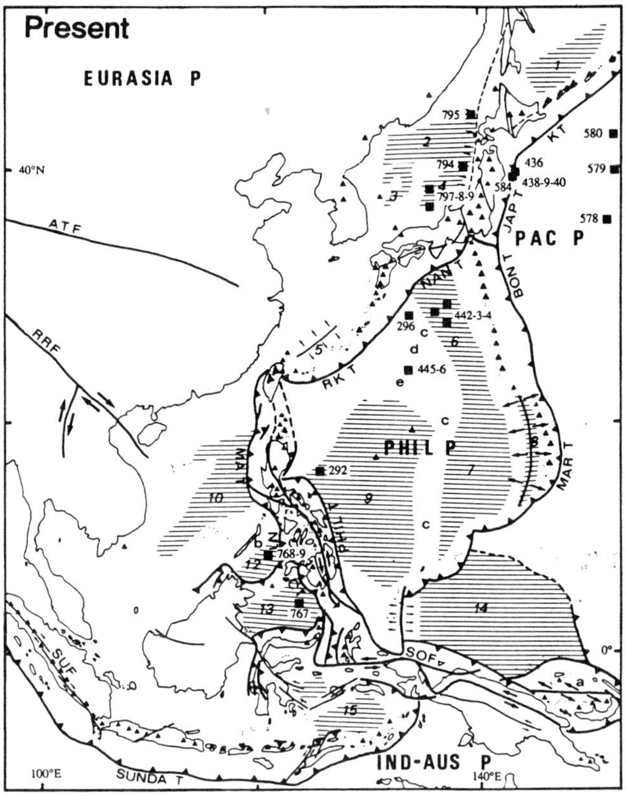

Fig. 1 . Tectonic map of the western Pacific (modified from Jolivet et al., 1989 and Rangin et al., 1990) with location of studied sites (full squares) and quaternary volcanism (full triangles):

- Numbers = marginal basins, Kuriles basin (1), Japan basin (2), Tsushima basin (3), Yamato basin (4), Okinawa basin (5), Shikoku Basin (6), Parece Vela Basin (7), Mariana basin (8), West-Philippine basin (9), South China Sea (10), Proto-South China Sea (11), Sulu Basin (12), Celebes Basin (13), Carolina basin (14), Banda Sea (15), Molucca Sea (16), Salomon Sea (17);
- Letters = Bismarck rift (a), Cagayan ridge (b), Palau-Kyushu ridge (c), Daïto ridge (d), Oki-Daito ridge (e);
- Acronyms = Altyn-Tagh fault (ATF), Bonin trench (BONT), Cotabato trench (CT), Japan trench (JAPT), Kuriles trench (KT), Nankai trench (NANT), Mariana trench (MART), Manilla trench (MAT), Negros trench (NT), Philippine trench (PHILT), Ryukyu trench (RKT), Red River fault (RRF), Sorong fault (SOF), Sumatra fault (SUF), Sunda trench (SUNDAT).

evolved and are calc-alkaline in composition. For this reason, the analyses from drill sites on the Pacific margin of the Tohoku arc show mainly arc tholeiites.

Volcanic activity of the Tohoku arc (Figure 4) is absent in the early Miocene (22-18 Ma at site 439); a major volcanic event occurred during middle Miocene times (17-13 Ma at sites 438 and 439); a volcanic quiescence followed during the late Miocene (12-6 Ma at sites 438 and 584); and a major

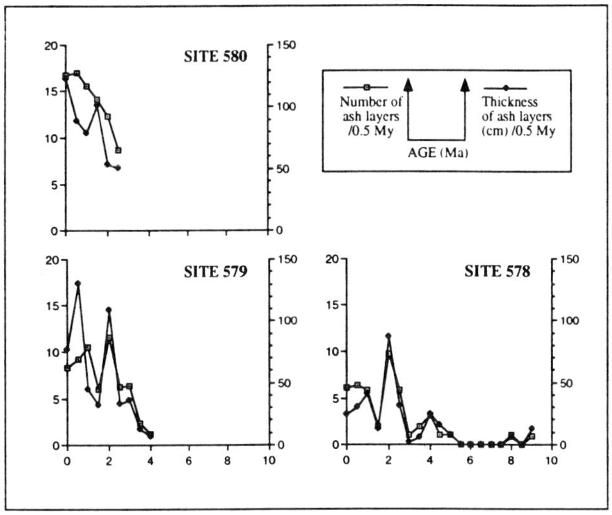

Fig. 2. Volcanic content through times of Leg 86 sites (NW Pacific basin).

Pliocene-Quaternary volcanic episode occurred with a Pliocene peak and a minimum around 1.5 Ma.

2.1.3. *Legs 127 and 128, Japan Sea.* Sites drilled during legs 127 and 128 (Table 3) are situated in the northeastern (sites 794 and 795) and southwestern parts (797, 798 and 799) of the Japan Sea (Figure 1). The geochemical analyses of volcanic glasses carried out in these sites define two major chemical clusters (Figure 5).

The first cluster contains differentiated glasses of dacitic to rhyolitic types ($68.0<SiO_2<76.0\%$) with high alkali contents ($5.0<Na_20 + K_20<10.0\%$). To the southeast, acidic glasses are associated with basic and intermediate types, which define a calc-alkaline differentiation trend. In the northeast, these calc-alkaline volcanic products are attributed to the inner Tohoku arc where volcanic materials are more alkalic than on the Pacific side [*Miyashiro*, 1974; *Aramaki and Ui*, 1982; *Okamura*, 1987]. In the southwest, the volcanic material is inferred from geochemical comparisons with Quaternary ash layers issued from major calderas in this area (Figure 5) to originate mainly from the Seinan arc. However, for these southwestern sites, a contribution of volcanic products from the Tohoku arc is possible. The second cluster of geochemical data is represented by Quaternary andesitic glasses ($56.0<SiO_2<64.0\%$). The analyses show very high alkali contents ($11.0<Na_20 + K_20<14.0\%$) and a negative trend between SiO_2 and alkalis. Glasses are undersaturated in silica including the presence of both olivine and nepheline in the norms. These glasses do not compare to any arc volcanic products in Japan. Differentiation trend and alkalic composition are assumed to correspond to intra-plate volcanic activity within the Japan Sea. For the southwestern sites, the most likely origin is the Ullung island, as shown by the chemical correlation

TABLE 3. Main characteristics of sites drilled in the Japan Sea

Leg-Site number	Water depth (m)	Recovery (%)	Thickness of sediments (m)	Oldest sediments	Total number of ash layers
127-794	2811	61	550	Mioc <>	122
127-795	3300	59	680	Mioc <>	139
127-797	2862	75	560	Mioc <	137
128-798	903	100	517	Plioc <	47
128-799	2084	94	1084	Mioc <	81

TABLE 4. Main characteristics of sites drilled in the Shikoku Basin

Leg-Site number	Water depth (m)	Recovery (%)	Thickness of sediments (m)	Oldest sediments	Total number of ash layers
31-296	2920	51	1087	Oligo <	90
58-442	4635	49	290	Mioc <>	37
58-443	4372	52	457	Mioc <>	48
58-444	4843	46	292	Mioc <>	120
58-445	3377	69	892	Eoc <>	56
58-446	4952	47	420	Eoc <	21

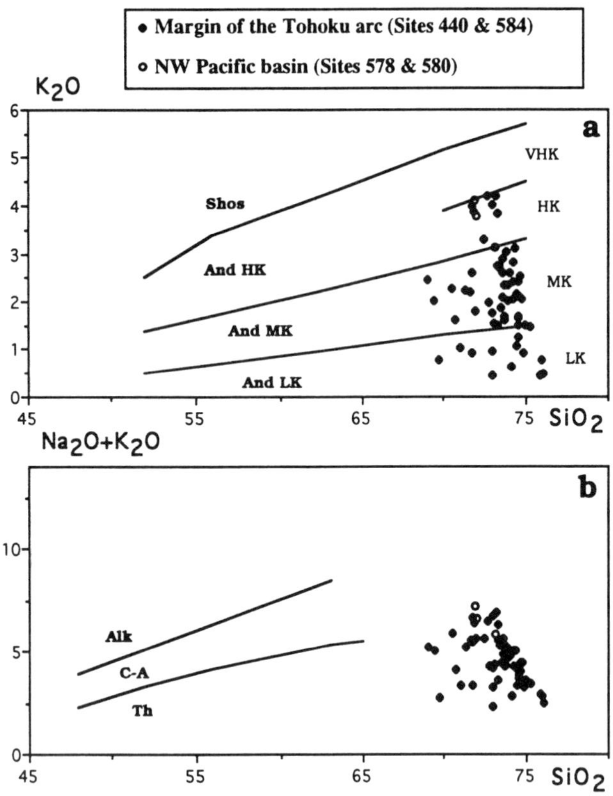

Fig. 3. (a) SiO_2-K_2O and (b) SiO_2-K_2O+Na_2O of ash layers sampled in Legs 57 (site 440), 86 (sites 578 and 580) and 87A (site 584); shoshonites (Shos), andesites (And), very high potassic (VHK), high potassic (HK), mid potassic (MK), low potassic (LK), alkalic (Alk), calc-alkalic (C-A) and tholeiitic (Th).

with U-Oki tephra derived from a recent eruption of Ullung volcano [*Moriwaki et al.*, 1983]. As a consequence, for these sites, the Quaternary record of arc volcanism needs to be distinguished from this intra-plate volcanic influence.

Tephrochronological study is useful to confirm the previous results obtained from sites on the Pacific margin. Chemical results show that the activity of the Tohoku arc is recorded in all sites in the Japan Sea. The southwestern sites, however, also contain volcanic products from southwest Japan. The volcanic content of Japan Sea sites (Figure 6) shows a first major early-middle Miocene episode (19-14 Ma at site 797), a late Miocene quiet volcanic period with a hiatus (14-6 Ma at sites 794 and 797) except for the northern part where site 795 shows two minor pulses and moderate Pliocene activity followed by major Quaternary activity.

2.1.4. *Legs 31 and 58, Shikoku Basin.* Sites drilled during legs 31 and 58 (Table 4) are in the Shikoku Basin proper (sites 296, 442, 443 and 444) and on the Daito ridge (445 and 446) on the southwestern edge of the basin (Figure 1). The geochemical analyses carried out on Neogene-Quaternary ash layers (Figure 7) sampled in sites 442, 443 and 444 allow us to characterize volcanic materials deposited in the Shikoku Basin.

The middle Miocene glasses can be divided into two chemical groups. The first contains rhyolitic glasses ($70.0<SiO_2<76.0\%$) with high sodium and low potassium contents (at about 3.5% Na_2O and 1% K_2O). These volcanic glasses are arc-tholeiite in affinity, and probably originated in the Bonin arc. This is supported by the above-average magnesia content of rhyolitic glasses (0.9% MgO for 72% SiO_2), compositions which favor a boninitic series [*Gill*, 1981]. The second chemical group shows a continuous calc-alkalic series from andesitic to dacitic glasses

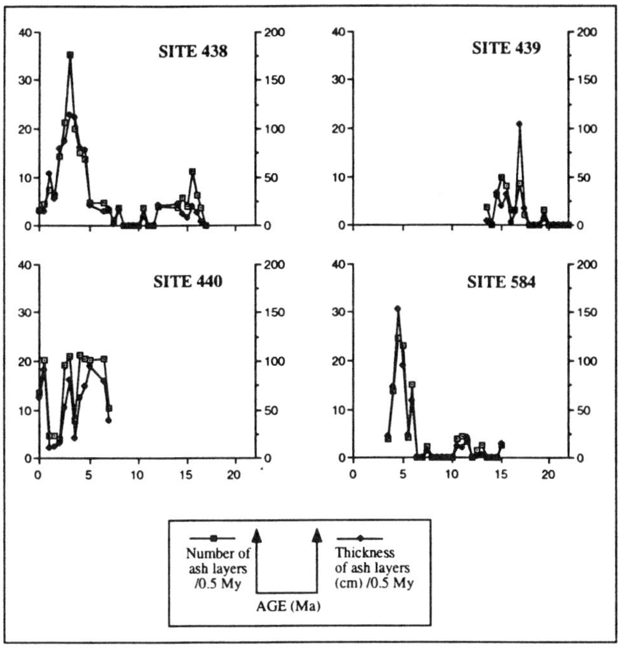

Fig. 4. Volcanic content through times of Legs 57 and 87A sites (Pacific margin of NE Japan).

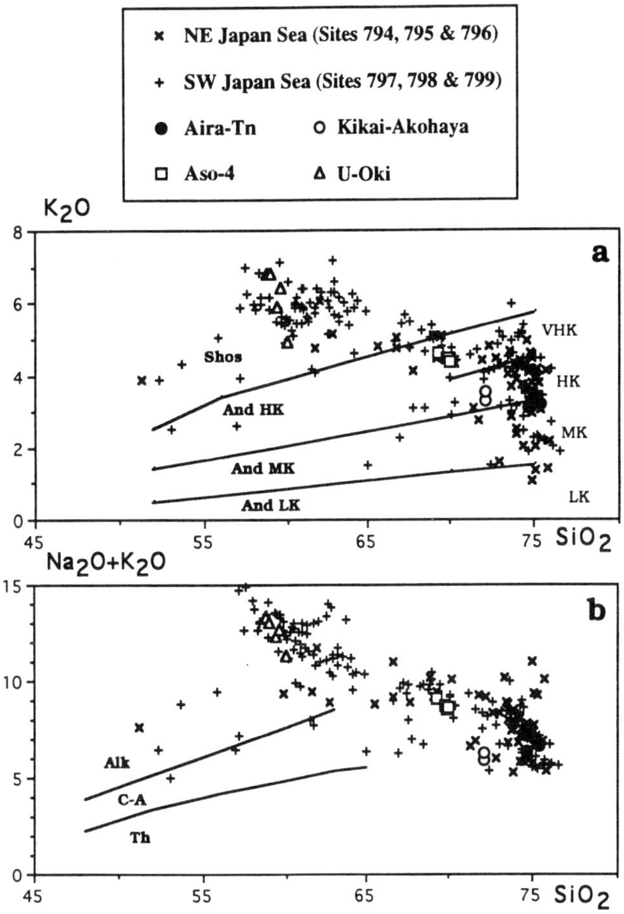

Fig. 5. (a) SiO_2-K_2O and (b) SiO_2-K_2O+Na_2O of ash layers sampled in Legs 127 (sites 794, 795, 796 and 797) et 128 (sites 798 and 799) and chemical analyses of four major tephra originated from Kyushu calderas (Aira, Aso and Kikai-Akohaya) and from Ullung island (U-Oki) (Machida, 81; Moriwaki et al., 1983; Furuta et al., 1986; Cambray et al., 1990); shoshonites (Shos), andesites (And), very high potassic (VHK), high potassic (HK), mid potassic (MK), low potassic (LK), alkalic (Alk), calc-alkalic (C-A) and tholeiitic (Th).

($56.0<SiO_2<71.0\%$; $8.0<Na_2O + K_2O<10.0\%$ for dacitic glasses). The chemical compositions and age indicate that these middle Miocene volcanic glasses certainly originated from the Setouchi volcanic province in the outer zone of southwest Japan [*Tatsumi and Ishizaka*, 1982]. As a consequence, sources of middle Miocene volcanogenic sedimentation in the Shikoku Basin are both the Bonin arc (arc tholeiite or possibly boninite) and the calc-alkaline volcanoes of the Seinan arc. Two types of basaltic glasses occur in late Miocene sediments of the Shikoku Basin (Figure 7). Between 12 and 10 Ma, the basaltic glasses ($51.0<SiO_2<56.0\%$) show a low potassium content ($0.2<K_2O<0.4\%$) and a low sodium content ($1.9<Na_2O<2.3\%$), which favor an arc-tholeiite affinity and can likely be attributed to sources in the Bonin arc. Between 10 and 8 Ma, the basaltic glasses were very undifferentiated ($47.0<SiO_2<48.0\%$), with high alkalic content ($4.0<Na_2O + K_2O<5.0\%$) and low iron content ($7.6<FeO<8.0\%$). Norms are silica-undersaturated and contain both olivine and nepheline. With such characteristics, these volcanic materials are not related to an arc setting but rather to oceanic intraplate volcanism. This hypothesis is enforced by high TiO_2 contents ($1.6<TiO_2<1.8\%$), which are larger than the limit currently admitted ($<1.3\%$) for arc volcanism [*Gill*, 1981]. These volcanic inputs are therefore assumed to originate from off-ridge volcanism in the seamounts of Kinian Ridge [*Klein, Kobayashi et al.*, 1978; *Furuta and Arai*, 1979; 1980]. In conclusion, the lower late Miocene sediments recorded a basaltic volcanism of tholeiitic affinity coming from the Bonin arc and from 10 Ma, the volcanic input mainly originated from off-axis seamounts located in the center of the Shikoku Basin. During Pliocene and Quaternary times, volcanogenic sedimentation in the Shikoku Basin was mainly rhyolitic in composition ($70.0<SiO_2<75.0\%$). Significant potassium variation ($1.7<K_2O<4.3\%$) indicates coexistence of arc-tholeiite and calc-alkalic series. In this case, calc-alkalic glasses characterize volcanic sources from southwest Japan, whereas the arc-tholeiitic ash layers originated either from the Seinan-Ryukyu arc or from the Bonin arc. The similar compositions of these two potential sources do not allow further chemical inference.

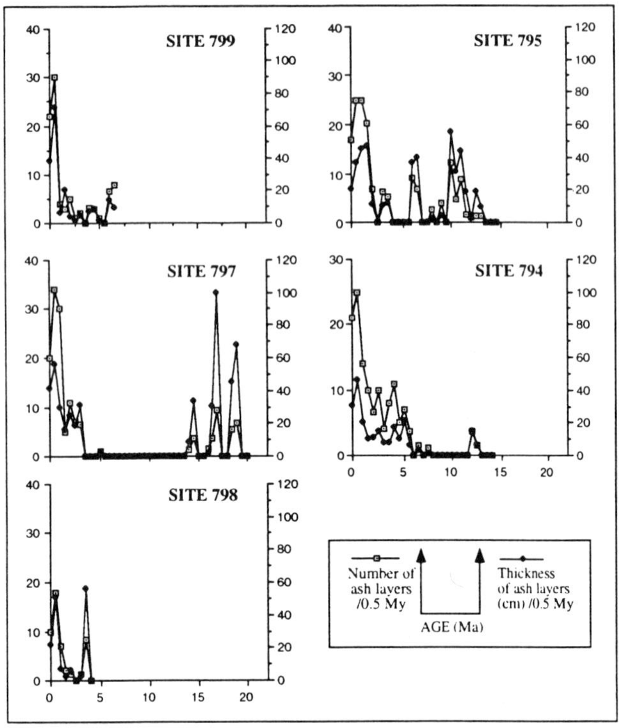

Fig. 6: Volcanic content through times of Legs 127 and 128 sites (Japan Sea).

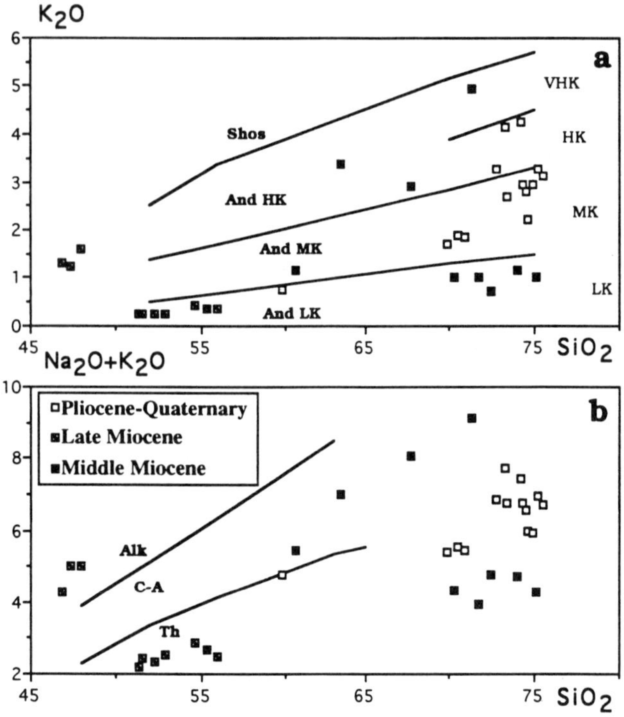

Fig. 7. (a) SiO_2-K_2O and (b) SiO_2-K_2O+Na_2O of ash layers sampled in Leg 58 (sites 442, 443 and 444); shoshonites (Shos), andesites (And), very high potassic (VHK), high potassic (HK), mid potassic (MK), low potassic (LK), alkalic (Alk), calc-alkalic (C-A) and tholeiitic (Th).

In its present geographic setting, the north Philippine Sea received volcanic products from both the east (the Bonin arc) and the west (the Seinan-Ryukyu arc). Forty million years ago, a different situation was due to the complex geodynamic evolution of this area. To determine the possible sources of the oldest volcanic episodes, it is important to reconstruct the past position of sites in relation to active volcanic arcs at given time periods. The first volcanic pulse is recorded (Figure 8) between 40 and 37 Ma (site 445). During Eocene times, spreading of the West Philippine basin was nearly complete and the Philippine Sea plate was located farther south than at present. The paleo-location of site 445 (Figure 9a) was near the Palau-Kyushu arc, and the Eocene ash probably originated from the activity of this ancient arc. The second volcanic pulse was recorded at site 296 between 30 and 24 Ma, with a major peak just before cessation of the activity. This late Oligocene pulse is also lightly marked at sites 445 and 446. Before the opening of the Shikoku Basin, the still active Palau-Kyushu arc was the most probable source for this upper Oligocene volcanic ash. The two sites (Figure 9b) were also near the arc, supporting this hypothesis. The third volcanic pulse is well expressed in most of the sites between 16 and 14 Ma. Based on data from site 296, the duration of this pulse was only 2 or 3 m.y., but drilling in the Shikoku Basin did not reach sediments old enough to confirm this. The geochemical analyses of ash layers recovered in the Shikoku Basin indicate two volcanic sources for middle Miocene times (Seinan-Ryukyu and Bonin arcs; Figure 7) and paleolocation of sites at this period (Figure 9c,d) confirms these. The same volcanic pulse is recorded at site 797 in the South Japan Sea, on the other side of the Seinan volcanic arc (Figure 6). The late Miocene is characterized by a hiatus of arc volcanism at most sites except between 12 and 10 Ma at sites 443 and 444 in the Shikoku Basin. Because this episode was only detected in the Shikoku Basin, it likely originated from the Bonin arc (Figure 9c). This is supported by geochemical analyses discussed above. The analyses also show that the peak of activity between 10 and 8 Ma, mainly recorded at site 444, was caused by eruption of off-ridge seamounts in the center of the Shikoku Basin. Deposition of ash in the Pliocene and Quaternary is well expressed at most of the sites. A maximum of deposition is observed from 2.5 Ma to present time. No data from the geochemical analyses allow us to infer the origin of this volcanic episode. Given the variation in total alkali contents of glasses, we assume that the Bonin arc, the Ryukyu arc, and the southern active part of the Seinan arc on Kyushu could equally have produced this recent volcanic sedimentation (Figures 1, 9e,f).

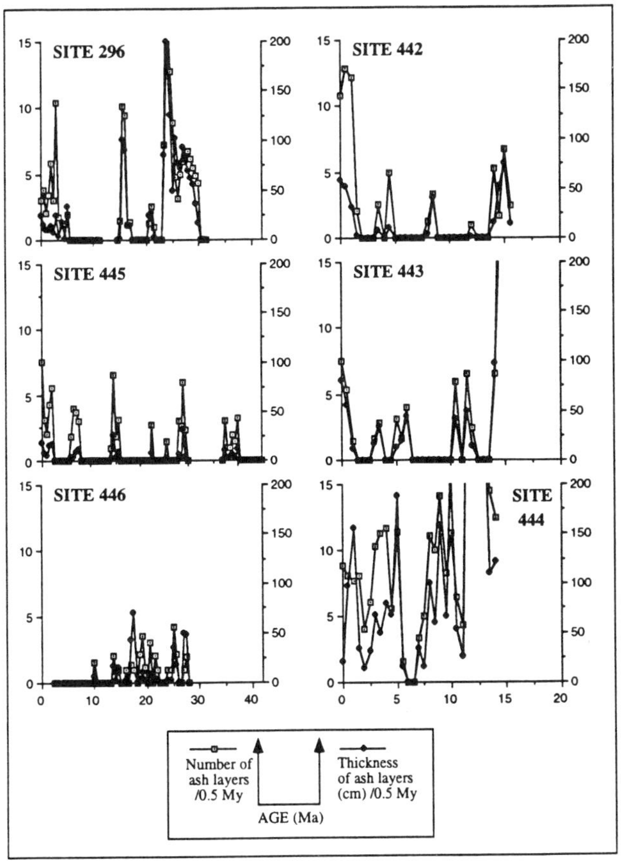

Fig. 8. Volcanic content through times of Legs 31 and 58 (Shikoku Basin).

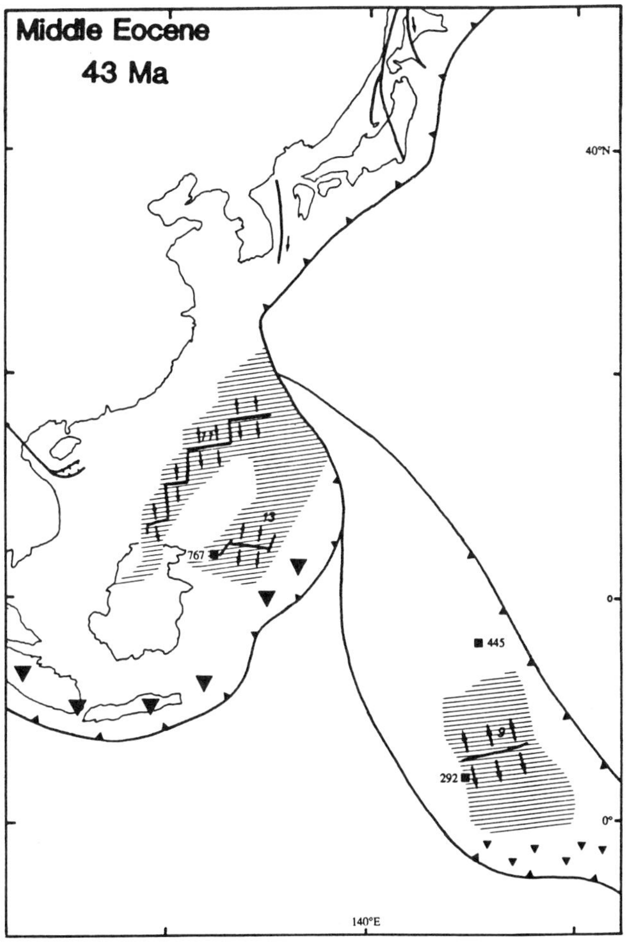

Fig. 9a. Paleogeodynamic reconstruction of the western Pacific (see Figure 1 for legend, modified from Jolivet et al., 1989 and Rangin et al., 1990) with location of studied sites (full squares) and Quaternary volcanoes (full triangles); Middle Eocene, West-Philippine basin (9), Proto-South China Sea (11) and Celebes Basin (13).

2.1.5. *Legs 31 and 124, Philippine archipelago.* From Luzon to Mindanao, the Philippine archipelago was formed by festoons of island arcs situated between the Philippine Sea plate and the Eurasiatic margin (Figure 1). Presently, active subduction is responsible for numerous volcanoes. Sites drilled during legs 31 and 124 (Table 5) are located (Figure 1) in the West Philippine basin (site 292), in the Sulu Basin (sites 768 and 769), and in the Celebes Basin (site 767). In these basins, high geothermal gradients coupled with active circulation of hydrothermal fluids induces the rapid alteration of ash layers [*Desprairies et al.*, 1991]. As a consequence, reliable geochemical analyses of volcanic glasses are limited to the Pliocene and the Quaternary and show two main chemical clusters (Figure 10).

The first is characterized by calc-alkalic rhyolitic glasses ($70.0 < SiO_2 < 76.0\%$) with a moderate to high potassium content ($2.0 < K_2O < 4.0\%$; $5.0 < Na_2O + K_2O < 8.0\%$). The second group contains shoshonitic basaltic to andesitic glass ($52.0 < SiO_2 < 66.0\%$) with both high potassium contents ($2.0 < K_2O < 5.6\%$ positively correlated to the silica contents) and high soda content ($4.0 < Na_2O < 5.0\%$). These two main chemical groups are not uniformly distributed in the cores. In the Celebes Sea (site 767), the recent volcanic ashes are mainly rhyolitic in composition with a calc-alkalic affinity, whereas, on Cagayan ridge (site 769), the ashes are mostly basaltic to andesitic shoshonites. An intermediate geographic location, site 768 in the Sulu Sea, contains both types of volcanic ash. In consequence, the source of calc-alkalic ashes was likely to the south, whereas shoshonitic volcanic products evidently came from the north, possibly from Luzon [*De Boer et al.*, 1980].

Given both the small areal extent of basins and the density of volcanic sources, it appears that most of the volcanic arcs in the area can have contributed to volcanogenic sedimentation in these sites. To identify the origin of each

TABLE 5. Main characteristics of sites drilled in the Philippine basin

Leg-Site number	Water depth (m)	Recovery (%)	Thickness of sediments (m)	Oldest sediments	Total number of ash layers
31-292	2943	55	367	Eoc <>	52
124-767	4906	79	739	Eoc >	65
124-768	4384	65	1047	Mioc <	84
124-769	3643	85	370	Mioc <	89

volcanic event, it is necessary to consider the older ash occurrences, the complex geodynamic evolution of this region (Figure 9), and the dating performed on land [*Wolfe*, 1981; *Bellon and Rangin*, 1990, 1991]. An initial volcanic pulse occurred at site 292 in sediments between 40 and 37 Ma (Figure 11). In the Celebes Sea (site 767), the amplitude of this pulse is smaller but activity was persistent from the middle Eocene to the Miocene with a major peak between 35 and 30 Ma [*Pubellier et al.*, 1991]. A middle Eocene reconstruction of location of the sites (Figure 9a) allows us to discuss the origin of the volcanic pulses in relation to the active volcanic arc in the neighborhood of the sites. For site 292 in the West Philippine basin, late Eocene volcanic activity originated from the arcs of the Philippine mobile belt and/or from the Palau-Kyushu arc. This is enforced by the identification of the same pulse at site 445 on Daito Ridge. In the Celebes Sea, the continuous and relatively weak volcanic activity from 42 and 24 Ma seen at site 767 mainly originated from the Sunda arc [*Pubellier et al.*, 1991]. The early Oligocene peak in activity is interpreted as volcanism of the proto Sulu arc, which crops out on Sabah [*Rangin et al.*, 1990]. The second pulse of volcanic activity at the location of site 292 occurred during late Oligocene times (30 to 24 Ma). In this case, it is difficult to assign this volcanism to either of the two arcs that bordered the Philippine Sea plate (Figure 9b). Evidence of this pulse to the north at sites 296, 445, and 446 favor an origin from the Palau-Kyushu arc.

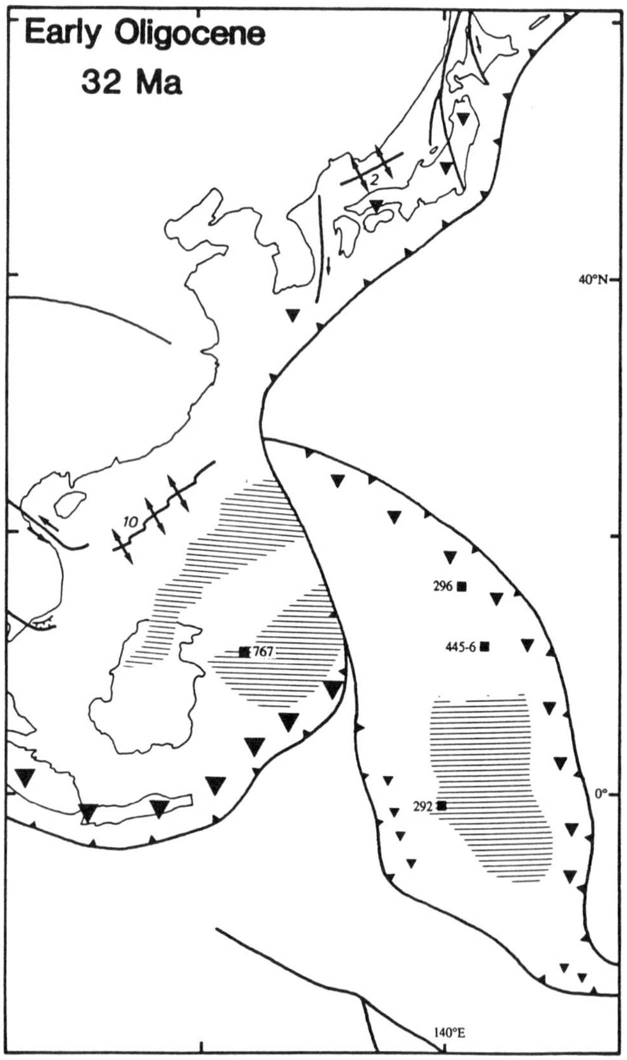

Fig. 9b. Early Oligocene, Japan basin (2) and South China Sea (10).

In late early Miocene times, thick pyroclastic tuffs were deposited at sites 768 and 769, in the Sulu Basin and on the Cagayan ridge respectively. Given the location of the two sites and the limited spatial dispersal of these pyroclastic deposits, the most likely source is the Cagayan arc, the leading arc split by the spreading of the Sulu Basin (Figures 9c,d). Other pyroclastic tuffs occur in the Sulu Basin (site 768) in sediments of about 10 Ma. At this time, volcanic activity originated either from the Sulu arc cropping out in Zamboanga peninsula (western Mindanao) or from the West Philippine arc at Panay, which are the two closest volcanic sources (Figure 9e). In the West Philippine basin (site 292), there is a hiatus between lower and middle Miocene sediments, and the first pulse of more recent volcanic activity was 12-10 Ma. During this interval (Figure 9d), combined northward motion and rotation of the Philippine Sea plate brought together the present area of sites drilled on the Philippine Sea plate and the Eurasiatic margin. In consequence, the late Miocene volcanic pulse recorded at site 292 may have originated from the arc volcanic activity in the Philippine mobile belt. Because of the early Miocene opening of the Parece Vela Basin, the active Mariana arc was too distant to noticeably influence the volcanogenic sedimentation at site 292 located in the western part of the Philippine Sea.

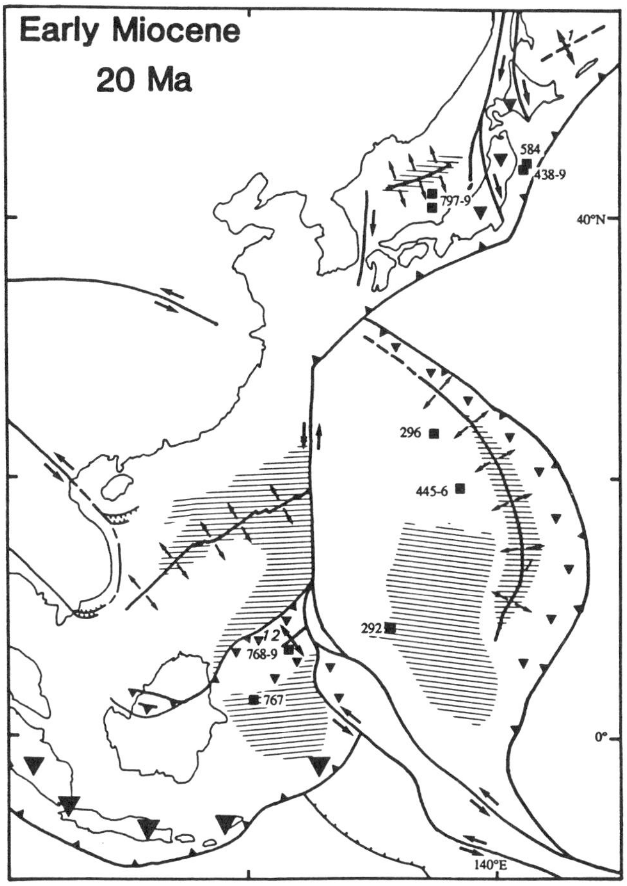

Fig. 9c. Early Miocene, Kuriles basin (1), Parece Vela Basin (7) and Sulu Basin (12).

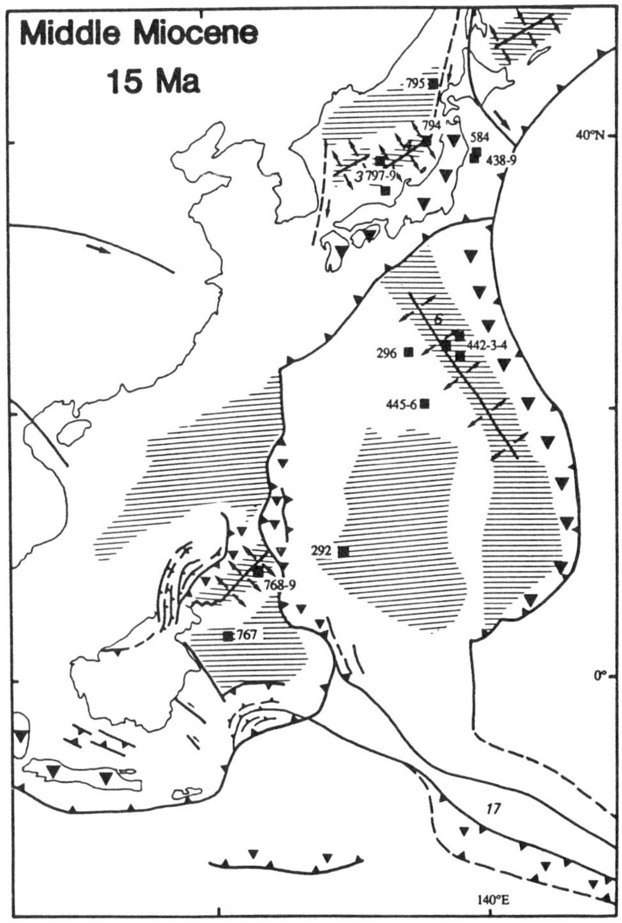

Fig. 9d. Middle Miocene, Tsushima basin (3), Yamato basin (4) and Shikoku Basin (6).

In both the Celebes and Sulu Basins, significant volcanic activity occurred in late Miocene (8-7 Ma) to middle Pliocene (4-3 Ma). Numerous volcanic sources were present around the Celebes and Sulu Seas to provide the late Miocene volcanogenic sedimentation (Figures 9d,e). Geochemical analyses allow us to characterize the origin of the volcanic products within the two basins (Figure 10). In the Celebes Sea, volcanic glasses are moderately potassic and calc-alkalic in composition and originated along the immature Sangihe volcanic arc located in the southeastern edge of the Celebes Sea. In the Sulu Sea, volcanic glasses are very highly potassic and shoshonitic in composition. The most likely volcanic sources were the Negros and the Sulu arcs. The late Pliocene was marked by a noticeable decrease in the volcanic deposits at all sites. This pattern may reflect quiescence of the volcanic activity of the surrounding arcs. Alternatively, many ash beds may be intensely altered by the hydrothermal circulation at shallow levels that occurred within the sediments at this time [*Desprairies et al.*, 1991]. Volcanic deposition at all sites resumed near the Pliocene-Quaternary boundary to reach a present maximum. This recent spurt of volcanic activity is well expressed at all sites and is particularly intense at sites 767 in the Celebes Sea and 292 in the West Philippine Basin. In the Sulu Sea, the Quaternary volcanic inputs were from the aforementioned arcs (mainly the Sulu and Negros arcs). A recent volcanic peak in the Celebes Sea is related to the Cotabato arc and the Philippine arc. In these cases, major element compositions are too similar to allow any kind of correlation or identification of volcanic sources.

2.1.6. *Summary*. Figures 12 and 13 summarize episodes of volcanic activity for the western Pacific volcanic arcs. Each pulse of volcanism recorded at a particular site is assigned to a volcanic source(s). At most sites, the record is limited to the Neogene and Quaternary, but several sites provide a tephrochronologic record dating from the Eocene. Accordingly, two time scales are used to display the data. Figure 12 displays the Neogene activity of the following arcs: Tohoku (northern Honshu), Seinan (southern Honshu and Kyushu)-Ryukyu, Sangihe (Celebes) and Cagayan arcs. Figure 13 displays the activity over the last 40 m.y. of the

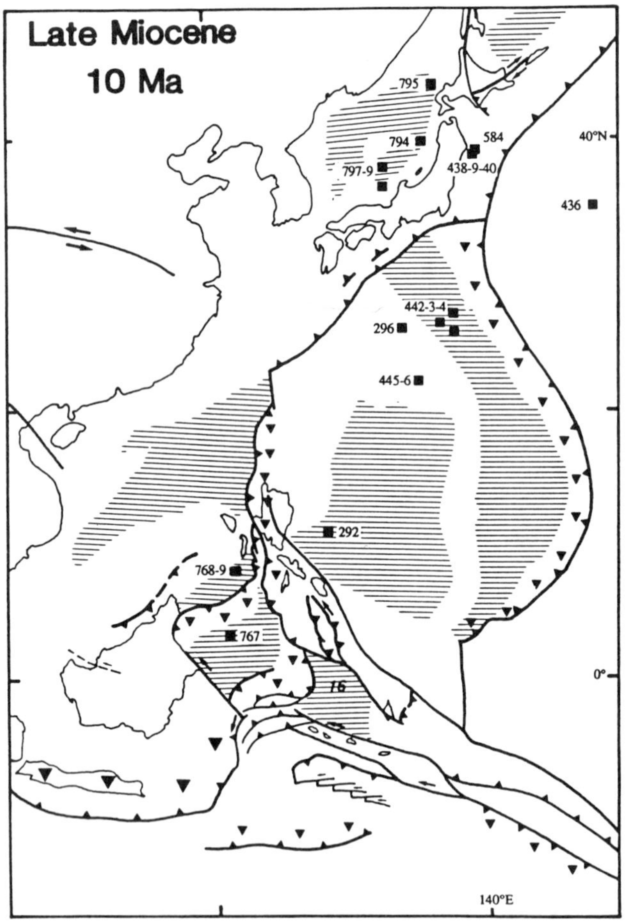

Fig. 9e. Late Miocene, Molucca Sea (16).

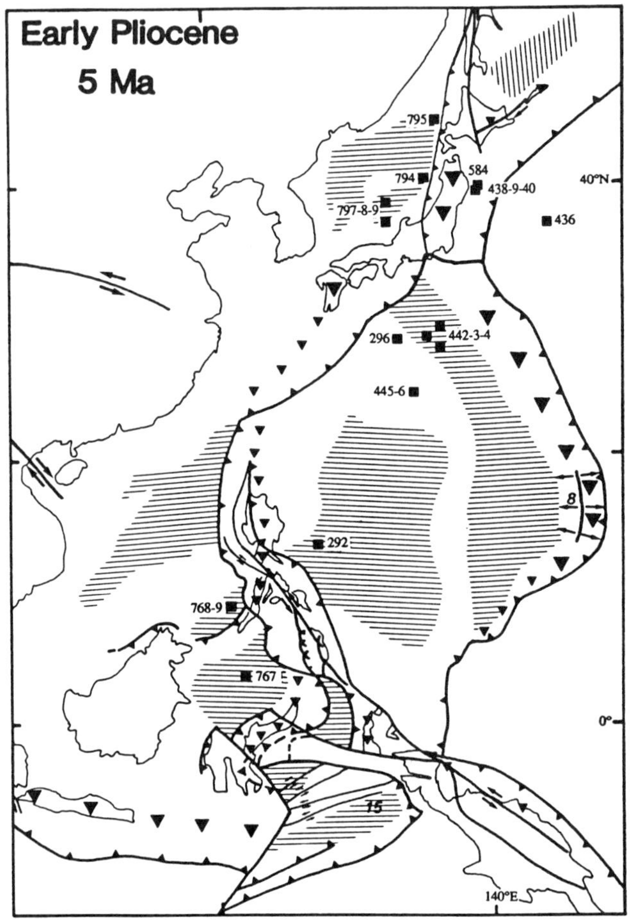

Fig. 9f. Early Pliocene, Mariana basin (8) and Banda Sea (15).

Palau-Kyushu and Bonin arcs, the Philippine mobile belt, the Sulu arc, and the Sunda arc.

2.2. *On Land Volcanism and Tectonic Setting of Arcs*

2.2.1. *The Tohoku arc.* In northeastern Japan, the Tohoku arc is formed by the subduction of the Pacific plate under the Eurasiatic plate (Figure 14). The arc formed on a continental fragment and has an outer non-volcanic arc and a wide inner volcanic arc. After the middle Eocene change in direction of the Pacific plate, magmatic activity of the Tohoku arc reached a maximum in late Oligocene-early Miocene times with a huge amount of andesitic and dacitic tuffs called the Green Tuff [*Sugimura et al.*, 1963]. Bimodal volcanism (basalt and rhyolite) continued until the middle Miocene (about 13 Ma). After a quiet period between 13 and 8 Ma, activity resumed with acidic and very explosive volcanism and formed huge polygenetic calderas. From about 2 Ma, volcanic activity became andesitic and less explosive [*Yagi et al.*, 1963; *Matsuda et al.*, 1967; *Konda et al.*, 1981; *Fujioka*, 1985].

During the Eocene, the Tohoku arc was part of the Eurasiatic continental margin, before the opening of the Japan Sea. The timing of this opening, based on different models, is not well established [*Otofuji et al.*, 1985; *Kimura and Tamaki*, 1986; *Tosha and Hamano*, 1988; *Jolivet et al.*, 1991]. Recent data from ODP leg 127 showed that the oceanic basins in the Japan Sea were partly opened by early Miocene times and their expansion continued until the middle Miocene [*Pisciotto et al.*, 1990]. The last stages of opening were coeval with a late Miocene uplift of the Tohoku arc [*Ninkovich and Robertson*, 1975]. During late Pliocene times, a convergent zone was initiated along the eastern side of the Japan Sea [*Tamaki et al.*, 1992]. The timing of this event is dated by the 1300 m uplift of the Okushiri ridge starting from 1.8 Ma [*Pisciotto et al.*, 1990].

The evolution of stress orientation in the southern part of the Tohoku arc is reconstituted from the framework of

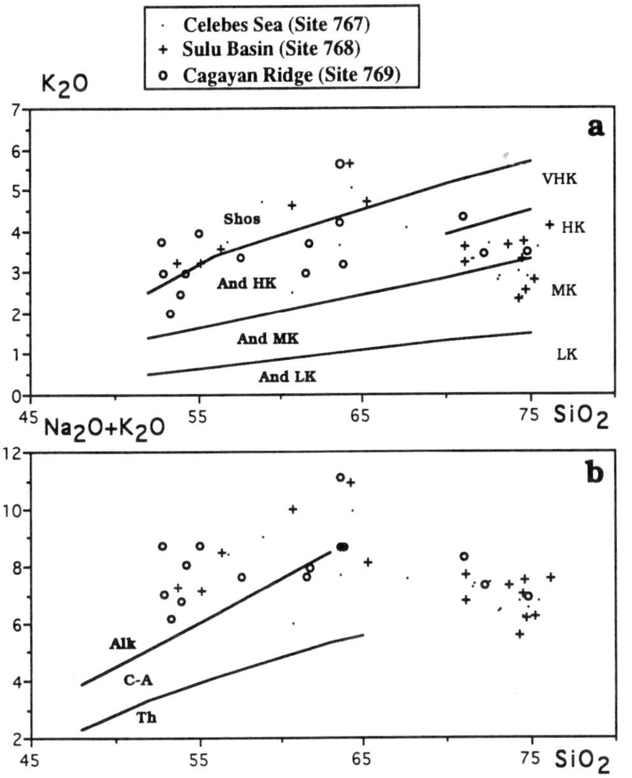

Fig. 10. (a) SiO_2-K_2O and (b) SiO_2-K_2O+Na_2O of ash layers sampled in Leg 124; shoshonites (Shos), andesites (And), very high potassic (VHK), high potassic (HK), mid potassic (MK), low potassic (LK), alkalic (Alk), calc-alkalic (C-A) and tholeiitic (Th).

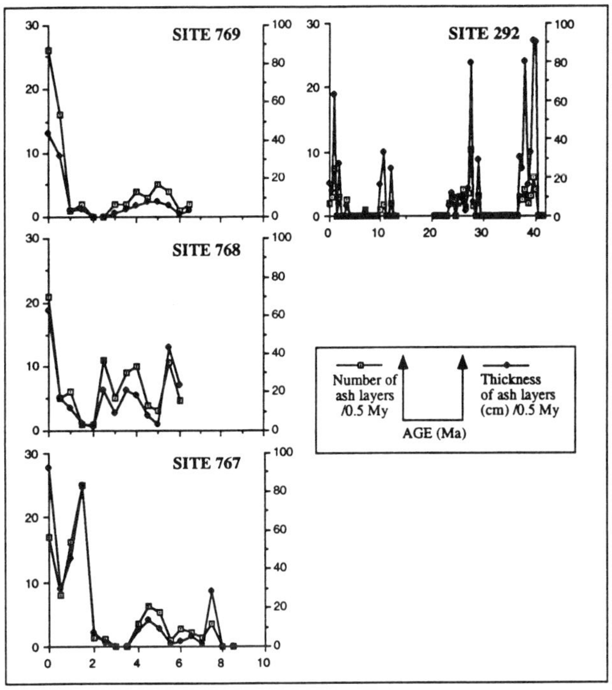

Fig. 11. Volcanic content through times of Legs 31 and 124 sites (Celebes and Sulu Seas).

orientation of combined faults and dikes (Figure 15). The lower Cenozoic is characterized by a NNW orientation of principal stress ($\sigma 1$) which corresponds to the direction of convergence of the Pacific plate before the major change in plate motion responsible for the Hawaiian-Emperor bend. During the early Miocene, the extensional stress field in the inner zone was related to the rifting and opening of the Japan Sea. After that, the effect of the Japan Sea opening decreased and a compressional stress field affected the Pacific side of the arc. This grew toward the inner part of the arc during the late Miocene. During the Pliocene and Quaternary, compressional stress affected the entirety of the Tohoku arc with $\sigma 1$ WNW in orientation, parallel to the convergence of the Pacific plate. From these data, the tectonic evolution of the Tohoku arc was considered to have been controlled successively by the effects of Pacific plate subduction and of the Japan Sea backarc opening. This stress field evolution is in accordance with data measured on faults [*Otsuki*, 1988; *Jolivet and Huchon*, 1989] and allows us to conclude that a right lateral tensional stress field was active on the Tohoku arc until the early late Miocene. Afterwards, the stress field was dominated by a E-W compression.

2.2.2. *The Seinan arc*. In contrast to the Tohoku arc, volcanism of the Seinan arc of southern Honshu and Kyushu, and its extension to the Ryukyu arc (Figure 14) is less abundant and restricted to middle Miocene and Quaternary times. Middle Miocene volcanism is distributed along the inner part of the arc, on Shikoku and Kyushu islands. Volcanic series associated with plutonic complexes give a range of ages between 17 and 12 Ma with a clear concentration of ages at 14 Ma [*Nakada and Takahashi*, 1979; *Tatsumi and Ishizaka*, 1982; *Tatsumi*, 1986]. Quaternary volcanic activity was mainly located on Kyushu and landward of the middle Miocene volcanism. Quaternary volcanism is characterized by calc-alkalic rhyolites and dacites, which erupted from large monogenetic calderas in an area of intense geothermal activity [*Aramaki and Ui*, 1982; *Kagiyama*, 1983].

Because of northward migration of the Philippine Sea plate (Figures 9b,c), subduction under southwest Japan began early-middle Miocene times [*Karig and Angevine*, 1985; *Seno*, 1989; *Hibbard and Karig*, 1990]. Afterwards, the subduction process was marked by a possible late Miocene cessation and a renewal during late Pliocene times. This hypothesis is based on the presence of Pliocene turbidites in the Shikoku Basin [*Karig and Angevine*, 1985] and by the fact that the present slab is not deeper than 80 km beneath southwest Japan as shown by the seismic data [*Seno*, 1989; *Notsu et al.*, 1990]. Furthermore, the late

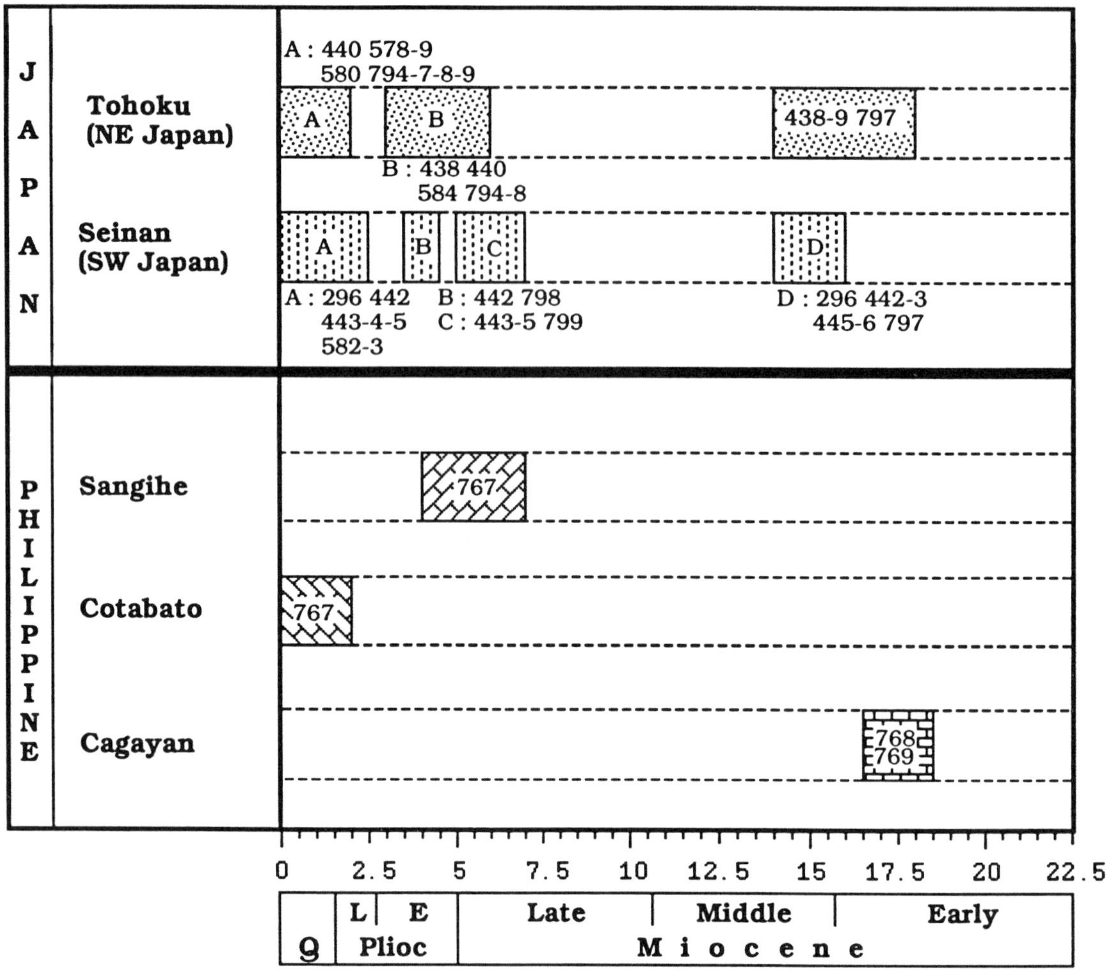

Fig. 12. Volcanic activity of the western Pacific arcs from Miocene to Present times, reconstructed from the volcanic content of the deep sea drillings (Numbers associated with each episode indicate the label of sites used to determine the episode).

Pliocene renewal of subduction was associated with a change in the direction of motion of the Philippine Sea plate [*Angelier and Huchon*, 1986]. The tectonic evolution of southwest Japan was influenced by the subduction of the oceanic crust of the Shikoku Basin under the margin. The particular buoyancy and the prominent topographic features of this young oceanic crust are factors in disruption of the subduction process. Thus, the Miocene tectonics of the arc are interpreted to reflect the occurrence in the trench of prominent features carried by the subducting plate. For example, intense thrusting in the early Miocene that affected the Shimanto belt is attributed to the subduction-collision of a microblock [*Fabbri*, 1989]. Again, the last stage of the Takachiho orogenesis at the middle-late Miocene boundary is interpreted as a consequence of closure between the Nankai margin and the Palau-Kyushu remnant arc with the Bonin active arc, both carried by the buoyant oceanic crust of the Shikoku Basin [*Charvet and Fabbri*, 1987]. Seinan arc extension to the Ryukyu arc is split by the Okinawa trough in the south and a narrow rift basin in the north [*Letouzey and Kimura*, 1986].

2.2.3. *The Bonin arc.* In southeastern Japan, the Bonin arc is formed by subduction of the Pacific plate beneath the Philippine Sea plate (Figure 14). The complex structure of this oceanic arc is composed, from east to west, of a non-volcanic submerged outer arc split by a forearc basin and a narrow volcanic arc comprising a single line of volcanoes [*Honza et Tamaki*, 1985]. On land, history of volcanic activity is difficult to estimate because the emerged part of the Bonin arc is restricted to a few islands. Therefore, marine sediments are helpful in reconstructing the volcanic episodicity of this arc (Figure 13). On the islands, the first evidence of arc volcanism is at the middle-late Eocene boundary [*Shiraki et al.*, 1978], and that is followed by

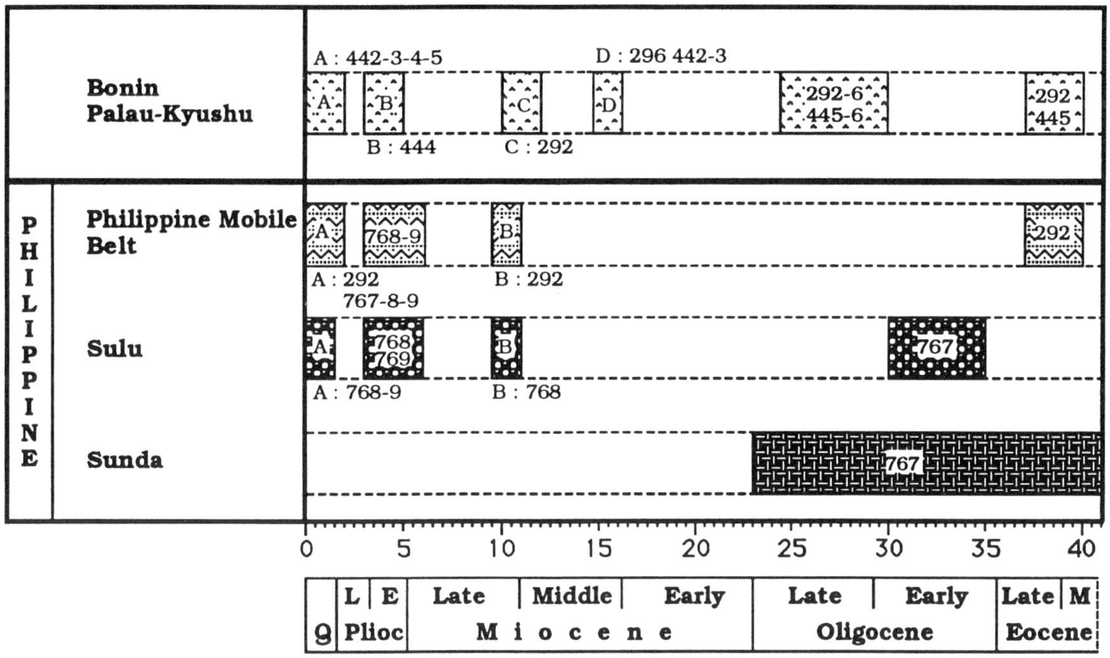

Fig. 13. Volcanic activity of the western Pacific arcs from Eocene to Present times, reconstructed from the volcanic content of the deep sea drillings (Numbers associated with each episode indicate the label of sites used to determine the episode).

upper Oligocene magmatism [*Kaneoka et al.*, 1970]. The Neogene volcanic arc built along a new front 100 km west of the old one [*Honza and Tamaki*, 1985] and was accompanied by opening of the forearc basin [*Taylor et al.*, 1990b]. The results of leg 126 show that the modern volcanic arc formed during Oligocene time and continued until 27 Ma [*Taylor et al.*, 1990a]. Arc volcanism resumed during the middle Miocene with a maximum between 13 and 10 Ma [*Scott et al.*, 1980; *Fujioka et al.*, 1989; *Taylor et al.*, 1990a]. There were a few volcanic events during the late Miocene [*Jones*, 1973; *Fujioka et al.*, 1989]. Then, Pliocene-Quaternary activity reached a maximum as shown by volcanic content of marine sediments recovered during leg 126 [*Fujioka et al.*, 1989; *Taylor et al.*, 1990a].

Convergence between the Pacific and the Philippine Sea plates is assumed to have begun middle Eocene times, at the time of the major kinematic reorganization of the Pacific plate [e.g., *Gordon and Jurdy*, 1986]. This is evidenced by older volcanic rocks collected on the Palau-Kyushu remnant arc [*Karig*, 1975; *Klein and Kobayashi*, 1979; *Meijer et al.*, 1983]. This incipient subduction was probably initiated along a transform fault [*Hilde and Lee*, 1984; *Jolivet et al.*, 1989]. Then, the evolution of this intra-oceanic arc was marked by its splitting into two marginal basins dated by the magnetic anomalies. First, between 30 Ma (An 10) and 17 Ma (An 5D-E) the Parece Vela Basin opened [*Mrozowski and Hayes*, 1979]. Farther north, backarc opening occurred belatedly and the Shikoku Basin accreted between 25 Ma (An

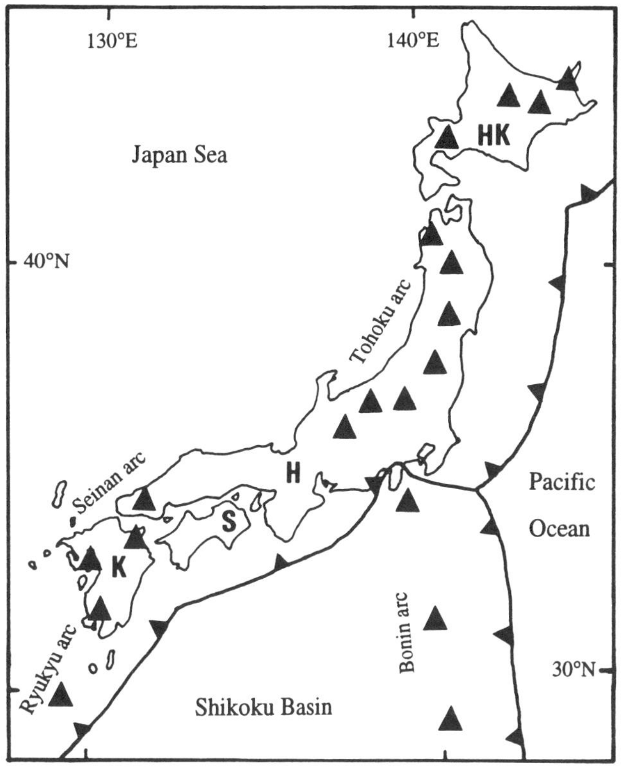

Fig. 14. Simplified tectonic and volcanic map of Japan, Quaternary arc volcanism (triangles), Honshu (H), Hokkaido (HK), Kyushu (K) and Shikoku (S).

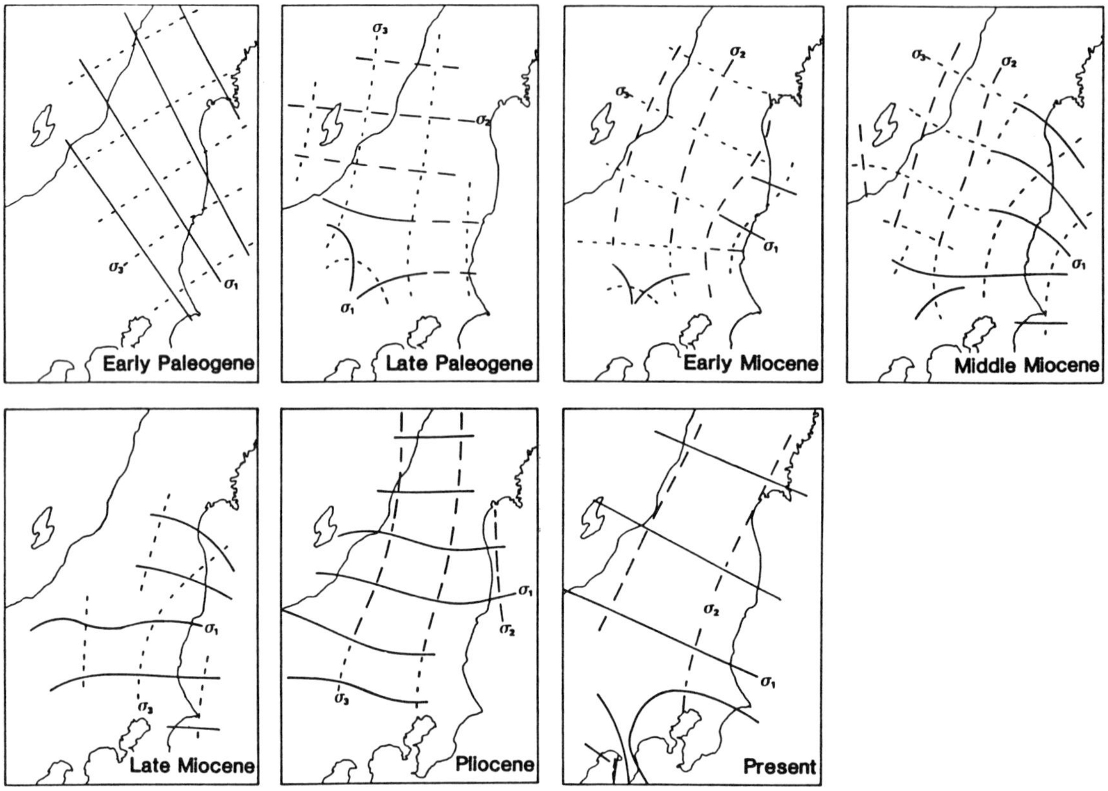

Fig. 15. Evolution of Cenozoic and Quaternary stress field in southern part of the Tohoku arc (modified from Kuwahara, 1982; Takeuchi, 1985; 1986; Tsunakawa, 1986).

6C) and 14 Ma (An 5B) [*Klein et al.*, 1978; *Chamot-Rooke et al.*, 1987]. The Palau-Kyushu ridge, which is the remnant arc of this system, began to subside at about 28 Ma [*Klein and Kobayashi*, 1979]. This age confirms the opening of the backarc basins from the late Oligocene. After the accretion in the basins, large-scale extension still prevailed into the middle Miocene, as evidenced by structural sedimentary rocks recovered during leg 126 [*Taylor et al.*, 1990b]. Afterwards, the upper Miocene sediments show only faults in the forearc basin whereas Pliocene and Quaternary formations are crossed by well expressed normal faults [*Taylor et al.*, 1990a]. These clear evidences of a large scale extension are assumed to be related to the first stage of rifting of the Bonin arc, initiated during late Pliocene times at about 3 Ma [*Taylor et al.*, 1990b]. In the southern part of this arc system, K/Ar dating of volcanic rocks recovered in the Mariana trough indicates that rifting began 8 Ma [*Crawford et al.*, 1986]. As a consequence, the recent evolution of this arc system can be considered as a recurrence of the late Oligocene splitting of the arc, with the identical time discrepancy between the south and the north.

2.2.4. *The Philippines*. The Philippine mobile belt, which is wedged between the Philippine Sea plate and the Eurasiatic margin, is the dominant feature in this region and contains important volcanic series linked to the Philippine arc in the east and to, from north to south, the Bataan, Negros and Cotabato arcs in the western part. Other arcs are located on the Eurasiatic margin and are, from north to south, the Cagayan ridge, the Sulu arc, the Sangihe arc, the north arm of Sulawesi arc and the Sunda arc (Figure 16). The duration of arc volcanism along these arcs together with geographic locations of dated volcanic formations are summarized on a sketch (Figure 17).

1. Philippine mobile belt. The distribution and ages of volcanic series of the Philippine mobile belt allow us to define three major arc systems. They are, namely, the old Philippine arc, the recent Philippine arc, and the arcs along the western edge. The Philippine arc formed in two episodes—one during the Paleogene and the other during Pliocene-Quaternary times [*Wolfe*, 1981, 1988]. On land, the old arc is interpreted to have formed by subduction of the Indian plate under the Philippine Sea plate (Figures 9a,b,c), and ranges in age from 37 to 23 Ma [*Bellon and Rangin*, 1990]. Since the late Pliocene, the recent arc has been active as a result of the newly born subduction of the West Philippine basin [*Knittel et al.*, 1988; *Bischke et al.*, 1990]. On the western edge of the Philippine mobile belt, the succession of volcanic arcs is clearly linked to subduction of

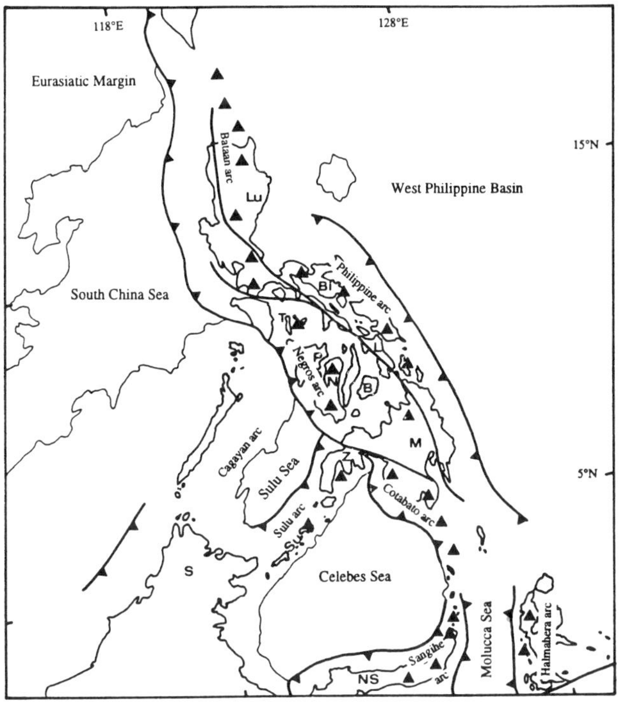

Fig. 16. Simplified tectonic and volcanic map of Philippine, Quaternary arc volcanism (triangles); Bohol (B), Bicol (Bi), Leyte (L), Luzon (Lu), Mindanao (M), Negros (N), North Arm of Sulawesi (NS), Panay (P), Sabah (S), Sulu islands (Su), Tablas (T) and Zamboanga (Z).

marginal basins on the Eurasiatic margin, such as the Bataan arc to the south China Sea, the Negros arc to the Sulu Basin, and the Cotabato arc to the Celebes Sea. The Bataan arc has been active since the late Miocene [*Defant et al.*, 1988; *Maleterre*, 1989], the Negros arc since the Miocene-Pliocene boundary [*Mitchell et al.*, 1986].

2. Sulu arc and Cagayan ridge. The Sulu arc has had several periods of volcanism. From evidence in Sabah (Borneo), a proto Sulu arc was active during the Eocene and especially during the Oligocene [*Rangin et al.*, 1990]. The Cagayan ridge, interpreted as the leading arc [*Rangin and Silver*, 1991], was active from 23 to 15 Ma [*Bellon and Rangin*, 1990]. These oldest activities of this arc system are interpreted to be related to subduction of a proto-south China Sea [*Holloway*, 1982]. Recent volcanic activity of the Sulu is due to the subduction of the Sulu Basin as observed on land in formations of middle-late Miocene and Pliocene-Quaternary ages [*Bellon and Rangin*, 1990].

3. North arm of Sulawesi and Sangihe arcs. The recent volcanic activity of these two arcs began in the latest Miocene [*Rangin and Silver*, 1991; *Wolfe*, 1981]. The north arm of Sulawesi volcanism, which is due to the Celebes Basin subduction, ceased by the early late Pliocene, whereas continuing volcanism on the present Sangihe arc is related to subduction of the Molucca Sea.

4. Sunda arc. The Sunda arc formed by the subduction of the Indian plate beneath the Eurasiatic margin. It has been active since the Eocene with a possible Miocene volcanic hiatus dated between 19 and 11 Ma [*Ninkovich and Donn*, 1976; *Bellon et al.*, 1990].

3. DISCUSSION

Before proceeding with discussion of the tephrochronologic data, it is important to discuss the consequence of arc submarine drowning on the distal dispersion of volcanic products. In fact, the data used in this paper are established from the expression of volcanic activity in marine sediments. Thus a minimum of volcanic productivity in the record can correspond either to real variations in arc volcanism or immersion of the volcanic sources due to arc subsidence. For instance, variations of volcanic components in leg 60 (Mariana arc transect) were interpreted in term of eustatic processes [*Hussong and Uyeda*, 1981], i.e., low ash content in sediments is interpreted in term of arc immersion due to subsidence of the basin and its sides during backarc spreading. In this case, lateral transport of volcanic material after at least the partial immersion of the arc is assumed to be limited [*Beccaluva et al.*, 1986].

The subsidence explanation for reduced volcanism is not confirmed by the tephrochronologic data from the Bonin and the Tohoku arcs. The Quaternary sediments of the Shikoku Basin record important volcanic activity that originated partly from the Bonin arc. This arc is at present mainly under water. Results of leg 126 off the Bonin arc also show a high density of ash layers in recent sediments [*Fujioka et al.*, 1989]. The middle Miocene volcanic activity of the Tohoku arc described on land is well expressed by sediments of both the Pacific margin and the Japan Sea, despite the maximum of the transgression event that occurred in Japan at about 15 Ma [*Matsuda et al.*, 1967]. As a consequence, the main events of arc volcanism seems to be well expressed in marine sediments independently of arc immersion. We conclude that the effects of arc immersion filter only the background volcanism of the arc, which is represented by disseminated glasses within the marine sediments. Density of thick ash layers observed in the sedimentary cores corresponds to episodes of powerful arc eruptions issued from subaerial stratovolcanoes. In this case, the dispersion of the volcanic products is not influenced by the average level of arc above or below sea-level.

3.1. *Marginal Basins Opening and Arc Volcanism*

The volcanic activity of the Tohoku arc expressed in sites both on the Pacific margin of Japan and in the Japan Sea (Figure 12) shows a dominant Pliocene pulse and two minors during the middle Miocene and the Quaternary (Figure 18). The major feature expressed by the data is the virtual absence of volcanic activity during the late Miocene. The major volcanic pulses expressed by marine sediments

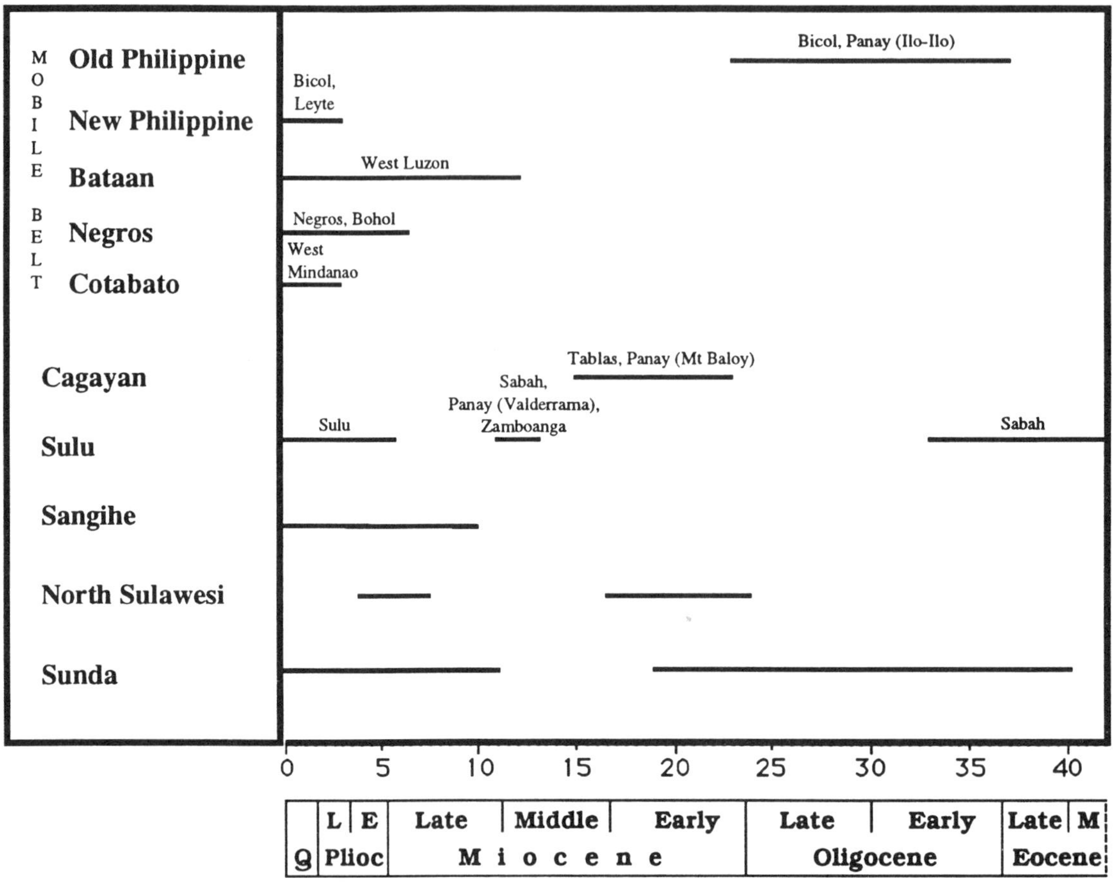

Fig. 17. Compilation of on land dating of volcanic formations for the arcs of the Philippine (see references in text).

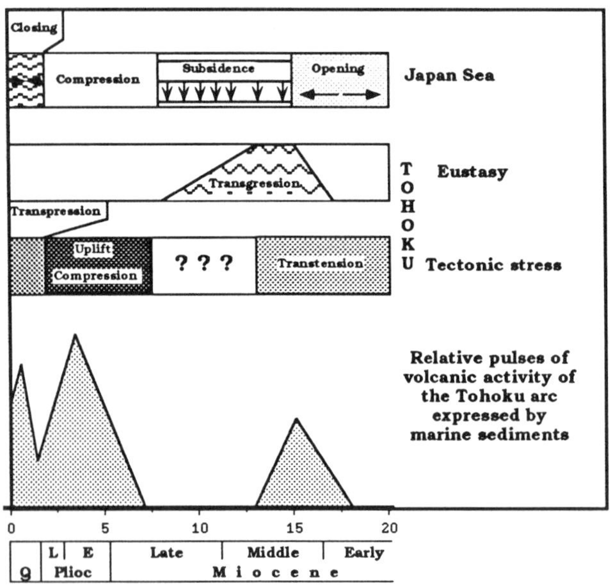

Fig. 18. Comparison of tectonic and volcanic evolution of the Tohoku arc (NE Japan); arrows indicate the main tectonic movements.

are respectively connected to the last stage of spreading in the Japan Sea (middle Miocene times) and to the uplift of the Tohoku arc from 8 Ma (Figure 18). Furthermore, the decrease in volcanic activity at the Pliocene-Quaternary boundary is coeval with the incipient convergence along the western side of the Tohoku arc. Besides the tephrochronologic data, one may underline the correspondence of the Green tuff formation and the rifting of the Japan Sea at the Oligocene-Miocene boundary.

The tephrochronologic data processed from sites located in the Philippine Sea (legs 31 and 58) show two old episodes, i.e., the late Eocene and the late Oligocene (Figure 13). Two more recent episodes are shown by middle Miocene and Pliocene-Quaternary sediments of the Shikoku Basin. The four volcanic episodes correlate with key periods of the arc system evolution (Figure 19). The late Eocene volcanic episode that lasted 3 m.y. associated with initiation or re-initiation of the subduction of the Pacific plate under the Philippine Sea plate. The late Oligocene episode occurred during the first stage of accretion in the Parece Vela and Shikoku Basins. The climax of this episode clearly occurred at 28 Ma in the southern basin (site 292) and at 25 Ma in the northern Shikoku Basin (site 296). The cessation of

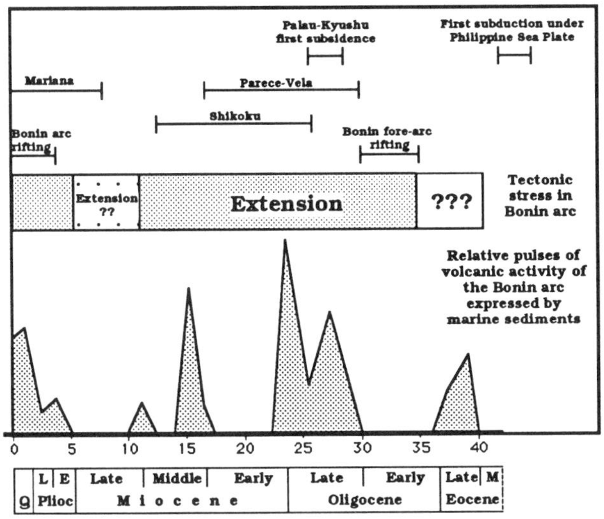

Fig. 19. Comparison of tectonic and volcanic evolution of the Bonin arc (SE Japan).

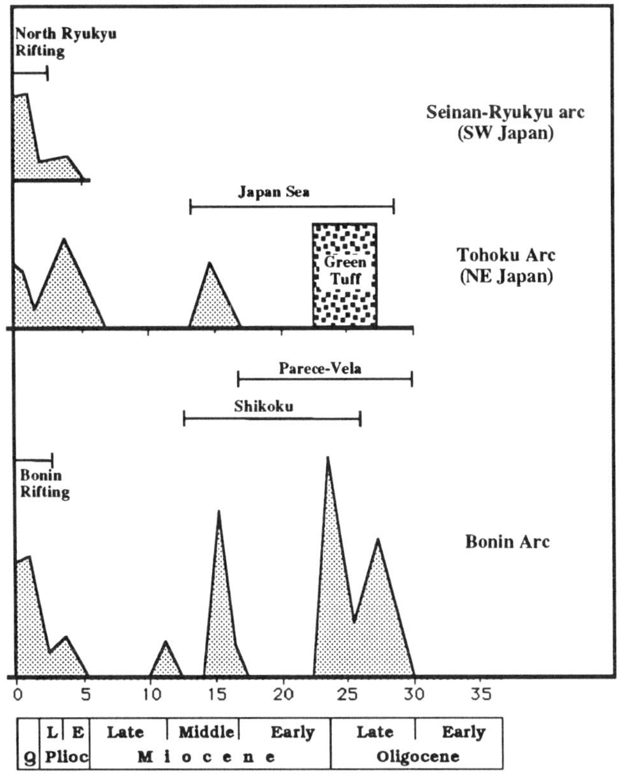

Fig. 20. Comparison between the relative volcanic episodicity of the Japanese arcs and the opening of marginal basins.

accretion in the Shikoku Basin was synchronous with resurgent middle-late Miocene volcanic activity on the Bonin arc. Finally, the present rifting of the Bonin arc is associated with an intense Pliocene and Quaternary volcanism.

The temporal relationship between arc volcanism and the opening of marginal basins is an important but still controversial problem. From the interpretations of the drilling data in the southern Philippine Sea, *Karig* [1983] suggested that the episodes of marginal basin expansion and intense arc volcanism are nearly synchronous. In support of this hypothesis, Karig noted that backarc expansion in the present arc system is associated with arc volcanism (Bonin-Mariana, Ryukyu, New-Hebrides, Tonga-Kermadec). Opposed to *Karig*, the data of legs 59 and 60 [*Scott et al.*, 1980; *Scott and Kroenke*, 1980; *Hussong and Uyeda*, 1981] suggest that, for the Mariana arc system, arc volcanism diminished during episodes of backarc opening and especially during the rifting. This hypothesis is also discussed by *Honza* [1983] who postulated that an alternation between arc volcanism and backarc opening is explained by material partitioning in the mantle due to convection cells. A compilation of the western Pacific arc systems showed that arc volcanism generally ceased after episodes of rifting [*Klein*, 1985]. Volcanism resumes after spreading, as during Miocene times in the Mariana [*Sartori*, 1980] and Bonin arcs [*Mrozowski and Hayes*, 1979]. Finally, drilling results from legs 125 and 126 show that there is no universal temporal relationship between arc volcanism and backarc opening, but that arc segments go through cycles of arc volcanism controlled by backarc opening [*Taylor*, 1992]. Given these controversial hypotheses, it is important to outline, step by step, the possible bearing on this discussion of the tephrochronological data originated from northwestern Pacific arcs (Figure 20).

1. Rifting. In the evolution of arc systems studied around Japan, the chronology of rifting stages is well established and the rifting periods correlate with arc volcanic episodes. During the late Oligocene, the volcanic arc located on the eastern edge of the Philippine Sea plate underwent a diachronous splitting that led to the opening of the Parece Vela Basin and then of the Shikoku Basin. This period is marked by an intense arc volcanic activity reaching, to the south, a maximum at 28 Ma (site 292) and, to the north, at 24 Ma (site 296). This chronological difference coincides with the spatially diachronous rifting of these two basins (Figure 21) and shows that rifting is accompanied by intense arc volcanism. This feature is also observed in the rifting of the Japan Sea at the Oligocene-Miocene transition. On land, the Tohoku arc is covered by a huge andesitic outpouring called the Green Tuff formation. During the Quaternary, sites in the north Philippine Sea were fed by abundant volcanic input coming from the Bonin and Ryukyu arcs, which were both undergoing rifting. In summary, all these examples favor intense arc volcanism during rifting.

2. Spreading. In contrast to rifting, spreading corresponds to quiet intervals of arc volcanism. The Bonin arc is marked by an early Miocene volcanic hiatus synchronous with the spreading of the Shikoku Basin. During oceanic accretion in

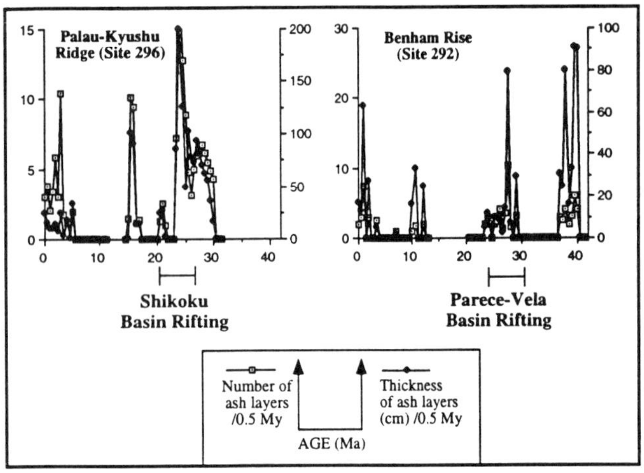

Fig. 21. Comparison of the volcanic activity recorded in the north and south part of the Bonin-Mariana arc system and the riftings.

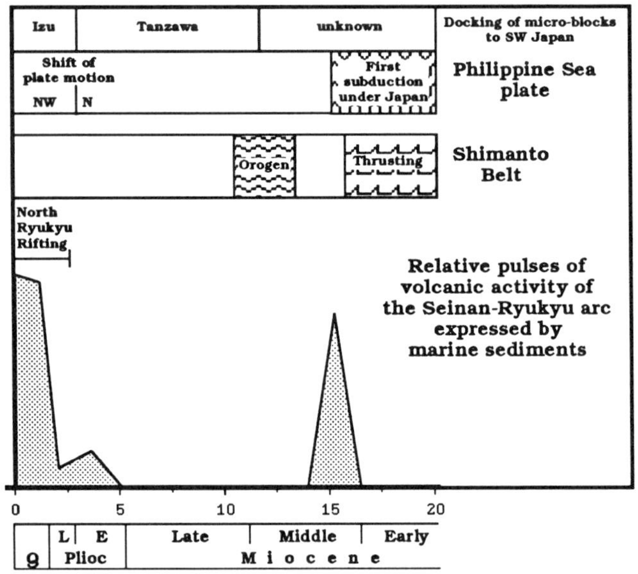

Fig. 22. Comparison of tectonic and volcanic evolution of the Seinan-Ryukyu arc (SW Japan).

the Japan Sea, the Tohoku arc after the Green Tuff episode shows little evidence of volcanism in the lower Miocene deep-sea sediments. Compared to the Bonin arc, Quaternary volcanism of the Mariana arc was less intense whereas the accretion of backarc crust was already active in the Mariana trough. Also, the southern part of the Ryukyu arc, which is presently without active volcanoes, was active during opening of the Okinawa trough [*Sibuet et al.*, 1987].

3. Post-expansion. Both for the Bonin and the Tohoku arcs, the tephrochronologic data express a middle Miocene renewal of arc volcanism. This renewal occurred just before or during the cessation of accretion in the adjacent marginal basins.

In conclusion, one may assume that the studied arcs went through the following cycle: (1) Intense volcanism during the rifting, (2) minimum volcanism during the spreading, and (3) renewal of volcanism during the last stages and cessation of spreading. This cycle is not universal, since a few particular examples do not conform, but it does apply to large segments of ancient and modern volcanic arcs in the Western Pacific region.

To understand correspondence between arc volcanism and backarc opening, it is important to accurately study the stress evolution across the arc. For the Tohoku arc, stress evolution compared to volcanic episodicity (Figure 18) shows the middle Miocene volcanic pulse correlates with the period of transtensional stress of the Tohoku arc. The volcanic lull recorded in the upper Miocene sediments corresponds to the transition between extensional and compressional stress fields. Since the Pliocene, where the volcanic input from the Tohoku arc was at a maximum, a compressional stress field affected the entirety of the Tohoku arc. Contrary to the Tohoku arc, the volcanic episodicity of the Bonin arc is always associated with extension. The waning of volcanic activity observed in the lower Miocene marine sediments correlates with spreading of the Shikoku Basin and was marked on the arc by a steady deformation stage. The late Miocene volcanic hiatus correlates with reduction of extensional stress on the arc, as evidenced by rare faults in sedimentary rocks of this age.

For the Seinan-Ryukyu arc, tephrochronology of ash beds from sites located both in the Shikoku Basin and in the Japan Sea shows two main pulses of volcanism, one during middle Miocene and other during Pliocene-Quaternary times (Figure 22). The short but well expressed middle Miocene volcanic pulse occurred just between two compressional deformation events defined in the Shimanto belt that are related to subduction disruption. As a consequence, this volcanic pulse can be interpreted as the result of either a release of compressional stress or a renewal of the subduction process. The first explanation is related to the mechanism of magma ascent whereas the second is related to the genesis of magma. The two mechanisms can have acted in concert. The main feature expressed by the tephrochronologic data is the large gap of volcanic activity during the late Miocene. This volcanic hiatus is synchronous with the late Miocene cessation of subduction. Quaternary volcanism thus originated from an area where renewed subduction was effective, i.e., where the downgoing slab was deep enough to produce the magmas erupted at Kyushu and the Ryukyu arc.

At a regional scale, the origin of arc volcanism depends on magma genesis, magma ascent and eruption [*Honda and Uyeda*, 1983; *McGeary et al.*, 1985]. The present worldwide tectonic framework shows that subduction is associated with the production of magma, except in a few cases such as southwest Japan [*Karig and Angevine*, 1985; *Seno*, 1989] and central Peru [*Nur and Ben-Avraham*, 1981; *Cande*, 1983; *Barberi et al.*, 1988]. For these two exceptions, lack of volcanism is related to the locking of the buoyant oceanic plate beneath the margin [*Sacks*, 1984]. The zone of magma genesis thus appears to be commonly present above the subducted slab. Convergence between two lithospheric plates can be considered as a continuous process through time [*Hilde et al.*, 1977; *Engebretson et al.*, 1985; *Gordon and Jurdy*, 1986]. Thus, one may wonder why there are ever areas without volcanism over subducting slabs. For the Western Pacific region such a question is particularly pertinent. Since Eocene times, there has been no noticeable variation either in rate or in direction of Pacific plate motion that can explain the breakup of arcs and the associated episodicity of volcanism. Also, the Quaternary subduction process of the Pacific Plate along the Eurasiatic margin cannot explain, to the north, the compressional stress field on Tohoku arc and, to the south, the extensional environment in the Bonin arc. As a consequence, it seems that the evolution of arcs is mainly controlled by the tectonic status of margins whereas subduction is only responsible for magma genesis.

To understand the relationship between the tectonic evolution of the arc substratum and volcanic episodes recorded in the deep-sea sediments, two opposite hypotheses can be proposed. The upward thrust of low density magma can induce fracturing of the crust and eruption on the surface [*Armstrong et al.*, 1969; *Barrash and Venkatakrishnan*, 1982]; alternatively, the final ascent of magma is driven to the surface along preexisting open faults or fractures in the crust. The first hypothesis might explain the triggering process of eruption from a magma body at shallow level, but it is not realistic to assume that the upward thrust of magma can explain the ascent and eruption of magma. In that case, the volcanism rate on surface should have been constant through times.

At the scale of volcanoes, numerous structural studies demonstrate that the triggering of eruptions is influenced by the stress environment. If the stress field induced by the hydrostatic pressure of magma is higher than the local stress field of the rocks, the rocks will fracture and magma will ascent to the surface. During the 1980 eruption of Mount St. Helens, the magma outpourings took place in a local extensional environment inside a major strike-slip zone [*Guffanti and Weaver*, 1988]. The example of the caldera El-Borracho in northern Mexico demonstrated that early Miocene thick flows of rhyolitic ash were erupted in a regional extensional stress field [*Deshler et al.*, 1986]. As a consequence, the more the stress field can induce fracturing, the easier are the conditions for magma to erupt. This is true for volcanoes, but needs to be considered for a whole volcanic arc. Similarly, at a regional scale, the status of a crustal stress field is the principal factor controlling ascent through the crust of magma formed in the upper mantle. The ascent of magma depends on preexisting subvertical fracturing in the crust (Figure 23). Theoretically, such fracturing appears along the plane formed by $\sigma 1$ and $\sigma 2$, respectively the principal and intermediate stresses. Fracturing is high angle when $\sigma 1$ (extension) is vertical, and subvertical when $\sigma 2$ (strike-slip) is vertical. Once the rock is fractured, it can be a conduit for magma if the hydrostatic pressure of the magma is higher than $\sigma 3$, the least principal stress. With $\sigma 3$ vertical (compression), fracturing is low angle and the ascent of magma is difficult. As a consequence, a compressional stress field can prevent eruptions and may induce the deep cooling of magma in sills (Figure 23). At a global scale, numerous authors are in favor of such a model to explain the nature and the episodic activity of arc volcanism [*Uyeda and Kanamori*, 1979; *Nakamura and Uyeda*, 1980; *Hildreth*, 1981; *Uyeda*, 1982; *McGeary et al.*, 1985; *Arculus*, 1987]. We wish to know to what extent this is true for the Western Pacific arcs.

From the foregoing volcanic and tectonic comparisons of Western Pacific arcs, there is no simple rule applicable to the relationship between stress field and arc volcanism. The Tohoku and Bonin arcs show that Pliocene volcanism occurred in a compressional and in an extensional stress field respectively. The conjunction of compression and arc volcanism is not in accordance with the theoretical model. One possibility to reconcile the theory with the example is to postulate a strike-slip component in the compressional stress field. In such a case, the strike-slip component can exist as soon as $\sigma 1$ or $\sigma 2$ are not horizontal. This will explain the occurrence of arc volcanism in a compressional stress field. On one hand, these two examples show that episodes of arc volcanism can occur in varied tectonic environments. On the other hand, it appears that the arc tectonics may have a critical influence on the nature of arc volcanism. The Tohoku arc experienced (1) middle Miocene bimodal volcanism (rhyolites and basalts) in an extensional environment, (2) large Pliocene rhyodacitic calderas forming in a compressional stress field, and (3) Quaternary andesitic volcanism accompanying release of the compressional stress field induced by the incipient subduction along the eastern side of the Japan Sea. This tectonic control may be explained by the transit times of the magma body in the crust. The more the compressional stress field prevails, the less easy is the ascent of magma, leading to more differentiation (2). In a different tectonic environment, vertical fractures and faults make the outpouring of juvenile magma easier (1 and 3).

We have compared the stress evolution of the Tohoku and Bonin arcs, and the episodicity of arc volcanism as recorded in marine sediments. The results show that the late Miocene volcanic hiatus at both arcs correlates with a major change in the tectonic environment. On the Tohoku arc, volcanic waning takes place between clear episodes of Miocene extension and Pliocene-Quaternary compression (Figure 18). These respectively correspond to the last stage of opening and to the first stage of closing of the Japan Sea. On the

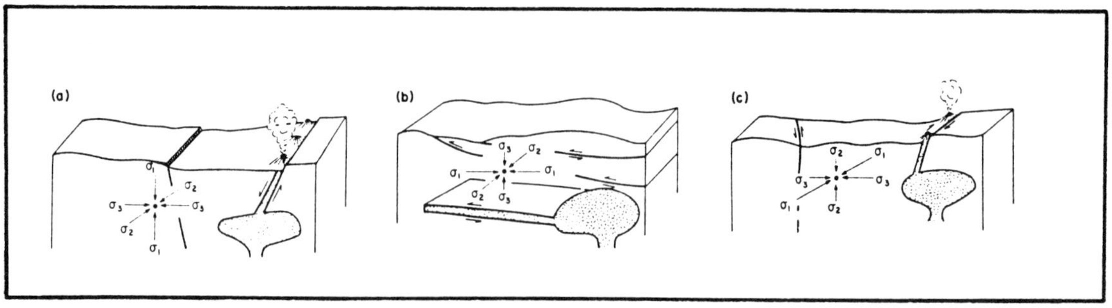

Fig. 23. Sketch of stress field orientations in the crust with the resulting deformation and fracturing susceptible to drive the magma; (a) extension; (b) compression; (c) strike-slip (from Cas et Wright, 1987).

Bonin arc, the late Miocene volcanic lull is associated with quiet tectonic activity except minor normal faulting in the inner part of the forearc basin [*Taylor et al.*, 1990b]. Just before and after this period, the last stage of opening of the Shikoku Basin and rifting of the Bonin arc, respectively, are associated with clear extensional features on the arc. These two examples show that the late Miocene volcanic hiatus is linked to a reorganization of the stress field, in one case by a shift from extension to compression and, in the other case, by a decrease of tectonic activity within an extensional environment. The time delay between these periods of active deformation is assumed to be characterized by an instability of the stress field lasting several million years. This stress instability induces a limited deformation of the arc substratum. As a consequence, mantle derived magma without access through the crust (fractures and faults) was not able to reach the surface. This could explain the observed volcanic hiatus.

In conclusion, tectonic changes in arc evolution may control volcanic episodicity recorded in marine sediments in the western Pacific. A major consequence of this hypothesis is a that local subduction has little influence on arc volcanic episodicity. Previous authors have assumed that subduction trends, depending on the nature and age of subducted oceanic crust, correlate with arc volcanism [*Helwig and Hall*, 1974; *Vallier*, 1974; *Scheidegger and Kulm*, 1975; *Ninkovich and Donn*, 1977]. Indeed, *Kennett and Thunell* [1975, 1977] concluded that pulses of arc volcanism may represent a sensitive amplifier of changes in subduction rates. Based on studies in the eastern Pacific, *Rea and Scheidegger* [1979] concluded that a high subduction rate may be associated with an increase in frictional heating and hence with more dehydration in the zone of magma genesis. As a consequence, partial melting is enhanced, and the increased magma production leads to a high rate of volcanic eruption. On the contrary, it appears here, that even if subduction is directly related to magma genesis and to migration of water-satured, low-density melt in the upper mantle, arc volcanism is mostly controlled by the tectonic status of the shallow arc substratum.

3.2. *Arc Volcanism and Collisions, the Philippine Case*

The tephrochronologic data expressed by the drilling results in the Philippine region are gathered on a sketch (Figure 24), which also shows the interpreted origin of each peak in volcanism. By comparing the ages of volcanic series on land with the data recorded in the basins, one can notice a partial discrepancy particularly for the oldest episodes. During the Pliocene and Quaternary, there is a good agreement between on-land data and data recorded within the basins. Site 767 in the Celebes Sea records the late Miocene-early Pliocene activity of the Sangihe and north arm of the Sulawesi arcs and the Quaternary activity of the Cotabato arc. Sites 768 and 769 in the Sulu Sea record the Pliocene-Quaternary activity of the Sulu and Negros arcs. Site 292 in the west Philippine basin records Quaternary activity of the Philippine arc. In all the sites, a systematic hiatus is expressed close to the Pliocene-Quaternary boundary (Figure 24).

For the oldest times, the volcanic frequency recorded in the sediments of the basins partially covers the arc activity dated on land. The volcanic episodes recorded in the basins correspond to either the end of the volcanic period described on land, such as the lower Miocene pyroclastics from the Cagayan ridge (sites 768 and 769), or the beginning of the period (site 292) such as the late Eocene episode on the old Philippine arc and that of the late Miocene on the Bataan arc. Other possible correlation are not so marked. The weak episode recorded in the Celebes Sea from the late Eocene to the early Miocene (site 767) correlates with activity on the Sunda arc [*Pubellier et al.*, 1991]. Two volcanic episodes correlate with the multi-period activity of the Sulu arc, the first one during the early Oligocene (site 767) and the second one during the late Miocene, expressed by pyroclastic deposits at site 768. The discrepancy between on-land and marine volcanic data can be interpreted in term of alteration, because the alteration process is able to fully remove evidence of volcanism in deep-marine sediments [*Vallier*, 1974; *Kelts and McKenzie*, 1976; *Hein et al.*, 1978; *Grechin et al.*, 1980; *Gardner et al.*, 1985]. In the Sulu and Celebes

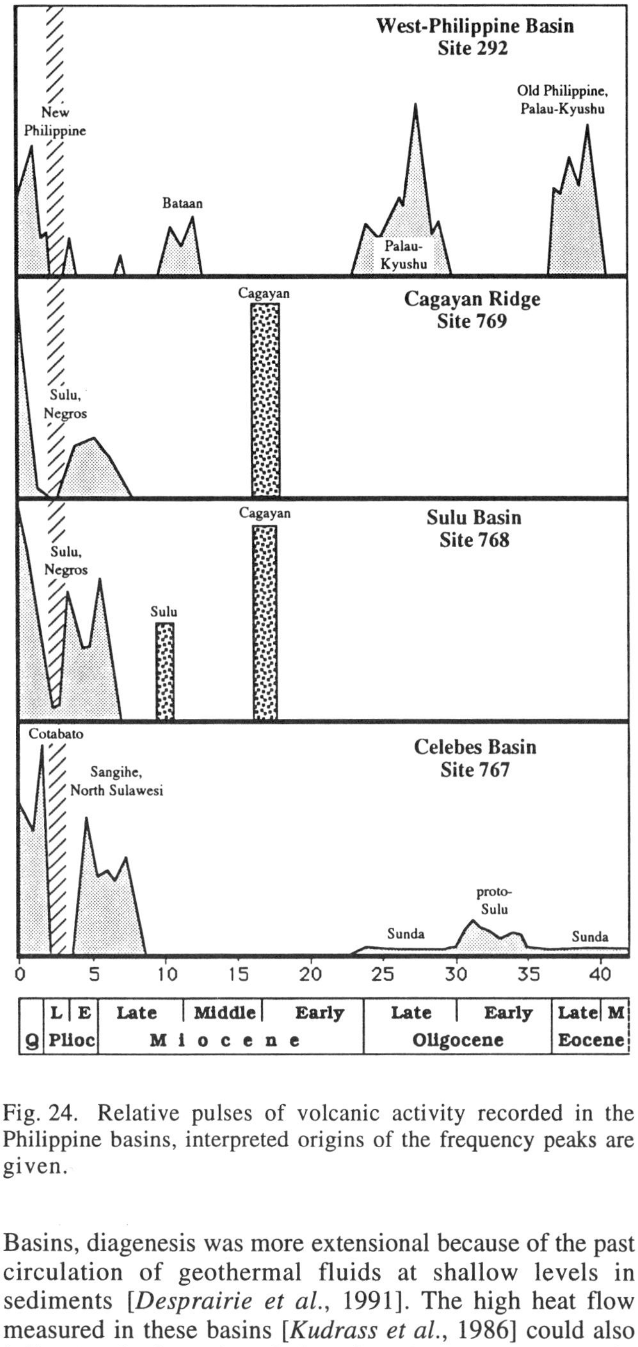

Fig. 24. Relative pulses of volcanic activity recorded in the Philippine basins, interpreted origins of the frequency peaks are given.

Basins, diagenesis was more extensional because of the past circulation of geothermal fluids at shallow levels in sediments [*Desprairie et al.*, 1991]. The high heat flow measured in these basins [*Kudrass et al.*, 1986] could also influence the intensity of alteration. As a consequence, in these basins it seems that diagenesis acted as a selective filter on the evidence for arc volcanism, i.e., only the maximum pulses of arc volcanism are detected in the marine sediments older than Pliocene times. The aim of the following discussion is to stress the origin and the geodynamic meaning of these major episodes of arc volcanism.

The Cenozoic evolution of the Philippine region is controlled by a complex collision between the Eurasiatic, Australian-Indian and Philippine Sea plates. During Paleogene times, an extensional environment was responsible for the thinning of the two continental margins and for the opening of marginal basins. During the Miocene, incipient collision between the Eurasian and Indian plates induced closure of marginal basins along a north-south direction. At the same time, the northern motion of the Philippine Sea plate decreased. Juxtaposition with the Eurasiatic margin also induced closure of marginal basins along a northwest-southeast direction. As a consequence, two roughly orthogonal directions of crustal shortening were active in the Philippine area since the Miocene. A collision may have a dual effect on arc volcanism: on one hand the head-on juxtaposition of two plates may block subduction; on the other hand tectonic closure of basins at margins initiates subduction. As a consequence, the tephrochronologic data recorded in the Philippine basins can be mainly used to detect the initiation and the cessation of arc volcanism and, thus to specify the chronology of collisions.

In the Celebes Basin (site 767), a background volcanic event attributed to the Sunda arc was recorded in upper to lower Miocene sediments. This volcanic output is expressed by regularly distributed ash layers and by disseminated glass in the sediments. From 23 Ma, the entire volcanic component in the sediments vanishes. This event is interpreted as a cessation of activity on the Sunda arc or at least on the eastern segment of this arc, which fringed the Celebes Basin in early Miocene times (Figure 9c). This event is explained by a complex interaction in the Miocene between the Philippine Sea plate, and the Eurasiatic and Australian margins. Particularly, it corresponds to the collision of the Sula micro-blocks with the eastern Sunda margin. This Sunda micro-block was scraped out of the Australian margin and pushed against the Eurasiatic margin by the northerly oblique motion of the Philippine Sea plate [*Silver et al.*, 1983; *Daly et al.*, 1991]. The consequences of this collision are observed in central Sulawesi and eastern Borneo by middle Miocene folding and thrusting [*Kundig*, 1956; *Letouzey et al.*, 1990]. In this example of collision, it appears that arc volcanism ceased several million years before the first phases of compressional deformation were recorded on land.

The volcanic episodicity of the Sulu arc system established by the tephrochronologic data can be separated into four distinct episodes. (1) The first is in the lower Oligocene sediments of the Celebes Basin (site 767). (2) The second corresponds to the lower-middle Miocene pyroclastic deposits (18-16 Ma) at sites 768 and 769, originated from Cagayan ridge. These thick pyroclastic flows, which contain lithic and glassy angular fragments, are turbiditic in origin and are assumed to be synchronous with arc volcanism. (3) The third episodes marked by thin upper Miocene pyroclastic tuffs (10 Ma) in the Sulu Basin (site 768). The tuffs are turbiditic in origin and are mixed with an important detrital component. These epiclastic tuffs are interpreted to be subsequent to the volcanic activity of the Sulu arc, because of the proportion of detrital components and the deep barrier effect of the active trench. (4) The fourth and final episode is represented by major Pliocene and Quaternary peaks in ash deposition (site 767 and 769).

(1) The early Oligocene activity of the Sulu arc can be interpreted as the result of subduction of the proto-south China Sea or the Celebes Basin (Figure 9b). Tilted blocks identified in seismic data of the northern margin of the Celebes Basin [*Hamilton*, 1979; *Rangin and Silver*, 1991] as well as magnetic patterns in the basin [*Weissel*, 1980; *Lee and McCabe*, 1986] contradict the hypothesis of northward subduction of the Celebes Sea. As a consequence, the early Oligocene volcanic episode may be explained by the subduction of the proto-south China Sea. This first stage of closure is not linked to collision of the Australian plate that occurred later, but it was contemporaneous with the rifting of the south China Sea [*Taylor and Hayes*, 1980; 1983].

(2) The late early Miocene episode of the Cagayan ridge was synchronous with both the end of arc volcanism and that of spreading in the Sulu Basin [*Silver et al.*, 1989]. This temporal relationship allows us to postulate that backarc opening of the Sulu Basin was caused by subduction of the Proto-south China Sea under the Cagayan arc, leaving the Sulu arc as a remnant arc. This ended with full closure of the basin. The resultant collision between the rifted margin of the south China Sea and the Cagayan arc is postdated by middle Miocene deformed sediments in the northwestern Sulu Basin [*Holloway*, 1982; *Rangin*, 1989]. The second episode of closure of the proto-south China Sea was associated to incipient collision in the south marked by the collision of the Sula micro-block and cessation of volcanism on the Sunda arc. The tephrochronological data thus support two periods of closure of the proto-south China Sea with a late Oligocene hiatus. This is supported by on-land data [*Bellon and Rangin*, 1990] and by the kinematic constraints that there was no wide basin in the Paleogene evolution of the southwest Pacific [*Jolivet et al.*, 1989].

(3 and 4) The late Miocene evidences of Sulu arc activity, consistent with but subsequent to arc volcanism, are related to the first episode of closing of the Sulu Sea. Subduction of this basin was a consequence of the first stage of the Australian plate collision with the Eurasiatic margin. This process resumed in the latest Miocene and was marked by important Pliocene and Quaternary activity of the Sulu arc as recorded in the Sulu Sea. This most recent episode is also expressed in Pliocene and Quaternary sediments of the Celebes Sea. These marine sediments record volcanism of the north arm of Sulawesi, which was caused by southward subduction of the Celebes Basin.

In summary, the effect of the collision between the Australian and the Eurasiatic margin is well expressed by the volcanic inputs to the various basins and allows us to establish a chronology for when they closed: 35-30 and 22-16 Ma for the Proto-south China Sea; 14-11, 7-4 Ma and Quaternary for the Sulu Basin; and 8-4 Ma and Quaternary for the Celebes Sea. Most of these episodes are directly related to crustal shortening induced by collision.

The recent volcanic episodicity of the Philippine mobile belt is well constrained by four sites distributed all along the islands. These allow us to discuss the timing of the conjunction of the Philippine Sea plate and the Eurasiatic margin. Late Eocene volcanism (40-37 Ma) in the west Philippine basin is correlated with the initiation of arc volcanism on the Philippine arc, which is related to subduction of the Indian-Australian plate beneath the Philippine Sea plate (Figures 9a-b). From the late Miocene, the tephrochronological data allow us to specify new arc volcanism all along the Philippine mobile belt. The late Miocene frequency peak (12.5-10 Ma) in the West Philippine Basin (site 292) corresponds to the initial activity of the Bataan arc (west of Luzon) formed by the incipient subduction of the south China Sea beneath the Philippine mobile belt. In the Sulu and Celebes Basins, the same Pliocene patterns are respectively related to the start of Negros and Sangihe arcs, resulting from subduction of the Sulu and Molucca Seas. After a brief hiatus at about 3 Ma, Quaternary sediments in all sites indicate resumed volcanic activity. This resulted from an important framework of recent arcs. These were (1) to the north (site 292), the Philippine arc formed by the newly born subduction of the West Philippine Basin under the Philippine mobile belt, (2) in the center (sites 768 and 769), the reactivated Negros arc, and (3) to the south (site 767), the Cotabato arc formed by the newly born subduction of the Celebes Basin beneath the Philippine mobile belt.

The effect of progressive collision of the Philippine Sea plate with the Eurasiatic margin is well expressed by the volcanic inputs to the basin. We may establish a chronology for the closing episodes of the marginal basins. The shortening is first accomplished to the north by subduction of the South China Sea beginning at 12.5 Ma (Bataan arc). Then, the effects of collision moved to the south with closing first of the Sulu Sea from 7 Ma (Negros arc) and then, the Celebes Sea from 3 Ma (Cotabato arc). Finally, the shortening was compensated to the east by subduction of the west Philippine basin from 3 Ma (Philippine arc). During the Quaternary, only the Sulu Basin was affected by N-S shortening along the Sulu trench. Except for this case, however, arc volcanism in the Philippine region was essentially associated with NW-SE shortening (e.g., the Bataan, Negros, Cotabato, and Philippine arcs). As a consequence, the Quaternary distribution of arc volcanism rather suggests the predominance of collision between the Philippine Sea plate and the Eurasiatic margin.

In the Philippine basins, the volcanic record is partially biased by the effect of alteration due to high heat flow and hydrothermal circulation. As a consequence, alteration diluted the volcanic record and only the major episodes of arc volcanism are expressed by the sedimentary data of the basin. From prior discussion, it appears that, despite alteration, the first stages of arc volcanism are particularly well expressed by the sediments of the basin, i.e., the first stage of arc volcanism corresponds to a sudden and abundant volcanic input well preserved in the sediments of the surrounding basins. This appears true for the old Philippine arc (the late Eocene of site 292), for the Cagayan arc (the early Oligocene of site 767), for the Bataan arc (the late Miocene of site 292), for the Negros arc (the late Miocene-early Pliocene of sites 768 and 769), for the Cotabato arc (the Quaternary of site 767) and, for the Philippine arc (the Quaternary of site 292). In these cases, one may wonder why there was such important volcanic activity recorded in the basins corresponding to the birth of volcanic arcs. However, the same feature is also observed for the end of volcanic activity of the Cagayan arc with the thick lower Miocene pyroclastic deposits in the Sulu Sea. In both cases, it is important to discuss the meaning of such episodes of arc volcanism. Two classes of hypotheses are available to explain the climax of arc volcanic activity. The first concerns the subducted materials and their influence on magmagenesis. The second concerns the migration of the magma in the crust and the consequences for differentiation. The two processes could act in combination.

The genesis of magma in a subduction zones is a complex problem for which it is difficult to assess the influence of the subducted components and particularly that of subducted sediments. Isotopic evidences for influence of sediments inherently is difficult to distinguish from those for crustal contamination of magma during their ascent. Sediments constitute at most a few percent of the mass of the source rock for arc volcanism and their parent magmas [e.g., 2%, *Marsh*, 1979], although sediments may contribute a larger percentage of some elements, e.g., Ba, Th and Pb [*Gill*, 1981]. Strontium isotopic data from andesitic lavas sampled on the New Hebrides arc show that a sedimentary component contributed to the genesis of magmas [*Briqueu and Lancelot*, 1982]. However simply the amount of subducted sediments may affect arc volcanism. In the Lesser Antilles, for instance, the subducted oceanic plate comprises a succession of ridges and grabens which are filled with thick sedimentary sequences. Observations on the arc showed that the subduction of these sediments corresponds to the period of most active segment of the volcanic arc [*Westercamp*, 1987; *Andreieff et al.*, 1988; *Bouysse and Westercamp*, 1988]. In the Aleutian arc, the absence of arc volcanism is linked to a segment of trench where sediments are offscraped and accreted; conversely, volcanic activity is located at places where sediments are subducting [*Kienle et al.*, 1983]. Regarding these examples, it seems that the subducted sediments may influence arc magmatism either by melting directly or by their content of water. The water introduced into a subduction zone may induce hydration and the partial melting of a large amount of mantle material, triggering important arc volcanism. In consequence, the bursts of arc volcanism during the first stages of subduction may be explained by the water contents of subducted material. Water in thick unconsolided sedimentary sequences may accumulate at the toes of passive margins. These will be subducted during the first stages of collision. This hypothesis may be postulated for the early Miocene volcanic activity of the Cagayan arc during the last stages of subduction, which corresponds to the final closing of the proto-south China Sea. During the incipient collision in the early Miocene, the basin was narrow and fed by much detritus. In such an environment, huge amounts of unconsolidated and hydrated sediments may have subducted during the last stages of basin closure.

Alternatively, we may consider the problem of magma ascent during a first trip to the surface following the onset of subduction. In fact, the transit time of magma should be longer if diapirs in the upper mantle and magma chambers in the crust are not connected. Longer transit time implicates more differentiated magma, more vigorous eruption on the surface, and therefore larger volumes of volcanic material susceptible to sedimentation in the surrounding basins. This model may apply to the end of the Cagayan arc volcanism, when transit time of magma was controlled by the incipient collision of the two sides of the basin.

4. CONCLUSION

Data from western Pacific ocean drilling allow us to define direct links between tectonic evolution and episodes of arc volcanism. From tephrochronological data recorded in the basins, backarc rifting is associated with intense arc volcanic activity that vanishes during backarc spreading and resumes during cessation of expansion in the marginal basin. This cycle is not universal but reflects an accurate examination of the Tohoku and Bonin arcs. Comparisons between the episodicity of arc volcanism and stress evolution of these arcs lead us to conclude that subduction only minimally influenced episodicity of arc volcanism. Instead, arc volcanism was mainly controlled by the deformation status of the arc substratum. For instance, the late Miocene volcanic hiatus both at the Tohoku arc and Bonin arcs occurred during the transition between two periods of deformation. In the Philippine area, the tephrochronologic data allow us to determine the births and the deaths of the volcanic arcs in relation to subduction closure of the marginal basins. These data are genuinely useful in considering the evolution of the complex collision between the Eurasiatic, the Indian-Australian and the Philippine Sea plates. In the Philippine basins, where diagenesis was so intense to alter the volcanic record, the discrepancy between marine and on-land data highlights the major volcanic pulses. These pulses are interpreted as mainly generated by subduction of a large amount of hydrated sediments or by the transit times of magma through the arc substratum. Explosive volcanic activity of arcs can therefore be interpreted either as an indicator of changing tectonics or as a marker of water content in subducted materials. As regards the processes which could have controlled volcanic episodicity of western Pacific arcs, it seems that shallow processes related to the tectonic state of the arc crust act in combination with deeper processes in the Wadati-Benioff zone, i.e., in the zone of partial melting. At this stage, no generalization is possible, because the controlling factors depend essentially on the regional geodynamic setting of each arc.

Acknowledgments. Many thanks are due to James H. Natland for his detailed review and helpful comments.We also would like to thank Seiya Uyeda and Renzo Sartori for reviewing this manuscript, and special thanks to Mariem Ayadi for the last editing.

REFERENCES

Andreieff, P., J. C. Baubron, and D. Westercamp, Histoire géologique de la Martinique, Petites Antilles, biostratigraphy (foraminiféres), radiochronologie (K-Ar), évolution volcano-structurale, *Géol. Fr., 2-3*, 39-70, 1988.

Angelier, J., and P. Huchon, Tectonics records of convergence changes in a collision area: the Boso and Muira peninsula, Central Japan, *Earth Planet. Sci. Lett., 81*, 397-408, 1986.

Aramaki, S., and T. Ui. Japan, in *Andesites: Orogenic Andesites and Related Rocks*, edited by R. S. Thorpe, pp. 259-292, John Wiley & Sons, New-York, 1982.

Arculus, R. J., The significance of source versus process in the tectonic controls of magma genesis, *J. Volcanol. Geotherm. Res., 32*, 1-12, 1987.

Armstrong, R. L., E. B. Ekren, E. H. Mc Kee, and D. C. Noble, Space-time relations of Cenozoic silicic volcanism in the Great Basin of the western United-States, *Am. J. Sci., 267*, 478-490, 1969.

Barberi, F., M. Coltelli, G. Ferrara, F. Innocenti, J. M. Navarro, and R. Santacroce, Plio-Quaternary volcanism in Ecuador, *Geol. Mag., 125*, 1-14, 1988.

Barrash, W., and R. Venkatakrishnan, Timing of late Cenozoic volcanic and tectonic events along the western margin of the north American plate, *Geol. Soc. Am. Bull., 93*, 977-989, 1982.

Beccaluva, L., G. Serri, and J. Dostal, Geochemistry of volcanic rocks from the Mariana, Yap and Palau trenches bearing on the tectono-magmatic evolution of the Mariana trench, arc and backarc, in *Development in Geotectonics, the Origin of Arcs*, vol. 21, edited by F. C. Wezel, pp. 481-508, Elsevier, Amsterdam, 1986.

Bellon, H., R. C. Maury, R. Soeria-Atmadja, and M. Polve, Tertiary and Quaternary volcanism in Java, Indonesia, K-Ar isotopic dates and their constraints (abstract), p. 9, in *Orogenesis in Action, Tectonics and Processes in the West Equatorial Pacific Margin,* edited by R. Hall, G. Nichols, and C. Rangin, Geological Society of London, London, 1990.

Bellon, H., and C. Rangin, K-Ar isotopic dating of magmatic activity in Sabah (Borneo) and the Central-southern Philippine archipelago: geodynamic implications (abstract), p. 8, in *Orogenesis in Action, Ttectonics and Processes in the West Equatorial Pacific Margin,* edited by R. Hall, G. Nichols, C. Rangin, , Geological Society of London, London, 1990.

Bellon, H., and C. Rangin, Geochemistry and isotopic dating of Cenozoic volcanic arc sequences around the Celebes and Sulu seas, in *Proceedings ODP, Scientific results*, vol. 124, C. Rangin, E. Silver, M. T. Von Breymann, pp. 321-338, Ocean Drilling Program, College Station, TX, 1991.

Bischke, R. E., J. Suppe, and R. Del Pilar, A new branch of the Philippine fault system from aeromagnetic and seismic data, *Tectonophysics, 183*, 243-265, 1990.

Bouysse, P., and D. Westercamp, Effets de la subduction de rides océaniques sur l'évolution d'un arc insulaire: l'exemple des Petites Antilles, *Géol. Fr., 2-3*, 3-38, 1988.

Briqueu, L., and J. R. Lancelot, Sr isotopes and K, Rb, Sr balance in sediments and igneous rocks from the subducted plate of the Vanuatu (New Hebrides) active margin, *Geochim. et Cosmochim. Acta, 47*, 191-200, 1982.

Cadet, J. P., and K. Fujioka, Neogene volcanic ashes and explosive volcanism: Japan trench transect, leg 57, DSDP, in *Init. Repts. DSDP*, 56-57, E. Honza, et al., pp. 1027-1041, U.S. Govt. Printing Office, Washington, D.C., 1980.

Cadet, J. P., A. Pouclet, and H. Cambray, Les cendres volcaniques sous-marines du Pacifique Oriental, enregistrement du volcanisme explosif de l'Amérique Centrale (large du Mexique et du Guatemala: legs 66, 67 et 84 du DSDP), *Bull. Soc. Géol. Fr., 8 VI, 1*, 37-47, 1990.

Cambray, H., and J. P. Cadet, Testing global synchronism in peri-Pacific arc volcanism, *J. Volcanol. Geotherm. Res., 63*, in press, 1994.

Cambray, H., J. P. Cadet, and A. Pouclet, Ash layers in deep-sea sediments as tracers of arc volcanic activity: Japan and Central America as cases of study, *The Island Arc, 2,* 72-86, 1993.

Cambray, H., F. Guichard, K. Fujioka, A. Pouclet, and J. P. Cadet, Tephrochronologic study of two piston cores off Japan, *Bull. Volcanol. Soc. Japan, II, 35, 3*, 313-322, 1990.

Cande, S. C., Nazca-south America plate interaction 80 MY to Present (abstract), *EOS Trans. AGU, 64*, 865, 1983.

Cas, R. A. F., and J. V. Wright, *Volcanic successions, modern and ancient*, 528 pp., Allen & Unwin, London, 1987.

Chamot-Rooke, N., V. Renard, and X. Le Pichon, Magnetic anomalies in the Shikoku Basin, a new interpretation, *Earth Planet. Sci. Lett., 83*, 214-223, 1987.

Charvet, J., and Fabbri, O., Vue générale sur l'orogenèse Shimanto et l'évolution tertiaire du Japon sud-ouest, *Bull. Soc. Géol. Fr., 8, III, 6*, 1171-1188, 1987.

Crawford, A. J., L. Beccaluva, G. Serri, and J. Dostal, Petrology, geochemistry and tectonic implications of volcanic dredged from the intersection of the Yap and Mariana trenches, *Earth Planet. Sci. Lett., 80*, 265-280, 1986.

Daly, M. C., M. A. Cooper, I. Wilson, D. G. Smith, and B. G. D. Hooper, Cenozoic plate tectonics and basin evolution in Indonesia, *Mar. Petrol. Geol., 8*, 2-21, 1991.

De Boer, J. Z., L. A. Odom, P. C. Ragland, F. G. Snider, and N. R. Tilford, The Bataan orogene: Eastward subduction, tectonic rotations and volcanism in the Western Pacific (Philippines), *Tectonophysics, 67*, 251-282, 1980.

Defant, M. J., J. Z. De Boer, and D. Oles, The western Central Luzon volcanic arc, the Philippines, two arcs divided by rifting?, *Tectonophysics, 145*, 305-317, 1988.

Deshler, R. M., J. D. Hoover, and J. R. Dyer, El Borracho caldera, a 17 MY rhyolotic edifice in the eastern Basin and Range, Chihuahua, Mexique, *Geol. Soc. Am. Abst. with Programs, 18, 6*, 584, 1986.

Desprairie, A., M. Riviere, and M. Pubellier, Diagenetic evolution of Neogene volcanic ashes (Celebes and Sulu Seas), in *Proc. ODP, Sci. Results*, 124, C. Rangin, E. Silver, M. T. Von Breymann, et al., pp. 489-503, Ocean Drilling Program, College Station, TX, 1991.

Donn, W. L., and D. Ninkovich, Rate of Cenozoic explosive volcanism in the north Atlantic ocean inferred from deep-sea cores, *J. Geophys. Res., B10*, 5455-5460, 1980.

Engebretson D. C., A. Cox, and R. G. Gordon, Relative motions between oceanic and continental plates in the Pacific Basin, *Geol. Soc. Amer. Spec. Paper., 206*, 59 pp., 1985.

Fabbri, O., Contribution à l'étude de la marge asiatique: la chaîne tertiaire de Shimanto (Japon SW); Rôles de la subduction et de la collision dans la tectogenèse, Thèse de doctorat d'Université, Université d'Orléans, 286 pp., 1989.

Fujioka, K., Synthesis of Neogene explosive volcanism of the Tohoku arc, deduced from the marine tephra drilled around the Japan trench region, DSDP, legs 56, 57 and 87A, in *Init. Repts. DSDP, 87*, H. Kagami, D. E. Karig, W. T. Coulbourn, et al., pp. 703-721, U.S. Govt. Printing Office, Washington, D.C., 1985.

Fujioka, K., B. Taylor, et al., Arc volcanism and rifting, *Nature, 342*, 18-20, 1989.

Furuta, T., and F. Arai, Petrographic and geochemical property of tephra in DSDP cores from the north Philippines Sea, in *Init. Repts. DSDP, 58*, G. DeV. Klein, K. Kobayashi, et al., pp. 617-627, US Government Printing Office, Washington, 1979.

Furuta, T., and F. Arai, Petrographic properties of tephra, leg 56, DSDP, in *Initial reports of the Deep-Sea Drilling Project*, vol. 56-57, E. Honza, et al., pp. 1043-1048, U.S. Govt. Printing Office, Washington, D.C., 1980.

Furuta, T., K. Fujioka, and F. Arai, Widespread submarine tephra aroud Japan, petrographic and chemical properties, *Mar. Geol., 72*, 125-146, 1986.

Gardner, J. V., C. S. Nelson, and P. A. Baker, Distribution and character of pale green laminae in sediment from Lord Howe rise: a probable late Neogene and Quaternary tephrostratigraphy record, in *Init. Repts. DSDP, 90*, J. P. Kennett, C. C. Von Der Borch, et al., pp. 1145-1159, U.S. Govt. Printing Office, Washington, D.C., 1985.

Gill, J. B., *Orogenic Andesites and Plate Tectonics*, 390 pp., Springer Verlag, Berlin-Heidelberg, 1981.

Gordon, R. G., and D. M. Jurdy, Cenozoic global plate motions, *J. Geophys. Res., 91, B12*, 12389-12406, 1986.

Grechin, V. I., A. R. Niem, R. O. Mahood, V. A. Alexandrova, and B. A. Sakharov, Neogene tuffs, ashes and volcanic breccias from offshore California, DSDP, leg 63: sedimentation and diagenesis, in *Init. Repts. DSDP, 63*, R. S. Yeats, B. U. Haq, et al., pp. 631-657, U.S. Govt. Printing Office, Washington, D.C., 1980.

Guffanti, M., and G. S. Weaver, Distribution of late Cenozoic volcanic vents in the Cascade range; Volcanic arc segmentation and regional tectonic considerations, *J. Geophys. Res., 93, B6*, 6513-6529, 1988.

Hamilton, W. B., Tectonics of the Indonesian region, *U. S. Geol. Survey Prof. Paper, 1078*, 345 pp., 1979.

Heath, G. R., L. H. Burckle, et al., *Init. Repts. DSDP, 86*, 804 pp, U.S. Govt. Printing Office, Washington, D.C., 1985.

Hein, J. R., D. W. Scholl, and J. Miller, Episodes of Aleutian ridge explosive volcanism, *Science, 199*, 137-141, 1978.

Helwig, J., and G. A. Hall, Steady states trenches, *Geology, 2*, 309-316, 1974.

Hibbard, J. P., and D. E. Karig, Alternative plate model for the early Miocene evolution of the southwest Japan margin, *Geology, 18*, 170-174, 1990.

Hilde, T. W. C., and C. S. Lee, Origin and evolution of the West Philippine basin: a new interpretation, *Tectonophysics, 102*, 85-104, 1984.

Hilde, T. W. C., S. Uyeda, and L. W. Kroenke, Evolution of the western Pacific and its margin, *Tectonophysics, 38*, 145-165, 1977.

Hildreth, W., Gradient in silicic magma chambers: implications for lithospheric magmatism, *J. Geophys. Res., 86, B11*, 10153-10192, 1981.

Holloway, N. H., The stratigraphy and tectonic relationship of Reed Bank, North Palawan and Mindoro to the Asian mainland and its significance in the evolution of the South China sea, *Am. Assoc. Petrol. Geol. Bull., 66*, 1357-1383, 1982.

Honda, S., and S. Uyeda, Thermal process in subduction zones, a review and preliminary approach and the origin of arc volcanism, in *Arc Volcanism: Physics and Ttectonics*, edited by D. Shimozuru, I. Yokoyama, pp. 117-140, Terrapub, Tokyo, 1983.

Honza, E., Evolution of arc volcanism related to marginal sea spreading and subduction at trench, in *Arc Volcanism: Physics and Ttectonics*, edited by D. Shimozuru, I. Yokoyama, pp. 177-189, Terrapub, Tokyo, 1983.

Honza, E., and K. Tamaki, The Bonin arc, in *the ocean basins and margins: the Pacific ocean*, vol. 7A, edited by A. E. M. Nairn, F. G. Stehli, S. Uyeda, pp. 459-502, Plenum Press, New York, 1985.

Horn, D. R., M. N. Delach, and B. M. Horn, Distribution of volcanic ash layers and turbidites in the north Pacific, *Geol. Soc. Amer. Bull., 80*, 1715-1724, 1969.

Huang, T. C., N. D. Watkins, and D. M. Shaw, Atmospherically transported volcanic glass in deep-sea sediments: volcanism in sub-antarctic latitudes of the south Pacific during late Pliocene and Pleistocene time, *Geol. Soc. Amer. Bull., 86*, 1305-1315, 1975.

Hussong, D., and S. Uyeda, Tectonic processes and the history of the Mariana arc: a synthesis of the results of DSDP, leg 60, in *Init. Repts. DSDP, 60*, D. Hussong, S. Uyeda, et al., pp. 909-929, U.S. Govt. Printing Office, Washington, D.C., 1981.

Ingle, J. C., K. Suyehiro, et al., Introduction, background and principal results of leg 128 of the Ocean Drilling Program, Japan Sea, in *Proc. ODP, Init. Repts., 128*, J. C. Ingle, K. Suyehiro, M. T. Von Breymann, et al., pp. 5-38, Ocean Drilling Program, College Station, TX, 1990.

Jolivet, L., and P. Huchon, Crustal-scale strike-slip deformation in Hokkaido, northern Japan, *J. Struct. Geol., 11, 5*, 509-522, 1989.

Jolivet, L., P. Huchon, J. P. Brun, X. Le Pichon, N. Chamot-Rooke, and J. C. Thomas, Arc deformation and marginal basin opening: Japan Sea as a case of study, *J. Geophys. Res., 96, B3*, 4367-4384, 1991.

Jolivet, L., P. Huchon, and C. Rangin, Tectonic setting of western Pacific marginal basins, *Tectonophysics, 160*, 23-47, 1989.

Jolivet, L., and K. Tamaki, Neogene kinematics in the Japan

Sea region and volcanic activity of the NE Japan arc, in *Proc. ODP, Init. Repts., 127-128, part 2*, K. Pisciotto, K. Tamaki, et al., pp. 1311-1331, Ocean Drilling Program, College Station, TX, 1992.

Jones, E. J. W., Volcanic glass in abyssal clays sampled at DSDP Leg 20 drilling sites, northwest Pacific, in *Init. Repts. DSDP, 20*, B. C. Heezen, I. D. Mac Gregor, et al., pp. 389-402, U.S. Govt. Printing Office, Washington, D.C., 1973.

Kagami, H., D. E. Karig, W. T. Coulbourn, et al., *Init. Repts. DSDP, 87*, 985 pp., U.S. Govt. Printing Office, Washington, D.C., 1985.

Kagiyama, T., Thermal activity of volcanoes in the Japan arc, nature and geological meaning, in *Arc Volcanism: Physics and Tectonics*, edited by D. Shimozuru, I. Yokoyama, pp. 13-27, Terrapub, Tokyo, 1983.

Kaneoka, I., N. Isshiki, and S. Zashu, K-Ar ages of the Izu-Bonins islands, *Geochem. J., 4*, 53-60, 1970.

Karig, D. E., Basin genesis in the Philippine Sea, in *Init. Repts. DSDP, 31*, D. E. Karig, J. C. Ingle, et al., pp. 857-879, U.S. Govt. Printing Office, Washington, D.C., 1975.

Karig, D. E., Temporal relationships between backarc basin formation and arc volcanism with special reference to the Philippine Sea, in *The Tectonic and Geologic Evolution of Southeast Asian Seas and Islands, part 2, Geophys. Monogr. Ser.*, vol. 27, edited by D. E. Hayes, pp. 318-325, Am. Geophys. Un., Washington, D. C., 1983.

Karig, D. E., and C. L. Angevine, Geologic constraints on subduction rates in the Nankai trough, in *Init. Repts. DSDP, 87,* H. Kagami, D. E. Karig, W. T. Coulbourn, et al., pp. 789-796, U.S. Govt. Printing Office, Washington, D.C., 1985.

Karig, D. E., J. C. Ingle, et al., *Init. Repts. DSDP, 31*, 927 pp., U.S. Govt. Printing Office, Washington, D.C., 1975.

Kelts, K., and J. A. Mc Kenzie, Cretaceous volcanogenic sediments from the Line island chain: diagenesis and formation of K feldspar, DSDP Leg 33, hole 315A and Site 316, in *Init. Repts. DSDP, 33*, S. O. Schlanger, E. D. Jackson, et al., pp. 789-831, U.S. Govt. Printing Office, Washington, D.C., 1976.

Kennett, J. P., A. R. Mc Birney, and R. C. Thunell, Episodes of Cenozoic volcanism in the circum-Pacific region, *J. Volcanol. Geotherm. Res., 2*, 145-163, 1977.

Kennett, J. P., and R. C. Thunell, Global increase in Quaternary explosive volcanism, *Science, 187*, 497-503, 1975.

Kennett, J. P., and R. C. Thunell, On explosive Cenozoic volcanism and climatic implications, *Science, 193*, 1231-1234, 1977.

Kienle, J., S. E. Swanson, and H. Pulpan, Magmatism and subduction in the eastern Aleutian arc, in *Arc Volcanism: Physics and Tectonics*, edited by D. Shimozuru, I. Yokoyama, pp. 191-224, Terrapub, Tokyo, 1983.

Kimura, G., and K. Tamaki, Collision, rotation and backarc spreading: the case of the Okhotsk and Japan Seas, *Tectonics, 5*, 389-401, 1986.

Klein, G. D., The control of depositional depth, tectonic uplift and volcanism on sedimentation processes in the backarc basins of the western Pacific ocean, *J. Geol., 93*, 1-25, 1985.

Klein, G. D., and K. Kobayashi, Geological summary of the north Philippines Sea, based on DSDP Leg 58 results, in *Init. Repts. DSDP, 58*, G. D. Klein, K. Kobayashi, et al., pp. 951-961, U.S. Govt. Printing Office, Washington, D.C., 1979.

Klein, G. D., K. Kobayashi, et al., Off-ridge volcanism and sea-floor spreading in the Shikoku Basin, *Nature, 273*, 746-748, 1978.

Klein, G. D., K. Kobayashi, et al., *Init. Repts. DSDP, 58*, 1022 pp, Washington, U.S. Govt. Printing Office, Washington, D.C., 1980.

Knittel, U., M. J. Defant, and M. J. Raczek, Recent enrichment in the source region of arc magmas from Luzon island, Philippines: Sr and Nd isotopic evidence, *Geology, 16*, 73-76, 1988.

Konda, T., H. Kurozawa, and Y. Oba, Cenozoic volcanism of north-east Japan and its implications for Japanese islands arc evolution, I, major element study (abstract), in *IAVCEI Symposium on Arc volcanism*, pp. 189-190, Tokyo and Hakone, 1981.

Kudrass, H. R., M. Wiedicke, P. Cepeck, H. Kreuser, and P. Muller, Pre-Quaternary rocks dredged from the south China Sea (Reed Bank area) and Sulu Sea, during Sonne cruise in 1982-1983, *Mar. Petrol. Geol., 3*, 19-30, 1968.

Kundig, E., Geology and ophiolite problems of east Celebes, *Verh. K. Ned. Geol. Mijnbouwkd. Genoot., 16*, 210-235, 1956.

Kuwaraha, T., Late Cretaceous to Pliocene fault systems and corresponding regional tectonic stress fields in the southern part of northeast Japan, Scientific Report, 111 pp., University of Tsukuba, 1982.

Lee, C. S., and R. McCabe, The Banda-Celebes-Sulu Basins: a trapped piece of Cretaceous Eocene oceanic crust ?, *Nature, 322*, 51-53, 1986.

Letouzey, J., and M. Kimura, The Okinawa trough: genesis of a backarc basin developping along a continental margin, *Tectonophysics, 125*, 209-230, 1986.

Letouzey, J., P. Werner, and A. Marty, Fault reactivation and structural inversion, backarc and intraplate compressive deformations, example of the eastern Sunda shelf (Indonesia), *Tectonophysics, 183*, 341-362, 1990.

Machida, H., Tephrochronology and Quaternary studies in Japan, in *Tephra studies*, edited by S. Self, R. S. J. Sparks, pp. 161-191, D. Reidel Publ. Co., Dordrecht, 1981.

Maleterre, P., Contribution à l'étude géologique de la frontière occidentale de la plaque de la mer des Philippines, Thèse de Doctorat d'Université, 315 pp., Université de Bretagne Occidentale, 1989.

Marsh, B. D., Island arc volcanism: volcanic arc ringing the Pacific owe their origin to magma generated in a mysterious fashion when tectonic plates are consumed by the inner earth, *Am. Scientist, 67*, 161-172, 1979.

Matsuda, T., K. Nakamura, and A. Sugimura, Late Cenozoic orogeny in Japan, *Tectonophysics, 4, 4-6*, 349-366, 1967.

McGeary, S., A. Nur, and Z. Ben-Avraham, Spatial gaps in arc volcanism: the effect of collision or subduction of oceanic plateaus, *Tectonophysics, 119*, 195-221, 1985.

Meijer, A., M. Reagan, H. Ellis, M. Shafiqullah, J. F. Sutter, P. E. Damon, and S. Kling, Chronology of volcanic events in the eastern Philippine Sea, in *The Tectonic and Geologic Evolution of Southeast Asian Seas and Islands, part 2, Geophys. Monogr. Ser.*, vol. 27, edited by D. E. Hayes, pp. 349-359, Am. Geophysl Un., Washington, D. C., 1983.

Mitchell, A. H. G., F. Hernandez, and A. P. Dela Cruz, Cenozoic evolution of the Philippines archipelago, *J. Southeast Asian Earth Sci., 1, 1*, 3-22, 1986.

Miyashiro, A., Volcanic rock series in island arcs and active

continental margins, *Am. J. Sci., 274*, 321-355, 1974.
Moriwaki, H., T. Furuta, and F. Arai, Tephrochronology in the Sea of Japan and the northeast China Sea, in *Preliminary report of the Hakuho Maru cruise, KH 82-4, IPOD*, edited by K. Kobayashi, pp. 160-170, Ocean Research Institut, Tokyo, 1983.
Mrozowski, C. L., and D. E. Hayes, The evolution of the Parece Vela Basin, eastern Philippines Sea, *Earth Planet. Sci. Lett., 46*, 49-67, 1979.
Nakada, S., and M. Takahashi, Regional variation in chemistry of the Miocene intermediate to felsic magmas in the outer zone and the Setouchi province of southwest Japan, *J. Geol. Soc. Japan, 85*, 571-582, 1979.
Nakamura, K., and S. Uyeda, Stress gradient in arc-backarc region and plate subduction, *J. Geophys. Res., 85, B5*, 6419-6428, 1980.
Ninkovich, D., and W. L. Donn, Explosive Cenozoic volcanism and climatic implications, *Science, 194*, 899-906, 1976.
Ninkovich, D., and W. L. Donn, Cenozoic explosive volcanism related to east and southeast Asian arcs, in *Island Arcs, Deep-Sea Trenches and Backarc Basins*, edited by M. Talwani, W. C. Pitman III, pp. 337-347, Am. Geophys. Un., Washington, D. C., 1977.
Ninkovich, D., and J. H. Robertson, Volcanogenic effects on the rates of deposition of sediments in the Northwest Pacific Ocean, *Earth Planet. Sci. Lett., 27*, 127-136, 1975.
Notsu, K., Y. Arakawa, and T. Kobayashi, Strontium isotopic characteristics of arc volcanic rocks at the initial stage of subduction in western Japan, *J. Volcanol. Geotherm. Res., 40*, 181-196, 1990.
Nur, A., and Z. Ben-Avraham, Vocanic gaps and the consumption of aseismic ridges in south America, *Geol. Soc. Amer. Mem., 154*, 729-740, 1981.
Okamura, S., Geochemical variation with time in the Cenozoic volcanic rocks of southwest Hokkaido, Japan, *J. Volcanol. Geotherm. Res., 32*, 161-176, 1987.
Otofuji, Y., T. Matsuda, and S. Nohda, Paleomagnetic evidences for the Miocene counter clockwize rotation of the northeast Japan, rifting process of the Japan arc, *Earth Planet. Sci. Lett., 75*, 265-277, 1985.
Otsuki, K., Neogene stress fields of northeast Honshu arc and dynamic conditions of plate boundaries, in *Tectonics of Eastern Asia and Western Pacific Continental Margin*, pp. 96-97, DELP symposuim, Tokyo, 1988.
Pisciotto, K., K. Tamaki, et al., Les premiers forages de la croûte océanique de la mer du Japon; résultats préliminaires du Leg ODP 127, *C. R. Acad. Sci. Paris, 311, II*, 837-844, 1990.
Pouclet, A., H. Cambray, J. P. Cadet, J. Bourgois, and P. De Wever, Volcanic ash from Leg 112 off Peru, in *Proc. ODP, Sci. Results, 112*, E. Suess, R. Von Huene, et al., pp. 465-480, Ocean Drilling Program, College Station, TX, 1990.
Pubellier, M., P. Spadea, A. Pouclet, R. Solidum, A. Desprairie, and H. Cambray, Correlations of tephra in Celebes and Sulu Basins, constraints in geodynamics, in *Proc. ODP, Sci. Results, 124,* C. Rangin, E. Silver, M. T. Von Breymann, et al., pp. 459-465, Ocean Drilling Program, College Station, TX, 1991.
Rangin, C., The Sulu Sea, a backarc basin setting within a Neogene collision zone, *Tectonophysics, 161*, 119-141, 1989.
Rangin, C., M. Pubellier, L. Jolivet, et al., A simple model for the tectonic evolution of the southeast Asia and Indonesia region for the past 43 my, *Bull. Soc. Géol. Fr., 8, VI, 6*, 889-905, 1990.
Rangin, C., and E. A. Silver, Neogene tectonic evolution of the Celebes-Sulu Basins: new insights from Leg 124 drilling, in *Proc. ODP, Sci. Results, 124*, C. Rangin, E. Silver, M. T. Von Breymann, et al., pp. 51-63, Ocean Drilling Program, College Station, TX, 1991.
Rangin, C., E. A. Silver, M. T. Von Breymann, et al., *Proc. ODP, Init. Repts., 124*, 916 pp., Ocean Drilling Program, College Station, TX, 1990.
Rea, D. K., and K. F. Scheidegger, Eastern Pacific spreading rate fluctuation and its relation to Pacific area volcanic episodes, *J. Volcanol. Geotherm. Res., 5*, 135-148, 1979.
Sacks, I. S., Subduction geometry and magma genesis, in *Studies in Geophysics: Explosive Volcanism, Inception, Evolution and Hazards*, pp. 34-46, National Academy Press, Washington, D.C., 1984.
Sartori, R., Tectonic significance of the sedimentary column at Site 449 (western Parece Vela Basin), DSDP, Leg 59, in *Init. Repts. DSDP, 59*, L. Kroenke, R. Scott, et al., pp. 609-614, U.S. Govt. Printing Office, Washington, D.C., 1980.
Scheidegger, K. F., and L. D. Kulm, Late Cenozoic volcanism in the Aleutian arc: information from ash layers in the northeastern gulf of Alaska, *Geol. Soc. Am. Bull., 86*, 1407-1412, 1975.
Scientific Party, *Init. Repts. DSDP 56 and 57 (Part 1 and 2)*, 1417 pp., U.S. Govt. Printing Office, Washington, D.C., 1980.
Scott, R. B., and L. W. Kroenke, Evolution of backarc spreading and arc volcanism in the Philippine Sea: interpretation of Leg 59 DSDP results, in *The Tectonic and Geologic Evolution of Southeast Asian Seas and Islands, Geophys. Monogr. Ser.*, vol. 23, edited by D. E. Hayes, pp. 283-291, Am. Geophys. Un., Washington, D.C., 1980.
Scott, R. B., L. W. Kroenke, G. Zakariadze, and A. Sharaskin, Evolution of the south Philippines Sea: DSDP, Leg 59 results, in *Init. Repts. DSDP*, vol. 59, L. Kroenke, R. Scott, et al., pp. 803-815, U.S. Govt. Printing Office, Washington, D.C., 1980.
Seno, T., Philippine Sea plate kinematics, *Modern Geol., 14*, 87-97, 1989.
Shaw, D. M., N. D. Watkins, and T. C. Huang, Atmospherically transported volcanic glass in deep-sea sediments: theoritical considerations, *J. Geophys. Res., 79, 21*, 3087-3094, 1974.
Shiraki, K., N. Kuroda, and S. Maruyama, Evolution of the Tertiary volcanic rocks in the Izu-Mariana arc, *Bull. Volcanol., 41*, 548-562, 1978.
Sibuet, J. C., J. Letouzey, F. Barbier, J. Charvet, J. P. Foucher, T. W. C. Hilde, M. Kimura, C. Lin-Yun, B. Marsset, C. Miller, and J. F. Stephan, Backarc extension in the Okinawa trough, *J. Geophys. Res., 92*, 14041-14063, 1987.
Silver, E. A., R. Mc Caffrey, and R. B. Smith, Collision, rotation and the initiation of subduction in the evolution of Sulawesi, Indonesia, *J. Geophys. Res., 88*, 9407-9418, 1983.
Sugimura, A., T. Matsuda, K. Chinzei, and K. Nakamura, Quantitative distribution of late Cenozoic volcanic materials in Japan, *Bull. Volcanol., 26*, 125-140, 1963.
Takeuchi, A., On the episodic vicissitude of tectonic stress field of the Cenozoic northeast Honshu arc, Japan, in *Formation of Active Ocean Margin*, edited by N. Nasu, pp. 443-465,

Terrapub, Tokyo, 1985.

Takeuchi, A., Oligocene/Miocene rotational block movement and paleostress field of Japan, *J. Geomag. Geoelect.*, *38*, 495-511, 1986.

Tamaki, K., K. Pisciotto, J. Allan, et al., *Proc. ODP, Init. Repts*, *127*, 844 pp., Ocean Drilling Program, College Station, TX, 1990.

Tamaki, K., K. Suyehiro, J. Allan, J. C. Ingle, and K. A. Pisciotto, Tectonic synthesis and implications of Japan Sea ODP drilling, in *Proc. ODP, Sci. Results*, *127-128*, part 2, K. Pisciotto, K. Tamaki, et al., pp. 1333-1348, Ocean Drilling Program, College Station, TX, 1992.

Tatsumi, Y., Formation of volcanic front in subduction zones, *Geophys. Res. Lett.*, *13*, *8*, 717-720, 1986.

Tatsumi, Y., and K. Ishisaka, Origin of high magnesian andesites in the Setouchi volcanic belt, southwest Japan: Petrographical and chemical characteristics, *Earth Planet. Sci. Lett.*, *60*, 293-304, 1982.

Taylor, B., Rifting and the volcanic-tectonic evolution of the Izu-Bonin-Mariana arc, in *Proc. ODP, Sci. Results*, *126*, B. Taylor, K. Fujioka, et al., pp. 627-651, Ocean Drilling Program, College Station, TX, 1992.

Taylor, B. et al., Correlation and comparison of backarc sites, in *Proc. ODP, Init. Repts.*, *126*, B. Taylor, K. Fujioka, et al., pp. 415-419, Ocean Drilling Program, College Station, TX, 1990a.

Taylor, B. et al., Correlation and comparison of forearc sites, in *Proc. ODP, Init. Repts.*, *126*, B. Taylor, K. Fujioka, et al., pp. 407-413, Ocean Drilling Program, College Station, TX, 1990b.

Taylor, B., and Hayes, D. E., The tectonic evolution of the south China basin, in *The Tectonic and Geologic Evolution of Southeast Asian Seas and Islands, Geophys. Monogr. Ser.*, vol. 23, edited by D. E. Hayes, pp. 89-104, Am. Geophys. Un., Washington, D.C., 1980.

Taylor, B., and D. E. Hayes, Origin and history of the south China basin, in *The Tectonic and Geologic Evolution of Southeast Asian Seas and Islands, part 2, Geophys. Monogr. Ser.*, vol. 27, edited by D. E. Hayes, pp. 23-26, Am. Geophys. Un., Washington, D.C., 1983.

Tosha, T., and Y. Hamano, Paleomagnetism of Tertiairy rocks from the Oga peninsula and the rotation of the northeast Japan, *Tectonics*, *7*, 653-662, 1988.

Tsunakawa, H., Neogene stress field of the Japanese arcs and its relation to igneous activity, *Tectonophysics*, *124*, 1-22, 1986.

Uyeda, S., Subduction zones: an introduction to comparative subductology, *Tectonophysics*, *81*, 133-159, 1982.

Uyeda, S., and H. Kanamori, Backarc opening and the mode of subduction, *J. Geophys. Res.*, *84*, *B3*, 1049-1061, 1979.

Vallier, T. L., Volcanogenic sediment and their relation to landmass volcanism and sea floor continents movements, western Indian ocean, Leg 25, DSDP, in *Init. Repts. DSDP*, *25*, edited by R. S. J. Simpson, R. Schlich, pp. 515-542, U.S. Govt. Printing Office, Washington, D.C., 1974.

Vogt, P. R., Global magmatic episodes: new evidence and implications for the steady state mid-oceanic ridge, *Geology*, *7*, 93-98, 1979.

Weissel, J. K., Evidence for Eocene oceanic crust in the Celebes Basin, in *The Tectonic and Geologic Evolution of Southeast Asian Seas and Islands, Geophys. Monogr. Ser.*, vol. 23, edited by D. E. Hayes, pp. 37-47, Am. Geophys. Un., Washington, D.C., 1980.

Westercamp, D., Magma generation in the Lesser Antille: geological constraints, *Tectonophysics*, *149*, 145-153, 1987.

Wolfe, J. A., Philippine geochronology. *J. Geol. Soc. Philipp.*, *35*, *1*, 1-30, 1981.

Wolfe, J. A., Arc magmatic and mineralization in North Luzon and its relationship to subduction at the East Luzon and Manilla trenches, *J. Southeast Asian Earth Sc.*, *2*, 79-93, 1988.

Yagi, K., Y. Kawano, and K. Aoki, Types of Quaternary volcanic activity in northeastern Japan, *Bull. Volcanol.*, *26*, 223-235, 1963.

H. Cambray, LMF, ODP Institut Méditerranéen de Technologie, 13451 Marseille cedex 20, France.

M. Pubellier, LGS, UPMC Paris VI, T26 E1, BP 129, 4 Place Jussieu, 75252 Paris cedex 05, France.

L. Jolivet, Dpt de Géologie, ENS, 24 rue Lhomond, 75005 Paris, France.

A. Pouclet, LGP, Université d'Orléans, BP 6759, 45067 Orléans cedex, France.

Evolution of the Lau Basin—Insights from ODP Leg 135

James W. Hawkins

Geological Research Division, Scripps Institution of Oceanography, La Jolla, California, 92093

Subduction at the Tonga Trench gives rise to arc volcanism and plays an active role in forming the Lau (backarc) Basin. Ocean Drilling Program drilling, together with GLORIA imagery, and intensive studies of backarc spreading centers have given us new insight into the evolution of the Lau Basin. Many of the features and processes appear to be common to other western Pacific backarc basins. Among the more significant new findings are recognition that the basin opened in a two-stage process. Initially, crustal extension and rifting formed small sub-basins that were partly filled with basaltic flows and sediments. The second stage of opening involves seafloor spreading from propagating rifts. Although the magma source is in a supra-subduction zone setting, the petrology of the crust indicates that the predominant source of melt has been mantle similar to the MORB-source. Evidence for mixing with melts, on a small scale, shows a contribution from a subduction component (i.e., similar to melts of arc-source mantle). As the basin opened, arc-composition, intra-basin edifices formed alongside the main production of MORB-like crust. A highly heterogeneous source is implied. Isotope data show that two distinct MORB-source mantle components were involved. Older units came from a "Pacific" source, whereas modern basalts come from mantle with an "Indian" source character. Parts of the northern basin carry the isotope signature of OIB source and helium isotope data suggest an influx like that of the Samoan "plume-source".

INTRODUCTION

Crustal recycling from mid-ocean ridges into subduction zones has been a major factor in earth differentiation and crustal evolution since Archaean time. The main site of formation of new oceanic crust is at the global mid-oceanic ridge system. New crust is also formed at convergent plate margins in suprasubduction zone (SSZ) settings where mantle-derived melts are transferred to the arc, forearc, and backarc systems, and oceanic sediments and fragments of oceanic crust are accreted. Most of the rock series known as ophiolites probably have their origin in a SSZ environment. Thus, even though arc-backarc-forearc systems may constitute only one-tenth of the annual volume of new oceanic crust formed, they may be fundamentally the most important sources of crust that eventually is accreted to existing continental nuclei.

The western Pacific basin is rimmed by an array of arc-trench-backarc systems, and inactive "fossil" remnants of older systems, that mark the location of active, or former, convergent plate margins. *Alfred Wegener* [1929] first called attention to the patterns of these systems and speculated that they had formed by extensional processes that rifted the east Asian margin. The first discussions of the geology of these systems in the light of modern data appeared in the early 1970s [cf., *Karig*, 1970, 1971; *Packham and Falvey*, 1971; *Moberly*, 1972]. We now know that many of the western Pacific convergent margin systems, especially those younger than about 40 Ma, formed in intra-oceanic settings and have very minor or no remnants of continental crust.

The recently concluded western Pacific drilling program focused on convergent margin systems. The results have given new insight to the development, especially the early stages, of the extension, rifting and magma-genesis in SSZ settings. This paper summarizes the geology of the Lau Basin and synthesizes our present understanding of backarc

Active Margins and Marginal Basins of the Western Pacific
Geophysical Monograph 88

basin evolution. In particular it evaluates the data in terms of the insights gained from drilling on the Lau Basin - Tonga arc transect of Ocean Drilling Program (ODP) leg 135. Three major problems related to SSZ evolution are addressed. These are (1) the tectonic style and the dynamic processes important in their early evolution, (2) the nature of the mantle sources for backarc basin magmas, and (3) the petrogenetic relations between backarc, arc and forearc magmatism in the SSZ environment.

GEOLOGIC SETTING

Convergent plate margins appear to be the locus of crustal shortening. However, the geology of most western Pacific SSZ systems indicates that the entire region between remnant arc and outer trench swell has geologic features indicative of extension as shown in Figure 1.

Two major geologic settings for western Pacific backarc basins may be recognized. These may represent two different stages in maturity of a common process. The main focus of this discussion will be the *intra-oceanic* settings where Pacific plate lithosphere is subducted beneath an intra-oceanic volcanic island arc chain. A forearc basin separates the trench from the volcanic arc and there is an extensional backarc basin and a remnant arc on the "lee" side of the volcanic arc. The Mariana and Tonga Trench convergent margins are type examples of these immature, intra-oceanic systems. They lack evidence for remnants of continental crust and are inferred to be entirely oceanic in origin. Tectonic erosion of the inner trench wall appears to have been important in both systems. A second type of backarc basin or marginal sea comprises seafloor that includes fragments of extended continental crust as well as newly formed seafloor of oceanic character. Examples of this type of basin, such as the Okinawa Trough [*Sibuet et al.*, 1987] and Sea of Japan [*Pisciotto et al.*, 1992] show a progression from extended, rifted and thinned continental crust, to extensional basins floored with basalt, to basaltic crust formed by seafloor spreading. The rifted continental crust has sediment filled basins separated by intervening horsts of continental crust. This type of extensional margin may be summarized as a continent/ocean convergent margin that failed in extension rather than developing into an Andean type margin. The evolution of these extensional systems will not be discussed here.

This summary is concerned with the geology of the Lau Basin but the inferences drawn about its evolution may be important for the North Fiji Basin, Mariana, and Izu-Bonin systems as well.

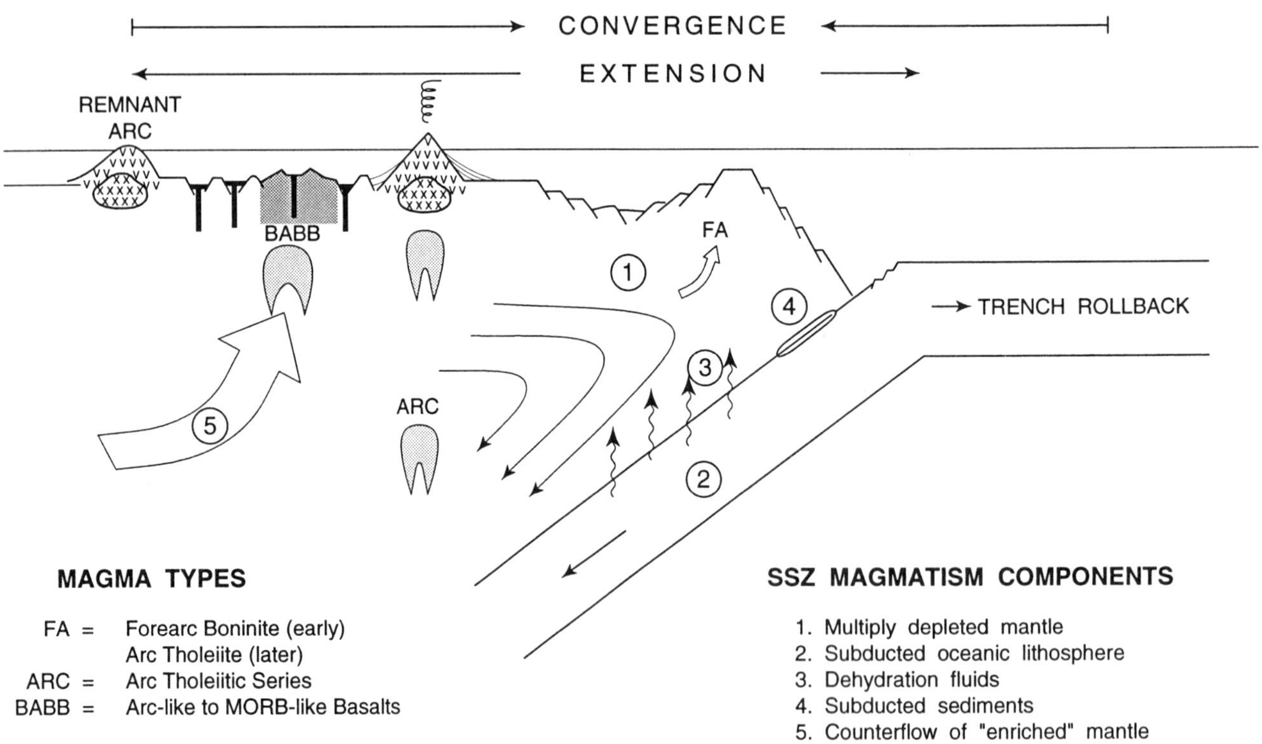

Fig. 1. Schematic cross-section of an intra-oceanic convergent margin showing the remnant arc (RA), backarc basin, active arc and forearc (FA). Sites of magmatism are shown by arrows and shaded crescentic patterns. Potential components that may contribute to SSZ magmatism are numbered.

TECTONIC CONTROLS AND PHENOMENA

Tectonic Setting

SSZ arcs and backarc basins are situated above the inclined seismic zone (Figure 1). The seismic zone trace (Figure 2) is presumed to mark the location of the subducted oceanic lithosphere but not necessarily its trajectory. The western Pacific seismic zones typically lie about 150 km beneath the arc volcanoes and from 200 to 300 km beneath the axial zone of the backarc basin *[Isacks and Barazangi*, 1977; *Giardini and Woodhouse*, 1984, 1986]. Exceptions to this generalization are found at the Mariana Trench and the North Fiji Basin. The seismic zone is about 150 km beneath the trace of the Mariana arc, but it steepens sharply and does not lie beneath the axial zone of the Mariana Trough [*Isacks and Barazangi*, 1977]. The (backarc) spreading axes of the North Fiji Basin lie between the opposing seismic zones of the New Hebrides and Tonga Trenches but there is no indication of a deep seismic zone beneath the axial ridges [*Isacks and Barazangi*, 1977]. This suggests that backarc spreading does not require the presence of a subducted slab *beneath* the spreading system. However, the spatial association between arc-backarc systems and zones of plate convergence argue for a cause/effect relation between subduction and the tectonic/petrologic processes that form crust in SSZ settings.

The mantle above the Tonga subduction zone (Figure 2) exhibits strong seismic wave attenuation (low Q) which is attributed to high temperature mantle and perhaps partial melting [*Barazangi and Isacks*, 1971]. There are no data to show whether or not this is common to all backarc basins, but it may be a fundamental factor both for the crustal extension mechanism and for controlling melt chemistry in the region of extension. The Tonga-Lau system has high heat-flow as shown by measurements [*Sclater et al.*, 1972] and by the general high-standing seafloor. Axial zones of spreading ridges in the Lau Basin rise to about 2200 meters depth and the average basin depth is on the order of 2500 to 3000 meters [*Hawkins*, 1974]. This differs from the Mariana

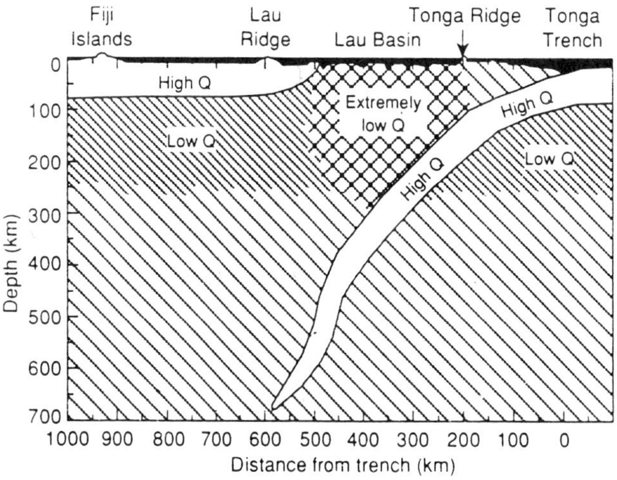

Fig. 2B. Cross-section from the Fiji Islands to Tonga Trench region showing inferred and extrapolated extent of high and low attenuation regions in the uppermost mantle. Note the extremely low Q (high attenuation) mantle beneath the Lau Basin, [*Barazangi and Isacks*, 1971].

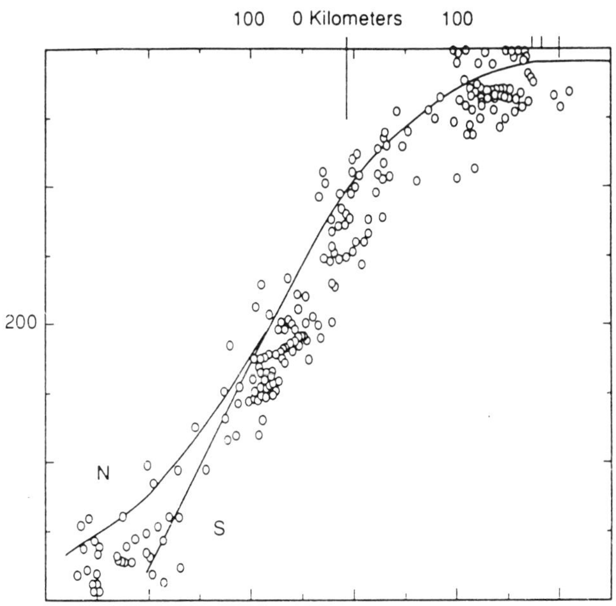

Fig. 2A. Composite cross section of earthquake hypocenters for the Tonga subduction zone. The "0 kilometer" mark on the horizontal axis designates the Tofua arc. Vertical lines are projected locations of the trench axis. Scale is in kilometers [*Isacks and Barazangi*, 1977].

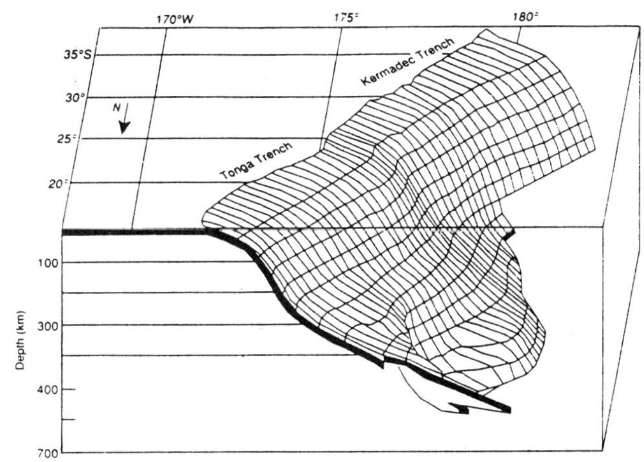

Fig. 2C. View to the south of a grid representing the upper surface of the Wadati-Benioff zone of the Tonga-Kermadec subduction zone. The projection is slightly distorted in that it does not take into account the Earth's sphericity [*Billington*, 1980].

Trough in which much of the axial ridge is deeper than 3500 meters and depths of 4500 meters are common in off-axis areas [*Hawkins et al.*, 1990]. This may reflect a contrast in the existence or scale of mantle upwelling under the Mariana Trough. Detailed surveys of the North Fiji Basin show highly variable ridge crest bathymetry on the <3 Ma axial ridge that ranges from depths less than 2000 meters near the North Fiji triple junction to more than 2900 meters depth on the southern extremity of the axial zone [*Lafoy et al.*, 1990]. The North Fiji Basin axial ridge is not located above the trace of a seismic zone.

The well-defined Tonga seismic zone (Figure 2) has a flexure or imbrication at about 525 to 575 km depth [*Billington*, 1980; *Louat and Dupont*, 1982]. Reconstructing the subduction history of this inflection suggests that there may be a connection between the time of initiation of this inflection and the beginning of opening of the Lau Basin [*Hawkins*, 1994]. The imbrication may have caused an increase in the velocity of subduction and started backarc extension [*Furlong et al.*, 1982].

On a global scale, convergent plate margins are zones of lithosphere shortening between remnant arcs and the incoming oceanic plates. Thermal controls on mantle rheology, the dynamic response of the mantle to stresses set up by subduction, and the geometry of the subduction system all must play a role in causing the SSZ crust to deform by regional extension. Several factors may be important in causing extension. Mantle upwelling above the inclined seismic zone, and seaward "rollback" of the trench axis [*Elsasser*, 1971; *Uyeda and Kanamori*, 1979], probably are major controls. Mantle diapirism above the seismic zone may be a passive response to the seaward rollback or it may be driven by the subduction process. In addition to generating magmas as a result of decompression melting, the rising diapirs displace the mantle wedge seaward. This must exert a downward force on the incoming lithosphere plate. Rollback may be related to the age, thickness, velocity and convergence vector of the subducted plate. Rollback may be possible only if there is a lack of transfer of stress from the incoming plate to the overriding plate [*Kanamori*, 1977]. The Mariana and Tonga Trenches are examples of subduction systems for which there is little or no evidence for significant stress transfer. They are characterized by earthquakes having relatively low stress-drop and typically these are of lesser magnitude than high-stress drop earthquakes [*Kanamori*, 1977]. Sediment supply to the trench, either seafloor sediments or detritus from nearby sources, also may be important in determining whether or not there is stress transfer.

The Initiation and Early Evolution of Backarc Basins

The western Pacific basin preserves a record of more than 40 m.y. of arc evolution and backarc basin extension that has been contemporaneous with subduction throughout most of Cenozoic time. The locations of the major western Pacific trenches (i.e., Izu-Bonin, Mariana, Yap-Palau, Vitiaz, Tonga, Kermadec) have migrated eastward. A sequential array of arcs and backarc basins formed more or less concentric with the trench systems and tracked their migration. Remnants of arc and backarc systems that constitute many of the terranes accreted to western North America suggest that in Paleozoic and early Mesozoic time the eastern Pacific basin may have looked much like the present day western Pacific. Is there a pattern of evolution to be discerned in this wealth of information?

The initiation of plate convergence at intra-oceanic trench systems, and the more or less synchronous backarc basin extension, are consequences of plate reorganization. These are inferred to be lithosphere responses to controls exerted by mantle rheology and global scale perturbations in mantle convection. The oldest arc-backarc systems in the northwestern Pacific, for example the Paleocene/Eocene age Palau-Kyushu Ridge and the West Philippine Sea, reflect the major change in plate motion at about 43 Ma recorded by the bend in the Emperor-Hawaii seamount chain. However, subduction and backarc basin generation may have been preceded by an earlier phase of *rifting and extension* that developed the boninitic magmas of the Izu-Bonin-Mariana system that are at least as old as 48 Ma [*Stern and Bloomer,* 1992].

The sequence of tectonic styles for southwestern Pacific arc systems parallels those of the northwestern Pacific but is even more complex. The beginning of evolution of these systems closely followed the beginning of seafloor spreading between Australia and Antarctica at about 83 Ma [*Cande and Mutter*, 1982]. The Tasman Sea began to open in late Cretaceous time (86 Ma) by rifting of the Lord Howe Rise from eastern Australia [*Hayes and Ringis*, 1973]. There are widely distributed Eocene boninitic series rocks extending from Luzon (then in the southern hemisphere) to Fiji that may reflect extensional tectonism. This early stage of rifting in these arc systems may be a continuation of earlier extension that caused continental breakup. Arguably, southwestern Pacific arc and backarc basin formation related to subduction began in Paleocene time; the New Caledonia Basin probably is one of the first expressions of subduction-related backarc basin development. There are many data, discussed in a later section, that demonstrate the role of Cenozoic subduction in the development of an array of arc and backarc systems that resemble those of the northwestern Pacific.

PETROLOGY OF BACKARC BASINS

Backarc basin crust ranges in composition from MORB-like basalts to rocks showing variable affinity to arc tholeiite. MORB-like varieties appear to be the most prevalent type erupted on modern backarc axial ridge systems but most backarc basalts are not truly equivalent to MORB in all respects. In this paper data for Lau Basin samples will be used to develop a general model for backarc petrology. Sample locations are shown in Figures 3 and 4 and data are in Tables 1-4. Major element data are compared with "true" MORB and with Mariana Trough backarc basin basalts in Figures 5 and 6.

SSZ magmas show variable depletions relative to MORB in high field-strength elements (HFSE), especially Nb and Ta. However, they have rare earth element (REE) abundances and chondrite normalized patterns typical of MORB; ratios such as $(La/Sm)_N$ are less than 1 in least fractionated samples. Many also show a variable enrichment relative to MORB in large ion lithophile elements (LILE) although this is not always the case. Even so, LILE/LILE ratios such as K/Rb, K/Ba, Ba/La commonly are MORB-like rather than arc-like. These relative LILE enrichments and HFSE depletions characteristic of backarc basin basalts are shown in Figure 7, which shows data for some of the least fractionated samples. The relative depletion in HFSE, especially for Nb and Ta, and enrichment in LILE is considered to be a reflection of their derivation from SSZ sources. These characteristic SSZ signatures are believed to be imparted by the multiply depleted MORB-source mantle (Figure 1) that has been variably enriched with subduction-derived fluids distilled from the altered oceanic crust and, possibly, from subducted sediments [e.g., *Pearce et al.*, 1984]. In addition, the SSZ setting is influenced by counterflow of "new" mantle which is driven into the mantle wedge region by subduction and trench roll-back [e.g., Hawkins et al., 1984; Hawkins and Melchior, 1985]. During the early evolution of SSZ convergent margins the effect of counterflow may be minimal. The Lau Basin data suggest that it may become the most important source as some backarc basins mature [e.g., *Hergt and Hawkesworth*, 1994; *Ewart et al.*, 1994a; *Hawkins*, 1994]. The question remains as to whether or not counterflow plays a role in backarc basins such as the North Fiji Basin where there is no subjacent seismic zone or the Mariana Trough where the trace of the subducted slab is not directly below the axial spreading ridge.

Perhaps one of the most distinctive features of backarc magmas is their elevated volatile content. The high volatile content is expressed in the relatively high vesicularity of nearly all samples of backarc crust. The elevated volatile contents are shown by data for the Mariana Trough [*Garcia et al.*, 1979; *Newman et al.*, 1989]; East Scotia Sea [*Muenow et al.*, 1980]; and Sumisu Rift, [*Ikeda and Yuasa*, 1989; *Hochstaedter et al.*, 1990a b]. *Dick* [1982] inferred high volatile contents for the highly vesicular rocks of the Shikoku Basin. Lau Basin basalts all have water contents that tend to be high relative to MORB [*Hawkins and Melchior*, 1985; *Hawkins et al.*, 1990]. Water is the major constituent of Mariana Trough glasses, as determined by measurement, but CO_2 is also important although the ratio H_2O/CO_2 is high [*Garcia et al.*, 1979]. There are limited data for combined H_2O, CO_2, and S in Lau Basin basaltic glasses; the reported range is 0.42-1.55%, 0-91 ppm, and 310-1880 ppm respectively [*Nilsson*, 1993; *Nilsson and Peach*, 1993]. The most likely source of these volatiles is subducted oceanic lithosphere.

The elevated water content of backarc basalts, and by implication their mantle source, is reflected in the common tendency for many backarc basalts to have higher Al_2O_3 and lower FeO* and TiO_2 than MORBs with comparable MgO [see Figures 5 and 6, and *Fryer et al.*, 1981; *Sinton and Fryer*, 1987]. These compositional features are attributed to the suppression or delay of plagioclase crystallization and to the early formation of Fe-Ti oxides. Many, but not all, backarc samples also show elevated Na_2O in comparison to MORB. This may be related to a relatively higher Na content in the source (related to the subduction component?), relatively lesser extent of melting, or to melt /conduit interaction over an ascent path relatively longer than for MORB settings.

Island arc magma systems commonly show evidence that their magma sources have been influenced by altered ocean crust, subducted sediments, or fluids derived from them. Not surprisingly, some backarc magma systems also carry at least some evidence of subduction component signature, (e.g., elevated water content). However, the LILE enrichments in Figure 7 are far less than for typical arc samples. The LILE generally show relative enrichments in the order Rb>K>Ba>La; the few data for Cs commonly show that it is more enriched than Rb. This hierarchy is expected from the mobility in aqueous solutions predicted from ionic potential and is verified experimentally [*Tatsumi et al.*, 1986; *Tatsumi*, 1989]. Dehydration of the subducted crust provides a source of hydrous fluids enriched in elements with high ionic potential. This contribution probably is made at depths less than about 90 km (approximate maximum depth limit of amphibole stability). The dehydrated sediment residue may be subducted to deeper levels and eventually may melt and contribute its trace element signature to mantle-derived melts [*Cousens et al.*, 1994].

The LILE enrichment thus is a measure of the contribution of a subduction component to the SSZ source, whereas the HFSE depletion is the mark of the multiply-depleted MORB-source mantle of the SSZ wedge.

GEOLOGY OF THE LAU BASIN

Introduction

The Lau Basin, like other western Pacific backarc basins, is but one element of an integrated SSZ tectonic/petrologic system that includes the subduction zone, active volcanic arc and the substrate upon which it is built, a remnant arc and the backarc basin [*Hawkins*, 1974]. The backarc basin

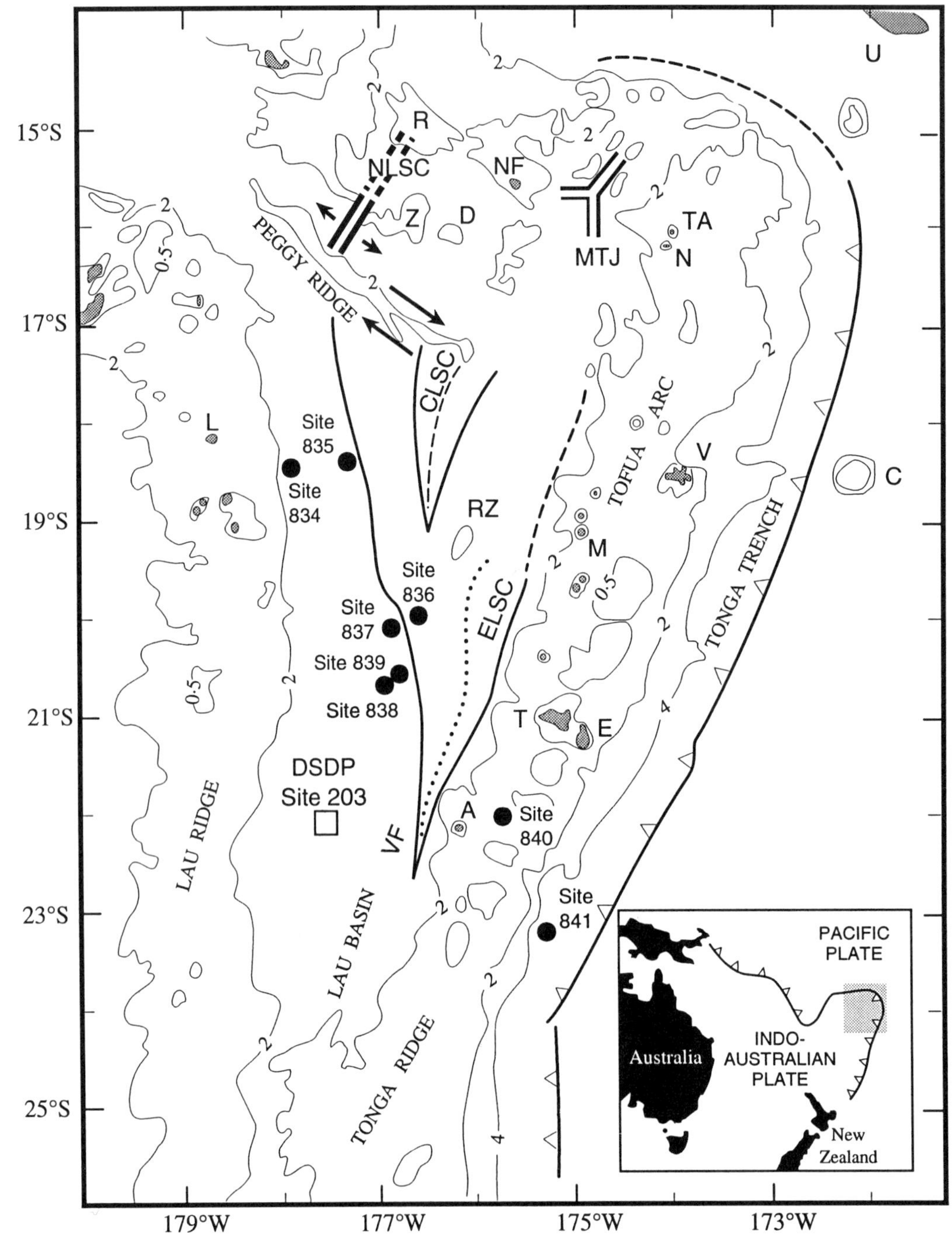

Fig. 3. Lau Basin and Tonga Trench-Arc system showing locations of ODP Leg 135 sites 834-841 and DSDP site 203. Major features shown are Mangatolu triple junction (MTJ); central Lau spreading center (CLSC), eastern Lau spreading center (ELSC), relay zone (RZ), and Valu Fa ridge (VF). Islands and shoals are Ata (A), Capricorn Seamount (C), Donna Seamount (D), 'Eua (E), Lakemba (L), Metis Shoal (M), Niuataputopu (N), Niuafo'ou (NF), Rochambeau Bank (R), Tongatapu (T), Tafahi (Ta), Upolu, Western Samoa (U), Vava'u (V), Zephyr Shoal (Z). Contour interval in kilometers.

(Figure 3) is a trapezoidal area of seafloor that separates the 1100 km long Tonga Ridge from the Lau Ridge (remnant arc). The Tonga Ridge is a composite feature comprising an eastern belt of uplifted blocks of Tertiary platform carbonates that form shoals and islands, and a western belt of young volcanoes, the Tofua arc, that form seamounts, shoals and islands. The eastern belt of platform carbonates overlie a crystalline basement comprising middle Eocene to

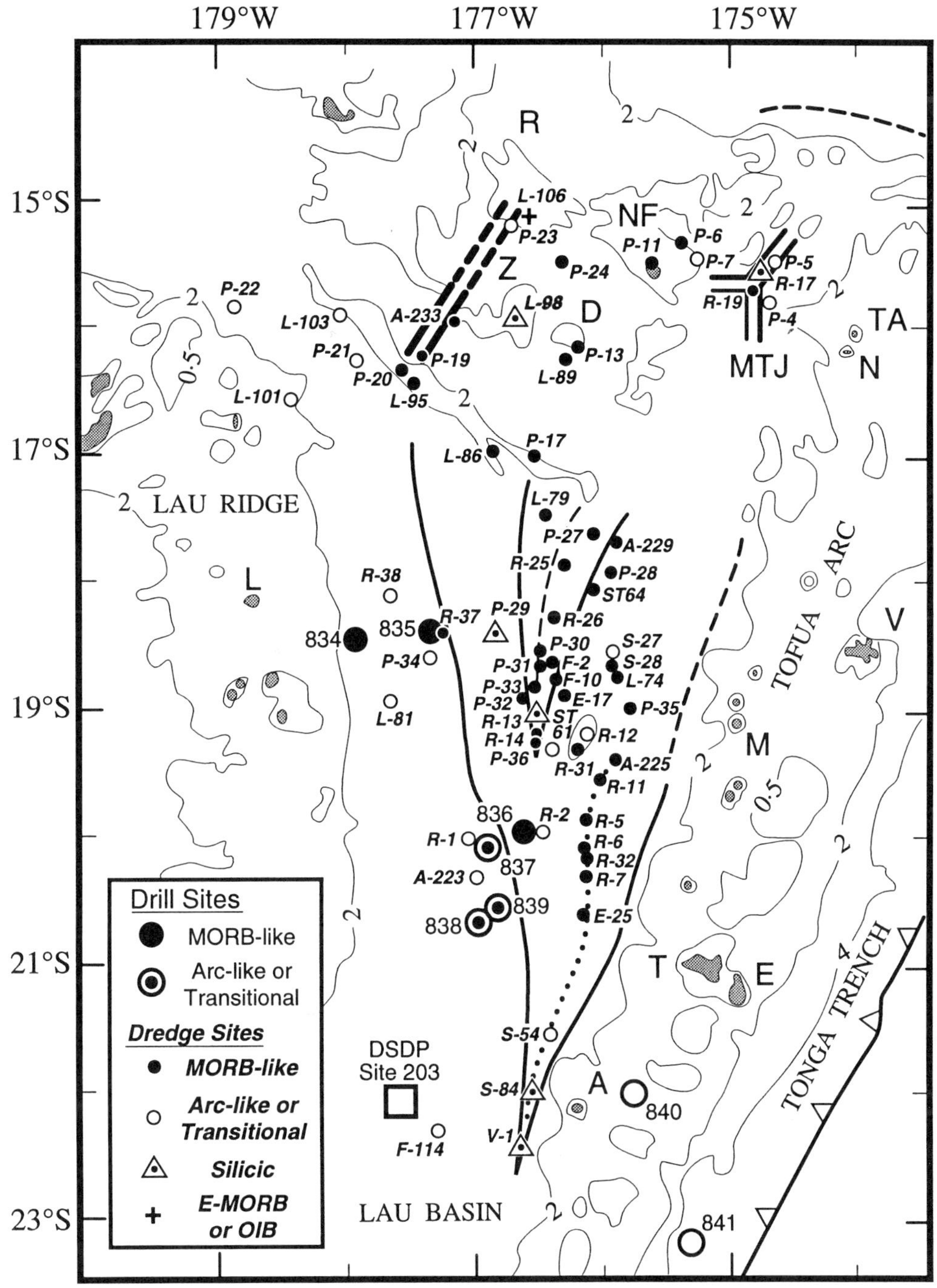

Fig. 4. Locations of dredge sites referred to in text and on tables. Dredge site acronyms are abbreviated as: A=ANT, L=123 [*Hawkins and Melchior*, 1985]; P=PPTU [*Hawkins*, 1988]; R=RNDB [*Hawkins*, 1989]; E= [*Ernewein et al., in* press]; F=[*Frenzel et al.*, 1990]; S=[*Sunkel*, 1990]; V=[*Vallier et al.*, 1991]. Numbers shown are the dredge site numbers.

late Miocene arc-composition volcanic and plutonic rocks. The oldest rocks known from the Tonga Ridge are 46 to 40 Ma beach boulders of arc tholeiitic hypersthene gabbro exposed on the island of 'Eua [*Duncan et al.*, 1985]. Rhyolitic tuffs, breccias and flows, with textures indicating subaerial or shallow water eruption, and dated at 44±2 Ma, were drilled at ODP site 841 on the Tonga Ridge some 225 km south of 'Eua [*Bloomer et al.*, 1994]. Miocene age volcanic and volcaniclastic rocks on the Tonga Ridge are known from island exposures, dredged samples, and drill core

TABLE 1. Representative Analyses, ELSC, CLSC, and Relay Zone

Location	ELSC							
Site	L-74	A-225	P-29	R-5	R-6	R-7	R-11	R-37
Sample	123	ANT	PPTU	RNDB	RNDB	RNDB	RNDB	RNDB
	74-1 GL	225-1 GL	29-3 GL	5-3 GL	6-1 GL	7-2 GL	11-1 GL	37-1RK
SiO_2	50.24	53.59	72.96	51.44	48.37	52.58	50.62	49.36
TiO_2	0.76	1.05	0.34	0.9	0.85	2.11	1.02	0.67
Al_2O_3	15.8	14.56	12.08	15.44	15.38	13.43	15/18	15.75
FeO*	9.13	11.55	4.56	9.39	7.95	14.76	9.54	8.73
MnO	0.21	0.21	0.13	0.19	0.17	0.21	0.18	0.27
MgO	7.39	5.27	0.12	8.02	6.21	3.99	7.51	8.49
CaO	13.37	10.22	2.21	12.87	11.01	8.55	12.83	12.53
Na_2O	2.1	2.34	2.23	1.91	2.09	3.24	1.9	3.05
K_2O	0.11	0.17	1.18	0.03	0.16	0.14	0.13	0.14
P_2O_5	0.09	0.11	0.02	0.06	0.11	0.2	0.12	0.08
Sum	99.2	99.07	95.83	100.26	99.21	99.21	99.04	99.07
Mg#	62.9	48.3	5.4	62.9	61.6	35.6	61.8	66.6
CaO/Al_2O_3	0.85	0.702	0.18	0.83	0.72	0.64	0.85	0.8
Trace Elements (ppm)			*					
Cr	300	50		337			342	
Ni	94	33	3	110	62	8	103	138
Co		90		43			43	
V	282	359	13	307	277	502	338	240
Zr	40	45	201	53	50	127	57	43
Y	21	26	82	27	31	55	26	18
Nb	<3		1	1	1	1	1	
Hf				1.16			1.55	
Ta				<.05			0.08	
Rb	2	0.5	31	1	5	4	6	2
Ba	10.8	32.5	143	4	18	27	8	
Sr	166	98	113	69	118	95	111	185

GL=glass data
other data for aphyric rocks
*trace elements from aphyric rock

TABLE 1. (continued)

Location	ELSC						
Site	E-25	S-54	S-84	V-1	ODP 836 A	ODP 836 B	ODP 836 B
Sample	Darwin-33	Sonne	Sonne	Lee	Unit 3 RK	Unit 4 RK	Unit 5 RK
	25-5-1	GC 54-1	KD 84-1	1-5	3H7/54-55	3R2/32-38	7R2/56-62
SiO_2	51.3	52.4	65.5	56.69	47.13	49.94	48.08
TiO_2	0.96	0.7	0.64	1.38	0.69	0.79	0.72
Al_2O_3	15.93	15.77	13.4	14.57	19.18	16.11	16.21
FeO*	9.28	8.07	6.69	11.51	8.08	8.27	8.13
MnO	0.17	0.16	0.16	0.22	0.16	0.12	0.14
MgO	6.5	7.78	1.34	3.73	8.88	7.68	9.07
CaO	11.29	12.5	4.54	7.65	13.03	14.85	13.78
Na_2O	2.11	1.79	4.2	3.36	1.9	1.96	2.33
K_2O	0.17	0.2	0.73	0.52	0.01	0.15	0.01
P_2O_5	0.08	0.08	0.14	0.16	0.14	0.06	0.14
Sum	97.79	99.45	97.34	99.2	99.2	99.93	98.59
Mg#	58.9	66.4	31.3	39.9	69.3	65.6	69.6
CaO/Al_2O_3	0.71	0.79	0.34	0.52	0.68	0.92	0.85
Trace Elements (ppm)							
Cr	116	216	16	4		318	
Ni	55	69	4	25	111	139	120
Co	35	38	12				
V	376	258	54	350	216		157
Zr	47	36	141	81	47	53	44
Y	23	21	58	36	19	18	20
Nb	0.69	<5	5		2	1	2
Hf	1.41			2.3			
Ta	0.04						
Rb	2.8	<3	12	8.6	0.1	3	2
Ba	40.7	44	120	96	17	14	19
Sr	119	116	128	168	140	288	319

GL=glass data
other data for aphyric rocks

TABLE 1. (continued)

Location				CLSC			
Site	ELSC	ELSC	ELSC	A-229	L-79	P-31	P-33
Sample	Mg#>60	Mg# 50-59	Mg# <50	ANT	123	PPTU	PPTU
	(Mean 20)	(Mean 7)	(Mean 2)	229 GL	79-1 GL	31-1 GL	33-1-GL
SiO_2	50.95	52.79	52.79	48.11	49.33	49.51	50.45
TiO_2	0.95	0.92	1.85	0.88	1.94	1.64	1.93
$Al_{22}O_3$	15.41	15.13	14.08	17.35	13.93	16	14.11
FeO*	9.36	10.47	13.62	9.02	13.27	9.76	13.29
MnO	0.18	0.19	0.21	0.17	0.28	0.18	0.25
MgO	7.52	6.52	4.28	8.73	6.64	7.56	5.41
CaO	12.64	11.47	8.9	12.56	11.74	11.33	10.17
Na_2O	2.03	1.92	3.03	2.59	2.59	2.66	2.85
K_2O	0.08	0.12	0.16	0.03	0.09	0.42	0.08
P_2O_5	0.08	0.08	0.19	0.05	0.14	0.15	0.13
Sum	99.22	99.61	99.09	99.49	99.95	99.21	98.67
Mg#	62.2	56.1	39.2	67	51.2	61.4	45.5
CaO/Al_2O_3	0.82	0.758	0.632	0.72	0.84	0.71	0.72
Trace Elements							
Cr						198	20
Ni						68	37
Co						46	44
V						364	455
Zr						68	128
Y						31	59
Nb						2	1
Hf						1.84	3.57
Ta						0.081	0.16
Rb						1	1
Ba						7	2
Sr						86	114

TABLE 1. (continued)

Location	CLSC				
Site	R-13	R-14	R-25	R-26	ST-64
Sample	RNDB	RNDB	RNDB	RNDB	STO
	13-2 GL	14-1 GL	25-1-GL	26-2 GL	64-3 GL
SiO_2	60.41	51.28	50.96	50.25	48.58
TiO_2	1.62	2.39	1.05	0.94	0.7
Al_2O_3	12.93	12.63	14.65	15.86	16.85
FeO*	12.16	16.1	10.6	9.79	9.54
MnO	0.22	0.25	0.2	0.19	0.19
MgO	1.99	4.64	7.6	8.25	8.93
CaO	6.15	9.12	12.76	13.06	12.77
Na_2O	3.91	3.12	2.26	2.05	2.42
K_2O	0.42	0.14	0.03	0.07	0.01
P_2O_5	0.32	0.2	0.07	0.08	0.03
Sum	100.13	99.87	100.17	100.53	100.02
Mg#	25.2	37.1	59.5	63.3	66.3
CaO/Al_2O_3	0.48	0.72	0.87	0.82	0.76
Trace Elements (ppm)					
Cr			303	325	430
Ni7	7	18	116	149	189
Co				43	70
V	147	469	332	323	202
Zr	441	165	45	51	33
Y	138	64	30	2929	20
Nb			1	2	2
Hf			1.89		
Ta			0.04		
Rb	10	3	0.1	0.3	0.5
Ba	75	31	2	4	6.2
Sr	86	91	49	63	62

*trace elements from aphyric rock

TABLE I. (continued)

Location	CLSC				Relay Zone		
Site					R-12	R-31	R-31
Sample	Mg#> 60 (Mean 5)	Mg# 50-59 (Mean 17)	Mg# 40-49 (Mean 10)	Mg# <40 (Mean 7)	RNDB 12-6 GL	RNDB 31-1 GL	RNDB 31-4 GL
SiO_2	49.15	52.04	50.81	53.76	53	51.44	49.25
TiO_2	0.99	1.24	1.94	2.07	0.57	1.21	0.79
Al_2O_3	16.97	14.46	13.41	13.01	16.2	15.31	17.36
FeO*	9.19	10.7	14.1	14.44	7.95	10.61	8.65
MnO	0.18	0.2	0.26	0.24	0.14	0.22	0.17
MgO	8.48	6.71	5.63	3.82	6.89	6.28	9.1
CaO	12.14	11.37	10.21	8.34	12.28	11.61	12.82
Na_2O	2.57	2.53	2.88	3.27	1.54	2.54	2.2
K_2O	0.11	0.14	0.12	0.26	0.32	0.08	0.03
P_2O_5	0.07	0.1	0.16	0.28	0.1	0.07	0.03
Sum	99.88	99.67	99.55	99.56	99	99.37	100.41
Mg#	65.4	56.2	45	35.2	64	54.8	68.3
CaO/Al_2O_3	0.715	0.786	0.761	0.641	0.758	0.758	0.738
Trace Elements (ppm)							
Cr							334
Ni					67	55	163
Co							44
V					271	326	212
Zr					66	83	68
Y					17	31	24
Nb					1	3	3
Hf							1.14
Ta							<0.05
Rb					6	1	1
Ba					29	7	3
Sr					197	122	93

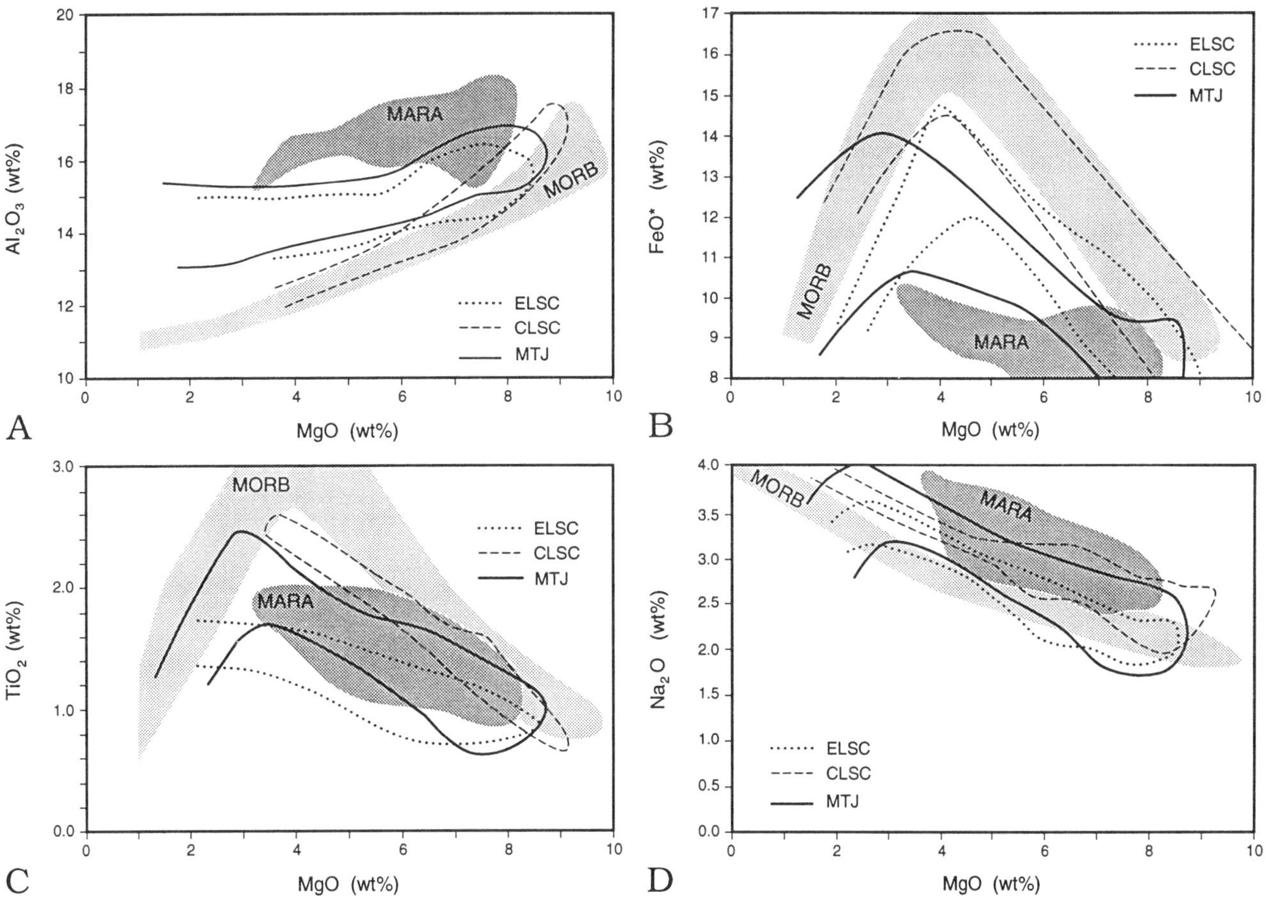

Fig. 5. Variation diagrams showing major elements vs. MgO. Fields shown are mid-ocean ridge basalt (MORB), [after *Hochstaedter et al.*, 1990a]; and Mariana Trough glasses (MARA), [*Hawkins et al.*, 1990]. Other fields are eastern Lau spreading center (ELSC); central Lau spreading center (CLSC); Mangatolu triple junction (MTJ). Data sources are discussed in the text.

data. Arc tholeiitic andesitic flows (33-31 Ma) and andesitic to silicic-andesite dikes (19-17 Ma) are found on 'Eua [*Duncan et al.*, 1985; *Hawkins and Falvey*, 1985]. ODP site 841 penetrated arc tholeiitic basalt and basaltic andesite dikes or sills intrusive into distal facies of late Miocene volcaniclastic turbidites [*Parson et al.*, 1992]. The interpretation is that the Eocene to late Miocene igneous rocks of the Tonga Ridge are parts of the former Lau Ridge volcanic arc and its forearc that were rifted eastward in the opening of the Lau Basin.

The Lau Ridge was an active volcanic arc from at least mid-Miocene time (14 Ma) until early Pliocene time(1.5-2.5 Ma) [*Gill*, 1976; *Cole et al.*, 1985; *Woodhall*, 1985]. The main rock types exposed on the islands of the Lau Ridge are basaltic andesite and andesite members of tholeiitic and calc-alkaline magma series; both low-K and high K-magma series are represented. An important aspect of Lau Ridge volcanism is that it continued during the initial stages of opening of the Lau Basin. A rifting stage is recognized on the Lau Ridge that coincides with the beginning of extension and rifting in the Lau Basin [*Whelan et al.*, 1985]. Isotopic data for the Lau Ridge magma series contrast with Tofua arc (modern) volcanoes in having relatively lower values for $^{87}Sr/^{86}Sr$ (0.7030-0.7033) which *Gill* [1976] attributes to a lesser subduction component than is present in the Tofua arc source.

The modern (Pleistocene to Recent) volcanoes of the Tofua arc have erupted arc tholeiitic series lavas that are mainly basalt and low-K basaltic andesite with lesser amounts of dacite and low-K rhyolite [*Bryan et al.*, 1972; *Ewart and Bryan*, 1972; *Ewart et al.*, 1973, 1977; *Bryan*, 1979a; *Ewart*, 1979; *Ewart and Hawkesworth*, 1987]. *Ewart and Hawkesworth* [1987] proposed that the arc series magmas were derived from a compositionally variable source in the SSZ mantle wedge that includes the residue from previous melting of an N-MORB source and less depleted MORB-source mantle. These sources have both been re-enriched with a subduction component. Tofua arc Nd and Sr

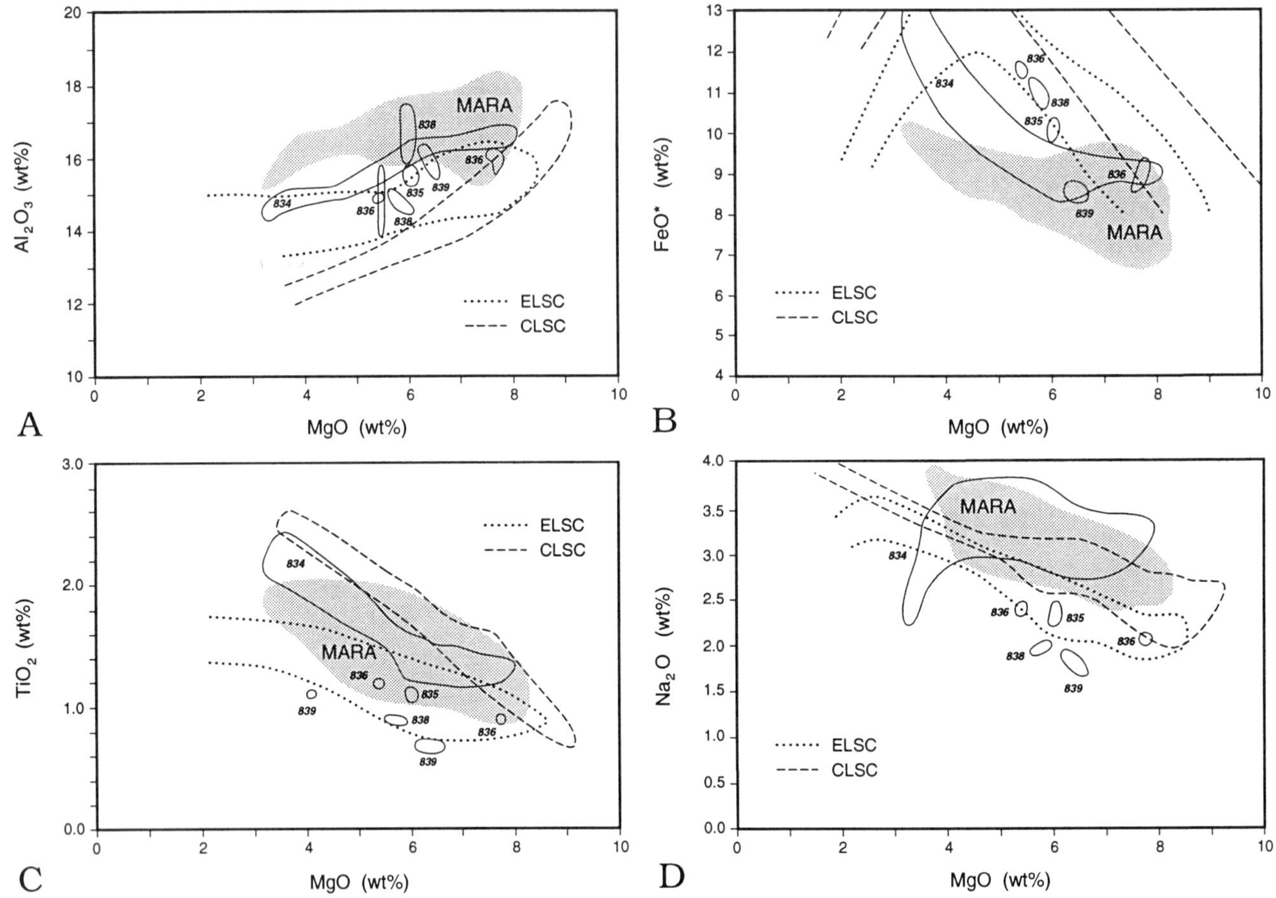

Fig. 6. Variation diagrams showing major elements vs. MgO. Numbered fields are for ODP leg 135 drill sites [*Hawkins*, 1994]. Other fields are as in Figure 5.

isotopic ratios lie within the mantle array, but the Sr isotope chemistry suggests that the arc lavas have more of a subduction component in them than is recognized in the Lau Ridge volcanic series (e.g., $^{87}Sr/^{86}Sr$ = 0.70361-0.70399, *Ewart and Hawkesworth*, 1987). The Pb isotope data for the Tofua arc vary along strike; samples from the northern islands of Tafahi and Niuatoputapu are significantly more radiogenic in ^{206}Pb, and to a lesser extent in ^{208}Pb, than the other volcanic islands. *Ewart and Hawkesworth* [1987] conclude that the evidence is more suggestive of mixing of OIB and MORB sources than MORB and sediment. The petrologic interpretations for the Tofua arc parallel the interpretations for the origin of Lau Basin backarc magmas.

The Lau Basin proper (Figure 3) has several constituent parts which include the active ridges, older extended and rifted crust with basin-range structure and sub-basins partly filled with lavas, seamounts, and uplifted blocks of older oceanic crust. Seafloor spreading forms new MORB-like crust at several sites [*Hawkins et al.*, 1970; *Sclater et al.*, 1972; *Hawkins*, 1974, 1976, 1977]. The eastern Lau spreading center (ELSC), and central Lau spreading center (CLSC), constitute two overlapping segments each the result of southward propagating rifts that originated from the trace of a major transform fault within the basin. In addition to the main north-south axis of spreading, there is a small ridge segment north of Peggy Ridge, the northwestern Lau spreading center (NLSC), and a three-limbed ridge system in the northeastern part of the basin, the Mangatolu triple junction (MTJ).

Tectonic Evolution

ODP drilling at six backarc basin sites, and sonar images made with GLORIA [*Parson et al.*, 1990], give us new insight to Lau Basin evolution. The most important finding is that the Lau Basin opened in two tectonic styles or stages. An initial phase of rifting was followed by seafloor spreading on propagating rifts. Initially, the *forearc* to the Lau Ridge was stretched, the crust was thinned, and a series of small, north-south trending, sub-basins opened. The crustal attenuation and rifting was accompanied by

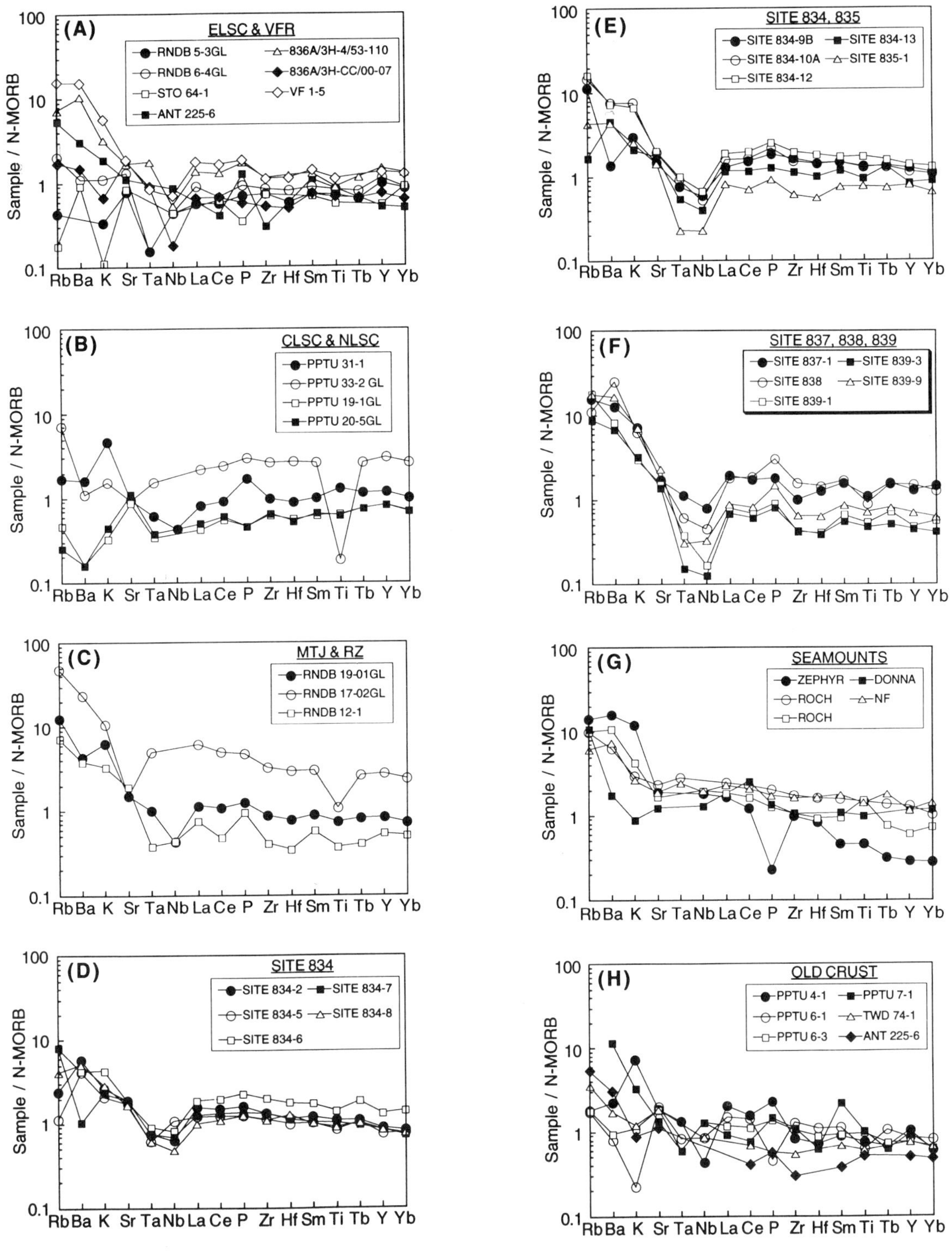

Fig. 7. Trace elements normalized to N-MORB values of *Sun and McDonough* [1989]. Note the relative depletion in Nb and Ta and enrichments in LILE that characterize SSZ magmas. A. Eastern Lau spreading center (ELSC) and Valu Fa ridge (VF). B. Central Lau spreading center (CLSC) and northwestern Lau spreading center (NLSC). C. Mangatolu triple junction (MTJ) and relay zone (RZ) between ELSC and CLSC. D., E., and F. are ODP drill sites [*Hawkins et al.*, 1994]. G. Seamounts in northern basin; ROCH is Rochambeau Bank, and NF is Niuafo'ou island. H. Older crust near spreading centers. PPTU samples are from northeastern Lau Basin near MTJ [*Hawkins*, 1988]. TWD 74-1 lies east of the CLSC and ANT 225-6 lies west of the ELSC [*Hawkins*, 1976; *Hawkins and Melchior*, 1985].

magmatism that partly filled the sub-basins as they opened. The sub-basins appear to have opened sequentially from west to east as indicated by the eastward younging of the *end* of magmatic activity in each. The intervening horsts may be largely remnants of old forearc crust, but at least some of them may be basaltic crust formed during the extension and rifting. Drill core data from ODP leg 135 show that basin rifting had begun by about 6 Ma [*Hawkins*, 1994]. The individual basins were sites of magmatic activity for periods as long as 1.8 Ma as indicated by interbeds of lava and fossiliferous sediments at ODP site 834 [*Parson et al.*, 1992; *Hawkins*, 1994]. The backarc basin magmatism was bimodal in composition; arc composition volcanoes formed within the basin contemporaneously with eruption of backarc basin basalts.

The initial rifting probably occurred mainly on the forearc of the Lau Ridge. There is no evidence that fragments of the active Lau Ridge volcanic arc were rifted away to the east. Clearly the initial rifting must have been close to the Lau Ridge arc edifice because some of the rifted material recognized on the present Tonga Ridge includes igneous rock that must have been formed at the Lau Ridge volcanic arc. These include distal facies of arc-derived volcaniclastic rocks and small intrusive dikes or sills. Pre-Oligocene crystalline basement and its coralline cover were rifted as well. The latter may represent part of the basement of the forearc to the Lau Ridge. It is important to note that the Lau Ridge continued as an active volcanic arc for at least 4 m.y. after rifting began. It is also important to emphasize that the present Tofua arc may be a much younger feature, superposed on or intrusive into older crust. It is not a petrogenetic descendant of the Lau Ridge.

The second stage of opening was due to seafloor spreading from a propagating rift that was initiated at about 4.5 to 5 Ma [*Parson and Hawkins*, 1994]. The propagator started from the trace of the Peggy Ridge (PR). Geometric and morphologic features argue that at that time the PR was a transcurrent fault probably with left-slip displacement. Subsequent to the beginning of spreading the PR has marked the location of a transform fault for which earthquake first-motion data indicates a near vertical right-slip fault plane [*Eguchi*, 1984; *Hamburger and Isacks*, 1988]. The southward propagating rift laid down crust from the ELSC giving rise to a coherent pattern for magnetic anomalies that presently extends as far south as about 22°50'S. A second propagator began on the Peggy Ridge at about 1.5 to 1.2 Ma and has overtaken the first. The second propagator formed the CLSC which overlaps the ELSC at about 19°22'S; between the two is a complicated relay zone (RZ) having pseudofaults, nodal basins and small discontinuous ridges (Figure 3).

Presently the Lau Basin is opening at about 124-144 mm/yr as determined by GPS measurements between Lakemba on the Lau Ridge and Vava'u [*Bevis et al.*, 1993; F. Taylor, personal communication, 1994]. The ELSC rift tip is estimated to have propagated southward at about 120 mm/yr and the CLSC probably at a similar rate [*Parson and Hawkins*, 1994]. Our new understanding of the two-stage opening of the Lau Basin helps to resolve the long-standing debate about the interpretation of magnetic anomalies in the Lau Basin [e.g., *Lawver et al.*, 1976; *Weissel*, 1977; *Lawver and Hawkins*, 1978]. The symmetric anomalies that were originally recognized [*Lawver et al.*, 1976] are associated with the ELSC and CLSC. The contentious anomalies are those from the sub-basins of the western Lau Basin which lack symmetric counterparts and cannot fit into a seafloor spreading interpretation. The postulated patterns with a triple junction [*Weissel*, 1977] simply do not exist. Detailed discussions of the morphology, tectonic elements, and tectonic evolution of the Lau Basin are given elsewhere [*Hawkins*, 1974; *Hawkins*, 1994; *Parson and Hawkins*, 1994].

Petrology and Chemistry

Introduction. There is a very large amount of data on the petrology of the Lau Basin; most of it comes from studies of the ELSC and CLSC. Results from ODP leg 135 complement these data with studies of crust that floors several sub-basins in the western extensional basin (WEB) morpho-tectonic province. This discussion will focus mainly on the petrology and chemistry of least fractionated rocks from the major spreading systems and compares them with drill site data. The goal is to define the Lau Basin magma series and the implications to the SSZ sources. More detailed discussions are presented elsewhere in references cited.

Eastern and central Lau spreading centers. The combined ELSC and CLSC extend for about 600 km through the Lau Basin in a generally north-south direction ending at the Valu Fa ridge segment of the ELSC. This ridge system has served as the main locus for new backarc crust formed by seafloor spreading for about 4.5 to 5 m.y. The extensive data base for the ELSC and CLSC offers insight to the petrologic evolution and the mantle sources for the melts that have erupted during spreading [*Hawkins*, 1974, 1976, 1977, 1988, 1989; *Hawkins and Melchior*, 1985; *Jenner et al.*, 1987; *Sunkel*, 1990; *Boespflug et al.*, 1990; *Vallier et al.*, 1991; *Falloon et al.*, 1994; *Ernewein et al.*, 1994]. The ridges carry the compositional stamp of a predominantly MORB-like source with a varied SSZ overprint, however each ridge has its own distinctive isotopic and chemical signature. Representative data for the ELSC and CLSC are in Table 1, and drill site data are in Table 2. Sample locations are in Figure 4. The modern spreading axes all are situated above the Wadati-Benioff zone, which lies at about 250 km depth beneath the ELSC and CLSC. Two secondary spreading ridge systems, the Mangatolu triple junction (MTJ) and the northwestern Lau spreading center (NLSC) are respectively about 250 km, and from about 300 to 350 km, above the seismic zone. This implies that the major source of the melts must be in the SSZ mantle, but there is a general MORB-like character to all of the basaltic samples.

Local variations in chemistry that point to variable affinity with arc composition magmas are attributed to variable mixing with a subduction component [*Hergt and Farley*, 1994; *Hawkins*, 1994], to assimilation of older crust [*Hilton et al.*, 1993], or to proximity to the Tofua arc [*Ernewein et al.*, 1994].

Petrology and chemistry of the ELSC and CLSC. Rocks from the ELSC and CLSC ridge system exhibit textures and mineralogy similar to rocks of mid-ocean ridges. Essentially they are glasses, aphyric, and variably-phyric tholeiitic basalts. Mineralogically they comprise plagioclase, olivine, clinopyroxene and Fe-Ti oxides. Cr-spinel and Fe-Cu sulfides are trace constituents. Plagioclase is the main mineral constituent. Its compositional range is typical of MORB (e.g., An_{60} to An_{85}). Some highly calcic grains are found in some of the rocks with compositions transitional to arc chemistry (e.g., phenocryst cores as calcic as An_{90}). Plagioclase is present and abundant in all samples; as discussed below, elevated P H_2O in parental melts may have delayed onset of plagioclase crystallization but it did not completely suppress it. Clinopyroxene occurs as a phenocryst phase as well as a groundmass constituent in nearly all samples with Mg# < 60. Both ilmenite and Ti-magnetite are present as minor constituents in the groundmass in more evolved samples.

The only textural feature that differs from typical MORB is their highly vesicular nature. Vesicle contents ranging from 5% to 50% are common in samples dredged from ridge depths as great as 2600 m [e.g., *Hawkins*, 1976]. Implications of the high vesicularity on volatile content and its effect on magma-genesis are discussed below.

The crust formed at the active spreading centers is MORB-like in most essential respects although there are minor but distinctive chemical compositional features which set some of the samples apart from N-MORB *sensu stricto*. The chemistry of representative rock types from the CLSC and ELSC axial ridges, and ELSC crust sampled at ODP site 836, is in Table 1.

Chemical distinctions between MORB and *some* of the axial ridge samples include elevated Al_2O_3, and lower FeO* and TiO_2 relative to MORB with comparable MgO (Figure 5). Both the ELSC and CLSC overlap with the MORB fields at high levels of MgO although the ELSC is displaced more towards the MARA (Mariana Trough basalt) field. The MARA data fields [*Hawkins et al.*, 1990] demonstrate element abundance patterns considered typical of the backarc basin basalt (BABB) signature [*Fryer et al.*, 1981]; they are clearly different from MORB especially for the more fractionated samples. The elevated Al may reflect higher water content and the lower Fe and Ti result from elevated fO_2 in parental melts. The more oxidizing environment relative to MORB may be due to higher water content. CLSC data are MORB-like but the ELSC is different. Samples with low MgO define trends shifted toward the Mariana Trough field for Al_2O_3. They lie parallel to the MORB field, but at a lower level, for FeO*. The TiO_2 field is subparallel to Mariana Trough but is depleted. The behavior of Al and Fe, and Ti in fractionated samples resembles the Mariana Trough (SSZ) trend [*Fryer et al.*, 1981; *Hawkins and Melchior*, 1985; *Hawkins et al.*, 1990]. The Na_2O field is MORB-like and ELSC glasses lack the relatively higher K content that distinguishes Mariana Trough (transitional to arc-type) basalts from MORB. Drill site data from ODP leg 135 are shown in Figure 6 and representative data are in Table 2.

Magmas that formed the northern part of the ELSC are mainly MORB-like tholeiitic basalt in terms of major and trace elements and element ratios [e.g., *Bryan et al.*, 1976; *Bryan*, 1979b; *Viereck et al.*, 1989]. Glasses range from relatively unfractionated (up to 8% MgO, Mg# 64, 0.9% TiO_2, <0.15% K_2O) to moderately fractionated (4-5% MgO and Mg# 43-50). There are lesser amounts of more strongly fractionated types including Fe-Ti enriched basalt (up to 2.1% TiO_2 and 14.8% FeO*). The range in composition of ELSC lavas is easily explained by low-pressure fractional crystallization of observed phenocryst assemblages. *Sunkel* [1990] described low-K tholeiitic basalts from the axial ridge south of 21°50' S which are depleted in HFSE, but enriched in LILE, relative to N-MORB (i.e., showing more of the SSZ component). Throughout its length, the ELSC shows low-level differences from MORB in trace elements. These are mainly subtle LILE enrichments and HFSE depletions that give variable similarity to arc tholeiitic magmas. These HFSE differences are seen in other backarc basins (e.g., Mariana Trough, *Hawkins and Melchior*, 1985; *Hawkins et al.*, 1990) and we consider it to be a reflection of the multiply depleted SSZ mantle source.

ODP site 836 is located close to the boundary between the western extensional basin province and probable ELSC crust. Five petrologic units are recognized. Units 1 and 2 (14-16 meters below seafloor, mbsf) are andesitic gravels probably derived from a nearby intra-basin seamount formed of arc-composition lavas. The other three are MORB-like basaltic units which lie in the data fields for ELSC on Figure 6; this supports the interpretation that they are genetically related to the ELSC and are some of the earliest ELSC crust formed at the latitude of the drill site. Their trace element and REE abundances also resemble the ELSC (Figures 7 and 8).

The southernmost part of the ELSC, south of 21° 20'S, is known as the Valu Fa ridge [*Morton and Sleep*, 1985]. This feature is at the propagating rift-tip and constitutes one of the youngest locations of backarc rifting in the Lau Basin. It is distinctive enough to deserve a separate discussion because it is capped with "oceanic andesite" having chemical and isotopic signatures resembling the nearby Tofua arc [*Vallier et al.*, 1991], an extensive hydrothermal field is found there, and because a small axial magma chamber has been recognized [*Morton and Sleep*, 1985]. The Valu Fa ridge (VFR) is a well defined, linear feature that extends for at least 165 km more or less parallel to the Tonga Ridge, and approaches to within about 40 km of the volcanic island of Ata in the Tofua arc. The VFR is propagating southward into older crust of uncertain composition. This crust may be equivalent to the rifted forearc terrane that forms the basin-range seafloor of the western Lau Basin.

The main rock types recovered from the VFR are highly vesicular, glass-rich blocky, tabular or bulbous andesites. The samples are mainly vitrophyric, or aphyric to sparsely phyric, fine grained rocks. Plagioclase is the main crystalline phase but it is rare. It forms both microphenocrysts (An_{87-77}) and microlites (An_{62}-An_{38}). Clinopyroxene is rare and there are very minor amounts of Fe-Ti oxides. The glasses are mainly low-K andesites with 54 to 60% SiO_2, 0.36-0.57% K_2O [*Davis et al.*, 1990; *Vallier et al.*, 1991]. Sunkel [1990] reported three samples of high silica dacitic glasses (65.5 to 69.8% SiO_2, 0.44 to 0.64% TiO_2, and 0.62 to 0.87% K_2O). Moderately fractionated basalt (MgO 5.66-6.07%) was described by *Davis et al.* [1990]. Representative analyses are in Table 1.

The axial magma chamber is unusual in that the top is about 2-3 km wide, flat, and relatively deep (3.5 km below seafloor) in comparison to those seen on mid-ocean ridges. The sharp well-defined seismic reflections indicate a large reflective coefficient. This may be related to the low density silica-rich melt (L. Dorman, personal communication, 1994). The flattened top may be related to the highly viscous siliceous magma.

ELSC magma chemistry appears to have remained fairly constant over at least 3 Ma as shown by the similarity between axial ridge samples and dredge samples from older parts of the ELSC. For example, using the model of *Parson and Hawkins* [1994], an approximate age of about 2 Ma can estimated for dredge site L-74 and about 3 Ma for dredge site R-37 (Figure 4). Rocks from these dredge sites are relatively unfractionated (Mg#=62.9 and 66.6 respectively). They have the Al, Fe, Ti characteristics of the Mariana Trough transitional basalts although they have low K_2O. They are similar in most respects to the modern lavas of the ELSC axial zone. The record of ELSC magmatism from about 3 Ma to the present suggests a fairly uniform long-term composition for the magma sources. The main variability in ridge chemistry seems to be expressed near the rift tip or as the rift approaches the Tofua arc (e.g. Ernewein et al., in press).

The CLSC formed by a propagating rift that penetrated crust originally formed by the ELSC (Figure 1). It was initiated at about 1.5 Ma. In general it comprises rock types like those for the ELSC and exhibits a similar range in composition. CLSC rocks have the chemical characteristics of MORB. Representative data are in Table 1. CLSC rocks range from relatively unfractionated basalts, represented by glasses with up to 8.8% MgO and Mg#=67.5, to more evolved Fe-Ti enriched basalt glasses with 4.8% MgO, Mg#=39.9, TiO=2.31%, and FeO*=14.7%. The trends on Figure 5 show that the CLSC samples are closer to MORB than are the ELSC samples, and overlap completely at high MgO. This may reflect a lower level of water in the source or a more extensive input from MORB-source mantle. Rocks from near the propagating rift tip include some low-K, high silica andesites (RNDB-13-2, Table 1) that probably are extreme fractionates formed as the rift penetrated older cold crust.

The compositional range for ELSC and CLSC samples may be explained in terms of low-pressure fractional crystallization of observed phenocryst assemblages. The spectrum of glass compositions for both the ELSC and CLSC lie close to the presumed low-pressure, anhydrous, multiply-saturated cotectic as projected from PL in the OL-CPX-QTZ diagram [*Hawkins and Allan*, 1994]. This supports the inference that fractionation may explain the compositional range.

Samples from the relay zone between the CLSC propagator and the dying north end of the ELSC have both transitional and MORB-like chemistry (Table 1). Two samples with similarly high Mg# are compared. Sample RNDB 12-1 shows the SSZ signature of depletion in HFSE, and elevated LILE (e.g. 0.3% K_2O, 24 ppm Ba), whereas RNDB 31-4 has HFSE somewhat depleted relative to MORB but lacks the elevated LILE (K_2O= 0.03%, 7 ppm Ba).

Least fractionated samples from the CLSC show only minor variation in trace element chemistry along strike whereas the ELSC shows much more variability. This may be related to different effects of crustal assimilation. The CLSC propagated southward into crust with broadly similar (MORB-like) composition that had been (recently) formed by the ELSC. The variations in sample chemistry along strike may be explained largely as a result of crystal fractionation and proximity to the cooled rift tip. The ELSC propagator encountered much older crust much of which probably is remnants of Lau Ridge forearc. Assimilation of this older crust may have been an important factor in developing more arc-like magmas. This idea is supported by helium isotope data [*Hilton et al.*, 1993] discussed in a separate section. An alternate explanation is that the north-south variation reflects increasing proximity to the trace of the Tofua arc [*Ernewein et al.*, 1994].

Mangatolu triple junction. Three well-defined axial ridges, which meet at a triple junction (Figure 3), have been identified in the northeastern part of the Lau Basin. Samples from the ridges include basaltic pillow fragments with glass-rich chill margins, and blocky fragments of silica-rich vitrophyric material of broadly andesitic composition. Fragments of a sulfide enriched chimney, comprising wurtzite and sphalerite, with lesser amounts of chalcopyrite and pyrite, in a matrix of amorphous silica and barite were dredged from the northeastern limb [*Hawkins*, 1986; *Hawkins and Helu*, 1986].

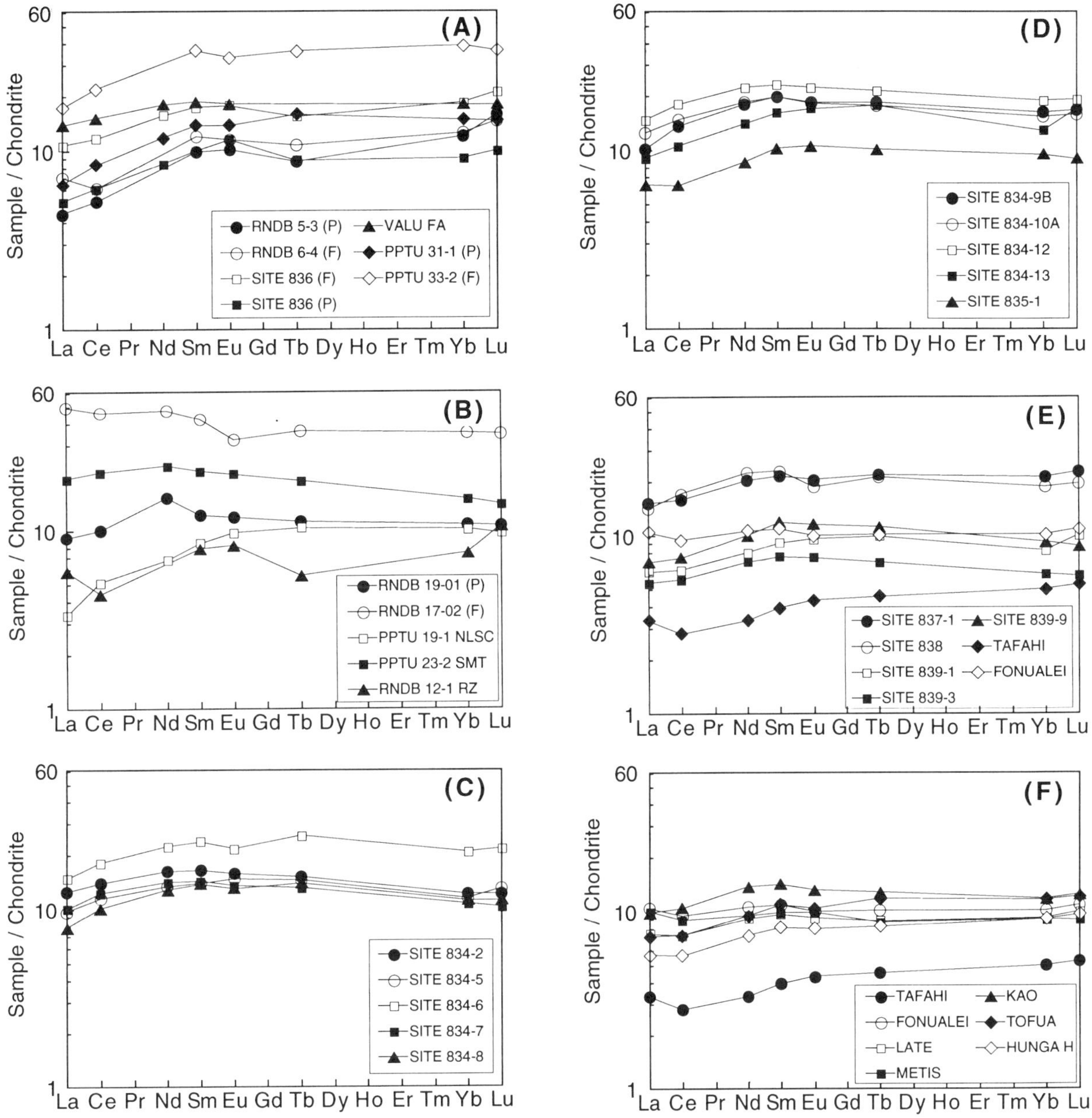

Fig. 8. Chondrite normalized REE data for representative rock types. Normalizing values from *Masuda et al.* [1973]. A. ELSC (RNDB 5-3, 6-4 and site 836); CLSC (PPTU 31-1, 33-2); and Valu Fa ridge. P=relatively primitive, F=fractionated. B. MTJ (RNDB 19-1, 17-02); NLSC (PPTU 19-1); Rochambeau Bank (PPTU 23-2) and relay zone (RNDB 12-1). C.ODP site 834 data; suffix is petrologic unit. D. ODP sites 834 and 835. Labels as for Fig. 7. E. Arc-like ODP leg 135 data and representative Tofua arc data. Labels as for Fig. 7. F. Representative Tofua arc data [*Ewart et al.*, 1994b].

Representative data for the range in rock types are in Table 3 and are summarized in Figures 7-12. They share many of the characteristics of the Valu Fa ridge samples for major and trace elements but in general are much more like the ELSC and CLSC lavas than to those of the Tofua arc. For example on the Harker diagrams (Figure 5), the trends for oxides of Al, Fe, Ti and Na are MORB-like although the field for Ti lies between those for the ELSC and CLSC; it coincides with the Mariana Trough field.

The REE and the extended REE plots (Figures 9 and 10) also resemble MORB as do some of the data shown on the Hf/3-Ta-Th diagram (Figure 11). Samples with the highest

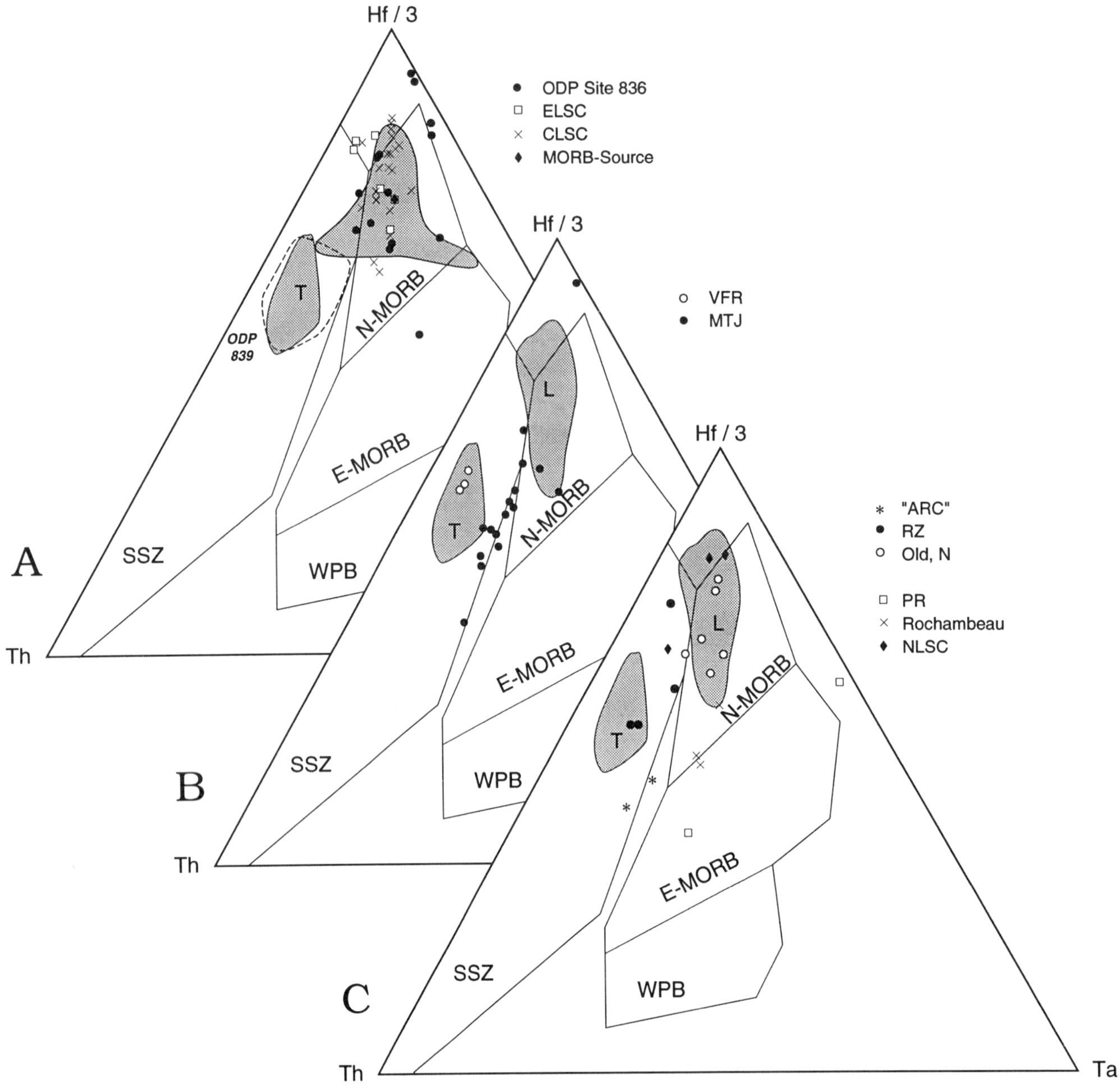

Fig. 9. Discriminant diagram for Hf-Ta-Th [*Wood*, 1980; *Wood et al.*, 1981]. WPB = within plate basalt. Other acronyms and labels as defined in the text. A. Axial ridge system data. Shaded field is for site 834 data [*Hawkins*, 1994]. T=Tofua arc, [*Ewart et al.*, 1994]. B. Valu Fa ridge and Mangatolu triple junction. Shaded field L=axial ridge data from Fig. 9A, T=Tofua arc. C. ARC= seafloor southeast of MTJ; Old, N=older crust northwest of MTJ; RZ=relay zone; PR=Peggy Ridge; NLSC=northwestern Lau spreading center.

Th levels lie close to, but not within, the SSZ field. Thus, even though the MTJ data plot close to the Tofua arc data, and one limb projects into it, the MTJ samples are like those of the NLSC and CLSC. Nilsson [1993] concluded that major element variations can be explained by extensive low-pressure fractionation of plagioclase, olivine, clinopyroxene, and titanomagnetite. Despite the proximity to the Tofua arc, the MTJ lavas have been derived from a different mantle source.

Western extensional basin. The presently active spreading centers create new crust from mantle-derived melts with little evidence for a subduction component. This new crust has replaced older crust that formerly was part of the forearc to the Lau Ridge. The Tonga Ridge includes fragments of the

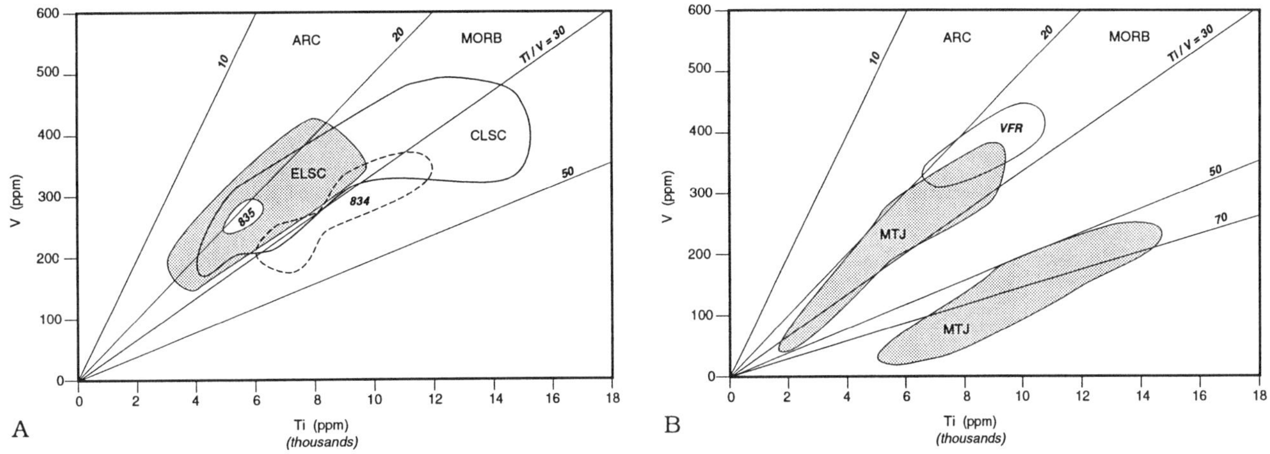

Fig. 10. Ti-V plot. ARC and MORB fields from *Shervais* [1982]. Field labels as for Fig. 7. ELSC field includes site 836 data.

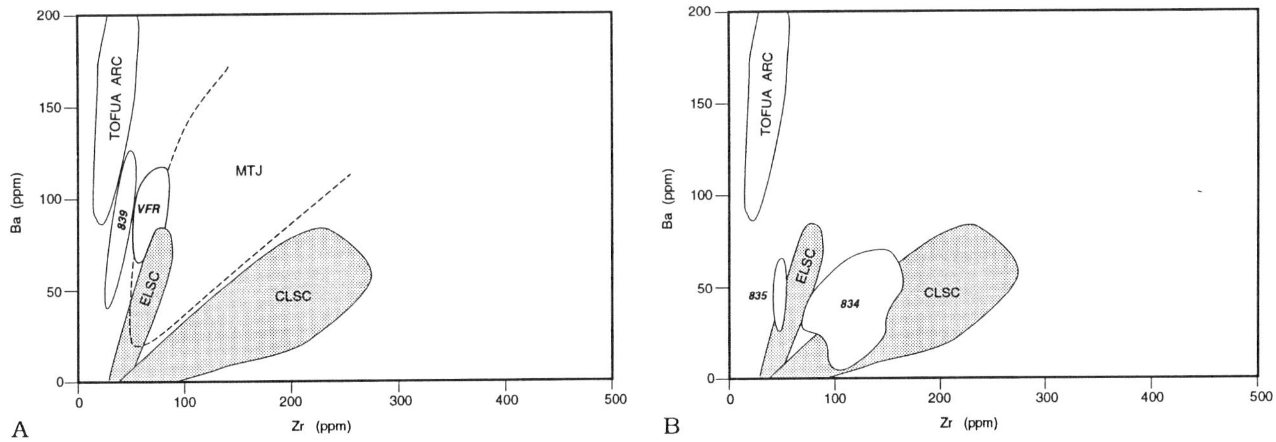

Fig. 11. Ba-Zr plot. ELSC field includes data for site 836. Field labels as for Fig. 7.

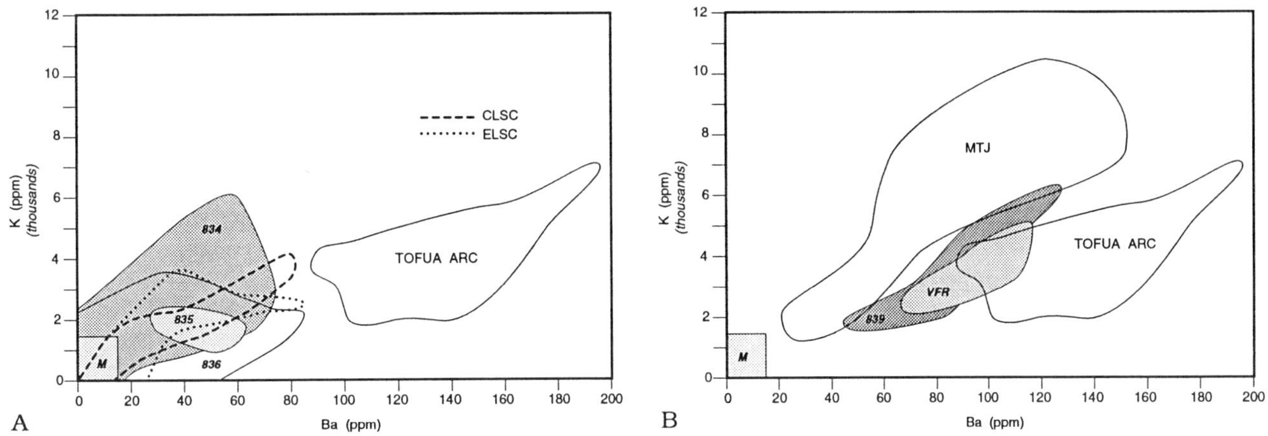

Fig. 12. K-Ba plot. M is field for fresh N-MORB [*Viereck et al.*, 1989]. Field labels as for Fig. 7.

outer edge of the Lau Ridge and its forearc as discussed previously. The broad region in the western Lau Basin that includes the extensional sub-basins is given the informal name western extensional basin (WEB). The WEB has horsts of older crust separated by grabens partly filled with younger basalt. Some of the older crust of the WEB may be remnants of the Lau Ridge volcanic arc, some of it may be formed of basalts erupted in the early stages of extension of what had been the forearc of the Lau Ridge, and, possibly, there may be trapped fragments of old Pacific lithosphere. None of the latter has been identified. Dredged samples from several ridges and scarps in the WEB are transitional between MORB and arc compositions [*Hawkins and Melchior*, 1985].

We know very little about the seafloor north and northeast of the Peggy Ridge, which may be comparable to the WEB. Rocks similar to those found in the WEB were dredged on the eastern edge of the Lau Basin from similar morphologic features. This suggests that they were formerly part of the same petrologic/tectonic province.

Crust of the WEB sub-basins was sampled at six drill sites on ODP leg 135 (Figure 3). All but one of these sites were in sub-basins that had opened during the early extension and rifting history of the basin. Three sites sampled MORB-like crust and three were on arc-like or transitional crust. One site (836) was on crust formed very early in the spreading history of the ELSC. A comprehensive discussion of the petrology of these drill sites is in the ODP leg 135 Initial Results and Scientific Results volumes [*Parson et al.*, 1992; *Hawkins et al.*, 1994].

Large areas of seafloor form ridges between the sub-basins drilled in the WEB, these are informally referred to here as "old crust." The name is also used for seafloor on the eastern side of the axial ridge system, in the northeastern part of the basin near the Mangatolu triple junction and for much of the region north of Peggy Ridge. *Hawkins and Melchior* [1985] proposed that the crust of the Lau Basin was compositionally zoned from a MORB-like central area on the axial ridge systems to more arc-like, or transitional to arc, composition crust on the outer margins. The "old crust" constitutes the outer margin. It is likely that this term includes a range in crustal ages; for most of it only a relative age can be assigned. It should be emphasized that the zoned pattern was based on analyses of glasses and fresh aphyric rocks and is not a reflection of alteration. The general pattern of the zonation, and locations of sample used to infer it, are in Figure 4. Representative analyses are in Table 3.

Dredged samples from ridges in the WEB include RNDB 1-1 from a ridge crest west of site 836, and RNDB 2-7, 2-8 (Table 3); the latter are from east of site 836 but resemble older ridges to the west. DSDP site 203 was drilled in a sub-basin; it did not reach basement but recovered a small basalt cobble deep in the core. Glass in the vitrophyric rim has transitional chemistry (e.g., 52% SiO_2, 0.7% TiO_2, 0.28% K_2O, Mg#=66.1), [*Fryer et al.*, 1981]. It is likely that it came from the adjacent intra-basin ridge. There is but one sample from the southern end of the WEB, F-114 on Figure 4, but it also shows the transitional chemistry of the postulated zonation. Rocks from dredge sites 123-81, PPTU-34 and ANT-223 (Table 3) are from crust which *Hawkins and Melchior* [1985] interpreted as transitional between MORB and arc. ANT-223 is from close to the western side of the ELSC crust about 100 km off the present axis and southwest of ODP site 836. Sample 123-81 is from the WEB northwest of site 837. Sample 123-101-2 is from the lower slopes of the Lau Ridge and represents either a part of the arc or some of the early formed backarc crust. Examples of old crust from the northeastern part of the basin, adjacent to the Mangatolu triple junction, are PPTU 4-1, 6-1 and 7-1. PPTU 4-1 and 7-1 have compositions transitional to arc chemistry whereas PPTU 6-1 is more MORB-like. Collectively these samples of "old crust" indicate that much of the earlier formed crust in the opening of the Lau Basin was transitional in composition between MORB and arc, as originally proposed by *Hawkins and Melchior* [1985]. However, some of the drill site data pose a problem for this simple model as discussed below.

Drill sites in the western extensional basin. MORB-like basalts were drilled at ODP sites 834, 835, and 836; representative data are in Table 2. Sites 834 and 835 are in extensional basins, partly filled with sediments, and site 836, discussed previously, is part of the earliest crust formed at the ELSC. It too is in an extensional basin like sites 834 and 835 that may represent a tectonic setting transitional from rifting to spreading. Age estimates from the paleontologic data indicate that site 834 basalts range in age from about 5.5 Ma, or older, to 3.8 Ma. We proposed that magmatism may have started at about 6 Ma [*Hawkins*, 1994]. The narrow sub-basin at site 834 was the locus of basalt eruption for more 1.7 Ma as it opened; the basaltic filling is interbedded with thin fossiliferous sedimentary layers that provide age control. At site 835, the age of the end of basaltic infilling is late Pliocene (3.5 Ma) as determined from the paleontologic age of overlying sediments. As for site 834, we do not know when volcanism began but it ended at about the time the ELSC propagator passed by on the east [*Parson and Hawkins*, 1994]. Site 836 basalts are intercalated with Pleistocene sediments. The basaltic crust is estimated to be from 0.8 to 0.64 m.y. old and it too coincides closely with the estimated time of passage of the ELSC propagator.

Thirteen petrologic units were identified at site 834 that range from a relatively "primitive" massive unit (Unit 7) having Mg#=61-64 to more fractionated units (e.g., Unit 12) having Mg#=34-36 and SiO_2=54%-56%. Identification of petrologic units was based on a combination of petrography, mineralogy, and chemistry as well as the presence of chilled margins and sedimentary interbeds. The basaltic rocks are MORB-like in composition but show some of the SSZ signature in major elements such as moderately elevated Al_2O_3 and Na_2O, and lower FeO* and TiO_2 relative to

MORB at a given value of MgO (Figure 6). As discussed for the axial ridges, this is considered to reflect the SSZ setting although there is a very strong MORB-like signature seen in the trace elements (Figures 9, 10, and 12) and isotopic data. All of the units sampled except Unit 7 are highly vesicular as are most other Lau Basin samples.

Site 834 lavas have trace element abundances and ratios that are more like the CLSC lavas than the ELSC even though they were erupted only about 1 m.y. before the ELSC propagator passed by to the east. Fields for element abundances are in variation diagrams (Figures 7-12). Trace element and isotope data and their bearing on petrogenesis are discussed in separate sections.

Many of the units show effects of fairly extensive fractionation and this implies that deeper, unsampled "remnant magma chambers" must exist. There is wide range in melt composition with depth and no correlation between depth and magma chemistry. For example, "primitive" Unit 7 was encountered from 214-287.6 meters below seafloor (mbsf) and is early Pliocene or older, whereas "evolved" Unit 12 was encountered from 363.5-407.9 mbsf. The deepest, oldest, unit sampled (Unit 13) at 423 mbsf is a fractionated unit with Mg#=57-58 and about 51% SiO_2. One of the youngest units (Unit 2) at 112 mbsf, has Mg#=52-55. It is clear that we did not drill to the bottom of the lava sequence and into the earliest melts erupted at site 834, but had encountered some relatively evolved rocks at the point where drilling had to be stopped.

The petrologic stratigraphy of the sequence cored shows cyclic eruptions of variably fractionated magmas and large variations in unit thickness. This may reflect episodic eruption, refilling, and melt storage in near surface magma chambers which experienced variable amounts of fractionation and mixing of melts. Variations in thicknesses (volumes) of each petrologic unit on the scale of tens of meters may reflect differences in volume of parental melt production. This in turn may be a response to non-uniform mantle upwelling. Several factors may have been involved such as episodic increases in the flux of thermal energy to the mantle source; episodic pulses of trench rollback or rate of subduction; or perturbations in the convective overturn or mixing of SSZ mantle/MORB-source mantle.

An unusual aspect of site 834 samples is their high sodium content, relative to other incompatible elements, at a given MgO content. For example, relatively "primitive" Unit 7 glasses have 2.9-3.1% Na_2O; this could be the result of a small extent of melting. Incompatible element abundances such as Zr (90 ppm), Ba (about 15 ppm), TiO_2 (1.35%), and K_2O (<0.08%) all seem consistent with fairly extensive melting and resemble N2-MORB [*Viereck et al.*, 1989]. The thickness (74 m) of this homogeneous unit indicates a large volume of melt. The source for site 834 magmas may have been relatively enriched in Na, or the magmas had extensive interaction with the conduits, or both. If we postulate that the backarc mantle source were homogeneous with respect to Na, the data would suggest that site 834 magmas represent somewhat lesser extent of melting than at the other sites. However, other incompatible elements require that the difference be small. Assimilation or reaction between Na-enriched mantle, and extensive melts from a massive upwelling mantle diapir seems more likely and would be consistent with the mantle counterflow and diapirism proposed in Figure 1.

Hawkins and Melchior [1985] proposed that the basin is zoned from crust transitional to arc composition on the east and west to a central part with more MORB-like crust. Sample sites supporting this are shown in Figure 4. Site 834 was selected to sample some of the oldest crust in the basin and indeed the cores include the oldest Lau Basin crust yet identified. Site 834 lavas were expected to be arc-like because of their location and relative age. The hypothesis was tested and it appears to have failed because some of the most MORB-like rocks in the basin were sampled at site 834. A similar situation was found at the Sumisu Rift where MORB-like lavas were erupted in the early rifting stages [*Hochstaedter et al.*, 1990a,b]. Lau Basin dredged samples clearly support the zonation model but the drill cores do not. Is there a reasonable explanation?

Site 834 lavas were erupted into the slowly opening sub-basin over a time span of more than 1.7 Ma. Relatively unfractionated basalts (e.g., Unit 7) form thick petrologic units that have low-partition-coefficient trace element abundances consistent with moderate extent melting of a depleted source [*Hawkins*, 1994]. We did not reach the base of new crust at site 834 and have no evidence of the chemistry of the lavas erupted first. However, glass samples from Unit 13, the deepest petrologic unit sampled, carry the depleted SSZ signature both in terms of LILE and HFSE. For example, sample 59R2/52-55 has Rb=1.46 ppm, Ba=29.4 ppm, Nb=0.95 ppm, Zr=79 ppm, Hf=2.14 ppm [*Hergt and Farley*, 1994] and Table 2. The lack of K, Rb, Ba enrichment relative to MORB differs from the values we would have expected if these were the transitional basalts mapped elsewhere in the western Lau Basin. However, Unit 13 has Al, Fe, Ti, and Na values relative to MgO that lie in the MARA (Mariana Trough basalt) field (Figures 5 and 6) and thus are not truly equivalent to MORB. These lavas must have come from depleted SSZ mantle that lacked the LILE enrichment. Some site 834 magmas were mixed with arc-source magmas as indicated by mineral data. Intricately zoned cores of plagioclase in some units are interpreted as the result of mixing between arc and MORB-type melts [*Bryan and Pearce*, 1994]. Strongly zoned Cr-spinels also support the idea of magma mixing [*Allan*, 1994].

The lack of LILE enrichment may be because the SSZ mantle source for site 834 escaped infiltration of subduction-derived fluids. A more likely answer is that the earliest eruptions (small amounts of melting ?) effectively stripped out the incompatible elements and the large volume of melt subsequently erupted to fill the basin at site 834 may have come largely from a multiply depleted MORB-like mantle source.

Site 835 data are limited to one moderately fractionated basaltic petrologic unit. Like the site 834 rocks, it is MORB-like in most respects as seen in Figures 6-11 and the data lie within the fields for ELSC rocks. An affinity to SSZ magmas is shown by the moderate LILE enrichment and the depletion in Nb and Ta (Figure 7E). Comparison with the ELSC data is relevant because the ELSC propagator tip passed by to the east at about the time (3.6 Ma) that site 835 lavas were erupted. Lower Na_2O and TiO_2 relative to MgO than for site 834 may indicate relatively larger extent of source melting, a source more depleted in Na and Ti, less interaction with conduits, or some combination of these factors. Site 835 lavas resemble the ELSC in Al_2O_3 and FeO* and have even less of the Mariana (MARA) signature. This may reflect a melt source less hydrous, less oxidizing, than the Mariana Trough source.

To summarize, rocks from WEB drill sites 834 and 835 resemble those from the active ridges in terms of major and trace element chemistry which suggests a similar source. They appear to be derived largely from MORB-source mantle with very minor chemical influence (mainly water) from a subduction component. Rocks from sites 837, 838 and 839 are more arc-like (Table 2 and Figures 7-12) than those from sites 834, 835, and 836; some are closely similar to modern Tofua arc tholeiites [*Ewart et al.*, 1973, 1977; *Ewart and Hawkesworth*, 1987; *Ewart et al.*, 1994a,b; *Hawkins and Allan*, 1994]. All three sites were drilled in small, narrow, partly sedimented basins, elongated in a northeasterly direction. It was assumed that they were backarc crust because the sites are all less than 90 km from the present axial zone of the ELSC; their crustal ages are estimated to be about 2 Ma. These correspond closely to the time of nearest approach of the ELSC propagating tip. Their arc-like composition was unexpected in view of their location although some similar rocks had been dredged from nearby high-standing ridges (e.g., RNDB 1-1, 2-7, 2-8, Table 3).

Site 837 and 838 igneous rock samples are moderately differentiated arc-like basalts and basaltic andesites. They are gravel clasts probably derived from a nearby seamount. No glasses were recovered at site 837; bulk rock analyses have 54%-56% SiO_2, and 3%-3.4% MgO. These are closely similar to Valu Fa ridge samples [*Jenner et al.*, 1987; *Vallier et al.*, 1991] and to island arc tholeiites in general. FeO* and Al_2O_3 are like ELSC lavas, but Na_2O and TiO_2 resemble Mariana Trough data. Glasses from site 838 comprise a bimodal basalt and low-K rhyolite suite with 52%-53% SiO_2 and about 6% MgO as one mode, and 69-73% SiO_2 and 0.2% MgO as the other. The basalts all resemble Mariana Trough glasses in Al_2O_3 content but are more like the ELSC for FeO*, TiO_2 and Na_2O (Figure 6). The basalts resemble rocks from the relay zone between the ELSC and CLSC and other Lau Basin crust transitional between MORB and arc chemistry.

Site 839 lavas are among the more unusual samples collected by drilling on leg 135. Some samples resemble those from sites 837 and 838, whereas others are even more arc-like. The samples include two-pyroxene basalts and basaltic andesite. Their SiO_2 ranges from 50%-55% and MgO from 4.4%-15%. The samples with more than 9% MgO are not liquid compositions as they have abundant phenocrysts of olivine and clinopyroxene as well as phenocrysts of chrome-spinel. *Allan* [1994] showed that the strongly zoned olivine and chrome-spinel phenocrysts support magma mixing as an explanation for some of the site 839 lavas. Site 839 is considered to be an arc-composition edifice that formed within the Lau Basin at about 1.9 Ma. It was cut off and abandoned on the west side of the ELSC as the rift tip passed by.

The melts erupted at sites 837, 838, and 839 all were formed from mantle with a robust arc-source signature even though they are contemporary with the MORB-like crust formed at the nearby ELSC. If site 839 lavas had been collected on the Tonga Ridge, they would have been considered to be part of the Tofua arc magma series. However, even though the arc signature is strong, the site 839 samples lack the high Al_2O_3 seen in the Mariana Trough or in typical arc lavas. The site 839 source may have been less hydrous than the Tofua arc SSZ source. The mantle source must have remained heterogeneous, retaining arc-source SSZ domains, even while the MORB-source mantle influx was feeding melts to the ELSC propagating rift. Although this is unusual, it is not unique inasmuch as a young, arc composition, volcanic edifice, aligned with other edifices having typical "backarc basin basalt" chemistry, is on the crest of the axial ridge in the Mariana Trough [*Hawkins et al.*, 1990].

Peggy Ridge. Peggy Ridge is a well-defined linear feature that rises to depths of about 1250 m. It has rough topography with many irregular high-standing features separated by deep, probably fault controlled, canyons. The linear shape suggests that it could have been a spreading axis but this is not supported by the magnetic data [*Sclater et al.*, 1972] or by the range of magma types collected by dredging. The ridge marks the location of a zone of right-slip faulting accommodating plate motion between the CLSC and the northern part of the Lau Basin. In some places it may be a leaky transform fault [*Parson and Tiffin*, 1993], but the few samples from the ridge do not support this proposal.

The morphology and the variety of dredged rock types suggest that the ridge comprises uptilted blocks of old crust much of it like rocks from the horsts of the WEB. The westernmost end, near the Lau Ridge, has transitional basalt (e.g., 123-103-3, Table 3) similar to that which forms much of the high areas of the western basin. These have element abundances and ratios suggesting an affinity to BABB or Mariana Trough type lavas [*Hawkins and Melchior*, 1985]. One site on the western part of the ridge yielded very primitive, highly depleted basalt that has many of the characteristics of a basaltic komatiite or a variety of boninite. Rocks and glasses from this dredge site (e.g., 123-

95-1, Table 3) represent extensive melting, up to 30%, of a depleted mantle source. The trace element data, especially REE, Cr, and Ni [*Hawkins and Melchior*, 1985] are consistent with a boninitic character. These samples have many of the chemical aspects of high Ca-boninites, but their relatively high Al_2O_3 (15-16%) argues against that classification. However, CaO/Al_2O_3 (0.9-1.0) is comparable to many high-Ca boninites [*Crawford et al.*, 1989]. Irrespective of the name assigned, these samples resemble early-formed melts in many SSZ settings and may represent some of the first magmas erupted as the Lau Basin opened. It is unlikely that they are derived from magma leakage on the transform fault because of the large amount of melting, and unfractionated composition, implied by their chemistry.

The middle and eastern parts of the ridge are composed of MORB-like basalt (e.g., 123-86-1, PPTU 17-1, PPTU 17-2, Table 3). These resemble fractionated basalts from site 834 (e.g., Unit 13) and the axial ridges. They have relatively high TiO_2 (e.g., 2%), FeO* (e.g., 11-12%), and Na_2O (e.g., 3%), but their calculated abundances at 8% MgO [*Klein and Langmuir*, 1987] are about 1.6%, 9.7%, and 2.6% respectively. The parental magmas may have formed by relatively smaller extent of melting of a source similar to that for site 834 lavas.

Peggy Ridge may comprise crustal blocks forced up by transpression on the northwest trending transform fault. Partial support for this suggestion comes from the west to east variation in chemistry from arc-like, to boninitic, to MORB-like. The disposition of these magma types mimics the regional zonation seen in the western Lau Basin.

The postulated NLSC abuts the Peggy Ridge and puts an overprint of young MORB composition lavas on the older crust. Samples from the intersection are PPTU-19-1, PPTU 20-5 (Table 3)

Seamounts. North and east of Peggy Ridge are a number of seamounts and the island of Niuafo'ou. All are shown on Figure 3 and some representative chemical data are in Table 4. Each feature has different geology which is briefly summarized here.

Donna Seamount is a very prominent high standing feature that rises more than 2300 m above its surroundings to about 888 m depth. It is somewhat elongated in an east-west direction. The summit is highly irregular and it may never have reached wave base. The south flank drops into a deep basin from which diabasic/ micro-gabbroic textured rocks were dredged [*Hawkins*, 1976]. The morphology suggests that the seamount may be very large uplifted uptilted crustal block rather than a "point source" magma leak. The age of this seamount is not known but all indications are that it is one of the older features in the basin.

Rock mineralogy and chemistry (e.g., 123-89-4, Table 4) is typical of the MORB-like basalts found throughout the basin. The calculated Na and Fe at 8% MgO (2.4% and 9.9% respectively) are typical of other MORB-like basalts from this part of the basin. The REE pattern is typical for MORB [*Hawkins and Melchior*, 1985] and other trace elements also indicate derivation from a MORB-source mantle. There is no evidence that Donna Seamount formed at the same time as the eruptions at site 834, but the source composition and petrogenesis for the two sites were nearly identical.

Rochambeau Bank is elongated in a northwesterly direction; it sits atop a broad 2000 m deep platform and rises nearly 1500 meters above it. Rochambeau Bank offers several contrasts with Donna Seamount; its form and the dredged samples suggest a point-source volcanic origin. These are relatively more enriched in alkalis and HFSE than most other Lau Basin rocks (e.g., 123-106-1, PPTU 23-2, Table 4) and are quite different from samples of the NLSC which extends toward Rochambeau Bank. In some respects the major element chemistry is arc-like but the enriched HFSE suggest an ocean island basalt (OIB) affinity. Ti, Zr, Hf, Ta, Nb all are enriched relative to other Lau Basin rocks. *Volpe et al.*, [1988] considered that Rochambeau Bank samples were E-MORB on the basis of Nd and Sr isotope ratios. *Poreda* [1985] and *Poreda and Craig* [1993] reported $^3He/^4He$ of 11.0 and 14.1 × R_A for Rochambeau Bank samples and a ratio of 22.0 × R_A for a sample from the nearby seafloor. They proposed that these ratios represent contributions from a Samoan plume component that has been entrained under the northern Lau Basin.

Niuafo'ou island is a shield volcano that was active as recently as the mid-1940s. It represents a melting anomaly as it rises more than 1300 m above the surrounding seafloor which itself is part of a broad pedestal, elongated in a northwesterly direction, and rising about 500 m above its surrounding seafloor. Samples from Niuafo'ou are moderately fractionated basalts, SiO_2=49%-50%, MgO=5.1-7.2%, and TiO_2=1.4-1.75% [*Reay et al.*, 1974]. These authors observed that the Ni content is low (34-60 ppm) for this MgO range, and in this respect resemble arc lavas, whereas the Ba is higher than typical arc values (35-68 ppm). However, the Ba enrichment is not unusual for Lau Basin lavas. The mean value for Ti/V (47.8), and all data, lie within the broad field for MORB and are unlike typical arc values [*Shervais*, 1982]. Isotope data (Sr, Nd, Pb) all show that Niuafo'ou lavas are different from Tofua arc samples. *Ewart and Hawkesworth* [1987] conclude that they are from a distinct magma source that has an ocean island basalt (OIB) signature in spite of the general chemical similarity to MORB. *Poreda and Craig* [1993] showed that gases collected in the crater lake Vai Kona on Niuafo'ou have $^3He/^4He$ 7.8 × R_A. This MORB-like ratio is typical of the surrounding seafloor and contrasts with ratios up to 22 × R_A measured from samples collected near Rochambeau Bank. The Samoan "plume" component proposed by *Poreda* [1985] is found at Rochambeau Bank some 100 km to the northwest. The "plume" component and the OIB component must be decoupled. *Hilton et al.* [1993] explain the helium data as the result of crustal assimilation as discussed in a separate section.

Zephyr Shoal rises to water depths of less than 700 m

TABLE 2. Representative Analyses, ODP Leg 135 Core Samples

Location Site Sample	WEB ODP 834 A Unit 1 12XCC/13-15	ODP 834 B Unit 7 mean (17)	ODP 834 B Unit 7 GL 29R1/07-12	ODP 834 B Unit 12 56R1/07-15	ODP 834 B Unit 13 GL 59R2/16-24	ODP 835 Unit 1 GL mean (27)	ODP 835 Unit 1 6R1/32-37
SiO_2	49.78	50.2	50.17	52.2	51.18	52.05	50.35
TiO_2	1.4	1.32	1.42	2.11	1.33	1.08	1.09
Al_2O_3	17.51	16.33	16.33	15.71	16.14	15.52	16.47
FeO*	8.47	8.78	8.67	12.41	9.08	9.99	8.46
MnO	0.15	0.17	0.2	0.2	0.19	0.19	0.16
MgO	5.06	7.73	7.73	3.4	6.1	6.12	7.67
CaO	12.71	11.88	11.87	8.58	11.41	11.14	12.9
Na_2O	3.08	3	3.08	3.87	2.98	2.35	1.82
K_2O	0.26	0.1	0.09	0.73	0.14	0.21	0.28
P_2O_5	0.15	0.12	0.11	0.21	0.1	0.12	0.11
Sum	99.48	99.63	99.67	100.78	98.66	98.77	100.23
Mg#	55	64.1	64.6	36	57.9	55.7	65
CaO/Al_2O_3	0.73	0.73	0.73	0.55	0.71	0.72	0.78
Trace Elements (ppm)			*				
Cr	267			0			179
Ni	53		102	12			78
Co							
V			227				
Zr	120		102	146	79		44
Y	28		29	40	27		22
Nb	2		3	1	0.95		1
Hf							
Ta							
Rb	5		3	13	1.46		5
Ba	41		22	60	29		42
Sr	245		176	195	169		130

GL=glass data
all other data are for aphyric rocks

* major elements on glass
trace elements, aphyric rock

TABLE 2. (continued)

Location	WEB						
Site	ODP 837	ODP 837	ODP 838	ODP 839 A	ODP 839 B	ODP 839 B	P-23
Sample	Unit 1	Unit 1	Unit 1 GL	Unit 1	Unit 2	Unit 3 GL	PPTU
	mean (5)	4$1/36-41	9H2/95-117	24X1/00-06	18R1/54-60	22R1/19-23	23-2
SiO_2	55.16	55.23	52.95	53.03	54.92	53.21	48.88
TiO_2	1.34	1.33	0.92	0.65	0.9	0.7	1.89
Al_2O_3	15.34	15.34	15	15.06	16.08	15.9	14.94
FeO*	11.7	11.67	11.19	9.04	10.1	8.71	11.35
MnO	0.2	0.2	0.23	0.17	0.18	0.17	0.19
MgO	3.5	3.56	5.64	9	4.64	6.43	6.63
CaO	8.08	8.12	10.72	11.66	10.08	12.24	11.79
Na_2O	2.87	3.06	1.93	1.35	1.9	1.8	3.1
K_2O	0.69	0.75	0.35	0.23	0.56	0.32	0.26
P_2O_5	0.21	0.23	0.09	0.07	0.12	0.09	0.17
Sum	99.09	99.5	99.01	101.26	100.62	99.57	99.2
Mg#	38	38.5	50.8	67.1	48.5	60.2	54.5
CaO/Al_2O_3	0.53	0.53	0.96	0.77	0.63	0.77	0.79
Trace Elements (ppm)							
Cr				618	7		238
Ni		5		123	1		115
Co							47
V		314					346
Zr		79		32	49		127
Y		35		16	21		41
Nb		5		1	0		
Hf							3.27
Ta							0.375
Rb		14		5	8		7
Ba		80		78	85		50
Sr		165		148	191		215

TABLE 3. Representative Analyses, Peggy Ridge, NLSC and MTJ

Location	Peggy Ridge			NLSC			MTJ	
Site	L-95	L-86	L-103	P-19	P-20	P-17	R-17	R-19
Sample	123	123	123	PPTU	PPTU	PPTU	RNDB	RNDB
	95-12 GL	86-1	103-1 GL	19-1 GL	20-5 GL	17-1	17-2 GL	19-1 GL
SiO_2	50.42	49.66	52.84	49.17	48.19	50.12	59.98	50.89
TiO_2	0.34	2	1.75	0.82	0.79	1.39	1.37	0.94
Al_2O_3	15.75	14.65	15.37	17.15	17.41	15.12	15.16	15.93
FeO*	8.59	11.88	9.46	8.72	9.11	10.79	10.5	8.53
MnO	0.16	0.29	0.19	0.18	0.18	0.21	0.23	0.17
MgO	9.78	6.77	4.72	9.51	8.98	6.94	1.91	7.52
CaO	14.67	11.54	9.35	12.75	12.42	11.82	5.62	13.1
Na_2O	0.99	3.01	3.35	2.11	2.34	2.99	3.89	2.14
K_2O	0.01	0.16	1.01	0.03	0.04	0.16	0.97	0.21
P_2O_5	0.02	0.18	0.5	0.04	0.04	0.07	0.42	0.11
Sum	100.73	100.14	98.54	100.62	99.69	99.24	100.05	99.54
Mg#	70	53.9	50.5	69.1	66.9	56.9	27.2	64.4
CaO/Al_2O_3	0.93	0.79	0.61	0.74	0.71	0.78	0.37	0.82
Trace Elements (ppm)								
Cr	670	700		741	470	113	39	309
Ni	271	380		312	221	56	6	90
Co	75			56	53	47	15	41
V	252			212	215	367	51	226
Zr	7			44	48	120	238	64
Y	11			20	21	53	78	24
Nb	3			1		1	9	1
Hf				1.15	1.08	3.53	6.09	1.59
Ta				0.045	0.045	0.07	0.65	0.133
Rb	0.153	2.9		0.3	0.1	0.8	27	7
Ba	1.16	30		1	1	10	147	27
Sr	12	109		79	100	120	135	133

TABLE 4. Representative Analyses, Old Crust, Seamounts, and Forearc Interior

Location	Old Crust							
Site	A-223	F-114	L-81	L-101	P-4	P-6	P-7	P-34
Sample	ANT	Sonne	123	123	PPTU	PPTU	PPTU	PPTU
	223-1 GL	114-GL	81-GL	101-4	4-1 GL	6-1 GL	7-1 GL	34-1 GL
SiO_2	50.89	50.88	51.57	55.97	49.72	48.22	50.08	49.18
TiO_2	1.32	0.83	1.11	1.04	0.96	0.79	1.28	0.79
Al_2O_3	17.1	18.84	16.03	14.45	16.17	17.46	15.48	15.58
FeO*	8.01	9.34	8.38	10.35	9.13	9.29	10.38	10.91
MnO	0.17	0.21	0.18	0.21	0.19	0.19	0.18	0.2
MgO	6.35	4.59	6.23	4.4	6.64	9.02	6.41	6.97
CaO	11.06	11.39	11.77	8.87	12.66	12.52	12.22	13.38
Na_2O	3.23	1.89	3.03	2.81	2.31	2.28	2.71	1.9
K_2O	0.38	0.46	0.43	0.61	0.66	0.02	0.3	0.17
P_2O_5	0.23	0.14	0.2	0.13	0.2	0.04	0.13	0.08
Sum	98.79	98.59	98.92	98.84	98.64	99.83	99.17	99.08
Mg#	61.9	50.1	60.9	46.6	59.8	66.6	55.9	56.7
CaO/Al_2O_3	0.647	0.6	0.73	0.61	0.78	0.72	0.79	0.86
Trace Elements (ppm)				*				
Cr	640	74		510	681	177	222	
Ni	332	34		193	257	114	52	133
Co	225	31		85	52	50	32	
V	262	212		297	249	250	244	279
Zr	76	64		38	62	95	81	34
Y	26			29	25	24	26	21
Nb	2					2		1
Hf					1.49	2.28	1.27	
Ta					0.175	0.11	0.078	
Rb	6			6	18	1	6	3
Ba	46.3	12		34.9	69	17	61	12
Sr	186	281		174	224	191	222	244

* from aphyric sample 123-101-2

TABLE 4. (continued)

Location				Donna Smt.	Rochambeau Bank		Zephyr Sh.	Forearc
Site	R-1	R-2	R-2	L-89	L-106	P-23	L-98	ODP 841
Sample	RNDB	RNDB	RNDB	123	123	PPTU	123	Unit 1
	1-1 RK	2-7 GL	2-8 GL	89-4 GL	106-1 GL	23-2	98-4RK	Average
SiO_2	49.33	55.86	56.43	50.53	48.77	48.88	62.99	51.52
TiO_2	0.51	1.2	1.25	1.26	1.6	1.89	0.471	1.05
Al_2O_3	11.89	14.95	14.14	14.3	16.09	14.94	14.34	16.33
FeO*	9.07	11.32	12.19	11.97	10.34	11.35	5.32	11.4
MnO	0.16	0.22	0.22	0.25	0.2	0.19	0.09	0.24
MgO	12.91	3.87	3.72	6.73	7.22	6.63	3.62	4.67
CaO	12.29	8.32	8.14	11.72	12.36	11.79	5.32	9.59
Na_2O	1.83	2.91	2.41	2.88	2.85	3.1	3.65	3.09
K_2O	0.35	0.24	0.29	0.08	0.32	0.26	0.88	0.28
P_2O_5	0.13	0.15	0.14	0.12	0.17	0.17	0.07	0.16
Sum	98.47	99.2	99.11	99.84	99.92	99.21	96.75	98.33
Mg#	72.8	41.2	38.5	53.5	58.9	54.5	58.2	45.6
CaO/Al_2O_3	1.03	0.557	0.576	0.82	0.77	0.79	0.37	0.59
Trace Elements (ppm)		*	*					
Cr				150	255	238		20
Ni	306	17	10	45	122	72		20
Co				50	106	47		
V	197	435	404	334	287	355		376
Zr	33	98	93	77	105	127		67
Y	13	38	37	32	61	31		25
Nb	3	1	2	3	5	7		1
Hf						3.27		
Ta						0.375		
Rb	9	6	5	16	5.6	5		2.7
Ba		55	55	11.1	66.6	40		157
Sr	298	131	137	90	230	215		234

* from aphyric rock samples

from depths of about 2200 m. Its top is capped with corals and reefal limestone [*Sclater et al.*, 1972; *Hawkins*, 1974, 1976, 1988]. This seamount is anomalous in that it has arc-composition rocks but, it is out in the middle of the northern Lau Basin. It is a hybrid of low-K rhyolite glass and phenocrysts that give it an overall dacite composition [*Hawkins*, 1985], (sample 123-98-4, Table 4). The extremely low REE patterns and general depletion in HFSE [*Gill*, 1976] suggest that Zephyr Shoal lavas may have been derived from a fractionated boninitic parental magma that mixed with crystal residue from other melts.

Petrologic Evolution of the Lau Basin

The early studies of backarc basin crust emphasized the similarity to MORB [*Hawkins et al.*, 1970; *Sclater et al.*, 1972; *Hart et al.*, 1972; *Hawkins,* 1976], although the volatile and alkali metal contents of some samples were recognized as being anomalously high [*Hart et al.*, 1972; *Fryer et al.*, 1981]. The term backarc basin basalt (BABB) was proposed for these "enriched" rocks. Some backarc basin samples show similarity to arc lavas [e.g., *Gill*, 1976]. The recently completed studies of ODP leg 135 samples [*Hawkins et al.*, 1994] give further evidence that Lau Basin crust ranges in element and isotopic chemistry from MORB to arc compositions. MORB-source mantle has played a major role in Lau Basin magma evolution but there has been mixing to varied extent with an arc-source component [e.g., *Hergt and Farley*, 1994]. These observations imply that many of the processes and magma sources that control Tofua arc chemistry [*Ewart and Hawkesworth*, 1987] are important for the Lau Basin.

Mid-ocean ridge basalts form a distinct class of rocks depleted in magmaphilic elements (LILE and HFSE) relative to estimates for the upper mantle. These inferred depletions are further supported by isotopic signatures (e.g., the Sm-Nd, Rb-Sr, and U-Pb systems). The mantle wedge in supra-subduction zone (SSZ) settings must have been depleted even more than the MORB-source and is considered to be residual from the extraction of MORB (i.e., it is a multiply-depleted mantle source). Magmas of intra-oceanic convergent margins must be evaluated in terms of the relative contributions of the SSZ mantle, more "normal" MORB-source mantle, other mantle such as an OIB-source, and a subduction component.

The nature of the SSZ mantle source. The chemical complexity of the SSZ mantle source for arc and backarc magmas is the result of previous melting, counterflow of "new" mantle into the source region, and addition of a subduction component (Figure 1). The subduction component may include aqueous fluids enriched in LILE derived from the subducted ocean crust; sediment, either directly or through an intervening fluid phase; or by the addition of partial melts (e.g., plagiogranite) derived from subducted amphibolitized ocean crust [*Pearce et al.*, 1992]. Trace element and isotope data offer the most useful tracers of these components.

HFSE are considered to be immobile under low temperature hydrous conditions [*Ludden and Thompson*, 1979; *Pearce and Norry*, 1979], thus they are not likely to be added to the mantle wedge by subduction-derived fluids. Their abundances in island arc volcanics (IAV), and backarc basin crust, are determined by abundances in the SSZ mantle and the melting process. This makes them useful tracers of the mantle source composition when least fractionated samples are evaluated.

IAV magma series typically are depleted in HFSE relative to N-MORB [e.g., *Arculus and Powell*, 1986]. They are characterized by a marked relative depletion in Nb and Ta [*Dupuy et al.*, 1982; *Briqueu et al.*, 1984; *Kelemen et al.*, 1990] on plots of trace elements normalized to MORB (Figure 7). Other HFSE show variable depletions; for example some IAV series have a lesser depletion in Zr than others.

The HFSE depletion seen in most intra-oceanic IAV series has been explained by some authors as the result of previous depletion of their mantle source, that is, the mantle source is multiply-depleted MORB-source mantle [*Hawkins et al.*, 1984; *Woodhead et al.*, 1993]. Other explanations include, retention of HFSE in a mantle mineral (e.g., a titanate) [*Briqueu et al.*, 1984; *Ryerson and Watson*, 1987], more extensive melting of MORB-source mantle [*Dupuy et al.*, 1982], or exchange reactions between melt and mantle enroute to the surface [*Kelemen*, 1986; *Kelemen et al.*, 1990]. *Kelemen et al.* [1990] showed that melts derived from a relatively more fertile mantle may react with overlying depleted mantle and become depleted in HFSE relative to MORB. This may be especially significant because long transit of small volumes of melts through large volumes of previously depleted mantle (e.g., SSZ mantle) may cause the strong HFSE depletions seen in IAV. HFSE bulk distribution coefficients (D_O) in depleted mantle depend largely on the amount of residual pyroxene. D_O for HFSE in clinopyroxene-poor mantle increases in the series Ta and Nb < Zr < Hf < Ti.

Subduction-derived "dehydration fluids" would enrich the mantle wedge in LILE but not the HFSE or HREE abundances because the latter are immobile in aqueous fluids. However, partial melts of subducted (amphibolitized) basaltic crust could impregnate the SSZ mantle with "immobile" elements [*Pearce et al.*, 1992]. Distribution coefficients [e.g., *Kelemen et al.*, 1990] suggests that small amounts of melt would be enriched selectively in the relative order LREE > Ta and Nb > Zr > Ti > Hf > HREE.

The discussion of HFSE partitioning in the genesis of IAV magmas applies to backarc magma-genesis as well. Plots of trace element abundances for Lau Basin basalts having Mg# > 60, normalized to N-MORB (Figure 9) show many characteristics of IAV. Nb, in particular, is generally at or below XRF detection limits of about 1-2 ppm. Nearly all Nb data obtained by ICP-MS or by spark-source emission spectrography range from 0.4-1 ppm; Ta ranges from 0.01-0.2 ppm [*Hergt and Hawkesworth*, 1994; *Ewart et al.*, 1994a]. Tofua arc data are Nb = 0.5-1.4 ppm and Ta = 0.03-0.09 ppm [*Ewart et al.*, 1994a]. Collectively, the data in Figure 7 support a multiply-depleted MORB-source mantle for the source of Lau Basin basalts. There is no indication of enrichment in Ti or Zr that might be due impregnation of the SSZ mantle with melts from the subducted basaltic crust. We cannot rule out a possible role for these melts but, if they were a factor, they did not impart a recognizable signal.

The REE data for the Lau Basin samples (Figure 8) include both MORB-like patterns, with La/Sm_{ef} <1 and some with arc-like La/Sm_{ef} =1 or >1. The variable LREE enrichment is typical of SSZ settings. The HREE are unremarkable in that they resemble MORB or show slight relative depletion. These depleted samples thus resemble low-K tholeiites of the Tofua arc [*Gill*, 1981; *Ewart and Hawkesworth*, 1987].

Thorium is generally considered to be an HFSE but it could be mobilized in subduction fluids. The Hf/3-Ta-Th diagram (Figure 9) indicates that nearly all of the Lau Basin samples have MORB-like ratios for these elements. Data for ELSC, CLSC, and sites 834, 835, and 836 all lie in the MORB field. Many of the data are shifted toward Hf on these diagrams mainly because of the low levels of Ta and Th. Very low concentrations of Th, generally < 0.2 ppm, argues for a minimal contribution from sediment or sediment-derived fluids. This plot was devised to evaluate the sources for ophiolites and to discriminate between SSZ and MORB provenances. The spread in data show that this is not a conclusive discriminant diagram inasmuch as the Lau Basin clearly is in an SSZ setting but the crust largely exhibits a MORB signature. The few data that plot in the SSZ field are either from old crust in the northeastern part of the basin, close to the Tofua arc; from the Valu Fa ridge, where the ELSC is propagating into older (forearc ?) crust; or from the relay zone separating the ELSC and CLSC. In each place, the mantle source may retain fragments of the SSZ mantle that formed arc-like melts. In contrast, the Mangatolu triple junction extends from the backarc basin into the Tofua arc. At the latitude of the MTJ, the Tofua arc may be as old as 3 Ma [*Tappin et al.*, 1994]; it has erupted depleted melts as seen on Tafahi Island [*Ewart and Hawkesworth,* 1987]. Data for the MTJ trend across the MORB field towards the SSZ field on Figure 11 but do not reach it. Possibly, the reason for the minimal arc signal in MTJ lavas, in spite of their location, is because the original "arc component" in the SSZ mantle wedge has long since been stripped out and has been replaced by MORB-source mantle.

Collectively, the HFSE data demonstrate that the Lau Basin mantle source is dominated by multiply-depleted MORB-source mantle with local remnant domains of less depleted mantle, or zones that have been replenished by counterflow with more fertile MORB-source mantle. There is little or no evidence for enrichment with partial melts of subducted crust. The HFSE shed no light on the role of infiltration of subduction-derived aqueous fluids.

The covariation of Ti with V, and Ti/V ratios have been proposed as useful discriminants for recognizing the original tectonic setting of magma series now preserved in ophiolites [*Shervais*, 1982] but are used here (Figure 10) to illustrate the relative oxygen fugacity of arc and backarc magmas. The crystal/liquid partition coefficients for V (K_D) are sensitive to oxygen fugacity and range from >1 to <1; higher fO_2 causes K_D for V to decrease. Inasmuch as the bulk K_D for Ti is almost always <1, the relative enrichment or depletion of V relative to Ti is strongly influenced by fO_2 [*Shervais*, 1982]. Other factors include the original composition of the source, extent of melting, and effects of fractional crystallization.

MORB-series magmas are characterized by Ti/V ranging from 20-50 and reflect derivation under conditions of relatively low fO_2, whereas arc-series magmas have Ti/V ranging from 10-20 reflecting more oxidizing environments [*Shervais*, 1982]. Backarc basins display a range in Ti/V that include both arc-like and MORB-like ratios. This is well illustrated in the data for the Lau Basin (Figure 10). For clarity, the arrays of data for each of the morpho-tectonic provinces are shown on separate plots.

Sites 834 and 835 are MORB-like although each defines a separate field. Nearly all of the CLSC data lie within the MORB field. Some samples have Ti/V as low as 15, whereas Ti-rich samples (>12,000 ppm) have ratios between 30 and 50. Site 836 data lie within the ELSC field and are not shown separately. The ELSC data lie on the plot between ratio lines of 15 and 30. The lower ratios straddle the boundary designated to separate MORB from arc. Thus it could be argued that the Ti-V relations indicate a transitional character for some CLSC and most of the ELSC samples. Data from Valu Fa ridge lie in the MORB field even though many HFSE data give them an arc affinity (Figure 7-9). Mangatolu triple junction data suggest that there may be two classes of magma types. Many of the data lie in the MORB field and overlap with Valu Fa data. There is also a broadly scattered field for high Ti, low V samples that extends into the field for alkali basalts [*Shervais*, 1982]. Many of the low V samples (e.g., < 100 ppm) are from highly fractionated samples that likely lost both Ti and V as a result of magnetite fractionation but additional V was lost to another phase. Fractionation of either clinopyroxene or amphibole or both would preferentially remove V relative to Ti.

A distinctive feature of N-MORB is their very low levels of K, Rb, Cs, Ba, and LREE [*Kay et al.*, 1970; *Bryan et al.*, 1976; *Viereck et al., 1989*]. N-MORB also have low ratios of $^{87}Sr/^{86}Sr$ [*Hedge and Peterman*, 1970; *Hart et al.*, 1974]. Abundances of these trace elements (LILE), and isotope ratios, increase with low-temperature alteration. K, Rb, and Cs increase much more readily than Ba, Sr, or LREE [*Hart*,

1969; *Hart and Nawalk*, 1970; *Staudigel and Hart*, 1983; *Hart and Staudigel*, 1989]. Unlike isotope ratios, abundances and ratios of trace elements can be modified by fractional crystallization. Inasmuch as the LILE are highly incompatible elements, one expects to find positive correlations between K, Rb, Cs, Ba and La, and SiO_2 or with HFSE such as Zr. However, the difference in abundances of alkalis and alkaline-earth elements between least-fractionated samples of IAV and MORB indicates that the IAV are relatively enriched in these elements. Equally important are the high ratios of Ba/La, alkalis/La, and an overabundance of Cs in IAV; these cannot be explained by crystal fractionation. They are considered to be signatures of IAV sources or signatures that arise during the melting process [*White and Patchett*, 1984]. Similar chemical signatures are seen in many backarc basin basalts, especially those erupted on relatively youthful features such as the Valu Fa ridge [*Jenner et al.*, 1987; *Vallier et al.*, 1991] and the ELSC. Most backarc basin basalts typically have abundances of K and Sr not much different from MORB. However, they are characterized by relatively high levels of Ba, and variable, but commonly high, Rb and La relative to MORB with comparable abundances of Mg. Many of the axial ridge glasses from the Mariana Trough and Lau Basin as well as site 834 samples are relatively enriched in Na (Figure 7). These observations suggest that the backarc mantle source has been enriched selectively in a "subduction component".

The LILE are likely constituents of fluids derived from dehydration of subducted crust [*Tatsumi et al.*, 1986; *Tatsumi*, 1989]. The LILE also are highly mobile in hydrous low-temperature environments and would increase without a parallel increase in HFSE in altered rocks. LILE vs. HFSE should show scatter in altered rocks. LILE enrichment in apparently unaltered glass in SSZ settings is strong evidence that they reflect a subduction fluid component in the melt source. Fractionation of these melts should produce linear correlations between LILE and an HFSE. For example if Ba correlates well with Zr, it can be argued that the Ba is a primary feature of the melt rather than an effect of alteration.

Lau Basin data for Ba and Zr (Figure 11) show good linear correlation. Ba/Zr differs for different features with the lower ratios at CLSC (<0.5) and ELSC (1) and higher ratios for VFR (1.5) and site 839 (2). The Tofua arc has the highest ratios (2-4). These changes may reflect relative increase of the subduction component, Ba, in different settings.

We might also look for correlation between Ba and other cogeners such as K, Rb, and La to evaluate an enriched melt source rather than alteration to explain overall enrichments in this group of elements. The Lau Basin data for K and Ba are in Figure 12. Some of these fields are broader, especially MTJ, Tofua arc and site 834, suggesting that some alteration was involved (e.g., uptake of K). The main spreading centers show fairly well defined, but separate, linear trends. Again, these data suggest that there are primary differences in the mantle sources with varied amounts of a subduction component.

Ewart and Hawkesworth [1987] noted that the average K/Ba for Tonga (Tofua) arc lavas is about 30 and contrast this with values of 90 for fresh MORB and about 60 for altered MORB. They proposed that a subduction fluid component has been added to the source of the arc lavas. Lau Basin basalts range from 30 - 90 encompassing both MORB source ratios and those for a subduction component. The most arc-like ratios are from the Valu Fa ridge and parts of the ELSC and CLSC. This may be a reflection of their proximity to the Tofua arc. However, the Mangatolu triple junction actually cuts into the arc trend yet it has ratios ranging mainly from 50 - 70. The highest K/Ba ratios (as high as MORB values of 90) are from the CLSC, from part of the Mangatolu triple junction, and areas of "old" crust. For comparison, Mariana Trough axial ridge glass samples have K/Ba ranging from 50 to 90 [*Hawkins et al.*, 1990].

Cesium is an especially sensitive tracer of alteration, and is considered to be highly mobile in aqueous fluids [*Tatsumi et al.*, 1986]. This mobility makes it potentially useful as an indication of a contribution to the SSZ mantle from subduction fluids. Cs is enriched in IAV relative to MORB together with the other alkali metals. The limited data for Cs in Lau Basin samples requires caution in attempting to determine how important subduction fluids have been in petrogenesis. Fresh, unaltered MORB has about 0.0086 ppm Cs and Rb/Cs ranges from about 43-140 [e.g., *Hart and Staudigel*, 1989] whereas Aleutian IAV have 0.7-2 ppm Cs and Rb/Cs ranging from 24-30 [*Morris and Hart*, 1983]. There are few data for the Tofua arc but a Cs range of 0.04-0.8 ppm and Rb/Cs ranging from 10-95 is reported by *Ewart et al.* [1994a]. The decrease in Rb/Cs of arc magmas can be attributed to input of sediment or sediment-derived fluids [e.g., *Ben Othman et al., 1989*]. The Lau Basin data are limited but most samples have MORB-like ratios. ODP site 834 glasses have Cs abundances ranging form 0.01-0.36 ppm and Rb/Cs is about 50 [*Hergt and Hawkesworth*, 1994; *Ewart et al.*, 1994]. ODP site 836 also has MORB-like values with Cs= 0.01-0.05 ppm and Rb/Cs=90-180. The CLSC has < 0.08-0.15 ppm Cs and Rb/Cs=37-77. The Valu Fa ridge has some of the most arc-like rocks; Cs data are very limited but those available are MORB-like (Cs=0.17-0.19 ppm and Rb/Cs=39-45). More data may change the picture, but the data available indicate that introduction of Cs by subduction fluids or sediments has been very limited.

A signature of subducted sediment ? It is highly probable that at least some of the seafloor sediments delivered to subduction zones are introduced to the mantle and influence SSZ magma chemistry. The Tonga Trench presently appears to have minimum sediment at the subduction interface but we do not know what may have been subducted. At DSDP site 595, outboard of the Tonga Trench, there is about 70 m of pelagic clay; the K/T boundary is at 20 mbsf [*Menard et al.*, 1987]. Possibly some of the subducted material included volcaniclastic debris shed from alkalic basalts of the Louisville Seamount Chain [*Hawkins et al.*, 1987]. The most useful tracers of sediment input to magma-genesis are abundances and ratios of the LILE, especially Ba and Sr, Cs (discussed above), the LREE, Th, Pb, B, Be, and isotope ratios such as $^{18}O/^{16}O$, $^{87}Sr/^{86}Sr$, $^{10}Be/^{9}Be$, and the Pb isotopes. Negative Ce anomalies, expressed as the Ce/Ce* ratio, are considered by some to be indicative of sediment input [e.g., *Hole et al.*, 1984]. With the exception of Ce anomalies, nearly all of these element signatures may be explained equally well by transport in a fluid phase rather than by direct incorporation of sediment [see discussions in *Ewart and Hawkesworth*, 1987; *Tatsumi*, 1989; *Woodhead*, 1989]. The argument for extensive *sediment* involvement seems weak and unnecessary although it is likely that dehydration fluids may play an important role. An effective way to evaluate this is with isotope data.

Isotope data. There are extensive Sr and Nd isotope data for Lau Basin rocks; these are complemented by a lesser number of Pb isotope data (Figures 13-15) and some data for He and O. Some arc magmas have ^{10}Be which give important implications to the role of subducted sediment in magma-genesis. There are no data for the Lau Basin, but backarc basins are an unlikely place to find it in view of the short half-life for ^{10}Be and the depth (transit time) to which the Be carrying phase would need to be subducted. The following summary presents some of the petrogenetic constraints from isotope data inferred by other workers.

Ewart and Hawkesworth [1987] showed that the Tofua (Tonga) arc and Kermadec arc volcanic series have Sr and Nd isotope data that lie in the mantle array. They concluded that the Pb isotope data for the Tonga-Kermadec arcs allow a subduction component but there are no unambiguous data for it. Isotope data for Southern Tofua arc volcanoes may represent mixtures between MORB and about 2% sediment, whereas OIB sources could give rise to the Pb isotope ratios observed throughout the region. They propose that the subduction component could have been dehydration fluids derived from subducted sediment. *Meijer* [1976] studied Mariana arc volcanoes and concluded that the Pb isotope data permitted no more than 1 to 2 % sediment input. He cautioned that the constraints depend critically on the assumption that the sediment data used are appropriate to what is actually incorporated. *White and Patchett* [1984] pointed out that other assumptions such as using depleted MORB-source mantle or altered MORB as end-members are critical in the modelling. They showed that mixes of sediment and depleted MORB-source mantle were adequate to explain the observed Nd and Pd data for the Mariana arc, but the Sr isotope ratios derived from the modelling were lower than observed values.

Magma-genesis in backarc basins presents problems similar to those for island arc volcanic series. IAV have slightly more radiogenic $^{207}Pb/^{204}Pb$ and Sr isotope ratios than MORB. Many backarc basin basalts also are more radiogenic. IAV lavas tend to have more radiogenic Sr at a given level of $^{143}Nd/^{144}Nd$ than comparable MORB [e.g., *DePaolo and Wasserburg*, 1977] although *White and Patchett* [1984] argue that the field of MORB is broad enough to include most of the IAV data. The IAV are, nonetheless, shifted to the right and at the lower Nd end of the MORB field. This characteristic is shared by many backarc basalts and is observed in Lau Basin axial ridge samples [*Volpe et al.*, 1987, 1988, 1990; *Loock et al.*, 1990].

It has long been recognized that the Lau Basin has anomalously high $^{87}Sr/^{86}Sr$ relative to MORB and to other western Pacific arc and backarc systems [*Hawkins*, 1976; *Gill*, 1976; *Stern*, 1982; *Hawkins and Melchior*, 1985;

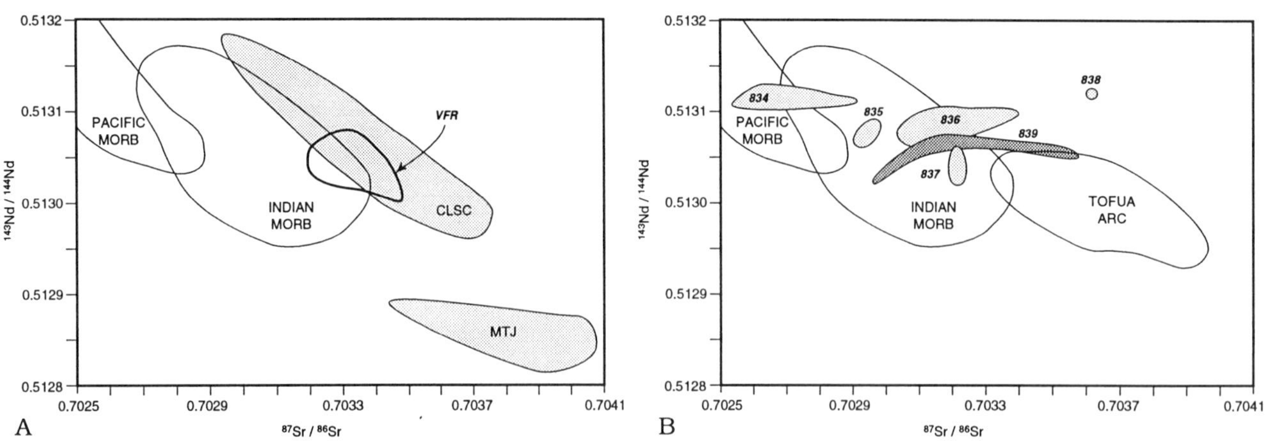

Fig. 13. Nd-Sr isotope data. Tofua arc field from Ewart and *Hawkesworth* [1987]. Pacific and Indian MORB fields and ODP data from *Hergt and Hawkesworth* [1994]. CLSC and ELSC fields from *Volpe et al.* [1988], *Boespflug et al.* [1990]; and *Loock et al.* [1990].

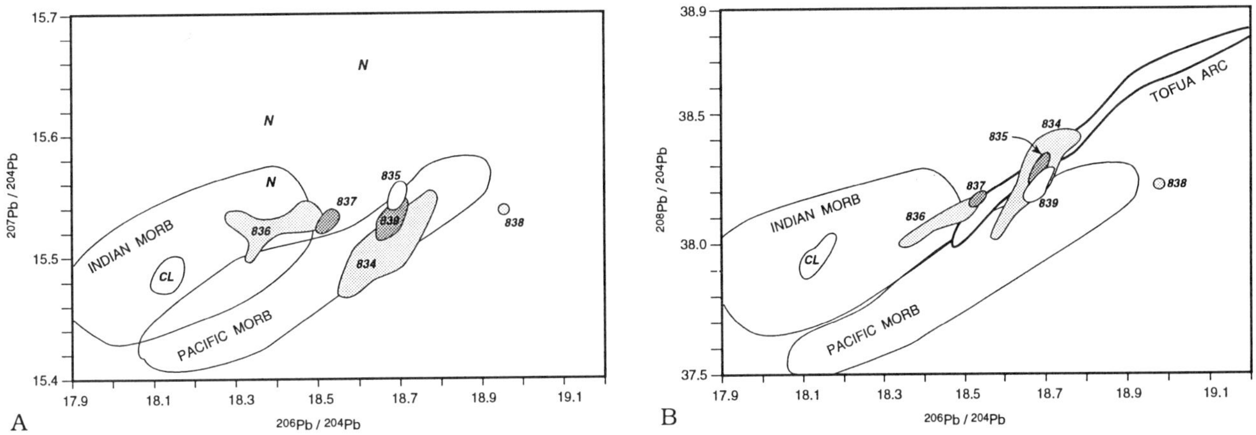

Fig. 14. Pb isotope data. N= Niuafo'ou. Sources as for Figure 13.

Volpe et al., 1988; *Loock et al.*, 1990]. Although most of the Sr isotope ratios are higher than typical MORB values, the Nd isotope ratios lie in the MORB field [*Carlson et al.*, 1978; *Volpe et al.*, 1988] (Figure 13). The isotope data, together with incompatible trace element abundances, have been used to suggest that several distinct mantle sources are required to explain the observed isotopic range of the active ridges and seamounts [e.g., *Volpe et al.*, 1988; *Loock et al.*, 1990; *Boespflug et al.*, 1990; *Hergt and Hawkesworth*, 1994].

Volpe et al. [1988] showed that Nd and Sr isotope data for basalt glasses from the CLSC and seamounts in the northern Lau Basin require multiple mantle sources. They proposed (1) a MORB-like component depleted in incompatible elements; (2) a component similar to an E-MORB or OIB source enriched in incompatible elements; and (3), a distinct OIB-like component having high $^{87}Sr/^{86}Sr$ and low $^{143}Nd/^{144}Nd$. The Nd-Sr relations are more like MORB from the Indian Ocean than to the East Pacific Rise although all data are shifted to Sr isotope ratios slightly higher than Indian MORB.

The Valu Fa ridge [*Jenner et al.*, 1987; *Vallier et al.*, 1991] is the youngest segment of the ELSC. It is at the propagating rift tip where it enters older oceanic crust at the southern end of the Lau Basin. The isotope data give a mixed signal. Sr and Nd isotopic data are similar to Lau Basin axial ridge samples, whereas the Pb isotopic data trend toward arc values. *Jenner et al.* [1987] proposed that the trace element and isotopic data indicate a minor, but significant, contribution from the subducted slab. *Vallier et al.* [1991] conclude that VFR lavas are isotopically similar to the (Tofua arc) volcanic island of Ata that is 40 km to the east of the VFR. However, Ata is anomalous in the Tofua arc in having a MORB-like character. The VFR has $^{87}Sr/^{86}Sr$ ratios lower than most of the other Tofua arc volcanoes and thus resembles the backarc. The VFR ratios for $^{143}Nd/^{144}Nd$ isotope data are higher than those of the Tofua arc; they too are more like the backarc data. The high U/Th for VFR is

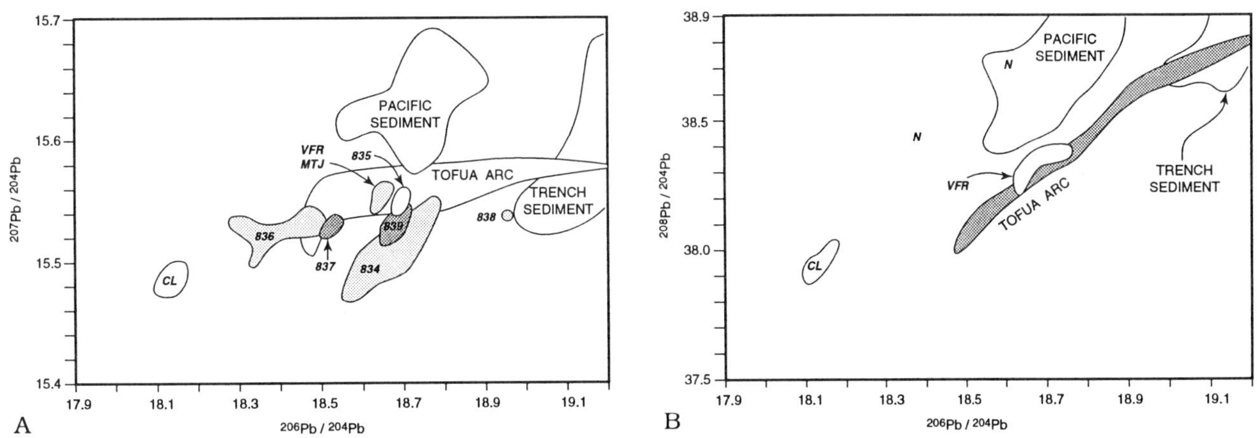

Fig. 15. Pb isotope data. Sources as for Figure 13.

arc-like as are the high ratios for $^{238}U/^{230}Th$, $^{230}Th/^{232}Th$, and $^{226}Ra/^{230}Th$ [*Vallier et al.*, 1991].

Loock et al. [1990] compared isotopic data for the CLSC, the Mangatolu triple junction in the northeastern Lau Basin, and Valu Fa ridge. Their data further support the interpretation that glasses from much of the northern Lau Basin and CLSC have the Indian Ocean signature ($^{87}Sr/^{86}Sr$=0.70309 to 0.70318 and $^{143}Nd/^{144}Nd$=0.51309 to 0.51312). The Mangatolu triple junction has ratios of 0.7040 and 0.51283, respectively; the VFR ratios are 0.70320 to 0.70339 and 0.51303 to 0.51306, respectively. *Loock et al.* [1990] and *Boespflug et al.* [1990] suggested that the arc-like isotope data for the VFR could be explained by mixing between *Pacific* MORB and either a Tongan arc component or a flux of trace elements from the subducted slab. *Hergt and Hawkesworth* [1994] note that the VFR is the only part of the modern Lau Basin ridge system lacking an Indian MORB mantle source signature. Inasmuch as it is the southernmost and youngest part of the axial rift system, it is likely that the influx of Indian-source mantle has not yet reached that far south.

Helium and argon isotope data offer some of the more intriguing insights to Lau Basin mantle sources. The axial ridge system has MORB-like $^{3}He/^{4}He$ ratios of about 8 to 8.6 x atmospheric ratio (R_A) and $^{40}Ar/^{36}Ar$ much greater (up to 4900) than the atmosphere [*Poreda*, 1985; *Poreda and Craig*, 1993; *Hilton et al.*, 1993]. This offers further support for derivation of these magmas from a depleted MORB-source mantle as suggested by trace elements and other radiogenic isotopes. Seamounts and seafloor rocks in the northern part of the basin have R_A ranging from 6.9 to 22.1. *Poreda* [1985] interpreted the high ratios (11.0 to 22.1 R_A) near Rochambeau Bank (Figure 1) as the mark of a "Samoan plume" component entrained above the inclined seismic zone. Data for the nearby island of Niuafo'ou are 8.1, whereas the surrounding seafloor ranges from 7.5 to 7.8 R_A [*Poreda and Craig*, 1993; *Hilton et al.*, 1993]. MTJ data are for two samples, 6.9 and 7.7 R_A. *Poreda*'s data [1985] also showed some very low ratios on the Peggy Ridge that range from 10.4 to 1.15 R_A; the lowest value being at the northwest end close to the Lau Ridge. *Hilton et al.* [1993] present data for the Valu Fa ridge that range from 7.65 to 1.12 R_A. These are interpreted as showing the effects of mixing with a source enriched in radiogenic helium; the argon isotope ratios are similar to atmospheric values [*Hilton et al.*, 1993]. They interpret the lower ratios as the result of assimilation of older Lau Basin crust, near the surface, by previously degassed magmas. There are limited data for the large area in the Lau Basin but the pattern of ratios suggests that MORB-like ratios occur on the axial ridges, "plume-like" ratios are at seamounts in the northern basin, and ratios significantly lower than MORB are on the margins of the basins where the oldest crust is located. This pattern parallels that for LILE as proposed by *Hawkins and Melchior* [1985].

Oxygen isotope data are limited to a collection of samples from the Peggy Ridge and the region around the CLSC [*Pineau et al.*, 1976]. They all are similar to MORB values (e.g., del ^{18}O +5.7 to +6 and indicate equilibration with mantle source material.

Prior to ODP leg 135, there was evidence that mantle magma sources for the axial ridges are dominated by Indian Ocean MORB, propagating rifts in the southern and northeastern parts of the basin have a strong signature of an arc component mixed with Pacific MORB, and the northern Lau Basin has an OIB component mixed with Indian MORB [*Volpe et al.*, 1988; *Jenner et al.*, 1987; *Loock et al.*, 1990]. This pattern is supported by newer data from which the best evidence for each of these components is the Pb isotope data, but it is also is implicit in the Sr and Nd isotopes (Figure 15) and [*Hergt and Hawkesworth*, 1994]. These interpretations, based largely on samples from the modern spreading centers, and other sites less than 1 m.y. old, did not consider sources important in the early history of the basin.

Among the most significant results of ODP leg 135 are the isotope studies that bear on mixing of multiple mantle components in the evolution of Lau Basin crust [*Hergt and Hawkesworth*, 1994]. They proposed that magmas erupted at the modern spreading centers are derived from a recent influx of mantle having the isotopic signature of Indian plate mantle. The Indian source has displaced an older source having the isotopic signature of the Pacific Plate. The Pb data suggest two different mixing trends. These require an "arc" component and a component from either a Pacific MORB-source or an Indian MORB-source. *Hergt and* for the trace of the arc component. These are (1) magma mixing between upwelling mantle and arc magmas during the early stages of extension; (2) low but significant infiltration of the backarc magma source by slab derived flux; and (3) contamination of the source by arc lithosphere during extension and upwelling. They propose that the "arc-like" component has low Ce/Pb, low Nb/U and relatively radiogenic Sr and Pb. The Pb composition of the "arc" end-member is more radiogenic than most of the data from the Tofua arc but overlaps with Tafahi and Niuatoputapu, two of the northernmost Tofua arc volcanoes, and with old Tonga arc crust such as the Eocene rocks of 'Eua. No data are available from the Lau Ridge to make a comparison for Pb isotopes but, arguably, 'Eua may be equivalent. The N-MORB component is defined by Unit 7 at site 834, which has relatively unradiogenic Pb and Sr like the Pacific N-MORB mantle sources. The trace element data for glasses from this unit have a smooth MORB-like normalized pattern [*Hergt and Farley*, 1994; *Hawkins*, 1994].

At ODP site 834, basalts erupted in the earliest stage of Lau Basin extension were cored. Their isotopic composition probably represents that for the "original" SSZ mantle. The Nd isotope data are remarkably uniform (0.513109 to 0.513125) and are similar both to MORB and to the modern Lau spreading centers. Sr data range from 0.702544 to

0.702958 (i.e., lower than most of the Lau Basin basalt). Although several samples are more radiogenic than MORB, they do not reach values typical of the modern Lau spreading centers (e.g., 0.7033). Site 834 samples have the most MORB-like Nd and Sr values reported to date for the Lau Basin. Good correlations exist between Sr and Pb isotope compositions. The Pb isotope data define linear trends indicating mixing that project into the field for *Pacific* MORB. This is quite remarkable in view of the widespread evidence elsewhere in the backarc basin for an Indian mantle source. Unit 7, at mid-depth in the hole, is the most MORB-like unit and has the least radiogenic Pb. The Pb isotope ratios become progressively more radiogenic upward to the youngest flow studied (Unit 2). Units 13 (deepest) to 8 also show a similar upward increase from less to more radiogenic; Unit 8 and Unit 2 are very similar. Units 7 and 8 offer the greatest contrast in Pb isotope ratios. The variations with depth, and the apparent cyclicity of the variations, are interpreted as the result of mixing of different proportions of MORB-like and arc-like melts [*Hergt and Farley*, 1994; *Hergt and Hawkesworth*, 1994]. For example, these authors propose that the variation between Unit 7, the most MORB-like sample, and Units 2 and 8, the most radiogenic samples, may be explained by mixing between "essentially pristine MORB" and up to 50% of the "arc" end-member. The timing and nature of the mixing process are fairly well constrained and require mixing of mafic melts rather than melting of a heterogeneous source [*Hergt and Farley*, 1994].

Lavas from site 837 and site 839 have a strong arc signature in terms of mineralogy and chemistry (e.g., they have two-pyroxene basaltic andesite and andesite). They also have arc isotopic characteristics (Figures 13-15). Site 837 samples are similar to the modern Tofua arc, and Pb isotope ratios lie on a trend between Indian mantle and arc. Site 839 Pb isotope ratios are confined to a relatively small field, suggesting that they come from a well-mixed and homogeneous source. The Pb isotope ratios lie on a trend between arc and the Pacific mantle and thus resemble site 834 samples. The isotopic signatures for site 839 resemble those for the modern Tofua arc volcanic island Ata as well as those for arc-derived intrusive dikes at forearc site 841 [*Bloomer et al.*, 1994]. The arc rocks of site 839 probably formed at an arc edifice that was cut off and isolated from the Tonga Ridge by the southward propagation of the ELSC [*Ewart et al.*, 1994b; *Hawkins*, 1994].

Site 836 sampled crust formed early in the history of the ELSC. The propagating rift tip would have passed close to the site at about the time (0.6 to 0.8 Ma) that the site 836 crust was formed [*Parson and Hawkins*, 1994]. Probable igneous basement at site 836 has a strong Indian mantle isotopic signature and resembles rocks of the present axial ridges. Samples from volcanic breccias or gravels are more radiogenic than the presumed basement and are similar to arc-like site 837 samples. The breccias probably came from a nearby seamount.

Collectively, the data of Hergt and Hawkesworth, together with data of *Loock et al.* [1990], suggest that a remnant of "old Pacific" asthenosphere has been available under the easternmost part of the Lau-Tonga system for at least the past 6 m.y. The remnant of Pacific mantle must have migrated eastward since about 5.5 Ma, perhaps in response to the eastward rollback of the Tonga Trench. This inference helps explain the Pacific MORB-source as evidenced in the oldest drill site samples (site 834) as well as its presence under the Valu Fa ridge [*Loock et al.*, 1990]. This depleted source mixed with an Indian MORB-source, an arc component, or both, to form magmas of the modern spreading centers (e.g., CLSC, ELSC, and Valu Fa ridge). *Hergt and Hawkesworth* [1994] propose that an influx of Indian MORB-source mantle under the Lau Basin progressively replaced Pacific MORB-source mantle within the past 5.5 m.y. They postulate that the influx was coupled with the southward propagation of the ELSC. The rift propagation may have been the result or the cause of clockwise rotation of the Tonga Ridge and its underlying mantle. They note that the complex mixture of Indian, Pacific, and arc sources at the Valu Fa rift tip may be explained as a consequence of this process. Another possibility is that the "Pacific" signature is derived from the subducted slab [*Loock et al.*, 1990].

Mixing between MORB-source mantle and other components are not unique to the Lau Basin. For example the Mariana Trough has both a MORB and an arc component [*Volpe et al.*, 1987, 1990]. They proposed a heterogeneous source rather than mixing of melts as proposed for the Lau Basin. The northern Mariana Trough narrows to a propagating rift tip penetrating the Mariana arc; mixing between MORB and either an arc component or a mantle metasomatized by fluid from the subduction zone has been suggested [*Stern et al.*, 1990]. Nascent backarc basins behind the Shichito Ridge, in the Izu-Ogasawara arc, are interpreted as showing evidence for mixing between E-MORB and island arc *magmas* rather than derivation from heterogeneous magma sources [*Ikeda and Yuasa*, 1989]. However, *Hochstaedter et al.* [1990a, 1990b] invoke mixing relations between sources having E-MORB and a subduction component for the Sumisu Rift of the Izu-Ogasawara arc system.

The North Fiji Basin differs from many backarc basins in that the present spreading axes are not situated above the seismic zone, and there is little to suggest that their subjacent mantle could have been infiltrated by subduction-derived fluids. The actively spreading well-developed ridges erupt N-type MORB; crust formed by incipient rifting, on short ridge segments between transform faults, and on seamounts are LILE-enriched alkalic or transitional (BABB) basalts. Melt sources for the N-type MORB are envisaged as massive upwellings of depleted asthenospheric mantle. Two component mixing of melts or the MORB-source and variable small amounts of melt derived from lithosphere enriched in incompatible elements has been proposed [*Price et al.*, 1990]. These authors question the existence of BABB as a magma type and suggest that they are hybrids. They also question the role or necessity of subduction-derived components in explaining backarc magmas. They explain the high volatile content of backarc magmas as a contribution from the alkalic melts, however this fails to explain the elevated volatiles of many of the BABB that are otherwise MORB-like in composition.

The question of whether the isotopic heterogeneity of backarc magmas is caused by selective melting of separate domains comprising a heterogeneous source, or by mixing with injections of different batches of melt, may be closely tied to the dynamics of backarc opening. Injection of new batches of MORB melt to mix with another type of mantle or melt (e.g., SSZ mantle) could be a consequence of mantle counterflow, upwelling, and melting in the extensional region above the subduction zone. Magmas should evolve toward more MORB-like systems as the "other" component is melted, mixed with, and overwhelmed by the influx of MORB-source mantle. Alternatively, a passive dynamic response may be visualized in which heterogeneous mantle is progressively melted with the more fusible (enriched) domains being consumed first. This too would lead to crust dominated by less enriched rocks. The second sequence of events would lead to a linear progression in composition rather than cyclicity and would not necessarily have N-MORB crust as the evolutionary end-point. The Lau Basin data suggest that the first model is a more likely explanation for Lau Basin evolution [*Hergt and Farley*, 1994].

Evidence for Intra-Basin Arc Magmatism

The main focus of most backarc basin petrologic work has been on the MORB-like magmas erupted on the axial ridges and seamounts. The drilling results gave information about the buried crust under small rift sub-basins but in addition gave us a record of arc-composition volcaniclastic material in the sediment column overlying the basaltic basement. With a remnant arc and an active arc on either side of the basin, arc detritus in the sediment column seemed unremarkable at first. However, textural and compositional studies of the detritus provided one of the more important discoveries of ODP leg 135 [*Bednarz and Schmincke*, 1994; *Clift and Dixon*, 1994; Parson et al., 1994]. The drill cores have abundant arc-composition detritus; textures of the shards indicate minimum transport distances. The volcanic debris must have come from ephemeral, arc-composition, volcanism at numerous centers that formed *within* the Lau Basin as it opened. These volcanic centers were a major source of the volcaniclastic sediment deposited in the sub-basins and were active synchronously with the nearby eruption of MORB-like lavas that formed much of the backarc basin crust.

The backarc basin drill sites, as well as the two drilled on the Tonga Ridge, display similar patterns of sedimentation. Late Miocene sediments are the oldest recovered, whereas the youngest are Holocene. Basal sediment ages are different at each site. There is a general trend of eastward younging from site 834, alongside the Lau Ridge, to site 836 on some of the oldest ELSC crust. Except for site 836, the sediments grade upward from a lower sequence of volcaniclastic turbidites interbedded with hemipelagic clayey nannofossil mixed sediments, and with nannofossil clays, into an overlying sequence of hemipelagic clayey nannofossil oozes which locally contain calcareous turbidites [*Rothwell et al.*, 1994; *Parson et al.*, 1994]. These authors interpret the volcaniclastic turbidites as being derived from nearby arc fragments or constructs within the basin rather than from adjacent island arcs. The Pliocene proximal volcaniclastic sediments are interpreted as having been derived from intra-basinal seamounts. Pleistocene sediments are mainly hemipelagic. Site 836, drilled on old ELSC crust, is distinctive in having only 20 m of sediment overlying presumed igneous basement. The lower half of this sediment sequence comprises more than 50% volcaniclastic material. The glass shards, comprise a limited compositional range (54.3 to 58.1 % SiO_2) which they interpret as the products of fractional crystallization of the MORB-like basaltic magmas that formed the basement at site 836 and the nearby ELSC. Similar basaltic andesite has been dredged from the high standing ridges near site 836 (e.g., RNDB 15-02, Table 3).

Rothwell et al. [1994] base their interpretation of a local source for the Pliocene volcaniclastic sediments on the presence of fresh, delicate, needle-like glass shards that could not have survived long distance transport. Neither the Lau Ridge nor the Tofua arc are considered to be likely sources. The chemistry of the glass shards indicates that nearly all have low-K arc tholeiitic compositions ranging from basaltic to rhyolitic. Some shards from sites 834, 835, and 837 have medium-K arc chemistry. Glasses from sites 834 and 835 span the full range from basalt to rhyolite, whereas site 836 has a fairly unimodal composition (basaltic andesite, clustered around 55% SiO_2). Site 838 glasses also have a unimodal concentration of values ranging from 50% to 55% SiO_2. Bimodal suites are found at sites 837 and 839 where basaltic andesites and rhyolites represent maxima. The compositionally similar vitrophyres of Zephyr Shoal, described earlier, are examples of an intrabasinal source like those inferred from the sedimentologic data.

Site 836 glasses are distinct in having trace element ratios similar to those of the nearby ELSC (e.g., MORB-like ratios for Ba/Zr of 0.9 to 1.4), in contrast to the much higher, arc-like values (3.9 to 10), found at the other sites. In general, all of the trace element data for glasses from the backarc sites have broadly similar patterns when normalized to N-MORB. These patterns show the SSZ magma signature, that is, moderate to strong enrichment in LILE, strong depletion in Nb, and minor depletion in other HFSE [*Clift and Dixon*, 1994; *Rothwell et al.*, 1994].

The provenance of the widespread volcaniclastic units is

important to understanding the magmatic evolution of the Lau-Tonga system. The only epiclastic unit that can be tied to the Lau Ridge is a 3.3 Ma bed at site 834 that closely matches the chemistry of the Korobasaga Volcanic Group on the Lau Ridge. The Korobasaga Group was erupted between 4.5 and 2.5 Ma [*Whelan et al*, 1985] and was a potential source of sediment. The limited record of Lau Ridge volcanic material in the drill cores may be due to differences in eruptive style or geographic extent of activity, sediment entrapment in the basin west of site 834, a different pattern of dispersal, a more limited locus of volcanism, or some combination of these factors. The relative unimportance of the Lau Ridge as a source of volcanic detritus for infilling the sub-basins may be related to the fact that the Korobasaga Group is limited to six islands, forming a 35-km-wide belt, in the northern 200 km of the Lau Group [*Cole et al.*, 1985; *Woodhall*, 1985].

The Tofua arc probably was not a robust system until about 1 Ma. It may have evolved over 3 m.y. by southward propagation of volcanic centers [*Parson and Hawkins*, 1994; *Hawkins*, 1994]. There is an extensive record of Miocene volcanism on the Tonga Ridge but the early Miocene beds probably had a source on the Lau Ridge. In latest Miocene time the most likely sources were small and ephemeral (?) intrabasinal eruptive sites [*Clift and Dixon*, 1994; *Rothwell et al.*, 1994]. Zephyr Shoal (Table 4, Figure 3) is an example of such a feature. The low-K arc-series chemistry of many of the Tonga Ridge volcaniclastic units matches the compositions found in the sedimentary units at the backarc drill sites; this supports the idea of derivation from intrabasin volcanoes.

The age of opening and sedimentation of the basins appears to young eastward. The time of transition from deposition of volcaniclastic sediments to sediments dominated by clayey nannofossil ooze varies from site to site but also tends to young eastward. The eastward age progression in timing of the locally derived voluminous eruptions of arc-like material suggests that there was a wave of arc volcanic activity that "tracked" the eastward retreat of the Tonga forearc as it followed the rollback of the trench. This is a previously unrecognized aspect of backarc basin evolution. Intrabasinal arc edifices erupted alongside magma leaks that form MORB-like backarc basin crust. The arc volcanic front, or zone, has been stabilized for about 1 Ma or less and constitutes the present Tofua arc. Well dated volcaniclastic units help constrain the longevity of the intrabasinal arc-like eruptive centers. For example, site 835 records nearby activity from 3.5 to 2.9 or 3.0 Ma. At site 837, proximal sources laid down deposits from 2.1 to 0.34 Ma; at site 838, proximal sources formed volcaniclastic beds from 1.92 to 0.50 Ma; at site 839, the range is from 1.73 to 0.37 Ma [*Bednarz and Schmincke*, 1994]. These "intrabasinal arc" constructs probably had eruptive life spans on the order of 1 to 2 m.y. years before becoming inactive as the arc front migrated eastward.

Forearc Magmatism

Forearcs are generally considered to be cold amagmatic sites of sediment accumulation. This view has been overturned with the discovery by ODP drilling of arc-derived, magmatic bodies intrusive into forearc units of volcaniclastic turbidites. In the Mariana forearc (ODP site 781) Pleistocene age (1.68±0.37 Ma) basalt has intruded late Pliocene sediments [*Marlow et al.*, 1993]. A sill of arc tholeiitic basaltic andesite, intrusive into 15 Ma sediments, was cored in the Bonin forearc at ODP site 793 (*Taylor*, 1991]. At ODP site 841 on the Tonga forearc eleven separate intrusives of arc tholeiitic basalt and basaltic andesite were cored. They intruded distal facies of late Miocene volcaniclastic turbidites that must have had their source on the Lau Ridge volcanic arc. Reconstruction of the geology indicates that site 841 was at least 75 km east of the volcanic arc at that time [*Parson et al.*, 1992; *Bloomer et al.*, 1994]. The intrusives have arc-like chemistry as indicated by enrichment in LILE, depletion in HFSE especially Nb and Ta, and depletion in LREE relative to MORB. Element ratios such as Y/Zr, Ba/Zr, Ba/La, La/Zr and Sr/Nd all resemble late Miocene age arc rocks of the Lau Ridge [*Bloomer et al.*, 1994]. Some representative data are in Table 4. Isotope data also support an arc origin [*Hergt and Hawkesworth*, 1994]. The Sr ratios all are slightly higher than Lau Ridge ratios but are less radiogenic than those of the Tofua arc; Nd ratios are slightly higher than those of the Lau Ridge. Pb isotopes lie in the Pacific MORB array for $^{207}Pb/^{204}Pb$ vs. $^{206}Pb/^{204}Pb$ and slightly above it on the plot of $^{208}Pb/^{204}Pb$ vs. $^{206}Pb/^{204}Pb$. No Lau Ridge Pb isotope data are available for comparison.

The source of the magmas must have been at Lau Ridge magma systems but the transport of melts into the forearc poses thermal and structural problems. These are discussed by *Bloomer et al.* [1994].

A MODEL FOR BACKARC EVOLUTION

Introduction

The origin and evolution of the Lau Basin, and other backarc basins, must be controlled by kinematic responses of SSZ crust and upper mantle to oceanic plate subduction and the consequent mantle counterflow, upwelling, and melting. Extension of supra-subduction zone lithosphere, and formation of new crust by emplacement of melts derived from the SSZ mantle are expressions of the kinematic response. The magma sources include SSZ mantle, a subduction component, and "new" mantle brought in by convective overturn. The model presented here helps explain Lau Basin evolution and probably applies to the Mariana Trough. With some modifications it may be suitable for other western Pacific backarc basins. The style of deformation, and related magmatic activity, resemble those

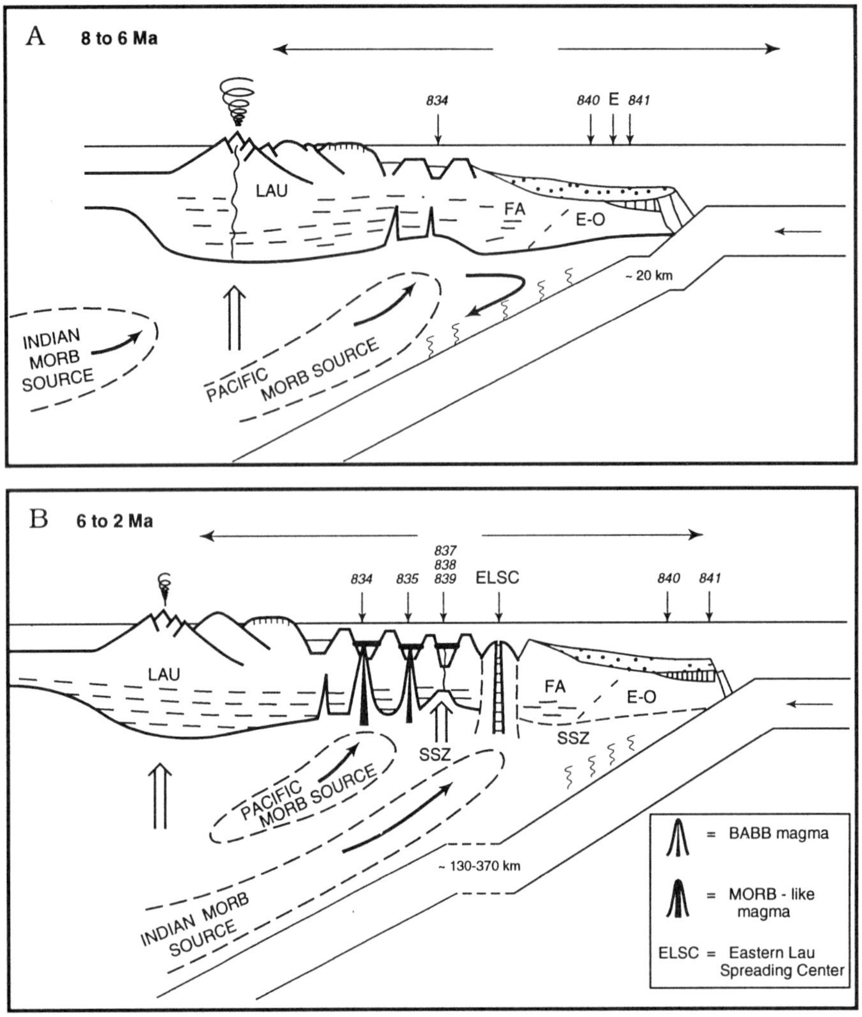

Fig. 16. A model for the evolution of the Lau Basin. Sections are projections on to latitude 20°S. A. Active Lau Ridge volcanism is contemporary with rifting of the forearc and the beginning of crustal extension to form the WEB. Magmatic activity has not yet filled the extensional basin but upwelling of Pacific mantle is setting the stage for it. Note the insertion of the flexure or imbrication in the subducted lithosphere. FA is remnants of rifted forearc, E-O is the rhyolitic complex at ODP site 841 and the mafic crust exposed on 'Eua (E). Numbers are ODP drill sites. B. This shows effects of extension; upwelling of Pacific mantle is the magma source for WEB sub-basins. The southward propagating ELSC is fed by Indian MORB-source mantle. Not shown are the ephemeral intrabasin arc-composition volcanoes that formed progressively from west to east. C. Present configuration with the Tofua arc and ELSC both active. The southward propagating CLSC lies north of this section. The vertical dimension in all sketches is exaggerated to show details of crustal geology.

that occurred in the extensional provinces in continental settings such as the Basin-Range Province of western North America. Backarc basin evolution also may be a small scale replica of the early stages of formation of the major ocean basins.

The Model

Opening of the Lau Basin has involved two modes of extension. Initially, rifting and extension formed small sub-basins separated by high standing blocks; subsequently, seafloor spreading began and continues today. The rifting and extension phase formed new crust within the sub-basins. The deep structure of these sub-basins is not known, but it is likely to be different from "true" deep sea floor in that "new" basaltic crust may overlie attenuated remnants of older crust (e.g., fragments of old arc or forearc). Extension was accompanied by extrusion of both arc-like and MORB-like magmas that formed new crust. The proximity to active island arcs, and the similarity in many respects between arc

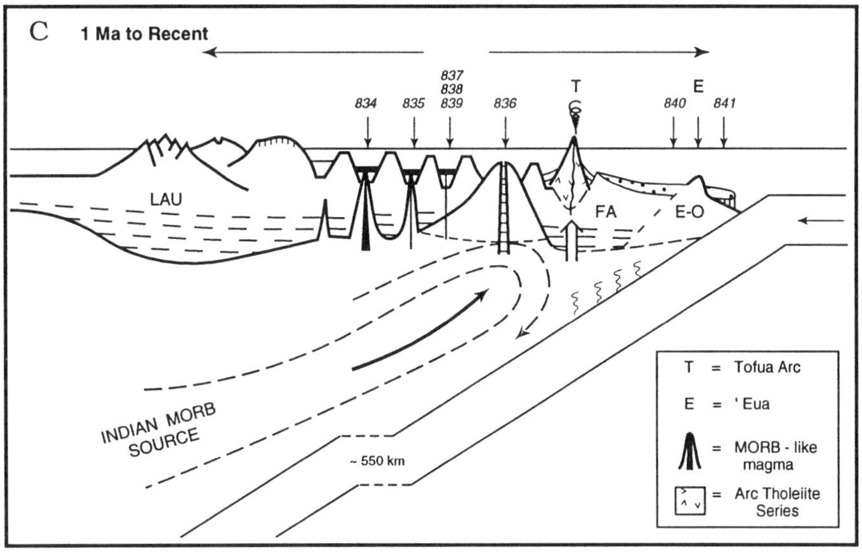

Figure 16. (continued)

and backarc magmas, suggests similarities in petrogenesis. Extreme rifting and extension ultimately separated all remnants of the older crust and began the second phase of opening in which rifts were filled with new ocean crust that directly overlies its mantle source. The second phase of opening was promoted by propagating rifts. Studies of other western Pacific backarc basins suggest a common evolutionary model in which crustal extension is followed by rift propagation and seafloor spreading [*Stern et al.*, 1984; *Parson et al.*, 1990; *Tamaki et al.*, 1992; *Taylor*, 1992; *Fryer and Pearce*, 1992; *Parson et al.*, 1992; *Parson and Hawkins*, 1994.

The model shown in Figure 16 is an updated version of earlier models by the author [*Hawkins et al.*, 1984; *Hawkins and Melchior*, 1985]. It is based on rock and sediment chemistry, geophysical data, ages, interpretations of the regional geology from the numerous contributions cited elsewhere in the text, and extensive data collected by the author. *Stern and Bloomer* [1992] proposed a similar interpretation of the early history of the Izu-Bonin-Mariana system and used it to explain the origin of ophiolite remnants in the California Coast Range. Interpretations of data from ODP drilling in the Sea of Japan, the Izu-Bonin system and the Mariana arc all have led to broadly similar, independently derived models [*Tamaki et al.*, 1992; *Taylor*, 1992; *Fryer and Pearce*, 1992; *Pearce et al.*, 1992].

A discussion of the initial conditions in the southwestern Pacific at the beginning of Cenozoic time is beyond the scope of this paper, but there is evidence for a vast zone of extension and upwelling along the 4400 km trace of the West Melanesian-Solomon-Vitiaz trench and arc system; details are discussed elsewhere [*Kroenke*, 1984; *Hawkins*, 1994]. A similar mega-arc is required to explain the evolution of the Palau-Kyushu Ridge [*Mattey et al.*, 1980; *Stern and Bloomer*, 1992]. A profound change in upper mantle rheology, perhaps accompanied by a change in relative plate motion, must have played a role in forming these major geologic features. Mantle upwelling and crustal extension promoted the development of boninitic magma systems early in arc history and at relatively shallow mantle levels. This has been invoked for the Izu-Bonin arc [*Pearce et al.*, 1992; *Stern and Bloomer*, 1992] and for the Mariana forearc [*Bloomer and Hawkins*, 1987; *Stern et al.*, 1991]. Boninite magmas were followed by, and perhaps overlapped with, eruption of arc tholeiitic series melts. There is a major problem in explaining the mismatch in the age of initiation of the western Pacific Eocene arc systems and the timing of the relative change in motion of the Pacific Plate (about 43 Ma) as inferred from the age of the bend of the Emperor-Hawaii seamount chain. Evidence for volcanism at least 5 to 6 m.y. earlier is found on the Palau-Kyushu Ridge and in the Bonin Islands [*Taylor*, 1992, and references therein]. Thus, it is difficult to relate *initiation* of arc volcanism to the change in motion even though plate convergence has been important subsequently. The boninitic and early arc tholeiitic stages of an arc require melting of a suitable mantle source at relatively shallow depths [*Stern et al.*, 1991; *Pearce et al.*, 1992]; plate convergence may not be required, and, in fact, lithosphere extension could well have been the tectonic control on early boninitic magmatism. Admittedly this is speculative, but an early phase of extension could initiate the boninitic magmatism and set up crustal buoyancy differences that could be exploited by subsequent plate motion changes to cause subduction. A similar model was proposed by *Stern and Bloomer* [1992] and a less refined version was presented by *Hawkins et al.* [1984]. Different, and equally plausible models, presented by *Pearce et al.* [1992] include subduction of a spreading ridge beneath young lithosphere. Whatever the mechanism for its initiation, the earliest stages of Eocene arc development in the northwestern Pacific resulted in an arc that was up to 300 km wide and 3000 km long formed of boninite and arc tholeiite [*Taylor*, 1992]. A comparable arc developed concurrently in the southwestern Pacific.

Considering the present Papua-Vanuatu-Fiji-Lau-Tonga region, there is evidence for an Eocene Vitiaz arc or West Melanesian arc on the northeastern margin of the Indian Plate [*Kroenke*, 1984; *Colley and Hindle*, 1984; *Colley et al.*, 1986, 1992, 1993; *Gill*, 1987; *Wharton et al.*, 1992] that may have been as much as 200 km wide and nearly 2000 km long. This ancestral arc includes 46 Ma arc tholeiitic gabbro and gabbro-norite, pillow basalts, low-K rhyolite, and boninite. Although these ages are close to the inferred age for the Emperor-Hawaii bend, they are hard to reconcile with a cause-effect relationship. We would not expect an instant response of magmatism to the beginning of subduction. It is probable that there was a delay of a few million years between the onset of magma separation and upwelling from a mantle source and the beginning of eruption. The silicic rocks are voluminous (?), highly differentiated melts, that must have had mafic precursors. A further delay is likely to allow the extreme fractionation. Whether these rocks are related to extension or are the result of subduction, they constitute some of the oldest crust of the West Melanesian or Vitiaz arc. Evidence for Oligocene arc magmatism is widespread and includes rocks on Fiji, 'Eua, and Vanuatu, as well as New Britain and New Ireland [*Gill*, 1987]. These were no doubt a response to subduction after the change in Pacific Plate motion. Opening of the South Fiji Basin as a backarc system appears to have been a response to convergence at the Vitiaz Trench in Oligocene time. Remnants of the Eocene-Oligocene arc are shown under the forearc on Figure 16 and labeled as "E-O".

The Miocene Lau Ridge volcanic arc (Figure 16A) probably faced the present Tonga Trench, however we have no control on its actual orientation and it is likely that it has undergone some rotation. It is uncertain whether or not there was a break in arc volcanism in the early Miocene; neither outcrops nor drill site data offer an indication for continuity. The Eocene-Oligocene arc probably formed part of the forearc to the Lau Ridge as evidence from 'Eua and other Tongan islands indicates that Miocene dikes and volcaniclastic material were emplaced into, or deposited on, the older forearc rocks. Nearly all of the rocks of the forearc have an affinity to arc magmas and this suggests that the underlying crust is largely arc material. Some evidence exists for older MORB-like crust under the forearc in the occurrence of basaltic clasts in the low-K rhyolite at site 841 [*Bloomer at al.*, 1994]. These are not unequivocal evidence for trapped older deep seafloor inasmuch as MORB-like rocks are among the earliest magma types formed in arc rifting [*Taylor et al.*, 1992]. MORB-like lavas in the Mariana forearc were reported by *Johnson and Fryer* [1990] who suggested that they may represent injection of magma during forearc rifting. It is likely that much, or all, of the forearc is underlain by crust formed in an island arc system. The SSZ source for the Lau Ridge volcanic arc is considered to be depleted harzburgite that had been metasomatised by fluid derived from subducted, altered ocean crust [e.g., *Pearce et al.*, 1984; *Tatsumi et al.*, 1986; *Tatsumi,* 1989]; a minor role may have been played by subducted sediments. The initial depleted mantle source carried the isotope signature of the source of Pacific MORB. Late in the evolution of the Lau Ridge, at about 6 Ma, it began to be rifted, and at about the same time crustal extension began to form rift basins in the forearc to the Lau Ridge. An imbrication or a flexure of the subducted plate may have formed at the trench at about 7 to 8 Ma [*Billington*, 1980; *Louat and Dupont*, 1982], as shown schematically in Figure 16. This flexure may have been either a cause or an effect of extension. The present depth of the flexure, between latitudes 20 and 23°S, is about 525 to 575 km (Figure 16). Opening of the Lau Basin included an extension and rifting phase followed by seafloor spreading. Extension and rifting formed the basins drilled at all of the backarc sites except for site 836, which is on crust formed by seafloor-spreading. Site 834 basalts appear to contradict the compositional zonation of the Lau Basin proposed by *Hawkins and Melchior* [1985] and shown on Figure 4. They should be transitional in composition like the rocks dredged from nearby ridges. Some site 834 petrologic units resemble Mariana Trough basalts in major elements and HFSE but they lack LILE enrichment. A possible explanation may be that the long-term melting that partly filled the sub-basin at site 834 thoroughly depleted the source in the LILE-enriched SSZ component and tapped the depleted MORB-source component which dominated the later stages of melting. The rocks we drilled recorded only the products of the later stages of melting.

Site 834 lavas came mainly from a "Pacific" MORB source as shown by *Hergt and Hawkesworth* [1994], whereas sites 837 to 839 had an SSZ source with some mixture of the Pacific source. Site 836 magmas, and the modern spreading centers, were derived from an "Indian" MORB source [cf. *Volpe et al.*, 1988]. The two-dimensional sketch does not show the complexity of the probable mixing pattern. The influx of the Indian MORB source would have swept down the axis of the basin from the north in the wake of the propagating rift tip of the ELSC. The trace of a "Samoan plume component" in the northern part of the basin [*Poreda*, 1985] offers further support for this idea. The Pacific MORB source either was consumed, bypassed, or overwhelmed by the Indian source as the basin evolved. The final phase of evolution included the initiation and southward propagation of a second rift system at about 1 to 2 Ma that formed the CLSC. Mixing of MORB-like and arc magmas has been important throughout the history of opening of the basin [*Hergt and Farley*, 1994]. The development of a new volcanic arc (Tofua arc) on the outboard edge of the Lau Basin probably occurred within the past 3 m.y. and has propagated from north to south down the axis of the Tofua arc. A complexity not shown because of limitations of the scale of the sketches is the seaward march of the small, intrabasinal, arc composition volcanoes that contributed to the infilling of the rift basins. These may have been no less significant at any time than the scattering of arc volcanoes that constitute the present Tofua arc. Thus, the long-standing question of whether or not there was synchronous backarc and arc magmatism has an indefinite answer in the Lau Basin. Both phenomena have been operative more or less continuously *within* the basin as it opened. However, a well-defined arc on the Tonga Ridge probably did not appear until after about 6 Ma of backarc spreading.

The Lau-Tonga system serves as a model for understanding relict convergent margin rocks found as terranes accreted to continental margins and as ophiolite series assemblages. It seems likely that eventually plate reorganization will lead to the amalgamation of the arc and backarc systems of the western Pacific and their eventual accretion to a continent. There is ample evidence in areas of composite terranes such as the Philippine archipelago for amalgamation of arc-backarc-forearc systems like the Lau-Tonga system [e.g., *Hawkins and Evans*, 1983; *Hawkins et al.*, 1985; *Florendo*, 1994]. There is an increasing general acceptance of the idea that many ophiolites are remnants of SSZ magma systems. The leg 135 transect, together with the other western Pacific ODP studies, provide important new data and insight to understanding the geology of components involved in continental evolution through terrane accretion.

SUMMARY

The Lau Basin has had a complex history that is part of the broader geologic evolution of the Tonga convergent plate margin. To understand Lau Basin history properly it has to be considered in the context of what happened on the adjacent remnant and active volcanic arcs. A brief synopsis is as follows:

(1) The Lau Ridge was an active volcanic island arc from at least 14 Ma to 2.5 Ma. The volcanism was in response to subduction at an ancestral Tonga Trench that may have had the same general orientation as the present one but the trench axis probably lay 100 km or more to the west. Volcanic activity younger than 2.5 Ma appears to have been related to extensional stresses to the rear of the arc and backarc rather than to frontal arc volcanism.

(2) The geologic history of the Lau Ridge may extend back to late Eocene time as suggested by the occurrence of Eocene arc volcanic rocks on Viti Levu, Fiji Islands, and Eocene arc plutonic rocks on 'Eua, in the Tonga forearc. Leg 135 drilling at site 841 also revealed pre-late Eocene and late Miocene arc series volcanic rocks on the Tonga Ridge. All of these rocks are interpreted as remnants of a former "Melanesian island arc" that included what are now parts of the Lau Ridge, Tonga Ridge, Fiji Platform, and New Hebrides-Vanuatu Ridge. The orientation of the trench related to this ancestral arc is not known but it probably faced to the east or the northeast.

(3) The Lau Basin began to open in latest Miocene time (about 6 Ma) as the eastern part of the Lau Ridge and its forearc were first extended and then rifted away to the east. Fragments of the Lau Ridge and its older basement are found on the Tonga Ridge.

(4) There has been more or less continuous arc volcanism in the Lau-Tonga region since late Eocene time. Lau Ridge volcanism overlapped with the time of initial crustal extension that formed the Lau Basin, but during the last 2.5 Ma it has diminished and been supplanted by seafloor spreading in the Lau Basin. Within the last 1 m.y. the Tofua arc has become a major feature on the Tonga Ridge. The Tofua arc may have developed sequentially from north to south, over about 3 m.y., in consort with the rotational opening of the Lau Basin.

(5) Lau Basin crust is largely floored by basaltic rocks having MORB mineralogy and many MORB-like chemical and isotopic characteristics. The Central Lau and Eastern Lau spreading centers are well-defined axial ridge systems formed mainly of N-MORB type basalts. Two other areas of active spreading are on the NLSC and MTJ.

(6) Although dominated by MORB-like chemistry, the crust of the Lau Basin, like other backarc basins, is heterogeneous. On a regional scale, the Lau Basin is compositionally zoned from older rocks having boninitic and "transitional to arc" compositions to younger MORB-like rocks on the spreading centers. However, the sub-basin drilled at site 834 contradicts this simple pattern; the oldest crust at this site has MORB chemistry. Seamounts in the northern Lau Basin include rocks having an ocean island basalt composition component; dacitic rocks of Zephyr Shoal may be silicic differentiates of boninite magma series. Rocks of the Valu Fa ridge and the Mangatolu triple junction range from basaltic to basaltic andesite; the VFR has a strong arc signature and probably comes from a mixed source whereas the MTJ has few arc characteristics. ODP sites 834, 835, and 836 although not on a true "flow line" in the sense of progressive accretion of new crust, sampled the evolution of magma types over a time span of about 5 m.y. The sites are located, respectively, on the oldest crust drilled (about 5.6 Ma to 6 Ma, latest Miocene), intermediate age crust (about 3.4 Ma, late Pliocene), and the youngest crust drilled (about 0.64 to 0.8 Ma, middle Pleistocene). Site 836 lies about 50 km west of the axis of the ELSC which is formed of MORB-like tholeiite. The ELSC becomes transitional from MORB to arc tholeiite southward towards the Valu Fa ridge. Sites 837, 838, and 839 form a cluster, about 125 km west of the active ELSC, on crust estimated to be 1.9, 2.0, and 2.1 Ma, respectively. Site 839 samples are arc-like in chemistry in spite of their proximity to the ELSC.

Acknowledgments. This synthesis has drawn extensively on the results of ODP leg 135 and contributions from shipmates that are in the Scientific Results Volume for leg 135. The author's work in the Lau Basin has been supported by several research awards from the National Science Foundation in particular from awards OCE 84-15472, OCE 87-16660, JOI JSC 5-89, and Texas A&M 20556. I thank Jo Griffiths, Evelyn Hegemier, and Fred Florendo for making the illustrations. Jim Natland and Jill Karsten are thanked for their reviews and suggestions that have improved the original manuscript. Lindsay Parson, my co-chief scientist on leg 135, provided valuable insight to our new understanding of Lau Basin evolution through his interpretations of GLORIA imagery.

REFERENCES

Allan, J. F., Cr-spinel in depleted basalts from the Lau Basin backarc, ODP Leg 135; petrogenetic history from Mg-Fe crystal-liquid exchange, in *Proc. ODP, Sci. Results, 135*, J. W. Hawkins, L .M. Parson, J. F. Allan, et al., pp. 565-584, Ocean Drilling Program, College Station, TX., 1994.

Arculus, R. J., and R. Powell, Source component mixing in the regions of arc magma generation, *J. Geophys. Res., 91*, 5913-5926, 1986.

Barazangi, M., and B. Isacks, Lateral variations of seismic-wave attenuation in the upper mantle above the inclined earthquake zone of the Tonga Island Arc: deep anomaly in the upper mantle, *J. Geophys. Res., 76*, 8493-8516, 1971.

Bednarz, U., and H. U. Schmincke, Composition and origin of volcaniclastic sediments in the Lau Basin (Southwest Pacific), Leg 135 (Sites 834-839), in *Proc. ODP, Sci. Results, 135*, J. W. Hawkins, L. M. Parson, J. F. Allan, et al., pp. 51-74, Ocean Drilling Program, College Station, TX, 1994.

Ben Othman, D., W. M. White, and J. Patchett, The geochemistry of marine sediments, island arc magma genesis, and crust-mantle recycling, *Earth Planet. Sci. Lett., 94*, 1-21, 1989.

Bevis, M., F. W. Taylor, B. E. Schutz, and S. Calmant, Geodetic observations of convergence and back-arc spreading in the SW Pacific (1988-1992), (abstract) *EOS, Trans. AGU, 74*, 60, 1993.

Billington, S., The morphology and tectonics of the subducted lithosphere in the Tonga-Kermadec-Fiji region from seismicity and focal mechanism solutions, Ph.D. thesis, Cornell Univ., Ithaca, NY, 1980.

Bloomer, S. H., and J. W. Hawkins, Petrology and geochemistry of boninite series volcanic rocks from the Mariana Trench, *Contrib. Mineral. Petrol., 97*, 361-377, 1987.

Bloomer, S. H., A. Ewart, J. Hergt, and W. B. Bryan, Geochemistry and origin of igneous rocks from the outer Tonga forearc, Site 841, in *Proc. ODP, Sci. Results, 135*, J. W. Hawkins, L. M. Parson, J. F. Allan, et al., pp. 625-646, Ocean Drilling Program, College Station, TX, 1994.

Boespflug, X., L. Dosso, H. Bougault, and J. L. Joron, Trace element and isotopic (Sr, Nd) geochemistry of volcanic rocks from the Lau Basin, *Geol. Jahrb., 92*, 503-516, 1990.

Briqueu, L., H. Bougault, and J. L. Joron, Quantification of Nb, Ta, Ti, and V anomalies in magmas associated with subduction zones: petrogenetic implications, *Earth Planet. Sci. Lett., 68*, 297-308, 1984.

Bryan, W. B., Low K_2O dacite from the Tonga-Kermadec island arc: petrography, chemistry, and petrogenesis, in *Trondhjemites, Dacites, and Related Rocks*, edited by F. Barker, pp. 581-600, Elsevier, Amsterdam, 1979a.

Bryan, W. B., Regional variation and petrogenesis of basalt glasses from the FAMOUS area, Mid-Atlantic Ridge, *J. Petrol., 20*, 293-325, 1979b.

Bryan, W. B., and T. H. Pearce, Plagioclase zoning in selected lavas from Holes 834B, 839B, and 841B, in *Proc. ODP, Sci. Results, 135*, J. W. Hawkins, L. M. Parson, J. F. Allan, et al., pp. 543-556, Ocean Drilling Program, College Station, TX, 1994.

Bryan, W. B., G. D. Stice, and A. E. Ewart, Geology, petrography, and geochemistry of the volcanic islands of Tonga, *J. Geophys. Res., 77*, 1566-1585, 1972.

Bryan, W. B., G. Thompson, F. A. Frey, and J. S. Dickey, Inferred geologic settings and differentiation in basalts from the Deep Sea Drilling Project, *J. Geophys. Res., 81*, 4285-4304, 1976.

Cande, S. C. and J. C. Mutter, A revised identification of the oldest sea-floor spreading anomalies between Australia and Antarctica, *Earth Planet. Sci. Lett., 58*, 151-160, 1982.

Carlson, R. W., J. D. Macdougall, and G. W. Lugmair, Differential Sm/Nd evolution in oceanic basalts, *Geophys. Res. Lett., 5*, 229-232, 1978.

Clift, P., and J. E. Dixon, Variations in arc volcanism and sedimentation related to rifting of the Lau Basin, Southwest Pacific, in *Proc. ODP, Sci. Results, 135*, J. W. Hawkins, L. M. Parson, J. F. Allan, et al., pp. 23-50, Ocean Drilling Program, College Station, TX, 1994.

Cole, J. W., J. B. Gill, and D. Woodhall, Petrologic history of the Lau Ridge, Fiji, in *Geology and Offshore Resources of Pacific Island Arcs - Tonga Region*, edited by D. Scholl and T. Vallier, pp. 379-414, Circum-Pacific Council for Energy and Mineral Resources, Houston TX, 1985.

Colley, H., and W. H. Hindle Volcano-tectonic evolution of Fiji and adjoining marginal seas, in *Marginal Basin Geology*: edited by B. P. Kokelaar and M. F. Howells, pp. 151-162, Blackwell, Oxford, 1984.

Colley, H., H. W. Hindle, and B. Hathaway, Early arc and basinal igneous activity on Viti Levu, Fiji, (abstract) IAVCEI Congress, New Zealand, 140, 1986.

Colley, H., M. R. Wharton, and B. Hathaway, Magmatism and initial arc formation in Fiji, *Ancient Volcanism and Modern Analogues*, (abstract), IAVCEI General Assembly, Canberra, Australia, 22, 1993.

Cousens, B. L., J. F. Allan, and M. P. Gorton, Subduction-modified pelagic sediments as the enriched component in back-arc basalts from the Japan Sea: Ocean Drilling Program Sites 797 and 794, *Contrib. Mineral. Petrol.*, in press, 1994.

Crawford, A. J., T. J. Falloon, and D. H. Green, Classification, petrogenesis and tectonic setting of boninites, in *Boninites and Related Rocks*, edited by A. J. Crawford, pp. 2-49, Unwin Hyman, London, 1989.

Davis, A. S., D. A. Clague, and J. L. Morton, Volcanic glass compositions from two spreading centers in Lau Basin, southwest Pacific Ocean, *Geol. Jahrb., 92*, 481-501, 1990.

DePaolo, D. J., and G. J. Wasserburg, The sources of island arcs as indicated by Nd and Sr isotope studies, *Geophys. Res.*

Lett., 4, 465-468, 1977.

Dick, H. J. B., The petrology of two back arc basins of the northern Philippine Sea, *Am. Jour. Sci., 282*, 644-700, 1982.

Duncan, R. A., T. Vallier, and D. A. Falvey, Volcanic episodes on 'Eua, Tonga, in *Geology and Offshore Resources of Pacific Island Arcs - Tonga Region.* edited by D. Scholl and T. Vallier, pp. 281-290, Circum-Pacific Council for Energy and Mineral Resources, Houston TX, 1985.

Dupuy, C., J. Dostal, G. Marcelot, H. Bougault, J. L. Joron, and M. Treuil, Geochemistry of basalts from central and southern New Hebrides arc: implications for their source rock composition, *Earth Planet. Sci. Lett., 60*, 207-225, 1982.

Eguchi, T., Seismotectonics of the Fiji Plateau and Lau Basin, *Tectonophysics, 102*, 17-32, 1984.

Elsasser, W. M., Sea-floor spreading as thermal convection, *J. Geophys. Res., 76*, 1101-1112, 1971.

Ernewein, M., J. A. Pearce, S. H. Bloomer, L. M. Parson, B. J. Murton, and L. E. Johnson, Geochemistry of Lau Basin volcanic rocks: influence of ridge segmentation and arc proximity, in *Volcanism Associated with Extension at Consuming Plate Margins*, edited by J. Smellie, Geol. Soc. Spec. Publ., London, in press, 1994.

Ewart, A., A review of the mineralogy and chemistry of Tertiary-Recent dacitic, latitic, rhyolitic, and related salic volcanic rocks, in *Trondhjemites, Dacites, and Related Rocks*, edited by F. Barker, pp. 13-22, Elsevier, Amsterdam, 1979.

Ewart, A., and W. B. Bryan, Petrography and geochemistry of the igneous rocks from 'Eua, Tongan Islands, *Geol. Soc. Am. Bull., 83*, 3281-3298, 1972.

Ewart, A. E., and C. J. Hawkesworth, Pleistocene to Recent Tonga - Kermadec arc lavas: interpretation of new isotopic and rare earth data in terms of a depleted mantle source model, *J. Petrol., 28*, 495-530, 1987.

Ewart, A., W. B. Bryan, and J. B. Gill, Mineralogy and geochemistry of the younger Tongan Islands, S.W. Pacific, *J. Petrol., 14*, 429-466, 1973.

Ewart, A., R. N. Brothers, and A. Mateen, An outline of the geology and geochemistry, and the possible petrogenetic evolution of the volcanic rocks of the Tonga-Kermadec-New Zealand island arc, *J. Volcanol. Geotherm. Res., 2*, 205-250. 1977.

Ewart, A., W. B. Bryan, B. W. Chappell, and R. L. Rudnick, Regional geochemistry of the Lau-Tonga arc and back-arc systems, in *Proc. ODP, Sci. Results, 135*, J. W. Hawkins, L. M. Parson, J. F. Allan, et al., pp. 385-426, Ocean Drilling Program, College Station, TX, 1994a.

Ewart, A., J. Hergt, and J. W. Hawkins, Major, trace element and Pb, Sr, and Nd isotope geochemistry of Site 839 basalts and basaltic andesites: implications for arc volcanism, in *Proc. ODP, Sci. Results, 135*, J. W. Hawkins, L. M. Parson, J. F. Allan, et al., pp. 519-532, Ocean Drilling Program, College Station, TX, 1994b.

Falloon, T. J., A. Malahoff, L. P. Zonenshain, and Y. Bogdanov, Petrology and geochemistry of back-arc basin basalts from Lau Basin spreading ridges at 15, 18, and 19°S, *Mineral. Petrol.*, in press, 1994.

Florendo, F. F., Tertiary arc rifting in northern Luzon, Philippines, *Tectonics*, in press, 1994.

Frenzel, G., R. Muhe, and P. Stoffers, Petrology of the volcanic rocks from the Lau Basin, southwest Pacific, *Geol. Jahrb., 92*, 395-479, 1990.

Fryer, P., J. Sinton, and J. A. Philpotts, Basaltic glasses from the Mariana Trough, in *Init. Repts. DSDP, 60*, D.M. Hussong, and S. Uyeda, et al., pp. 601-610, U.S. Govt. Printing Office, Washington D.C., 1981.

Fryer, P., and J. A. Pearce, J.A., Introduction to the scientific results of Leg 125, in *Proc. ODP, Sci. Results, 125*, P. Fryer, J. A. Pearce, L. B. Stokking, et al., pp. 3-14, Ocean Drilling Program, College Station, TX, 1992.

Furlong, K. P., D. S. Chapman, and P. W. Alfeld, Thermal modeling of the geometry of subduction with implications for the tectonics of the overriding plate, *J. Geophys. Res., 87*, 1786-1802, 1982.

Garcia, M., N. W. K. Liu, and D. W. Muenow, Volatiles from submarine volcanic rocks from the Mariana Island Arc and Trough, *Geochim. Cosmochim. Acta, 43*, 305-312, 1979.

Giardini, D., and J. H. Woodhouse, Deep seismicity and modes of deformation in Tonga subduction zone, *Nature, 307*, 505-509, 1984.

Giardini, D., and J. H. Woodhouse, Horizontal shear flow in the mantle beneath the Tonga Arc, *Nature, 319*, 551-555, 1986.

Gill, J. B., Composition and age of Lau Basin and ridge volcanic rocks: Implications for evolution of an interarc basin and remnant arc, *Geol. Soc. America Bull., 87*, 1384-1395, 1976.

Gill, J. B., *Orogenic Andesites*, Springer-Verlag, Berlin, 390 pp., 1981.

Gill, J. B., Early geochemical evolution of an oceanic island arc and backarc: Fiji and the South Fiji Basin, *J. Geol., 95*, 589-615, 1987.

Hamburger, M. W., and B. L. Isacks, 1988. Diffuse back-arc deformation in the southwestern Pacific, *Nature, 332*, 599-604.

Hart, S. R., 1969. K, Rb, Cs contents and K/Rb, K/Cs ratios of fresh and altered submarine basalts, *Earth Planet. Sci. Lett., 6*, 295-303.

Hart, S. R. and A. J. Nawalk, 1970. K, Rb, Cs, and Sr relationships in submarine basalts from the Puerto Rico Trench, *Geochim. Cosmochim. Acta., 34*, 145-155.

Hart, S. R. and H. Staudigel, Isotopic characterization and identification of recycled components, in *Crust/Mantle Recycling at Convergence Zones*, edited by S. R. Hart and L. Gulen, pp. 15-20, Kluwer Academic Publishers, 1989.

Hart, S. R., W. A. Glassley, and D. E. Karig, Basalts and seafloor spreading behind the Mariana Island arc, *Earth Planet. Sci. Lett., 15*, 12-18, 1972.

Hart, S. R., A. J. Erlank, and E. J. D. Kable, Sea floor basalt alteration: some chemical and strontium isotopic effects, *Contrib. Mineral. Petrol., 44*, 219-230, 1974.

Hawkins, J. W., Geology of the Lau Basin, a marginal sea behind the Tonga Arc, in *The Geology of Continental Margins*, edited by C. A. Burk, and C. L. Drake, pp. 505-520, Springer-Verlag, New York, 1974.

Hawkins, J. W., 1976. Petrology and geochemistry of basaltic rocks of the Lau Basin, *Earth Planet. Sci. Lett., 28*, 283-297.

Hawkins, J. W., Petrological and geochemical characteristics of marginal basin basalts, in *Island Arcs, Deep Sea Trenches, and Back Arc Basins*, edited by M. Talwani, and W. C.

Pitman, III, pp. 355-365, Maurice Ewing Series, 1, AGU, Washington, D.C., 1977.

Hawkins, J. W., Low-K rhyolitic pumice from the Tonga Ridge, in *Geology and Offshore Resources of Pacific Island Arcs - Tonga Region*, edited by D. Scholl and T. Vallier, pp. 171-178, Circum-Pacific Council for Energy and Mineral Resources, Houston TX, 1985.

Hawkins, J. W., "Black smoker" vent chimneys, Lau Basin, *EOS, 67*, 430-431, 1986.

Hawkins, J. W., Cruise Report - PAPATUA expedition, Leg 04, R/V *Thomas Washington SIO Ref. Ser.*, 88-14, 1988.

Hawkins, J. W., Cruise Report - ROUNDABOUT expedition, Legs 14, 15, R/V *Thomas Washington SIO Ref. Ser.*, 89-13, 1989.

Hawkins, J. W. Petrologic synthesis, ODP Leg 135 - Lau Basin transect, in *Proc. ODP, Sci. Results, 135*, J. W. Hawkins, L. M. Parson, J. F. Allan, et al., pp. 879-908, Ocean Drilling Program, College Station, TX, 1994.

Hawkins, J. W., and J. F. Allan, Petrologic evolution of the Lau Basin, sites 834-839, in *Proc. ODP, Sci. Results, 135*, J. W. Hawkins, L. M. Parson, J. F. Allan, et al., pp. 427-470, Ocean Drilling Program, College Station, TX, 1994.

Hawkins, J. W. and C. A. Evans, Geology of the Zambales Range, Luzon, Philippines: ophiolite derived from an island-arc backarc basin pair, in *The Tectonic Evolution of Southeast Asian Seas and Islands*: (pt. 2), *Geophys. Monogr. Ser.*, vol 27, edited by D. Hayes, pp. 124-138, AGU, Washington, D.C., 1983.

Hawkins, J. W., and D. A. Falvey, Petrology of andesitic dikes and flows from 'Eua, Tonga, in *Geology and Offshore Resources of Pacific Island Arcs - Tonga Region*, edited by D. Scholl and T. Vallier, pp. 269-280, Circum-Pacific Council for Energy and Mineral Resources, Houston TX, 1985.

Hawkins, J. W., and S. Helu, Polymetallic sulphide deposit from "black-smoker" chimney: Lau Basin, (abstract), *EOS, Trans. AGU, 67*, 378, 1986.

Hawkins, J. W., and J. T. Melchior, Petrology of Mariana Trough and Lau Basin basalts, *J. Geophys. Res., 90*, 11431-11468, 1985.

Hawkins, J. W., J. G Sclater, and G. Hohnhaus, Petrologic and geophysical characteristics of the Lau Basin Ridge: a spreading center behind the Tonga Arc, (abstract), *Geol. Soc. Am. Ann. Mtg, 2*, 71, 1970.

Hawkins, J. W., S. H. Bloomer, C. A. Evans, and J. T. Melchior, Evolution of intra-oceanic arc-trench systems, *Tectonophysics, 102*, 174-205, 1984.

Hawkins, J. W., G. F. Moore, R. Villamor, C.A. Evans, and E. Wright, Geology of the composite terranes of east and central Mindanao, in *Tectonostratigraphic Terranes of the Circum-Pacific Region*, edited by D. G. Howell, pp. 437-463, Circum-Pacific Council for Energy and Mineral Resources, Earth Sci. Ser. No. 1, Houston TX, 1985.

Hawkins, J. W., P. F. Lonsdale, and R. Batiza, Petrologic evolution of the Louisville Seamount chain, in *Seamounts, Islands, and Atolls, Geophys. Monogr. Ser.*, vol. 43, edited by B. H. Keating, P. Fryer, R. Batiza, and G. Boehlert, pp. 235-254, AGU, Washington, D.C., 1987.

Hawkins, J. W., P. F. Lonsdale, J. D. Macdougall, and A. M. Volpe, Petrology of the axial ridge of the Mariana Trough backarc spreading center, *Earth Planet. Sci. Lett., 100*, 226-250, 1990.

Hawkins, J. W., L. M. Parson, J. F. Allan, and Leg 135 Scientific Party, New insight to the evolution of arc-backarc systems, results of Ocean Drilling Program (ODP) Leg 135, Lau-Tonga transect. (abstract) *EOS, Trans. AGU, 72*, 541, 1991.

Hawkins, J. W., L. M. Parson, J. F. Allan, et al, *Proc. ODP, Sci. Results, 135*, 975 pp., Ocean Drilling Program, College Station, TX, 1994.

Hayes, D., and J. Ringis, Seafloor spreading in the Tasman Sea, *Nature, 243*, 545-458, 1973.

Hedge, C. E., and Z. E. Peterman, The strontium isotopic composition of basalts from the Gorda and Juan de Fuca Rises, northeastern Pacific Ocean, *Contrib. Mineral. Petrol., 27*, 114-120, 1970.

Hergt, J. M. and K. N. Farley, Major, trace element, and isotope (Pb, Sr, and Nd) variations in Site 834 basalts: implications for the initiation of backarc opening, in *Proc. ODP, Sci. Results, 135*, J. W. Hawkins, L. M. Parson, J. F. Allan, et al., pp. 471-486, Ocean Drilling Program, College Station, TX, 1994.

Hergt, J. M., and C. J. Hawkesworth, The Pb, Sr, and Nd isotopic evolution of the Lau Basin: implications for mantle dynamics during the back-arc opening, in *Proc. ODP, Sci. Results, 135*, J. W. Hawkins, L. M. Parson, J. F. Allan, et al., pp. 505-518, Ocean Drilling Program, College Station, TX, 1994.

Hilton, D. R., K. Hammerschmidt, G. Loock, and H. Friedrichsen, Helium and argon isotope systematics of the central Lau Basin and Valu Fa ridge: Evidence of crust/mantle interaction in a back-arc basin, *Geochim. Cosmochim. Acta, 57*, 2819-2841, 1993.

Hochstaedter, A. G., J. B. Gill, M. Kusakabe, S. Newman, M. Pringle, B. Taylor, and P. Fryer, Volcanism in the Sumisu Rift, I. Major element, volatile, and stable isotope geochemistry, *Earth Planet Sci. Lett., 100*, 179-194, 1990a.

Hochstaedter, A. G., J. B. Gill, and J. D. Morris, Volcanism in the Sumisu Rift, II. Subduction and non-subduction related components, *Earth Planet Sci. Lett., 100*, 100-195, 1990b.

Hole, M. J., A. D. Saunders, G. F. Marriner, and J. Tarney, Subduction of pelagic sediments: implications for the origin of Ce-anomalous basalts from the Mariana Islands, *J. Geol. Soc. London, 141*, 453-472, 1984.

Ikeda, Y., and M. Yuasa, Volcanism in nascent back-arc basins behind the Shichito Ridge and adjacent areas in the Izu-Ogasawara arc, northwest Pacific: evidence for mixing between E-type MORB and island arc magmas at the initiation of back-arc rifting, *Contrib. Mineral. Petrol., 101*, 377-393, 1989.

Isacks, B. L. and M. Barazangi, Geometry of Benioff zones: lateral segmentation and downwards bending of the subducted lithosphere, in *Island Arcs, Deep Sea Trenches, and Back Arc Basins*, edited by M. Talwani, and W.C. Pitman, III, pp. 94-114, Maurice Ewing Series, 1, AGU, Washington, D.C., 1977.

Jenner, G.A., P. A. Cawood, M. Rautenschlein, and W. M. White, 1987. Composition of back-arc basin volcanics, Valu Fa ridge, Lau Basin: evidence for a slab-derived component in their mantle source, *J. Volcanol. Geotherm. Res., 32*, 209-222, 1987.

Johnson, L. E., and P. Fryer, The first evidence for MORB-like lavas from the outer Mariana forearc: geochemistry, petrography and tectonic implications, *Earth Planet. Sci. Lett.*, 100, 304-316, 1990.

Kanamori, H., Seismic and aseismic slip along subduction zones and their tectonic implications, in *Island Arcs, Deep Sea Trenches and Back-arc Basins*, edited by M. Talwani and W.C. Pitman III, pp. 163-174, Maurice Ewing Ser. 1, AGU, Washington, D.C., 1977.

Karig, D.E., Ridges and basins of the Tonga-Kermadec Island arc system, *J. Geophys. Res., 75*, 239-254, 1970.

Karig, D.E., Origin and development of marginal basins in the western Pacific, *J. Geophys. Res., 76*, 2542-2561, 1971.

Kay, R. W., N. Hubbard, and P. Gast, Chemical characteristics and origins of oceanic ridge volcanic rocks, *J. Geophys. Res., 75*, 1585-1613, 1970.

Kelemen, P. B., Assimilation of ultramafic rocks in subduction related magmatic arcs, *J. Geol., 94*, 829-843, 1986.

Kelemen, P. B., K. T. M. Johnson, R. J. Kinzler, and A. J. Irving, High-field strength element depletions in arc basalts due to mantle-magma interaction, *Nature, 345*, 521-524, 1990.

Klein, E. M., and C. H. Langmuir, Global correlations of ocean ridge basalt chemistry with axial depth and crustal thickness, *J. Geophys. Res., 92*, 8089-8115, 1987.

Kroenke, L., *Cenozoic Development of the Southwest Pacific*, CCOP/SOPAC, Tech. Bull., No. 6. U.N. Econ. Soc. Comm. Asia Pac., Suva, Fiji. 122 pp., 1984.

Lafoy, Y., J. M. Auzende, P. Ruellan, P. Huchon, and E. Honza, The 16°40' S triple junction in the North Fiji Basin (SW Pacific), *Mar. Geophys. Res., 12*, 285-296, 1990.

Lawver, L., J. W. Hawkins, and J. G. Sclater, Magnetic anomalies and crustal dilation in the Lau Basin, *Earth Planet. Sci. Lett., 33*, 27-35, 1976.

Lawver, L. A., and J. W. Hawkins, Diffuse magnetic anomalies in marginal basins: their possible tectonic and petrologic significance, *Tectonophysics, 45*, 323-339, 1978.

Loock, G., W. F. McDonough, S. L. Goldstein, and A. W. Hoffman, Isotopic composition of volcanic glasses from the Lau Basin, *Marine Mining, 9*, 235-245, 1990.

Louat, R., and J. Dupont, Seismicite'de l'arc des Tonga-Kermadec, in *Equipe de Geologie-Geophysique du Centre ORSTOM de Noumea, contribution a' l'etude geodynamique du Sud-Ouest Pacifique*, Trav. Doc. 147 ORSTOM (Nouvelle Caledonie'): 299-317, 1982.

Ludden, J. N., and G. Thompson, An evaluation of the behavior of the rare earth elements during the weathering of sea-floor basalt, *Earth Planet. Sci. Lett., 43*, 85-92, 1979.

Marlow, M. S., L. E. Johnson, J. A. Pearce, P. Fryer, L. G. Pickthorn, and B. J. Murton, Pleistocene volcanic rocks in the Mariana forearc revealed by drilling at Site 781, in *Proc. ODP, Sci. Results 125*, P. Fryer, J. A. Pearce, L. B. Stokking, et al., pp. 293-312, Ocean Drilling Program, College Station, TX, 1992.

Masuda, A., N. Nakamura, and T. Tamaka, Fine structure of mutually normalized rare-earth element patterns of chondrites, *Geochim. Cosmochim. Acta., 37*, 239-248, 1973.

Mattey, D. P., N. G. Marsh, and J. Tarney, The geochemistry, mineralogy, and petrology of basalts from the west Philippine and Parece Vela Basins and from the Palau-Kyushu and West Mariana Ridges, DSDP Leg 59, in *Init. Repts. DSDP, 59*, L. Kroenke and R. Scott, et al., pp. 753-800, U.S. Govt. Printing Office, Washington, D.C., 1980.

Meijer, A., Pb and Sr isotopic data bearing on the origin of volcanic rocks from the Mariana island-arc system, *Bull. Geol. Soc. America, 87*, 1358-1369, 1976.

Menard, H. W., J. Natland, T. H. Jordan, J. A. Orcutt, *Init. Repts. DSDP*, 91, U.S. Govt. Printing Office, Washington, D.C., 1987.

Moberly, R., Origin of lithosphere behind island arcs with reference to the western Pacific. *Geol. Soc. Am. Mem., 132*, 35-55, 1972.

Morris, J. D., and S. R. Hart, Isotopic and incompatible element constraints on the genesis of island arc volcanics from Cold Bay and Amak Island, Aleutians, and implications for mantle structure, *Geochim. Cosmochim. Acta., 47*, 2015-2-30, 1983.

Morton, J. and N. H. Sleep, Seismic reflections from a Lau Basin magma chamber. in *Geology and Offshore Resources of Pacific Island Arcs - Tonga Region,* edited by D. Scholl and T. Vallier, pp. 441-453, Circum-Pacific Council for Energy and Mineral Resources, Houston TX, 1985.

Muenow, D. W., N. W. K. Liu, M. O. Garcia, and A. D. Saunders, Volatiles in submarine volcanic rocks from the spreading axis of the East Scotia Sea backarc basin, *Earth Planet. Sci. Lett., 47*, 272-278, 1980.

Newman, S., Water and carbon dioxide contents in the basaltic glasses from the Mariana Trough, (abstract), *EOS, Trans. AGU, 70*, 1387, 1989.

Nilsson, K., Oxidation state, sulfur speciation, and sulfur concentration in basaltic magmas: examples from Hess Deep and the Lau Basin, Ph.D. thesis, Scripps Institution of Oceanography, U. Calif.-San Diego, La Jolla, 1993.

Nilsson, K., and C. L. Peach, Sulfur speciation, oxidation state, and sulfur concentration in backarc magmas, *Geochim. Cosmochim. Acta, 57*, 3807-3813, 1993.

Nilsson, K., F. Florendo, and J. W. Hawkins, Petrology of a nascent triple junction, northeastern Lau Basin, (abstract), *EOS, Trans. AGU, 73*, 1389, 1989.

Packham, G. H., and D. A. Falvey, An hypothesis for the formation of marginal seas in the western Pacific, *Tectonophysics, 11*, 79-109, 1971.

Parson, L. M., and D. L. Tiffin, Northern Lau Basin: backarc extension at the leading edge of the Indo-Australian plate, *Geo-Marine Lett., 13*, 107-115, 1993.

Parson, L. M., and J. W. Hawkins, Two-stage ridge propagation and the geological history of the Lau backarc basin. in *Proc. ODP, Sci. Results, 135*, J. W. Hawkins, L. M. Parson, J. F. Allan, et al., pp. 819-828, Ocean Drilling Program, College Station, TX, 1994.

Parson, L. M., J. A. Pearce, B. J. Murton, and RRS *Charles Darwin* Scientific Party, 1990. Role of ridge jumps and ridge propagation in the tectonic evolution of the Lau back-arc basin, SW Pacific, *Geology, 18*, 470-473, 1990.

Parson, L. M., J. W. Hawkins, J. F. Allan, et al., *Proc. ODP, Init. Repts., 135*, 1230 pp., Ocean Drilling Program, College Station, TX, 1992.

Parson, L. M., R. G. Rothwell, and C. J. MacLeod, Tectonics and sedimentation in the Lau Basin, southwest Pacific, in *Proc. ODP, Sci. Results, 135*, J. W. Hawkins, L. M. Parson, J. F. Allan, et al., pp. 9-22, Ocean Drilling Program,

College Station, TX, 1994.

Pearce, J. A., and M. J. Norry, Petrogenetic implications of Ti, Zr, Y, and Nb variations in volcanic rocks, *Contrib. Mineral. Petrol., 69*, 33-47, 1979.

Pearce, J. A., S. J. Lippard, and S. Roberts, Characteristics and tectonic significance of supra-subduction zone ophiolites. in *Geology of Marginal Basins*, edited by P. Kokelaar, and M. Howells, pp. 77-94, Geol. Soc. London. Spec. Publ., 16, 1984.

Pearce, J. A., S. R. van der Laan, R. J. Arculus, B. J. Murton, T. Ishii, D. W. Peate, and I. J. Parkinson, Boninite and harzburgite from Leg 125 (Bonin-Mariana forearc): a case study of magma genesis during the initial stages of subduction, in *Proc. ODP, Sci. Results, 125*, P. Fryer, J. A. Pearce, L. B. Stokking, et al., pp. 623-659, Ocean Drilling Program, College Station, TX, 1992.

Pineau, F., M. Javoy, J. Hawkins, and H. Craig, Oxygen isotope variations in marginal basin and ocean ridge basalts, *Earth Planet. Sci. Lett., 28*, 299-307, 1976.

Pisciotto, K. A., J. C. Ingle, M. T. von Breymann, J. Barron, et al., *Proc. ODP, Sci. Results, 127/128*, Pt 1, 776 pp., Ocean Drilling Program, College Station, TX, 1992.

Poreda, R., Helium-3 and deuterium in backarc basin basalts: Lau Basin and Mariana Trough, *Earth Planet. Sci. Lett., 76*, 244-254, 1985.

Poreda, R. and H. Craig, He and Sr isotopes in the Lau Basin mantle: depleted and primitive mantle components, *Earth Planet. Sci. Lett., 119*, 319-329, 1993.

Price, R. C., L. E. Johnson, and A. J. Crawford, Basalts of the North Fiji Basin: the generation of backarc basin magmas by mixing of depleted and enriched mantle sources, *Contrib. Mineral. Petrol., 105*, 106-121, 1990.

Reay, A., J. M. Rooke, R. C. Wallace, and P. Whelan, Lavas from Niuafo'ou Island, Tonga, resemble ocean-floor basalts, *Geology, 2*, 605-606, 1974.

Rothwell, R. G., U. Bednarz, R. Boe, P. Clift, R. A. Hodkinson, J. K. Ledbetter, C. E. Pratt, and S. Soakai, Sedimentation and sedimentary processes in the Lau Basin: results from Leg 135 of the Ocean Drilling Program, in *Proc. ODP, Sci. Results, 135*, J. W. Hawkins, L. M. Parson, J. F. Allan, et al., pp. 829-842, Ocean Drilling Program, College Station, TX, 1994.

Ryerson, F. J., and E. B. Watson, Rutile saturation in magmas; implications for Ti-Nb-Ta depletion in island arc basalts, *Earth Planet. Sci. Lett., 86*, 225-239, 1987.

Sclater, J. G., J. W. Hawkins, J. Mammerickx, and C. G. Chase, Crustal extension between the Tonga and Lau Ridges; petrologic and geophysical evidence, *Geol. Soc. Am. Bull., 83*, 505-518, 1972.

Shervais, J. W., Ti-V plots and the petrogenesis of modern and ophiolitic lavas, *Earth Planet. Sci. Lett., 59*, 101-118, 1982.

Sibuet, J. C., J. Letouzey, F. Barbier, J. Charvet, J. P. Foucher, T. Hilde, M. Kimura, C. Ling-Yun, B. Marsset, C. Muller, and J. F. Stephan, Back arc extension in the Okinawa Trough. *J. Geophys. Res., 92*, 14091-14112, 1987.

Sinton, J., and P. Fryer, Mariana Trough lavas from 18°N: implications for the origin of back arc basin basalts, *J. Geophys. Res., 92*, 12782-12802, 1987.

Sleep, N., and M. N. Toksoz, Evolution of marginal basins, *Nature, 233*, 548-550, 1971.

Staudigel, H., and S. R. Hart, Alteration of basaltic glass; mechanisms and significance for the ocean's crust-seawater budget, *Geochim. Cosmochim. Acta, 47*, 337-350, 1983.

Stern, R. J., Strontium isotopes from circum-Pacific intraoceanic island arcs and marginal basins: regional variations and implications for magmagenesis, *Geol. Soc. Am. Bull., 93*, 477-486, 1982.

Stern, R. J. and S. H. Bloomer, Subduction zone infancy: examples from the Izu-Bonin-Mariana and Jurassic California arcs, *Geol. Soc. Am. Bull., 104*, 1621-1636, 1992.

Stern, R. J., N. C. Smoot, and M. Rubin, Unzipping of the Volcano Arc, Japan, *Tectonophysics, 102*, 153-174, 1984.

Stern, R. J., P. N. Lin, J. D. Morris, M. C. Jackson, P. Fryer, S. H. Bloomer, and E. Ito, Enriched back-arc basin basalts from the northern Mariana Trough: implications for the magmatic evolution of back-arc basins, *Earth Planet. Sci. Lett., 100*, 210-225, 1990.

Stern, R. J., J. Morris, S. H. Bloomer, and J. W. Hawkins, The source of the subduction component in convergent margin magmas: trace element and radiogenic isotope evidence from Eocene boninites, Mariana forearc, *Geochim. Cosmochim. Acta, 55*, 1467-1481, 1991.

Sun, S.-S. and W. F. McDonough, Chemical and isotopic systematics of oceanic basalts: implications for mantle composition and processes, in *Magmatism in the Ocean Basins*, edited by A. D. Saunders, and M. J. Norry, pp. 313-346, Geol. Soc. Spec. Publ. 42, Blackwell, London, 1989.

Sunkel, G., Origin of petrological and geochemical variation of Lau Basin lavas (SW Pacific), *Mar. Mining, 9*, 205-234, 1990.

Tamaki, K., K. Suyehiro, J. Allan, J. C. Ingle, and K. A. Pisciotto, Tectonic synthesis and implications of Japan Sea ODP drilling, in *Proc. ODP, Sci. Results, 127/128*, Pt. 2, K. Tamaki, K. Suyehiro, J. Allan, M. McWilliams, et al., pp. 1333-1348, Ocean Drilling Program, College Station, TX, 1992.

Tappin, D. R., T. R. Bruns, and E. L. Geist, Rifting of the Tonga/Lau Ridge and formation of the Lau backarc basin: evidence from Site 840 on the Tonga Ridge, in *Proc. ODP, Sci. Results 135*, J. W. Hawkins, L. M. Parson, J. F. Allan, et al., pp. 367-372, Ocean Drilling Program, College Station, TX, 1994.

Tatsumi, Y., Migration of fluid phases and genesis of basalt magmas in subduction zones, *J. Geophys. Res. 94*, 4697-4707, 1989.

Tatsumi, Y., D. L. Hamilton, and R. W. Nesbitt, Chemical characteristics of fluid phase released from a subducted lithosphere and origin of arc magmas: evidence from high pressure experiments and natural rocks. *J. Volcan. Geotherm. Res. 29*, 293-309, 1986.

Taylor, P. W., The geology and petrology of Niuafo'ou Island, Tonga: subaerial volcanism in an active back-arc basin, M.Sc. thesis, Macquarie Univ., 1991.

Taylor, B., 1992. Rifting and the volcanic-tectonic evolution of the Izu-Bonin-Mariana Arc, in *Proc. ODP, Sci. Results 126*, B. Taylor, K. Fujioka, et al., pp. 627-651, Ocean Drilling Program, College Station, TX, 1992.

Uyeda, S., and H. Kanamori, Back-arc opening and the mode of subduction. *J. Geophys. Res. 84*, 1049-1061, 1979.

Vallier, T. L., R. M O'Conner, D. W. Scholl, A. J. Stevenson, and P. Quinterno, Petrology of rocks dredged from the landward slope of the Tonga Trench: implications for middle Miocene volcanism and subsidence of the Tonga Ridge forearc, in *Geology and Offshore Resources of Pacific Island Arcs - Tonga Region* edited by D. Scholl and T. Vallier, pp. 109-120, Circum-Pacific Council for Energy and Mineral Resources, Houston TX, 1985.

Vallier, T. L., G. A. Jenner, F. A. Frey, J. B. Gill, A. M. Volpe, J. W. Hawkins, J. D. Morris, P. A. Cawood, J. L. Morton, D. W. Scholl, M. Rautenschlein, W. M. White, R. W. Williams, A. J. Stevenson, and L. D. White, Subalkaline andesite from Valu Fa ridge, a back-arc spreading center in the southern Lau Basin; petrogenesis, comparative chemistry, and tectonic implications, *Chem. Geol. 91*, 227-256, 1991.

Viereck, L. G., M. F. J. Flower, J. Hertogen, H. U. Schmincke, and G. A. Jenner, The genesis and significance of N-MORB sub-types, *Contrib. Mineral. Petrol. 102*, 112-126, 1989.

Volpe, A. M., J. D. Macdougall, and J. W. Hawkins, Mariana Trough basalts (MTB): trace element and Sr-Nd isotopic evidence for mixing between MORB-like and Arc-like melts. *Earth Planet. Sci. Lett. 82*, 21-254, 1987.

Volpe, A. M., J. D. Macdougall, and J. W. Hawkins, Lau Basin basalts (LBB): trace element and Sr-Nd isotopic evidence for heterogeneity in back-arc basin mantle, *Earth Planet. Sci. Lett, 90*, 174-186, 1988.

Volpe, A. M., J. D. Macdougall, G. W. Lugmair, J. W. Hawkins, and P. F. Lonsdale, Fine-scale isotopic variation in Mariana Trough basalts: evidence for heterogeneity and a recycled component in backarc basin mantle, *Earth Planet. Sci. Lett. 100*, 251-264, 1990.

Wegener, A., *Die Entstehung der Kontinente und Ozeane*, (translated by J. Biram, 1969, as *The Origin of Continents and Oceans*), Dover, New York, 246 pp., 1929.

Weissel, J. K., Evolution of the Lau Basin by the growth of small plates, in *Island Arcs, Deep Sea Trenches, and Back Arc Basins*, edited by M. Talwani, and W. C. Pitman, III, pp. 429-436, Maurice Ewing Series, 1, AGU, Washington, D.C., 1977.

Wharton, M. R., B. Hathaway, and H. Colley, Volcanism associated with extension in the Vitiaz island arc, Fiji, (abstract) Symposium on Volcanism Associated with Extension at Consuming Plate Margins, Geological Society of London, 1992.

Whelan, P. M., J. B. Gill, E. Kollman, R. Duncan, and R. E. Drake, Radiometric dating of magmatic stages in Fiji. in *Geology and Offshore Resources of Pacific Island Arcs - Tonga Region*, edited by D. Scholl and T. Vallier, pp. 415-440, Circum-Pacific Council for Energy and Mineral Resources, Houston TX, 1985.

White, W. M., and J. Patchett, Hf-Nd-Sr isotopes and incompatible element abundances in island arcs: implications for magma origins and crust-mantle evolution, *Earth Planet. Sci. Lett. 67*, 167-185, 1984.

Wood, D. A., The application of a Th-Hf-Ta diagram to problems of tectonomagmatic classification and to establishing the nature of crustal contamination of basaltic lavas of the British Tertiary volcanic province, *Earth Planet. Sci. Lett. 50*, 151-162, 1980.

Wood, D. A., N. G. Marsh, J. Tarney, J. L. Joron, J. Cotten, and M. Treuil, Geochemistry of igneous rocks recovered from a transect across the Mariana Trough, Arc, Fore-arc, and Trench, Sites 453 through 461, Deep Sea Drilling Project Leg 60, in *Init. Repts. DSDP 60*, D.M. Hussong, S. Uyeda, et al., pp. 611-646, U.S. Govt. Printing Office, Washington, D.C., 1981.

Woodhall, D., Geology of the Lau Ridge. in *Geology and Offshore Resources of Pacific Island Arcs - Tonga Region*, edited by D. Scholl and T. Vallier, pp. 351-378, Circum-Pacific Council for Energy and Mineral Resources, Houston TX, 1985.

Woodhead, J. D., Geochemistry of the Mariana arc (western Pacific) source compositions and processes, *Chem. Geol. 76*, 1-24, 1989.

Woodhead, J., S. Eggins, and J. Gamble, High field strength and transition element systematics in island arc and back-arc basin basalts: evidence for multi-phase melt extraction and a depleted mantle wedge, *Earth Planet. Sci. Lett. 114*, 491-504, 1993.

James W. Hawkins, Geological Research Division, Scripps Institution of Oceanography, La Jolla, CA, 92093-0220, USA

The Indian Ocean-Type Isotopic Signature in Western Pacific Marginal Basins: Origin and Significance

Rosemary Hickey-Vargas

Department of Geology, Florida International University, Miami, FL, 33199, USA

Janet M. Hergt

School of Earth Sciences, The University of Melbourne, Parkville, Victoria, 3052, Australia.

Piera Spadea

Institution di Scienze della Terra, Universita Udine, Viale Ungerhia, I-33100, Italy

Basalts forming the floor of the western Pacific marginal basins sampled by DSDP and ODP legs and dredging cruises over the past 20 years have isotopic characteristics like those of basalts erupted at Indian Ocean spreading ridges, distinct from those of basalts from Pacific Ocean ridges. The basins, which include the Yamato and Japan Basins of the Sea of Japan, the West Philippine, Shikoku and Parece Vela Basins and Mariana Trough of the Philippine Sea Plate, the South China Basin, the Celebes Sea, the Sulu Sea and the Lau Basin, range in age from >60 to 0 Ma and have differing tectonic histories. The origin of the Indian Ocean-type isotopic signature in basalts from these basins has been investigated by examining and comparing geochemical patterns found within each basin in the light of the tectonic processes, such as subduction and continental rifting, involved in basin development. Results suggest that the basins tapped a shared asthenospheric mantle source that is related to the source of Indian Ocean ridge basalts, but which is modified by local processes in some cases. Based on this conclusion, it is likely that the Indian Ocean asthenospheric domain extends beneath the western Pacific marginal basins. Flow of Indian Ocean-type asthenosphere into this region may have occurred in two stages: initial flow to the north of Australia, followed by displacement to the northwest, north, northeast, and southeast during the rapid northward motion of the Australian continent after 50 Ma.

1. INTRODUCTION

Results from the ocean floor dredging and drilling cruises that explored and sampled the marginal basins of the western Pacific Ocean over the past 20 years have transformed our concepts of basin formation, tectonics, and geochemistry. Geochemical studies of basalts recovered from marginal basins allow us to characterize and interpret the upper mantle asthenospheric sources of the basin magmas and the influence of basin-related tectonic processes, such as subduction and continental rifting, on the underlying asthenosphere. A long-established goal of such research has been to determine if marginal basin basalts, particularly backarc basin basalts from active spreading centers, are systematically different from basalts generated at the major oceanic spreading ridges (e.g., *Hart et al.*, 1972; *Fryer et al.*, 1981; *Hawkins and Melchior*, 1985]. As continuing research on mid-ocean ridges has shown that significant geochemical differences exist among MORBs from the different ocean basins on a regional scale [e.g., *Hamelin et al.*, 1985; *Ito et al.*, 1987; *White et al.*, 1987],

Active Margins and Marginal Basins of the Western Pacific
Geophysical Monograph 88

a further objective has been to determine if marginal basin magmas are related to a specific type of MORB [e.g., *Volpe et al.*, 1988; *Hochstaedter et al.*, 1990a,b; *Loock et al.*, 1990]. The result that has emerged from these latter studies is that basalts from all western Pacific marginal basins for which isotopic data have been collected to date (Figure 1a): the Japan Basin, the Yamato Basin, the West Philippine Basin, the Shikoku and Parece Vela Basins, the Mariana Trough and Sumisu Rift, the South China Basin, the Celebes Basin, the Sulu Basin and the Lau Basin, have the characteristics of Indian Ocean-type MORB. In this paper, we summarize the evidence for this feature and examine hypotheses for its origin as a widespread characteristic of western Pacific marginal basins.

2. MARGINAL BASIN SITES SAMPLED

Figure 1a shows the location of marginal basins of the western Pacific, and Figures 1b and c show the location of western Pacific dredging and drilling sites where igneous basement rocks have been recovered and analyzed for their isotopic composition.

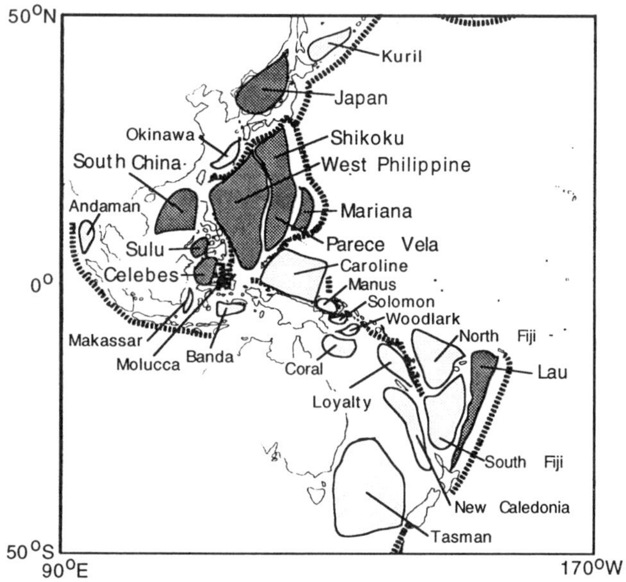

Fig. 1a. Map showing the location of Western Pacific marginal basins, from *Tamaki et al.* [1992]. Basins discussed in this paper are darkly shaded.

2.1. Philippine Sea Plate

The Philippine Sea Plate comprises the West Philippine Basin [37 to >60 Ma; *Scott et al.*, 1981; *Hilde and Lee*, 1984], the Shikoku and Parece Vela Basins [31-17 Ma; *Mrozowski and Hayes*, 1979; *Scott et al.*, 1981], and the active Mariana Trough [6 Ma-present; *Hussong and Uyeda*, 1981](Figure 1b). Incipient arc rifting occurs at the Sumisu Rift in the northern Izu-Bonin Ridge [*Taylor et al.*, 1990]. Sites in these basins were drilled on Deep Sea Drilling Project (DSDP) legs 6, 31, 58, 59 and 60 and ODP (Ocean Drilling Program) legs 125 and 126. Igneous rocks were recovered from six sites in the West Philippine basin (Figure 1b), sampling the eastern basin floor near the Central Basin ridge at sites 290 and 447, the western basin floor at site 291 and the Benham Rise (site 292), and the northwest corner near the Daito-Oki Daito ridges at sites 294 and 446. Igneous rocks of the Shikoku and Parece Vela Basins were drilled at sites 442, 443 and 444, and site 449 and 450 (Figure 1b). The Mariana Trough was drilled at site 453 and 454 and the Sumisu Rift at sites 790 and 791. The young lavas of these latter areas have also been sampled in detail by dredging and submersible [*Taylor et al.*, 1990; *Hochstaedter et al.*, 1990a; *Stern et al.*, 1990; *Hawkins and Melchior*, 1985].

Tectonic interpretations of these areas were summarized in the reports of the DSDP legs [*Karig*, 1975; *Klein and Kobayashi*, 1980; *Scott et al.*, 1981; *Hussong and Uyeda*, 1981], but recent controversy has emerged concerning the origin of the West Philippine Basin [*Jolivet et al.*, 1989; *Hall et al.*, 1994b]. This basin may be either a trapped section of ocean floor, centered on an extinct mid-ocean ridge segment [*Uyeda and Ben Avraham*, 1972; *Jolivet et al.*, 1989; *Wells*; 1989], or a backarc basin [*Karig*, 1975; *Seno and Maruyama*, 1984; *Hall et al.*, 1994b]. The Parece Vela and Shikoku Basins and Mariana Trough are backarc basins that formed successively along a convergent margin with the Pacific Plate that was producing arc volcanics as early as 45-50 Ma [*Taylor*, 1992]. The Sumisu Rift appears to be the site of incipient interarc spreading and may represent an early stage of basin formation [*Taylor et al.*, 1990].

2.2. Celebes and Sulu Seas

These basins were explored on ODP leg 124, and igneous rocks were drilled at sites 767 and 770 in the Celebes Sea, sites 768 in the Sulu Sea and sites 769 and 771 on the Cagayan ridge (Figure 1b). Results of leg 124 show that the Celebes Sea is mid-Eocene in age, but its tectonic origin is disputed [*Silver and Rangin*, 1991a and b; *Rangin and Silver,* 1991]. Sedimentological evidence suggesting an open ocean setting favors an origin as a section of trapped ocean crust, as proposed for the West Philippine Basin. Ocean crust derived from an ancestral Indian Ocean spreading ridge [*Lee and McCabe,* 1989],

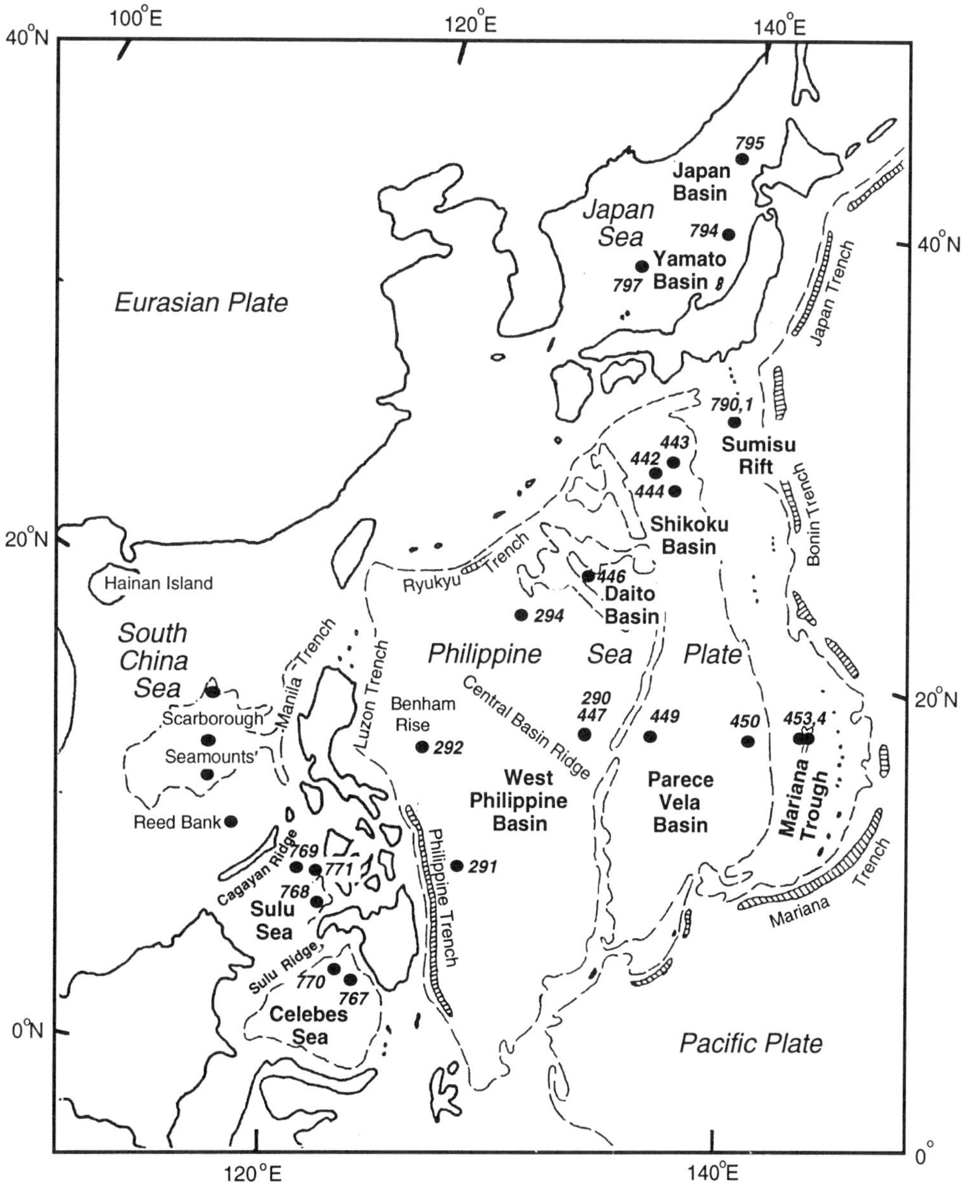

Fig. 1b. Map of northwestern Pacific marginal basins showing the location of samples discussed in this paper, including DSDP/ODP drilling sites (numbered dots) in the Sea of Japan, Philippine Sea Plate, Sulu and Celebes Basins, and dredge sites (dots) in the South China Sea.

fragments of Tethyan ocean floor [*Audley-Charles et al.*, 1988] and fragments of a larger, now subducted, Molucca Sea plate [*Silver and Rangin,* 1991b] have all been suggested as possible origins. Alternatively, the basin may have formed by stretching and rifting of the Eurasian continental margin by collisional extrusion, as proposed for the South China Basin to the north [*Tapponier et al.*, 1982; *Silver and Rangin*, 1991b; Figure 1b].

The early Miocene Sulu Sea is more likely a backarc basin that formed behind an island arc marked by the Cagayan ridge on the north edge [*Silver and Rangin*, 1991a]. Results of leg 124 make less probable alternative origins either by backarc spreading behind the Sulu Ridge to the south [*Rangin*, 1989] or by rifting of the Eurasian continental margin [*Rangin et al.*, 1990].

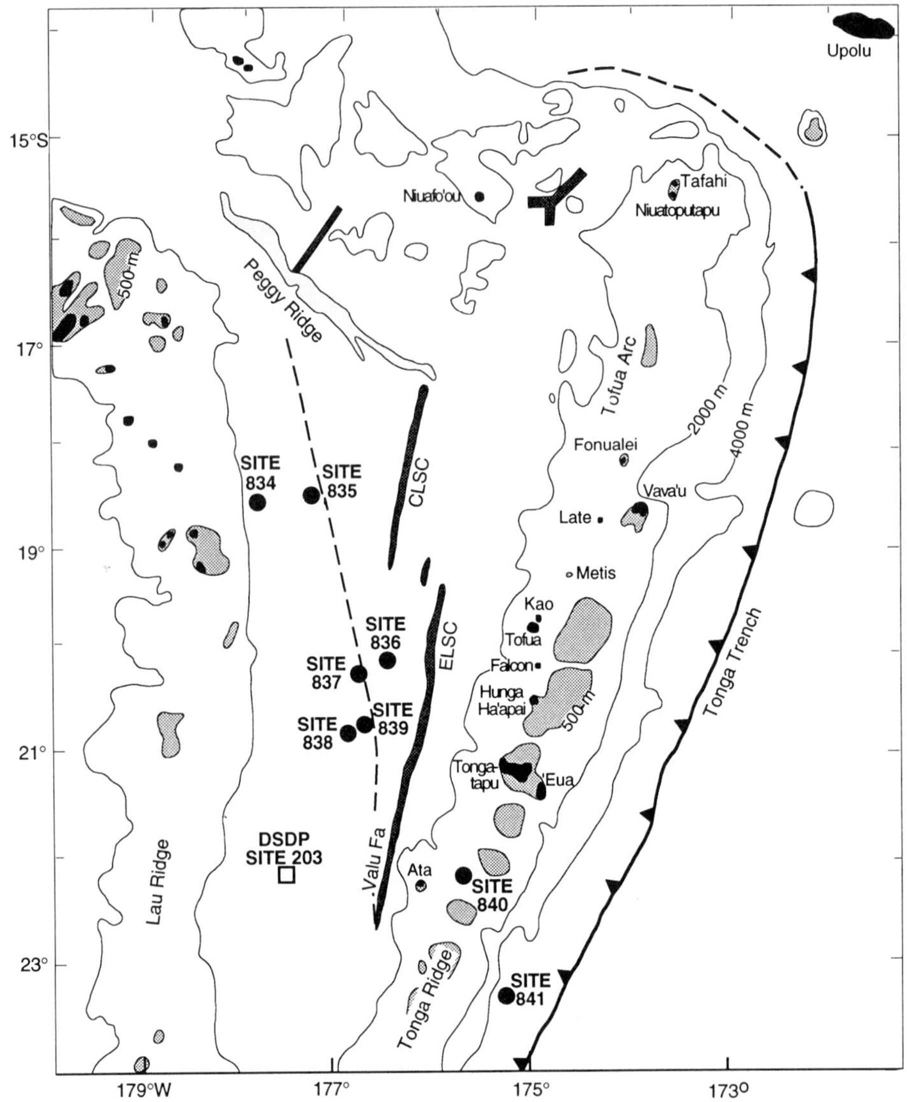

Fig. 1c. Map of the Lau Basin region showing the location of ODP drilling sites. After *Hergt and Hawkesworth* [1994].

2.3. South China Basin

The South China Basin has not been sampled by drilling on any DSDP or ODP leg, but dredge samples were recovered from seamounts that formed following basin opening on cruises of the R/V *Vema* [*Taylor and Hayes*, 1983] and *Sonne-II* [*Kudrass et al.*, 1986]. Basin opening occurred between 32 and 17 Ma [*Taylor and Hayes,* 1980; 1983], probably as a response to southeastward extrusion of southeast Asia during collision of the Indian and Eurasian plates [*Tapponier et al.*, 1982; *Silver and Rangin*, 1991b]. Post-spreading [<14 Ma, *Wang et al.*, 1984] alkaline volcanism occurred on basin floor crust, forming seamounts such as the Scarborough and Reed Bank seamounts (Figure 1b), at isolated continental fragments within the basin, and in adjacent areas of China and southeast Asia, such as Hainan island (Figure 1b).

2.4. Sea of Japan

Igneous rocks from the Sea of Japan were drilled on DSDP legs 127 and 128 at sites 794 and 797 in the Yamato Basin and site 795 near the edge of the Japan Basin (Figure 1b). Previously, igneous samples were dredged from seamounts in the Yamato basin during the DELP 1985 and KH 1984 cruises [*Kimura et al.*, 1987; *Kaneoka et al.*, 1990]. The tectonic history of the Sea of Japan was summarized by *Tamaki et al.* [1992]. The Sea of Japan area formed by rifting of the Eurasian continental margin beginning about 32 Ma. Major basin forming processes were crustal thinning and seafloor spreading within the Japan and Yamato basins, which ceased about 10 Ma. The Japan basin is probably underlain by ocean crust, whereas the Yamato basin, which opened later, may include fragments of rifted arc crust or possibly be underlain by stretched continental basement.

2.5. Lau Basin

The Lau Basin was drilled during ODP leg 135 at sites 834-839 (Figure 1c). Previously, igneous rocks had been dredged from the northeast, northern and southern parts of the basin [*Hawkins*, 1976; *Jenner et al.*, 1987; *Sunkel*, 1990]. The Lau basin is a backarc basin associated with the active Tofua island arc. It originated by rifting of the older Tonga arc, which is marked to the west of the basin by the remnant Lau ridge and to the east by sections of the Tonga ridge, to the immediate west of which the active arc developed [e.g., *Hawkins*, 1974; *Gill*, 1976]. Rifting began about 6 Ma, and spreading has occurred along short ridge segments referred to as the Central Lau and East Lau spreading centers (CLSC and ELSC on Figure 1c) and Valu Fa ridge (Figure 1c) [e.g., *Lawver and Hawkins*, 1978; *Parson et al.*, 1992].

3. ISOTOPIC RESULTS

Table 1 summarizes sources of geochemical data for each of the basins, together with information on the location of sampling sites and the age and tectonic origin ascribed each basin. An important consideration in isotopic studies of basin basalts, especially for the older basins, is the influence of post-eruptive alteration by seawater. Some authors have dealt with this effect by restricting samples to fresh-appearing glasses and mineral separates [*Volpe et al.*, 1987; 1988; 1990; *Loock et al.*, 1990; *Hickey-Vargas*, 1991; *Hergt and Hawkesworth, 1994*], whereas other have used acid leaching to test for and remove elements added by seawater exchange [*Hochstaedter et al.*, 1990a,b; *Cousens and Allan*, 1992; *Tu et al.*, 1992; *Spadea et al.*, 1991; *Serri et al.*, 1991]. In either case, alteration limits the number of samples appropriate for isotopic analysis and the isotopic coverage of the basins discussed is clearly incomplete, both in terms of number of sites, and number of samples per site. Considerable future work is needed in this area.

The western Pacific marginal basins described above have differing proposed tectonic origins: trapping of ocean floor generated at a mid-ocean ridge (Celebes and West Philippine basins), back arc and interarc spreading (Sulu, Shikoku-Parece Vela, Mariana-Sumisu, and Lau Basins), and spreading associated with continental rifting (Sea of Japan, South China Basin, and possibly the Celebes Sea). These different processes can influence the geochemical characteristics of the mantle sources of basin magmas and produce geochemical complexity among the basins, as large-scale asthenospheric sources are influenced by local processes. In discussing the isotopic results, we focus in turn on isotopic trends that (1) are similar to those of large-scale MORB domains, (2) are shared by different basins, and (3) are unique to individual basins.

3.1. Philippine Sea Plate

3.1.1. West Philippine Basin. Based on major and trace element characteristics, basalts recovered from the West Philippine Basin (WPB) are MORB-like (site 447) and oceanic island basalt-like (OIB-like) (Daito basin site 446 and Benham Rise site 292) [*Marsh et al.*, 1980; *Mattey et al.*, 1981; *Hickey-Vargas*, 1991]. Two MORB-like basalts have Sr and Nd isotope ratios that overlap with those of MORB from both the Pacific and Indian Oceans (Figure 2a). Three OIB-like basalts analyzed for Sr and Nd isotopes have lower $^{143}Nd/^{144}Nd$ and higher $^{87}Sr/^{86}Sr$, and the five analyzed for Pb isotopes have higher $^{206}Pb/^{204}Pb$, $^{207}Pb/^{204}Pb$ and $^{208}Pb/^{204}Pb$ than the site 447 samples. Both MORB-like and OIB-like WPB basalts have higher $^{208}Pb/^{204}Pb$ and lower $^{143}Nd/^{144}Nd$ for a given $^{206}Pb/^{204}Pb$ than Pacific MORB (Figure 2c) and most northern hemisphere OIB. Their values overlap with data for basalts from Indian Ocean spreading ridges or extend from these values toward the EMII and HIMU OIB mantle end-members of *Hart* [1988].

3.1.2. Parece Vela and Shikoku Basins. Trace element features of samples from five DSDP sites [*Marsh et al.*, 1980; *Mattey et al.*, 1981; *Hickey-Vargas*, 1991] are similar to those of normal to enriched MORB [N-MORB and E-MORB, e.g., Wood, 1979]. Sr and Nd isotope ratios are like those of MORBs from the major oceans and Pb isotope ratios have the Indian Ocean imprint, much like basalts from WPB site 447 (Figure 2).

3.1.3. Mariana Trough and Sumisu Rift. Basalts dredged from the Mariana Trough have trace element and isotopic features that form a well-defined continuum between MORB-like and island arc basalt end-member compositions [*Volpe et al.*, 1987; *Stern et al.*, 1990; *Volpe et al.* 1990]. The most MORB-like basalts were recovered south of 18°

TABLE 1. Summary of Western Pacific Marginal Basins Having Isotopic Data

Basin	Age (Ma)	Tectonic Setting	Sites with isotopic data for basement DSDP/ODP (Leg-Site)	Sites with isotopic data for basement Dredges	Number of analyses	Isotopic data sources
Japan Basin	32-10	Continental Rift	128-795		2	Cousens and Allan (1992)
Yamato Basin	32-10	Continental Rift	128-794		5	Cousens and Allan (1992)
			128-797		5	"
				several	7	Tatsumoto and Nakamura (1991)
West Philippine	>60-37	Trapped seafloor	31-292		2	Hickey-Vargas (1991)
			58-446		3	"
			59-447		5	"
					1	Cohen and O'Nions (1982a)
Shikoku Basin	25-17	Back arc	58-442		2	Hickey-Vargas (1991)
			58-443		3	"
			58-444		2	"
Parece Vela Basin	31-18	Back arc	59-449		2	Hickey-Vargas (1991)
			59-450		2	"
					1	Cohen and O'Nions (1982a)
Mariana Trough	6-0	Back arc	none			
				several	7	Volpe et al. (1987)
					12	Stern et al. (1990)
					47	Volpe et al. (1990)
Sumisu Rift	0	Incipient	126-790		1	Gill et al. (1992)
		Back arc	126-791		4	"
				several	21	Hochstaedter et al. (1990a,b)
South China Basin	32-14	Pull apart	none			
				several	12	Tu et al. (1992)
Celebes Basin	50	Trapped seafloor	124-767		1	Serri et al. (1991) and
			124-770		11	Spadea et al (unpubl. info.)
Sulu Basin	20	Back arc	124-768		12	Spadea et al. (1992; unpubl. info.)
Lau Basin	6-0	Back-arc	135-834		25	Hergt and Hawkesworth (1994)
			135-835		6	"
			135-836		13	"
			135-837		3	"
			135-838		1	"
			135-839		16	"
				several	5	Jenner et al. (1987)
					23	Volpe et al. (1988)
					9	Loock et al. (1990)

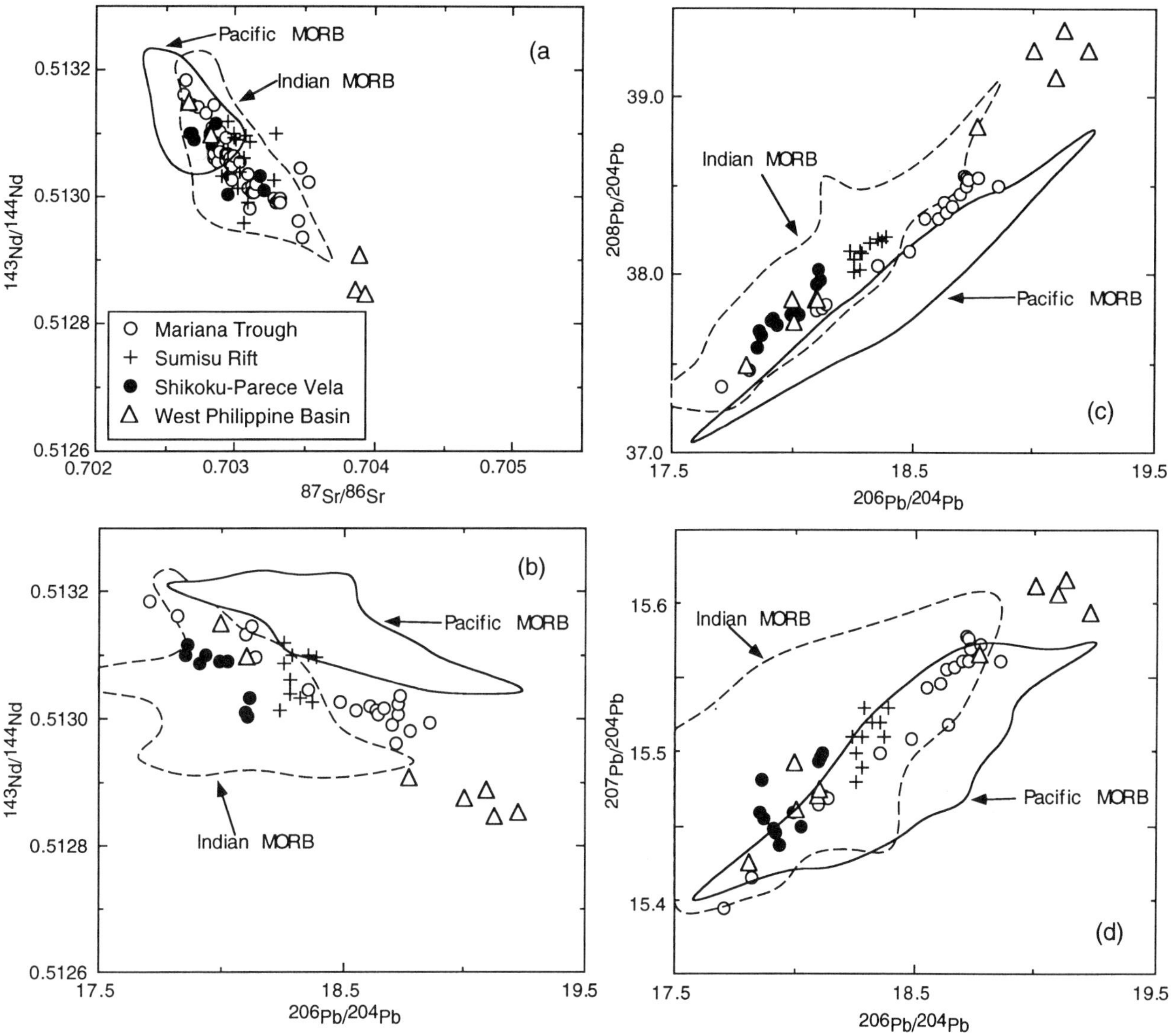

Fig. 2: Isotopic characteristics of volcanic rocks from Philippine Sea Plate marginal basins compared with Pacific and Indian Ocean MORB. Data for marginal basins from *Cohen and O'Nions* [1982a], *Volpe et al.* [1987; 1990], *Stern et al.* [1990], *Hochstaedter et al.* [1990a,b], and *Hickey-Vargas* [1991]. MORB fields are constructed with data from *Church and Tatsumoto* [1975], *Tatsumoto* [1978], *Cohen et al.* [1980], *Sun* [1980], *Cohen and O'Nions* [1982b], *White and Hofmann* [1982], *Dupre and Allegre* [1983], *Hamelin et al.* [1984], *Hamelin et al.* [1985], *Hamelin and Allegre* [1985], *Macdougall and Lugmair* [1986], *Price et al.* [1986], *Ito et al.* [1987], *White et al.* [1987], *Dosso et al.* [1988], *Klein et al.* [1988], *Hanan and Schilling* [1989], *Mahoney et al.* [1989, 1992, 1994], and *Pyle et al.* [1992]. Samples from ridge segments near hotspots that show plume isotopic characteristics are not used in the construction of the fields. All data used in Figures 2 through 8 are normalized, when possible, to mutually consistent values for Eimer and Amend standard $SrCO_3$ ($^{87}Sr/^{86}Sr$ =0.7080); NBS 981 ($^{87}Sr/^{86}Sr$=0.71025); BCR-1 ($^{143}Nd/^{144}Nd$=0.51264); La Jolla Nd standard ($^{143}Nd/^{144}Nd$=0.51185); and Pb-isotope values for NBS 981 [*Catanzaro et al.*, 1968].

N [*Volpe et al.*, 1990], whereas most dredge samples from the northern trough have geochemical characteristics transitional to basalts from the adjacent arc. The MORB-like basalts have the highest $^{143}Nd/^{144}Nd$ and lowest $^{87}Sr/^{86}Sr$ of Philippine Sea Plate basins (Figure 2a), and $^{208}Pb/^{204}Pb$ ratios are elevated, like those of Indian Ocean MORB and older Philippine Sea Plate basin lavas (Figure 2c). The arc end-member of the trends is like Mariana island arc basalts, and has higher $^{87}Sr/^{86}Sr$, lower $^{143}Nd/^{144}Nd$ and higher $^{206}Pb/^{204}Pb$, $^{207}Pb/^{204}Pb$ and $^{208}Pb/^{204}Pb$ than the southern Mariana Trough samples. The arc end-member has higher $^{143}Nd/^{144}Nd$ and lower $^{208}Pb/^{204}Pb$ for a given $^{206}Pb/^{204}Pb$ than the trend defined by WPB MORB and OIB-like basalts (Figures 2b and c).

Sumisu Rift basalts have trace element features similar to E-MORB with some island arc basalt-like characteristics [*Hochstaedter et al.*, 1990b]. Like other MORB-like basalts from the Philippine Sea Plate, Sr and Nd isotope ratios coincide mainly with those of Indian Ocean MORB, and Pb-isotope ratios plot in the field of Indian Ocean MORB (Figure 2). Sr, Nd and Pb isotope ratios plot as a cluster within the array of Mariana Trough basalts. Displacement of the cluster from the MORB-like end-member is toward either Mariana island arc basalt or WPB OIB-like basalts.

3.2. Celebes and Sulu Seas

Based on major and trace element features, *Serri et al.* [1991], *Spadea et al.* [1991] and *Smith et al.* [1991] classified basalts from the floor of the Celebes Sea (sites 767 and 770) as N-MORB, lavas from the Cagayan ridge (sites 769 and 771) as island arc tholeiites and calc-alkaline rocks, and basalts from the Sulu Sea (site 768) as backarc basin basalts with transitional N-MORB to island arc character. Isotopic data for samples from these basins were obtained by *Spadea et al.* [unpublished information]. With the exception of two samples that have probably suffered post-eruptive alteration of $^{87}Sr/^{86}Sr$, Sr, Nd and Pb isotope ratios in Sulu basin basalts coincide with values for Indian Ocean MORB (Figure 3). Celebes Basin basalts also plot within or very close to the field of Indian Ocean MORB, at the lower range of $^{143}Nd/^{144}Nd$, but intermediate $^{206}Pb/^{204}Pb$, that is associated with the influence of near-ridge plumes like Amsterdam-St Paul [*Dosso et al.*, 1988]. Interestingly, the arc-like Cagayan ridge lavas have Pb and Sr isotope ratios like those of the adjacent basins, but lower $^{143}Nd/^{144}Nd$ (Figure 3).

3.3. South China Basin

No basin floor (i.e., syn-spreading) basalts from the South China Basin have been analyzed for isotopes; however, *Tu et al.* [1992] report data for two presumed basin floor samples from ophiolites in the Philippine Islands and use these in their interpretations. Nd and Pb isotopes ratios for the two samples plot within the field defined by Indian Ocean MORB, at lower $^{206}Pb/^{204}Pb$, $^{207}Pb/^{206}Pb$, $^{208}Pb/^{204}Pb$ and higher $^{143}Nd/^{144}Nd$ than those of samples from the Sulu and Celebes Basins (Figure 3). High $^{87}Sr/^{86}Sr$ in these samples (Figure 3a), like the Sulu Sea basalts, probably reflects post-eruptive interaction with seawater [*Tu et al.*, 1992]. Isotopic data for alkaline basalts from the post-spreading Scarborough and Reed Bank seamounts (<14 Ma), also reported by *Tu et al.* [1992], are displaced toward lower $^{143}Nd/^{144}Nd$ and higher $^{206}Pb/^{204}Pb$ than those of the basin floor basalts, and most have high $^{208}Pb/^{204}Pb$ and $^{207}Pb/^{204}Pb$ for a given $^{206}Pb/^{204}Pb$ (Figure 3c and d), similar to values for OIB-like basalts from the West Philippine Basin. Some samples have unusually high $^{207}Pb/^{204}Pb$ for a given $^{206}Pb/^{204}Pb$ and $^{208}Pb/^{204}Pb$. *Tu et al.* [1992] found that acid leaching of some samples diminished $^{207}Pb/^{204}Pb$ values, so this feature may be attributable to post-eruptive alteration by seawater. The seamount basalts also overlap in isotopic composition with alkali basalts from nearby Hainan Island [*Tu et al.*, 1991; *Flower et al.*, 1992].

3.4. Sea of Japan

Basalts from ODP sites on the floor of the Yamato and Japan Basins were analyzed by *Cousens and Allan* [1992], and dredge samples from seamounts in the Yamato basin and alkali basalts from islands in the southwestern Japan Sea were analyzed by *Tatsumoto and Nakamura* [1991]. Based on trace element patterns, *Allan and Gorton* [1992] classified the ODP samples as N-MORB to E-MORB (Yamato basin site 794 and lower site 797), and island arc tholeiitic (upper site 797) to transitional (Sea of Japan site 795). The highest $^{143}Nd/^{144}Nd$ are found in the highly incompatible element-depleted arc rocks from upper units of site 797. These rocks have higher $^{87}Sr/^{86}Sr$ than MORB (Figure 4a), which may be either a primary characteristic, similar to many arc rocks, or a result of post-eruptive alteration by seawater. Other Yamato and Japan Basin floor basalts have lower $^{143}Nd/^{144}Nd$ and higher $^{87}Sr/^{86}Sr$ than upper site 797, and most are outside the range of MORBs and similar to arc and oceanic island basalts. Only the enriched Yamato Basin seamount basalts fall within or close to the MORB fields on Figures 4a-d, and in each case they plot with Indian Ocean MORBs. The lowest $^{143}Nd/^{144}Nd$ and highest $^{87}Sr/^{86}Sr$ values are found in alkali basalts from the southwestern islands, which have $^{143}Nd/^{144}Nd$ of 0.51246-0.51260 and $^{87}Sr/^{86}Sr$ of 0.7045-

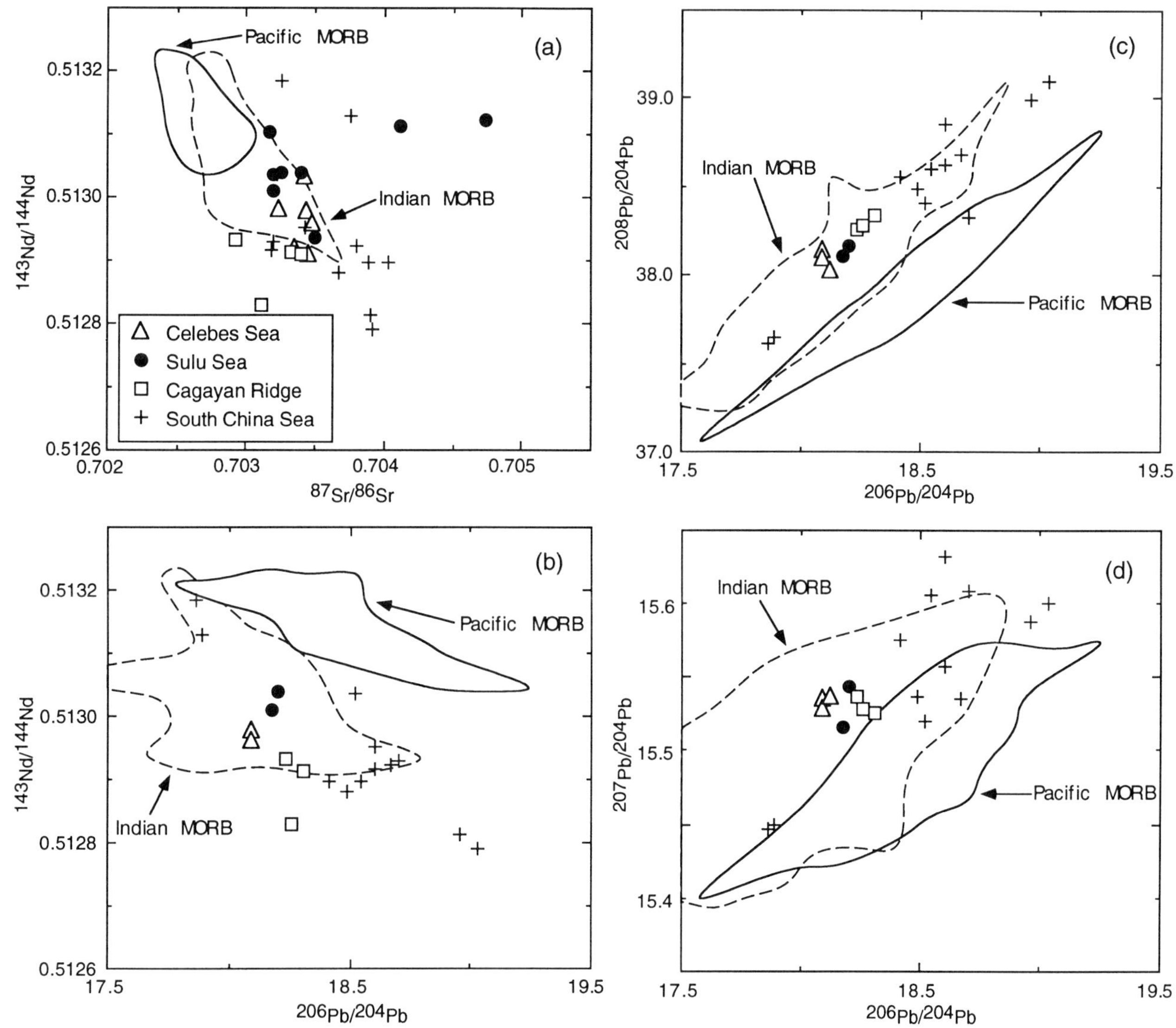

Fig. 3. Isotopic characteristics of volcanic rocks from the South China, Sulu and Celebes Basins. Data from *Tu et al.* [1992]; *Spadea et al.* [1991] and *Spadea et al.* [in press]. MORB fields are constructed as in Figure 2.

0.7059 (only one sample plotted in Figures 4a and b). All Yamato and Japan Basin volcanics, including arc-like, MORB-like, and incompatible element enriched rocks have $^{208}Pb/^{204}Pb$ ratios higher than those of Pacific Ocean MORB, and similar to values for Indian Ocean MORB (Figure 4c). $^{207}Pb/^{204}Pb$ values are also generally like those of Indian Ocean MORB, with the exception of two Yamato Basin seamount samples which plot at lower $^{207}Pb/^{204}Pb$ than the Pacific MORB field. Alkali basalts from the southwestern islands have exceptionally high $^{208}Pb/^{204}Pb$ for a given $^{206}Pb/^{204}Pb$ and these plot above the range for Indian Ocean MORB (Figure 4c).

3.5. Lau Basin

Isotopic data for ODP sites in the Lau basin were reported by Hergt and Hawkesworth [1994] and data for earlier dredge samples were reported by *Jenner et al.* [1987], *Volpe et al.* [1988] and *Loock et al.* [1990]. This area has been studied in some detail and several geochemical trends have been noted (Figure 5). First, volcanics that occur where the spreading center converges with the arc (Valu Fa Ridge, Fig 1c) have trace element and isotopic features with strongly arc-like characteristics [*Jenner et al.*, 1987; *Volpe et al.*, 1988]. Such features were also found in samples

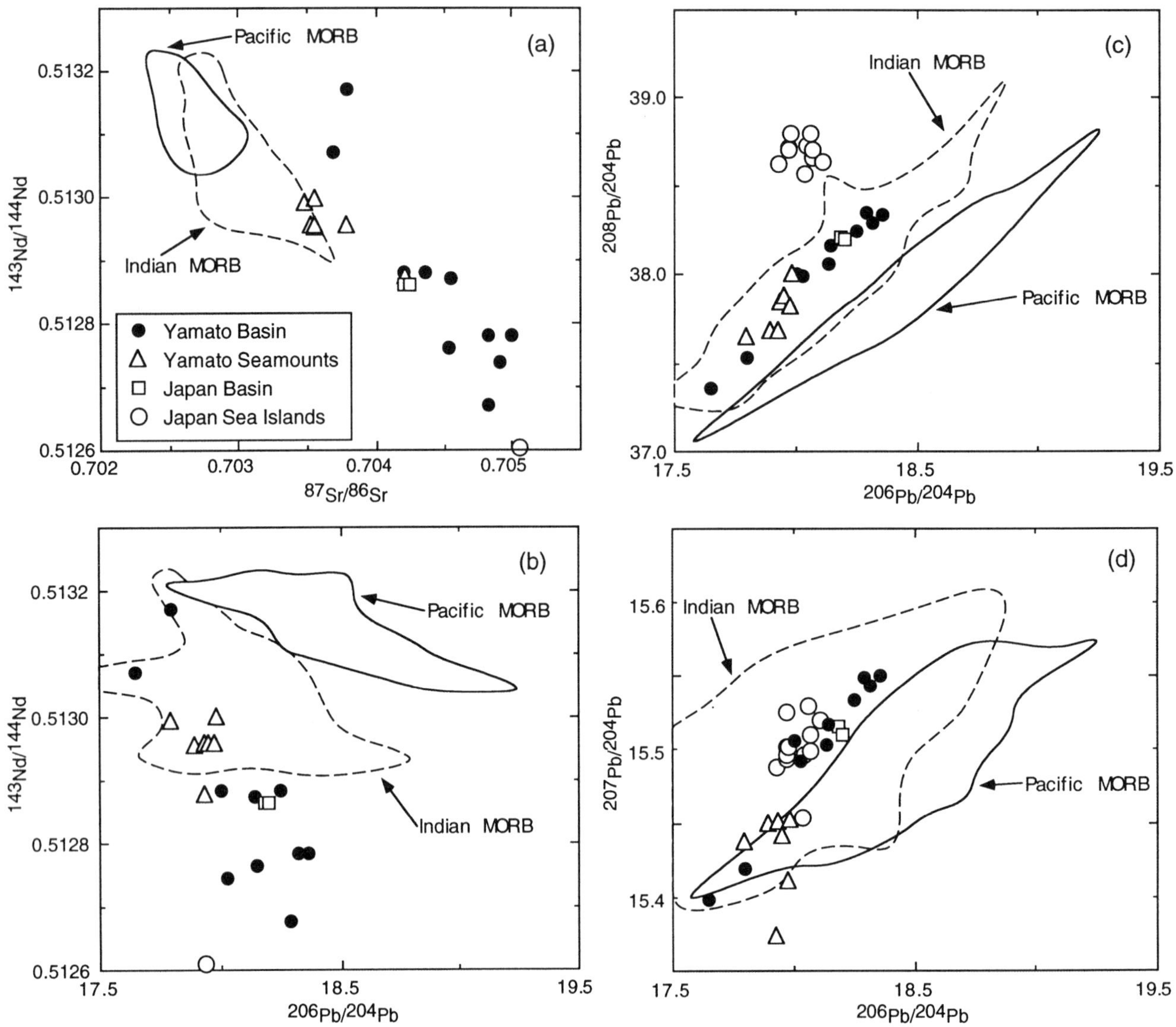

Fig. 4. Isotopic characteristics of volcanic rocks from the Sea of Japan. Data from *Tatsumoto and Nakamura* [1991] and *Cousens and Allan* [1992]. MORB fields are constructed as in Figure 2.

from ODP sites 835, 838 and 839. Well defined transitional MORB-arc basalt isotopic trends were found in ODP samples from sites 834, 836, and 837 [*Hergt and Hawkesworth, 1994*]. Remarkably, trends formed by the site 836 and 837 samples define a MORB end-member of Indian Ocean-type composition and trends formed by the site 834 samples define a MORB end-member of Pacific Ocean-type composition (Figures 5a and c). On Pb-isotope diagrams (Figures 5c and d), both trends converge at a common end-member similar in isotopic composition to some Tonga and Tofua arc basalts [the slab-derived flux end-member of *Hergt and Hawkesworth, 1994*]. The Pacific and Indian Ocean-type MORB end-members are distinguished from one another by lower $^{87}Sr/^{86}Sr$, very slightly higher $^{143}Nd/^{144}Nd$ and significantly higher $^{206}Pb/^{204}Pb$ at similar $^{207}Pb/^{204}Pb$ and $^{208}Pb/^{204}Pb$ in the Pacific type. The relative age of sites sharing a similar latitude indicates that the Indian Ocean-type MORB source was tapped more recently than was the Pacific Ocean-type MORB source [*Hergt and Hawkesworth, 1994*]. Compared with the MORB end-members, the slab-derived flux end-member, which is inferred to be material transferred to the mantle from the subducted crust [*Hergt and Hawkesworth, 1994*], has higher $^{87}Sr/^{86}Sr$ for a given $^{143}Nd/^{144}Nd$ and higher

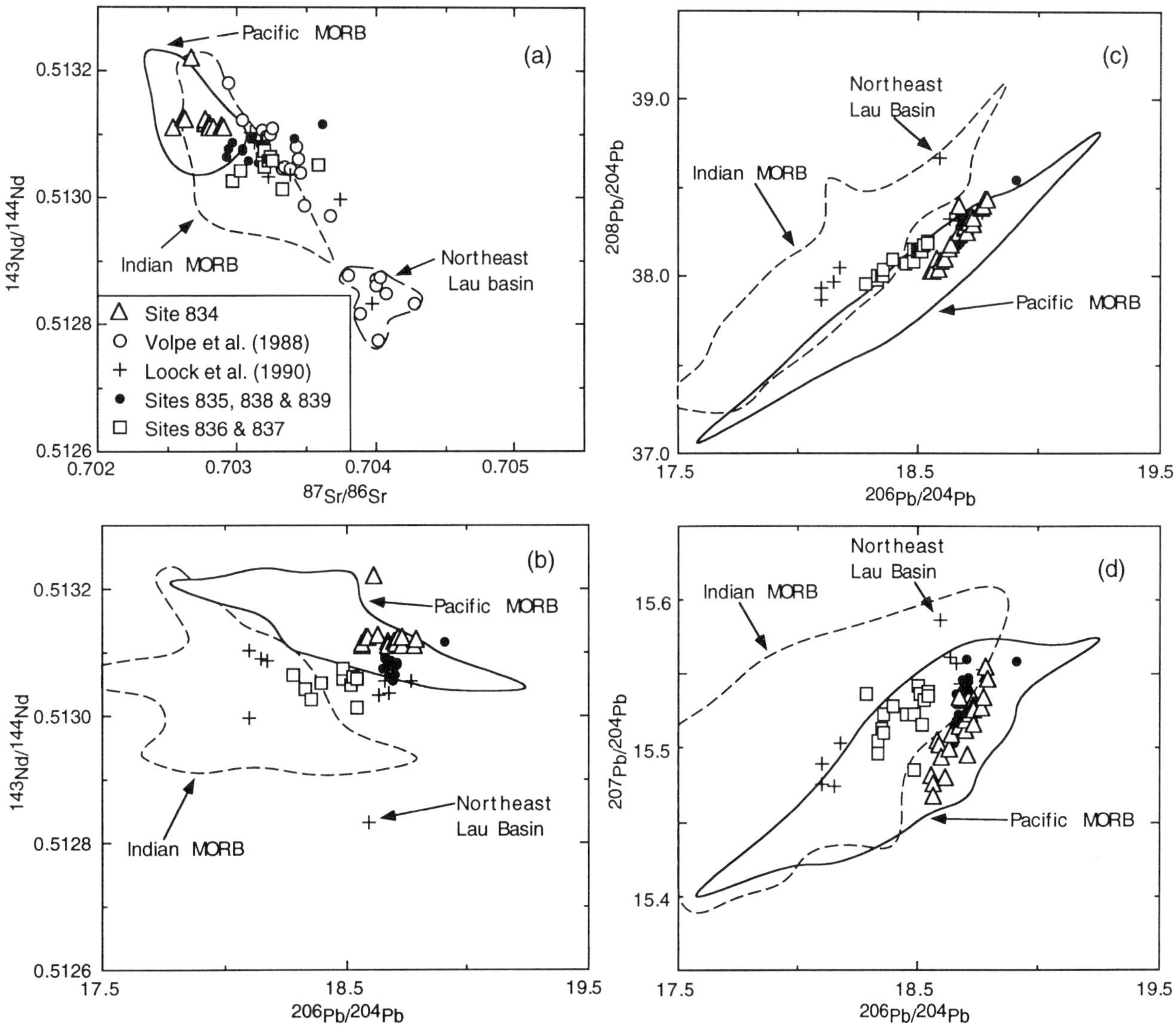

Fig. 5. Isotopic characteristics of volcanic rocks from the Lau basin. Data from *Hergt and Hawkesworth [1994]*; *Volpe et al* [1988]; *Loock et al.* [1990]. MORB fields are constructed as in Figure 2.

$^{206}Pb/^{204}Pb$. Like many of the site 836 and 837 basalts, dredge samples from the central and northern Lau basin studied by *Loock et al.* [1990] have the Indian Ocean-type isotopic signature. Nine samples from the northeastern Lau basin form a cluster with lower $^{143}Nd/^{144}Nd$ and higher $^{87}Sr/^{86}Sr$ than values for MORBs and one sample analyzed for Pb isotopes has higher $^{207}Pb/^{204}Pb$ and $^{208}Pb/^{204}Pb$ than those of all other Lau basin basalts (Figures 5c and d), a feature that was attributed by *Volpe et al.* [1988] and *Loock et al.* [1990] to influence from the nearby Samoan hotspot.

4. EVIDENCE FOR AN INDIAN OCEAN-TYPE ISOTOPIC SIGNATURE IN WESTERN PACIFIC MARGINAL BASINS

Over the past ten years, geochemical research on basalts from active mid-ocean spreading ridges has shown that there are clearly defined differences between MORBs from the different ocean basins [e.g., *Hamelin et al.*, 1985; *Ito et al.*, 1987; *White et al.*, 1987]. The most pronounced difference is between the Atlantic and Pacific Ocean ridges (northern and central Mid-Atlantic Ridge, Juan de Fuca and Gorda

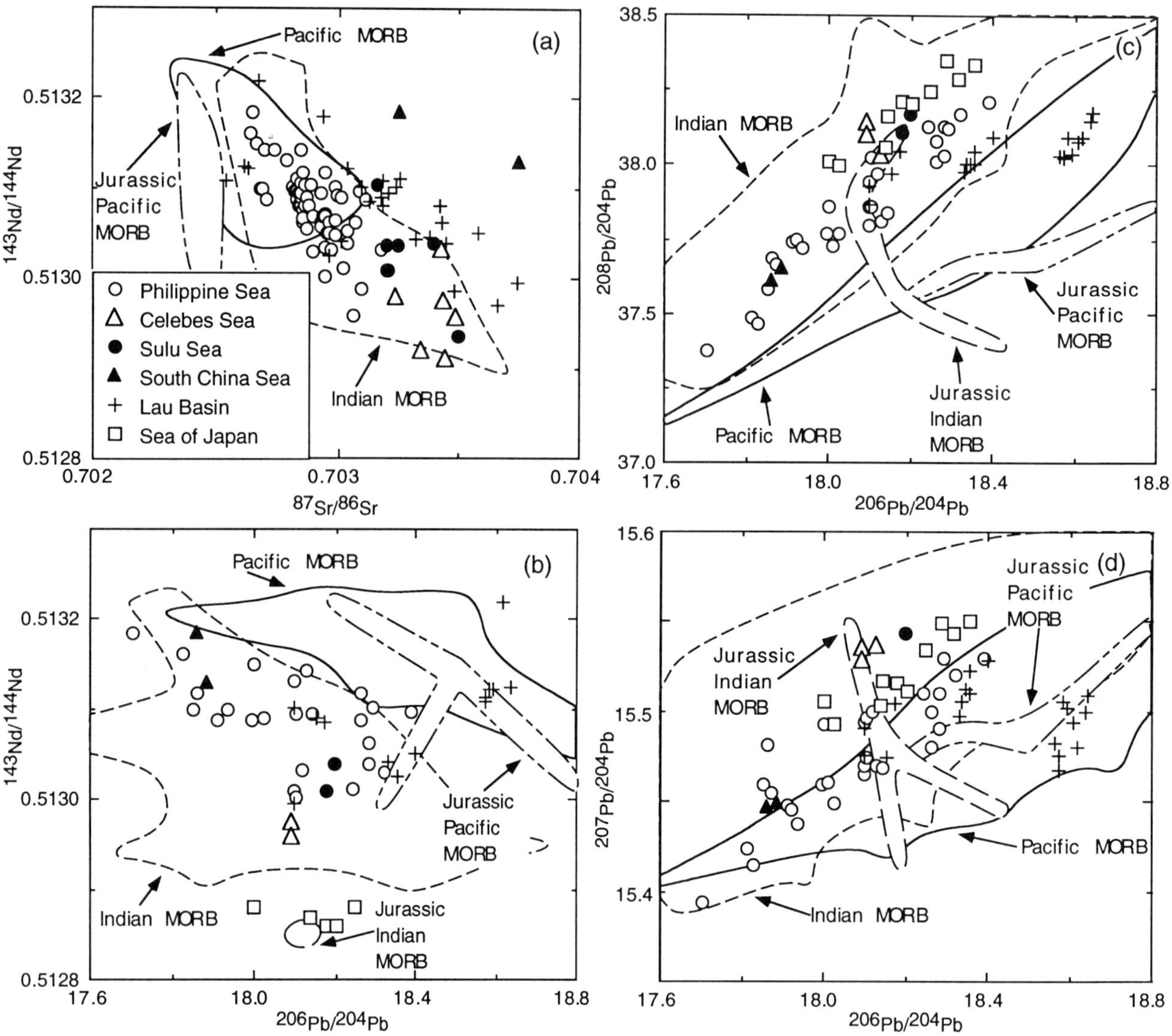

Fig. 6. Isotopic patterns in MORB-like Western Pacific marginal basin basalts compared with large-scale MORB reservoirs. Criteria for classification as MORB-like are, roughly, chondrite normalized (La/Yb) < 2, La/Nb < 2, Ba/La < 10, Sm/Nd > 0.3 with some samples lacking trace element data included on the basis of geographic association. MORB fields are constructed as in Figure 2. Also shown are fields for Jurassic Pacific MORB from ODP site 801 [*Castillo et al.*, 1992] and Jurassic Indian Ocean MORB from ODP site 865 [*Ludden and Dionne*, 1992]. For the Jurassic MORB, Nd and Sr isotope ratios are corrected for in situ decay of ^{147}Sm and ^{87}Rb.

Ridges, East Pacific Rise, and Southeast Indian Ridge east of 126°E) and ridges of the Indian Ocean (Central, Southwest and Southeast Indian Ridges west of 126°E, and Carlsberg Ridge). Isotopic differences include higher ^{87}Sr/^{86}Sr for a given ^{143}Nd/^{144}Nd, and higher ^{207}Pb/^{204}Pb and ^{208}Pb/^{204}Pb for a given ^{206}Pb/^{204}Pb for Indian Ocean ridges, and generally lower ^{206}Pb/^{204}Pb (Figure 6). Transitions from one MORB type to the other occur along the Southeast Indian Ridge at the Australian-Antarctic Discordance at 126°E [*Klein et al.*, 1988; *Pyle et al.*, 1992] and along the western Southwest Indian Ridge between 17-26°E [*Mahoney et al.*, 1992]. The transition along the Southeast Indian Ridge is remarkably abrupt [40 km; *Pyle et al.*, 1992], in contrast to the transition along the Southwest Indian Ridge, which is gradual, with interspersed segments having anomalous enriched MORB [*Mahoney et al.*, 1992].

One of the most interesting geochemical findings of dredging and drilling in the western Pacific marginal basins is predominance of an Indian MORB-type isotopic signature. Of the basins sampled, only the Lau Basin includes basalts with Pacific Ocean isotopic character, and these are replaced at the major basin spreading center by basalts with Indian Ocean-type MORB features between the period 6 Ma and present. A complete picture of the isotopic characteristics of western Pacific marginal basins is clearly yet to be established, because, in addition to the low sample coverage for some of the basins studied to date, isotopic data for basin floor basalts from the Banda and Molucca Seas, which mark the Pacific-Eurasian plate margin together with the Sulu, Celebes, and Philippine Seas, and the numerous basins that together with the Lau Basin form the Pacific-Indo-Australian plate margin (Figure 1a), are not yet available. However, based on the existing data, it appears at least that areas west of the major western Pacific active convergent margins—the Japan, the Izu-Mariana, and the Tonga-Kermadec arc-trench systems—are underlain by Indian Ocean-type MORB mantle sources. An interesting point in support of a regional presence for Indian Ocean-type asthenospheric mantle is that new data for lavas sampled by ODP leg 134 in the North Aoba Basin, adjacent to the North Fiji Basin (Figure 1a) on the Pacific Plate side of the New Hebrides trench, also exhibit Indian Ocean-type isotopic characteristics [*Briqueu et al.*, 1994].

Figure 6 highlights data for WPMB basalts that are MORB-like (enriched or normal) in their trace element characteristics. An important feature of the data, together with the Indian Ocean signature, is that, with the exception of the Yamato Basin site 794, lower site 797 and Lau Basin site 834, isotopic ranges defined by the different basins overlap. This overlap is remarkable given the wide range of ages for the basalts (Paleocene-recent), and their past and present areal coverage. Overall, the basins studied currently encompass 4.5×10^6 km^2, with large sections probably subducted [*Seno and Maruyama*, 1984; *Silver and Rangin*, 1991b; *Hall et al.*, 1994b]. Their isotopic similarity points to, but does not prove, a single, long-lived mantle reservoir giving rise to the basin basalts.

Many of the marginal basins studied also contain OIB-like basalts: the Sea of Japan, West Philippine Basin, South China Basin, and Lau Basin. Like the MORB-like basalts, the OIB-like basin basalts follow isotopic trends that are consistent with OIB end-member compositions recognized globally (Figure 7). For the Sea of Japan, alkaline basalts from the southwestern islands plot between the MORB fields and a region between the OIB end-members EMI and EMII defined by *Hart* [1988]. OIB-like basalts from the West Philippine Basin (Benham Rise and Daito Basin) and South China Basin (Scarborough and Reed Bank seamounts) diverge from the MORB field toward the region between the EMII and HIMU end-members. Enriched Lau Basin basalts [Type II of *Volpe et al.*, 1988] are displaced toward EMII. Therefore, all of the globally recognized OIB mantle sources are represented within the western Pacific marginal basins.

5. ORIGIN OF THE INDIAN OCEAN-TYPE ISOTOPIC SIGNATURE IN WESTERN PACIFIC MARGINAL BASINS

Two competing hypotheses for the isotopic similarity of West Pacific marginal basin basalts and Indian Ocean MORB are (1) that their upper mantle sources are or were once part of a single, relatively homogeneous reservoir; or (2) that their sources formed independently by similar processes. In this section we examine evidence for and against each of these hypotheses.

5.1. Association with the Indian Ocean Asthenospheric Domain

5.1.1. *The age and extent of distinct Pacific and Indian asthenospheric domains.* Because the western Pacific marginal basins vary in age, and because the West Philippine and Celebes Basins are thought to be trapped crust formed at major ocean spreading centers, an important consideration in assessing whether the sub-Indian Ocean and sub-western Pacific asthenosphere were once physically continuous is the age and past areal extent of distinct Pacific and Indian MORB asthenospheric domains. For the Pacific Ocean, the older seafloor sampled has been mainly anomalous—i.e., the Cretaceous Manihiki and Ontong Java Plateaus and the Nauru Basin. These areas have strong plume related geochemical signatures that, together with geologic evidence, suggest an origin at massive near-ridge hotspots [*Mahoney*, 1987; *Mahoney and Spencer*, 1991; *Castillo et al.*, 1991]. Recent samples of Pacific crust east of the Mariana Trough from ODP leg 129 also revealed material affected by plume activity, and seamounts in this region can be tracked back to known hotspots in the South Pacific [*Castillo et al.*, 1992; *Staudigel et al.*, 1991]. However, in the lower section of ODP leg 129 site 801, Jurassic Pacific seafloor was sampled that has isotopic features like those of the modern Pacific ridges, including low $^{87}Sr/^{86}Sr$ and $^{208}Pb/^{204}Pb$ (Figures 6a and c). This finding establishes the long term existence of the Pacific reservoir, but does not constrain its extent. Tectonic reconstructions [*Uyeda and Ben Avraham*, 1972, *Seno and Maruyama*, 1984; *Jolivet et al.*, 1989] show several

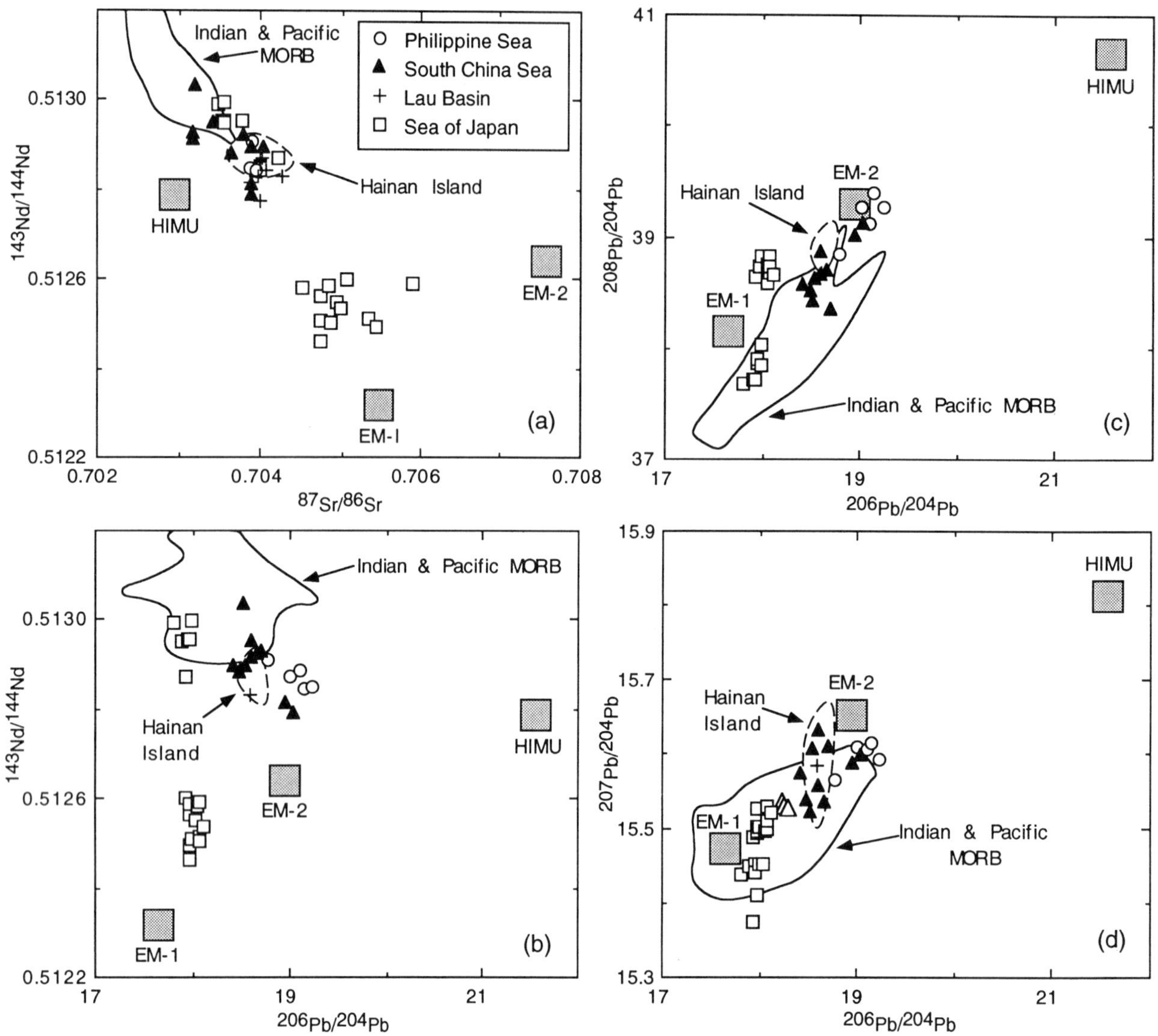

Fig. 7. Isotopic patterns in OIB-like Western Pacific marginal basin basalts compared with the globally recognized OIB mantle end-members of *Hart* [1988]. Criteria for classification as OIB-like are, roughly: chondrite normalized (La/Yb) > 3, La/Nb <1.5, Ba/La < 10, Sm/Nd < 0.3. For the Lau basin, Type II basalts are included in this category [*Hawkins and Melchior,* 1985]. Data for continental alkaline basalts from Hainan Island, China are also shown [*Tu et al.*, 1991]. MORB fields are constructed from sources listed for Figure 2.

different "Pacific" ridges (Kula-Pacific, South Pacific-Pacific, Northern New Guinea-Pacific) in existence between the Jurassic and early Tertiary. Thus the identity of the ridge producing the seafloor at site 801, which tapped Pacific Ocean-type asthenosphere, and its location in the Jurassic is unknown, except that it may have migrated northward over South Pacific hotspots in the Cretaceous [*Castillo et al.*, 1992]. For the Indian Ocean, older seafloor has also been sampled mainly at strongly plume-influenced plateau areas such as the Kerguelen Plateau and Ninetyeast Ridge, and Mascarene Plateau and Chagos-Maldive Ridge [e.g., *Weis et al.*, 1992]. Examination of isotopic data for these areas shows the presence of a depleted MORB-source mixing end-member [*White et al.*, 1990; *Weis and Frey*, 1991; *Weis et al.*, 1992] that could be similar to that tapped at modern Indian Ocean ridges, but the data do not rule out a different MORB-source end-member. Similarly, limited data for Indian Ocean basalts from the Jurassic Argo abyssal plain (ODP site 865) northwest of Australia reported by *Ludden and Dionne* [1992] show both Indian Ocean and Pacific

Ocean-type isotopic signatures (Figure 6), suggesting that the Indian Ocean ridge forming seafloor at this site at about 155 Ma may have tapped both Indian and Pacific-type asthenospheric domains. Overall, these very sparse data suggest that distinct MORB types existed as early as the Jurassic, considerably before the formation of the West Philippine and Celebes Basins, but the boundaries of the asthenospheric domains are not well constrained. Therefore, it is possible that the Paleocene-Eocene West Philippine and the Celebes Basins formed in an area underlain by the Indian Ocean asthenospheric domain. However, this does not preclude their being trapped crust generated within the region (in a hotspot reference frame) of the present-day Pacific Ocean basin, since the present and ancient eastward extent of the Indian Ocean domain, and hence the geochemical characteristics of ancient Pacific ridges, is unknown.

5.1.2. *Barriers to asthenospheric flow in the western Pacific region.* A second consideration in evaluating the concept of a shared, rather than independently produced, asthenospheric source for western Pacific marginal basins is the existence of barriers to asthenospheric flow in the western Pacific region. The exact physical nature of the asthenospheric domains that give rise to MORBs and their boundaries is not well established. In the model proposed by *Klein et al* [1988], who first documented the sharpness of the Indian-Pacific isotopic boundary along the Australian-Antarctic Discordance at 126°E, the domains are separate convective units, with boundaries formed where there is downwelling in adjacent cells. Assuming such convection units are limited to the upper mantle, asthenospheric flow at present beneath western Pacific marginal basins at present must blocked and confined by the numerous subducting oceanic lithospheric plates (Figure 1a) and the subcontinental mantle lithosphere of Australia. However, many tectonic reconstructions show a more open configuration in the past. For example, up until 50 Ma, when the rate of northward motion of Australia away from Antarctica increased, there was a significant oceanic gap between the northern border of the Australian continent and southernmost Asia [*Wells*, 1989; *Jolivet et al.*, 1989; *Royer et al.* 1992]. Between 80 and 50 Ma, a now extinct, rapidly spreading Indian Ocean ridge propagated to the northeast of Australia, separating India and Australia [e.g., *Royer et al.*, 1992], and, in the reconstruction of *Wells* [1989], this ridge continues north of Australia and into the Pacific Ocean basin. Remarkably, both early [e.g., *Louden*, 1977] and recently collected paleomagnetic data for the Western Philippine Basin [*Haston and Fuller*, 1991; *Hall et al., 1994a* and b] indicate that it migrated from south of the equator northward over the past 50 Ma. According to *Hall et al.* [1993b] the Philippine Sea Plate at 50 Ma was located in the present position of northern Australia and New Guinea, consistent with an origin adjacent to the eastern Indian Ocean basin, and its southern margin underwent collision with northern New Guinea at 25 Ma. Paleomagnetic data for the Celebes Basin also suggest that it originated close to its present position north of Australia, although comparison of spreading rates for the West Philippine and Celebes Basins suggest that they were not formed by the same ridge [*Silver and Rangin*, 1991b]. Nevertheless, geologic, geochronologic, and paleomagnetic data suggest both basins formed in an oceanic region centered on the present position of northern Australia at a time when this region was occupied by an Indian Ocean ridge.

Continuity of Indian Ocean asthenosphere with mantle beneath the basins formed by rifting of the Eurasian continental margin, like the Japan and South China Seas, is more problematic. Prominent subduction barriers to mantle flow into this region, such as the Indonesian subduction system have existed from 150-175 Ma [*Gill*, 1981]. Spreading in both basins commenced about 32 Ma, considerably after the formation of the West Philippine and Celebes Basins, at a time when Australia and the Philippine Sea Plate were both moving northward. Flow of asthenosphere into the opening South China Basin from the east, possibly caused by displacement by the northward moving Philippine Sea Plate, could have occurred. However, such asthenosphere would soon be isolated from the sub-Philippine Sea Plate asthenosphere by the development of the Manila, Philippine, and Luzon trenches over the period 14-3 Ma [*Jolivet et al.*, 1989; *Rangin et al.*, 1990]. Asthenospheric flow into the opening Sea of Japan at 32 Ma is more difficult to explain, although migration from the south, either from the South China or West Philippine Basins is possible.

The Lau basin is the easternmost of numerous basins that developed along the northeast and eastern Australian margin west of the convergent boundary with the Pacific Plate (Figure 1a). The pronounced bending of this convergent margin north of the Lau Basin took place relatively recently due to collision of the Ontong Java Plateau at about 8 Ma [e.g., *Kroenke et al.*, 1986]. Flow into this region could have occurred from the west and northwest as the northward migration of Australia displaced the intervening asthenosphere. In order to test this possibility, many other marginal basins—North Fiji, South Fiji, Solomon, Loyalty, and New Caledonian—(Figure 1a) must be studied for isotopes.

5.1.3. *Origin of the Indian Ocean asthenospheric domain by contamination and mantle flow.* Resolution of a common versus separate origin for the Indian Ocean-type signature in western Pacific marginal basins also rests in part on the origin of the Indian Ocean asthenospheric domain itself. Extensive dredging studies of the Indian Ocean spreading ridges together with recent ODP exploration of Indian Ocean plateaus have greatly contributed to our understanding of this problem. In simplest terms, the isotopic characteristics of Indian MORB can been modelled by two types of contamination of the depleted upper mantle reservoir tapped by Atlantic and Pacific MORB [e.g., *Hamelin et al.*, 1985]. One contaminant produced the elevated $^{208}Pb/^{204}Pb$, $^{207}Pb/^{204}Pb$ and $^{87}Sr/^{86}Sr$, and low $^{206}Pb/^{204}Pb$, characteristic of many Indian Ocean MORB. Subsequent homogenization of the upper mantle has apparently erased evidence, in the form of mixing arrays, for this contamination event, although Indian Ocean MORB close in isotopic composition to this end-member are notably heterogeneous [Figure 6b; *Mahoney et al.*, 1992]. A second type of contamination has led to a correlated array of increasing $^{206}Pb/^{204}Pb$ ratios and decreasing $^{143}Nd/^{144}Nd$ (Figure 6b). A number of sources for these contaminants have been proposed. Deep mantle plumes rising through the asthenosphere and feeding the numerous DUPAL [*Hart*, 1984] oceanic islands in the Indian Ocean region were proposed by *Castillo* [1988] for both contaminants, whereas *Mahoney et al.*, [1989; 1992] considered active plume contamination likely for only the second type. Contamination by delaminated fragments of the sub-Gondwana lithosphere during and after continental breakup at approximately 200 Ma was favored as a source for the first type of contaminant by *Mahoney et al.* [1989]. Based on geochemical results for MORB of the Southwest Indian Ridge, *Mahoney et al.* [1992] proposed that this contaminant was the degraded subcontinental lithosphere produced by contact with a series of massive thermal plume heads, marked by the present-day Marion, Crozet, and Kerguelen hotspots, at the beginning of Gondwana breakup. An early hypothesis for the origin of the plumes of the DUPAL anomaly [*Hart*, 1984], favored by *Staudigel et al.* [1991] following *Anderson* [1982], was that long-term subduction along the Gondwanan margin prior to breakup focused the accumulation of subducted materials in the deep mantle beneath this region.

If the Indian Ocean asthenospheric domain originated by contamination caused by a specific plume or plumes, the age and areal extent of the Indian Ocean asthenospheric domain may be restricted by the age of that plume and the area reachable through mantle flow [*Mahoney et al.*, 1992]. For example, based on an origin at the Marion, Crozet, and Kerguelen hotspots, and a maximum age of 200 Ma for the Marion hotspot, *Mahoney et al.* [1992] calculated asthenospheric flow rates of 24 mm/year to explain the present areal extent of the Indian Ocean domain to the northeast, and 15 mm/year by flow to the east. If the domain in fact extends to the presently active Sumisu Rift and Lau Basin, higher average asthenospheric flow rates of about 50 and 55 mm/year, respectively, are required. Within the context of this model, a plausible flow mechanism is initial rapid motion of the contaminated asthenosphere northeast through the gap between northern Australia and Southern Eurasia prior to 50 Ma, possibly associated with rapid spreading in the Indian Ocean between 80 and 43 Ma, as proposed by *Mahoney et al.* [1992], followed by displacement of this asthenosphere to the northwest, north, northeast and southeast resulting from the rapid northward migration of Australia between 50 Ma and the present. The older sections of the Philippine Sea Plate and proto-Celebes Sea may have been generated over Indian Ocean asthenosphere north of Australia, whereas the other, younger basins have tapped a displaced asthenospheric reservoir, disconnected from the present Indian Ocean domain by the collision of northern Australia with the Indonesian subduction system.

5.2 Origin of the Indian Ocean-type Isotopic Signature by Local Processes

An alternative hypothesis is that the Indian Ocean type isotopic signature was generated separately within each basin by basin related processes such as subduction and continental rifting. Evidence in support of this concept is that the same types of contaminants present in many marginal basins—subcontinental lithosphere and subducted materials—have been proposed as end-members for the Indian Ocean asthenospheric domain [e.g., *Mahoney et al.*, 1989; 1992; *Staudigel et al.*, 1991]. In order to evaluate the origin of the Indian Ocean-type isotopic features of western Pacific marginal basins basalts by processes operating separately within each basin, we look in detail at the isotopic effects of such local processes as revealed by the data.

5.2.1. *Local effects of continental rifting.* Basins that probably formed by continental rifting are the Sea of Japan and South China Basin. A major common geochemical feature of these basins is the presence of OIB-like seamount or island basalts. Alkaline basalts from the southwestern Sea of Japan islands have very low $^{143}Nd/^{144}Nd$, relatively low $^{206}Pb/^{204}Pb$ but high $^{208}Pb/^{204}Pb$ that is outside the range of compositions of basalts from the other marginal basins (Figure 7). *Tatsumoto and Nakamura* [1991] interpreted this trend as the result of syn-rifting contamination of the asthenosphere by local ancient subcontinental lithosphere including an ancient subduction component—i.e., subcontinental margin lithosphere. As shown by Figures 4 and 7, the Yamato Basin seamounts have Sr, Nd and Pb isotopic compositions which could result from similar subcontinental lithospheric contamination of either an Indian Ocean or Pacific Ocean MORB asthenospheric end-member with $^{206}Pb/^{204}Pb$ of about 17.8. In contrast, Yamato and Japan Basin floor basalts have Pb-isotope compositions that cannot be produced in this way because $^{207}Pb/^{204}Pb$ and $^{206}Pb/^{204}Pb$ ratios are too high for a given $^{208}Pb/^{204}Pb$ (Figure 4c and d). On the basis of this feature, together with trace element evidence, *Cousens and Allan* [1992] and *Cousens et al.* [1993] proposed that the DUPAL isotopic characteristics of these basalts were caused by admixture of a residual component from subducted Pacific Plate sediment into the backarc asthenosphere. It is notable that on Pb-isotope diagrams (Figures 4c and d), the sediment contamination trend extends from an end-member within the field of Indian Ocean MORB at low $^{206}Pb/^{204}Pb$. In the South China Basin, data for enriched basalts from the Scarborough and Reed Bank seamounts diverge from MORB fields (Figure 7) toward low $^{143}Nd/^{144}Nd$, high $^{87}Sr/^{86}Sr$, but high $^{206}Pb/^{204}Pb$, unlike values for the southwestern Japan Sea Islands. Isotopic data for these OIB-like basalts overlap with those for the West Philippine Basin, which is not rift related, as well as Hainan Island (Figure 7), which is sited on continental crust. For the post-spreading South China Basin seamounts and Hainan island, *Tu et al.* [1991; 1992] and *Flower et al.* [1992] proposed as mantle sources a sediment-contaminated Indian Ocean MORB-type asthenosphere that then interacts with subcontinental mantle lithosphere at Hainan island and other continental sites. Two possible origins for the Indian Ocean-type asthenosphere suggested by *Tu et al.* [1991, 1992] were either transport from the Gondwana margin in the late Paleozoic with the South China block [*Lin et al.*, 1985] or contamination by subcontinental lithosphere disturbed by the Indo-Eurasian collision. In summary, for rift-related western Pacific marginal basins, the Sea of Japan and South China Basin, there are isotopic signals of contamination of the asthenosphere by sub-continental mantle lithosphere, subducted sediment and subduction modified lithosphere, as well as possible plume-type enriched mantle sources. In addition to these, there is evidence for an Indian Ocean-type asthenospheric source, on which the contamination trends are superimposed. However, as in the case for Indian Ocean MORBs, there is little evidence, in the form of mixing arrays, that the identified contaminants interacted with Pacific Ocean-type asthenosphere to locally produce the Indian Ocean-type MORB end-member.

5.2.2 *Local effects of subduction.* The best examples of the local effects of subduction are the Lau Basin and Mariana Trough, where extensive sampling of fresh lavas has revealed basalts with transitional MORB-arc isotopic and trace element characteristics [*Volpe et al.*, 1988; *Stern et al.*, 1990; *Volpe et al.*, 1990; *Hergt and Hawkesworth, 1994*; *Hergt and Nilsson*, 1994]. Isotopic patterns formed by these transitional basalts are highlighted in Figure 8. Trends formed separately by Mariana Trough basalts, Lau Basin basalts from sites 836 and 837, and Lau basin site 834, reach highest $^{87}Sr/^{86}Sr$, $^{206}Pb/^{204}Pb$, $^{207}Pb/^{204}Pb$, and $^{208}Pb/^{204}Pb$ within or close to the field of compositions defined by basalts from nearby island arcs. One of the clearest features of these trends is that the MORB-like end-members, which have lowest $^{87}Sr/^{86}Sr$ and lowest $^{206}Pb/^{204}Pb$, $^{207}Pb/^{204}Pb$ and $^{208}Pb/^{204}Pb$, are distinct: Pacific Ocean-type MORB for Lau Basin site 834 (Figures 5 and 8), and Indian Ocean-type MORB for Lau Basin site 836 and 837 and the Mariana Trough (Figures 2, 5 and 8). It is notable that the Indian Ocean-type MORB end-members do not plot in a position intermediate between Pacific Ocean-type MORB and arc basalt (e.g., Figure 8c), therefore, it is unlikely that the subduction component sampled by the arc basalts is responsible for the Indian Ocean imprint. More importantly, as shown by *Hergt and Hawkesworth [1994]* and *Hergt and Nilsson* [1994], the different arcs also have isotopic features that reflect the nature of the adjacent basin mantle: e.g., both the modern Mariana and Tofua arcs, formed adjacent to backarc basins with Indian Ocean-type MORB features, have lower $^{143}Nd/^{144}Nd$ than the older Tonga arc, which formed when Lau basin volcanism was of Pacific Ocean MORB-type. For the Tonga-Lau-Tofua system, Pb-isotopes are especially distinctive: $^{206}Pb/^{204}Pb$ ratios are higher in the older arc, consistent with the concurrent Pacific Ocean Pb-isotope signature in the Lau basin (Figure 8c). A similar influence of asthenospheric composition on island arc lavas was noted by *Mukasa et al.* [1987] for arc lavas from the eastern and western Philippine volcanic belts. Both belts have a clear DUPAL signature in Pb-isotopes, and in the western arc the effect of sediment subduction was clearly superimposed on this feature [*Mukasa et al.*, 1987].

5.2.3. *Summary.* Although sampling and isotopic analysis of western Pacific marginal basins is incomplete, evidence drawn from the existing data supports the contention that the Indian Ocean-type isotopic signature in western Pacific marginal basins results from their derivation from the same Indian Ocean asthenospheric domain that supplies modern Indian Ocean ridges, rather than the separate formation of an Indian Ocean type signature in each marginal basin by local processes. Geochemical evidence in favor is (1) the overall isotopic similarity of the marginal basins basalts to the Indian Ocean ridge basalts; (2) the isotopic similarity of the basins to one another, suggesting derivation from a single long-lived and relatively well-mixed reservoir; (3) that isotopic trends within the basins

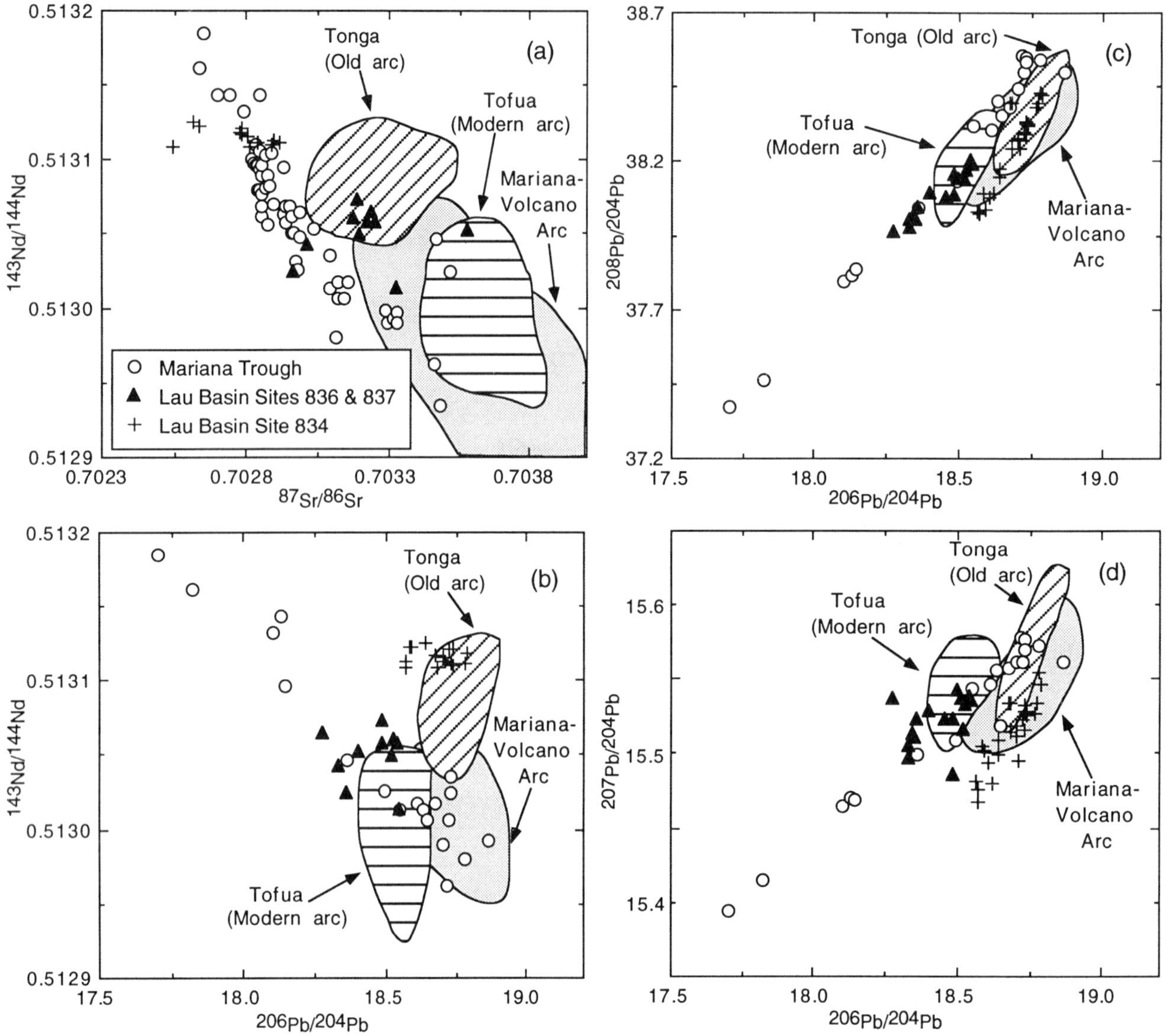

Fig. 8. MORB-like back arc basin basalt to island arc basalt isotopic trends for the Mariana Trough-Mariana arc system, and Lau Basin-Tofua and Tonga arc system. Transitional MORB-island arc basalts (excluded from Figure 6) have La/Nb > 1.5; Ba/La >15. Arc fields from *Hergt and Hawkesworth [1994]*, *Hergt and Nilsson* [1994]; *Ewart and Hawkesworth* [1987]; *Lin et al.* [1990] and *Woodhead* [1989].

toward OIB, local continental or subduction end-members originate within and are superimposed on the Indian Ocean MORB field, rather than some other MORB end-member, and that such end-members often are not isotopically appropriate to produce the Indian Ocean signature.

6. IMPLICATIONS FROM BASIN EVOLUTION FOR UPPER MANTLE DYNAMICS

Our conclusion of a common origin for the mantle sources of western Pacific marginal basins and the Indian Ocean asthenospheric domain can provide some useful information about both the history of the basins and the nature of geochemical provinces in the upper mantle. For the modern basins: Lau Basin, Mariana Trough and Sumisu Rift, it is apparent that the Indian Ocean asthenospheric domain extends to their current geographic position. For the Lau basin, it is also clear that the boundary is rather close since it was transited by the basin spreading center [*Hergt and Hawkesworth, 1994*]. The present position of the Mariana Trough and Sumisu Rift is the westernmost limit for the location of the boundary in the northern hemisphere, since there are no active spreading ridges immediately to the east to indicate the nature of the underlying mantle. To the extent that the paleopositions of older West Philippine and Celebes Basins are known, based on paleomagnetic and sedimentological studies, the past position of the Indian Ocean asthenospheric domain can be defined. Based on our hypothesis of two-stages of asthenospheric flow, Indian Ocean asthenosphere had reached the gap between Australia and southern Eurasia by 60 Ma, to form the West Philippine and Celebes Basins. Later flow westward into the South China Sea, and northeastward into the Shikoku and Parece Vela Basins had taken place by about 32 Ma, and to the southeast into the Lau Basin by only < 6 Ma.

Early models for the West Philippine Basin were that it was trapped Pacific Ocean crust [*Uyeda and Ben Avraham*, 1972], and *Lee and McCabe* [1989] proposed that the Celebes Sea was trapped Indian Ocean crust. Isotopic studies of these areas promised to help resolve these issues based on the differing composition of modern Indian and Pacific Ocean ridge basalts. On the basis of the isotopic data, it is clear that both basins formed over Indian Ocean-type asthenosphere at about the same time. In contrast, differences in spreading rates and change of latitude with time between the two basins indicate that they are unlikely to be fragments formed by the spreading system [*Silver and Rangin*, 1991b]. These observations demonstrate that isotopic data may be helpful in delineating the paleoposition of ancient marginal basin crust, but only to the extent that we know the past extent of the Indian and Pacific asthenospheric domains, and the location of ancient spreading ridges relative to them.

A last point is that, although the Indian Ocean asthenospheric domain may extend to the east of convergent margins bounding the Sumisu Rift, Mariana Trough, and Lau Basin, it is plausible that there is a relationship between the position of these convergent margins, most of which formed concurrently at 45-50 Ma [*Taylor*, 1992], and the boundary of the asthenospheric domains. For example, if the boundary is a major zone of downflow in the upper mantle as suggested by *Klein et al.* [1988], then one would predict a propensity for subduction zones to form there or to become stationed there over time. Similarly, if the Indian Ocean isotopic signature was spread by asthenospheric flow, these convergent margins may have served as subduction barriers and contributed to northward and southward deflection of asthenospheric flow. Since there are no spreading ridges in the western Pacific, this concept is difficult to test. Possibly, the identification of the MORB-source asthenospheric end-members for basaltic magma erupted at active western Pacific hotspots, like Samoa and the Caroline Islands, could pin down the eastward extent of the Indian Ocean domain.

7. PROBLEMS FOR FUTURE RESEARCH

Numerous western Pacific marginal basins have not been sampled or studied for their isotopic composition. Since the basins have varying ages and tectonic settings, each basin contributes uniquely to the overall picture of mantle sources along the past and present Pacific-Eurasian and Pacific-Indo-Australian plate margins. It is important to test the conclusion that the Indian Ocean asthenospheric domain is pervasive in this area at present and to define its position in the past. Studies of these basins will improve our understanding of origin, age and boundary of the Pacific and Indian Ocean asthenospheric domains.

To date, detailed isotopic studies of volcanic rocks have centered on the young basins, for the obvious reasons of sample freshness and exposure. While these are of obvious importance, detailed geochemical studies of the older basins, especially the West Philippine and Celebes Basins, which are oceanic, are a crucial part of the overall picture of the causes of isotopic variations in the major spreading ridges.

Finally, detailed studies of older ocean floor, like that sampled on legs 123 and 129, add to our knowledge of the composition of sections of the upper mantle not currently tapped by spreading ridges, and help to define the past position and extent of the Pacific and Indian Ocean asthenospheric domains. Research on ancient ocean crust will also help to resolve the issue of trapped ocean crust as an origin for the Celebes and West Philippine Basins, and improve our understanding of the tectonic evolution of the western Pacific Ocean.

Acknowledgments. The authors thank John Mahoney and Brian Cousens for thoughtful and thorough reviews which contributed greatly to the improvement of this manuscript. RHV acknowledges financial support of this research from NSF grants OCE: 8707448 and 8923070, and a FIU College of Arts & Science mini-grant. JMH would like to acknowledge the support of a NERC Special Topic grant, and the encouragement of Steve Eggins in this work. Finally, we thank the participants in the JOI/USSAC symposium on western Pacific Marginal Basins for inspiring and informative discussions.

REFERENCES

Allan, J. F. and M. P. Gorton, Geochemistry of igneous rocks from Legs 127 and 128, Sea of Japan, in *Proc. ODP, Sci. Results, 127/128, Pt. 2*, K. Tamaki, K. Suyehiro, J. Allan, M. McWilliams, et al., pp. 905-929, Ocean Drilling Program, College Station, TX, 1992.

Anderson, D. L., Hotspots, polar wander, Mesozoic convection and the geoid, *Nature*, *297*, 391-393, 1982.

Audley-Charles, M. G., P. D. Ballantyne, and R. Hall, Mesozoic-Cenozoic rift-drift sequence of Asian fragments from Gondwanaland, *Tectonophysics 155*, 317-330, 1988.

Briqueu, L., C. Laporte, A. J. Crawford, T. Hasenaka, P. E. Baker, and M. Coltorti, Temporal magmatic evolution of the Aoba Basin, Central New Hebrides island arc: Pb, Sr and Nd isotopic evidence for the coexistence of two mantle components beneath the arc, in *Proc. ODP, Sci. Results, 134*, H. G. Green, J. -Y. Collot, L. B. Stokking, et al., Ocean Drilling Program, College Station, 1994, in press.

Castillo, P. R., The Dupal anomaly as a trace of the upwelling lower mantle, *Nature*, 336, 667-670, 1988.

Castillo, P. R., R. W. Carlson, and R. Batiza, Origin of the Nauru basin igneous complex, Sr, Nd and Pb-isotope and REE constraints, *Earth Planet. Sci. Lett, 103*, 200-213, 1991.

Castillo, P. R., P. A. Floyd, and C. France-Lanord, Isotope geochemistry of Leg 129 basalts: implications for the origin of the widespread Cretaceous volcanic events in the Pacific, in *Proc. ODP, Sci. Results*, 129, R. L. Larson, Y. Lancelot, et al., pp. 405-413, Ocean Drilling Program, College Station, TX, 1992.

Catanzaro, E.J., T. J. Murphy, W. R. Shields, and E. L. Garner, Absolute isotopic abundance ratios of common, equal atom, and radiogenic lead isotope standards, *J. Res. Natl. Bur. Stand., 72A*, 261-267, 1968.

Church, S. E., and M. Tatsumoto, Lead isotope relations in oceanic ridge basalts from the Juan de Fuca-Gorda Ridge area, N.E. Pacific Ocean, *Contrib. Mineral. Petrol., 53*, 253-279, 1975.

Cohen, R. S. and R. K. O'Nions, Identification of recycled continental material in the mantle from Sr, Nd and Pb isotopic investigations, *Earth Planet. Sci. Lett., 61*, 73-84, 1982a.

Cohen, R. S., and R. K. O'Nions, The lead, neodymium and strontium isotopic structure of ocean ridge basalts, *J. Petrol.*, 23, 299-324, 1982b.

Cohen, R. S., N. M. Evensen, P. J. Hamilton, and R. K. O'Nions, U-Pb, Sm-Nd and Rb-Sr systematics of mid-ocean ridge basalt glasses, *Nature, 23*, 149-152, 1980.

Cousens, B. L. and J. F. Allan, A Pb, Sr and Nd isotopic study of basaltic rocks from the Sea of Japan, Legs 127/128, in *Proc. ODP, Sci. Results, 127/128, Pt. 2*, K. Tamaki, K. Suyehiro, J. Allan, M. McWilliams, et al., pp. 805-817, Ocean Drilling Program, College Station, TX, 1992.

Cousens, B. L., J. F. Allan, and M. P. Gorton, Subducted sediments and mantle source enrichment for backarc basin basalts, Japan Sea, *EOS Trans. AGU, 74*, 675, 1993.

Dosso, L., H. Bougault, P. Beuzart, J.-Y. Calvez, and J.-L. Joron, The geochemical structure of the Southeast Indian Ridge, *Earth Planet. Sci. Lett., 88*, 47-59, 1988.

Dupre, B., and C. J. Allegre, Pb-Sr isotope variation in Indian Ocean basalts and mixing phenomena, *Nature, 303*, 142-146, 1983.

Ewart, A. and C. J. Hawkesworth, The Pleistocene-Recent Tonga-Kermadec arc lavas: interpretation of new isotopic and rare earth data in terms of a depleted source model, *J. Petrol., 28*,495-530, 1987.

Flower, M. F. J., M. Zhang, C.-Y. Chen, K. Tu and G. Xie, Magmatism in the South China Basin 2. Post-spreading Quaternary basalts from Hainan Island, south China, *Chem. Geol., 97*, 65-87, 1992.

Fryer, P., J. M. Sinton, and J. A. Philpotts, Basaltic glasses from the Mariana Trough, in, *Init. Repts. DSDP, 60*, D. M. Hussong, S. Uyeda, et al., pp. 601-610, U.S. Govt. Printing Office, Washington, D.C., 1981.

Gill, J. B., Composition and age of the Lau Basin and Ridge volcanic rocks: Implications for evolution of an interarc basin and remnant arc, *Geol. Soc. Amer. Bull., 87*, 1384-1395, 1976.

Gill, J. B., *Orogenic Andesites and Plate Tectonics*, Springer Verlag, Berlin, 390 pp., 1981.

Gill, J. B., C. Seales, P. Thompson, A. G. Hochstaedter, and C. Dunlap, Petrology and geochemistry of Plio-Pleistocene volcanic rocks from the Izu arc and Sumisu Rift, in *Proc. ODP, Sci. Results, 126*, B. Taylor, K. Fujioka, et al., pp. 383-404, Ocean Drilling Program, College Station, TX, 1992

Hall, R., J. Ali, and C. D. Anderson, Tertiary motion of the Philippine Sea Plate: Paleomagnetic evidence from Eastern Indonesia, *Earth Planet. Sci. Lett.*, in press 1994a.

Hall, R., J. R. Ali, C. D. Anderson, and S. J. Baker, Rotation of the Philippine Sea Plate: Evidence from Eastern Indonesia, *Geology*, 1994b.

Hamelin, B., and C. Allegre, Large-scale regional units in the depleted upper mantle revealed by an isotope study of the South-West Indian Ridge, *Nature, 315*, 196-200, 1985.

Hamelin, B., B. Dupre, and C. J. Allegre, Lead-strontium isotopic variations along the East Pacific Rise and the Mid-Atlantic Ridge: a comparative study, *Earth Planet. Sci. Lett., 67*, 340-350, 1984.

Hamelin, B., B. Dupre and C. J. Allegre, Pb-Nd-Sr isotopic data of Indian Ocean ridges: new evidence of large-scale mapping of mantle heterogeneities, *Earth Planet. Sci. Lett., 76*, 288-298, 1985.

Hanan, B. B., and J.-G. Schilling, Easter microplate evolution, Pb isotope evidence, *J. Geophys. Res., 94*, 7432-7448, 1989.

Hart, S. R. , The DUPAL Anomaly: a large scale isotopic anomaly in the southern hemisphere, *Nature, 309*, 753-756, 1984.

Hart, S. R., Heterogeneous mantle domains: signatures, genesis and mixing chronologies, *Earth Planet. Sci. Lett., 90*, 273-296, 1988.

Hart, S.R., W. E. Glassley, and D. E. Karig, Basalts and seafloor spreading behind the Mariana Island arc, *Earth Planet. Sci. Lett., 15*, 12-18, 1972.

Haston, R. B., and M. Fuller, Paleomagnetic data from the Philippine Sea plate and their tectonic significance, *J. Geophys. Res., 96*, 6073-6098, 1991.

Hawkins, J. W., Geology of the Lau Basin, a marginal sea behind the Tonga Arc, in *Geology of the Continental Margins,* edited by C. Burke and C. Drake, pp. 505-520, Springer-Verlag, 1974.

Hawkins, J. W., Petrology and geochemistry of basaltic rocks of the Lau basin, *Earth Planet. Sci. Lett., 28*, 283-298, 1976.

Hawkins, J. W., and J. T. Melchior, Petrology of Mariana Trough and Lau Basin basalts, *J. Geophys. Res., 90*, 11431-11468, 1985.

Hergt, J. M., and C. J. Hawkesworth, The Pb, Sr and Nd isotopic evolution of the Lau basin: implications for mantle dynamic during backarc opening, in *Proc. ODP, Sci. Results, 135*, L. M. Parson, J. W. Hawkins, J. Allan, et al., pp. 505-517, Ocean Drilling Program, College Station, TX, 1993.

Hergt, J. M. and K. Nilsson Farley, Major, trace element and isotope (Pb, Sr and Nd) variations in Site 834 basalts: implication for the initiation of backarc opening, in *Proc. ODP, Sci. Results, 135*, L. M. Parson, J. W. Hawkins, J. Allan, et al., pp. 471-485, Ocean Drilling Program, College Station, TX, 1994.

Hickey-Vargas, R., Isotope characteristics of submarine lavas from the Philippine Sea: implications for the origin of arc and basin magmas of the Philippine Plate, *Earth Planet. Sci. Lett., 107*, 290-304, 1991.

Hilde, T. W. C., and C. Lee, Origin and evolution of the West Philippine Basin: a new interpretation, *Tectonophysics, 102*, 85-104, 1984.

Hochstaedter, A. G., J. B. Gill, M. Kusakabe, S. Newman, M. Pringle, B. Taylor and P. Fryer, Volcanism in the Sumisu Rift, I. Major element, volatile, and stable isotope geochemistry, *Earth Planet. Sci. Lett., 100*, 179-194, 1990a.

Hochstaedter, A.G., J. B. Gill, and J. D. Morris, Volcanism in the Sumisu Rift, II. Subduction and non-subduction related components, *Earth Planet. Sci. Lett., 100*, 195-209, 1990b.

Hussong, D. M., and S. Uyeda, Tectonic processes and the history of the Mariana arc: a synthesis of the results of DSDP Leg 60, in *Init. Repts. DSDP, 60*, D. M. Hussong, S. Uyeda, et al., pp. 909-929, U.S. Govt. Printing Office, Washington, D.C., 1981.

Ito, E., W. M. White, and C. Gopel, The O, Sr, Nd and Pb isotope geochemistry of MORB, *Chem. Geol., ii*, 157-176, 1987.

Jenner, G.A., P. A. Cawood, M. Rautenschlein, and W. M. White, Composition of backarc basin volcanics, Valu Fa Ridge, Lau Basin: Evidence for a slab-derived component in their mantle source, *J. Volcanol. Geotherm. Res.*, 92, 209-222, 1987.

Jolivet, L., P. Huchon, and C. Rangin, Tectonic setting of western Pacific marginal basins, *Tectonophysics, 160*, 23-47, 1989.

Kaneoka, I., K. Notsu, Y. Takigami, K. Fujioka, and H. Sakai, Constraints on the evolution of the Japan Sea based on $^{40}Ar/^{39}Ar$ ages and Sr isotope ratios for volcanic rocks of the Yamato Seamount chain in the Japan Sea, *Earth Planet. Sci. Lett., 97*, 211-225, 1990.

Karig, D. E., Basin genesis in the Philippine Sea, in *Init. Repts. DSDP, 31*, D. E. Karig, J. C. Ingle Jr., et al., pp. 857-879, U.S. Govt. Printing Office, Washington, D.C., 1975.

Kimura, M., T. Matsuda, H. Sato, I. Kaneoka, H. Tokuyama, S. Kuramoto, A. Oshida, K. Shimamura, K. Tamaki, H. Kinoshita, and S. Uyeda, Report on DELP 1985 cruises in the Japan Sea, Part VII: Topography and geology of the Yamato Basin and its vicinity, *Bull. Earthquake Res. Inst., 62*, 447-483, 1987.

Klein, E., C. H. Langmuir, A. Zindler, H. Staudigel, and B. Hamelin, Isotope evidence of a mantle convection boundary at the Australian-Antarctic Discordance, *Nature, 333*, 623-629, 1988.

Klein, G. and K. Kobayashi, Geological summary of the North Philippine Sea, based on DSDP Leg 58 results, in *Init. Repts. DSDP*, 58, G. deVries Klein, K. Kobayashi, et al., pp. 951-961, U.S. Govt. Printing Office, Washington, D.C., 1980.

Kroenke, L. W., J. M. Resig, and P. A. Cooper, Tectonics of the southeastern Solomon Islands: foımation of the Malaita Anticlinorium, in *Geology and Offshore Resources of Pacific Island Arcs—Central and Western Solomon Islands,* edited by J. J. Vedder, K. S. Pound, and S. W. Boundy, pp. 109-116, Circum-Pacific Council Energy Mineral Resources, Earth Science Series, *4*, Houston, TX, 1986.

Kudrass, H. R., M. Wiedicke, P. Cepek, H. Kreuzer, and P. Muller, Mesozoic and Cenozoic rocks dredged from the South China Sea (Reed Bank area) and Sulu Sea, and their significance for plate tectonic reconstruction, *Mar. Pet. Geol., 3*, 19-30, 1986.

Lawver, L., and J. W. Hawkins, Diffuse magnetic anomalies in marginal basins, their possible petrologic and tectonic significance, *Tectonophysics, 43*, 323-339, 1978.

Lee, C. S., and R. McCabe, The Banda-Celebes-Sulu basins: a trapped piece of Cretaceous-Eocene oceanic crust? *Nature*, 322, 51-54, 1989.

Lin, J., M. Fuller, and W. Zhang, Preliminary Phanerozoic polar wander paths for the North and South China blocks, *Nature, 313*, 444-449, 1985.

Lin, P. N., R. J. Stern, J. Morris, and S. H. Bloomer, Nd-and Sr-isotopic compositions of lavas from the northern Mariana and southern Volcano arcs: implications for the origin of island arc melts, *Contrib. Mineral. Petrol., 105*, 381-392, 1990.

Loock, G., W. F. McDonough, S. L. Goldstein, and A. W. Hofmann, Isotopic compositions of volcanic glasses from the Lau basin, *Marine Mining, 9*, 235-245, 1990.

Louden, K. E., Paleomagnetism of DSDP sediments, phase shifting of magnetic anomalies and rotations of the West Philippine basin, *J. Geophys. Res., 82*, 2989-3002, 1977.

Ludden, J. N. and B. Dionne, The geochemistry of ocean crust at the onset of rifting in the Indian Ocean, *in Proc. ODP, Sci. Results, 123*, F. M. Gradstein, J. N. Ludden, et al., pp. 791-799, Ocean Drilling Program, College Station, TX, 1992.

Macdougall, J. D., and G. W. Lugmair, Sr and Nd isotopes in basalts from the East Pacific Rise: significance for mantle heterogeneity, *Earth Planet. Sci. Lett., 77*, 273-284, 1986.

Mahoney, J. J., An isotopic survey of Pacific oceanic plateaus: implications for their nature and origin, in *Seamounts, Islands, Atolls, Geophys. Monogr. Ser.*, vol. *43*, edited by B. H. Keating, P. Fryer, R. Batiza, and G. W. Boehlert, pp. 207-220, AGU, Washington, D.C., 1987.

Mahoney, J. J., and K. J. Spencer, Isotopic evidence for the origin of the Manihiki and Ontong Java oceanic plateaus, *Earth Planet. Sci. Lett., 104*, 196-210, 1991.

Mahoney, J., J. H. Natland, W. M. White, R. Poreda, S. H. Bloomer, R. L. Fisher, and A. N. Baxter, Isotopic and geochemical provinces of the western Indian Ocean spreading centers, *J. Geophys. Res., 94*, 4033-4052, 1989.

Mahoney, J., A. P. LeRoex, Z. Peng, R. L. Fisher, and J. H. Natland, Southwestern limits of Indian Ocean ridge mantle and the origin of low $^{206}Pb/^{204}Pb$ mid-ocean ridge basalt: isotope systematics of the central Southwest Indian Ridge (17°-50°E), *J. Geophys. Res., 97*, 19771-19790, 1992.

Mahoney, J. J., J. M. Sinton, M. D. Kurz, J. D. Macdougall, K. J. Spencer, and G. W. Lugmair, Isotope and trace element characteristics of a super-fast spreading ridge: East Pacific rise, 13°-23°S, *Earth Planet. Sci. Lett., 121*, 173-193, 1994.

Marsh, N.G., A. D. Saunders, and J. Tarney, Geochemistry of basalts from the Shikoku and Daito basins, DSDP Leg 58, in *Init. Repts. DSDP*, 58, G. DeVries Klein, K. Kobayashi, et al., pp. 805-842, U.S. Govt. Printing Office, Washington, D.C., 1980.

Mattey, D. P., N. G. Marsh, and J. Tarney, The geochemistry, mineralogy and petrology of basalts from the West Philippine and Parece Vela basins and from the Palau-Kyushu and West Mariana Ridges, DSDP Leg 59, in *Init. Repts. DSDP, 59*, L. Kroenke, R. Scott, et al., pp. 753-797, U.S. Govt. Printing Office, Washington, DC, 1981.

Mrozowski, C. L. and D. E. Hayes, The evolution of the Parece Vela Basin, eastern Philippine Sea, *Earth Planet. Sci. Lett.*, *46*, 49-67, 1979.

Mukasa, S. B., R. McCabe, and J. B. Gill, Pb-isotopic compositions of volcanic rocks in the West and East Philippine island arcs: presence of the Dupal isotopic anomaly, *Earth Planet. Sci. Lett.*, 84, 153-164, 1987.

Parson, L. M., J. W. Hawkins, and P. A. Hunter, Morphotectonics of the Lau Basin seafloor, implications for the opening history of backarc basins, in *Proc. ODP, Init. Repts., 135*, L. M. Parson, J. W. Hawkins, J. Allan, et al., pp. 81-82, Ocean drilling Program, College Station, TX, 1992.

Pyle, D. G., D. M. Christie, and J. J. Mahoney, Resolving an isotopic boundary within the Australian-Antarctic Discordance, *Earth Planet. Sci. Lett., 112*, 161-178, 1992.

Price, R.C., A. K. Kennedy, M. Riggs-Sneeringer, and F. A. Frey, Geochemistry of basalts from the Indian Ocean Triple junction: implications for the generation and evolution of Indian Ocean ridge basalts, *Earth Planet. Sci. Let.*, *78*, 379-396, 1986.

Rangin, C., The Sulu Sea, a backarc basin setting within a Neogene collision zone, *Tectonophysics 161*, 119-141, 1989.

Rangin, C., and E. A. Silver, Neogene tectonic evolution of the Celebes-Sulu basins: new insights from Leg 124 drilling, in *Proc. ODP, Sci. Results*, 124, E. A. Silver, C. Rangin, M. T. von Breymann, pp. 51-63, Ocean Drilling Program, College Station, TX, 1991.

Rangin, C., L. Jolivet, M. Pubellier, and the Tethys Pacific Working Group, A simple model for the tectonic evolution of Southeast Asia and Indonesia regions for the past 43 Ma., *Bull. Soc. Geol. Fr.*, *6*, 887-905, 1990.

Royer, J. Y., J. G. Sclater, D. T. Sandwell, S. C. Cande, R. Schlich, M. Munschy, J. Dyment, R. L. Fisher, R. D. Muller, M. F. Coffin, P. Patriat, and H. W. Bergh, Indian Ocean Plate reconstructions since the Late Jurassic, in *Synthesis of Results from Scientific Drilling in the Indian Ocean, Geophys. Monogr. Ser.*, vol. 70, edited by R. A. Duncan, D. K. Rea, R. B. Kidd, U. von Rad, and J. J. Weissel, pp. 471-475, AGU, Washington, D.C., 1992.

Scott, R. B., L. Kroenke, G. Zakariadze, and A. Sharaskin, Evolution of the South Philippine Sea: DSDP Leg 59 results, in *Init. Repts. DSDP, 59,* L. Kroenke, R. Scott, et al., pp. 803-816, U.S. Govt. Printing Office, Washington, D.C., 1981.

Seno, T., and S. Maruyama, Paleogeographic reconstruction and origin of the Philippine Sea, *Tectonophysics, 102*, pp. 53-84, 1984.

Serri, G., P. Spadea, L. Beccaluva, L. Civetta, M. Coltorti, J. Dostal, F. Sajona, C. Vaccaro, and O. Zeda, Petrology of igneous rocks from the Celebes Sea Basement, in *Proc. ODP, Sci. Results, 124*, E. A. Silver, C. Rangin, M. T. von Breymann, et al., pp. 271-296, Ocean Drilling Program, College Station, TX, 1991.

Silver, E. A., and C. Rangin, Leg 124 Tectonic synthesis, in *Proc. ODP, Sci. Results*, *124*, E. A. Silver, C. Rangin, M. T. von Breymann, et al., pp. 3-9, Ocean Drilling Program, College Station, TX, 1991a.

Silver, E. A., and C. Rangin, Development of the Celebes Basin in the context of Western Pacific marginal basin history, in *Proc. ODP, Sci. Results*, *124*, E. A. Silver, C. Rangin, M. T. von Breymann, et al., pp. 39-49, Ocean Drilling Program, College Station, TX, 1991b.

Smith, T. E., C. H. Huang, and F. G. Sajona, Geochemistry and petrogenesis of the volcanic rocks from Holes 768 and 767, Sulu Sea, in *Proc. ODP, Sci. Results*, *124*, E. A. Silver, C. Rangin, M. T. von Breymann, et al., pp. 297-310, Ocean Drilling Program, College Station, TX, 1991a.

Smith, T.E., C. H. Huang, and F. G. Sajona, Geochemistry and petrogenesis of the volcanic rocks from the Celebes Sea, in *Proc. ODP, Sci. Results*, *124*, E. A. Silver, C. Rangin, M. T. von Breymann, et al., pp. 311-320, Ocean Drilling Program, College Station, TX, 1991b.

Spadea, P., L. Beccaluva, L. Civetta, M. Coltorti, J. Dostal, F. Sajona, G. Serri, C. Vaccaro, and O. Zeda, Petrology of basic igneous rocks from the floor of the Sulu Sea, in *Proc. ODP, Sci. Results*, 124, edited by E. A. Silver, C. Rangin, M. T. von Breymann, et al., 251-269, Ocean Drilling Program, College Station, TX, 1991.

Staudigel, H., K.-H. Park, M. Pringle, J. L. Rubenstone, W. H. F. Smith, and A. Zindler, The longevity of the South Pacific isotopic and thermal anomaly, *Earth Planet. Sci. Lett.*, *102,* 24-44, 1991.

Stern, R. J., P. N. Lin, J. D. Morris, M. C. Jackson, P. Fryer, S. H. Bloomer, and E. Ito, Enriched backarc basin basalts from the northern Mariana Trough: implications for the magmatic evolution of backarc basins, *Earth Planet. Sci. Lett.*, *100*, 210-225, 1990.

Sun, S.-S., Lead isotopic study of young volcanic rocks from mid-ocean ridges, ocean islands and island arcs, *Phil. Trans. R. Soc. Lond., A 297*, 409-445, 1980.

Sunkel, G., Origin of the petrological and geochemical variations of Lau Basin lavas (SW Pacific), *Marine Mining*, *9*, 205-234, 1990.

Tamaki, K., K. Suyehiro, J. Allan, J. Ingle Jr., and K. A. Pisciotto, Tectonic synthesis and implications of Japan Sea ODP Drilling, in *Proc. ODP, Sci. Results, 127/128,* K. Tamaki, K. Suyehiro, J. Allan, M. McWilliams, et al., p. 1333-1348, Ocean Drilling Program, College Station, TX, 1992.

Tapponier, P., G. Peltzer, A. Y. LeDain, R. Armigo, and P. Cobbold, Propagation extrusion tectonics in Asia: new insights from simple experiments with plasticine, *Geology*, *7*, 611-616, 1982.

Tatsumoto, M., Isotopic composition of lead in oceanic basalt and its implication to mantle evolution, *Earth Planet. Sci. Lett.*, *38*, 63-87, 1978.

Tatsumoto, M. and Y. Nakamura, DUPAL anomaly in the Sea of Japan: Pb, Nd and Sr isotopic variations at the eastern Eurasian continental margin, *Geochim. Cosmochim. Acta*,

55, 3697-3708, 1991.

Taylor, B., Rifting and the volcanic-tectonic evolution of the Izu-Bonin-Mariana arc, in *Proc. ODP, Sci. Results,* 126, B. Taylor, K. Fujioka, et al., pp. 627-651, Ocean Drilling Program, College Station, TX, 1992.

Taylor, B., and D. E. Hayes, The tectonic evolution of the South China Basin, in *The Tectonic and Geologic Evolution of Southeast Asian Seas and Islands, Geophys. Monogr. Ser.*, vol. 23, edited by D.E. Hayes, pp. 89-104, AGU, Washington, D.C., 1980.

Taylor, B., and D. E. Hayes, Origin and history of the South China Basin, in *The Tectonic and Geologic Evolution of Southeast Asian Seas and Islands, Geophys. Monogr. Ser.*, vol. 27, pp. 25-56, edited by D.E. Hayes, AGU, Washington, D.C., 1983.

Taylor, B., G. Brown, P. Fryer, J. Gill, A. G. Hochstaedter, H. Hotta, C. H. Langmuir, M. Leinen, A. Nishimura, and T. Urabe, ALVIN-SeaBeam studies of the Sumisu Rift, Izu-Bonin Arc, *Earth Planet. Sci. Lett., 100,* 127-147, 1990.

Tu, K, M. Flower, R. Carlson, M. Zhang and G. Xie, Sr, Nd and Pb isotopic compositions of Hainan basalts (South China): Implications for a subcontinental DUPAL source, *Geology, 19*, 567-569, 1991.

Tu, K., M. J. F. Flower, R. Carlson, G. Xie, C-Y. Chen, and M. Zhang, Magmatism in the South China Basin 1. Isotopic and trace-element evidence for an endogenous Dupal mantle component, *Chem. Geol., 97*, 47-63, 1992.

Uyeda, S. and Z. Ben Avraham, Z., Origin and development of the Philippine Sea, *Nature,* 240, 176-178, 1972.

Volpe, A. M., J. D. Macdougall, and J. W. Hawkins, Mariana Trough basalts (MTB): trace element and Sr-Nd isotopic evidence for mixing between MORB-like and arc-like melts, *Earth Planet. Sci. Lett., 82,* 241-254, 1987.

Volpe, A. M., J. D. MacDougall, and J. W. Hawkins, Lau Basin Basalts (LBB): trace element and Sr-Nd isotope evidence for heterogeneity in back arc basin mantle, *Earth Planet. Sci. Lett., 90*, 174-186, 1988.

Volpe, A. M., J. D. Macdougall, G. W. Lugmair, J. W. Hawkins and P. Lonsdale, Fine-scale isotopic variation in Mariana Trough basalts: evidence for heterogeneity and a recycled component in backarc basin mantle, *Earth Planet. Sci. Lett., 100*, 251-264, 1990.

Wang, M., D. Wu, I. Liang, and A. Yin, Some geochemical characteristics of basalts from the South China Sea, *Geochimica, 4*, 332-340, 1984.

Weis, D., and F. A. Frey, Isotope geochemistry of Ninetyeast Ridge basalts: Sr, Nd and Pb evidence for the involvement of the Kerguelen hotspot, in *Proc. ODP, Sci. Results, 121*, J. Weissel, J. Pierce, E. Taylor, J. Alt, et al., pp. 591-610, Ocean Drilling Program, College Station, TX, 1991.

Weis, D., W. M. White, F. A. Frey, R. A. Duncan, M. R. Fisk, J. Dehn, J. Ludden, A. Saunders, and M. Storey, The influence of mantle plumes in the generation of Indian oceanic crust, In: *Synthesis of Results from Scientific Drilling in the Indian Ocean, Geophys. Monogr. Ser.*, vol. 70, edited by R. A. Duncan, D. K. Rea, R. B. Kidd, U. von Rad, and J. J. Weissel, pp. 57-89, AGU, Washington, D.C., 1992.

Wells, R. E., Origin of the oceanic basalt basement of the Solomon Island arc and its relationship to the Ontong-Java Plateau—insights from Cenozoic plate motions, *Tectonophysics, 165*, 219-235, 1989.

White, W. M., and A. W. Hofmann, Sr and Nd isotope geochemistry of oceanic basalts and mantle evolution, *Nature, 296*, 821-825, 1982.

White, W. M., A. W. Hofmann, and H. Puchelt, Isotope geochemistry of Pacific mid-ocean ridge basalt, *J. Geophys. Res., 92*, 4881-4893, 1987.

White, W. M., M. C. Cheatham, and R. A. Duncan, Isotope geochemistry of Leg 115 basalts and inferences on the history of Reunion mantle plume, in *Proc. ODP, Sci. Results,* 115, R. A. Duncan, J. Backman, L. C. Peterson, et al., pp. 53-61, Ocean Drilling Program, College Station, TX, 1990.

Wood, D. A., A variably veined suboceanic upper mantle—genetic significance for mid-ocean ridge basalts from geochemical evidence, *Geology, 7*, 499-503, 1979.

Woodhead, J. D., Geochemistry of the Mariana arc (western Pacific): Source composition and processes, *Chem. Geol., 76*, 1-24, 1989.

Coexistence of Indian and Pacific Oceanic Upper Mantle Reservoirs Beneath the Central New Hebrides Island Arc

Anthony J. Crawford

Department of Geology, University of Tasmania, Hobart, Tasmania, Australia

L. Briqueu and C. Laporte

Laboratoire de geochemie isotopique, Departement des Sciences de la Terre, Universite Montpellier II, France

T. Hasenaka

Department of Petrology and Mineralogy, Faculty of Science, Tokohu University, Aoba, Japan

Throughout most of its history (30 Ma-present), the New Hebrides intra-oceanic island arc has produced lavas with Pb-Nd-Sr isotopic signatures defining arrays interpreted to reflect mixing between a Pacific-type MORB mantle source and a more radiogenic component derived from dehydration of subducted Pacific oceanic crust. Modern volcanoes in the southern section of the arc are still show Pacific-type Pb-Nd-Sr isotopic values. However, collision of an Eocene intra-oceanic arc (the d'Entrecasteaux zone) with the central part of the New Hebrides arc ~2-3 Ma produced a dramatic shift in isotopic signatures of lavas in this part of the arc. Post-collision arc lavas have Pb-Nd-Sr isotopic compositions reflecting mixing between a mantle source like that of Indian Ocean mid-ocean ridge basalts and a slab-derived component very close to that involved in arc magmatism away from the collision zone. This isotopic shift is attributed to an eastward deflection (50-100 km) of the site of arc magmatism in the collision zone, such that dehydration of the slab led to magma generation not from old Pacific-type arc lithosphere as in most of the arc, but from ascending diapirs of asthenospheric mantle providing backarc basin-type basalts to the embryonic backarc troughs within the New Hebrides arc. These backarc troughs are absent in the collision zone. Two models are proposed for the existence of Indian-type asthenosphere in the western Pacific. The first assumes that Indian-type mantle formed within the Indian Ocean domain migrated north and south of Australia into the western Pacific. The second hypothesis proposes that the isotopic characteristics of Indian-type asthenospheric mantle are formed by interaction of typical Pacific-Atlantic-type MORB asthenosphere with delaminated subcontinental lithosphere. Indian-type asthenosphere formed in this way beneath the thinned margins of eastern Asia and the continental fragments east of Australia migrated a short distance eastward to ascend and partially melt whenever extension in western Pacific arc settings leads to the formation of backarc basins.

INTRODUCTION

To begin to understand the dynamics of the solid-Earth convection system, and address the debate relating to whole-mantle convection versus layered-mantle convection, we need detailed information on the distribution, composition, and longevity of the mantle reservoir sources of basaltic magmas erupted at the Earth's surface. Perhaps the least complex of the various mantle isotope reservoirs currently recognized [*Hart* 1988] is the global depleted asthenospheric source of mid-ocean ridge basalts (MORB). However, even this simplest source shows regional variation, with the source of Indian Ocean MORB being compositionally

Active Margins and Marginal Basins of the Western Pacific
Geophysical Monograph 88

distinct in terms of key Pb-Nd-Sr isotopic ratios from the sources of Pacific and Atlantic MORB [*Hamelin and Allegre*, 1985; *Ito et al.*, 1987; *White et al.*, 1987; *Mahoney et al.*, 1989, 1992; *Hickey-Vargas et al.*, this volume].

Klein et al. [1988] showed that the boundary between these two mantle isotopic domains lies on the Southeast Indian Ridge south of Australia, within a morphologically and geophysically anomalous ridge segment referred to as the Australian-Antarctic Discordance (AAD). Amplifying an idea proposed by *Alvarez* [1982], *Klein et al.* [1988] argued that one interpretation of the significance of this mantle isotope boundary could be that the Pacific asthenosphere commenced convecting westward into the Indian Ocean mantle isotope domain once the Southern Ocean opened enough to allow mantle communication between the continental roots of Antarctica and the South Tasman Rise (southernmost continental crust of Australia), around 35-40 Ma. The concept of Pacific asthenosphere invading westward has been strongly supported by (i) *Pyle et al.* [1992], who showed that 4 m.y.-old crust just off-ridge immediately east of the AAD has an Indian Ocean isotopic signature, whereas adjacent zero-age crust carries a distinct Pacific isotopic signature, and (ii) by *Lanyon and Crawford* [unpublished information], who show that 36-66 m.y.-old Southern Ocean basaltic crust northeast of the AAD also shows Indian Ocean isotopic signatures. These data imply a broad V-shaped 'front' of Pacific Ocean asthenosphere driving westward into the opening Southern Ocean, with the nose of the invading mantle presently located within the AAD.

Pyle et al. [1993] reported that for the one location from which basaltic crust of the Tasman Sea has been sampled, at DSDP hole 283 on 60-70 m.y.-old crust, isotopic data show the basalts to be Pacific mantle-type. It is presumed, therefore, that asthenosphere with a Pacific Ocean isotopic fingerprint existed to the east of the eastern margin of Gondwana (formed in this region by the then-contiguous continental blocks of Australia and Antarctica) prior to and during the opening of the Tasman Sea (~80-52 Ma). Furthermore, data summarized by *Pyle et al.* [1993] and our own data [*Lanyon*, 1993; *Lanyon and Crawford,* unpublished information] indicate that basalts produced over the last 15 m.y. at the eastern end of the Southeast Indian Ridge between Macquarie Island and the Balleny Islands (including those lavas exposed on Macquarie Island, and drilled at DSDP sites 274, 278 and 279A) were generated from Pacific asthenosphere contaminated by the HIMU Balleny plume.

Given the long-lived presence of Pacific asthenospheric upper mantle east of Australia, it is therefore surprising that basalts with clear Indian Ocean-type isotopic signatures dominate all the Tertiary and Recent backarc basins of the west Pacific region, from the Sea of Japan down to the Lau Basin [*Hickey-Vargas et al.*, this volume; *Nohara et al.*, 1994]. In fact, the only occurrence of Pacific MORB-type basalt isotopic ratios is for Lau Basin lavas from ODP site 834, for which the isotopic data extend from a Pacific-type end-member towards a more radiogenic Sr- and Pb-rich component generally regarded as being derived from the subducted slab. As shown by *Hergt and Hawkesworth* [1994], with progressive opening of the Lau Basin, the Pacific asthenospheric source was rapidly replaced by a mantle source with distinct Indian Ocean-type isotopic signatures, essentially identical to those defined by basalts from the other west Pacific backarc basins.

In the accompanying paper, *Hickey-Vargas et al.* focus on the Indian Ocean-type mantle signatures in backarc basin basalts from the western Pacific. Herein, we provide further evidence for the existence of Indian Ocean asthenosphere to the east of Australia, this time in post-2 m.y.-old picritic high-K arc lavas from several major island arc shield volcanoes in the central section of the New Hebrides island arc in Vanuatu. As reported by *Briqueu et al.* [1994], until 2-3 Ma, arc volcanoes in the central and southern part of this arc had Pb-Sr-Nd isotopic compositions suggesting mixing between Pacific asthenospheric mantle and a slab-derived component. However, following collision of an Eocene island arc (the d'Entrecasteaux Zone) with the central section of the New Hebrides island arc some 2-3 Ma, the source of subsequent arc magmas switched to an Indian Ocean-type asthenosphere source mixed with a slab-derived component, whereas those in the southern part of the arc, away from the collision zone, have maintained Pacific-type mantle sources.

The existence of this band of Indian Ocean-type upper mantle to the east of Australia, and its apparent isolation from its 'parent' Indian Ocean mantle by the extensive Tasman Sea-Antarctic SW Pacific province, characterized by Pacific-type isotopic signatures, is intriguing. The mechanism by which it has maintained its identity in the convecting upper mantle, particularly considering the strong evidence that the Pacific asthenosphere is migrating westward into the Southern Ocean, requires explanation. In this paper, we discuss in more detail the significance of the available isotope data for the New Hebrides island arc and adjacent areas, and we address the question as to how upper mantle with an Indian Ocean-type asthenosphere signature came to exist in this region, apparently isolated from its 'parent' mantle in the Indian Ocean.

REGIONAL PLATE TECTONIC SETTING

The New Hebrides island arc (Figures 1,2) is part of a narrow chain of Tertiary to Recent volcanic island arcs extending from Papua New Guinea and the Solomon Islands, through Fiji, Tonga, and the Kermadec Islands, to New Zealand. The Vanuatu segment of the New Hebrides island arc extends from the Torres Islands in the north to the islands of Matthew and Hunter in the south, and is a 700 km-long, partly emergent ridge with an average width of 200 km overlying a steeply eastward-dipping subducted slab that is relatively continuous along the length of the arc, and marks the site of subduction of the Australia-India plate

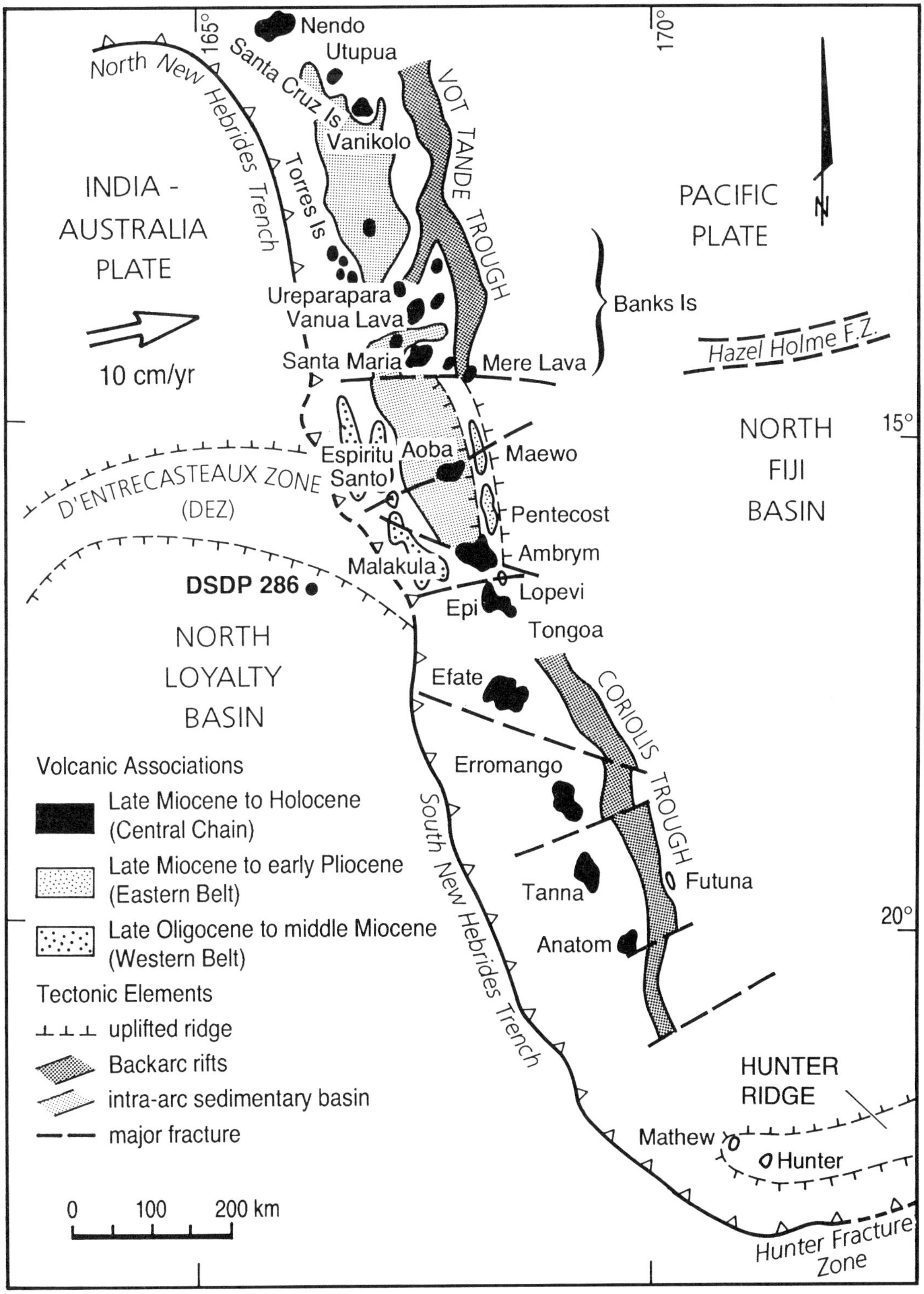

Fig. 1. Simplified tectonic map of the New Hebrides island arc showing western belt, central chain volcanoes, and eastern belt islands, the d'Entrecasteaux zone colliding with arc, and the narrow, embryonic backarc troughs (modified from *Greene et al.*, 1988).

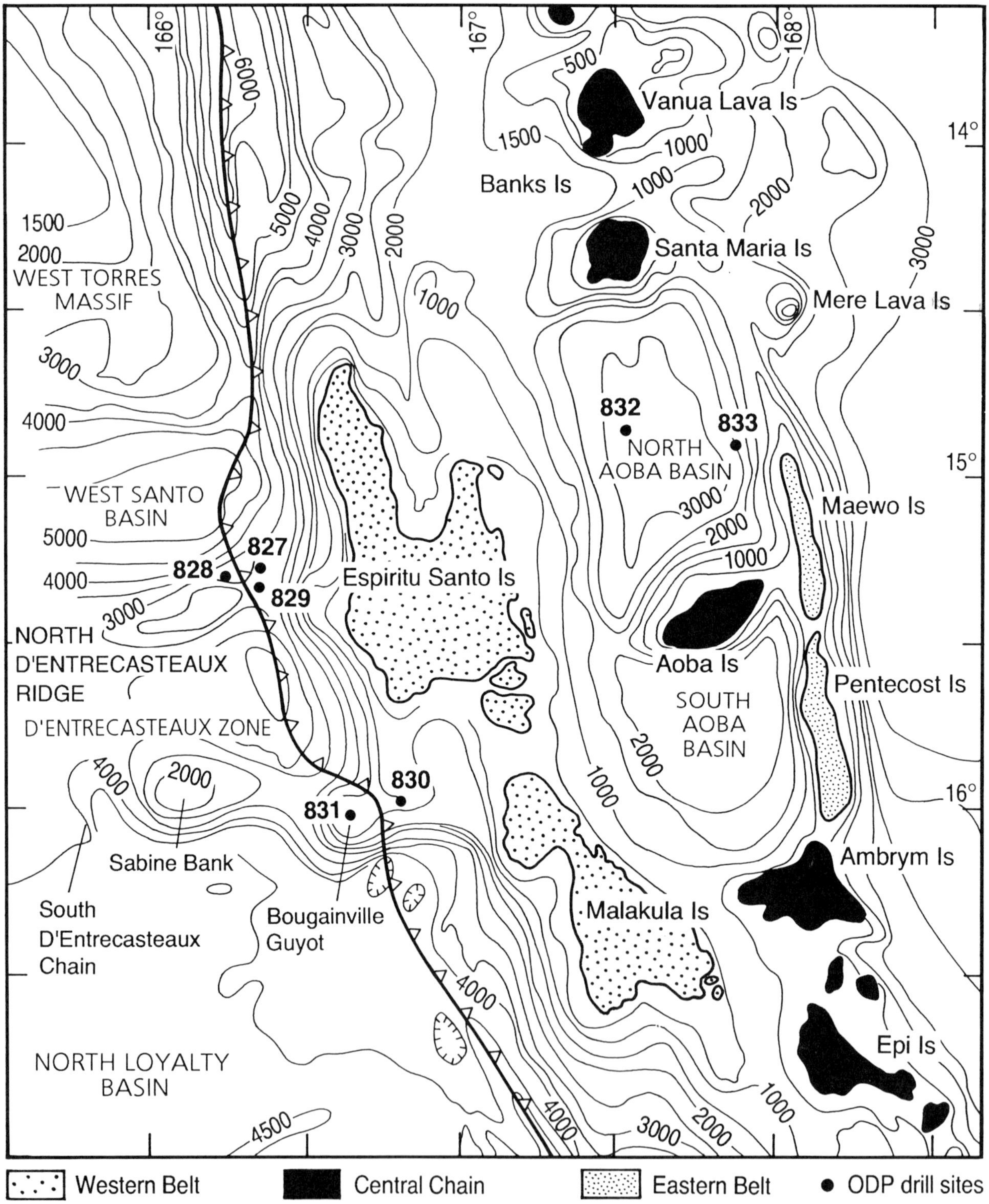

Fig. 2. Detailed map of the d'Entrecasteaux zone collision showing the major volcanoes in the collision zone, and the location of ODP Leg 134 Sites 832 and 833 in the north Aoba basin.

beneath the Pacific plate (Figure 1). The New Hebrides Trench, however, which lies west of, and runs parallel to, the New Hebrides island arc over much of its length, is absent opposite the large Espiritu Santo and Malakula islands.

Immediately west of the New Hebrides Trench, at depths of ~1,000-5,000 m, is the north Loyalty Basin (Figure 2), which is floored by oceanic crust that has been dated as middle Eocene at DSDP site 286 [*Andrews et al.,* 1975]. The Loyalty ridge, together with the ridge bearing the island of New Caledonia, passes northwestward into the d'Entrecasteaux zone (DEZ), an arcuate horst-and-graben structure some 100 km wide. Opposite Espiritu Santo and Malakula, the DEZ is colliding with, and being subducted beneath, the Pacific plate and is responsible for the shoaling of the trench in this region. Here, the twin-spined east-west-trending DEZ is made up of the north d'Entrecasteaux ridge (Figure 2) of backarc basin or forearc oceanic-type crust [*Maillet et al.* 1986; *Collot et al.* 1985; 1992], the smooth-bottomed central d'Entrecasteaux basin (to 4,000 m deep), and a line of seamounts, the south d'Entrecasteaux chain, which ODP leg 134 drilling showed to be Eocene intra-oceanic arc volcanoes [*Baker et al.*, 1994]. North of the DEZ are the west Santo basin and the west Torres massif (Figure 2), which, from the few data available, are considered to be, respectively, an old trench-like feature [*Collot et al.* 1985] and an oceanic plateau with intermediate crustal thickness.

East of the New Hebrides island arc lies the North Fiji Basin, a marginal sea no older than middle to late Miocene [*Pelletier et al.*, 1993; *Malahoff et al.*, 1982]. This basin is separated from the older South Fiji Basin of Oligocene age [*Malahoff et al.* 1982] by the northeast-trending Hunter fracture zone and Hunter Ridge. Complex multi-stage opening of the North Fiji Basin [*Auzende et al.* 1988, *1995*] and backarc extensional troughs [*Recy et al.*, 1990; *Monjaret et al.*, 1991; *Maillet et al.*, 1995] accompanied rotation of the New Hebrides arc.

The direction of relative plate convergence is N75°E along most of the New Hebrides island arc, and the present convergence rate varies from 16 cm/yr at 11°S to 12 cm/yr at 20°S [*Louat and Pelletier,* 1989]. A minimum convergence rate (9 cm/yr) occurs where the DEZ is colliding with the central part of the arc near 15°S. Oblique extension generally occurs along the eastern (backarc) flank of the main part of the New Hebrides island arc, except at the latitude of the DEZ collision, where backarc compression is active. A well-defined Benioff zone dips steeply to the east, extending to 350 km downdip along much of the arc, although the slab is only ~220 km or less long south of Anatom island in the southern part of the arc [*Louat and Pelletier,* 1989]. Along the main part of the arc, Quaternary-Recent arc volcanoes of the central chain define the current volcanic front and lie at 100-200 km above the subducted slab.

TECTONIC HISTORY

The tectonic evolution of the New Hebrides island arc has been greatly influenced by the opening of the North Fiji Basin. Before the opening of this basin, probably 10-12 Ma [*Auzende et al.*, 1988], a single arc (the Vitiaz arc) existed between the Indian-Australian and Pacific plates, along the Vitiaz trench. Parts of the the New Hebrides arc, the Fiji Islands, and the Lau and Tonga ridges all formed part of this long Vitiaz arc. The Late Oligocene to Middle Miocene volcanics of the western belt islands (Espiritu Santo, Malakula and the Torres group) of the New Hebrides arc is believed to have been produced during this west-dipping subduction. Collision of the Ontong-Java plateau with the northern end of the Vitiaz arc, perhaps around 10 Ma, forced a reversal of the subduction polarity and eventual opening of the North Fiji Basin [*Chase,* 1971; *Falvey,* 1978; *Carney and Macfarlane,* 1982].

Espiritu Santo island in the western belt developed probably in the late Oligocene to early Miocene, with extensive submarine volcanism between 25 and 15 Ma [*Macfarlane et al.*, 1988]. The oldest rocks include basaltic and andesitic lavas and volcanic breccias, including pyroxene- and hornblende-phyric calc-alkaline andesites and basalts. Early to middle Miocene composite and zoned intrusions of basalt, andesite, microdiorite, and minor gabbro are also not uncommon on Espiritu Santo. There is no post-middle Miocene magmatism on Espiritu Santo, and by late Miocene to early Pliocene time (11-4 Ma) the western belt had emerged as a landmass. Final rapid uplift of the western belt took place 2-0 Ma, in response to the continuing collision and subduction of the DEZ.

The eastern belt is an uplifted submarine ridge also produced mainly by collision of the DEZ and is exposed on Pentecost and Maewo islands. Voluminous late Miocene-Pliocene basaltic to andesitic submarine volcanism with medium- to high-K affinity [*Macfarlane et al.*, 1988] took place along the eastern belt, and later at the eastern edge of the central chain (5.8 Ma). Vertical uplift in the early to late Pliocene (4-2 Ma) exposed the eastern belt and may have been associated with emplacement of basement complex ophiolites on Pentecost island. Volcanism on Maewo and Pentecost ceased after late Pliocene.

Although volcanism as old as late Miocene is known from the central chain islands, it is limited to the northern and southernmost parts of the arc. Most central chain volcanism is less than 3 Ma [*Macfarlane et al.*, 1988]

THE D'ENTRECASTEAUX ZONE COLLISION

As noted above, the central section of New Hebrides island arc is in collision with a major submarine ridge, the DEZ. This collision commenced further south (around 17°S) ~2-3 Ma, and is sweeping northward along the trench at about

3.2-4 cm/yr [*Taylor et al.*, 1994]. In the collision zone, the old Oligo-Miocene volcanic islands of Espiritu Santo and Malakula are being rapidly uplifted and eroded [*Taylor et al.*, 1994], and a major flexural basin, the Aoba basin (Figure 2), has formed to the east of these islands [*Collot et al.*, 1985; *Falvey and Greene*, 1988; *Greene et al.*, 1994]. The Aoba basin is bordered to the north by the large active volcanic island of Santa Maria (Gaua), and to the south by the immense caldera island of Ambrym. The basaltic shield volcano of Aoba in the central part of the basin divides the Aoba basin into the north Aoba basin and south Aoba basin (Figure 2).

The north Aoba basin, with a maximum fill thickness of 4.9 km [*Holmes*, 1988], should preserve a good sedimentary record of post-early Miocene volcanism on nearby islands or submarine volcanic centers [*Greene et al.*, 1994]. Seismic reflection studies [*Fisher et al.*, 1994] have demonstrated a uniform stratification in the north Aoba basin, and *Fisher et al.* [1988] correlated the acoustic basement of the north Aoba basin with early and middle Miocene volcanics presently exposed on Espiritu Santo island, less than 100 km to the west of ODP site 832. A major objective for drilling the two deep ODP holes in the north Aoba basin was to recover a full stratigraphic section that would provide a sedimentological and volcanic record of the tectonic history (including a hypothetical reversal of subduction polarity; *Falvey*, 1978; *Carney and Macfarlane*, 1982; *Macfarlane et al.*, 1988; *Greene et al.*, 1988; 1994] of this arc-backarc basin system.

NEW HEBRIDES ISLAND ARC MAGMATISM

A compositional transition exists in modern New Hebrides island arc lavas from the central to the southern part of the arc [*Gorton*, 1974, 1977; *Roca*, 1978; *Briqueu*, 1985; *Javoy et al.*, 1986; *Eggins*, 1989, 1993; *Monzier et al.*, 1993]. In the southern segment of the arc, away from the influence of the DEZ collision, arc lavas are dominantly low- to medium-K basalts and andesites. *Girod et al.*, 1979; *Briqueu et al.*, 1984; *Eggins*, 1993], and picritic compositions are voluminous, especially on the islands of Aoba, Ambrym, and Santa Maria. This geochemical transition is precisely delineated by Sr isotopes [*Briqueu*, 1985; *Javoy et al.*, 1986; *Peate et al.*, 1991]. Lavas from arc volcanoes from Santa Maria to Efate (Figure 1) show significantly higher Sr isotope ratios (0.7036-0.7043) than those of the southern volcanic islands of Erromango, Tanna, and Anatom (0.7030-0.7032).

We have recently reported detailed petrological-geochemical [*Hasenaka et al.*, 1994] and isotopic studies [*Briqueu et al.*, 1994] of the extensive lavas, tephras and sills drilled at ODP sites 832 and 833 in the upper crustal section of the north Aoba basin. Coupled with published and unpublished studies of the major volcanoes of Aoba [*Eggins*, 1993] and Santa Maria [M. Monzier, J-P. Eissen and C. Robin; pers. comm. 1993] behind the current collision zone, and Epi Island further to the south, close to the site of initial impact of the DEZ [*Crawford et al.*, 1988; *Barsdell and Berry*, 1990], these data allow more precise documentation of the compositional (including isotopic) variability of arc magmatism in this region.

ODP HOLES 832 AND 833

Background

Lithostratigraphic units for ODP holes 832B and 833B in the north Aoba basin were defined on the basis of core descriptions and petrographic observations; full details are given by *Hasenaka et al.* [1994]. Volcanic rocks are present as clasts/fragments in volcanic breccias or sedimentary rocks in both holes, and a sill complex occurs at the base of hole 833B. Here we summarize the distribution and significance of the various magmatic rocks recovered during leg 134 drilling in the north Aoba basin as a prelude to a more detailed discussion of the implications for the nature and distribution of the mantle sources of these rocks.

Compositional-Stratigraphic Relationships

On the basis of compositional characteristics, four magma suites were recognized in holes 832B and 833B. The oldest suite is represented by the low- to medium-K calc-alkaline basaltic to andesitic clasts from the late middle Miocene epiclastic lava breccias constituting litho-stratigraphic unit VII in hole 832B. The calc-alkaline nature and middle Miocene minimum age of the unit VII clasts in hole 832B both suggest that these were derived from Espiritu Santo island, some 90 km to the west.

In hole 832B, lithostratigraphic units II, III and IV are all lava breccias of Pleistocene-late Pliocene age, whereas unit VI is a late Miocene lava breccia. All are low Al_2O_3 (13-16%) medium- to high-K tholeiitic to transitional alkaline basalts with FeO*/MgO values mainly from 1 to 2. The spreads of compositions defined by basalts (<51% SiO_2 and >6% MgO) from these units on plots of SiO_2 vs TiO_2, K_2O, P_2O_5, FeO*, Ce, Y and Zr are all at high angles to typical crystal fractionation trends. This probably reflects a spread in the primary magma compositions, with higher-pressure, lower-degree partial melts having lower SiO_2, higher FeO*, higher K_2O, and higher high field strength elements (HFSE0, compared to lower-pressure, higher-degree melts [*Hasenaka et al.*, 1994].

Late Pliocene lithostratigraphic unit III in hole 833B contains lava clasts belonging to a low-K basaltic to andesitic suite that has virtually flat rare earth elements (REE0 patterns. These clasts are strikingly similar petrographically and compositionally to lavas from Mere Lava volcano, some 75 km north-northeast (Figure 2). Mere Lava is a young stratocone volcano astride the western rim of the juvenile Northern (Jean Charcot or Vot Tande) Trough that is currently splitting the backarc side of the northern section of the New Hebrides island arc [*Charvis and Pelletier*, 1989].

Hasenaka et al. [1994] suggested that the unit III clasts are either derived from an early phase of Mere Lava magmatism, or perhaps more likely, from a closer, compositionally very similar volcano in a similar tectonic setting to Mere Lava along the western margin of the central belt. Uplift of the Maewo-Pentecost ridge since the Pliocene has clearly disrupted development of the juvenile backarc trough in this central part of the arc (Figure 1).

Despite their compositional differences, epiclastic volcanic breccias constituting lithostratigraphic unit IV in hole 832B and unit III in hole 833B are probably temporally correlated. They may have been deposited in response to the major DEZ collision in the latest Pliocene to earliest Pleistocene, which is marked by a major angular unconformity and biostratigraphic hiatus [*Fisher et al.*, 1994].

A sill complex intrudes Pliocene sedimentary rocks at the base of hole 833B. It includes high-K to shoshonitic basalts and andesites that are strongly enriched in HFSE and large ion lithophile elements (LILE) relative to the other suites in holes 832B and 833B. Two sill samples have been K-Ar dated at 3.65 and 3.30 Ma [*Rex,* 1994]. They thus appear to predate the DEZ collision and were intruded prior to emplacement of the unit III volcanic breccia in this hole. The distinctive compositions of the evolved basaltic to andesitic sills in hole 833B are very similar to compositions of high-Ti series basalts and andesites on adjacent Aoba volcano [*Eggins,* 1993], although the sills are older by perhaps 3 m.y. than the Aoba lavas. Thus, relatively HFSE-enriched basaltic magmas have been generated intermittently during the last 4 m.y. in the north Aoba basin region. These are volumetrically subordinate to the less HFSE-enriched series, best represented by the low-Ti series on Aoba [*Eggins,* 1993] to the south of hole 833B, and similar lavas on Santa Maria volcano, at the northern margin of the north Aoba basin. Most of the hole 832B lithostratigraphic units VI, IV, III, and II are broadly correlated compositionally with the Aoba low-Ti series of *Eggins* [1993]. Preliminary trace element data for ashes from the upper sections of holes 832B and 833B [*Gerard and Person,* 1994] suggest that Aoba-type high-Ti series magmas are not represented among the ashes, which presumably derive in large part from nearby Santa Maria volcano.

ISOTOPIC DATA

Isotopic data for the ODP leg 134 drilled rocks and for selected samples from volcanic islands of the New Hebrides arc together with analytical procedures in *Briqueu et al.* [1994] and representative isotopic data for pre- and post-DEZ collision arc lavas are given in Table 1. These data are plotted in Figures 3 and 4. Below, we summarize the significance of these data and discuss their implications for mantle isotope sources in this region since Mid-Miocene.

On the basis of these new isotopic data, *Briqueu et al.* [1994] made the following key observations. A pronounced difference in isotopic compositions exists between basalts from the southern New Hebrides arc volcanoes of Tanna, Anatom, and Futuna, and those farther north in the DEZ collision zone. This is most evident for $^{87}Sr/^{86}Sr$ and $^{206}Pb/^{204}Pb$ (Figures 3, 4). The southern arc volcanoes have a mantle source interpreted as Pacific MORB-type, with $^{87}Sr/^{86}Sr$ < 0.7033 and $^{206}Pb/^{204}Pb$ values > 18.6. In contrast, the central arc volcanoes have much higher $^{87}Sr/^{86}Sr$ (< 0.7038), and $^{206}Pb/^{204}Pb$ values <18.2, and appear to be derived from a source that was isotopically very close to Indian Ocean MORB-source asthenosphere.

The hole 832B unit VII calc-alkaline clasts have quite high $^{87}Sr/^{86}Sr$ values (0.70395-0.70446), $^{143}Nd/^{144}Nd$ values of 0.51305 ± 0.00003, and $^{206}Pb/^{204}Pb$ = 18.80-18.88 (ave. 18.85), and are isotopically similar to clasts drilled at site 830, interpreted to have been derived from Miocene lavas exposed on nearby Espiritu Santo. Almost certainly, the unit VII clasts and site 830 clasts represent the isotopic composition of the upper mantle that provided arc magmas in the east-facing New Hebrides island arc, *before* postulated subduction polarity reversal in this region around 10 Ma. We note that despite the strong leaching procedure applied prior to isotopic analysis of these samples [*Briqueu et al.,* 1994], the $^{87}Sr/^{86}Sr$ values for the unit VII breccia clasts may be higher than primary values, due to significant recrystallization of these clasts at relatively high temperatures, following seawater alteration. If this is so, the unit VII calc-alkaline lava breccias become even closer isotopically to those from site 830. Whatever the case, the unit VII lava breccia clasts in hole 832B are isotopically unlike modern lavas from volcanoes in the central section of the arc, and unlike the lavas, lava breccias and ashes in the upper levels of both holes 832B and 833B (Figures 3, 4). They are isotopically very similar, however, to modern lavas from volcanoes in the southern section of the arc, away from the DEZ collision zone. This is most obvious for Pb and Nd isotopes, for which both groups have $^{206}Pb/^{204}Pb$ > 18.66 and $^{143}Nd/^{144}Nd$ >0.5130.

The early to middle Pliocene unit 5 sills at the base of hole 833B pre-date the widespread early Pleistocene DEZ collision-related angular unconformity in this part of the north Aoba basin. These sills have $^{206}Pb/^{204}Pb$ values (18.75- 18.79) that are (i) close to those for the basal hole 832B unit VII breccias (18.80-18.88), (ii) notably higher than values for the upper ashes and latest Pliocene volcanic breccias at both sites (ave. 18.375), and (iii) notably higher than modern lavas from volcanoes in this central part of the arc (Figure 4). Their low $^{87}Sr/^{86}Sr$ values (0.7027- .7032) are equivalent to those for the southern arc volcanoes (0.7030-0.7033), and are significantly lower than those for the unit VII breccia in hole 832B (0.7040-0.7045). Although this may reflect a primary difference, it is hard to avoid an alternative interpretation noted above, that the relatively high $^{87}Sr/^{86}Sr$ values for the clasts in the unit VII breccias are due to sea-water alteration and subsequent prehnite-pumpellyite facies recrystallization. The coherent and massive intrusive sills avoided such seawater interaction,

TABLE 1. Isotopic Data for Drilled Rocks from Holes 832 and 833 in the North Aoba Basin, and from Islands in the New Hebrides Island Arc. Data for Rocks Drilled During ODP Leg 134 are Reported in *Briqueu et al.* (1994).

Sample (Hole, Unit, mbsf)	$^{87}Sr/^{86}Sr$	$^{143}Nd/^{144}Nd$	$^{206}Pb/^{204}Pb$	$^{207}Pb/^{204}Pb$	$^{208}Pb/^{204}Pb$
832-I-0.26	0.70404		18.118	15.497	38.275
832-I-1.0	0.70403	0.51294	18.361	15.559	38.518
832-I-146	0.70378		18.613	15.598	38.639
832-IV-625.8	0.70266		18.530	15.529	38.097
832-IV-683.9	0.70346	0.51300	18.275	15.506	38.297
832-VII-952.8	0.70446		18.866	15.583	38.624
832-VII-972.1	0.70407	0.51302	18.796	15.570	38.564
832-VII-1039.3	0.70395		18.852	15.587	38.662
832-VII-1097.1	0.70408	0.51309	18.879	15.608	38.750
833-I-20.35	0.70366		18.332	15.515	38.402
833-I-28.6	0.70368	0.51292	18.233	15.536	38.392
833-I-40.2	0.70404		18.372	15.541	38.493
833-II-162.4	0.70409	0.51306	18.432	15.523	38.317
833-III-415.4	0.70377	0.51295	18.452	15.550	38.437
833-III-501.2	0.70436	0.51298	18.413	15.582	38.611
833-V-834.3	0.70306		18.761	15.562	38.505
833-V-866.2	0.70317	0.51306	18.751	15.515	38.344
833-V-905.7	0.70279		18.766	15.515	38.344
833-V-944.2	0.70270	0.51312	18.792	15.583	38.568
833-V-992.4	0.70318		18.761	15.544	38.421
SMaria 18	0.70395	0.51298	18.168	15.512	38.273
SMaria 22	0.70389	0.51299	18.193	15.506	38.253
SMaria 1	0.70405	0.51294	18.170	15.521	38.303
Epi 8	0.70397	0.51297	18.545	15.566	38.449
Epi 26	0.70363		18.472	15.553	38.519
Tanna 72	0.70309	0.51298	18.658	15.542	38.357
Anatom 42	0.70325	0.51304	18.805	15.554	38.435
	0.70305	0.51304	18.789	15.563	38.465
834-2-10-13	0.702841	0.51311	18.781	15.554	38.428
834-5-83-86	0.702782	0.51312	18.738	15.529	38.324
834-6-28-32	0.702626		18.643	15.509	38.177
834-6-35-37	0.702674	0.51312	18.617	15.480	38.093
834-6-96-104	0.702618	0.51313	18.638	15.500	38.147
834-6-7-13	0.702637	0.51312	18.584	15.505	38.095
834-7-107-109	0.702524		18.561	15.482	38.027
834-7-98-104	0.702542	0.51311	18.572	15.476	38.026
834-7-105-110	0.702544		18.605	15.494	38.081
834-8-77-83	0.702776	0.51312	18.790	15.546	38.425
834-10a-42-44	0.70278	0.51312	18.677	15.534	38.395
834-10a-45-47	0.702914	0.51311	18.732	15.533	38.328
834-11-37-40	0.702808	0.51311	18.680	15.514	38.245
834-12-142-144	0.702896	0.51311	18.712	15.495	38.243
834-12-9-15	0.702891	0.51311	18.738	15.525	38.320
834-13-111-118	0.702802	0.51312	18.701	15.518	38.275
839-1-84-87			18.684	15.539	38.261
839-1-100-104	0.703201		18.710	15.548	38.293
839-2-0-6		0.51306	18.674	15.519	38.192
839-2-0-5	0.703081	0.51306	18.689	15.536	38.247
839-3-15-19		0.51307	18.659	15.518	38.182

TABLE 1. Isotopic Data for Drilled Rocks from Holes 832 and 833 in the North Aoba Basin, and from Islands in the New Hebrides Island Arc. Data for Rocks Drilled During ODP Leg 134 are Reported in *Briqueu et al.* (1994).

Sample (Hole, Unit, mbsf)	$^{87}Sr/^{86}Sr$	$^{143}Nd/^{144}Nd$	$^{206}Pb/^{204}Pb$	$^{207}Pb/^{204}Pb$	$^{208}Pb/^{204}Pb$
839-3-38-41	0.70304	0.51307	18.654	15.514	38.168
839-3-53-56		0.51309	18.659	15.519	38.170
839-4-48-53	0.703093	0.51309	18.664	15.536	38.198
839-5-7-11	0.703175	0.51308	18.712	15.545	38.280
839-6-29-33	0.703043	0.51308	18.667	15.523	38.193
839-9-39-43	0.703173	0.51308	18.712	15.540	38.264
839-9-121-126	0.703159	0.51306	18.692	15.522	38.209
836-1-23-24	0.703189	0.51307	18.485	15.548	38.154
836-1-30-95	0.703237		18.458	15.523	38.075
836-2-53-110	0.70325	0.51306	18.485	15.523	38.086
836-3-0-7	0.702967	0.51303	18.357	15.523	38.041
836-4-54-58	0.703016	0.51304	18.330	15.497	37.974
836-4a-47-52	0.703578	0.51305	18.401	15.528	38.089
836-4b-20-25	0.702998		18.335	15.505	38.001
836-5-62-67	0.703236	0.51307	18.279	15.536	37.958
836-9m-97-102	0.703174	0.51306	18.530	15.533	38.170
837-1-59-63	0.703194	0.51305	18.519	15.516	38.142
837-1-86-92	0.703225	0.51301	18.547	15.535	38.189
828-3-89.8			18.608	15.560	38.485
828-4-95.7	0.70615		18.618	15.568	38.339
828-4-108.3	0.7043	0.51323	18.613	15.566	38.344
828-4-109	0.70488	0.51324	18.561	15.518	38.142
829-7-399	0.7045		18.727	15.550	38.248
829-16-533	0.7058		18.865	15.659	38.862
829-16-534	0.7032		18.602	15.609	38.500
829-16-552.7	0.70325	0.51307	18.569	15.601	38.474
829-16-552.7d	0.70306		18.582	15.627	38.563
829-16-581		0.51317	18.616	15.619	38.547
830-2-175	0.70379	0.00000	18.696	15.556	38.388
830-2-175.4	0.70331	0.00000	18.754	15.575	38.570
831-4a-739.5	0.70287	0.51308	18.500	15.499	38.048
831-4-766	0.70282	0.00000	18.511	15.526	38.123
Loyalty286-9.3-68.7	0.70335	0.51299	18.466	15.550	38.405
Loyalty286-27.2-21.2	0.70322	0.51312	18.523	15.496	38.041
Loyalty286-36.1-110	0.7038	0.51317	18.529	15.485	38.070

and have retained their primary Sr isotopic compositions, which therefore are representative of the pre-collisional upper mantle in this region.

At both sites 832 and 833, the ashes in the upper parts of the holes show essentially identical Pb, Nd and Sr isotopic compositions, with high Sr (>0.7036), low Nd (<0.5130) and generally low $^{206}Pb/^{204}Pb$ values (<18.4). These match the isotopic compositions of the adjacent modern arc volcanoes, especially nearby Santa Maria [Figure 4; Table 1, and *Briqueu*, 1985], but are strikingly different from volcanoes in the southern part of the arc, and from the high-^{206}Pb rocks from the pre-DEZ collision drilled in this central section of the arc. Note that the ash units in both holes show a general decrease in $^{206}Pb/^{204}Pb$ with decreasing age. It is reasonable to conclude (see also *Baker et al.*, 1994; *Gerard and Person,* 1994] that the ashes drilled in

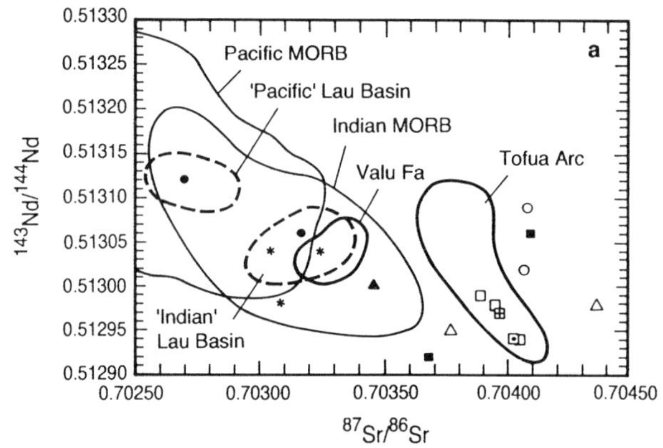

Fig. 3. (a) $^{143}Nd/^{144}Nd$ vs $^{87}Sr/^{86}Sr$ (a) relationships for ODP Sites 832 and 833 lavas and for New Hebrides island arc lavas from Santa Maria volcano in the collision zone, Epi volcano in the immediate post-collision zone, and for lavas from southern volcanoes (Tanna, Anatom and Futuna) away from the collision zone. Fields for Pacific and Indian MORB are from *Pyle et al.* [1992], and for Lau Basin lavas and Tofua arc from *Hergt and Hawkesworth* [1994].

the north Aoba basin are locally derived, basin-wide volcano-sedimentary units.

The latest Pliocene-earliest Pleistocene breccia horizons in both holes (832B unit IV and 833B unit III) are isotopically similar, and compositionally intermediate (Figures 3, 4) between the pre-DEZ collision unit VII lava breccias/unit V sills at the base of these holes, and the post-collision unit I Pleistocene-Recent ashes at both sites 832 and 833. They are, however, isotopically much closer to the unit I ashes than the pre-DEZ collision rocks. This breccia horizon is stratigraphically directly above a major early Pleistocene angular unconformity interpreted as resulting from the DEZ collision [*Fisher et al.*, 1994], and thus records the changing isotopic composition resulting from impact of the DEZ with the arc.

It is interesting to note that the $^{206}Pb/^{204}Pb$ values for Epi volcano (18.47-18.55; Table 1), located in the transition region between the collision-affected central section of the arc, and the unaffected southern arc volcanoes, are also transitional between $^{206}Pb/^{204}Pb$ values for lavas from the volcanoes to the north and south. This may reflect a 'recovery' of the Epi volcano mantle source subsequent to the early Pleistocene passage past this volcano of the colliding DEZ, so that present $^{206}Pb/^{204}Pb$ values are gradually increasing again towards values typical of the southern arc volcanoes unaffected by the DEZ collision.

DISCUSSION

Introduction

Our isotopic data, particularly for Pb (Figure 4), show that the older, pre-DEZ collision lavas and breccias drilled in the two north Aoba basin sites are compositionally close to the lavas from volcanoes in the southern part of the arc, remote from the DEZ collision. Both groups are interpreted to be derived from the regional, depleted Pacific-type sub-oceanic mantle with some amount of contamination by a slab-derived component. The latter may be hydrous or supercritical fluids which transport some of the Sr- and Pb-budget of the subducted slab (including both altered basaltic crust and sediment carapace) into the overlying mantle wedge. However, a similar source cannot be ascribed to the post-DEZ collision magmas in the central part of the arc, which appear to have been derived from typical Indian-type MORB-source mantle.

Nature and Location of the Low-$^{206}Pb/^{204}Pb$, High-$^{87}Sr/^{86}Sr$ Mantle Source

Subducted slab source? The anomalous Indian Ocean-type low-$^{206}Pb/^{204}Pb$, high-$^{87}Sr/^{86}Sr$ isotopic signature of the post-DEZ collision magmas might reasonably be expected to derive from the slab subducting beneath this section of

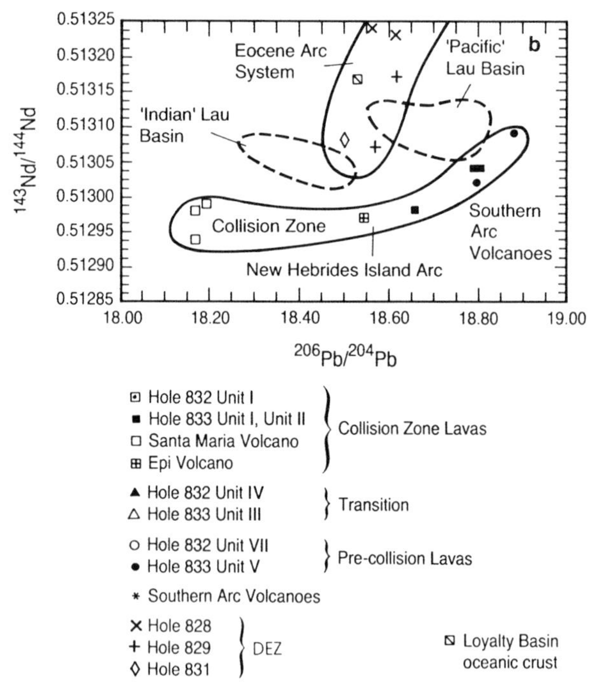

Fig. 3. (b) $^{143}Nd/^{144}Nd$ vs $^{206}Pb/^{204}Pb$ relationships for modern New Hebrides Island Arc volcanoes showing the transition to decreasing $^{206}Pb/^{204}Pb$ from the southern arc volcanoes towards the collision zone. Also shown are data for ODP drill sites on the DEZ (828, 831) and in the forearc adjacent to the collision zone(829), and a MORB sample from the North Loyalty Basin DSDP Site 286 (see Fig. 1), showing that the low $^{206}Pb/^{204}Pb$ component in the collision zone lavas is unlikely to be derived from the DEZ.

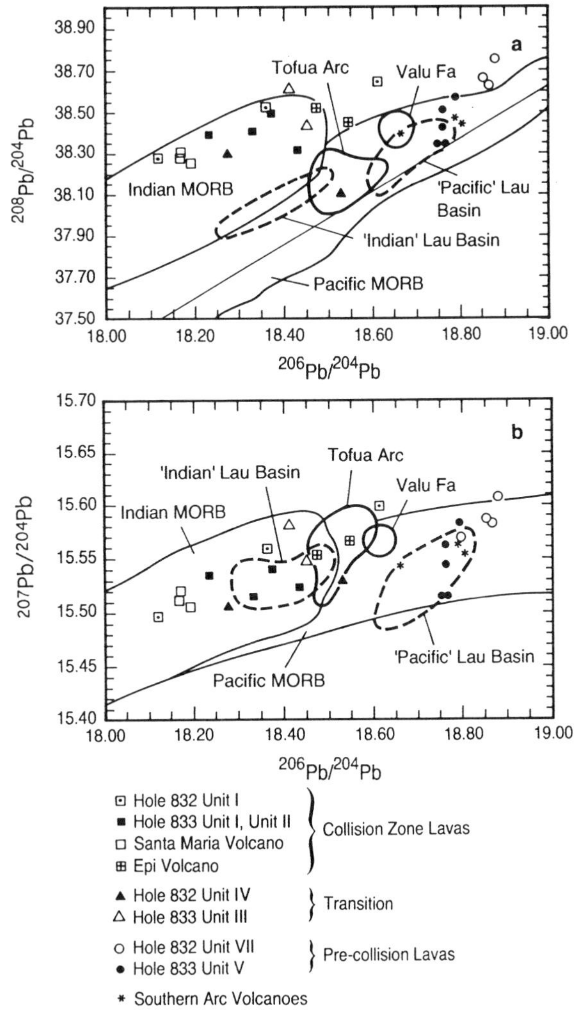

Fig. 4. (a) $^{206}8b/^{204}Pb$vs $^{206}Pb/^{204}Pb$ and (b) $^{207}Pb/^{204}PbPb$ vs $^{206}Pb/^{204}Pb$ for samples from Sites 832 and 833 (Table 2) island arc volcanoes (Table 3). The data clearly fall into a higher $^{206}Pb/^{204}Pb$ group (away from collision zone, and pre-collision lavas) and a lower $^{206}Pb/^{204}Pb$ group (collision zone post-collision lavas), a feature also showed by ODP drilled lavas in the Lau Basin [*Hergt and Hawkesworth*, 1994]. See text for discussion.

the arc, which is abnormal relative to the rest of the slab beneath the New Hebrides island arc in that it bears the DEZ. However, isotopic data for the DEZ [*Briqueu et al.*, 1994], including data for arc tholeiites (site 831) and the forearc(?) oceanic crust (sites 828 and 829) of this Eocene arc, are all *high-*^{206}*Pb*, typical of Pacific intra-oceanic arcs, and the southern arc volcanoes of the current New Hebrides arc. As is evident from Figure 4c, no part of the subducting oceanic crust in this region, including the DEZ, the Pacific MORB-like Loyalty Basin oceanic crust south of the DEZ (Figure 2, Table 1), nor the pre-subduction-reversal oceanic lithosphere in this region (as indicated by Middle Miocene clasts from holes 832B unit VII and 830), has Pb-Sr isotopic similarities to the mantle source of the post-collision lavas. The source of this anomalous Indian Ocean-type isotopic signature must therefore, be associated with the over-riding plate.

Recycled crust? Noting that there are no known old continental crustal blocks on the subducting plate, an alternative explanation for the origin of the 'abnormal' Indian Ocean asthenosphere-type source of post-collisional volcanics in the central New Hebrides island arc might involve recently recycled crustal materials. However, the isotopic signature of the 'Indian' mantle source is clearly much older than the time of collision of the DEZ, and our calculations using the DM growth curve [*Zartman and Haines*, 1988] show that *at least* 500 m.y. of mantle differentiation are necessary to produce the observed Pb isotope characteristics.

Two-layered mantle? In the southern part of the New Hebrides arc, where only Pacific-type mantle is contributing to the arc magmatism, the subducted slab is continuous, the top of the slab is only ~130 km below the arc volcanic islands of Tanna and Anatom, and the leading edge of the subducted slab may lie around 180 km depth [*Pascal et al.*, 1978; *Goula and Pascal,* 1979; *Louat et al.*, 1988]. Farther north, the length of the subducted slab increases sharply, and seismic events as deep as 350 km are recorded [*Louat et al.*, 1988]. In contrast to the relatively simple southern section of the arc, in the central section, the slab configuration is more complex. *Louat et al.* [1988] suggested, on the basis of earthquake foci in the central part of the arc, that the subducted slab may be broken, with a seismic gap from ~50-150 km depth, and *Prevot et al.* [1991] argued that this seismic gap represents a thermal anomaly of some 750°C, which alters the physical properties of the subducted slab beneath this region. In this region, the top of the slab is about 150-200 km beneath the arc volcanoes.

One simple hypothesis for these two contrasting mantle sources, pre- and post-DEZ collision, involves a two-layered mantle, with Pacific-type asthenosphere above Indian-type asthenosphere, the latter being tapped when magma generation is driven to deeper levels during DEZ collision. However, pre-DEZ collision Pliocene sills show distinctly high-K to shoshonitic, transitional alkaline compositions, that are strikingly similar to modern lavas of the high-Ti series on Aoba. Both these magmatic groups presumably evolved from parental magmas with very similar major element compositions, and were derived from partial melting at broadly similar levels beneath this central section of the arc. The sills, however, have isotopic signatures typical of the pre-collision magma grouping, with high $^{206}Pb/^{204}Pb$ and low $^{87}Sr/^{86}Sr$, whereas the Aoba lavas are low $^{206}Pb/^{204}Pb$ and high $^{87}Sr/^{86}Sr$. Whereas this demonstrates that the 'anomalous' low $^{206}Pb/^{204}Pb$ isotopic signature is not simply correlated with 'alkalinity' of the magma type, which in turn relates more or less to depth of magma segregation from source peridotite, it does not rule out the two-layered mantle hypothesis. It simply indicates that magmas with both isotopic signatures segregated from their source/residual mantles at broadly similar depths.

COMPARISON TO THE LAU BASIN

Isotopic data for backarc basin lavas drilled in the Lau Basin during ODP leg 135 [*Hergt and Hawkesworth,* 1994] show striking similarities to our data for arc lavas from the New Hebrides island arc. The oldest (5-6 m.y.-old) backarc-type oceanic crust in the Lau Basin at site 834 defines Pb-Nd-Sr isotopic arrays (Figures 3, 4), particularly for $^{208}Pb/^{204}Pb$ and $^{207}Pb/^{204}Pb$, that are very close to those defined by the pre-collision north Aoba basin lavas drilled during leg 134, and the southern arc volcanoes away from the collision zone. This trend corresponds to mixing between a more radiogenic Pb-enriched slab-derived component, and a MORB source that is clearly identified as having the Pacific isotopic signature. However, published data [*Loock et al.,* 1990; *Boespflug et al.,* 1990] and the new data of *Hergt and Hawkesworth* [1994] show that basaltic magmas erupted since 1 Ma at the modern spreading centres in the Lau Basin have isotopic signatures defining mixing trends between an *Indian MORB-type source* and the same slab-derived component that mixed with Pacific MORB source mantle to produce the older (e.g. leg 134) backarc basin basalts. For the Lau Basin, *Hergt and Hawkesworth* [1994] proposed a model in which pre-existing Pacific MORB-source asthenosphere located under the arc-backarc region 5-6 Ma was displaced by eastward invasion of Indian MORB-source asthenosphere, perhaps linked to oceanward rollback of the Tonga Trench and southward propagation of the backarc spreading centres.

This same change of mantle sources is evident in arc volcanoes in the central New Hebrides arc, following the collision of the DEZ with this part of the arc. Pre-collision arc lavas in the central part of the arc, and lavas from arc volcanoes away from the collision zone, define a mixing trend between Pacific MORB-type source mantle and aslab-derived component essentially identical in terms of Pb isotopes to the slab-derived component in the Lau-Tonga arc-backarc system [*Hergt and Hawkesworth,* 1994]. Furthermore, the nature of the slab-derived component, at least in terms of Pb-Sr-Nd isotopic ratios, has varied little over the life of the New Hebrides arc, despite the probable reversal of subduction polarity. Comparison of mixing trends between the Indian MORB-type source and the arc flux component for the Lau and New Hebrides occurrences (Figure 4) shows that the slab-derived component involved in producing the post-collision New Hebrides mixing trend was probably slightly more ^{207}Pb-rich, and significantly more ^{208}Pb-rich than that of the Lau Basin.

Thus, at the time of the collision of the DEZ with the central section of the New Hebrides arc, magma sources in this region shifted from Pacific- to Indian-type radiogenic isotopic compositions. At the same time, there was apparently little change in the composition of the slab-derived component that mixed with the MORB source mantle to create the broad mixing trends shown on the Pb-Pb isotope plots.

These observations led to an important deduction concerning the genesis of arc magmas. The arc platform extending along the length of the New Hebrides island arc from the Torres Islands in the north to Tanna island in the south (Figure 2) formed above the Vitiaz arc in pre-Miocene time from Pacific-type asthenosphere. Upon reversal of subduction polarity and the subsequent complex opening of the North Fiji Basin [*Auzende et al.*, 1995], this strip of Pacific upper mantle with its Oligo-Miocene carapace of arc volcanoes (exposed on Espiritu Santo, Malakula and the Torres group) eventually rotated clockwise to its present position. We can be sure that the mantle source beneath the modern southern New Hebrides arc volcanoes of Tanna, Anatom and Futuna still has isotopic characteristics of Pacific MORB asthenosphere. However, basalts erupting at the southern spreading segments of the North Fiji Basin (Figure 2) some 350-400 km to the east of these islands and on the same North Fiji Basin microplate, show Sr-Nd relationships (no Pb isotopic data are available) indicating an Indian Ocean asthenospheric-type source [*Nohara et al.,* 1994]. We have no isotopic data for North Fiji Basin crust located between the southern arc volcanoes and the active spreading ridge at this latitude. If, as we suspect, it also has Indian Ocean affinities, then it would imply that asthenospheric convection in the mantle wedge beneath these volcanoes has been so limited that the original Pacific mantle strip beneath these volcanoes has not convected away during opening of the North Fiji Basin. Rather, it has continued to yield arc magmas with mixing trends between Pacific mantle and a slab-derived component.

A similar situation occurs in the Lau Basin, where old arc volcanoes in the Tongan arc [e.g., 'Eua; *Hawkins*, 1994] and early backarc basin crust were derived from a source that was Pacific-type mantle modified by a slab-derived component. This Pacific mantle is presumably trapped as a slice along the arc, and probably now is frozen in the lithosphere attached to the base of the entire arc complex. Regional extension leading to backarc basin opening induced passive tapping first of the slab-derived fluid-metasomatized Pacific lithosphere, then of the invading Indian Ocean-type asthenospheric mantle.

For the New Hebrides arc, the old Pacific upper mantle has continued to provide arc magmas along the arc, despite upwelling Indian Ocean asthenosphere less than 400 km to the east, in the North Fiji Basin. Only when this steady-state scenario was interrupted by collision of the DEZ did the mantle source for volcanoes in the collision zone shift to Indian Ocean mantle. How did this happen?

MANTLE RESERVOIRS IN THE NEW HEBRIDES ARC: A MODEL

By analogy with other modern and inactive Tertiary western Pacific backarc basins [*Hickey-Vargas et al.*, this volume], we argue that Indian Ocean asthenospheric-type mantle has been welling up beneath the North Fiji Basin probably since at least 10 Ma. As a result of large scale readjustments along the India-Pacific plate boundary, the location of spreading centres in the North Fiji Basin has been variable and transient [*Auzende et al.*, 1988; *Huchon et al.*, in press]. Partly in response to regional stress changes, and partly to the decrease in convergence rate at the New Hebrides Trench from 16 cm/yr beneath the southern arc volcanoes to 9 cm/yr beneath the central islands affected by DEZ collision, the upper plate is not extending as rapidly in the collision zone as away from it to the north or south, where narrow backarc troughs (Figure 1) are developing behind the arc [*Maillet et al.*, 1995]. These troughs commenced formation around 5 Ma, and are interrupted in the collision zone of the DEZ. Accretion of new crust may have occurred in the northern backarc trough, but in the southern (Coriolis) trough, extension has probably involved only stretching of arc lithosphere, with no accompanying accretion of oceanic crust to the present time [*Maillet et al.*, in press]. Again by analogy with all other western Pacific backarc basins, we argue that extension in these New Hebrides backarc troughs has led to passive upwelling of Indian Ocean-type asthenosphere, and that continued extension will eventually lead to accretion of Indian Ocean-type mantle-derived basaltic crust in these troughs. We have commenced isotopic studies of lavas from the northern New Hebrides backarc trough to determine their (Indian versus Pacific) affinities [*Crawford, Eggins and Maillet*, unpublished information].

Collision of the DEZ with the arc some 2-3 Ma has produced a significant eastward shift (*Louat et al.*, 1988) of around 50 km of the arc axial chain volcanoes north of Efate (Figures 1, 2). We hypothesize that this shift has driven the site of arc magma generation (volcanoes here are 150-200 km above the slab) eastward, into the path of Indian -type asthenospheric mantle ascending beneath the incipient backarc troughs (Figure 5). Arc magma generation then occurs via the same slab-derived fluid component triggering magma production, not from the old arc block of Pacific lithosphere but from this deep, ascending Indian asthenospheric mantle. Away from the collision zone, for example beneath the southern volcanoes of Tanna and Anatom, arc magma generation remains spatially removed from the ascending Indian Ocean-type asthenosphere beneath the Coriolis Trough, and these arc lavas show no sign of the distinctive low-^{206}Pb Indian-type asthenospheric signal. Rather, they are produced from the old slice of proto-New Hebrides (Vitiaz) arc Pacific-type upper mantle trapped beneath the modern arc by opening of the North Fiji Basin.

WHY IS INDIAN-TYPE ASTHENOSPHERIC MANTLE TAPPED IN THE WESTERN PACIFIC REGION?

Extent and Origin of Indian-type Mantle in the Western Pacific?

Hickey-Vargas et al. [this volume] have demonstrated the extensive occurrence of Indian Ocean-type asthenospheric mantle beneath the western Pacific backarc basins from the Sea of Japan to the Lau Basin, and have argued convincingly that the regional homogeneity of this isotopic signal implies the existence of a single, long-lived and relatively well-mixed reservoir. This presumably was isolated from its parent Indian-type asthenosphere at some time following initiation of subduction along the Sunda arc system, which these authors take to have been around 150-175 Ma. The Indian asthenospheric signature is also recorded from the Xigaze MORB-type ophiolite in Tibet, thought to have formed in the northern part of the Indian Ocean around 110 Ma [*Gopel et al.*, 1984], and in post-South China Sea opening intra-plate basalts from Hainan (south China) [*Tu et al.*, 1991]. In contrast, basalts in the Aleutian arc [*von Drach et al.*, 1986], and in the Bransfield Strait backarc volcanoes [*Keller et al.,* 1991] have isotopic signatures reflecting mixing between Pacific-type asthenosphere and slab-derived component(s), suggesting that the Indian Ocean-type asthenosphere isotopic signature is not simply a process-related feature of modern subduction.

We are interested in knowing the original extent of Indian asthenospheric mantle. To some extent, this requires knowledge of the mode of origin of this unique mantle reservoir. However, it is beyond the scope of this paper to discuss the origin of the Indian Ocean asthenosphere isotopic composition. We simply note that several models have been proposed to account for this distinctive isotopic reservoir, via long-term (>500 m.y.) subduction beneath (especially the eastern margin of) Gondwana [*Dupre and Allegre,* 1983], and contamination of the original (Pacific-type?) MORB reservoir with either delaminated subcont-inental lithosphere, or plume-derived products associated with one (Kerguelen) or more Indian Ocean plumes [*Mahoney et al.*, 1989, 1992; *Storey et al.*, 1989, 1992; *Weis et al.*, 1992].

Models for the Origin of Indian-type Asthenosphere

Two broad classes of models can be considered in seeking an explanation of how the Indian-type mantle signature appears in western Pacific backarc basin basalts (and the central New Hebrides island arc volcanoes). The first of these involves active convection of Indian-type asthenospheric mantle out of the Indian Ocean domain into the western Pacific. The second follows the notion that the anomalous Indian Ocean MORB signature may be produced by interaction of typical Pacific-type MORB with delaminated Gondwana subcontinental lithosphere. If this is so, then interaction of Pacific MORB with similar delaminated(?) subcontinental lithosphere beneath the thinned margin of eastern Asia and the submerged continental blocks and ribbons east of Australia may produce Indian-type mantle when extensional events in the supra-subduction zone lithosphere draw up, and tap magmas from, this modified convecting asthenosphere.

Convection from the Indian Ocean domain. For the former model, once formed, Indian-type asthenosphere may have been squeezed eastward, north of Australia into the western Pacific as Southern Ocean opening commenced around 100 Ma. *Hickey-Vargas et al.* [this volume] suggest that subduction along the Sunda arc, which they took to have commenced at least 150 Ma, must have blocked this potential 'escape route' for Indian convecting asthenosphere to get into the western Pacific. However, it is more likely that subduction along this system commenced around 60 Ma [*Audley-Charles et al.*, 1988] rather than 150 Ma, since the old ages originally associated with this arc were for Sumatran rocks that almost certainly formed part of the southeast Asian mainland.

The simple notion of Indian-type asthenosphere convecting north of Australia into the western Pacific is compromised by other data. Basalts from ODP site 261 [*Gasparon,* 1993] and several of the analyzed samples from ODP site 265 [*Ludden and Dionne,* 1992], both in the easternmost Indian Ocean (Argo abyssal plain), have Pb isotopic signatures that are apparently of Pacific affinity. The Argo abyssal plain formed as part of Tethys (Tethys III) by rifting of 'Argoland' [now one of the terranes in eastern Asia; see *Ogg et al.,* 1992] from northwestern Australia around 155-160 Ma. If future isotopic studies show that the entire oceanic crust of the Argo abyssal plain indeed has a Pacific isotopic flavour, then it is difficult to understand how the western Pacific belt of Indian-type asthenospheric mantle has reached its present position, except via large-scale, deep westward convection *beneath* the Pacific-type lithosphere of the Argo abyssal plain.

The eastern limit of Indian-type asthenosphere beneath Australia is recorded by ~60 m.y.-old basalts from west of Tasmania [*Pyle et al.,* 1993] and northeast of the Australia-Antarctic Discordance in the Southern Ocean [*Lanyon and Crawford,* unpublished information]. By the time the Southern Ocean opened to allow circulation between the Pacific and Indian Oceans around 35-40 Ma, when the South Tasman Rise cleared Antarctica, Pacific asthenosphere commenced convecting westward at reasonable rates of 2.5-4 cm/yr into the Indian Ocean mantle domain [*Pyle et al.,* 1992], so that Indian asthenosphere probably never extended east of the Tasmania-South Tasman Rise continental roots. Available data [*Pyle et al.,* 1993] suggest that at least the southern Tasman Sea crust (ODP site 283; age ~60-70 m.y.) is derived from a Pacific-type mantle reservoir, as is the crust of the South Fiji Basin [*Briqueu et al.*, 1994], which probably formed as a backarc basin to the Vitiaz arc ~35-28 Ma [*Kroenke*, 1984]. Therefore, to arrive beneath the Lau and North Fiji backarc basins in the last 10 m.y., Indian-type asthenospheric mantle would have had to migrate from south of Tasmania around 40 Ma northeastward at ~10 cm/yr, passing *beneath* the Pacific MORB crust of the Tasman Sea, beneath the continental roots of the Lord Howe Rise and Norfolk-New Caledonia Ridges, at the same time sweeping just south of the plume conduits of the Tasmantid and Lord Howe seamount chains.

Su et al. [1992] showed that a striking low shear wave velocity anomaly exists in the upper mantle from 100-600 km depth extending from the Macquarie triple junction south of New Zealand north across the Tasman Sea and Antarctic southwest Pacific Ocean to the North Fiji Basin. At depths of 100-600 km, this is the most pronounced velocity anomaly on the Earth. Except for the major plume-related South Pacific thermal anomaly, which is also characterized by low velocities from 200-600 km, this Tasman anomaly is the only low velocity anomaly not clearly associated with spreading ridges. A controversial hypothesis we suggest is that this low velocity anomaly is associated with the convecting, northeastward-migrating Indian asthenosphere, escaping beneath shallower Pacific mantle that is invading westward into the Southern Ocean.

The absence beneath the Japan-West Philippine occurrence of Indian asthenospheric mantle of a deep low velocity zone such as exists southeast of Australia [*Su et al.,* 1992] may indicate that this mantle is no longer migrating from beneath eastern Asia due to effective cessation of the India-Asia collision.

Contamination of Pacific-type mantle by delaminated subcontinental lithosphere. The second model to account for the existence of Indian-type MORB mantle in the western Pacific involves minimum lateral migration of this mantle source. Rather, interaction between Pacific MORB mantle and delaminated subcontinental lithosphere (eastern Asia, and the rotated and thinned fragments of easternmost Gondwana, such as the Lord Howe Rise) produces mantle with the isotopic characteristics of Indian-type asthenosphere. This Indian-type asthenospheric mantle probably exists between the continental fragments and the mainly west-dipping subduction zones of the west Pacific, from the Kuriles to New Zealand. Arc magmas along the Pacific rim have isotopic compositions with a strong Pacific-type flavour, probably due to 'flooding' of the extremely low-Pb of MORB-source or MORB-residue upper mantle with a radiogenic Pb-rich component derived largely from dehydration of Pacific-type altered basalts in the subducted slab. Extension in the mantle wedge will induce ascent and eventual partial melting of the Indian-type asthenosphere, producing backarc basin lithosphere which is little- or unmodified by Pacific-type Pb-rich slab-derived fluids.

CONCLUSIONS

Until 2-3 Ma, magmas erupted along the New Hebrides arc

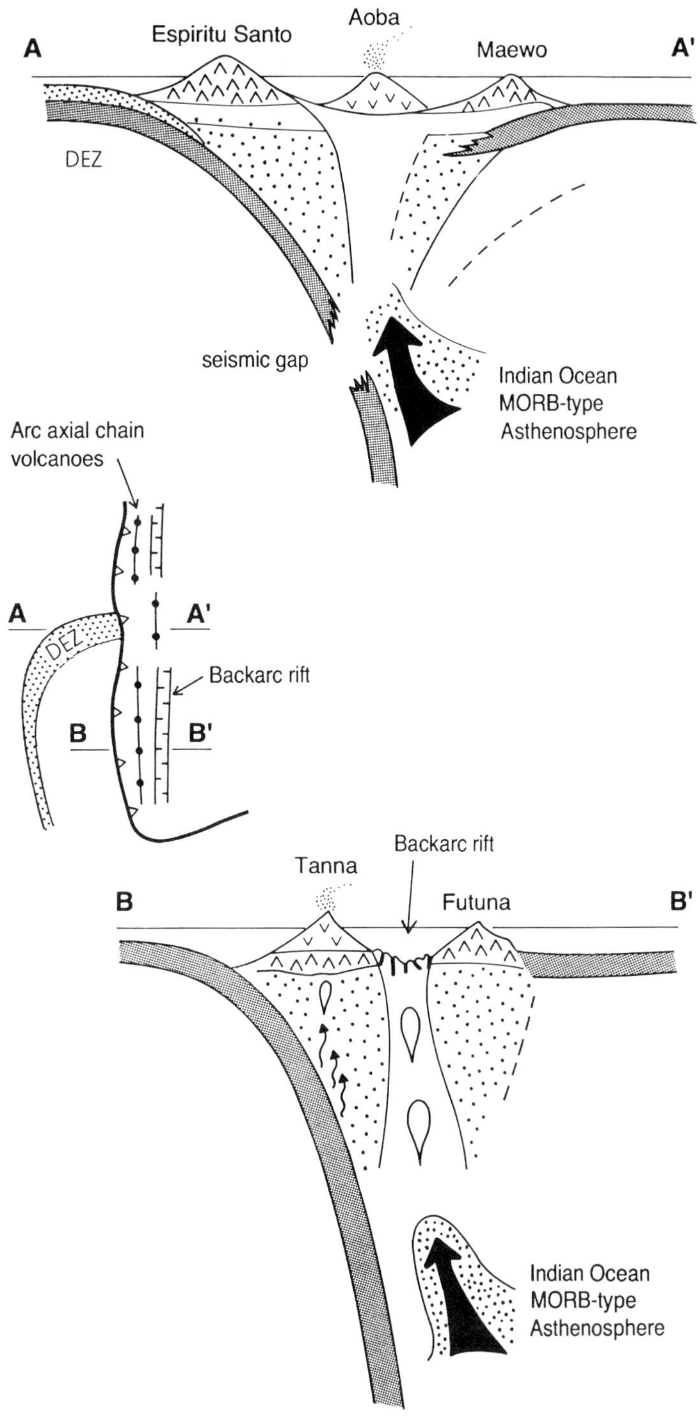

Fig. 5. Schematic sketch of New Hebrides Island Arc and colliding DEZ (inset), showing locations of cross-sections A-A' and B-B'. In A - A', collision of the DEZ has pushed the site of arc magma generation deeper, and to the east, so that slab dehydration and arc magma generation involves Indian Ocean-type MORB source mantle ascending beneath the embryonic backarc trough system in the arc. In contrast, in the southern part of the arc (B-B'), away from the collision zone, arc magma generation occurs at shallower levels and involves old Pacific Ocean MORB-type lithosphere; here, lavas lack the low $^{206}Pb/^{204}Pb$ component derived from the Indian Ocean -type MORB source.

showed isotopic compositions indicating derivation from a Pacific asthenospheric mantle source that had been contaminated by a slab-derived component. The same scenario still occurs today in the southern part of this arc, but collision around 2-3 Ma of an Eocene intra-oceanic arc (the d'Entrecasteaux zone) led to a profound change in the Sr-Nd-Pb isotopic fingerprint of the post-collision lavas in the collision-affected central part of the arc. The mantle source for the latter is typical of Indian Ocean-type asthenospheric mantle, with significantly lower $^{206}Pb/^{204}Pb$ (18.1-18.6) and higher $^{87}Sr/^{86}Sr$ (>0.7035) compared with pre-collision lavas (>18.75, <0.7033).

This replacement of a Pacific-type mantle source by an Indian-type reservoir mimics the situation recorded by *Hergt and Hawkesworth* [1994] for the opening of the Lau Basin. In both regions, blocks of old Pacific-type mantle are trapped in the sub-forearc and sub-arc, and continue to play an active role in arc magma genesis until either collision (New Hebrides) or backarc basin opening (Lau) interfere with arc magma generation processes, and lead to tapping of the Indian Ocean asthenosphere. For the Lau Basin, *Hergt and Hawkesworth* [1994] suggest that Indian-type asthenosphere invades the sub-arc mantle from the west. For the New Hebrides, we argue that collision of the DEZ with the arc has pushed the arc axial chain in the collision zone some 50 km to the east, so that it crosses the path of Indian Ocean-type asthenosphere ascending beneath the embryonic backarc troughs lying between the arc and the North Fiji backarc basin. These backarc troughs are absent in the collision zone, where magma genesis and melt segregation take place at deeper levels than, for example, in the backarc trough immediately north of the collision zone. Collision zone parental magmas are high-K transitional alkaline picritic basalts [*Eggins,* 1993].

The origin of the isotopically anomalous Indian Ocean-type asthenosphere remains poorly understood. If the Indian-type asthenosphere derives from the Indian Ocean domain, then the presence of Pacific oceanic lithosphere forming the crust of the Argo abyssal plain off northwestern Australia, and the Tasman Sea off southeastern Australia, seems to require that the Indian-type asthenosphere appearing over the last 40 m.y. in the backarc basins of the western Pacific region has been able to migrate eastward and then northward from the Southern Ocean beneath Pacific-type upper mantle (now lithosphere?) to its current location. Development, and continued existence of the circum-west Pacific subduction zones dipping towards Asia and Australia, from the Japan Trench to the Tonga Trench, probably limited the westward spread of Indian-type asthenosphere, and may be responsible for the steep slab dips in the areas where backarc spreading has been, or is still, active. The mechanism triggering eastward migration of the Indian-type asthenosphere is also unclear, but invasion of the Indian Ocean domain by the immense Kerguelen and Marion plumes around 115 Ma, and also by Pacific asthenosphere south of Australia since around 40 Ma, may have forced a gradual squeezing out of Indian asthenosphere around the easternmost margins of the Indian and Southern Oceans.

Alternatively, if, as has been postulated by others, the Indian-type asthenosphere evolved beneath Gondwana, it may also be possible that asthenospheric mantle similar to the Indian-type mantle reservoir formed beneath eastern Asia and eastern Gondwana during the last 1 Ga, to be squeezed out eastward from beneath Asia as a result of the Indian continent's collision with Asia, and to be drawn up during the complex extensional history of the microplates east of Asutralia. The extrusion model has been applied by *Tapponier et al.* [1986] to the east- and southeastward tectonic translation of continental blocks and terranes in southeast Asia. Whatever the case, the origin, distribution, and tracking of the anomalous Indian Ocean-type asthenosphere will remain a fertile source of further study.

Acknowledgments. We thank leg 134 co-chief scientists H. Gary Greene and Jean-Yves Collot for their efforts and contributions to the understanding of the New Hebrides island arc. Reviews by John Mahoney, Dave Christie and Jim Natland were very useful, as were discussion, Steve Eggins, and Jean-Philippe Eissen. We thank Rosemary Hickey-Vargas and Janet Hergt for a pre-print of their article published in this book. AJC gratefully acknowledges Gary Greene's enthusiastic support of his work in Vanuatu.

REFERENCES

Alvarez, W., Geological evidence for the geographical pattern of mantle return flow and the driving mechanism of continental drift, *J. Geophys. Res., 87*, 6697-6710, 1982.

Andrews, J. E., G. H. Packham, et al., Site 286, in *Init. Repts. DSDP, 30*, pp. 69-131, U.S. Govt. Printing Office, Washington, D.C., 1975.

Audley-Charles, M. G., P. D. Ballantyne, and R. Hall, Mesozoic-Cenozoic rift-drift sequence of Asian fragments from Gondwanaland, *Tectonophysics, 155*, 317-330, 1988.

Auzende, J-Y., B. Pelletier, and J-P. Eissen, The North Fiji Basin: geology, structure and geodynamic evolution, in *Backarc Basins: Tectonics and Magmatism*, edited by B. Taylor, Plenum Press (in press), 1994.

Auzende, J. M., Y. Lafoy, and B. Marsset, Recent geodynamic evolution of the North Fiji Basin (Southern Pacific), *Geology,* 16, 925-929, 1988.

Baker, P., M. Coltorti, L. Briqueu, T. Hasenaka, A. J. Crawford, and E. Condliffe, Petrology and composition of the volcanic basement of Bougainville Guyot, ODP Site 831, in *Proc. ODP, Sci. Results, 134,* H. G. Greene, J-Y. Collot, and L. B. Stokking, et al., Ocean Drilling Program, College Station, TX, (in press), 1994.

Barsdell, M. and R. Berry, The petrology and geochemistry of Western Epi, *J. Petrol., 31*, 747-777, 1990.

Boespflug, X., L. Dosso, H. Bougault, and J-L. Joron, Trace element and isotopic (Sr, Nd) geochemistry of the volcanic rocks from the Lau basin, *Geol. Jahrb., D 92*, 503-516, 1990.

Briqueu, L., Etude du magmatisme associe aux zones de

subduction a l'aide traceurs geochimiques multiples: elements en trace et rapports isotopiques $^{87}Sr/^{86}Sr$ et $^{143}Nd/^{144}Nd$, Ph.D. thesis, Univ. Montpellier II, 1985.

Briqueu L, H. Bougault and J-L Joron, Quantification of Nb, Ta, Ti and V anomalies in magmas associated with subduction zones: petrogenetic implications, *Earth Planet. Sci. Lett.*, *68*, 297-308, 1984.

Briqueu, L., C. Laporte, A. J. Crawford, T. Hasenaka, P. E. Baker, and M. Coltorti, Temporal magmatic evolution of the Aoba basin, Central New Hebrides island arc: Pb, Sr and Nd isotopic evidence for the coexistence of two mantle components beneath the arc, in *Proc. ODP, Sci. Results, 134,* H. G. Greene, J-Y. Collot, and L. B. Stokking, et al., pp. 393-401, Ocean Drilling Program, College Station, TX, 1994.

Carney, J. N., and A. Macfarlane, Geological evidence bearing on the Miocene to Recent structural evolution of the New Hebrides arc, *Tectonophysics*, *87*, 47-175, 1982.

Charvis, P., and B. Pelletier, The northern New Hebrides back-arc troughs: history and relation with the North Fiji Basin, *Tectonophysics*, *170*, 259-277, 1989.

Chase, C. G., Tectonic history of the Fiji plateau, *Bull. Geol. Soc. Am.*, *82*, 3087-3110, 1971.

Collot, J-Y, S. Lallemand, B. Pelletier, J-P. Eissen, G. Glacon, M. A. Fisher, H. G. Greene, J. Boulin, J. Daniel and M. Monzier, Geology of the d'Entrecasteaux New Hebrides arc collision zone: results from a deep submersible survey, *Tectonophysics, 212*, 213-241, 1992.

Collot, J-Y., J. Daniel, and R. V. Burne, Recent tectonics associated with the subduction/ collision of the d'Entrecasteaux Zone in the central New Hebrides, *Tectonophysics, 112*, 325-356, 1985.

Crawford, A. J., H. G. Greene and N. F. Exon, Geology, petrology and geochemistry of submarine volcanoes around Epi island, New Hebrides island arc, in *Geology and Offshore Resources of Pacific Island Arcs—Vanuatu Region,* edited by H. G. Greene and F. L. Wong, Circum-Pacific Council for Energy and Mineral Resources Earth Science Series, 8, Houston, TX, pp. 301-327 1988.

Dupre, B. and C. J. Allegre, Pb-Sr isotope variation in Indian Ocean basalts and mixing phenomena, *Nature, 303*, 142-146, 1983.

Eggins, S. M., Origin and differentiation of picritic arc lavas from Aoba (Ambae), Vanuatu, *Contrib. Mineral. Petrol., 114*, 79-100, 1993.

Eggins, S. M., The origin of primitive ocean island and island arc basalts, Ph.D. thesis, University of Tasmania, 1989.

Falvey, D. A., Analysis of paleomagnetic data from the New Hebrides, *Austr. Soc. Explor. Geophys. Bull.*, *9,* 117-123, 1978.

Falvey, D. A., and H. G. Greene, Origin and evolution of the sedimentary basins of the New Hebrides arc, in *Geology and Offshore Resources of Pacific Island Arcs—Vanuatu Region,* edited by H. G. Greene and F. L. Wong, Circum-Pacific Council for Energy and Mineral Resources Earth Science Series, 8, Houston, TX, pp. 413-442, 1988.

Fisher, M. A., D. A. Falvey, and G. L. Smith, Seismic stratigraphy of the summit basin of the New Hebrides island arc, in *Geology and Offshore Resources of Pacific Island Arcs—Vanuatu Region,* edited by H. G. Greene and F. L. Wong, Circum-Pacific Council for Energy and Mineral Resources Earth Science Series, 8, Houston, TX, pp. 201-224, 1988.

Fisher, M. A., H. G. Greene, E. L. Geist, and J.-Y. Collot, Seismic stratigraphy of the North Aoba Basin, central New Hebrides island arc, in *Proc. ODP, Sci. Results, 134,* H. G. Greene, J-Y. Collot, and L. B. Stokking, et al., pp. 59-69, Ocean Drilling Program, College Station, TX, 1994.

Gasparon, M., Origin and evolution of mafic volcanics of Sumatra (Indonesia): their mantle sources, and the roles of subducted oceanic sediments and crustal contamination, Ph.D. thesis, University of Tasmania, 1993.

Gerard, M., and A. Person, Composition and alteration of volcanic ash in the North Aoba basin, ODP Sites 832 and 833, in *Proc. ODP, Sci. Results, 134,* H. G. Greene, J-Y. Collot, and L. B. Stokking, et al., Ocean Drilling Program, College Station, TX, (in press), 1994.

Girod, M., R. Maury, and M. Barsdell, Coexistence de basalte sur-saturee et sous-saturees en silice aux nouvelles Hebrides; problem de leur relation genetique, *C.R. Acad. Sci. Ser., 2*, 288, 1355-1358, 1979.

Gopel, C., C. J. Allegre, and R-H Xu, Lead isotope study of the Xigaze ophiolite, Tibet; the problem of the relationship between magmatites and tectonites, *Earth Planet. Sci. Lett.*, *69*, 301-310, 1984.

Gorton, M. P., Geochemistry and geochronology of the New Hebrides, Ph.D. thesis, Australian National University, 1977.

Gorton, M. P., The geochemistry and origin of Quaternary volcanism in the New Hebrides, *Geochim. Cosmochim. Acta*, *41*, 1257-1270, 1977.

Goula, X., and G. Pascal, Structure of the upper mantle in the convex side of the New Hebrides arc, *Geophys. J. R. Astron. Soc.*, *58*, 145-167, 1979.

Greene, H. G., J-Y. Collot, M. A. Fisher, and A. J. Crawford, Neogene tectonic evolution of the New Hebrides island arc—A review incorporating ODp drilling results, in *Proc. ODP, Sci. Results, 134,* H. G. Greene, J-Y. Collot, and L. B. Stokking, et al., pp. 19-46, Ocean Drilling Program, College Station, TX, 1994.

Greene, H. G., A. Macfarlane, D. P. Johnson, and A. J. Crawford, Structure and tectonics of the central New Hebrides arc, in *Geology and Offshore Resources of Pacific Island Arcs—Vanuatu Region,* edited by H. G. Greene and F. L. Wong, Circum-Pacific Council for Energy and Mineral Resources Earth Science Series, 8, Houston, TX, pp. 377-412, 1988.

Hamelin, B., and C. J. Allegre, Large-scale regional units in the depleted upper mantle revealed by an isotopic study of the South-west Indian Ridge, *Nature, 315,* 196-199, 1985.

Hart, S.R., 1988. Heterogeneous mantle domains: signatures, genesis and mixing chronologies, *Earth Planet. Sci. Lett., 76,* 288-298

Hasenaka, T., A. J. Crawford, L. Briqueu, M. Coltorti, P. E. Baker, and A. Fujinawa, Magmatic evolution of the North Aoba intra-arc basin, ODP Sites 832 and 833, in *Proc. ODP, Sci. Results , 134*, H. G. Greene, J-Y. Collot, and L. B. Stokking, et al., Ocean Drilling Program, College Station, TX, (in press), 1994.

Hawkins, J. W., Petrologic synthesis: Lau Basin transect (Leg 135), in *Proc. ODP, Sci. Results, 135,* J. W. Hawkins, L. Parson, J. A. Allan, et al., pp. 879-905, Ocean Drilling Program, College Station, TX, 1994.

Hergt, J. M. and C. J. in *Proc. ODP Sci. Results, 135*, J. W. Hawkins, L. Parson, J. A. Allan, et al., pp. 505-517, Ocean Drilling Program, College Station, TX, 1994.

Holmes, M. L., Seismic refraction measurement in the summit basins of the New Hebrides arc, in *Geology and Offshore Resources of Pacific Island Arcs—Vanuatu Region,* edited by H. G. Greene and F. L. Wong, Circum-Pacific Council for Energy and Mineral Resources Earth Science Series, 8, Houston, TX, pp. 163-176, 1988.

Huchon, P., E. Gracia, E. Ruellan, M. Joshima, and J-M. Auzende, J-M., Kinematics of active spreading in the central North Fiji Basin (SW Pacific), *Mar. Geol., 116,* 69-87, 1994.

Ito, E., W. M. White, C. Gopel, The O, Nd and Pb isotope geochemistry of MORB, *Chem. Geol., 62,* 157-176, 1987.

Javoy, M., L. Briqueu, J. R. Lancelot, and M. Tatsumoto, Oxygen, strontium and neodymium investigations of Vanuatu intra-oceanic arc: structural control of magmatism, *Abstr. I.C.O.G.,* Cambridge, *177,* 1986.

Keller, R. A., M. R. Fisk, W. M. White, and K. Birkenmajer, Isotopic and trace element constraints on mixing and melting models of marginal basin volcanism, Bransfield Strait, Antarctica, *Earth Planet. Sci. Lett., 111,* 287-303, 1991.

Klein, E., C. H. Langmuir, A. Zindler, H. Staudigel, and B. Hamelin, Isotope evidence of a mantle convection boundary at the Australian-Antarctic Discordance, *Nature, 333,* 623-629, 1988.

Kroenke, L. W., Vanuatu and the Eastern Outer Solomon Islands-formation of the New Hebrides and Vitiaz Arcs and development of the North Fiji Basin, *UN ESCAP CCOP/SOPAC Technical Bull., 6,* 63-75, 1984.

Lanyon, R., The Balleny plume and the Australia-Antarctic Discordance, Ph.D. thesis, University of Tasmania, 1993.

Loock, G., W. F. McDonough, S. L. Goldstein, and A. W. Hofmann, Isotopic composition of basaltic glasses from the Lau Basin, *Marine Mining 9,* 235-245, 1990.

Louat, R., and B. Pelletier, Seismotectonics and present day relative plate motions in the New Hebrides North Fiji basin region, *Tectonophysics, 167,* 41-55, 1989.

Louat, R., M. Hamburger, and M. Monzier, Shallow and intermediate depth seismicity in the New Hebrides arc: constraints on subduction processes, in *Geology and Offshore Resources of Pacific Island Arcs—Vanuatu Region,* edited by H. G. Greene and F. L. Wong, Circum-Pacific Council for Energy and Mineral Resources Earth Science Series, 8, Houston, TX, pp. 329-356, 1988.

Ludden, J. N., and B. Dionne, The geochemistry of oceanic crust at the onset of rifting in the Indian Ocean, in *Proc. ODP, Sci. Results, 123,* F. M. Gradstein, J. N. Ludden, et al., Ocean Drilling Program, College Station, TX, pp. 791-799, 1992.

Macfarlane, A., J. N. Carney, A. J. Crawford, and H. G. Greene, Vanuatu: a review of the onshore geology, in *Geology and Offshore Resources of Pacific Island Arcs—Vanuatu Region,* edited by H. G. Greene and F. L. Wong, Circum-Pacific Council for Energy and Mineral Resources Earth Science Series, 8, Houston, TX, pp. 413-442, 1988.

Mahoney, J. J., A. P. Le Roex, Z. Peng, R. L. Fisher, and J. H. Natland, Southwestern limits of Indian Ocean Ridge mantle and the origin of low $^{206}Pb/^{204}Pb$ mid-ocean ridge basalts: isotope systematics of the Central Southwest Indian Ridge (17°-50°E), *J. Geophys. Res., 97,* 19771-19790, 1992.

Mahoney, J. J., J. H. Natland, W. M. White, R. Poreda, S. H. Bloomer, R. L. Fisher, and A. N. Baxter, Isotopic and geochemical provinces of the Western Indian Ocean spreading centers, *J. Geophys. Res., 94,* 4033-4052, 1989.

Maillet, P., E. Ruellan, M. Gerard, A. Person, H. Bellon, J. Cotten, J-L. Joron, S. Nakada, and R. C. Price, Tectonics, magmatism and evolution of the New Hebrides backarc troughs (SW Pacific), in *Backarc Basins: Tectonics and Magmatism,* edited by B.Taylor, Plenum Press (in press), 1994.

Maillet, P., M. Monzier, M. Selo, and D. Storzer, The d'Entrecasteaux zone (southwest Pacific); a petrological and geochronological reappraisal, *Mar. Geol., 53,*179-197, 1986.

Malahoff, A., R. H. Feden, and H. S. Fleming, Magnetic anomalies and tectonic fabric of marginal basins north of New Zealand, *J. Geophys. Res., 87,* 4109-4125, 1982.

Monjaret, M. C., H. Bellon, and P. Maillet, Magmatism of the troughs behind the New Hebrides island arc (RV Jean Charcot SEAPSO Cruise 2): K-Ar geochronology and petrology, *J. Volcanol. Geotherm. Res., 46,* 265-280, 1991.

Monzier, M., L. Danyushevsky, A. J. Crawford, H. Bellon, and J. Cotten, High-Mg andesites from the southern termination of the New Hebrides island arc, *J. Volcanol. Geotherm. Res. 57,* 193-217, 1993.

Nohara, M., K. Hirose, J-P. Eissen, T. Urabe, and M. Joshima, North Fiji Basin basalts and their magma sources: Part II, Br-Nd isotopic and trace element constraints, *Mar. Geol., 116,* 179-1 95, 1994.

Ogg, J. G., F. M. Gradstein, J.A. Dumoulin, M. Sarti, and P. Brown, Sedimentary history of the Tethyan margins of eastern Gondwana during the Mesozoic, in *Synthesis of Results from Scientific Drilling in the Indian Ocean, Geophys. Monog. Ser.,* vol. 70, edited by R. A. Duncan, D. K. Rea, R. B. Kidd, U. von Rad, and J. K. Weissel, pp. 203-224, AGU, Washington, D.C., 1992.

Pascal, G., B. L. Isacks, M. Baranzagi,. and J. Dubois, Precise relocations of earthquakes and seismotectonics of the New Hebrides Island Arc, *J. Geophys. Res., 83,* 4957-4973, 1978.

Peate, D. W., J. A. Pearce, and H. C. Colley, Geochemical variations along the Vanuatu Island Arc, south western Pacific, *Terra Abstr, 3,* 46, 1991.

Pelletier, B., Y. Lafoy, and F. Missegue, Morphostructure and magnetic fabric of the northwestern North Fiji Basin, *Geophys. Res. Lett., 20,* 1151-1154, 1993.

Prevot, R., S. W. Roecker, B. L. Isacks, and J. L. Chatelain, Mapping of low P wave velocity structures in the subducting plate of the central New Hebrides, southwest Pacific, *J. Geophys. Res., 96,* 19,825-19.842,1991.

Pyle, D. G., D. M. Christie, and J. J. Mahoney, Resolving an isotope boundary within the Australian-Antarctic Discordance, *Earth Planet. Sci. Lett.* 112, 161-178, 1992.

Pyle, D. G., R. A. Duncan, and J. J. Mahoney, Geochemical signature of southeast Indian and southwest Pacific seafloor: DSDP Legs 28 and 29 basement revisited (abstract), *Eos Trans. AGU, 74,* 632-633, 1993.

Recy, J., B. Pelletier, P. Charvis, M. Gerard, M. C. Monjaret, and P. Maillet, Structure, age et origine des fosses arriere-arc des Nouvelles-Hebrides (Sud-Ouest Pacifique), *Oceanologica Acta, 10,* 165-182, 1990.

Rex, D., K-Ar ages of volcanic rocks from ODP Leg 134, in *Proc. ODP, Sci. Results, 134,* H. G. Greene, J-Y. Collot and L. B. Stokking, et al., Ocean Drilling Program, College Station, TX, in press, 1994.

Roca, J. L., Contribution a l'etude petrologique et structurale des

Nouvelles Hebrides, These 3° cycle, Univ. Montpellier II, 1978.

Storey, M., A. D. Saunders, J. Tarney, I. L. Gibson, M. J. Norry, M. Thirlwall, P. Leat, R. N. Thompson, and M. A. Menzies, 1989. Contamination of Indian Ocean asthenosphere by the Kerguelen-Heard mantle plume, *Nature, 338,* 574-576.

Storey, M., R. W. Kent, A. D. Saunders, et al., Lower Cretaceous volcanic rocks on continental margins and their relationship to the Kerguelen Plateau, in *Proc. ODP, Sci. Results, 120,* S. W. Wise, R. Schlich, et al., pp. 33-53, Ocean Drilling Program, College Station, TX, 1992.

Su, W., R. L. Woodward, and A. M. Dziewonski, Deep origin of mid-ocean ridge seismic velocity anomalies, *Nature,* 360, 149-152, 1992.

Tapponier, P., G. Peltzer, and R. Armijo, On the mechanics of the collision between India and Asia, in *Collision Tectonics,* edited by M. P Coward and A. C. Ries, Geol. Soc. Lond. Spec. Publn., *42,* 313-345, 1986.

Taylor, F. W., T. M. Quinn, and C. G. Gallup, Quaternary plate convergence rates at the New Hebrides island arc from the chronostratigraphy of Bougainville Guyot (Site 831), in *Proc. ODP, Sci. Results, 134*, H. G. Greene, J-Y. Collot, and L. B. Stokking, et al., Ocean Drilling Program, College Station, TX, in press, 1994.

Tu, K., M. J. F. Flower, R. W. Carlson, M. Zhang, and G. Xie, Sr, Nd and Pb isotopic compositions of Hainan basalts (south China): implications for a subcontinental lithosphere Dupal source, *Geology*, 19, 567-569, 1991.

Von Drach, V., B. D. Marsh, and G. J. Wasserburg, Nd and Sr isotopes in the Aleutians: multi-component parenthood of island arc magmas, *Contrib. Mineral. Petrol., 92,* 13-34, 1986.

Weis, D., W. M. White, F. A. Frey, R. A. Duncan, M. R. Fisk, J. Dehn, J. N. Ludden, A. D. Saunders, and M. Storey, The influence of mantle plumes in the generation of Indian Ocean crust, in *Synthesis of Results from Scientific Drilling in the Indian Ocean, Geophys. Monog. Ser.*, vol. 70, edited by R. A. Duncan, D. K. Rea, R. B. Kidd, U. von Rad, and J. K. Weissel, pp. 57-90, AGU, Washington, D.C., 1992.

White, W. M., A. W. Hofmann, and H. Puchelt, Isotope geochemistry of Pacific mid-ocean basalts, *J. Geophys. Res.*, 92, 4881-4893, 1987.

Zartman, R. E., and S. M. Haines, The plumbotectonic model for Pb isotopic systematics among major terrestrial reservoirs. A case for bi-directional transport, *Geochim. Cosmochim. Acta*, *52*, 1327-1339, 1988.

A. J. Crawford, Department of Geology, University of Tasmania, GPO Box 252C, Hobart, Tasmania, 7001, Australia.

L. Briqueu and C. Laporte, Laboratoire de Géochemie Isotopique, SNRS, USTL, CP.066, 34095 Montpellier 2, France.

T. Hasenaka, Department of Petrology and Mineralogy, Faculty of Science, Tokohu University, Aoba, Sendai, Miyagi 980, Japan.

Origins of Saline Fluids at Convergent Margins

Jonathan B. Martin[1] and Miriam Kastner
University of California, Scripps Institution of Oceanography, La Jolla, CA, USA

Per Kr. Egeberg
Agder College, Department of Chemistry, Kristiansand, Norway

The compositions of pore and venting fluids at convergent margins differ from seawater values, reflecting mixing and diagenesis. Most significantly, the concentration of Cl^-, assumed to be a conservative ion, differs from its seawater value. Chloride concentrations could be elevated by four processes, although two, the formation of gas hydrate and ion filtration by clay membranes, are insignificant in forming saline fluids at convergent margins. During the formation of gas hydrate, the resulting Cl^--rich fluids, estimated to contain an average excess of ~140 mM Cl^- over seawater value, probably would be flushed from the sediment when the pore fluids vent to seawater. Ion filtration by clay membranes requires compaction pressures typical of >2 km burial depths. Even at these depths, the efficiency of ion filtration will be negligible because (1) fluids will flow through fractures, thereby bypassing clay membranes, (2) concentrations of clay minerals are diluted by other phases, and (3) during burial, smectite converts to illite, which has little capacity for ion filtration. A third process, mixing with subaerially evaporated seawater, elevates Cl^- concentrations to 1043 mM in forearc basins along the Peru margin. Evaporation of seawater, however, will be important only in limited geographic regions that are characterized by enclosed basins, arid climates, and permeable sediments. At the New Hebrides and Izu-Bonin margins, Cl^- concentrations are elevated to a maximum of 1241 mM. The process responsible for this increase is the alteration of volcanic ash to hydrous clay and zeolite minerals. Mass balance calculations, based on the decrease in $\delta^{18}O$ values to -9.5‰ (SMOW), suggest that the Cl^- concentrations could increase solely from the formation of smectite in a closed system. The diagenesis of volcanic ash also alters the concentrations of most dissolved species in addition to Cl^-. Depending on the volume of this altered fluid, it could influence seawater chemistry when vented from the sediment.

INTRODUCTION

At convergent margins, pore fluids influence all physical and chemical processes [e.g., see reviews by *von Huene and Scholl*, 1991; *Kastner et al.*, 1991; *Moore and Vrolijk*, 1992]. High pore pressures commonly drive fluids, which carry large amounts of dissolved solids, from the accreted sediments [*Bray and Karig*, 1985; *von Huene and Scholl*, 1991]. Fluids also transport significant quantities of heat, thereby influencing energy budgets and geothermal gradients. High pore pressures allow low-angle thrust faulting, which is common in convergent margin sediments [*Hubbert and Rubey*, 1959]. Fluid-solid interactions alter the strength of the sediments through the generation of authigenic minerals and cements, thereby influencing deformation styles. The important diagenetic reactions depend on the types of detrital, accreted, and subducted sediments and these reactions change pore fluids from their pristine composition, which approximates bottom seawater values. Depending on the magnitude of this alteration, and the quantity of fluids that vent from the sediment, mass fluxes from convergent margins may impact oceanic geochemical cycles of certain elements and isotopes [*Han*

[1]Current address: Department of Geology, University of Florida, 1112 Turlington Hall, PO Box 117340, Gainesville, FL 32611-7340.

Active Margins and Marginal Basins of the Western Pacific
Geophysical Monograph 88

and Suess, 1989; *Martin et al.*, 1991; *Martin,* 1993]. In addition, subduction zones are the principal route for material flux into the mantle [*Kay et al.*, 1978; *Morris and Hart*, 1983; *Morris and Tera*, 1989; *Morris et al.*, 1990]. Diagenetic and metamorphic reactions alter the composition, particularly concentrations of mobile elements such as Li, Be, B, Ba, and U, of the solid material carried into the mantle.

Pore fluids at convergent margins typically exhibit Cl^- concentrations that differ from seawater values [*Gieskes et al.*, 1989; *Kastner et al.*, 1991]. This observation is remarkable because, with the exception of halite diagenesis, Cl^- seems to be conservative in every common diagenetic reaction. Therefore, the Cl^- concentrations in pore fluids provide important constraints for identifying what physical and chemical processes control pore fluid compositions. Every convergent margin that has been thoroughly sampled contains fluids with Cl^- concentrations depleted below seawater value, but fluids with elevated Cl^- concentrations do not occur at all convergent margins [*Kastner et al.*, 1991]. Examples of elevated Cl^- concentrations and associated saline fluids have been discovered at the Peru margin [*Kastner et al.*, 1990; *Martin et al.*, 1993], the New Hebrides margin [*Martin*, 1994], the Izu-Bonin margin [*Egeberg et al.*, 1990; *Egeberg*, 1992] and the Nankai trough [*Gieskes et al.*, 1993; *Kastner et al.*, 1993]. In this paper, we identify processes that could be responsible for the elevated Cl^- concentrations and estimate the importance of each process to pore fluid compositions at convergent margins worldwide. In particular, we will describe examples of saline fluids found at the Peru, New Hebrides, and Izu-Bonin convergent margins and suggest possible origins.

POSSIBLE MECHANISMS FOR FORMATION OF SALINE FLUIDS

The early observation that fluids at convergent margin have Cl^- concentrations depleted below seawater values led to a variety of hypotheses regarding their origins. Possible processes include dissociation of gas hydrate, ion filtration by clay membranes, mixing with meteoric water, and dehydration of hydrous minerals [*Gieskes et al.*, 1989; 1990; *Kastner et al.*, 1990; 1991]. In contrast, formation of gas hydrate elevates Cl^- concentrations and ion filtration by clay membranes produces a residual fluid with elevated Cl^- concentrations. In addition, subaerial evaporation of seawater and dissolution of evaporites form brines that commonly have Cl^- concentrations exceeding seawater concentrations. Therefore, mixing between these brines and otherwise unaltered pore fluids would elevate Cl^- concentrations in marine pore fluids. Because convergent margins are associated with volcanism, their sediments are commonly volcanic ash-rich. Volcanic ash rapidly converts to hydrous mineral phases, which could reduce the water contents of pore fluids, thereby increasing Cl^- concentrations. The purpose of the following discussion is to describe the mechanistic aspects of each process and to provide examples where they have been found. If no examples exist, we suggest reasons for their absence.

Formation of Gas Hydrate

Gas hydrates are solid structures of water and gas that commonly form in continental and convergent margin sediments where pressures and temperatures are suitable for their stability [e.g. *Kvenvolden*, 1993]. The most common gas in natural hydrate is methane [*MacDonald*, 1990], although CO_2, H_2S, and heavier hydrocarbons up to pentane, can also substitute in the lattice. The ratio of water to gas in hydrate depends on the composition of gas that is incorporated into the structure, but saturated methane hydrate is characterized by a stoichiometry of $CH_4 5 \cdot 75H_2O$ [*Claussen*, 1951; *Barrer and Stuart,* 1957; *Glew*, 1962]. Rarely are all sites filled with gas molecules, but natural gas hydrate typically is greater than 80% saturated with methane [*Davidson*, 1973]. Gas hydrate is commonly observed in seismic profiles as bottom-simulating reflectors (BSR) which are estimated to correspond to the base of the hydrate stability field [*Shipley et al.*, 1979]. Bottom-simulating reflectors are especially common at convergent margins, an observation that *Hyndman and Davis* [1992] attributed to the supply of methane provided by fluid flow.

The occurrence of gas hydrate depends on the geothermal gradients in margin sediments but, assuming an average gradient of 40°C km^{-1}, *MacDonald* [1990] estimated that the thickness of the gas hydrate stability zone in continental shelves extends from near the sediment-water interface to ~500 meters below seafloor (mbsf). At greater water depths, such as those at many accretionary prisms, the stability zone will be even thicker because of increased pressure [e.g., *Kvenvolden and Barnard,* 1983]. Only in regions of high heat flow, for example at the Japan margin, will gas hydrate be restricted to shallow burial depths [*Gieskes et al.*, 1993; *Kastner et al.*, 1993].

When gas hydrate forms, it excludes most solutes from the structure, resulting in removal of water from the pore fluids and a corresponding increase in the pore fluid salinity. Although BSR's are common at convergent margins and actual gas hydrate has been recovered from sediments at some margins [*Kvenvolden and Kastner*, 1990;

Martin, 1993], none of the saline fluid that occurs at convergent margins has been shown to result from formation of gas hydrate. Based on mass balance calculations of the concentrations of water and organic carbon (OC), it is possible to estimate the maximum quantity of methane hydrate that can form in sediments at convergent margins, and thus the resulting increase in Cl^- concentration.

Von Huene and Scholl [1991] calculated that 1.9 km^3 of sediment and 1 km^3 of pore fluid annually enter convergent margins. Some of the fluid will be transported to mantle depths in pore space that remains open, but most of the fluid will be expelled to the oceans. Assuming that porosity decreases to a minimum of 10% in accreted sediments and 3% in subducted sediments, *von Huene and Scholl* [1991] estimate that ~0.9 km^3 yr^{-1} of fluid is expelled from convergent margins. Gas hydrate forms predominately at sedimented margins, including those with accretionary wedges, and therefore the following calculation is restricted to the amounts of organic carbon, sediment, and pore fluid that are accreted at convergent margins. *Von Huene and Scholl* [1991] estimated that 57% of the world's margins are accreting and thus ~0.5 km^3 yr^{-1} of the total accreted fluid is expelled from accretionary wedges. Assuming accreted sediment contains on average 0.5 wt% OC and is characterized by a density of 2.5 g cm^{-3}, ~1.1 X 10^{12} moles OC yr^{-1} are available to form gas hydrate. Considering the maximum limiting situation and assuming that all of the accreted OC is regenerated to methane and forms fully saturated hydrate, ~0.12 km^3 yr^{-1}, or ~25% of the total water that is venting will form gas hydrate.

Many unconstrained factors could alter this calculation of the fraction of water incorporated into gas hydrate. It is unlikely that all of the gas hydrate that forms is fully saturated, but assuming that a minimum of 80% of the methane sites are occupied, the amount of water that is trapped in gas hydrate will increase to only 0.15 km^3 yr^{-1}, or ~30% of the venting water. In addition, not all of the OC entering convergent margin systems is likely to convert to hydrocarbons and certainly not all of it will convert to methane. Although CO_2 and H_2S can substitute for methane in natural gas hydrate, they generally constitute a small fraction of the gas [*Kvenvolden and Barnard*, 1983; *Kvenvolden*, 1984; *Kvenvolden et al.*, 1984]. Organic carbon that converts to pentanes and heavier hydrocarbons will be excluded from the hydrate structure [*Carson and Katz*, 1942] and much carbonate will be incorporated into diagenetic minerals such as calcite and dolomite rather than gas hydrate. Therefore, even less than the 25% of the venting water is likely to be incorporated into gas hydrate.

Other than Cl^- concentrations, the increase in solute compositions would be difficult to detect considering the wide variety of diagenetic reactions and the resulting changes in composition common in convergent margin pore fluids. Assuming that no other reactions influence the Cl^- concentrations, no mixing occurs, the pore fluids are static during the formation of gas hydrates, and that 25% of the water is removed by gas hydrate formation, the pore fluid Cl^- concentrations would be elevated from seawater value to nearly 700 mM. Because of the ubiquitous flow at convergent margins, however, these Cl^--rich fluids would be flushed from the sediment, increasing the difficulty of observing the residual saline fluids [e.g., *Hyndman and Davis*, 1992]. Nonetheless, solid gas hydrate would remain as the high-Cl^- fluids vent, and its subsequent dissociation during sampling would dilute the Cl^- concentrations to less than seawater value.

Ion Filtration by Clay Membranes

Considerable experimental evidence and theoretical calculations show that ions are excluded and isotopes are fractionated when fluids are forced through clay minerals that have been compacted at high pressures [*Bredehoeft et al.*, 1963; *Coplen and Hanshaw*, 1973; *Hanshaw and Coplen*, 1973; *Kharaka and Berry*, 1973; *Graf*, 1982]. The proposal by *Bredehoeft et al.* [1963] that ion filtration by clay membranes forms the brines in the Illinois Basin has sparked a long-standing debate over the significance of this process in natural geological settings. Critics generally discount the importance of this process because most sedimentary sequences lack sufficient compaction pressures to reduce porosity of the clay minerals to values that would allow ion filtration [*Manheim and Horn*, 1968; *Carpenter*, 1978; *Lambert*, 1978]. In addition, the pore fluid pressures commonly are insufficient to drive the fluid through highly compacted clay minerals, particularly in sedimentary basins where flow is driven by density contrasts or gravity. At convergent margins, however, where tectonism develops elevated stress and rapidly thickened piles of sediments, the compaction of clay minerals and elevated pore pressures might reach levels sufficient to force pore fluids through the clay membranes.

Experimental information about the filtration capacity of clay membranes is derived by forcing fluids through clay discs that have been compressed under pressures up to ~70 MPa [*Coplen and Hanshaw*, 1973; *Hanshaw and Coplen*, 1973; *Kharaka and Berry*, 1973; *Haydon and Graf*, 1986]. The concentrations of solutes and isotopic ratios of the filtrate and residual fluids are then measured in order to quantify the extent of chemical alteration. These experiments have shown that the filtration efficiency (defined as the solute concentration in the effluent divided by the concentration in the residual fluid) correlates linearly and positively with the compaction pressure [*Kharaka and Berry*, 1973] and that the efficiency increases by ~25% between 20° and 180°C [*Haydon and Graf*, 1986]. Experiments also reveal that at laboratory temperatures and compaction pressures of 33 MPa, hydrogen and oxygen isotopes of water are fractionated by up to 2.5‰ and 0.8‰ respectively [*Coplen and Hanshaw*, 1973] and Na^+ concentrations increase by a factor of four [*Hanshaw and Coplen*, 1973]. Additional experiments were designed to test the effect that various clay mineralogies have on the filtration efficiency. Using Wyoming bentonite (92% montmorillonite), an illite sample (90% illite) compacted at pressures ranging from 14 to 69 MPa, and fluid with an initial composition similar to seawater, *Kharaka and Berry* [1973] reported that the efficiency of the filtration process varied for each ion. For the bentonite sample, Cl^- concentrations were ~18% and ~30% lower in the effluent at 14 and 28 MPa compaction pressure, respectively. Surprisingly, the illite sample showed little capacity to alter the ion concentration, regardless of the compaction pressure.

If the fluid with low Cl^- concentration that is common at convergent margins results from ion filtration, a residual fluid with high Cl^- concentrations would be retained on the opposite side of the clay membranes. The increased efficiency with compaction pressure and temperature [*Kharaka and Berry*, 1973; *Haydon and Graf*, 1986] indicates that the residual saline fluid would be trapped only at certain depths, which could be estimated with information about the compaction pressure of the sediments. Through finite element modeling constrained by seismic information and drilling results from the Deep Sea Drilling Project (DSDP) and Ocean Drilling Program (ODP), *Shi and Wang* [1988] and *Shi et al.* [1990] have calculated the distribution of effective stress at the Barbados and Peru margins (Figure 1). These authors define effective stress as P_t-P_p where P_t is the sum of the overburden pressure and tectonic stress and P_p is the pore fluid pressure. Therefore, the distribution of effective stress within accretionary wedge sediments corresponds to the compaction pressure used for experimental investigation of ion filtration by clay membranes, and should correspond to the efficiency of the process [e.g., *Kharaka and Berry*, 1973]. Although several different initial conditions were used to calculate effective stress, the results shown in Figure 1 are the maximum values obtained from the models [*Shi and Wang*, 1988; *Shi et al.*, 1990] and therefore the values that would produce the most extensive alteration by ion filtration. Based on the results from *Kharaka and Berry's* [1973] experiments using bentonite, Figure 1A shows the region at the Peru margin where the filtration efficiency should be >18%, the lowest efficiency obtained in their experiments. Although elevated temperatures with increased burial should increase the efficiency above the experimentally determined values, the geothermal gradients at the two margins are $\leq$ 50°C km^{-1}, and thus the temperature effects will be minimal.

At Peru, the burial depth required for ~18% efficiency in ion filtration is >1 km at conditions identical to the experimental results (Figure 1A). In addition, the process should become important only at distances >10 km arcward from the trench. At the Barbados margin, the effective stress values that were calculated by *Shi and Wang* [1988] do not increase to the compaction pressures required for 18% efficiency of membrane filtration (Figure 1B). Assuming that effective stress increases at a constant gradient with depth, the level where ion filtration efficiency would reach 18% occurs ~40 km arcward of the trench at burial depths of ~2 km. Fluids from these depths have not been sampled by drilling, and thus only the filtrate (i.e., fluid with low Cl^- concentrations) would have been collected. Fluids with Cl^- concentrations less than seawater values occur at the Peru slope and Barbados accretionary wedge, but based on their Br^-/Cl^- ratios and the isotopic compositions of the water, processes other than ion filtration by clay membranes are more important in controlling the Cl^- concentrations [*Gieskes et al.*, 1989; *Martin et al.*, 1993].

Results of calculations shown in Figure 1 neglect important sedimentological characteristics that would reduce the filtration efficiency of clay membranes. Flow through conduits such as faults and fractures would allow fluids to bypass the clay membranes, preventing chemical alteration. At the Peru and Barbados margins, depth profiles of Cl^- concentration suggest that flow through structural features is important [*Gieskes et al.*, 1989; 1990; *Kastner et al.*, 1990] and veins are common, also reflecting flow through fractures [*Kemp*, 1990; *Vrolijk and Sheppard*, 1991]. In addition, the greatest compositional changes in *Kharaka and Berry's* [1973] experiments resulted from a nearly pure montmorillonite, but natural sediment compositions are more complex. At Barbados and Peru, clay contents are diluted by a variety of other detrital and biogenic components such as biogenic opal (diatoms), volcanic ash, continental detritus, and organic matter. Smectite minerals (including montmorillonite) mostly convert to illite and mixed-layer illite-smectite when temperatures reach ~100°C [*Perry and Hower*, 1970]. Therefore, depending on the geothermal gradient, some

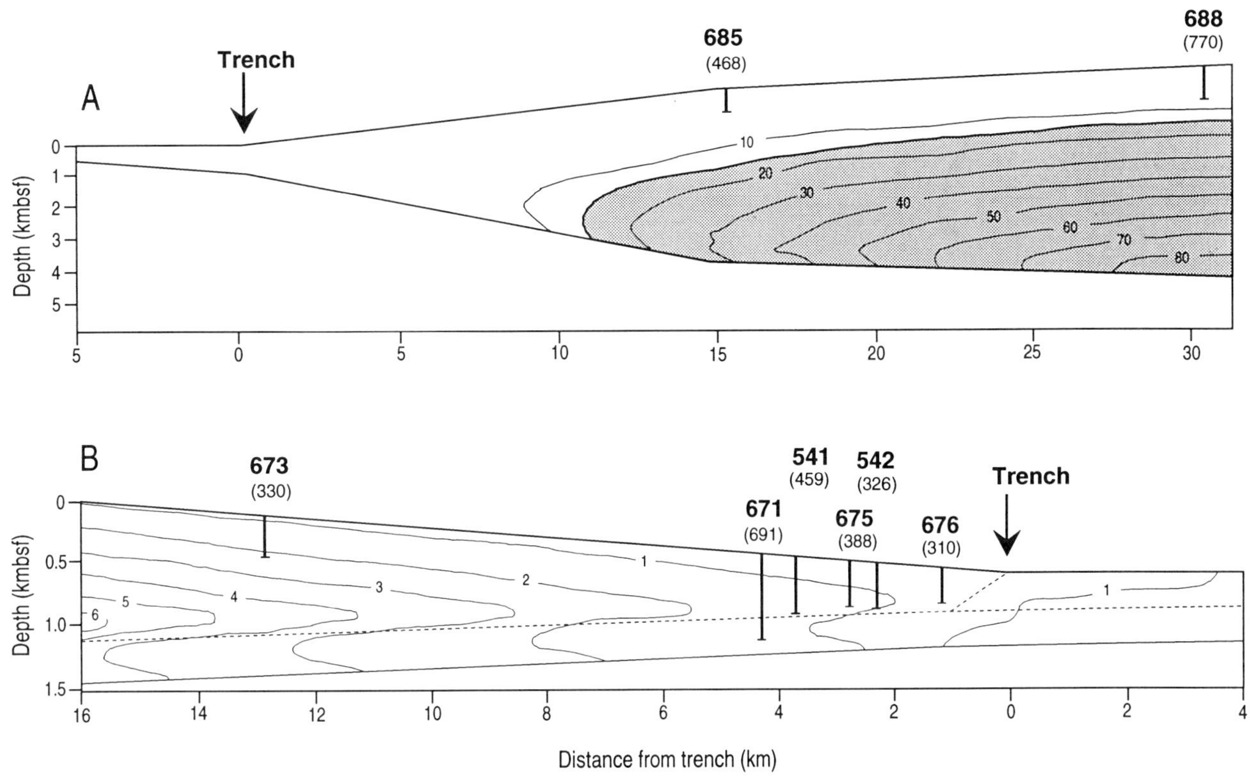

Fig. 1. Cross section of effective stress values (MPa) that were calculated with a finite element model for A. the Peru margin [*Shi et al.*, 1990] and B. the Barbados accretionary wedge [*Shi and Wang,* 1988]. In A, the shaded portion represents the region where membrane filtration efficiency could be >18% based on experimental results from *Kharaka and Berry* [1973]. Contour interval is 10 MPa. In B, a similar region would exist outside of the modelled portion of the accretionary wedge. Contour interval is 1 MPa. Also shown are the locations of the DSDP and ODP sites drilled during legs 78A, 110, and 112. The bold numbers represent the site numbers and the numbers in parentheses are the total depth drilled (mbsf).

smectite will have converted to illite at depths where compaction pressures are sufficient to produce ion filtration efficiency of 18% and following deeper burial, most smectite will convert to illite. Illite appears to be considerably less important than montmorillonite for ion filtration, reaching only <10% filtration efficiency at compaction pressures of ~69 MPa [e.g., *Kharaka and Berry*, 1973]. Therefore, even though sufficient pressures can be generated at convergent margins for ion filtration by clay membranes, the complex geology of these regions implies that this process does not elevate Cl^- concentrations above seawater values.

Seawater Evaporation and Halite Diagenesis

Concentrations of Cl^- that are elevated above seawater values commonly result from subaerial evaporation and the dissolution of halite. These processes could be important wherever vertical tectonics produce enclosed basins. Although deformation typically produces large vertical relief at convergent margins, most examples of marine pore fluids that are concentrated because of evaporation or dissolution of evaporites occur at passive margins: at the Mazagan Plateau, off the northwest coast of Morocco [*Gieskes et al.*, 1980; 1984], in the Mediteranean Sea [*McDuff et al.*, 1978], and in the Black Sea [*Manheim and Schug*, 1978]. Based on Na-Cl relationships, these elevated Cl^- concentrations form predominately from the dissolution of halite [*McDuff et al.*, 1978; *Gieskes et al.*, 1984], but elevated Cl^- concentrations in the pore fluid of the Black Sea sediments appear to be a relict fluid that formed by evaporation of the Black Sea during the Pleistocene [*Manheim and Schug*, 1978]. Among convergent margins, only the Peru margin contains pore fluids with Cl^- concentrations that appear to be caused by evaporations of seawater [*Suess et al.*, 1988b; *Kastner et al.*, 1990; *Martin*

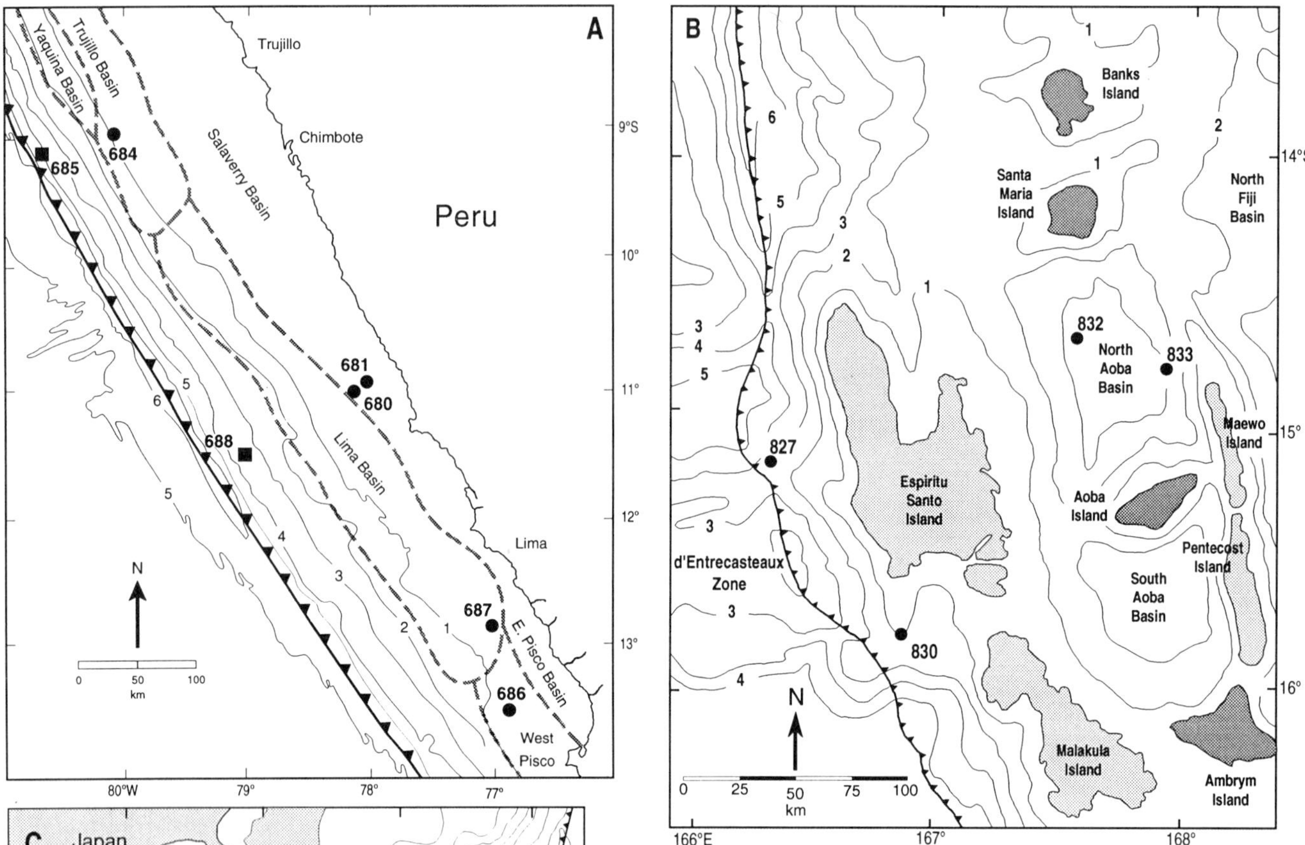

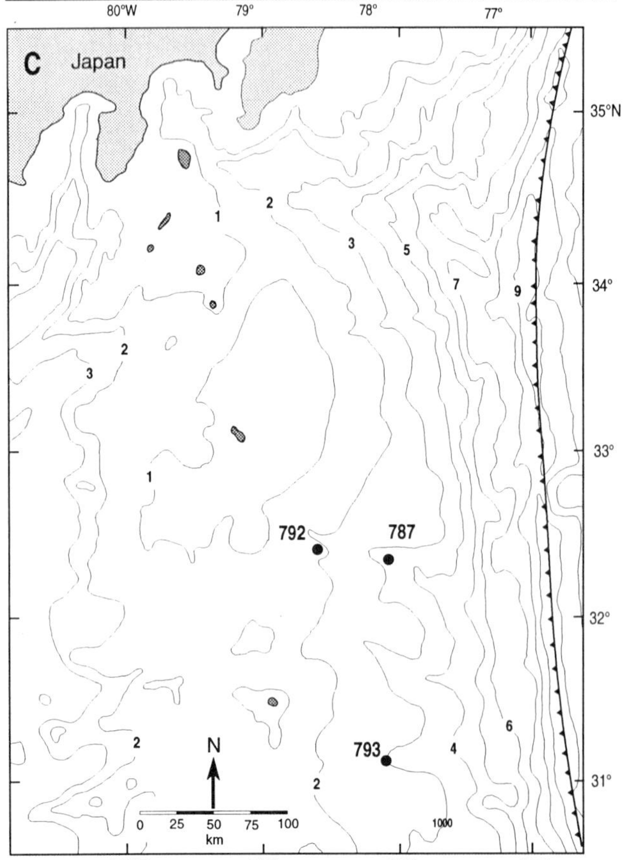

Fig. 2. Bathymetric maps of A. the Peru convergent margin [modified from *Suess et al.*, 1988], B. the central New Hebrides convergent margin [modified from *Collot et al.*, 1992] and C. the northern Izu-Bonin convergent margin [modified from *Taylor et al.*, 1990]. Locations of the drill sites discussed in this paper are shown. The heavy lines with barbs represent the trench axes. Contour interval is one kilometer. In B, the lightly shaded islands are non-volcanic and the darkly shaded islands are volcanic.

et al., 1993]. Therefore, the following section describes the distribution of this Cl^--rich fluid and suggests possible origins.

Geological Setting of Peru. In the early Mesozoic, a convergent tectonic regime that was characterized by subduction, magmatism, and tectonic erosion and accretion evolved at the Peru convergent margin (Figure 2A). The present day configuration of the margin was established in the Eocene [*Noble et al.*, 1979]. Benthic foraminifera indicate that the slope region has subsided at least 4 km over 40 m.y. An Oligocene to early Miocene regional unconformity marks this convergent margin and pre-Eocene continental crust extends seaward to 15-20 km east of the trench axis [e.g., *Hussong et al.*, 1976; *Hussong and*

TABLE 1. Sedimentological Data and Chloride Gradients, Peru Shelf[a]

Site	Water Depth[b] (m)	Sediment Age	Average Cl^- Gradient (mM/m)	Range Organic Carbon (wt. %)	Sedimentation Rate (m/My)
684	426	Mio.-Quat.	4.4	1.2-10.6	20-50
681	151	Quat.	3.4	0.8 - 4.0	80
680	253	L. Plio. - Quat.	2.5	2.9-10.0	25
687	307	Quat.	1.2	0.0-11.6	55
686	447	Quat.	0.5	0.5 - 5.0	160[c]

[a]Data taken from *Suess et al.* [1988a] and *Kastner et al.* [1990].
[b] Water depth at Hole A for Sites with more than one hole.
[c]Minimum value.

Wipperman, 1981; *Thornburg and Kulm*, 1981; *von Huene et al.*, 1985; *Miller et al.*, 1986; *Suess et al.*, 1988a]. Extensional faulting and tectonic erosion, related to the subduction of the Nazca Ridge, dominated this margin through late Miocene to early Pliocene. A relatively small accretionary complex has evolved north of ~10°S in the last 6-7 m.y., and at ~12°S only during the past 3 m.y. [*von Huene and Miller*, 1988].

The forearc basin sediments are situated beneath an intense upwelling system and consist predominately of diatomaceous organic carbon-rich mudstones, commonly exceeding 5 wt% OC. Despite the proximity to the Andes mountain belt, volcanic ash is not a major component of these sediments. Here sedimentation rates range from 20 to 80 m $m.y.^{-1}$, except for higher rates at Site 686 [Table 1; *Suess et al.*, 1988a]. The low oxygen bottom-water (<10 μM) at 200-1000 m water depth [*Burnett*, 1980; *Reimers and Suess*, 1983] is associated with the high OC flux to the sediments. In contrast, the OC content of the lower slope sediment, deposited at >100 m $m.y.^{-1}$, is generally ≤ 3 wt %, and the overlying bottom water at >3000 m water depth is oxygenated (~150 μM O_2).

Saline Pore Fluids in Forearc Basins of Peru. The forearc basins are characterized by pore fluids with Cl^- and Br^- concentrations that increase with depth [*Kastner et al.*, 1990; *Martin et al.*, 1993, respectively]. The maximum measured Cl^- value of 1043 μM at 196 mbsf at ODP Site 680 (Figure 3), is almost twice that of bottom seawater. It is important to stress, however, that the end-member brine was not recovered. The profiles of Cl^- concentration (Figure 3) indicate diffusional communication with a subsurface residual brine of marine origin. The geographic distribution of decreasing gradients of Cl^- and Br^- concentration from ODP site 684 to site 686 indicates that the brine's origin must be north of ODP site 684 (Figure 2A). The increasing extent of dilution of the brine from 9°S to ~13°S suggests that the brine has flowed or is flowing southward. Flow rates of approximately 3-4 cm yr^{-1} were estimated based on its Sr isotopic composition, assuming they reflect a minimum mid-Miocene age for the brine [*Kastner et al.*, 1990].

With only minor volcanic ash present and with the absence of gas hydrate at these shallow water depths of <450 m [*Kvenvolden and Kastner*, 1990], the elevated Cl^- concentrations of the pore fluids indicate that the brine originated either from the dissolution of subsurface evaporites or by evaporation of seawater and subsequent modification by fluid-sediment reactions and mixing with the pore fluids [*Kastner et al.*, 1990]. A similar dense brine is forming presently in the Bocana de Virrila, a 20-km long and narrow estuary at 6°S [*Morris and Dickey*, 1959; *Brantley et al.*, 1984].

Behavior of the "Conservative" and Related Solutes. Concentrations of Na^+ and Cl^- increase conservatively with evaporation of seawater up to the onset of halite precipitation at 10-11 times seawater concentration (Figure

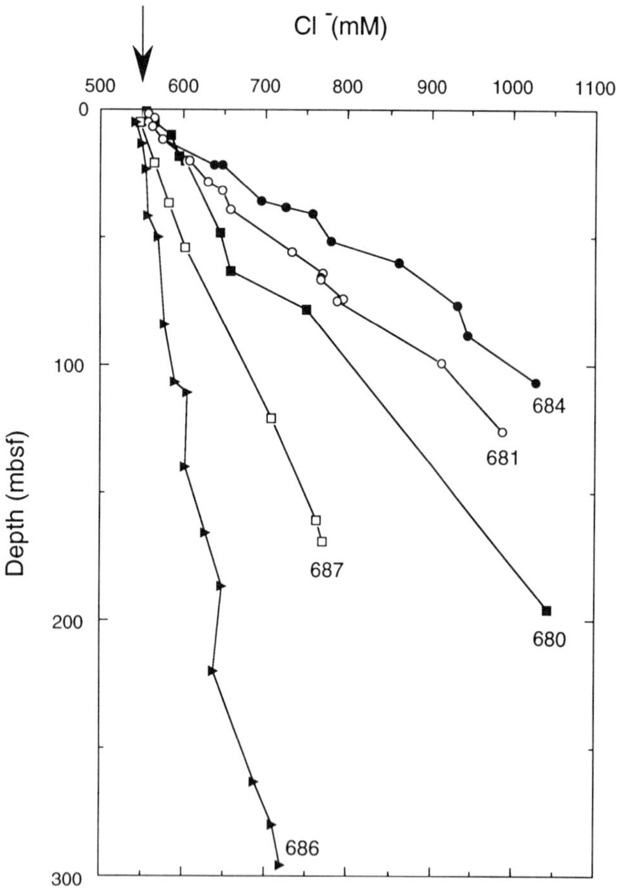

Fig. 3. Cl^- depth gradients at ODP drill sites in the Peru forearc shelf basins. Arrow represents seawater concentration. Site numbers are indicated at the bottom of each gradient. Modified from *Kastner et al.* [1990].

4A and B), K^+ and Mg^{2+} form highly soluble salts only at ≥70 times seawater concentration, and Br^- behaves conservatively up to ~90 times seawater concentration [e.g., *Usiglio*, 1849; *Braitsch*, 1971; *Holser*, 1979; *McCaffrey et al.*, 1987]. Accordingly, a pristine marine brine within the halite stability field should be enriched in Cl^-, Br^-, Na^+, K^+, and Mg^{2+}. Sulfate concentrations also should be enriched after a short negative excursion due to gypsum or anhydrite precipitation. Because the ratio of seawater Na^+ and Cl^- concentrations is 0.86, precipitation of halite (NaCl) lowers this ratio, while dissolution of halite increases it.

The evolution of Br^- and Na^+ versus Cl^- concentrations during seawater evaporation is plotted in Figure 4A and B along with the concentrations of Br^-, Na^+, and Cl^- in pore fluids of forearc basins of the Peru shelf in Figure 4C and D. The Na^+/Cl^- ratio of the Peru pore fluids indicates that the original brine formed by evaporation to ~19 times seawater concentration, but the Br^-/Cl^- ratio indicates only ~12 times seawater concentration. Because of limited sources and sinks for bromine in these sediments [*Martin et al.*, 1993], this discrepancy may be caused by addition of Na^+ or loss of Cl^- from the pore fluids, although the latter is less likely. Sodium could originate either from ion exchange sites in clay minerals or from dissolution of halite by the brine upon its flow southward. Because the pore fluid NH_4^+ concentrations are low at all sites except for site 686 [*Kastner et al.*, 1990], *Martin et al.* [1993] calculated that there is insufficient ion exchange capacity of the clay minerals to account for this large discrepancy. Consequently, Figure 4 suggests that the end-member brine feeding the saline fluids in the Peru forearc basins formed by seawater concentrated beyond the point of halite precipitation with minor additional modification from halite dissolution.

The brine systematically becomes more diluted with increasing distance from the point of origin by mixing with pore fluids. The non-linearity of the reactive Mg^{2+} versus the conservative Cl^- concentration plots, shown by *Kastner et al.* [1990], attests to the increased mixing of the original brine with pore fluids which are highly modified by intensive diagenesis of the organic-rich sediments of the Peru forearc basins. The oxygen isotope ratios of the pore fluids support this conclusion [*Martin*, 1993].

Oxygen Isotopes. Seawater that has been concentrated by 12 to 19 times by evaporation should have $\delta^{18}O$ values of +4-6‰ [SMOW; *Craig and Gordon*, 1965; *Lloyd*, 1966]. Assuming that the Cl^- concentrations in the Peru forearc are diluted by unaltered pore fluid with $\delta^{18}O$ = 0‰ (SMOW) from their original concentrations (of 12 to 19 times seawater value) to twice seawater value, the fluid with the highest Cl^- concentrations should have $\delta^{18}O$ values of +0.8 to +1.2‰. At the two most saline sites (680 and 684), the measured $\delta^{18}O$ values of the pore fluids range from +0.7 to -0.5‰ (SMOW). Interestingly, the deepest and most Cl^--rich samples have the lowest $\delta^{18}O$ values (Figure 5). Furthermore, depth profiles of the $\delta^{18}O$ values of the pore fluids do not show the same smooth gradients that typify the Cl^- profiles (e.g., Figure 3). This decoupling betweeen the depth profiles of $\delta^{18}O$ values and Cl^- concentration excludes the possibility that dilution by a source of isotopically light water, such as meteoric water from the Andes, lowers the $\delta^{18}O$ values and dilutes the Cl^- concentrations [e.g., *Martin*, 1993]. Extensive carbonate diagenesis, reflected in the Ca^{2+}, Sr^{2+}, and Mg^{2+} concentrations profiles [*Kastner et al.*, 1990], is most likely responsible for the depth inversion of the oxygen isotopic values.

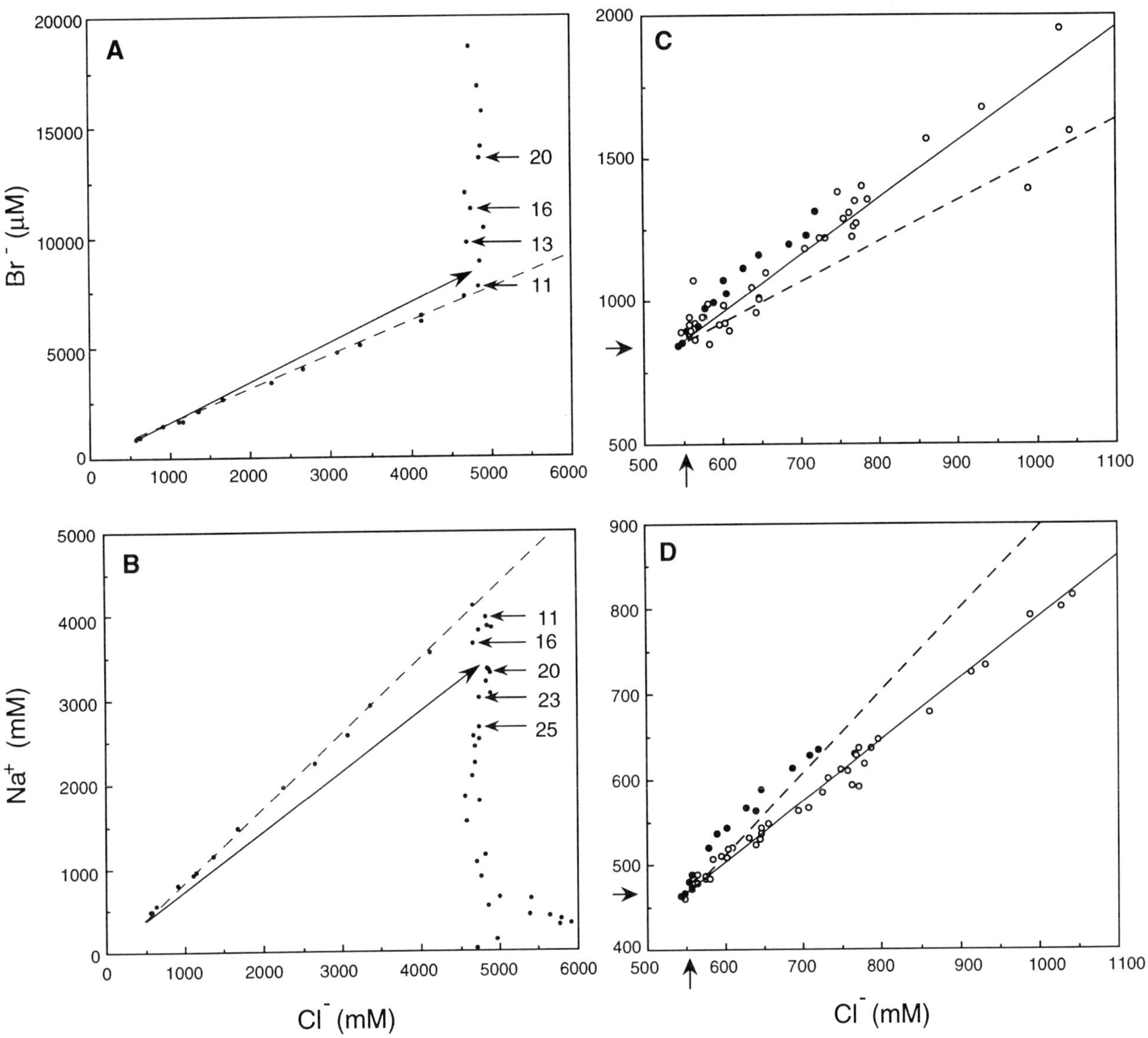

Fig. 4. A. Bromide versus Cl^- concentrations and B. Na^+ versus Cl^- concentrations from salt evaporating pans on Great Inagua, Bahamas; data from *McCaffrey et al.* [1987]. The change in slope of the data results from the onset of halite precipitation. The numbers indicate the extent of evaporation of the sample indicated by the short arrows. C. Bromide versus Cl^- concentrations and D. Na^+ versus Cl^- concentrations in pore fluids from the forearc basins of the Peru shelf. The solid lines represent the results of linear regressions of the data from all the shelf sites except Site 686 represented by the filled symbols. In A and B the long arrows indicate the extrapolation of regression correlations in C and D. The dashed lines indicate seawater Br^-/Cl^- and Na^+/Cl^- ratios. Modified from *Martin et al.* [1993].

Diagenesis of Volcanic Ash

Intense structural deformation common at convergent margins forms forearc, intra-arc, and backarc basins that fill rapidly with sediments characterized by a range of compositions. Sediment composition depends mostly on the proximity to continents and productivity in the overlying water column, which provide hemipelagic detritus and organic matter to the sediments, respectively. Because of ubiquitous volcanism, however, sediments at most convergent margins contain significant quantities of fresh volcanogenic material. This material is unstable in

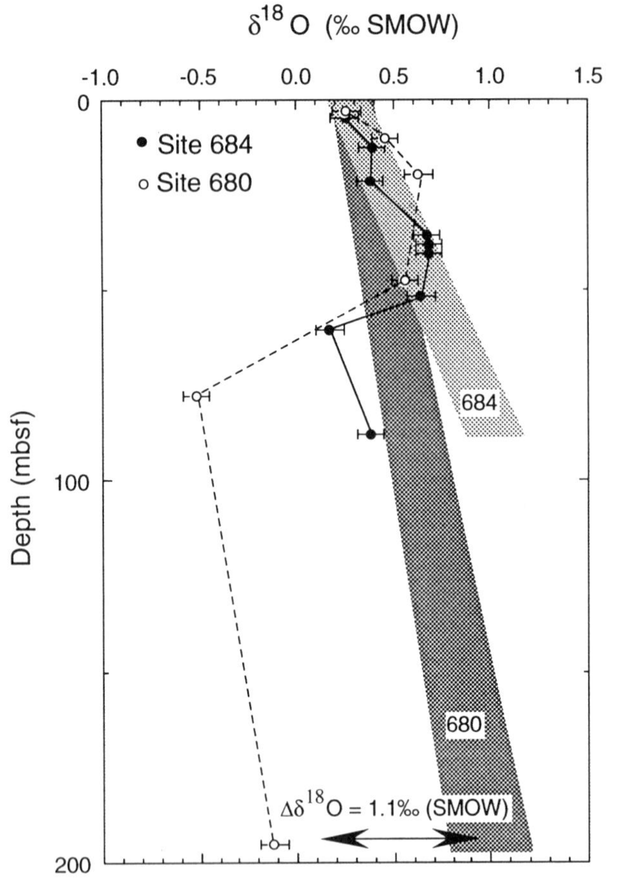

Fig. 5. $\delta^{18}O$ values (‰ SMOW) versus depth for pore water at sites 680 and 684, the two sites most enriched in Cl^- at the Peru shelf. The shaded regions show the $\delta^{18}O$ values (‰ SMOW) expected from mixing between a brine and Peru shelf waters. The end-member brine is assumed to be saturated with respect to halite, to have formed by evaporation from seawater at low humidities, and to be characterized by $\delta^{18}O$ values of +4 to +6‰ [SMOW; *Lloyd*, 1966]. The $\delta^{18}O$ values of the bottom water at the Peru shelf are assumed to be between +0.2 to +0.4‰ [SMOW; *Craig and Gordon*, 1965]. From *Martin*, 1993.

the marine environment and therefore reacts to form authigenic minerals. These reactions typically alter the compositions of the pore fluid. Striking examples of this phenomenon are found at the New Hebrides and Izu-Bonin margins (Figures 2B, 2C and 6) where the volcanogenic material is largely undiluted by other detrital phases [*Egeberg et al.*, 1990; *Egeberg*, 1992; *Martin*, 1994].

At the New Hebrides margin, the location of the island arc is controlled by subduction of the d'Entrecasteaux Zone [*Taylor et al.*, 1980, 1987; *F.W. Taylor*, 1992; *Collot and Fisher*, 1989], a bathymetric feature with elevations up to 3000 m above the surrounding seafloor [*Daniel and Katz*, 1981]. Uplift of Espiritu Santo and Malekula islands has separated the accretionary wedge from the Aoba basin (Figure 2B). In the accreted sediments, the geothermal gradients range between ~20 to 40°C km^{-1}, but because of volcanism in the Aoba basin, the temperature gradients are higher there, ranging up to ~67°C km^{-1}. Along the margin, the sediments are mixtures of two principal components, biogenic carbonate and volcanic material, either air-borne ash or slumped detritus. The content of organic carbon is low (0 to 0.5 wt%). Ash compositions are basaltic, with average SiO_2 contents of ~50 wt%, and average Ti/Al molar ratios of 0.13 [*Collot et al.*, 1992]. In the Aoba Basin, maximum sediment thickness is ~5 km and the sedimentation rates are rapid, ranging from 65 to >300 m $m.y.^{-1}$. These rapid sedimentation rates retard the diffusive loss from the sediments of solutes that are produced by diagenetic reactions.

Compared with the New Hebrides margin, the Izu-Bonin forearc has experienced little deformation since subduction of the Pacific lithosphere beneath the Philippine Sea plate began [*Honza and Tamaki*, 1985]. Thus a well-defined forearc basin, filled with volcaniclastic and hemipelagic sediments, has developed behind an outer-arc high (Figure 2C). The forearc basin formed by rifting in the early Oligocene [*B. Taylor*, 1992] in water depths of 3.5 km [*Kaiho*, 1992]. Variations in the faunal assemblages of benthic foraminifera suggest that there has been ~1 km of total basement uplift in the inner part of the forearc basin at site 792, and total basement subsidence of about 0.6 km in the central forearc basin at site 793 [*Kaiho*, 1992]. Within the forearc basin, information about the thermal regime is available only from site 792 where the geothermal gradient was measured to be 54°C km^{-1} [*Taylor et al.*, 1990]. The geothermal gradient is significantly less in the outer forearc at site 784 where it was measured to be only 23°C km^{-1} [*Fryer et al.*, 1990]. Oligocene sedimentation occurred at a rate of 250-300 m $m.y.^{-1}$ at sites 792 and 793 and 50 m $m.y.^{-1}$ at site 787. The basement rocks and volcanic ash are predominately andesite with average SiO_2 contents of 55 wt% and Ti/Al molar ratios of 0.06, although some dacite also was recovered, which is characterized by SiO_2 contents of 64 to 74 wt% and Ti/Al molar ratios of 0.02. The sediments consist of debris-flow deposits and turbidites produced by concurrent andesitic volcanism [*Hiscott and Gill*, 1992] and erosion of local highs [*Taylor et al.*, 1990]. The content of organic matter in the Oligocene section is extremely low (<0.03% organic carbon). Measured formation factors range up to 30 [*Taylor et al.*, 1990] suggesting that pore fluid mobility and diffusion have been restricted [*Egeberg et al.*, 1990].

Compositions of Pore Fluid and Mineral Stability. Compositions of pore fluids commonly reveal information about diagenesis and mineral formation and transformation reactions that otherwise could not be observed through studies of the solid phases. This important characteristic of pore fluids occurs because minor, or even imperceptible, amounts of diagenetic reactions cause large changes in the composition of the pore fluids. At both the New Hebrides and Izu-Bonin margins, concentrations of the major elements in the pore fluids differ greatly from seawater values (Table 2). The directions that these compositions change from seawater values are identical at the four sites shown in Figure 6, suggesting that a similar suite of diagenetic reactions occurs at both the New Hebrides and Izu-Bonin margins regardless of differences in the tectonic setting, sediment composition, and heat flow.

At sites 792, 793, 832, and 833 (Figure 2B and 2C), concentrations of Ca^{2+} and Cl^- increase above seawater values and concentrations of Na^+ and Mg^{2+} decrease below seawater values (Figure 6). Such changes in the concentrations of Ca^{2+} and Mg^{2+} typically result from the generation of smectite clay minerals from volcanic ash [*McDuff and Gieskes*, 1976; *Lawrence and Gieskes*, 1981]. Because authigenic carbonates, gypsum, and anhydrite are not common at either margin, reactions involving these minerals are unlikely to influence the Ca^{2+} and Mg^{2+} concentrations. In the Aoba Basin, smectite becomes increasingly more abundant with depth, suggesting that some fraction may be authigenic [*Martin*, 1994]. At the Izu-Bonin margin, smectite also is a common component in the sediments [*Taylor et al.*, 1990], although its concentration displays no systematic depth distribution. However, thermodynamic calculations based on Mg^{2+} and Fe^{2+} concentrations in the Izu-Bonin pore fluids reveal that smectite is a stable phase [*Egeberg,* 1992], implying that at least some of the observed smectite is authigenic. In the absence of other diagenetic reactions, evaporation would cause the concentrations of both Cl^- and Na^+ to increase until halite begins to precipitate (Figure 4). Therefore, the increase in Cl^- concentration (Figure 6) does not result from evaporation of seawater. Decreases in Na^+ concentration, however, could result from the formation of albite, analcite, clinoptilolite, or stilbite, all of which have been observed in the sediments by X-ray diffractometry [*Taylor et al.*, 1990; *Gerard and Person*, 1994; *Martin*, 1993].

Calculations of chemical activities reveal that at low to intermediate temperatures, stable phases include clinoptilolite, stilbite, and kaolinite, but with increased temperatures, stilbite becomes the dominant stable phase (Figure 7). These calculations were made using the algorithms described by *Egeberg* [1992] except that the pH values were estimated assuming carbonate saturation. This assumption avoids inaccuracies in the measured pH values caused by the loss of CO_2 from the pore fluids during the core recovery. In general, compositions of the pore fluids reflect stability of similar suites of minerals at both the New Hebrides and Izu-Bonin margins. The greatest difference between the two regions occurs at intermediate to elevated temperatures where pore fluids in the Aoba basin plot closer to the analcite and albite stability fields than at the Izu-Bonin margin. Nonetheless, at intermediate temperatures, all pore fluid compositions are close to the

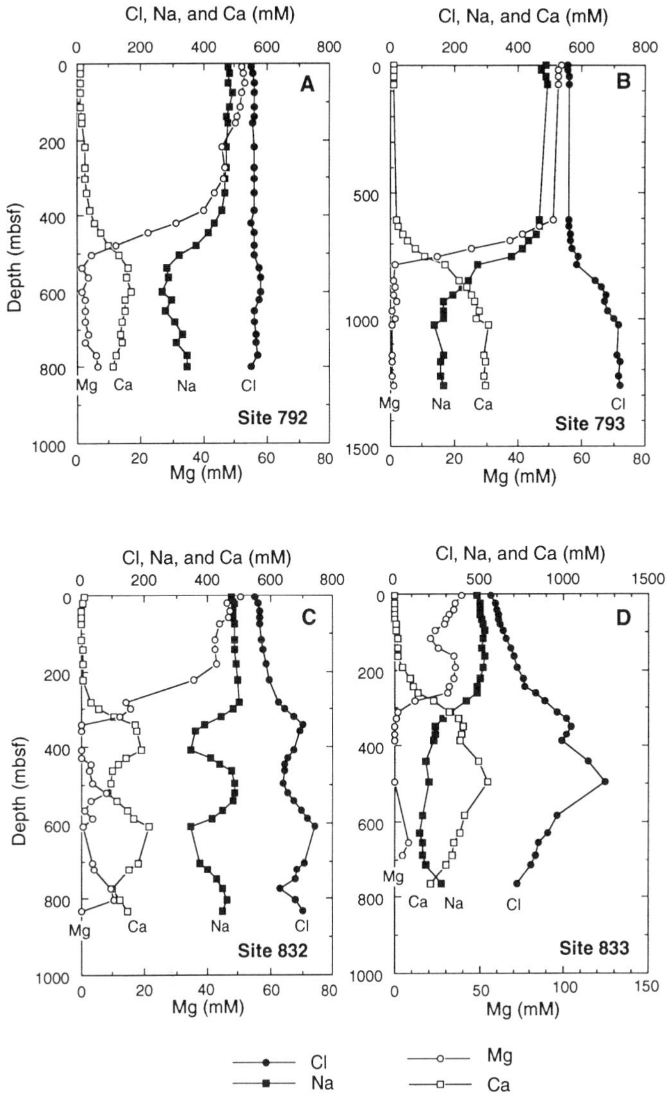

Fig. 6. Solute concentrations (Cl, Na, Mg, and Ca) versus depth at Izu-Bonin margin sites 792 (A) and 793 (B), and at the New Hebrides margin sites 832 (C) and 833 (D).

TABLE 2. Pore Fluid Compositions at Izu-Bonin and New Hebrides Margins[a]

Hole	Core and Section	Interval (cm)	Depth (mbsf)	Cl (mM)	Na (mM)	Mg (mM)	Ca (mM)	$\delta^{18}O$ (‰ SMOW)	δD (‰ SMOW)
792A	1H-5	140-150	7.5	550	476	52.2	9.4	0.0	-8.3
	3H-5	140-150	26.6	556	484	52.5	7.9	0.4	-4.9
	6H-2	140-150	50.2	561	480	53.0	9.2	0.4	-0.5
	9H-1	89-97	76.5	562	491	52.0	10.2	0.2	-0.8
792B	8X-3	140-150	112.4	561	485	51.8	11.0	0.2	-2.8
	11X-1	75-85	137.6	560	471	50.6	15.1	0.2	-2.9
792E	3R-2	60-70	157.0	557	478	50.2	16.5	0.3	-3.2
	9R-4	40-50	217.6	560	471	46.0	23.5	0.1	-2.9
	15R-2	115-125	273.2	559	472	46.6	23.0	0.0	-1.8
	18R-3	115-125	303.7	559	468	46.4	23.9	0.0	-3.2
	22R-2	140-150	340.9	559	468	43.6	28.8	0.2	0.6
	27R-1	27-37	386.6	560	457	40.0	36.9	0.0	-3.2
	30R-5	11-21	421.4	551	432	31.2	53.7	-0.2	-6.0
	33R-2	95-105	446.7	559	414	22.2	71.5	1.0	
	36R-4	67-77	478.4	560	375	12.0	98.5	-0.6	-3.0
	39R-2	115-125	504.9	562	321	4.3	130.6	-0.8	-7.2
	42R-4	140-150	537.2	578	281	1.3	159.0		
	45R-2	0-10	561.6	579	287	3.5	156.1		
	48R-7	140-150	599.1	579	270	1.7	169.3	-1.4	-8.0
	51R-2	140-150	620.1	577	296	2.4	153.1		
	54R-3	140-150	650.6	560	280	2.4	153.1		
	57R-5	140-150	682.6	563	306	2.5	143.1	-1.4	-3.4
	60R-6	140-150	713.0	568	331	3.4	135.6	-1.4	-2.5
	63R-3	138-150	736.9	567	313	2.2	143.8		
	66R-5	139-150	769.0	569	346	5.9	121.7	-1.2	-6.0
	69R-4	136-146	796.5	549	344	6.4	114.2	-1.6	-5.9
793A	1H-2	140-150	3.0	554	486	53.5	10.7	-0.2	-1.4
	3H-3	140-150	18.0	555	473	52.8	10.4	0.0	-0.3
	6H-1	140-150	43.5	561	490	52.8	9.5	0.4	-2.3
	9H-4	140-150	76.9	561	491	52.9	9.8	0.5	4.0
793B	3R-3	130-140	608.7	560	470	51.4	19.2	0.0	-1.5
	6R-3	0-10	636.4	560	469	46.8	24.7	0.0	-2.8
	9R-2	132-142	665.0	568	460	41.6	41.2	0.2	0.6
	12R-1	100-110	692.2	566	436	37.4	52.3	-0.1	1.8
	15R-3	117-127	724.3	569	413	25.4	80.2	-0.3	1.7
	18R-3	115-125	753.0	588	381	14.8	108.6	-0.7	1.2
	21R-6	133-143	786.6	586	271	1.5	171.5		
	27R-7	33-58	844.9	643	242	0.8	213.2	-2.6	-5.3
	30R-4	110-135	870.1	662	223	1.6	237.3	-3.2	-9.3
	33R-6	0-25	900.9	677	195	1.0	255.2	-3.5	-7.6
	36R-3	113-138	926.6	671	166	2.0	265.1	-3.7	-4.7
	40R-5	96-121	968.1	685	165	0.6	276.3		
	43R-5	120-150	997.1	700	166	1.5	268.1	-4.2	-3.9
	46R-3	116-141	1023.0	715	136	0.6	307.5		
	58R-4	76-106	1139.4	710	165	0.3	292.2		
	61R-3	56-86	1166.6	724	158	0.6	298.7	-4.9	-6.6
	67R-4	0-30	1225.5	717	156	0.5	293.9		-8.4
	70R-7	0-30	1259.0	722	164	1.2	296.8		-6.3
832A	1H-3	145-150	4.5	551	477	50.6	9.3		
	4H-3	145-150	23.0	560	486	46.1	3.0		

TABLE 2. (CONT.)

Hole	Core and Section	Interval (cm)	Depth (mbsf)	Cl (mM)	Na (mM)	Mg (mM)	Ca (mM)	$\delta^{18}O$ (‰ SMOW)	δD (‰ SMOW)
	6H-3	145-150	42.0	566	483	47.3	1.9		
	8H-2	144-150	59.5	568	484	46.7	2.3		
	10H-3	143-150	77.0	566	489	43.8	2.2		
	16H-3	145-150	117.5	573	489	42.3	2.3		
	19H-1	145-150	142.5	577	490	42.6	2.5		
	24H-2	145-150	180.7	583	495	42.7	3.4		
832B	9R-2	140-150	222.9	596	499	35.5	4.7		
	15R-2	58-68	280.0	622	501	14.2	30.6		
	17R-1	10-20	300.3	642	484	15.7	54.5		
	19R-1	140-150	319.5	673	443	12.1	102.3		
	21R-3	140-150	341.8	700	392	0.0	168.3		
	23R-2	16-31	358.2	692	363	0.0	177.3		
	28R-2	135-150	407.8	674	344	0.0	192.4		
	30R-4	138-150	428.0	653	410	0.0	140.8		
	32R-2	135-150	445.8	645	441	2.9	116.7		
	34R-1	135-150	462.9	645	477	2.5	97.4		
	37R-3	130-150	494.7	637	488	3.3	90.5		
	40R-1	123-135	520.7	655	488	7.8	85.1		
	42R-2	0-13	540.4	674	481	2.7	110.3		
	45R-1	130-150	569.0	698	447	0.9	148.4		
	47R-1	130-150	588.3	719	415	3.3	166.8		
	49R-CC	0-20	611.2	742	344	0.5	215.9		
	59R-3	130-150	706.3	708	374	3.2	180.2		
	61R-2	130-150	723.6	682	401	3.9	150.6		
	63R-6	0-20	747.3	678	431				
	66R-3	130-150	773.1	631	448	9.2	98.4		
	69R-4	0-20	802.3	680	462	10.1	122.8		
	72R-4	127-146	832.8	704	450	0.0	147.0		
833A	1H-3	145-150	4.5	568	488	39.7	4.3	-0.9	-4.5
	3H-3	145-150	23.5	596	506	35.5	2.6	-2.2	+2.5
	6H-1	145-150	40.7	603	503	34.3	3.0		
	8H-1	145-150	55.5	613	507	31.8	3.0	-2.7	+7.5
	10H-4	145-150	68.9	617	508	29.8	3.5		
	13H-1	145-150	79.5	620	518	29.1	4.8		
	16H-2	140-150	95.6	636	533	24.1	7.4	-3.5	+2.5
	21X-2	140-150	144.9	688	512	25.9	15.6	-4.2	-0.5
	23X-2	0-10	162.8	702	534	35.0	22.1		
	26X-1	92-102	191.2	726	521	35.5	43.1		
833B	5R-2	90-101	118.2	660	522	21.3	14.9		
	16R-1	140-150	223.5	761	500	35.1	88.7	-5.2	+2.5
	18R-1	141-150	242.3	771	485	31.7	109.6		
	20R-1	140-150	261.4	830	483	30.7	135.7		
	22R-2	0-10	280.8	884	419	11.5	227.9	-5.9	+0.5
	25R-1	95-105	309.2	956	329	1.6	316.7		
	27R-1	136-150	328.9	1014	284	0.7	373.5		
	29R-1	117-131	348.1	1042	237	0.0	399.5		
	31R-2	135-150	369.1	1015	242	0.0	389.0		
	33R-1	112-125	386.3	988	228	0.0	381.0	-7.1	+2.5
	38R-4	135-150	439.3	1141	187		491.6		
	46R-3	135-150	494.5	1241	197	0.0	548.5	-9.5	+8.5

TABLE 2. (CONT.)

Hole	Core and Section	Interval (cm)	Depth (mbsf)	Cl (mM)	Na (mM)	Mg (mM)	Ca (mM)	$\delta^{18}O$ (‰ SMOW)	δD (‰ SMOW)
	55R-4	10-30	583.8	960	163		407.6		
	60R-5	0-18	631.4	908	150		385.2		
	63R-2	0-15	656.4	851	166	8.1	349.4	-6.8	+2.0
	66R-4	0-15	688.3	828	168	4.4	336.9	-6.1	-2.5
	69R-2	0-15	713.7	805	186		299.3		
	74R-2	130-150	763.6	719	276		213.3		

[a]The concentration data from the Izu-Bonin margin (sites 792 and 793) are taken from *Egeberg* [1992] and the New Hebrides data (sites 832 and 833) are taken from *Martin* [1994].

stability fields of several Na-bearing minerals (Figure 7B), implying that only a small perturbation in the pore fluid composition will cause these minerals to precipitate [e.g., *Egeberg*, 1992], thereby depleting Na concentrations in the pore fluid.

Most of the Na-bearing minerals, as well as the other observed authigenic minerals such as smectite, contain water in structural, interlayer, and channel sites. Because Cl^- concentrations are not thought to be greatly altered when these minerals form, the increase in concentration of Cl^- (Figure 6) is likely to result from the partitioning of water from the pore fluids into these authigenic solid phases. Smectite is the most common of these hydrous minerals [*Taylor et al.*, 1990; *Martin*, 1994] and can incorporate up to ~20 wt% water into its structural and interlayer sites [*Newman and Brown*, 1987]. Therefore, the formation of smectite may be primarily responsible for increasing the Cl^- concentrations in the pore fluids (Figure 6). Because of the large isotopic fractionation between smectite and water, the formation of smectite should be reflected in the isotopic composition of the pore water.

Oxygen Isotopes and Extent of Alteration. The low geothermal gradients in the Aoba Basin and at the Izu-Bonin margin [*Taylor et al.*, 1990; *Martin*, 1994] indicate that temperatures are low, generally < 50°C where the Cl^- concentrations are maxima. Therefore, because of the inverse relationship between temperatures and fractionation factors, the isotopic composition of the pore water should deviate greatly from its pristine value. Although isotopic fractionation between pore water and volcanic ash is poorly constrained for hydrogen, empirical evidence suggests that the pore water $\delta^{18}O$ values decrease while δD values increase when smectite forms [*Lawrence and Gieskes*, 1981; *Lawrence and Taviani*, 1988]. At the New Hebrides margin, $\delta^{18}O$ values of the pore fluids decrease to -9.5‰ (SMOW), but the δD values are somewhat scattered, ranging from -4.5 to +8.5‰ (Table 2). These δD and $\delta^{18}O$ values correlate more or less inversely suggesting that the isotopic composition of the pore water is strongly influenced by the formation of smectite [Martin, 1994]. At the Izu-Bonin margin, the oxygen isotope ratios of the pore water are also depleted in ^{18}O, but most of the δD values are less than 0‰ (Table 2).

If the alteration from volcanic ash to authigenic smectite takes place in a closed system, the total amount of ^{18}O will not change regardless of the amount of smectite that forms. Thus

$$\frac{d(O_a R_a)}{dm} + \frac{d(O_w R_w)}{dm} + \frac{d(O_{ss} R_{ss})}{dm} + \frac{d(O_{si} R_{si})}{dm} = 0 \quad (1)$$

where O is the number of moles of oxygen and R is the $^{18}O/^{16}O$ ratio of the ash (a), water (w), structural sites in the smectite (ss), and interlayer sites in the smectite (si), and m is the number of moles of smectite formed. Because of the rapid exchange of water between the pore fluids and interlayer sites [*Savin and Epstein*, 1970], $R_{si} = R_w$. Assuming that the isotopic ratio of the unaltered ash remains constant when smectite forms,

$$\frac{dR_a}{dm} = 0.$$

The fractionation factor between the pore water and structural sites in the smectite is given by

$$\alpha = \frac{R_{ss}}{R_w}$$

which when rearranged and differentiated yields

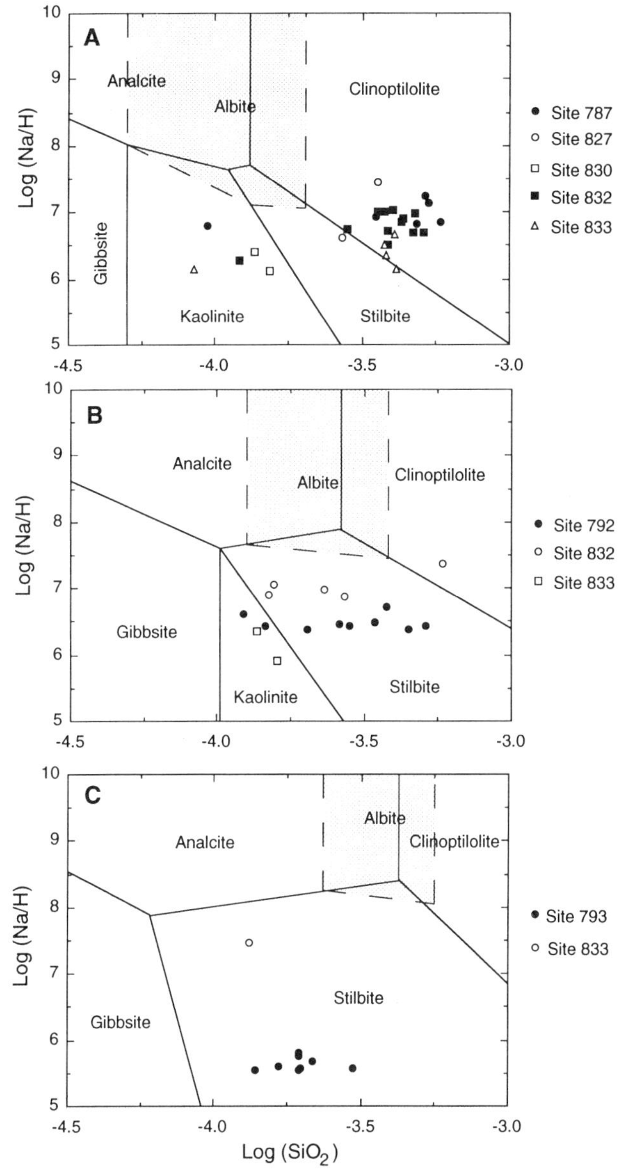

Fig. 7. Data from sites 787, 797, 793, 827, 830, 832, and 833 plotted on activity diagrams for the system Na_2O-Al_2O_3-SiO_2-H_2O at 15° (A), 40° (B), and 60°C (C). The phase boundaries of stilbite were estimated at the average $[Ca^{2+}]/[H^+]^2$ activity ratio of each temperature interval. The shaded region represents the stability field of albite.

$$\frac{dR_{ss}}{dm} = \alpha \frac{dR_w}{dm} \tag{2}$$

At any stage during the formation of smectite, the number of moles of oxygen in the pore water is given by a mass balance relationship

$$O_w = O_w^o + m\frac{dO_a}{dm} - m\frac{dO_{ss}}{dm} - m\frac{dO_{si}}{dm} \tag{3}$$

where O_w^o is the initial number of moles of oxygen in the pore water. Expanding equation 1, substituting equations 2 and 3, and rearranging the terms gives the ordinary differential equation

$$\frac{dR_w}{dm} + R_w(m)\left[\frac{f_1 + f_3 + af_3}{O_w^o + m(f_1 - f_3 + af_3)}\right] = R_a\left[\frac{f_1}{O_w^o + m(f_1 - f_3 + af_3)}\right] \tag{4}$$

where $f_1 = dO_a/dm$, the number of moles of oxygen released from ash dissolution per mole of smectite formed, and $f_3 = dO_{ss}/dm$, the number of moles of oxygen in the structural sites of smectite per mole of smectite formed. The analytical solution to equation 4 is given by

$$R_w(m) = R_w^o\left(\frac{O_w^o}{O_w^o + m(f_1 - f_3 + \alpha f_1)}\right) + mR_a\left(\frac{f_1}{O_w^o + m(f_1 - f_3 + \alpha f_1)}\right) \tag{5}$$

where R_w^o is the initial $^{18}O/^{16}O$ ratio in the pore water.

If most of the water lost from the pore fluids is incorporated into authigenic smectite, the increase in Cl^- concentration should reflect the number of moles of smectite formed. This formulation requires information about the amount of water that is contained by the smectite. Reported direct measurements of the water content of smectite based on a unit cell of $O_{10}(OH)_2$ range from ~4 to 8 moles of water per mole of smectite [*Newman and Brown*, 1987]. Based on the increase in interlayer spacing during smectite hydration, *Ransom and Helgeson* [1994] calculated the molar ratio of interlayer water in fully hydrated smectite to be ~4.5. At the in situ temperatures and pressures of the Aoba basin and Izu-Bonin margin, the smectite should be within ~10% of fully hydrated [*Colten-Bradley*, 1987; *Ransom and Helgeson*, 1994]. Therefore, assuming that the molar ratio of smectite to water in the smectite is 1/6,

$$m = \frac{O_w^o}{6}\left(1 - \frac{1}{r}\right) \tag{6}$$

where r is the ratio of the observed Cl^- concentration to the Cl^- concentration in seawater.

The final parameters required for the model calculations of $R_w(m)$ are values for f_1 and f_3. Because the dissolved Si concentrations are low [Egeberg, 1992; Martin, 1994] most of the Si released during ash dissolution must go into authigenic phases. Assuming that all Si is incorporated into smectite, values for f_1 and f_3 can be estimated based on the Si/O ratios of smectite and the volcanic ash. The Si/O molar ratio of ash at the Izu-Bonin margin and Aoba basin is 0.32± 0.02 [Taylor et al., 1990; Collot et al., 1992], while the Si/O molar ratio for smectite given by Newman and Brown [1987] is 0.36±0.04. Because $(Si/O)_{ash} \simeq (Si/O)_{smectite}$, $f_1 = f_3$. Assuming that all oxygen in each mole of smectite with a unit cell of $O_{10}(OH)_2$ isotopically equilibrates with the pore water, then $f_1 = f_3 = 12$. However, if not all of the oxygen-cation bonds in the volcanic ash are broken during the transformation from volcanogenic material to smectite, then not all of the structural oxygen would equilibrate with the pore water and $f_1 = f_3 < 12$.

In order to calculate $R_w(m)$ from equation 5, the temperature dependent relationship for α given by *Savin and Lee* [1988] was used, based on the geothermal gradients reported in *Taylor et al.* [1990] and *Martin* [1994]. R_w^o is assumed to be 0‰ SMOW, which is an absolute $^{18}O/^{16}O$ ratio of 0.0019934 [*Craig*, 1961]. Based on information given by *Taylor* [1968], R_a is assumed to be 6‰ (SMOW), which is an absolute $^{18}O/^{16}O$ ratio of 0.002005. It is unlikely that all of the smectite formed at its present depth and thus the temperature at which the fractionation occurred will be a function of the range of temperatures encountered during burial. If the rate of growth of smectite is more or less continuous, the temperature at which it formed will progressively increase with depth of burial. Therefore, as a first order approximation, the values for $R_w(m)$ were calculated assuming that all of the smectite formed at the average temperature the smectite had encountered during its burial. In Figure 8, results are shown assuming that different proportions (50%, 75%, and 100%) of the stuctural oxygen in the smectite have equilibrated with the pore fluid.

At both the New Hebrides and Izu-Bonin margins, the basin sites (793 and 833) show similar trends in the relationship between the model results and the observed data (Figure 8A and B). As the depth of burial increases, the model more closely approximates the observed $\delta^{18}O$ values only if an increasingly large proportion of the smectite oxygen is assumed to be in isotopic equilibrium with the pore water. At site 793, where fluid samples were collected above 700 mbsf, the Cl^- concentration is close to seawater value (Figure 6B), and thus there is little variability in the modelled $\delta^{18}O$ values (Figure 8A). The observed $\delta^{18}O$ values also do not change much over the upper 700 m, reflecting either little diagenetic reaction or diffusive exchange with the overlying seawater. From 700 mbsf to near the total depth of the site, the model best reproduces the observed $\delta^{18}O$ values assuming that 75% of the oxygen is in isotopic equilibrium. The deepest sample, however, suggests that all of the structural oxygen in the smectite has isotopically equilibrated with the pore water. At site 833 above 400 mbsf, the model and observed data are comparable when 50 to 75% of the oxygen in the smectite is in isotopic equilibrium with the pore water. However, below 400 mbsf, the best fit of the model reflects isotopic equilibrium of 75% of the oxygen in the smectite with the pore water (Figure 8B).

At site 827 in the New Hebrides accretionary wedge, oxygen isotopic composition of the pore water also appears to be dominated by the conversion of ash to smectite minerals (Figure 8C). Unlike the basin sites (793 and 833), however, the calculated $\delta^{18}O$ values best fit the observed data only by assuming that all of the oxygen in the smectite is in isotopic equilibrium. The differences between the basin and accretionary wedge sites probably result from differences in their tectonic setting, but the exact cause of these differences is difficult to ascertain from these data. Because fluid flow is common in accretionary wedges, this difference could reflect upward flow of highly altered pore fluids. Alternatively, because of elevated stress and pore fluid pressures, the ash may dissolve more completely allowing a complete equilibration of the smectite oxygen.

The results shown in Figure 8 imply that the observed variation in $\delta^{18}O$ values and Cl^- concentrations can be attributed to the formation of smectite clay minerals from volcanic ash. Formation of smectite undoubtably influences the concentrations of other solutes as well, especially the Ca^{2+} and Mg^{2+} concentrations. On the other hand, smectite generation is unlikely to be the sole control of the pore fluid compositions. Other hydrous minerals, particularly those zeolite and feldspar minerals that are thermodynamically stable (Figure 7), will also contribute to the pore-fluid compositions. Because the zeolites are hydrous, they also would alter Cl^- concentrations, while fractionating the oxygen isotopes. However, they compose a smaller portion of the sediments [*Taylor et al.*, 1990; *Martin*, 1994] and contain less water than the smectite. Thus, smectite generation is probably the principal cause of the formation of saline fluids found at the New Hebrides and Izu-Bonin margins.

The influence of smectite in elevating the Cl^- concentrations and fractionating the oxygen isotopes appears to be similar at both the New Hebrides and Izu-Bonin margins, even through they are characterized by

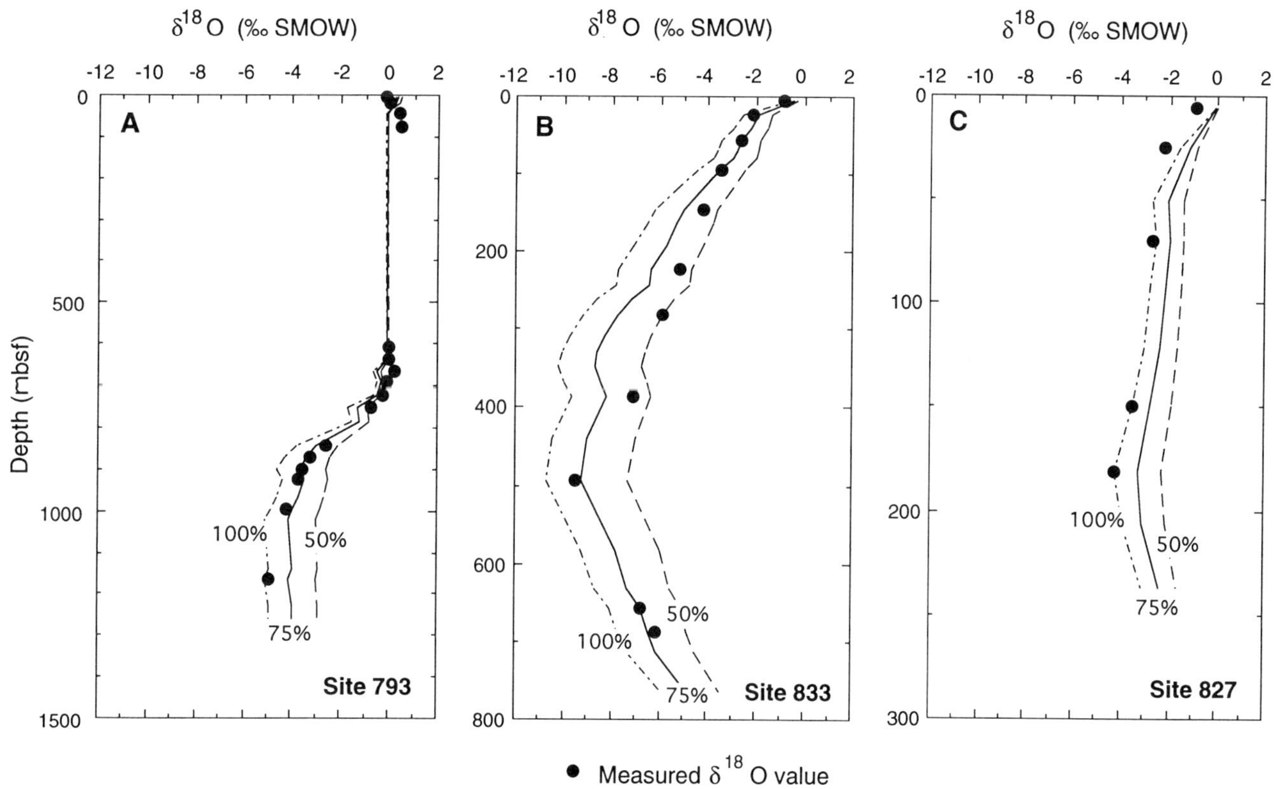

Fig. 8. Observed and modelled $\delta^{18}O$ values (‰ SMOW) for pore water at basin sites 793 (A) and 833 (B) and accretionary wedge site 827 (C). Analytical error in the measured value (0.1‰) is less than the symbol size. The modelled $\delta^{18}O$ values are calculated by assuming that smectite forms in a closed system, and are based on the temperature-dependent fractionation factor reported by *Savin and Lee* [1988]. The three different model calculations shown assume that 50%, 75%, and 100% of the structural oxygen in smectite is in isotopic equilibrium with the pore water.

chemically distinct volcanic material, and their sediment ages are significantly different. In addition, at the Izu-Bonin margin, the diagenesis of basement rocks could influence compositions of the pore fluids, but because of the sediment thickness at the Aoba Basin, the pore fluids are unlikely to be altered by similar reactions. The results from the basin sites (Figure 8A and B) also suggest that because not all of the oxygen in the smectite is in isotopic equilibrium with the pore water, some of the oxygen-cation bonds do not break during the transformation from volcanic ash to smectite. Therefore, the mechanism of smectite formation is not complete dissolution of the ash, followed by smectite precipitation.

CONCLUSIONS

At convergent margins, the compositions of pore fluids differ from seawater values, even for concentrations of those solutes, such as Cl^-, that appear to be unreactive. Because of insufficient compaction pressure at shallow burial depths and the conversion of smectite to illite at temperatures of ~100°C, Cl^- concentrations are unlikely to be altered during ion filtration by clay membranes. Instead, Cl^- concentrations change because of a variety of processes, including evaporation and the formation of gas hydrate and hydrous minerals, that selectively remove water from the pore fluids. Loss of water through evaporation requires specific geographic circumstances, primarily enclosed basins, an arid climate, and permeable conduits that allow flow of the evaporated fluids into the subsurface. Although an example occurs at the Peru margin, saline fluids generated by evaporation are unlikely to be volumetrically significant at convergent margins. In contrast, gas hydrate is common at convergent margins, but there are no examples of fluids with elevated salinities because of its formation. Saline fluids caused by the formation of gas hydrates would likely be flushed from the sediments and, therefore, gas hydrate will dilute fluids when they dissociate as they are buried. Volcanic ash is a common component of sediments at convergent margins. In special situations, such as at the New Hebrides and Izu-Bonin margins where there are few sources of sediments other than volcanic ash, its transformation to hydrous minerals such as clays and zeolites removes sufficiently large volumes of water from the pore fluids to increase their salinities. Volcanic ash is not concentrated enough to produce saline fluids at all margins. Nonetheless, because it is ubiquitous in convergent margin sediments, its diagenesis is an important control of the chemical and isotopic compositions of their pore fluids.

Of the various processes that are capable of generating saline fluids at convergent margins, only diagenesis of volcanic ash and formation of hydrous minerals are likely to affect chemical and isotopic fluxes to seawater. Processes such as evaporation and formation of gas hydrate are volumetrically insignificant. In addition, although these processes increase solute concentrations, they represent merely a recycling of seawater through the pore spaces with no addition of new material. In contrast, dissolution of volcanic ash and formation of authigenic minerals alter the composition of the pore fluids through addition and removal of mass. These fluids, when vented to seawater, thus represent a flux of new material to seawater. The absolute flux of the material, however, depends on the volumes of fluids that vent from convergent margins, a value that is poorly constrained. If only the accreted fluid vents [i.e., ~0.9 km^3 yr^{-1}; *von Huene and Scholl,* 1991], then there will be minimal impact on seawater composition. However, if there are external sources of fluids such as topographic flow from continental margins, or temperature-driven convection cells through the sediments, then larger volumes of fluid than the accreted fluid could vent from convergent margins, thereby providing a greater influence on seawater compositions.

Acknowledgments. JBM is grateful for the hospitality shown by Casey Moore, the Earth Sciences Department, and the Institute of Marine Sciences at the University of California, Santa Cruz while writing this paper. The manuscript has been improved by careful and thorough reviews by Russ McDuff and an anonymous reviewer. The research has been supported by NSF grant OCE92-03879 to MK.

REFERENCES

Barrer, R. M., and W. I. Stuart, Non-stoichiometric clathrate compounds of water, *Proc. Roy. Soc. London, Ser. A, 243,* 172-189, 1957.

Brantley, S. L., N. E. Møller, D. A. Crerar, and J. H. Weare, Geochemistry of a modern evaporite: Bocana de Virrila, Peru, *J. Sediment. Petrol., 54,* 447-462, 1984.

Braitsch, O., *Salt Deposits, Their Origin and Composition*, 297 pp., Springer-Verlag, Berlin, Heidelberg, 1971.

Bray, C. J., and D. E. Karig, Porosity of sediments in accretionary prisms and some implications for dewatering processes, *J. Geophys. Res., 90,* 768-778, 1985.

Bredehoeft, J. D., C. R. Blyth, W. A. White, and G. B. Maxey, Possible mechanism for concentrating brines in subsurface formations, *Am. Assoc. Petrol. Geol., 47,* 257-269, 1963.

Burnett, W. C., Apatite-glauconite associations off Peru and Chile. Paleoceanographic implications, *J. Geol. Soc. London, 137,* 757-764, 1980.

Carpenter, A. B., Origin and chemical evolution of brines in sedimentary basins, *Oklahoma Geological Surv. Circ. 79* , 60-77, 1978.

Carson, D. B. and D. L. Katz, Natural gas hydrates, *Petrol. Trans. Am. Inst. Mining Eng., 146,* 150-158, 1942.

Claussen, W. F., Suggested structures of water in inert gas hydrates, *J. Chem Phys., 19,* 259-260, 1951.

Collot, J-Y., and M. A. Fisher, Formation of forearc basins by collision between seamounts and accretionary wedges: An example from the New Hebrides subduction zone, *Geology, 17,* 930-933, 1989.

Collot, J-Y. H. G. Greene, L. B. Stokking, et al., *Proc. ODP, Init. Repts., 134*, Ocean Drilling Program, College Station, TX, 1992.

Colten-Bradley, V. A., Role of pressure in smectite dehydration-effects on geopressure and smectite-to-illite transformation, *Am. Assoc. Petrol. Geol. Bull., 71,* 1414-1427, 1987.

Coplen, T. B. and B. B. Hanshaw, Ultrafiltration by a compacted clay membrane- I. Oxygen and hydrogen isotopic fractionation, *Geochim. Cosmochim. Acta*, *37*, 2295-2310, 1973.

Craig, H., Standard for reporting concentrations of deuterium and oxygen-18 in natural water, *Science*, *133*, 1833-1834, 1961.

Craig, H., and L. I. Gordon, Deuterium and oxygen 18 variations in the ocean and the marine atmosphere, in *Stable Isotopes in Oceanographic Studies and Paleotemperatures,* Spoleto (Consiglio Nazionale de le Ricerche, Pisa), 9-130, 1965.

Daniel, J., and H. R. Katz, d'Entrecastaux zone, trench and western chain of the central New Hebrides island arc: Their significance and tectonic relationship, *Geo. Mar. Lett., 1,* 213-219, 1981.

Davidson, D. W., Clathrate hydrates, in *Water: A Comprehensive Treatise,* v. 2: edited by F. Franks, pp. 115-234, Plenum Press, New York, 1973.

Egeberg, P. K., Thermodynamic aspects of Leg 126 interstitial water, in *Proc. ODP, Sci. Results*, *126*, B. Taylor, K. Fujioka, et al., pp. 519-529, Ocean Drilling Program, College Station, TX, 1992.

Egeberg, P. K., and Shipboard Scientific Party Leg 126, Unusual composition of pore water found in the Izu-Bonin fore-arc sedimentary basin, *Nature*, *34*, 215-218, 1990.

Fryer, P., J. A. Pearce, L. G. Stokking, et al., *Proc. ODP, Init. Repts.,* 125, Ocean Drilling Program, College Station, TX, 1990.

Gerard, M., and A. Person, Hydrothermalism and diagenesis effects in the North Aoba Basin, in *Proc. ODP*, *Sci. Results*,

112, H. G. Greene, J-Y. Collot, et al., Ocean Drilling Program, College Station, TX, in press, 1994.

Gieskes, J. M., D. Graham, and R. Ellis, Interstitial-water studies, Deep Sea Drilling Project Sites 415 and 416, in *Init. Repts., DSDP, 50*, Y. Lancelot, E. L. Winterer, et al., pp. 691-694, U.S. Govt. Printing Office, Washington, D.C., 1980.

Gieskes, J. M., R. Johnston, and M. Boehm, Interstitial water studies, Leg 79, in *Init. Repts. DSDP, 79,* K. Hinz, E. L. Winterer, et al., pp. 825-836, U.S. Govt. Printing Office, Washington, D.C., 1984.

Gieskes, J. M., G. Blanc, et al., Hydrogeochemistry in the Barbados accretionary complex: Leg 110 ODP, *Palaeogeog. Palaeoclim. Palaeoecol., 71,* 83-96, 1989.

Gieskes, J. M., P. Vrolijk, and G. Blanc, Hydrogeochemistry, ODP Leg 110: An overview, in *Proc. ODP, Sci. Results, 110,* J. C. Moore, A. Mascle, et al., pp. 395-408, Ocean Drilling Program, College Station, TX, 1990.

Gieskes, J. M., T. Gamo, and M. Kastner, Major and minor element geochemistry of interstitial waters of Site 808- Nankai trough: An overview, in *Proc. ODP, Sci. Results, 131,* I. A. Hill, A. Taira, et al., pp. 387-396, Ocean Drilling Program, College Station, TX, 1993.

Glew, D. N., Aqueous solubility and the gas hydrates: the methane-water system, *J. Phys. Chem., 66,* 605-609, 1962.

Graf, D.L., Chemical osmosis, reverse chemical osmosis, and the origin of subsurface brines, *Geochim. Cosmochim. Acta, 46,* 1431-1448, 1982.

Han, M. W., and E. Suess, Subduction induced pore fluid venting and the formation of authigenic carbonates along the Cascadia continental margtin: implication for the global Ca-cycle, *Paleogeog., Palaeoclim., Palaeoecol., 71,* 97-118, 1989.

Hanshaw, B. B., and T. B. Coplen, Ultrafiltration by a compacted clay membrane-II. Sodium ion exclusion at various ionic strengths, *Geochim. Cosmochim. Acta, 37,* 2311-2327, 1973.

Haydon P. R., and D. L Graf, Studies of smectite membrane behavior: Temperature dependence, 20°-180°C, *Geochim. Cosmochim. Acta, 50,*115-121, 1986.

Hiscott, R. N., and J. B. Gill, Major and trace element geochemistry of Oligocene to Quaternary volcanoclastic sands and sandstones from the Izu-Bonin Arc, in *Proc. ODP, Sci. Results, 126,* B. Taylor, K. Fujioka, et al., pp. 467-485, Ocean Drilling Program, College Station, TX, 1992.

Holser, W. T., Mineralogy of evaporites, in *Marine Minerals; Reviews in Mineralogy, vol. 6,* edited by R. G. Burns, pp. 211-294, Mineral. Soc. Amer., New York, 1979.

Honza E., and K. Tamaki, Bonin Arc, in *The Ocean Basins and Margins: The Pacific Ocean (vol. 7),* edited by A. E. M. Nairn, F. G. Stehli, and S. Uyeda, pp. 459-499, Plenum Press, New York, 1985.

Hubbert, M. K., and W. M. Rubey, Role of fluid pressure in mechanics of overthrust faulting, *Geol. Soc. Am. Bull., 70,* 115-166, 1959.

Hussong, D. M., and L. K. Wipperman, Vertical movement and tectonic erosion of the continental wall of the Peru-Chile trench near 11°30'S latitude, in *Nazca Plate: Crustal Formation and Andean Convergence,* edited by L. D. Kulm, J. Dymond, E. J. Dasch, D. M. Hussong, and R. Roderick, Geol. Soc. Am. Mem. *154,* 509-524, 1981.

Hussong, D. M., P. B. Edwards, S. H. Johnson, J. F. Campell, and G. H. Sutton, Crustal structure of the Peru-Chile trench: 8°S to 12°S latitude, in *The Geophysics of the Pacific Ocean Basin and its Margin, Geophys. Monogr. Ser.*, vol. 19, edited by G. H. Sutton, M. H. Manghnani, and R. Moberly, pp. 71-86, AGU, Washington, D.C., 1976.

Hyndman, R. D., and E. E. Davis, A mechanism for the formation and paleobathymetry of the Izu-Bonin Arc, Legs 125 and 126, in *Proc. ODP, Sci. Results, 126,* B. Taylor, K. Fujioka, et al., pp. 285-310, Ocean Drilling Program, College Station, TX, 1992.

Kaiho, K, Eocene to Quaternary benthic foraminifers and paleobathymetry of the Izu-Bonin Arc, Legs 125 and 126, in *Proc. ODP, Sci. Results, 126,* B. Taylor, K. Fujioka, et al., pp. 467-485, Ocean Drilling Program, College Station, TX, 1992.

Kastner, M., H. Elderfield, J. B. Martin, E. Suess, K. A Kvenvolden, and R. E. Garrison, Diagenesis and interstitial water chemistry at the Peruvian continental margin-major constituents and strontium isotopes, in *Proc. ODP, Sci. Results, 112,* edited by E. Suess, R. von Huene, et al., pp. 413-440, Ocean Drilling Program, College Station, TX, 1990.

Kastner, M., H. Elderfield, and J. B. Martin, Fluids in convergent margins: What do we know about their composition, origin, role in diagenesis and importance for oceanic chemical fluxes? *Phil. Trans. Roy. Soc. London, Ser. A, 335,* 243-259, 1991.

Kastner, M., H. Elderfield, W. J. Jenkins, J. M. Gieskes, and T. Gamo, Geochemical and isotopic evidence for fluid flow in the western Nankai subduction zone, Japan, in *Proc. ODP, Sci. Results, 131,* I. A. Hill, A. Taira, et al., pp. 397-413, Ocean Drilling Program, College Station, TX, 1993.

Kay, R. W., S.-S. Sun, C. N. Lee-Hu, Rb and Sr isotopes in volcanic rocks from the Aleutian Islands and Pribilof Islands, Alaska, *Geochim. Cosmochim. Acta, 42,* 263-273, 1978.

Kemp, A. E. S., Sedimentary fabrics and variations in lamination style in Peru continental margin upwelling sediments, in *Proc. ODP, Sci. Results, 112,* E. Suess, R. von Huene, et al., pp. 33-42, Ocean Drilling Program, College Station, TX, 1990.

Kharaka, Y. K., and F. A. F. Berry, Simultaneous flow of water and solutes through geological membranes-I. Experimental investigation, *Geochim. Cosmochim. Acta, 37,* 2577-2603, 1973.

Kvenvolden, K. A., Comparison of marine gas hydrates in sediments of an active and passive continental margin, *Mar. Petrol. Geol., 2,* 65-71, 1984.

Kvenvolden, K. A., A primer on gas hydrates, in *The Future of Energy Gases,* edited by D. G. Howell, pp. 279-291, USGS Professional Paper 1570, 1993.

Kvenvolden, K., and L. Barnard, Hydrates of natural gas in continental margins, in *Studies in Continental Margin Geology, Mem. 34,* edited by J. Watkins, and C. Drake, pp. 631-640, Tulsa, OK, Am. Assoc. Petrol. Geol., 1983.

Kvenvolden, K. A., and M. Kastner, Gas hydrates of the Peruvian outer continental margin, in *Proc. ODP, Sci. Results, 112,* E. Suess, R. von Huene, et al., pp. 517-526, Ocean Drilling Program, College Station, TX 1990.

Kvenvolden, K. A., G. E. Claypool, C. N. Threlkeld, and E. D. Sloan, Geochemistry of a naturally occuring massive gas hydrate, *Org. Geochem., 6,* 703-706, 1984.

Lambert, S. J., Geochemistry of Deleware Basin groundwaters. *New Mexico Bur. Mines. Min. Res. Circ. 159,* 33-38, 1978.

Lawrence, J. R., and J. M. Gieskes, 1981, Constraints on water transport and alteration in the oceanic crust from the isotopic composition of pore water, *J. Geophys. Res., 86,* 7924-7934.

Lawrence, J. R., and M. Taviani, Extreme hydrogen, oxygen, and carbon anomalies in the pore waters and carbonates of the sediments and basalts of the Norwegian Sea: Methane and hydrogen from the mantle? *Geochim. Cosmochim. Acta, 52*, 2077-2083, 1988.

Lloyd, R. M., Oxygen isotope enrichment in seawater by evaporation, *Geochim. Cosmochim. Acta, 30*, 801-814, 1966.

MacDonald G. J., Role of methane clathrates in past and future climates, *Climatic Change, 16*, 247-281, 1990.

Manheim, F. T., and M. K. Horn, Composition of deeper subsurface waters along the Atlantic continental margin, *Southeastern Geol., 8*, 215-236, 1968.

Manheim, F. T., and D. M. Schug, Interstitial waters of black sea cores, in *Init. Repts. DSDP, 42B*, D. A. Ross, Y. P. Neprochnov, et al., pp. 637-651, U.S. Govt. Printing Office, Washington, D.C., 1978.

Martin, J. B., *Origins and Compositions of Fluids at Convergent Margins,* Ph.D. dissertation, 186 pp., University of California, San Diego, 1993.

Martin, J. B., Diagenesis and hydrology at the New Hebrides forearc and Aoba basin, in *Proc. ODP, Sci. Results, 134*, H. G. Greene, J-Y. Collot, L. B. Stokking, et al., Ocean Drillling Program, College Station, TX, in press, 1994.

Martin, J. B., M. Kastner, and H. Elderfield, Lithium: sources in pore fluids of Peru slope sediments and implications for oceanic fluxes, *Mar. Geol., 102*, 281-292, 1991.

Martin, J. B., J. M. Gieskes, M. Torres, and M. Kastner, Bromine and iodine in Peru convergent margin sediments and pore fluids: implications for fluid origins, *Geochim. Cosmochim. Acta, 57*, 4377-4389, 1993.

McCaffrey, M. A., B. Lazon, and H. D. Holland, The evaporation path of seawater and the coprecipitation of Br^- and K^+ contents of the halite, *J. Sed. Petrol., 57*, 928-939, 1987.

McDuff, R. E., and J. M. Gieskes, Ca and Mg profiles in DSDP interstitial waters, diffusion or reaction? *Earth Planet. Sci. Lett., 33*, 1-10, 1976.

McDuff, R. E., J. M. Gieskes, and J. R. Lawrence, Interstitial water studies Leg 42A, in *Init. Repts. DSDP, 42A*, K. S. Hsü, L. Montadent, et al., pp. 561-568, U.S. Govt. Printing Office, Washington, D.C., 1978.

Miller, J., D. Hussong, and R. von Huene, Peru continental margin, record section 3, in *Seismic Images of Modern Convergent Margin Tectonic Structure,* edited by R. von Huene, Am. Assoc. Petrol. Geol. Studies in Geol. Ser., *26,* 32-33, 1986.

Moore, J. C., and P. Vrolijk, Fluids in accretionary prisms. *Rev. Geophys., 30,* 113-135, 1992.

Morris, J. D., and S. R. Hart, Isotopic and incompatible element constraints on the genesis of island arc volcanics from Cold Bay and Amak Island, Aleutians and implications for mantle structure, *Geochim. Cosmochim. Acta, 47,* 2015-2030, 1983.

Morris, J. D., and F. Tera, ^{10}Be and ^{9}Be in mineral separates and whole rocks from volcanic arcs: Implications for sediment subduction, *Geochim. Cosmochim. Acta, 53*, 3197-3206, 1989.

Morris, J. D., W. P. Leeman., and F. Tera, The subducted component in island arc lavas: constraints from Be isotopes and B-Be systematics, *Nature, 344*, 31-36, 1990.

Morris, R. C., and P. A. Dickey, Modern evaporite deposition in Peru, *Am. Assoc. Petrol. Geol., 41*, 2467-2474, 1959.

Newman A. C. D., and G. Brown, The chemical constituents in clays, in *Chemistry of Clays and Clay Minerals,* edited by A. C. D. Newman, Mineralog. Soc. Monogr. 6, 2-166, John Wiley and Sons, New York, 1987.

Noble, D. C., E. H. McKee, and F. Megard, Early Tertiary "Incaic" tectonism, uplift, and volcanic activity, Andes of central Peru, *Geol. Soc. Am. Bull., 90,* 903-907, 1979.

Perry, E., and J. Hower, Burial diagenesis in Gulf Coast pelagic sediments, *Clays and Clay Minerals, 18,* 165-177, 1970.

Ransom, B., and H. C. Helgeson, A chemical and thermodynamic model of dioctahedral 2:1 layer clay minerals in diagenetic processes: Regular solution representation of interlayer dehydration in smectite, *Am. Jour. Sci.,* in press, 1994.

Reimers, C. E., and E. Suess, Spatial and temporal patterns of organic matter accumulation on the Peru continental margin, in *Coastal Upwelling*, edited by J. Theide and E. Suess, pp. 311-345, Plenum Publishing, New York, 1983.

Savin, S. M., and M. Lee, Isotopic studies of phyllosilicates, in *Hydrous Phyllosilicates (Exclusive of Micas)*, edited by S. W. Bailey, Rev. Mineral., *19,* 189-223, 1988.

Savin, S. M., and S. Epstein, The oxygen and hydrogen isotope geochemistry of clay minerals, *Geochim. Cosmochim. Acta, 34,* 43-63, 1970.

Shi, Y., and C-Y. Wang, Generation of high pore pressures in accretionary prisms: Inferences from the Barbados subduction complex, *J. Geophys. Res., 93,* 8893-8910, 1988.

Shi, Y., C-Y. Wang, and R. von Huene, Modeling of pore pressure, central Peru Margin, in *Proc. ODP, Sci. Results, 112,* E. Suess, R. von Huene, et al., pp. 663-670, Ocean Drilling Program, College Station, TX, 1990.

Shipley, T. H., M. H. Houston, R. T. Buffler, F. J. Shaub, K. J. McMiller, J. W. Ladd, and J. L. Worzel, Seismic evidence for widespread possible gas hydrate horizons on continental slopes and rises, *Am. Assoc. Petrol. Geol. Bull., 63*, 2204-2213, 1979.

Suess, E., R. von Huene, et al., *Proc. ODP, Init. Repts., 112*, 1015 pp., Ocean Drilling Program, College Station, TX, 1988a.

Suess, E., R. von Huene, and Leg 112 Shipboard Scientists, The Peru continental margin: Results of Leg 112 drilling, part II. History and geochemistry of depositional environment, *Geology, 16,* 939-943, 1988b.

Taylor, B., Rifting and the volcanic-tectonic evolution of the Izu-Bonin-Marian Arc, in *Proc. ODP, Sci. Results, 126*, edited by B. Taylor, K. Fujioka, et al., pp. 627-651, Ocean Drilling Program, College Station, TX, 1992.

Taylor, B, K. Fujioka, et al., *Proc. ODP, Sci. Results, 126*, Ocean Drilling Program, College Station, TX, 1990.

Taylor, F. W., Quaternary vertical tectonics of the central New Hebrides island arc, in *Proc. ODP, Init. Repts., 134*, J-Y. Collot, H. G. Greene, L. B. Stokking, et al., pp. 33-42, Ocean Drilling Program, College Station, TX, 1992.

Taylor, F. W., B. L. Isacks, C. Jouannic, A. L. Blon, and J. Dubois, Coseismic and Quaternary vertical tectonic movements, Santo and Malekula Islands, New Hebrides island arc, *J. Geophys. Res., 85,* 5367-5381, 1980.

Taylor, F. W., C. Frolich, J. Lecolle, and M. Strecher, Analysis of partially emerged corals and reef terraces in the central Vanuatu arc: Comparison of contemporary coseismic and non-seismic with Quaternary vertical movement, *J. Geophys. Res., 92,* 4905-4933, 1987.

Taylor, H. P., Jr., The oxygen isotope geochemistry of igneous rocks, *Cont. Mineral. Petrol., 19,* 1-71, 1968.

Thornburg, T. M., and L. D. Kulm, Sedimentary basins of the Peru continental margin: structure, stratigraphy and Cenozoic tectonics from 6°S to 16°S latitude, in *Nazca Plate: Crustal*

Formation and Andean Convergence, edited by L.D. Kulm, J. Dymond, E.J. Dasch, D.M. Hussong, and R. Roderick, Geol. Soc. Am. Mem. *154,* 393-422, 1981.

Usiglio, M. J., Etudes sur la composition de l'eau de la Mediterranée et sur l'exploitation des sel qu'ele continent, *Annu. Chim. Phys., 27,* 172-191, 1849.

von Huene, R., and J. Miller, Erosion and accretion along the southern Peru Margin, *Eos Trans. AGU, 69,* 1406, 1988.

von Huene, R., and D. W. Scholl, Observations at convergent margins concerning sediment subduction, subduction erosion, and the growth of continental crust, *Rev. Geophys., 29,* 279-316, 1991.

von Huene, R., L. D. Kulm, and J. Miller, Structure of the frontal part of the Andean convergent margin, *J. Geophys. Res., 90,* 5429-5442, 1985.

Vrolijk P. and S. M. F. Sheppard, Syntectonic carbonate veins from the Barbados accretionary prism (ODP Leg 110): record of palaeohydrology, *Sedimentology, 38,* 671-690, 1991.

Jonathan B. Martin and Miriam Kastner, Scripps Institution of Oceanography-0212, University of California, San Diego, La Jolla, CA 92093-0212, USA

Per Kr. Egeberg, Department of Chemistry, Agder College, Tordenskjoldsgate 65, 4601 Kristiansand, Norway

Sediment Diagenesis in Western Pacific Basins

Marta E. Torres

GEOMAR, Kiel, Germany

Kathleen M. Marsaglia

Department of Geological Sciences, The University of Texas at El Paso, TX

Jonathan B. Martin

Scripps Institution of Oceanography, La Jolla, CA

Richard W. Murray

Department of Earth Sciences, Boston University, MA

The tectonic environment of the western Pacific has created numerous forearc, intra-arc and backarc basins that are characterized by widely varying sediment compositions, oceanographic settings, and geothermal regimes. Because of common volcanism throughout the region, ash is a major component of the sediment where it is not diluted by continental detritus. Dissolution of glass, olivine, pyroxene and amphiboles occurs rapidly (within 1 Ma), followed by formation of a variety of clay and zeolite minerals. Simultaneously, pore fluid compositions change greatly: Ca^{+2}, Cl^- and Br^- concentrations increase whereas the concentration of Mg and values for the $\partial^{18}O$ of the pore water decrease. The formation of authigenic clay and zeolite minerals reduces permeability in the sediments and retards diffusive exchange with seawater, allowing large concentration gradients to develop in the pore fluids. In the Japan Sea, high productivity in the Middle Miocene produced silica rich deposits (app. 80% SiO_2), but nonetheless, the abundance of silica was insufficient to form thick sections of bedded chert. High productivity also provides a large source of barium to the sediments, which is remobilized and concentrated as barite deposits at the base of the sulfate reduction zone. Authigenic carbonate is common throughout the western Pacific. The shift from a predominance of rhodochrosite to dolomite reflects increasing bottom water oxygenation of the Japan Sea in the Middle Miocene. In the Sulu Sea, ^{13}C-depleted dolomites form from terrestrial organic matter deposited in turbidites.

INTRODUCTION

The western Pacific region is characterized by chains of island arcs and associated backarc and forearc basins (Figure 1). These basins undergo major tectonic processes and are subject to changing paleoceanographic conditions during their evolution. They vary in their proximity to active arcs or continental margins and experience a wide range of sediment accumulation rates [*Marsaglia et al.*, this volume; *Underwood et al.*, this volume]. Therefore, each basin records unique processes of diagenesis and mineralization. We have selected several locations where specific diagenetic processes can be best studied (Table 1). For example, thick sequences of arc-derived pyroclastic strata in intra-oceanic basins (e.g. Sulu Sea, Izu-Bonin forearc and Aoba basins) provide excellent opportunities to study diagenetic alteration

Active Margins and Marginal Basins of the Western Pacific
Geophysical Monograph 88

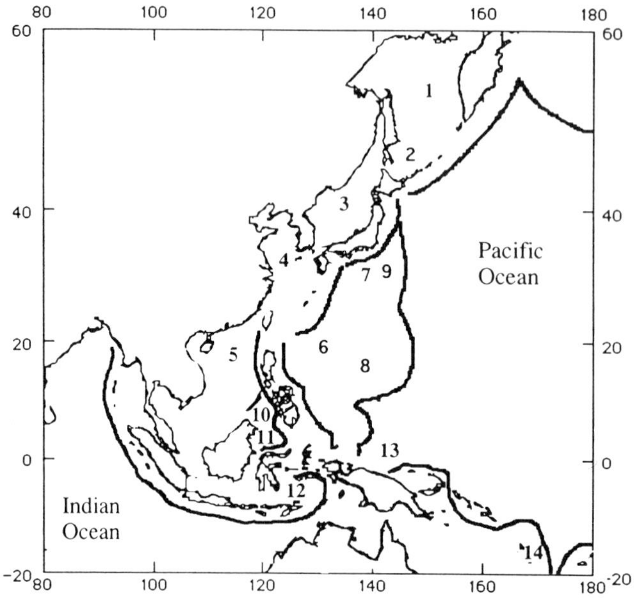

Fig. 1. Basins of the western Pacific region. (1) Sea of Okhotsk; (2) Kurile Basin; (3) Japan Sea; (4) East China and Yellow Seas; (5) South China Sea: (6) west Philippine basin; (7) Shikoku basin; (8) Parece Vela Basin; (9) Izu-Bonin forearc basin; (10) Sulu Sea; (11) Celebes Sea; (12) Banda Sea; (13) Caroline Basin; (14) Aoba Basin (New Hebrides). The solid lines indicate the approximate location of the trenches.

of volcaniclastic deposits and its effect on the physical and chemical characteristics of sediment and pore fluids. On the other hand, the transformation of silica phases and the formation of authigenic carbonates and barites can be well studied in sediments from the Japan Sea; a continental marginal basin that experienced wide extremes of temperature, productivity and circulation since the time of its formation [e.g., *Ingle*, 1975; *Ingle* et al., 1990].

BACKGROUND

The Celebes and Sulu Seas are deep restricted basins located between the island of Borneo to the west, and the Philippine Archipelago to the east. The Celebes Basin originated in the middle Eocene in an environment protected from significant volumes of terrigenous sediment. The Sulu Sea originated in a backarc setting in the late-early Miocene [*Rangin* et al., 1990]. Of particular interest are thick sequences of lower Miocene volcaniclastic deposits recovered from all sites drilled in the Sulu Sea. These thick, lithologically homogeneous sequences nicely document the nature of diagenetic changes that occur during burial of pyroclastic strata. Both the Celebes (sites 767 and 770) and Sulu Seas (sites 768, 769 and 771) show clear synchroneity of terrigenous turbidites in the latest middle Miocene, a time of very rapid sedimentation in both basins. A decrease in turbidites and an increase in the volcanic ash component in the late Miocene sequences from these basins have been associated with southward subduction of the Celebes Sea [*Rangin* et al., 1990].

Thick sequences of volcaniclastic material also were recovered from the forearc region of the Izu-Bonin arc [*Taylor* et al., 1990]. The Izu-Bonin forearc basin is part of the Izu-Bonin-Mariana system, a complex intra-oceanic arc-trench system which includes the trench, the forearc terrain, the active Izu-Bonin and Mariana arcs, the backarc basins, the Parece-Vela and Shikoku marginal basins and the Palau-Kyushu Ridge. The thickest stratigraphic unit in all sites drilled in the Izu-Bonin forearc basin (787, 792, and 793) is a lower to upper Oligocene turbidite succession derived from the contemporary Izu-Bonin intra-oceanic arc. Sedimentation rates at this time were less than 20 m $m.y.^{-1}$. The lower to middle Miocene sequences of the forearc are represented by fine grained, deep water carbonate-rich sediments and are characterized by low sedimentation rates (5 to 15 m $m.y.^{-1}$) and multiple unconformities. Sediments of late Miocene age were recovered only at site 792, and consist of sandy mudstones with nannofossil-rich intervals. The Pliocene and Pleistocene sections consist of mixtures of pumiceous and scoriaceous sand and gravel at sites 787 and 793, and of nannofossil-rich silty clay with minor resedimented volcanic ash at site 792 [*Taylor* et al., 1990].

Alteration of volcaniclastic sequences was also studied in the north Aoba basin, an intra-arc basin within the central New-Hebrides island arc (Vanuatu). The north Aoba basin sediments are primarily composed of volcanic ash and breccias with some biogenic carbonate. This basin experienced rapid accumulation of sediment, with rates ranging from 100 to 356 m $m.y.^{-1}$ [*Collot* et al., 1992].

The Japan Sea is a large backarc basin with water depths in excess of 3500 m, separated from the Pacific Ocean by shallow bathymetric sills. Its sedimentary sequences record the major tectonic and paleoceanographic changes during the evolution of the basin [*Tamaki* et al., 1990; *Ingle* et al., 1990]. Delta-front sands and siltstones rich in plant debris were deposited during the initial backarc rifting and basin subsidence in the early Miocene. Basin subsidence accelerated during the middle and late Miocene accompanied by widespread deposition of diatomaceous muds at bathyal depths. These deposits reflect increasing rates of primary productivity in the Japan Sea, which probably resulted from climatic cooling and a decrease in the flux of terrigenous sediments to the deepening basinal areas. The diatom flux decreased during the Pleistocene, indicating lower pro-

TABLE 1. Site Description

	Site	Location		Water Depth (m)	Oldest Sediment Depth (m)	Age
Celebes Sea[a]	767	4°47.50'N	123°30.21'E	4905	787	Middle Miocene
	770	5°08.69'N	123°40.11'E	4505	423	Early Oligocene
Sulu Sea[a]	768	8°00.04'N	121°31.18'E	4384	1047	Early Miocene
	769	8°47.12'N	121°17.69'E	3644	278	Late Miocene
	771	8°40.69'N	120°40.78'E	2859	234	Middle Miocene
Izu-Bonin[b]	787	32°22.50'N	140°44.64'E	3259	320	Late Oligocene
	788	30°55.37'N	140°00.22'E	1102	374	Early Pliocene
	792	32°23.96'N	140°22.79'E	1788	800	Late Oligocene
	793	31°06.33'N	140°53.27'E	2965	1404	Early Oligocene
Aoba Basin[c]	832	14°47.78'S	167°34.35'E	3089	846	Late Miocene
	833	14°52.56'S	167°52.78'E	2629	953	Early Pliocene
Japan Sea[d,e]	794	40°11.40'S	138°13.87'E	2811	414	Middle Miocene
	796	42°50.56'N	139°24.50'E	2623	465	Middle Miocene
	797	38°36.58'N	134°32.10"E	2865	553	Early Miocene
	798	37°03.85'N	134°08.00'E	900	518	Early Pliocene
	799	39°22.04'N	133°86.64'E	2072	1084	Early Miocene

Sources: [a]*Rangin et al.* [1990]; [b]*Taylor et al.* [1990]; [c]*Collot et al.* [1992]; [d]*Tamaki et al.* [1990); and [e]*Ingle et al.* [1990].

ductivity in the surface waters concurrent with a major progradation of submarine fans and deposition of coarse siliciclastic sequences.

VOLCANICLASTIC SEDIMENTS

The sand-sized grains of forearc and backarc basin sediments are predominantly volcanic glass fragments [e.g. *Lee and Klein*, 1986; *Packer and Ingersoll*, 1986; *Smith*, 1991; *Marsaglia*, 1992; *Marsaglia and Ingersoll*, 1992]. These volcaniclastic sands are prone to diagenetic alteration because of their reactivity and high surface area. As a result, ancient arc-related sandstones often exhibit a wide range of authigenic phases and complex paragenetic mineral sequences produced by grain dissolution, grain alteration, and cementation. High geothermal gradients along the arc axis and within the actively spreading backarc basins, appear to enhance these diagenetic reactions. The authigenic phases associated with alteration of primary volcaniclastic grains include clay minerals (mainly smectite and chlorite), carbonates, and a variety of zeolite minerals [*Lee and Klein*, 1986; *Smith*, 1991; *Marsaglia and Tazaki*, 1992; *Boggs and Seyedolali*, 1992].

Mineralogical Changes

In intraoceanic settings, the main source of sand-sized material is the volcanic arc [*Marsaglia* et al., this volume; *Underwood* et al., this volume]. Across a given arc segment, there is little difference between forearc and backarc composition of sands [*Marsaglia*, 1992; *Marsaglia and Ingersoll*, 1992]. Petrographic and compositional analysis of Izu-Bonin [*Marsaglia*, 1992] and Mariana (Marsaglia and Devaney, 1995) sands and sandstones show these lithologies to be composed of a mixture of felsic to mafic volcaniclastic grains.

Lee and Klein [1986]; *Klein and Lee* [1984]; *Lee* [1984]; and *Smith* [1991] studied diagenetic phases of older sediment sequences drilled in backarc settings. They reported an ubiquitous occurrence of smectite and chlorite

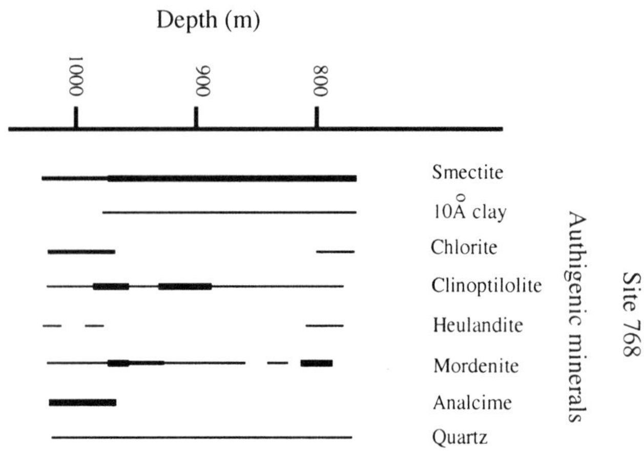

Fig. 2A. Depth distribution of authigenic minerals of Miocene pyroclastic strata from site 768. The width of the vertical lines is used to denotate the relative abundance of the minerals. Modified from *Smith* [1991].

cementation, followed by a wide array of zeolite minerals. *Lee and Klein* [1986] compared major diagenetic changes in sites drilled in several backarc basins in the western Pacific during the Deep Sea Drilling Project (DSDP). These authors showed that highly altered sediment sections (sites 445 and 446) accumulated slowly and experienced high heat

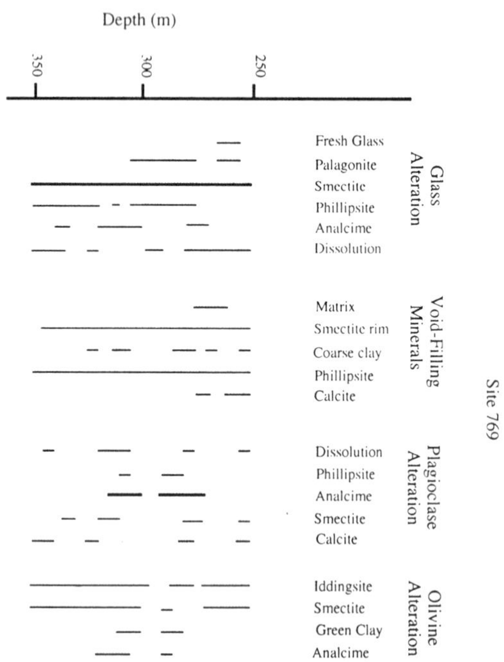

Fig. 2B. Depth distribution of authigenic minerals of Miocene pyroclastic strata from site 769. The width of the vertical lines is used to denotate the relative abundance of the minerals. Modified from *Smith* [1991].

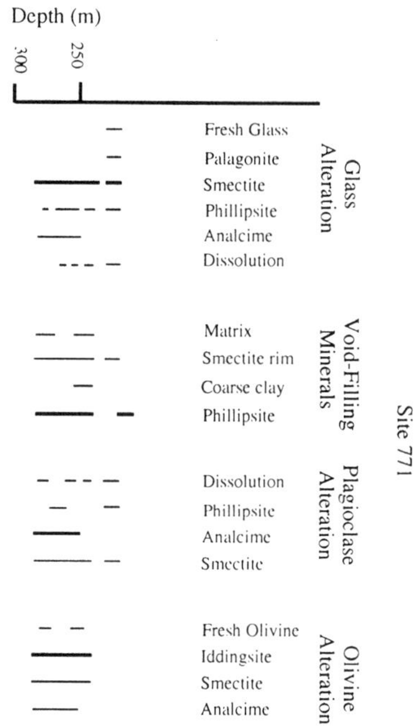

Fig. 2C. Depth distribution of authigenic minerals of Miocene pyroclastic strata from site 771. The width of the vertical lines is used to denotate the relative abundance of the minerals. Modified from *Smith* [1991].

flow associated with earliest rifting. In Miocene sandstones from the Sulu Sea, *Smith* [1991] reports a correlation between tuff composition and early diagenetic zeolite mineralogy: phillipsite was found in basaltic tuffs, whereas mordenite and clinoptilolite characterize the silicic tuffs. On the other hand, analcime occurs as a late-stage replacement mineral in the more deeply buried sections, independent of tuff composition (Figure 2).

Because *Lee and Klein* [1986] and *Smith* [1991] confined their studies to deeply buried sediments, their results provide little information on progressive changes with burial. *Marsaglia and Tazaki* [1992] document secondary porosity and cement relationships in loosely consolidated to well-indurated samples from forearc and backarc sequences ranging from Oligocene to Quaternary in age. They found that early diagenetic changes of volcaniclastic sandstones from Izu-Bonin basins are primarily characterized by dissolution of brown intermediate to mafic glass and precipitation of thin clay-mineral (smectite) rim cements (Figures 3 and 4). These two processes are followed by minor alteration of colorless felsic glass to clay minerals, and precipitation of zeolite (clinoptilolite) cements. Older forearc sections are pervasively altered to smectite and a diverse suite of zeolite minerals. The dissolution of glassy fragments began soon after burial (< 1 m.y.), followed by

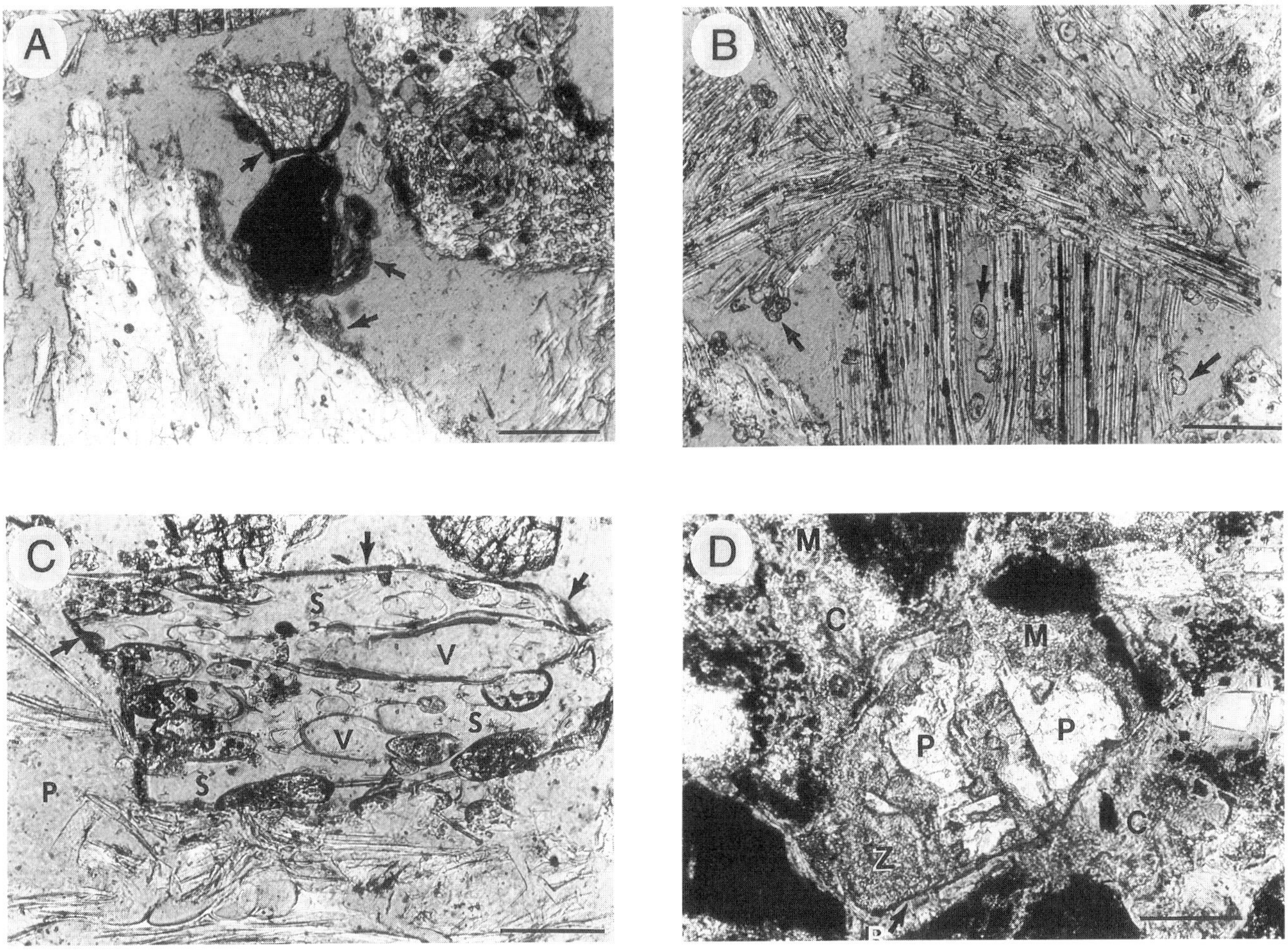

Fig. 3. Photomicrographs illustrating alteration of volcaniclastic sediments. A) Thick clay mineral rim cements (arrows) developed adjacent to opaque grain in sample 126-788C-19H-1, 69-71 cm. Clay mineral cements are dark gray and porosity is light gray (stained/etched portion of slide). B) Incipient cementation and compactional features in sample 126-788D-11R-1, 17-20 cm. Pumice fragment in center is bent around lower pumice fragment by compaction, incipient zeolite (phillipsite?) cementation occurs as small globular clusters (arrows) of submicroscopic crystals within interparticle pores (light gray) and intraparticle vesicles (unstained part of slide). C) Dissolved microlitic/vesicular brown (?) glass in sample 126-788C-19H-1, 69-71 cm. Primary interparticle porosity (P), primary intraparticle porosity (V - vesicles), and secondary intraparticle porosity (S - dissolved brown glass). Grain and vesicles are outlined by a thin discontinuous clay mineral (smectite?) rim (arrows) (stained/etched portion of slide). D) Plagioclase feldspar grain (P) partially altered to calcium-bearing zeolite (Z) in sample 126-793-19R-4, 47-51 cm. Interparticle cements include carbonate (C), mixed-stained zeolites (M) and chlorite rim (R) cements (stained/etched portion of slide). All photomicrographs were taken with plane-polarized light; scale bars equal 0.1 mm. Reproduced from *Marsaglia and Tazaki* [1992].

lithification and cementation after a few million years. In addition, olivine, pyroxene and amphibole also readily undergo dissolution and/or alteration at shallow burial depths prior to extensive cementation and lithification [e.g., *Lee and Klein*, 1986; *Marsaglia and Tazaki*, 1992; *Smith*, 1991]. The resulting secondary porosity is later filled with authigenic phases such as clay minerals, calcite, and zeolites. Plagioclase dissolution and replacement is, however, much more complex. Only minor dissolution occurs at shallow burial depths prior to extensive cementation [*Marsaglia and Tazaki*, 1992]. In older, deeply buried sections, feldspars are extensively dissolved and replaced by albite and/or zeolites (Figures 3 and 4).

Unlike intraoceanic backarc basins, the composition of

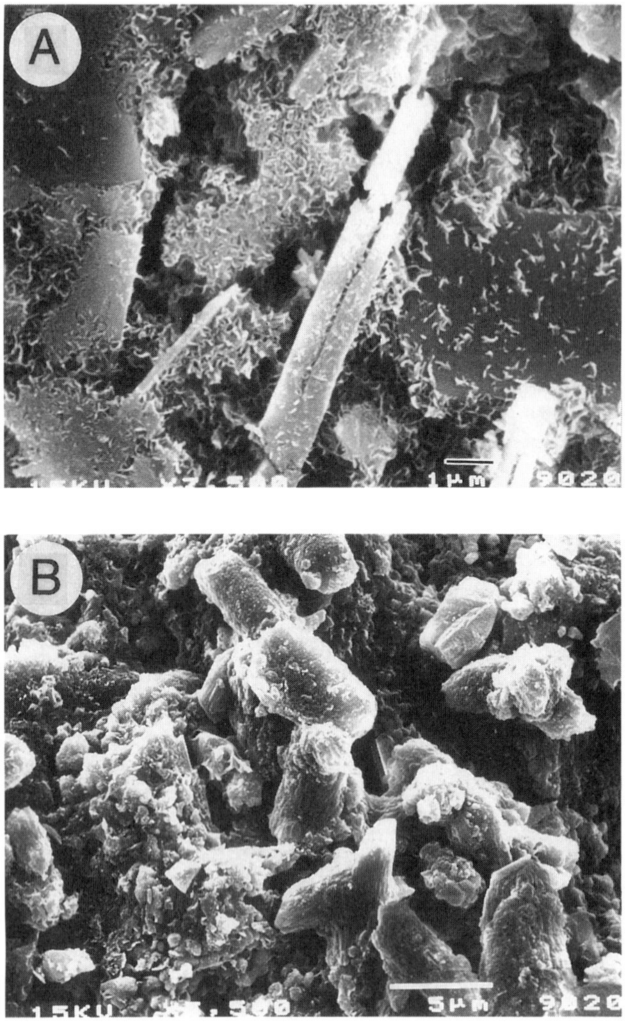

Fig. 4. Scanning electron microscope photographs. A) Irregular film of smectite covering surfaces of zeolite (phillipsite and wairakite), feldspar and calcite crystals from sample 126-787B-15R-1, 108-111 cm. B) Defaced surface of wairakite and feldspar covered by small particles of secondary minerals with smectite, from sample 126-792E-47R-2, 88-93 cm. Reproduced from *Marsaglia and Tazaki* [1992].

sand deposited in a continental margin basin, such as the Japan Sea, can be highly variable [*Packer and Ingersoll*, 1986; *Marsaglia and Ingersoll*, 1992]. These compositional variations control, in part, the history and intensity of diagenetic alteration in continental backarc sequences. *Boggs and Seyedolali* [1992] contrasted diagenetic changes in volcaniclastic sandstones from sites 796 (offshore Hokkaido) and 797 (Yamato basin) with the quartzo-feldspathic sandstones from site 799 located on the Yamato Rise. Diagenetic changes common to all three sites include the local replacement of volcanic lithic grains by zeolites, clay minerals, and carbonates. The mud matrix is replaced by mixed-layer clay minerals and chlorite at site 797, and plagioclase and feldspars are extensively albitized in samples from sites 797 and 799. According to *Boggs and Seyedolali* [1992], cementation within these sandstones is volumetrically less important than grain alteration. At site 796, cementation is limited to minor chlorite partly filling molds of glass shards; whereas, at site 797 minor clay mineral, zeolite, opal-CT, quartz, and albite cements are present. Site 799 is unique among these backarc sites because of the occurrence of kaolinite cement that is probably related to the abundance of K-feldspars in these sandstones. In contrast to the Miocene sandstones studied by *Boggs and Seyedolali* [1992], Miocene "Blue Tuff" intervals from sites 794 and 796 are extensively altered to smectite [*Barnes* et al., 1992].

Authigenic phases also undergo dissolution and replacement through time. For example, *Marsaglia and Tazaki* [1992] noted textural evidence for dissolution of zeolite cements in the Izu-Bonin forearc sections; and *Smith* [1991] found that smectite and celadonite cements were replaced by chlorite with depth. Late stage replacement of early formed zeolites (heulandite and clinoptilolite) by analcime is also commonly observed [*Smith*, 1991; *Boggs and Seyedolali*, 1992].

Effects on Physical Properties of the Sediments

Diagenetic formation of clay minerals and zeolites from original volcanic material has important effects on the physical properties of these deposits. However, only a few investigations have examined the physical properties in volcaniclastic sediments [*Bouma and Moore*, 1975; *Carlson* et al., 1980; *Carlson*, 1981]. Recently, *Dadey and Klaus* [1992] presented a comprehensive analysis of the effects of post-depositional processes on the physical properties of volcaniclastic sediments in the Izu-Bonin area. For example, at sites 792 and 793, sonic velocity values ranging from 2.5 to 4 km s^{-1} and formation factors that approach 30 reflect an increase in the amount of authigenic minerals and corresponding decrease in the amount of volcanic glass (Figure 5). *Dadey and Klaus* [1992] concluded that the cemented volcaniclastic rocks are composed of large grains of high porosity pumice welded to form an indurated rock. Although this sediment retains porosity values of approximately 40% (Figure 5), it is relatively impermeable to fluid flow, dissolved chemical species, and electrical energy.

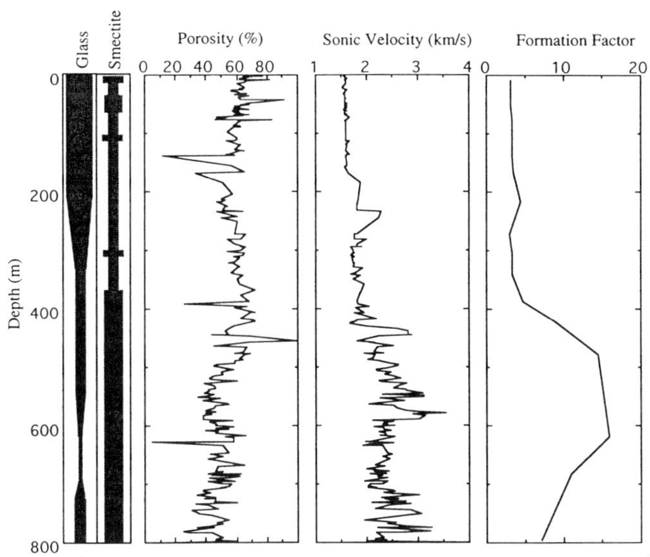

Fig. 5A. Porosity, sonic velocity and formation factors corresponding to volcaniclastic sediment strata from site 792. Data from *Taylor* et al. [1990].

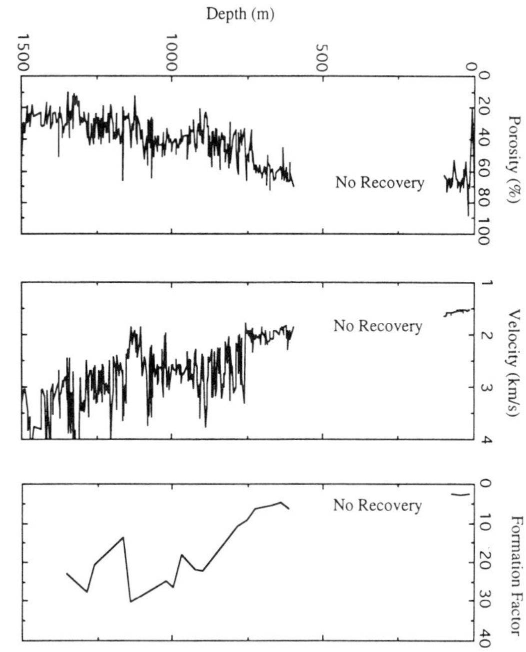

Fig. 5B. Porosity, sonic velocity and formation factors corresponding to volcaniclastic sediment strata from site 793. Data from *Taylor* et al. [1990].

Diagenetic Influence on Pore Fuid Composition.

The chemical composition of pore fluids associated with alteration of volcanic ash and Layer II basalts was extensively documented by deep sea drilling [*McDuff and Gieskes*, 1976; *Gieskes*, 1983; *Gieskes and Lawrence*, 1981; *Lawrence and Gieskes*, 1981]. As would be expected in western Pacific basins, the intense diagenetic alteration of unstable volcanic glass has greatly altered the composition of pore fluids of sediments from these basins [*Egeberg* et al., 1990; *von Breymann* et al., 1991; *Egeberg* 1992; *France-Lanord et al.*, 1992; *Martin*, 1994). Typical compositional changes in the pore fluids include a decrease in the concentration of Mg^{+2} and K^{+} and an enrichment of Ca^{+2}. Such trends are present in pore fluids from Celebes, Sulu, Izu-Bonin, and Aoba basins. Fluids from the Izu-Bonin forearc and Aoba basins display extremely large Ca^{+2} concentrations (up to 538 mM at site 833), as well as increased concentrations of typically conservative components, such as Cl- and Br- [Figure 6; *Egeberg* et al., 1990; *Martin*, 1994]. With exception of halite diagenesis, Cl^{-} is not directly involved in diagenetic reactions; therefore, changes in its concentration must reflect addition or removal of water from the pore fluids. In volcanic-rich sediments of the western Pacific, water is most likely removed through hydration of ash and formation of hydrous diagenetic minerals [*Martin* et al., this volume].

In addition to changes in the cation concentrations, alteration of volcanic glass depletes pore water in ^{18}O. At DSDP sites where alteration of Layer II basalts controls pore fluid chemistry, there is an inverse correlation between $\delta^{18}O$ values of the pore water and the dissolved calcium concentration [*Lawrence and Gieskes*, 1981]. A similar correlation was observed at sites 767 and 768 in the Sulu and Celebes Basins (Figure 7), suggesting that volcanic sequences underlying the sediment column control the calcium concentration and oxygen isotope distributions. At site 833 in the Aoba basin, a similar correlation occurs, although with a steeper slope than that observed at sites 767 and 768 (Figure 7).

In the Izu-Bonin forearc and Aoba basins, the large deviations in the compositions of pore fluids result from special sedimentological circumstances coupled with intense diagenesis [*Egeberg*, 1992; *Martin*, 1994; *Martin et al.*, this volume]. At sites 792 and 793, the most altered fluids occur at depths where high formation factors reflect low permeability zones (Figures 5 and 8). The low permeability appears to restrict the diffusive loss of solutes to the overlying seawater [*Egeberg*, 1992]. In contrast, the main factor responsible for elevated solute concentrations at the Aoba basin is the rapid accumulation of sediment. Sedimentation rates in the Aoba basin are as high as 356 m $m.y.^{-1}$ [*Collot et al.*, 1992]. Using this value for the rate of sedimentation, and assuming a diffusion coefficient of 2 x 10^{-6} cm^2 sec^{-1} [e.g., *Li and Gregory*, 1974], any changes

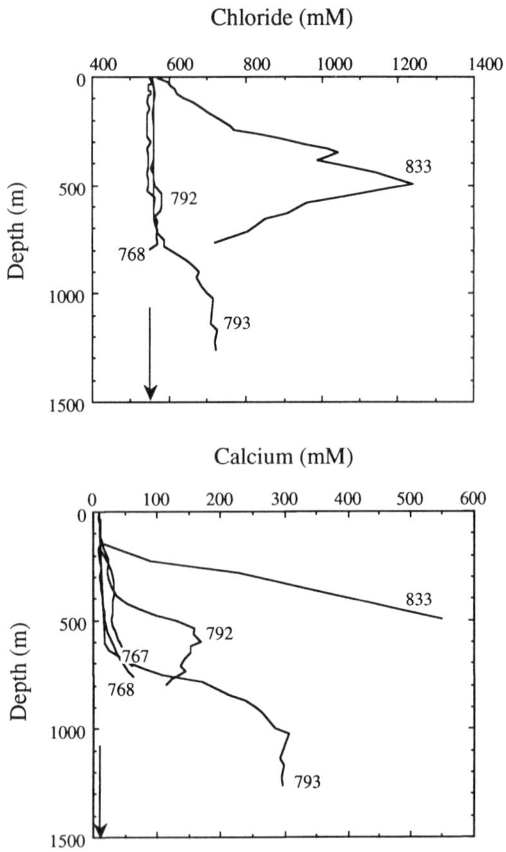

Fig. 6. Dissolved chloride and calcium depth profiles from sites where alteration of volcaniclastic sediments has modified the pore water composition. Note the large deviations in the calcium and chloride concentrations of samples from the Aoba (site 833) and the Izu-Bonin forearc (sites 792 and 793) basins, relative to samples form the Sulu (site 768) and Celebes (site 767) Seas. The arrows denote seawater concentrations.

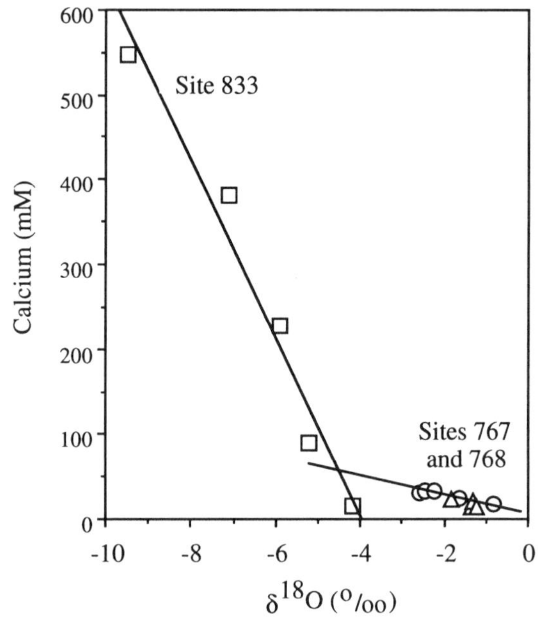

Fig. 7. Dissolved calcium concentration versus $\partial^{18}O$ values for pore water of deep sediment sequences from sites 767 (circles), 768 (triangles) and, site 833 (squares). Data for sites 767 and 768 are from *von Breymann* et al. [1991]; data for site 833 are from *Martin* [1994].

to fluid composition resulting from diagenesis at depths greater than 60 m will not be lost to the overlying seawater by ion diffusion.

Thermodynamic calculations using the composition of the pore waters and the stability of authigenic minerals (gypsum, zeolites, feldspars, smectite, chlorites and mica) in the Izu-Bonin forearc basin, show that many of these phases are in equilibrium with the pore fluids [*Egeberg*, 1992]. In the Aoba basin, similar thermodynamic calculations indicate that a suite of minerals comparable to that of the Izu-Bonin forearc basin is in equilibrium with the pore fluids [*Martin et al.,* this volume]. The sediment associated with the most altered fluid in the Aoba basin ranges in age from 1.89 to 0.45 Ma, indicating that alteration reactions and large changes in pore fluid composition can occur rapidly.

SILICA DIAGENESIS

During burial of marine sediment, biogenic silica undergoes several dissolution-reprecipitation phase transitions, the most important of which are the transformations of opal-A to opal-CT and of opal-CT to quartz. The rates and stratigraphic occurrences of these diagenetic transformations are dependent on time, temperature, pressure, host sediment composition, and organic matter content [e.g., *Murata and Larson*, 1975; *Kastner et al.*, 1977; *Kastner*, 1981; *Isaacs*, 1982; *Hein et al.*, 1981a, b; *Williams and Crerar*, 1985; *Williams et al.*, 1985; *Hinman*, 1990; *Monty et al.*, 1991]. Regional porosity reduction during burial occurs by compaction and dissolution/reprecipitation of the biogenic opal phase [e.g., *Isaacs*, 1981; *Compton*, 1991; *Nobes et al.*, 1992]; this is frequently accompanied by local dissolution of biogenic opal in the host sediment and transport of dissolved silica to chert nodules, lenses, and beds [*Cole*, 1917, and references therein; *Davis*, 1918; *Taliaferro and Hudson*, 1943; *Murata and Larson,* 1975; *Hein et al.*, 1981a, b; *Geeslin and Chafetz*, 1982; *Brueckner and Snyder*, 1985a, b; *Brueckner et al.*, 1987; *Tada*, 1991; *Murray et al.*, 1992b, c].

In the Japan Sea, the diagenetic phase transitions occur at depths of several hundreds of meters. The opal-A to opal-

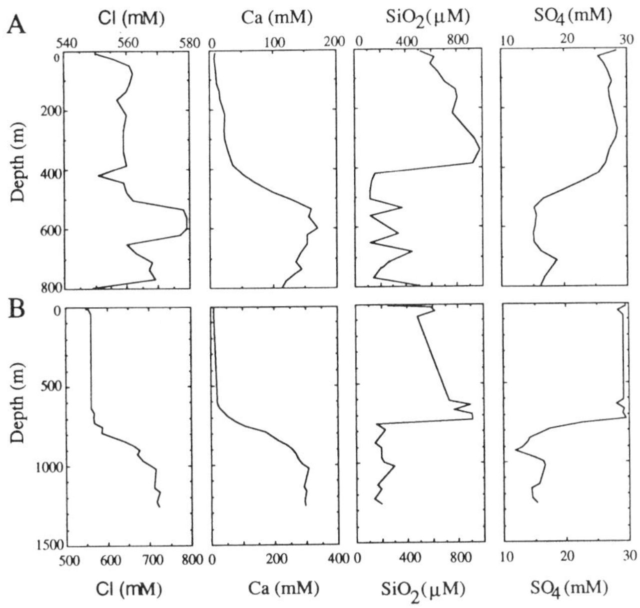

Fig. 8. Distribution of dissolved chloride, calcium, silica and sulfate in pore fluids from (A) site 792 and (B) site 793 [data from *Egeberg*, 1992]. The large magnitude of changes in the composition of these fluids arise from the development of low permeability horizons during alteration of volcaniclastic material illustrated in Fig. 5.

CT transformation occurs at temperatures ranging from 22 to 68°C; the conversion to quartz occurs from 44 to 64°C [*Murray et al.*, 1992a; *Pisciotto et al.*, 1992]. These burial depths and temperatures broadly agree with previous observations of marginal deposits such as the Monterey Formation and various DSDP sites [e.g., *Murata and Larson*, 1975; *Murata et al.*, 1977; *Pisciotto*, 1981]. Nevertheless, extremely early diagenetic mobilization of silica has been documented in Southern Ocean sediments [*Weaver and Wise*, 1973; *Botz and Bohrmann*, 1991; *Bohrmann et al.*, 1990, 1992; *Matheney and Knauth*, 1993], suggesting that under certain conditions chemical fractionations occur over ranges of temperature and pressure that are larger than those previously considered.

Chert nodules and porcellanites were recovered throughout much of the cored section in the Japan Sea. However, there is a paucity of thick sections of bedded chert in Neogene deposits of this marginal basin compared to other Mesozoic deposits [e.g., *Jenkyns and Winterer*, 1982; *Karl*, 1984; *Murray et al.*, 1991, 1992b]. It is useful to compare the Japan Sea to the Monterey Formation because of the close similarity of their depositional histories and stratigraphic successions [*Ingle*, 1981]. While specific parts of the Monterey Formation (and coeval deposits of western North America) display more pronounced bedding than others, bedded chert on the scale of hundreds of meters is rarely found in Monterey-aged North American sequences [e.g., the Claremont Formation; *Cerling*, 1976; *Pisciotto and Garrison*, 1981; *Murray et al.*, 1992b]. Potential reasons for the stratigraphic contrasts between bedded and non-bedded deposits have been discussed by *Jenkyns and Winterer* [1982] and *Hein and Karl* [1983]. These workers considered nodular and bedded chert as distinct chemical phenomena; however, recent work has shown that many aspects of the chemical histories of nodular and bedded chert may be similar [*Brueckner and Snyder*, 1985a; *Brueckner et al.*, 1987; *Murray et al.*, 1992b, c].

Murray et al. [1992b] showed that the bulk SiO_2 concentration in bedded chert sequences is always substantially higher than the bulk SiO_2 in non-bedded (but nodular chert bearing) stratigraphies. Mass balance calculations indicate that the formation of bedded chert occurs only in sequences containing enough labile biogenic SiO_2 to form entire stratigraphies of interlayered beds. Modern oozes and opaline Neogene deposits usually do not contain enough SiO_2 to supply the wholesale silicification of entire sequences through tens of meters of stratigraphic section [*Murray et al.*, 1992b]. Thus, the lack of bedded chert in certain siliceous sequences may not to be due to the absence of an appropriate chemical mechanism (e.g., biogenic opal dissolution, porewater SiO_2 transport), but rather seems to reflect that a vast majority of siliceous sequences do not contain enough of the key silica ingredient for bedded chert formation. Analogously, *Botz and Bohrmann*, [1991] proposed that the abundance of SiO_2, and the concomitant dilution of clay phases, are the dominant parameters controlling chert formation. Low temperature and shallow diagenetic formation of opal-CT chert requires substantial transport and reprecipitation of biogenic silica and apparently occurs only in Southern Ocean deposits, where the abundance of opal-A in the host sediment is extremely high [> 95 %; *Botz and Bohrmann*, 1991].

We are able to develop further this hypothesis by comparing the chemical composition of Japan Sea sediment and the Monterey Formation to bedded-chert sequences (Table 2). When comparing chemical compositions reported on a percentage basis, it is important to understand the statistical closure effects of compositional data [e.g., *Murray*, 1994]. For example, consider a rock with 80% SiO_2, i.e., 80 g of SiO_2 and 20 g of all other components. By doubling the SiO_2 content, the new sample will have 160 g SiO_2 in a total of 180 g, or a normalized composition of 89% SiO_2. Thus, a sample with 90% SiO_2 (dry weight) contains twice as much SiO_2 as a sample with

TABLE 2. Composition of Siliceous Sediments and Chert

	SiO_2	Opal
Japan Sea, Leg 127	80	70
Monterey Formation		
South Elwood field, offshore	77	33 (57)
Orcutt field, onshore	83	36 (56)
Siliceous member, outcrop	85.2	73
Bedded chert sequences		
Franciscan Assemblage (Mesozoic)	90.1	82
Claremont Formation (Monterey-aged)	92.0	80

All values are given in dry weight percent, parenthetical data indicate maximum values. For the Japan Sea, opal values are estimated from total silica and aluminum concentrations (opal = SiO_2 - (3 x Al_2O_3), using data from *Murray et al.*, [1992d]. South Elwood data from *Isaacs et al.* [1990a]. Orcutt field data from *Isaacs et al.* [1990b]. Monterey outcrop (Siliceous Member), Franciscan Assemblage and Claremont Formation data from *Murray et al.* [1992b].

80% SiO_2 on a dry weight basis. These calculations demonstrate that there are large differences between samples with only slightly different percentage values. The biogenic silica values for the Monterey Formation and the Japan Sea summarized in Table 2 are substantially smaller than those found in bedded-chert sequences in which the average composition of the entire sequence is commonly larger than 90% SiO_2 (dry weight). Therefore, even though the Japan Sea and Monterey sequences record deposition of elevated siliceous production throughout the Neogene, the overall absolute amount of biogenic silica is relatively small compared to that found in bedded-chert sequences. Notwithstanding the elevated siliceous productivity of the overlying water, the amount of terrigenous material in these areas diluted the concentration of the biogenic opal component below that required for the formation of bedded chert.

Effects on the Physical Properties of the Sediments and on the Composition of Pore-Fluids

Dissolved silica concentration-depth profiles (Figure 9) from sites drilled in the Japan Sea monitor the transitions from amorphous biogenic opal to progressively more ordered diagenetic silica phases. The transition from opal-A to opal-CT can be correlated with a distinct seismic reflector that results from a significant decrease in porosity across this diagenetic boundary [*Nobes* et al., 1992]. Figure 10 summarizes the porosity of sediments in a north-south transect across the Japan Sea. The high porosities typical of

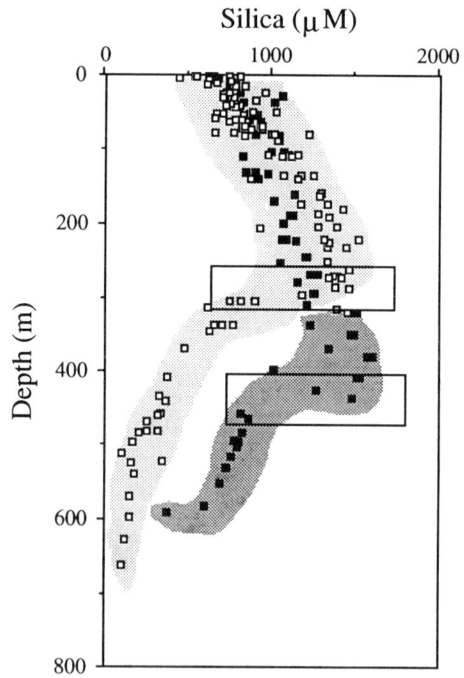

Fig. 9. Dissolved silica profiles for sites drilled in the Japan Sea. The profiles overlay each other to approximately 300 m. The deep basinal sites (open squares) experience the opal-A to opal-CT diagenetic transition at approximately 300 m, as evidenced by the sharp decrease in the dissolved silica profiles denoted by the upper rectangle. Sites 798 and 799 (closed squares) record the diagenetic transition at greater depth (lower rectangle), due to the more gradual geothermal gradient at these sites. Modified from *Murray et al.* [1992a].

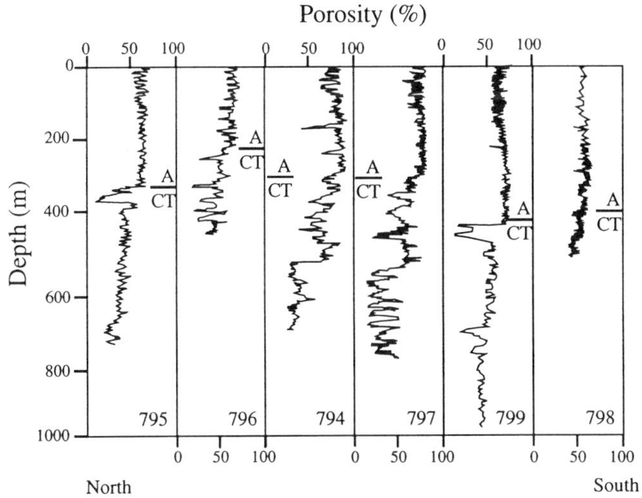

Fig. 10. Depth distribution of sediment porosity (%) on sites drilled in the Japan Sea. In all cases, there is a significant decrease in sediment porosity below the opal-A/opal-CT boundary. At Sites 799 and 795 the porosity drops quite sharply in the opal-A/opal-CT diagenetic horizon, before returning to more moderate values. Modified from *Nobes et al.* [1992].

sediments rich in biogenic silica are also associated with low wet bulk densities and low acoustic velocities, properties that are also distinctly different above and below the opal-A/opal-CT transition zone [*Nobes et al.*, 1992].

The opal-A/opal-CT transition profoundly affects interstitial water chemical profiles and divides each site into two diffusive regimes, with the upper section reflecting communication with the overlying seawater and the section below the transition affected dominantly by basement alteration reactions [*Murray et al.*, 1992a]. This effect is illustrated by the dissolved lithium profiles (Figure 11). The upper sediment section reflects a release of lithium which is followed by a removal at depth by alteration of volcanic ash and crystalline basement rocks. It is noteworthy that the most distinct discontinuity in pore water profiles occurs at sites 795 and 799, sites in which the silica diagenetic front results in the largest porosity barrier (Figure 10). The distinct discontinuity in the composition of the pore fluids at sites 795 and 799 is further illustrated by the distribution of dissolved chloride, calcium and strontium, and the strontium isotopic composition in pore fluids shown in Figure 12. The lithologic barrier associated with the opal-A/opal-CT transition efficiently isolate the fluids below the diagenetic horizon from exchange with the overlying interstitial waters.

DIAGENETIC CARBONATES

The Japan Sea experienced wide extremes of temperature, productivity and circulation since its formation [*Ingle*, 1975; *Ingle et al.*, 1990]. Thus, this marginal sea offers a good opportunity to study environmental changes that result from tectonically and eustatically controlled changes in basin geometry and gateways as well as from variations in global and local climatic patterns. For example, the distribution of diagenetic carbonates at site 799 (Figure 13) appears to result from changes in environmental factors at the site of deposition [*Matsumoto*, 1992]. *Pedersen and Price* [1982] showed that a high concentration of organic carbon is not critical for the generation of rhodochrosite, instead this mineral forms in regions characterized by a large supply of manganese oxides and coarse grain-size sediments. Accumulation of manganese oxides requires well-oxygenated bottom water, low productivity, and rapid sediment accumulation; conditions that are prevalent in the modern Japan Sea [*Ingle,* 1975; *Ingle et al.*, 1990]. The vertical distribution of rhodochrosite (Figure 13) may reflect a change in the degree of oxygenation of the bottom water

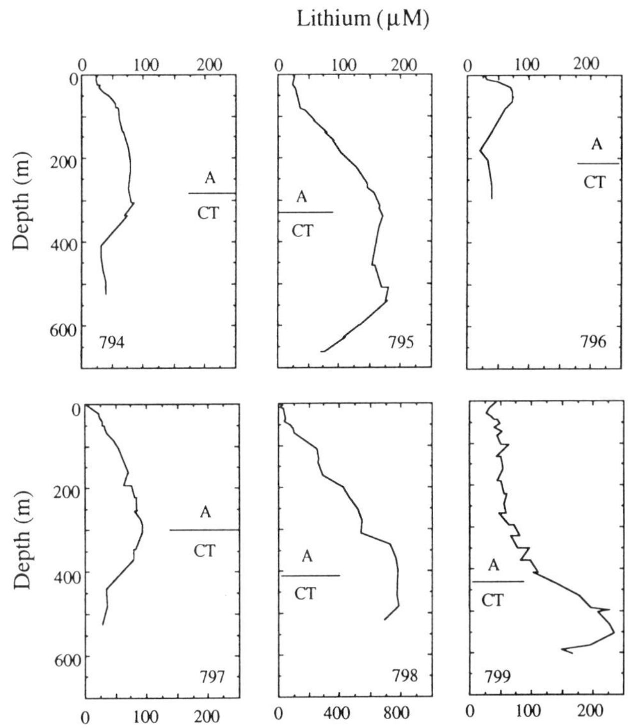

Fig. 11. Distribution of dissolved lithium in the pore fluids from sites drilled in the Japan Sea, showing the location of the opal-A/opal-CT diagenetic boundary. Modified from *Murray et al.* [1992a].

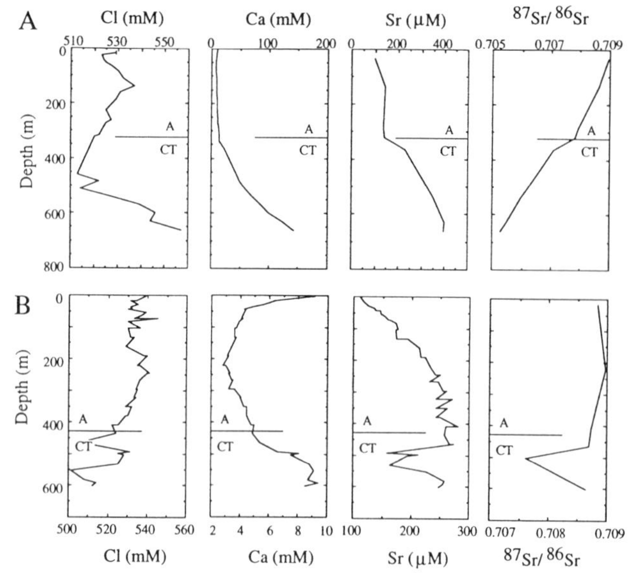

Fig. 12. Distribution of dissolved chloride, calcium and strontium, and strontium isotopic composition in pore fluids from (A) site 795 [data from *Brumsack et al.*, 1992], and (B) site 799 [data from *Sturz et al.*, 1992]. At both sites, the diagenetic transformation of opal-A to opal-CT results in a distinct discontinuity in the composition of the pore fluids.

from anoxic to well-oxygenated from the late Miocene to the present [*Matsumoto*, 1992].

Diagenetic dolomite, on the other hand, appears to require high productivity and elevated concentrations of organic carbon [*Garrison et al.*, 1984; *Baker and Burns*, 1985]. In the Middle Miocene, the Japan Sea experienced high levels of primary productivity [*Ingle et al.*, 1990] resulting in formation of diagenetic dolomite in the corresponding sediment sequences (Figure 13). Based on petrographic and isotopic analysis, *Matsumoto* [1992] concluded that protodolomite formed at shallow burial depths in a manganese-poor anoxic environment. This dolomite subsequently recrystallized at elevated temperatures below 400 m.

In contrast with typical diagenetic dolomite recovered from regions of high productivity [*Garrison et al.*, 1984; *Baker and Burns*, 1985], diagenetic dolomites were also found in sediments from the Celebes (site 767) and Sulu (site 768) basins [*Betzler*, 1991], areas of very low primary productivity. Nonetheless, these dolomites occur at depths associated with the onset of methane production and depletion of sulfate, where the source of the carbon is terrestrial organic matter deposited by turbidity currents in the Middle Miocene [*Bertrand* et al., 1991]. The rapid deposition of this continentally derived organic material led to the production of high levels of biogenic methane with $\delta^{13}C$ values as low as -74$^o/_{oo}$ [*von Breymann et al.*, 1991]. Sulfate reduction in sediments that overlie Miocene turbidites is restricted by their low organic carbon content, allowing the oxidation of upwardly diffusing CH_4. The distributions of total organic carbon, methane, and sulfate in the deep basin (sites 767 and 768) and on bathymetric highs (sites 769 and 770) illustrate this point (Figure 14). In the deep basin, sulfate reduction is driven by the oxidation of methane [*von Breymann et al.*, 1991). In contrast, on bathymetric highs where no organic-rich turbidites were deposited, there is no significant accumulation of methane, and no depletion of sulfate (Figure 14).

Oxidation of the ^{13}C-depleted upwardly diffusing methane by sulfate results in isotopically light CO_2. At sites 767 and 768, $\delta^{13}C$ values for dissolved ΣCO_2 are as low as -42 $^o/_{oo}$ at depths corresponding to the onset of methane accumulation [*von Breymann et al.*, 1991]. Dolomites from the Celebes and Sulu Seas record isotopic signatures imparted by the sulfate oxidation of methane, and have $\delta^{13}C$ values as low as -28.8$^o/_{oo}$ (Torres, unpublished data). Dolomites depleted in ^{13}C are typically formed in the sul-

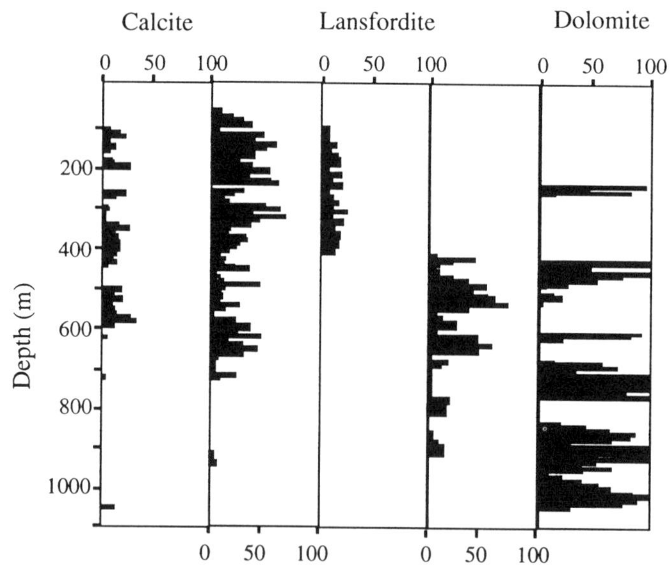

Fig. 13. Relative abundance of diagenetic carbonates at site 799 in the Japan Sea, estimated by intensity of X-ray diffraction peaks. Note that rhodochrosite is the dominant carbonate phase in the younger sequences, whereas dolomite occurs in deeper strata. This distribution is thought to represent changes in the conditions of the bottom water and in paleoproductivity regimes of this marginal basin. Modified from *Matsumoto* [1992].

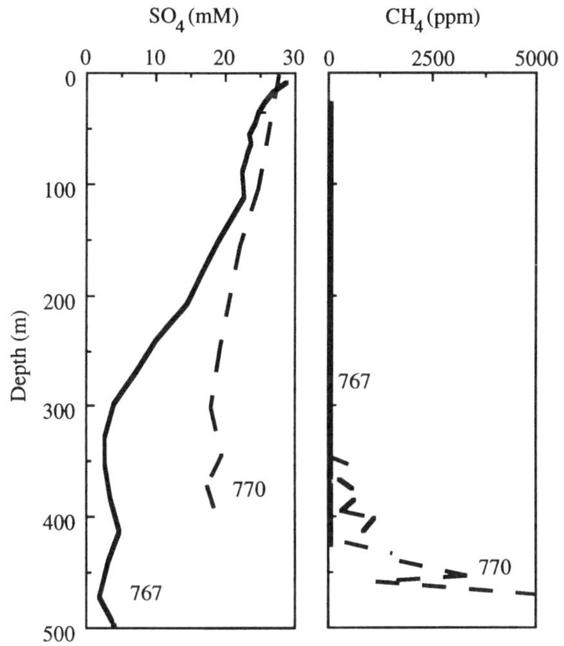

Fig. 14A. Sulfate and methane distributions in the Celebes sea. Site 767 (solid lines) is located in the deep basin; whereas, site 770 (dashed lines) is located on a bathymetric high and is isolated from the deposition of turbidites of terrigenous detritus rich in organic carbon. Thus, site 770 do not show accumulation of methane nor depletion of sulfate. From *von Breymann et al.* [1991].

fate reduction zone of sediments underlying highly productive environments. Examples of these deposits can be found in the Japan Sea [*Matsumoto,* 1992]; the Monterey formation [e.g., *Garrison and Graham,* 1984] and the Peru margin [*Kulm et al.*, 1984]. The Celebes and Sulu basins provide a contrasting scenario for the formation of ^{13}C-depleted dolomites, where the source of the organic carbon is terrestrial organic matter deposited in turbidites.

BARITE

Several diagenetic models have been proposed for the origin of non-volcaniclastic barites. For example, *Goldberg et al.* [1969] suggested that barite nodules from the coast of California originated in a coastal lagoon or a shallow hydrothermal environment. In contrast, *Dean and Schreiber* [1978] concluded that the barite at DSDP sites 369 and 370 deposits formed diagenetically in sediments exposed at the sediment-water interface during long hiatuses. Still another scenario is presented by *Jewell and Stallard* [1991]; they suggest that the bedded barite deposits of central Nevada formed beneath an anoxic basin in a paleo-upwelling zone.

A 2-cm barite layer was recovered from site 799 in the Japan Sea at 323.2 m, within a fairly homogeneous diatomaceous ooze of Pliocene age [*von Breymann et al.*, 1992]. Diatomaceous deposits have been related to enrichment of barium in marine sediments, because of precipitation of barium sulfate microcrystals in association with the biogenic component of the sediment [*Dehaires et al.*, 1980]. The Japan Sea experienced elevated siliceous productivity during the Middle Miocene [*Ingle*, 1975; *Ingle et al.*, 1990], which resulted in a large flux of biogenic barium to the sediments [*von Breymann et al.*, 1992]. In strongly reducing sediments sulfate reduction triggers remobilization of biogenic barium. Subsequently, barite reprecipitates in diagenetic fronts at the base of the sulfate reduction zone [Figure 15; *Brumsack*, 1986; *von Breymann et al.*, 1992, 1993]. In this scenario, the precipitation of barite involves sulfate ions that come, at least in part, from a reservoir that has been enriched in the heavy ^{34}S isotope by selective consumption of the lighter isotope during bacterial oxidation of organic matter [*Mizutani and Rafter*, 1973; *Goldhaber and Kaplan*, 1980]. Thus, the barites will be isotopically heavy relative to seawater. Such heavy

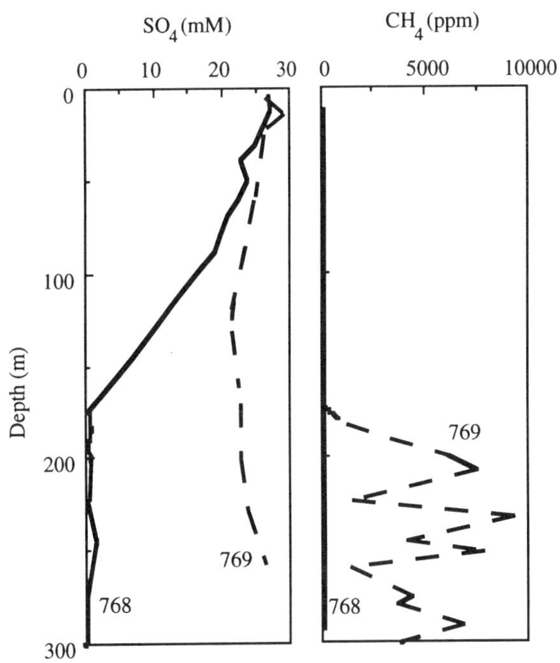

Fig. 14B. Sulfate and methane distributions in the Sulu sea. Site 768 (solid lines) is located in the deep basin; whereas, site 769 (dashed lines) is located on a bathymetric high and is isolated from the deposition of turbidites of terrigenous detritus rich in organic carbon. Thus, site 769 do not show accumulation of methane nor depletion of sulfate. From *von Breymann et al.* [1991].

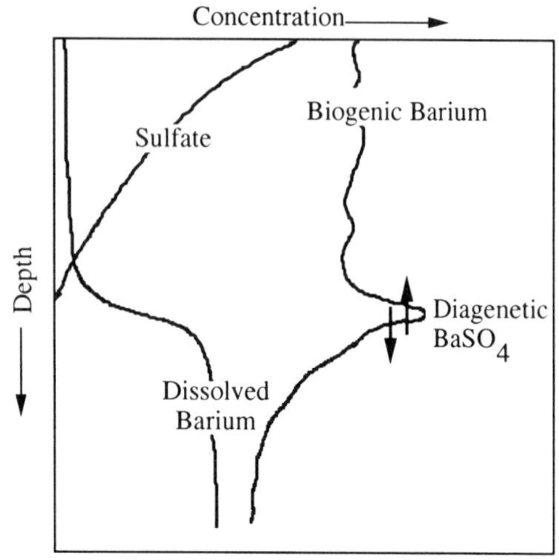

Fig. 15. Schematic diagram illustrating the precipitation of authigenic barite near the zone of sulfate depletion by upward migration of dissolved barium and downward diffusion of sulfate. From *von Breymann et al.* [1992].

barites (up to 84‰ CDT) were recovered by deep sea drilling in the Japan Sea and the Peru margin (Torres et al., unpublished information), and are analogous to the sediment-hosted barites that occur in a variety of localities around the world. These studies are very useful in elucidating paleo-depositional environments, and support a diagenetic model for the origin of the barite deposits enriched in ^{34}S.

Acknowledgments. The authors wish to thank the scientific members, technicians and crew of the JOIDES Resolution for the productive and effective working atmosphere through several years of drilling in the Western Pacific. RWM received financial support from USSAC. MET acknowledges support from the German Government (DFG-ODP) and from COAS, Oregon State University. We are grateful for the comments and suggestions provided by J. S. Compton and by an anonymous reviewer which significantly improved the manuscript.

REFERENCES

Baker, P. A., and S. J. Burns, Occurrence and formation of dolomite in organic-rich continental margin sediments, *Am. Asosoc. Petrol. Geol. Bull., 69,*1917-1930, 1985.

Barnes, D. A., P. Thy, and P. Renne, Sedimentology, phenocryst chemistry and age—Miocene "Blue Tuff": Sites 794 and 796, Japan Sea, in *Proc. ODP, Sci. Results, 127/128,* K. A. Pisciotto, J. C. Ingle, Jr., M. T. von Breymann, J. Barron, et al., pp. 115-130, Ocean Drilling Program, College Station, TX, 1992.

Betzler, C., Occurrence of dolomite at Site 768 (Sulu Sea), in *Proc. ODP, Sci. Results, 124,* E. A. Silver, C. Rangin, M. T. von Breymann, et al., pp. 527-530, Ocean Drilling Program, College Station, TX, 1991.

Bertrand, P., U. Berner, and E. Lallier-Verges, Organic sedimentation in Celebes and Sulu basins: type of organic matter and evaluation of organic carbon accumulation rates, in *Proc. ODP, Sci. Results, 124,* E. A. Silver, C. Rangin, M. T. von Breymann, et al., pp. 217-226, Ocean Drilling Program, College Station, TX, 1991.

Boggs, S. Jr., and A. Seyedolali, Diagenetic albitization, zeolitation, and replacement in Miocene sandstones, Sites 796, 797, and 799, Japan Sea, in *Proc. ODP, Sci. Results, 127/128,* K. A. Pisciotto, J. C. Ingle, Jr., M. T. von Breymann, J. Barron, et al., pp. 131-151, Ocean Drilling Program, College Station, TX, 1992.

Bohrmann, G., G. Kuhn, A. Abelmann, R. Gersonde, and D. Futterer, A young porcellanite occurrence from the Southwest Indian Ridge, *Mar. Geol., 92,* 155-163, 1990.

Bohrmann, G., V. Spieb, H. Hinze, and G. Kuhn, Reflector "Pc" a prominent feature in the Maud Rise sediment sequence (eastern Weddell Sea): occurrence, regional distribution and implications to silica diagenesis, *Mar. Geol.,106*, 69-87, 1992.

Botz, R., and G. Bohrmann, Low-temperature opal-CT precipitation in Antarctic deep-sea sediments: evidence from oxygen isotopes, *Earth Planet. Sci. Lett., 107,* 612-617, 1991.

Bouma, A. H. and J. C. Moore, Physical properties of deep sea sediments from the Philippine Sea and the sea of Japan, in *Init. Repts. DSDP, 31,* D. E. Karig, J. C. Ingle, Jr. et al., pp. 535-568, U. S. Govt. Printing Office, Washington, 1975.

Brueckner, H. K., and W. S. Snyder, Chemical and Sr-isotopic variations during diagenesis of Miocene siliceous sediments of the Monterey Formation, California, *J. Sediment. Petrol., 55,* 553-568, 1985a.

Brueckner, H. K., and W. S. Snyder, Structure of the Havallah sequence, Golconda Allochthon, Nevada, evidence for prolonged accretion in an accretionary prism. *Geol. Soc. Am. Bull., 96,* 1113-1130, 1985b.

Brueckner, H. K., W. S. Snyder, and M. Boudreau, Diagenetic controls on the structural evolution of siliceous sediments in the Golconda Allochthon, Nevada, U. S. A., *J. Struct. Geol., 9,* 403-417, 1987.

Brumsack, H. J., The inorganic geochemistry of Cretaceous black shales (DSDP Leg 41) in comparison to modern upwelling sediments from the Gulf of California, in *North Atlantic Paleoceanography,* edited by C. P. Summerhayes and N. J. Shackleton, pp. 447-462, London (Blackwells), 1986.

Brumsack, H. J., E. Zuleger, E. Gohn, and R. W. Murray, Stable and radiogenic isotopes in pore waters from Leg 127, Japan Sea, in *Proc. ODP, Sci. Results, 127/128,* K. A. Pisciotto, J. C. Ingle, Jr., M. T. von Breymann, J. Barron, et al., pp. 635-650, Ocean Drilling Program, College Station, TX, 1992.

Carlson, R. L., Acoustic properties of tuffaceous and calcareous sediments, Deep Sea Drilling Project Leg 60, in *Init. Repts. DSDP, 60,* D. M. Hussong, S. Uyeda, et al., pp. 803-804, U. S. Govt. Printing Office, Washington, 1981.

Carlson, R. L., N. I. Christensen, and R. H. Wilkens, Acoustic properties of volcaniclastic sediments recovered from the floor of the Philippine Sea, Deep Sea Drilling Project Leg 59, in *Init. Repts. DSDP, 59,* L. Kroenke, R. Scott, et al., pp. 5219-522, U. S. Govt. Printing Office, Washington, 1980.

Cerling, B., Diagenesis of Miocene bedded cherts, Berkeley Hills, California, M.S. thesis, 64 pp., University of California, Berkeley, 1976.

Cole, G. A. J., The rhythmic deposition of flint. *Geol. Mag., Decade VI,* 64-68, 1917.

Collot, J. Y., H. G. Greene, L. B. Stokking, et al., *Proc. ODP. Init. Reports, 134,* 1136 pp., Ocean Drilling Program, College Station, TX, 1992.

Compton, J. S., Porosity reduction and burial history of siliceous rocks from the Monterey and Sisquoc Formations, Point Pedernales area, California, *Geol. Soc. Am. Bull., 103,* 625-636, 1991.

Dadey, K. A., and A. Klaus, Physical properties of volcaniclastic sediments in the Izu-Bonin area, in *Proc. ODP, Sci. Results, 126,* B. Taylor, K. Fujioka, et al., pp. 543-550, Ocean Drilling Program, College Station, TX, 1992.

Davis, E. F., The radiolarian rocks of the Franciscan Group. Univ. Calif. Publ., Dept., *Geol. Bull., 11,* 235-432, 1918.

Dean, W. E., and B. C. Schreiber, Authigenic barite, Leg 41 Deep Sea Drilling Project, in *Init. Repts. DSDP, 41,* Y. Lancelot, E. Seibold, et al., pp. 915-931, U. S. Govt. Printing Office, Washington, 1978.

Dehaires, F., R. Chesselet, and J. Jedwab, Discrete suspended particles of barite and the barium cycle in the ocean, *Earth Planet. Sci. Lett., 49,* 528-550, 1980.

Egeberg, P. K., Thermodynamic aspects of Leg 126 interstitial waters, in *Proc. ODP, Sci. Results, 126,* B. Taylor, K. Fujioka, et al., pp. 519-530, Ocean Drilling Program, College Station, TX, 1992.

Egeberg, P. K., and the Leg 126 Shipboard Scientific Party, Unusual composition of pore waters found in the Izu-Bonin forearc sedimentary basins, *Nature, 344,* 215-218, 1990.

France-Lanord, C., A. Michard, and A. M. Karpoff, Major element and Sr isotope composition of interstitial waters in sediments from Leg 129: The role of diagenetic reactions, in *Proc. ODP, Sci. Results, 129,* R. L. Larson, Y. Lancelot, et al., pp. 267-281, Ocean Drilling Program, College Station, TX, 1992.

Garrison, R. E., and S. A. Graham, Early diagenetic dolomites and the origin of dolomite-bearing breccias, Lower Monterey Formation, Arroyo Seco, Monterey County, California, in *Dolomites of the Monterey Formation and Other Organic-Rich Units,* edited by R. E. Garrison, M. Kastner, and D. H. Zenger, pp. 87-102, Pacific Section SEMP, Los Angeles, California, 1984.

Garrison, R. E., M. Kastner, and D. H. Zenger (Eds.) *Dolomites of the Monterey Formation and Other Organic-Rich Units,* 215 pp. Pacific Section SEMP, Los Angeles, California, 1984.

Geeslin, J. H., and H. S. Chafetz, Ordovician Aleman ribbon cherts: an example of silicification prior to carbonate lithification, *J. Sed. Petrol., 52,* 1283-1293, 1982.

Gieskes, J. M., The chemistry of interstitial waters of deep-sea sediments: interpretation of deep-sea drilling data, in *Chemical Oceanography (vol.8),* edited by J. P. Riley, and R. Chester, pp. 222-269, Academy Press, London, 1983.

Gieskes, J. M., and J. R. Lawrence, Alteration of volcanic matter in deep sea sediments: evidence from the chemical composition of interstitial waters from deep-sea drilling cores, *Geochim. Cosmochim. Acta, 45,* 1687-1703, 1981.

Goldberg, E. D., L. K. Somayajulu, J. Galloway, I. R. Kaplan, and G. Faure, Differences between barites of marine and continental origins. *Geochim. Cosmochim. Acta 33,* 287-289, 1969.

Goldhaber, M. B., and I. R. Kaplan, Mechanisms of sulfur incorporation and isotope fractionation during early diagenesis in sediments of the Gulf of California, *Mar. Chem. 9,* 95-143, 1980.

Hein, J. R., and S. M. Karl, Comparisons between open-ocean and continental margin chert sequences, in *Siliceous Deposits in the Pacific Region,* edited by A. Iijima, J. R. Hein, and R. Siever, pp. 25-43, Elsevier, Amsterdam, 1983.

Hein, J. R., T. L. Vallier, and A. M. Allan, Chert petrology and geochemistry, Mid-Pacific Mountains and Hess Rise, Deep Sea Drilling Project Leg 62, in *Init. Repts. DSDP, 62,* J. Theide, T. L. Vallier, et al., pp. 711-748, U. S. Govt. Printing Office, Washington, 1981a.

Hein, J. R., C. Sancetta, and L. A. Morgenson, Petrology and geochemistry of silicified upper Miocene chalk, Costa Rica Rift, Deep Sea Drilling Project Leg 69, in *Init. Repts. DSDP, 69,* J. R. Cann, M.G. Langseth, et al., pp. 395-422, U. S. Govt. Printing Office, Washington, 1981b.

Hinman, N. W., Chemical factors influencing the rates and sequences of silica phase transitions: effects of organic constituents, *Geochim. Cosmochim. Acta, 54,* 1563-1574, 1990.

Ingle, J. C. Jr., Summary of late Paleogene-Neogene stratigraphy, paleobathymetry, and correlations, Philippine Sea and Sea of Japan region, in *Init. Repts. DSDP, 31,* D. E. Karig, J. C. Ingle, Jr., et al., pp. 837-855, U. S. Govt. Printing Office, Washington, 1975.

Ingle, J. C., Jr., Origin of Neogene diatomites around the North Pacific Rim, in *The Monterey Formation and Related Siliceous Rocks of California,* edited by R. E. Garrison, K. A. Pisciotto, C. M. Isaacs, and J. C. Ingle, Jr., pp. 159-179, Society of Economic Paleontologists and Mineralogists, Pacific Section, volume 15, 1981.

Ingle, J. C., Jr., K. Suyehiro, M. T. von Breymann, et al., *Proc. ODP, Init. Repts., 128,* 651 pp. Ocean Drilling Program, College Station, TX, 1990.

Isaacs, C. M., Porosity reduction during diagenesis of the Monterey Formation, Santa Barbara coastal area, California, in *The Monterey Formation and Related Siliceous Rocks of California,* edited by R. E. Garrison, K. A. Pisciotto, C. M. Isaacs, and J. C. Ingle, Jr., Society of Economic Paleontologists and Mineralogists, Pacific Section, volume 15, 1981.

Isaacs, C. M., Influence of rock composition on kinetics of silica phase changes in the Monterey Formation, Santa Barbara area, California, *Geol., 10,* 304-308, 1982.

Isaacs, C. M., J. H. Tomson, J. E. Jr. Taggart, and L. L. Jackson, Abundances of major elements and sedimentary components in Miocene and Early Pliocene cuttings from a well in the South Elwood Oil Field area, offshore Santa Barbara-Ventura Basin, Southern California, *Open-File Report 89-79,* 27 pp, U. S. Geological Survey, 1990a.

Isaacs, C. M., J. H. Tomson, K. C. Stewart, and L. L. Jackson, Abundances of major elements and sedimentary components of cuttings from the Foxen Mudstone and the Sisquoc, Monterey, and Point Sal Formations in the Union Hobbs 22 well, and preliminary comparison with cores from the Union Newlove 51 well, Orcutt Oil Field, onshore Santa Maria Basin, California, *Open-File Report 89-466,* 21 pp, U. S. Geological Survey, 1990b.

Jenkyns, H. C., and E. L. Winterer, Paleoceanography of Mesozoic ribbon radiolarites, *Earth Planet. Sci. Lett., 60,* 351-375, 1982.

Jewell, P. W. and R. F. Stallard, Geochemistry and paleoceanographic setting of central Nevada bedded barites, *J. Geology, 99,* 151-170, 1991.

Karl, S. M., Sedimentologic, diagenetic, and geochemical analysis of Upper Mesozoic ribbon cherts from the Franciscan Assemblage at the Marin Headlands, California, in *Franciscan Geology of Northern California,* edited by M. C. Blake, Jr., pp. 71-88, Society of Economic Paleontologists and Mineralogists, Pacific Section, Vol. 43, 1984.

Kastner, M., Authigenic silicates in deep-sea sediments: Formation and diagenesis, in *The Oceanic Lithosphere, The Sea Vol. 7,* edited by C. Emiliani, pp. 915-980, John Wiley and Sons, New York, 1981.

Kastner, M., J. B. Keene, and J. M. Gieskes, Diagenesis of siliceous oozes--I. Chemical controls on the rate of opal-A to opal-CT transformation--an experimental study, *Geochim. Cosmochim. Acta, 41,* 1041-1059, 1977.

Klein, D. de V., and Y. I. Lee, A preliminary assessment of geodynamic controls and depositional systems and sandstone diagenesis in backarc basins, western Pacific Ocean, *Tectonophysics, 102,* 119-152, 1984.

Kulm, L. D., E. Suess, and T. M. Thornburg, Dolomites in organic-rich muds of the Peru forearc basin: Analog to the Monterey Formation, in *Dolomites of the Monterey Formation and Other Organic-Rich Units,* edited by R. E. Garrison, M. Kastner, and D. H. Zenger, pp. 29-48, Pacific Section SEMP, Los Angeles, California, 1984.

Lawrence, J. R. and J. M. Gieskes, Constraints on water transport and alteration in the oceanic crust from the isotopic composition of the pore water, *J. Geophys. Res., 86,* 7924-7934, 1981.

Lee, Y. I., Isotopic aspects of thermal and burial diagenesis of sandstones at DSDP site 445, Daito Ridge, northwest Pacific Ocean, *Chemical Geology, 65,* 95-102, 1984.

Lee, Y. I. and G. de V. Klein, Diagenesis of sandstones in the backarc basins of the western Pacific Ocean, *Sedimentology, 33,* 651-675, 1986.

Li, Y. H. and S. Gregory, Diffusion of ions in seawater and in deep-sea sediments, *Geochim. Cosmochim. Acta, 48,* 47-62, 1974

Marsaglia, K. M., Provenance and petrography of volcaniclastic sands recovered on ODP Leg 126, the Izu-Bonin arc, in *Proc. ODP, Sci. Results, 126,* B. Taylor, K. Fujioka, et al., pp. 139-154, Ocean Drilling Program, College Station, TX, 1992.

Marsaglia, K.M., and K. A. Devaney, Tectonic and magmatic controls on backarc basin sedimentation: The Mariana region re-examined, in *Backarc Basins: Tectonics and Magmatism*, edited by B. Taylor, Plenum Press, NY, in press.

Marsaglia, K. M. and R. V. Ingersoll, A reassessment of magmatic-arc provenance: compositional trends in arc-related deep-marine sand and sandstone, *Geol. Soc. Am. Bull., 104,* 1637-1649, 1992.

Marsaglia, K. M. and K. Tazaki, Diagenetic trends in Leg 126 sandstones, in *Proc. ODP, Sci. Results, 126,* B. Taylor, K. Fujioka, et al., pp. 125-138, Ocean Drilling Program, College Station, TX, 1992.

Martin, J. B., Diagenesis and hydrology at the New Hebrides forearc and Aoba basin, in *Proc. ODP, Sci. Results, 134,* H. G. Greene, J. Y. Collot, L. Stokking, et al., in press, Ocean Drilling Program, College Station, TX, in press, 1994.

Matheney, R. K., and L. P. Knauth, New isotopic temperature estimates for early silica diagenesis in bedded cherts, *Geology, 21,* 519-522, 1993.

Matsumoto, R., Diagenetic dolomite, calcite, rhodochrosite, magnesite and lansfordite from Site 799, Japan Sea-Implications for depositional environments and the diagenesis of organic-rich sediments, in *Proc. ODP, Sci. Results, 127/128,* K. A. Pisciotto, J. C. Ingle, Jr., M. T. von Breymann, J. Barron, et al., pp. 75-96, Ocean Drilling Program, College Station, TX, 1992.

McDuff, R. E., and J. M. Gieskes, Calcium and magnesium profiles in DSDP interstitial waters: diffusion or reaction?, *Earth Planet. Sci. Lett., 33,* 1-10, 1976.

Mitzutani, Y., and A. Rafter, Isotopic behavior of sulfate oxygen in the bacterial reduction of sulfate, *Geochem. Jour., 6,* 183-191, 1973.

Monty, C. L. V., F. Westall, and S. van der Gaast, Diagenesis of siliceous particles in subantarctic sediments, Hole 699A, possible microbial mediation, in *Proc. ODP, Sci. Results, 129,* R. L. Larson, Y. Lancelot, et al., pp. 685-710, Ocean Drilling Program, College Station, TX, 1991.

Murata, K. J., and R. R. Larson, Diagenesis of Miocene siliceous shales, Temblor Range, California, *U. S. Geol. Surv. J. Res., 3,* 553-566, 1975.

Murata, K. J., I. Friedman, I., and J. D. Gleason, Oxygen isotope relations between diagenetic silica minerals in Monterey Shale, *Am. J. Sci., 277,* 259-272, 1977.

Murray, R. W., M. R. Buchholtz ten Brink, D. C. Gerlach, G. P. Russ, III, and D. L. Jones, Rare earth, major and trace elements in chert from the Franciscan Complex and Monterey Group, California: Assessing REE sources to fine-grained marine sediments, *Geochim. Cosmochim. Acta, 55,* 1875-1895, 1991.

Murray, R. W., H. J. Brumsack, M. T. von Breymann, A. A. Sturz, R. B. Dunbar, and J. M. Gieskes, Diagenetic reactions in deeply buried sediments of the Japan Sea: a synthesis of interstitial water chemistry results from Legs 127 and Leg 128, in *Proc. ODP, Sci. Results, 127/128,* K. A. Pisciotto, J. C. Ingle, Jr., M. T. von Breymann, et al., pp. 1261-1274, Ocean Drilling Program, College Station, TX, 1992a.

Murray, R. W., D. L. Jones, and M. R. Buchholtz ten Brink, Diagenetic formation of bedded chert: Evidence from chemistry of the chert-shale couplet. *Geology, 20,* 271-274, 1992b.

Murray, R. W., M. R. Buchholtz ten Brink, D. C. Gerlach, G. P. Russ, III, and D. L. Jones, Rare earth, major, and trace element composition of Monterey and DSDP chert and associated host sediment, assessing the influence of chemical fractionation during diagenesis. *Geochim. Cosmochim. Acta, 56,* 2657-2671, 1992c.

Murray, R. W., M. R. Buchholtz ten Brink, H. J. Brumsack, D. C. Gerlach, and G. P. Russ, III., Rare earth, major, and trace element composition of Leg 127 sediments, in *Proc. ODP, Sci. Results, 127/128,* K. A. Pisciotto, J. C. Ingle, Jr., M. T. von Breymann, J. Barron, et al., pp. 677-695, Ocean Drilling Program, College Station, TX, 1992d.

Murray, R. W., Chemical criteria to identify the depositional environment of chert: general principles and applications, *Sediment. Geol., 89,* in press, 1994.

Nobes, D. C., R. W. Murray, S. Kuramoto, K. A. Pisciotto, and P. Holler, Impact of silica diagenesis on physical property variations, in *Proc. ODP, Sci. Results, 127/128,* K. A. Pisciotto, J. C. Ingle, Jr., M. T. von Breymann, J. Barron, et al., pp. 3-32, Ocean Drilling Program, College Station, TX, 1992.

Packer, B. M. and R. V. Ingersoll, Provenance and petrology of Deep Sea Drilling project sands and sandstones from the Japan and Mariana forearc and backarc regions, *Sediment. Geol., 51,* 5-28, 1986.

Pedersen, T. F. and N. B. Price, The geochemistry of manganese carbonate in Panama Basin sediment, *Geochim. et Cosmochim. Acta, 46,* 59-68, 1982.

Pisciotto, K. A., Distribution, thermal histories, isotopic compositions and reflection characteristics of siliceous rocks recovered by the Deep Sea Drilling Project, in *The Deep Sea Drilling Project: A Decade of Progress,* edited by J. E. Warme, R. G. Douglas, and E. L. Winterer, pp. 129-147, Soc. Econ. Paleon. Mineral. Spec. Publ. 32, 1981.

Pisciotto K. A. and R. E. Garrison, Lithofacies and depositional environments of the Monterey Formation, California, in *The Monterey Formation and Related Siliceous Rocks of California,* edited by R. E. Garrison, K. A. Pisciotto, C. M. Isaacs, and J. C. Ingle, Jr., pp. 97-122, Society of Economic Paleontologists and Mineralogists, Pacific Section, volume 15, 1981.

Pisciotto, K. A., R. W. Murray, and H. J. Brumsack, Thermal history of Japan Sea sediments from isotopic studies of diagenetic silica and associated pore waters, in *Proc. ODP, Sci. Results, 127/128,* K. Tamaki, K. Suyehiro, J. Allan, M. McWilliams, et al., pp. 49-56, Ocean Drilling Program, College Station, TX, 1992.

Rangin, C., E. A. Silver, M. T. von Breymann, et al., *Proc. ODP, Init. Repts., 124,* 916 pp. Ocean Drilling Program, College Station, TX, 1990.

Smith, R. B., Diagenesis and cementation of lower Miocene pyroclastic sequences in the Sulu Sea, Sites 768, 769 and 771, in *Proc. ODP, Sci. Results, 124,* E. A. Silver, C. Rangin, M. T. von Breymann, et al., pp. 181-194, Ocean Drilling Program, College Station, TX, 1991.

Sturz, A. A., B. B. Hanan, and G. Emanuele, Geochemistry of sediments and interstitial waters from Oki Ridge and Kita-Yamato Trough, Japan Sea, in *Proc. ODP, Sci. Results, 127/128,* K. A. Pisciotto, J. C. Ingle, Jr., M. T. von Breymann, J. Barron, et al., pp. 605-622, Ocean Drilling Program, College Station, TX, 1992.

Tada, R., Compaction and cementation in siliceous rocks and their possible effect on bedding enhancement, in *Cycles and Events in Stratigraphy,* edited by G. Einsele, W. Ricken, and A. Seilacher, pp. 480-491, Springer, 1991.

Taliaferro, N. L., and F. S. Hudson, Genesis of the manganese deposits of the Coast Ranges of California, in *Manganese in California,* edited by P. Jenkins, pp. 217-275, State of California, Department of Natural Resources, Division of Mines, Bull 125, 1943.

Tamaki, K., K. A. Pisciotto, J. Allan, et al., *Proc. ODP, Init. Repts., 127,* 842 pp., Ocean Drilling Program, College Station, TX, 1990.

Taylor, B., K. Fujioka, et al., *Proc. ODP, Init. Repts., 126,* 1002 pp., Ocean Drilling Program, College Station, TX, 1990.

von Breymann, M. T., P. K. Swart, G. W. Brass, and U. Berner, Pore-water chemistry of the Sulu and Celebes Seas: Extensive diagenetic reactions at Sites 767 and 768, in *Proc. ODP, Sci. Results, 124*, E. A. Silver, C. Rangin, M. T. von Breymann, et al., pp. 203-215, Ocean Drilling Program, College Station, TX, 1991.

von Breymann, M. T., H. Brumsack, and K. C. Emeis, Depositional and diagenetic behavior of barium in the Japan Sea, in *Proc. ODP, Sci. Results, 127/128,* K. A. Pisciotto, J. C. Ingle, Jr., M. T. von Breymann, J. Barron, et al., pp. 651-665, Ocean Drilling Program, College Station, TX, 1992.

von Breymann, M. T., K. C. Emeis, and E. Suess, Water-depth and diagenetic constraints in the use of barium as a paleoproductivity indicator, in *Upwelling systems: Evolution Since the Early Miocene,* 15

C. P. Summerhayes, W. Prell, and K. Emeis, pp. 273-284, Geol. Soc. Spec. Pub. No 64, 1993.

Weaver, F. M., and Wise, S. W., Early diagenesis of deep sea bedded cherts, *Antarct. J.,* 228-230, 1973.

Williams, L.S., and D. A. Crear, Silica diagenesis, II. General mechanisms. *J. Sed. Petrol., 55,* 312-321, 1985.

Williams, L. A., G. A. Parks, and D. A. Crerar, Silica diagenesis, I. Solubility controls, *J. Sed. Petrol., 55,* 301-311, 1985.

Marta E. Torres, GEOMAR, 1-3 Wischhofstrasse, D-24148 Kiel, Germany. Current address: College of Oceanic and Atmospheric Sciences, Oregon State University, Ocean. Adm. Bld. 104, Corvallis, OR 97331-5503.

Kathleen M. Marsaglia, Department of Geological Sciences, The University of Texas at El Paso, El Paso, TX 79968, USA.

Jonathan B. Martin, Scripps Institution of Oceanography, La Jolla, CA, 92093-0208, USA. Current address: Department of Geology, University of Florida, 1112 Turlington Hall, PO Box 117340, Gainesville, FL 32611-7340.

Richard W. Murray, Department of Earth Sciences, Boston University, Boston, MA 02215, USA.

Serpentine Bodies in the Forearcs of Western Pacific Convergent Margins: Origin and Associated Fluids

Patricia Fryer[1], Michael Mottl[1], Lynn Johnson[2], Janet Haggerty[3], Steven Phipps[4], and Hirokazu Maekawa[5]

Deposits from serpentine seamounts on the forearc regions of the Mariana and Izu-Bonin convergent plate margins are very similar to some serpentine bodies found in exotic forearc terranes on land. Ideas regarding the formation of serpentine bodies have undergone a profound change as a consequence of seafloor mapping, submersible investigations, and scientific ocean drilling efforts in the western Pacific. These seamounts are of two end-member types, serpentine mud volcanoes and serpentine horst blocks. Various composites of these two end-member types exist. The seamounts are associated with faulting in these forearcs and fluids escaping from conduits or fissures within the edifices have a slab origin. The timing of formation of the serpentine is not known with certainty, but the presence of serpentine in sediments of middle Eocene age recovered from the Mariana forearc suggests that formation of sedimentary serpentine deposits took place at least since that time. Although care must be taken not to generalize the nature of the deposits observed on these seamounts to serpentinite bodies world-wide, it is clear that bodies of sedimentary serpentinite that originally formed in active convergence zones are far more abundant than previously thought. The study of forearc metamorphism and the fluids associated with these seamounts will help to explain the problem of mass chemical balance at convergent margins.

1. INTRODUCTION

Serpentine bodies forming in active tectonic provinces are of several types, but we confine our discussions to those formed in the supra-subduction zone region of convergent plate margins. Of the two types of interoceanic convergent plate margins the active accretionary margins have historically been more thoroughly investigated, but none of these so far have been shown to contain large serpentine bodies. The active nonaccretionary plate margins lack the thick overburden of accreted sediment wedges and thus provide a more direct means of understanding the processes involved in evolution of supra-subduction zone regions. Recent geophysical studies, sampling programs, submersible observations, and deep sea drilling of convergent plate margins in the western Pacific (Figure 1) have changed our models of formation of nonaccretionary interoceanic convergent plate margins [*Hussong and Fryer,* 1981, *Hussong and Uyeda*, 1981; *Bloomer,* 1982; *Taylor,* 1992; *Fryer,* 1992].

In addition to addressing the formation of serpentine bodies in forearcs, these models are important for helping to address broader issues of continental growth and mass balance at convergent plate margins. Accretion of exotic terranes to continental margins is one of the most widely accepted mechanisms by which continents grow and terranes formed in interoceanic convergent plate margins make up some of the larger and more coherent fragments of continental margins. Serpentine bodies are common elements of these terranes, but many have been interpreted as being metamorphosed during or after the process of obduction. The problem of mass balance at convergent margins is one of sums. The volume of volatile constituents

[1]Hawaii Institute of Geophysics and Planetology, University of Hawaii, Honolulu, Hawaii.

[2]Naval Research Laboratory, Washington, D.C.

[3]University of Tulsa, Tulsa, Oklahoma.

[4]Department of Geology, University of Pennsylvania, Philadelphia, Pennsylvania.

[5]Department of Earth Sciences, Faculty of Science, Koba University, Nada, Kobe 657 Japan

Active Margins and Marginal Basins of the Western Pacific
Geophysical Monograph 88

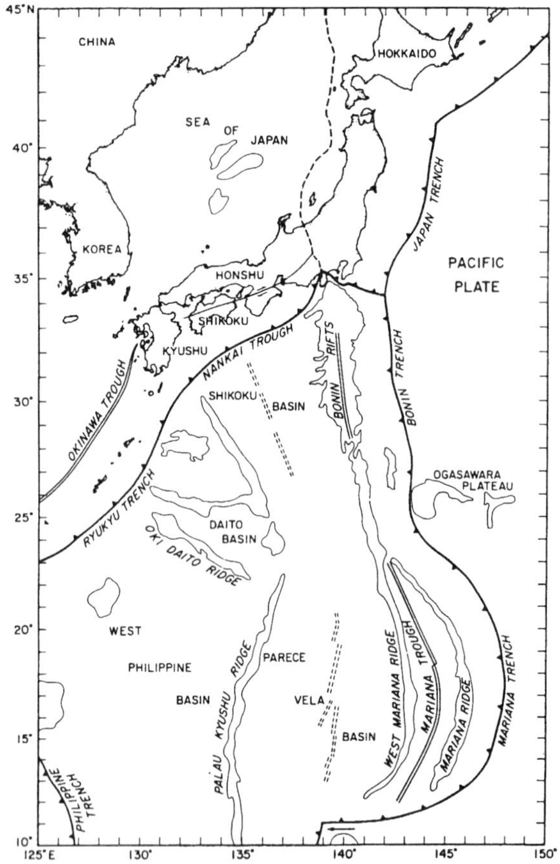

Fig. 1. Bathymetry and geologic features in the Philippine Sea region. Basins and ridges are outlined by the 4-km bathymetric contour, except for the Izu-Bonin arc, West Mariana Ridge, and Mariana arc, which are outlined by the 3-km contour. Barbed lines locate axes of trenches; medium double lines locate active spreading centers; dashed double lines locate relict spreading centers [*Fryer,* 1992].

lost through arc and backarc volcanism is far too small to account for the volumes of plate material subducted and thereby subjected to dehydration reactions. It is not the intent of this paper to address either of these larger issues in detail. However, if they are to be solved it is critical to understand the history of formation of supra-subduction zone lithosphere, the nature of interaction between that lithosphere and the fluids derived from the down-going plate, the degree and extent of metamorphism occurring there, and the nature of tectonic deformation occurring within the supra-subduction zone. This paper focuses principally on the geologic history of the serpentine bodies on the Mariana and Izu-Bonin outer forearc regions. It discusses the type of fluids associated with serpentine seamounts, briefly the potential for fluid flux through the forearc, and presents a comparison of serpentine bodies with fragments of those within forearc terranes exposed on land.

2. BACKGROUND

The most detailed information on serpentinite bodies in currently active forearc regimes has come from study of the Mariana and Izu-Bonin forearcs, at the boundary between the eastern Philippine Sea and the Pacific plate. These convergent plate margins formed through a series of subduction and rifting events since at least the Eocene [e.g., *Karig,* 1971]. On these forearc regions the serpentinite bodies comprise large seamounts located within about 100 km of the trench axis [*Hussong and Fryer,* 1985; *Fryer et al.,* 1985; *Fryer and Fryer,* 1987](Figure 2), and older, sediment draped seamounts are exposed on a narrow ridge on the inner trench wall of the Izu-Bonin system within 50 km of the trench axis [*Honza and Tamaki,* 1885; *Ishii*, 1985; *Fryer and Fryer,* 1987; *Horine et al.,* 1990] (Figure 3). The results of scientific ocean drilling (DSDP leg 60 and ODP legs 125 and 126) have shown that the process of metamorphism of the forearc and formation of serpentine seamounts has been active on the Mariana and Izu-Bonin convergent margins since initiation of subduction [*Fryer,* 1992].

The fluids involved in the metamorphism of the forearc lithosphere to form the serpentine bodies are presumably derived from the down-going plate through a series of dehydration reactions [*Fryer and Fryer,* 1987; *Peacock,* 1990; *Mottl,* 1992]. Fluid flux through the forearc region is thus controlled partly through metamorphism of supra-subduction zone materials and partly through venting associated with serpentine seamounts and fractures. The relative volumes of fluid that are involved in each process are still unknown, but this fluid flux is critical to control of the geochemical budget of the convergent margin. It is also important in the evolution of thermal structure.

Although numerous serpentine seamounts on the Mariana and Izu-Bonin forearcs have been imaged with seismic reflection profiles and have been sampled with dredging, only three on the Mariana forearc and one on the Izu-Bonin forearc have been studied in detail. The differences in each highlight the variability of the formation of these seamounts and the range of their relationships to the tectonic and metamorphic processes active on these forearcs. The three Mariana seamounts are Conical seamount located at 19°33'N, 146°40'E, Pacman seamount located at 19°15'N, 147°05'E, and an unnamed seamount located at 13°45'N, 146'E. The best studied of the Mariana seamounts are Conical and Pacman seamounts located near a large graben in the outer forearc at about 19°30'N (Figures 2 and 4). Dredging all three seamounts yielded serpentinized ultramafic rocks, metamorphosed mafic rocks, manganese crusts with adhering pelagic sediments, and some semilithified, vitric siltstones. SeaMARC II sidescan sonar images and bathymetry (Figure 5) of Conical and Pacman seamounts showed flow features of various types on their flanks [*Fryer et al.,* 1985; *Hussong and Fryer,* 1985; *Fryer and Fryer,* 1987]. Three sites on Conical seamount were drilled on

ODP leg 125, one at the summit, one on the east midflank region and one on the south midflank region.

The Izu-Bonin forearc seamount most thoroughly studied is Torishima forearc seamount, located at 30°57'N, 140°47'E, 45 km from the trench. Dredges from this seamount yielded serpentinized ultramafic rocks, manganese crusts with adhering pelagic sediments, and semilithified, vitric siltstones [*Ishii,* 1985; *Ishii et al.,* 1985; *Ishii et al.,* 1992]. Multichannel seismic profiling, detailed mapping and submersible observations show an edifice mantled with pelagic sediment and lacking evidence of active serpentine flows or fluid venting [*Horine et al.,* 1990; pers. comm., *K. Fujioka,* 1993]. Torishima forearc seamount was also drilled on ODP leg 125, one site on the lower north flank, and one on the lower west flank.

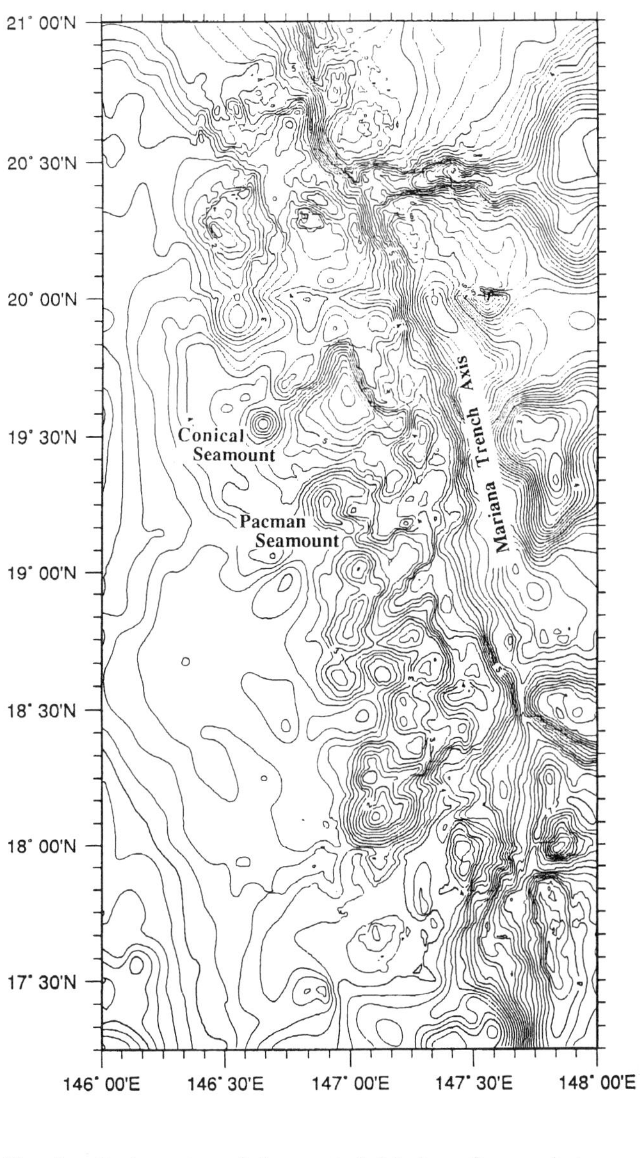

Fig. 2. Bathymetry of the central Mariana forearc between 17°15'N and 21°N. Contour interval 200 m.

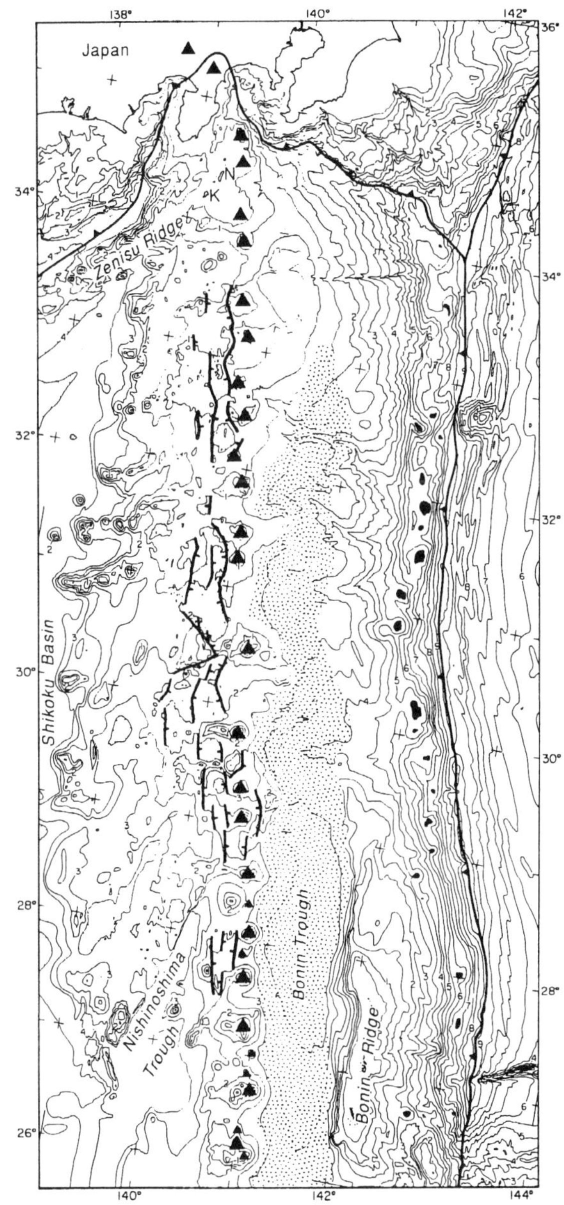

Fig. 3. Bathymetry of the Izu-Bonin forearc showing the major structural elements of the convergent margin. Solid lines with triangles indicate trench axes. Solid lines with hatchures indicate major bounding faults of rift graben. Shaded area indicates forearc rift basins. Triangles indicate active volcanoes and irregular-shaped solid markings indicate forearc serpentine seamount locations. Contours are in 500-m intervals [*Taylor,* 1992].

Fig. 4a. SeaMARC II sidescan image of the Conical and Pacman area. Dark shaded regions indicate high-backscatter portions of the seafloor [after *Hussong and Fryer*, 1985].

3. SERPENTINE SEAMOUNTS ON THE MARIANA AND IZU-BONIN FOREARCS

3.1. *Geologic Settings and Morphology*

Conical seamount lies at the apex of the large forearc graben (Figure 2) along at least two major forearc fault zones [*Newsom,* 1992] at a distance of about 80 km from the trench axis. It is the only one of these seamounts that has probably formed wholly as a consequence of the rise of fluids and fine-grained unconsolidated serpentine and clasts of metamorphosed mafic and ultramafic rock along one or more of these faults. The summit lies at a depth of about 3100 m. Conical seamount covers an area of approximately 700 km^2.

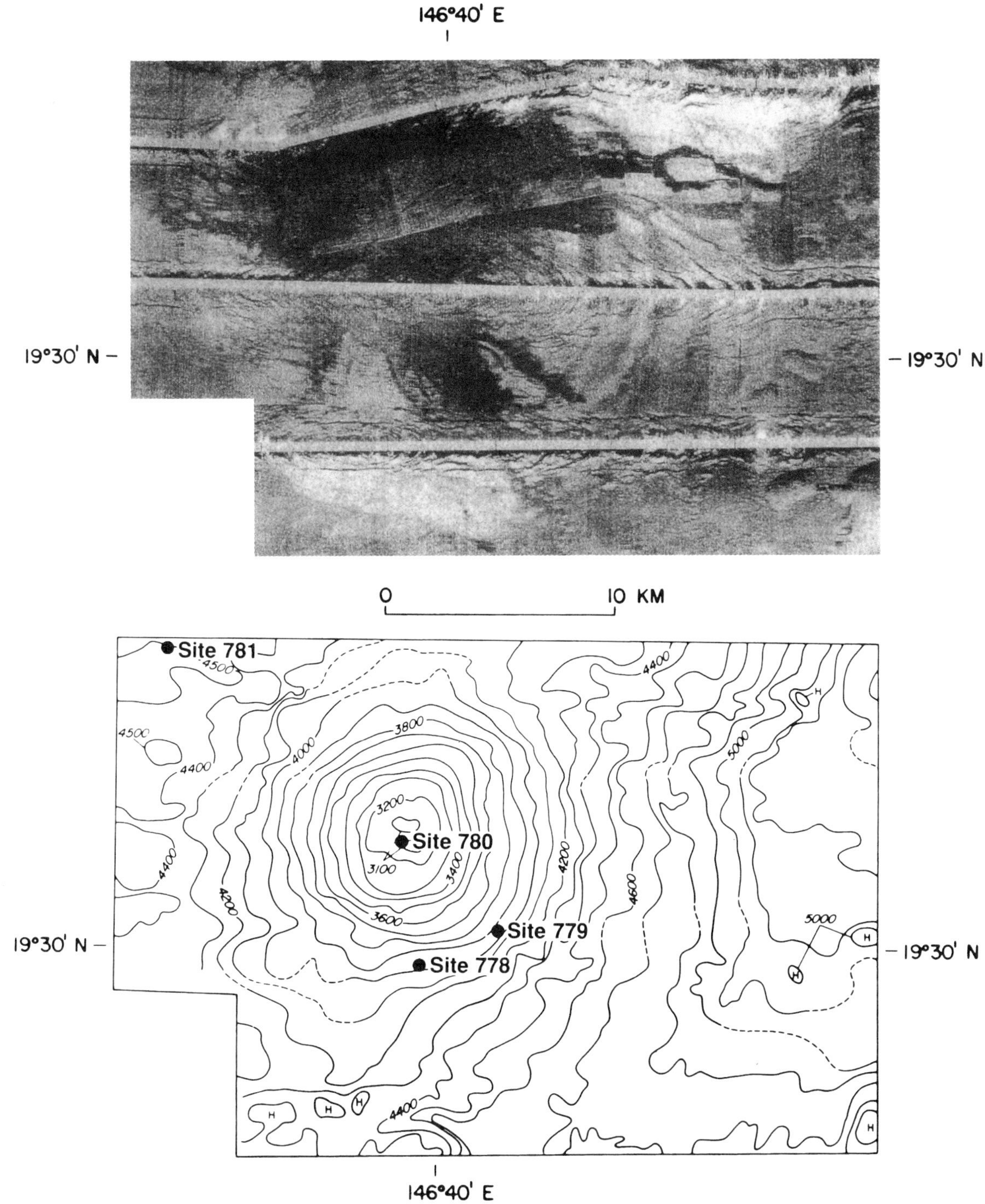

Fig. 4b. SeaMARC II sidescan sonar image (top) and bathymetry (bottom) of Conical seamount showing locations of leg 125 sites 778 through 781. Contours are in 100-m intervals [after *Fryer and Fryer*, 1987].

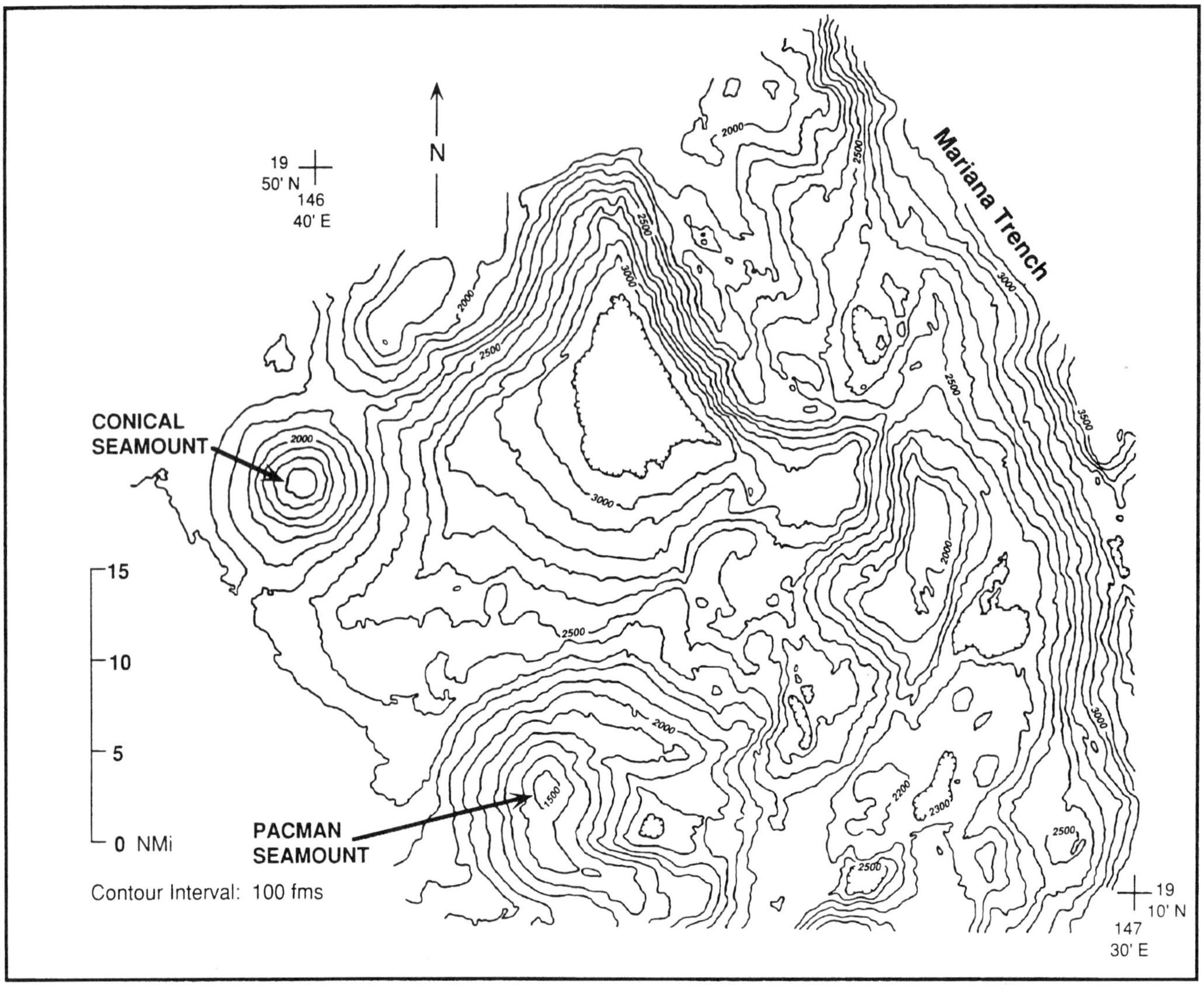

Fig. 5. Bathymetry of the region immediately surrounding Conical and Pacman seamounts showing the position of ODP site 781 and the Mariana Trench. Contour interval 100 fathoms [*Marlow et al.*, 1992]

The morphology of the seamount is different from any of the other well-studied seamounts. It is a smooth-sided, nearly circular edifice with a slightly flattened summit region. It alone shows long sinuous flow features on its flanks on sidescan imagery. Bottom photography and *Alvin* dive observations revealed that the flows are composed of unconsolidated fine-grained serpentine in the regions of high backscatter (dark regions) on the sidescan sonar image (Figure 4). These flows mantle all flanks of the seamount, extending for distances of up to 18 km from the summit. The total area covered by the sinuous flows is approximately 550 km^2.

By contrast Pacman seamount is a crescent-shaped edifice that lies about 45 km southeast of Conical seamount and covers an area of about 1,950 km^2. A single, large (~250 km^2), oval-shaped flow, showing moderately high backscatter on the sidescan sonar image, partially fills a graben on the east flank of Pacman seamount (Figure 4). A region of overlapping oval-shaped high backscatter patches on the southeast flank of the seamount, at the edge of the flank graben, is a region of small serpentine flows. No sinuous flows are apparent on the flanks of the seamount. A series of east-west oriented lineaments cross the southern summit region of the seamount, indicating possible recent fault activity along the southern half of the seamount.

The Mariana forearc seamount, at 13°47'N, 146°03'E, is one of three edifices located in a small cluster at the southern edge of a forearc graben. Thus, its setting is similar to that of Conical and Pacman seamounts. Its summit is about 3100 m deep. The seamount covers an area of about 400 km^2. Reflection profiles across the seamount show deformation of the flanks, possibly faulting. Although chrysotile muds were recovered adhering to dredged samples from the seamount, the lack of detailed bathymetry maps and imagery of the seamount makes it impossible to determine whether serpentine mud flows or debris flows are associated with this seamount.

Torishima forearc seamount is located on a lower forearc terrace of the Izu-Bonin inner trench slope (Figure 6) about 50 km from the trench axis. It is part of the 50 km-wide terrace on the lower part of the inner trench slope. The summit of the seamount is at about 4000 m and the area covered is approximately 100 km^2. Dredging and coring of the summit and upper slopes yielded serpentine muds and clasts of peridotite [*Ishii,* 1985; *Ishii et al.,* 1985]. The seamount is slightly asymmetrical. The western flank of the seamount is relatively smooth, but the eastern flank is cut by several ridges and valleys oriented northeast- southwest. Multichannel seismic data from the seamount show no coherent internal faulting [*Horine et al.,* 1990].

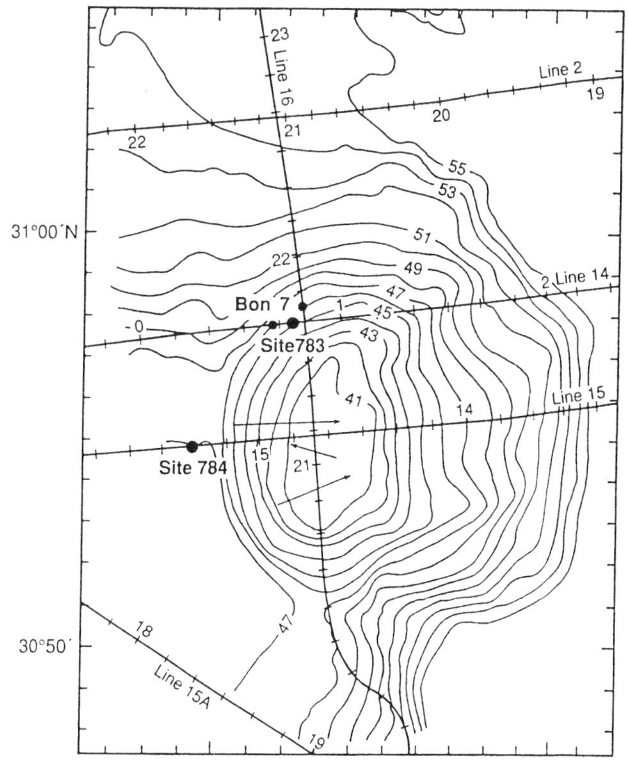

Fig. 6. Bathymetry of Torishima forearc seamount showing locations of leg 125 sites 783 and 784. Contours are in 100-m intervals. Ticked lines locate multichannel seismic profiles [*Horine et al.*, 1990].

3.2. *Internal Structure*

The structure of Conical seamount is inferred from drilling results and from seismic and gravity studies of the edifice [*Fryer et al.*, 1990; *Newsom*, 1992]. The drilling on Conical seamount shows that unconsolidated serpentine with entrained clasts persists to at least 300 m on the flank of the edifice. The seamount shows no coherent internal reflectors on seismic profiles [*Fryer et al.,* 1990], consistent with the interpretation that it consists of thick sequences of protruded serpentine flows. Gravity studies of the region are consistent with this interpretation. Unusually low densities of 2.0 to 2.2 g cm^{-3} for Conical seamount [*Newsom,* 1992] can best be explained if Conical seamount is a mud volcano composed of sequences of unconsolidated serpentine flows.

The structures and textures of the lithologic units recovered from the flank sites of the seamount are very similar to many structures observed in serpentine deposits on land. These textures provide an important insight into the mechanisms of emplacement and subsequent deformation within the edifice. Small-scale original layering within the matrix serpentine is apparently preserved locally, especially at the summit of the seamount. Foliation, either horizontal or oblique, is present especially around major shear bands in the cores (Figure 7). A thin, anastomosing cleavage, similar to scaley-clay cleavages (argille scagliose) has also developed, almost parallel to the foliation. On a microscopic scale, the deformation is not homogeneous and it is not possible to define a uniform field of deformation. Decimeter-size folds may affect the foliation. Anastomosing foliation planes and curving foliation planes around clasts define shear lenses (phacoids) on all scales. The presence of horizontal to gently dipping shear zones suggests that horizontal displacement took place on the flank of the seamount. These displacements can be related to gravitational instability of the upper part of the seamount. On the sidescan sonar image, there is a series of concentric ridges located on the lower flanks of the seamount (Figure 4) that suggest that the seamount has essentially collapsed on itself and flowed outward, forming a series of pressure ridges. The slightly flattened summit of the seamount supports this suggestion. Deformation within the serpentine matrix material is consistent with gentle flow under gravitational forces. It is likely that pore fluids present in the serpentine facilitate this type of deformation.

The *Alvin* submersible dives on Pacman seamount underscored the differences between a mud volcano type of serpentine seamount and a serpentine seamount generated by vertical tectonic processes. One dive on the oval-shaped serpentine flow that partially fills the eastern flank graben on the seamount showed broad undulations, most likely related to the concentric ridges observed on the sidescan sonar image of the flow (see Figure 4). The concentric ridges that outline the flow converge at a point on the eastern side of the summit that is probably the source of the flow. A dive there showed a sedimented surface and no signs of vent

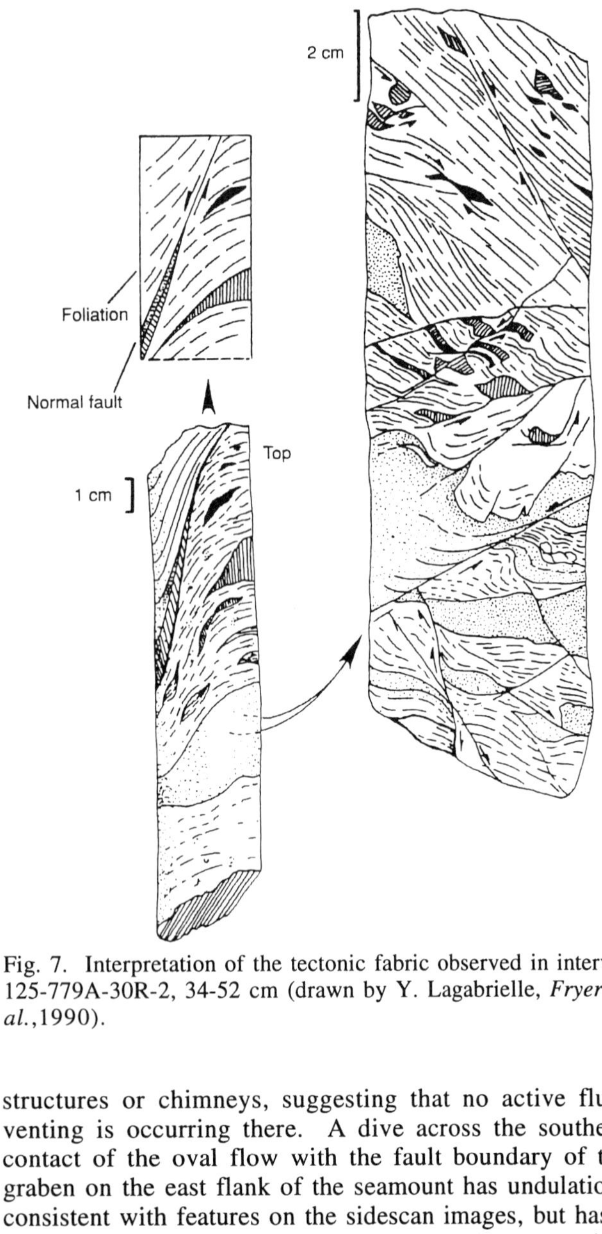

Fig. 7. Interpretation of the tectonic fabric observed in interval 125-779A-30R-2, 34-52 cm (drawn by Y. Lagabrielle, *Fryer et al.*,1990).

structures or chimneys, suggesting that no active fluid venting is occurring there. A dive across the southern contact of the oval flow with the fault boundary of the graben on the east flank of the seamount has undulations consistent with features on the sidescan images, but has a series of secondary drools of serpentine mud no more than 0.5 m in diameter, of varying lengths on the north-facing (downward dipping) contact with the fault boundary. These drools project down-slope like fingers and indicate reactivation of the surface of the oval flow as a consequence of recent activity on the fault that defines the southern wall of the graben. Above the fault contact the character of the exposure changes to a steeper slope exposing primarily serpentinized harzburgite, massive bastite. The bastite is highly fractured, and contains veins of chrysotile. This part of the slope lacks the voluminous unconsolidated serpentine muds characteristic of the deeper part of the slope. Dredge samples from this site show brecciation and infilling of fractures by secondary carbonate.

The lack of fresh serpentine flows, and the heavier sediment cover on Pacman seamount suggest that it is older than Conical seamount. The faulting that formed the graben on the east side of the seamount was probably localized, affecting only the eastern side of the edifice and may have been active recently. The paucity of serpentine muds at depths less than 3,500 m on the south wall of the graben and the massive character of the exposed bastite indicate that the upper portion of the fault exposure represents a horst block of metamorphosed forearc mantle. The suspected source area of the oval flow gives no indication of escaping fluids. A lack of fluid egress accompanying the emplacement of the flow could account for the apparently lower mobility of the flow, as compared with that of flows from Conical seamount. The oval flow was probably a debris flow, possibly generated as a consequence of local tectonic activity. It is unlikely that it represents material squeezed up from great depth, as is probable for the more fluid-charged flows emanating from Conical seamount.

Gravity studies of the region near Pacman seamount show a density of 2.4 to 2.5 g cm^{-3} for Pacman seamount and the range of seamounts that lie to the south, much higher than those of the mud volcano, Conical seamount *Newsom,* 1992]. These densities are consistent with the interpretation that Pacman seamount is a faulted block of uplifted serpentinized peridotite. The dive observations also suggest that at least the upper portion of the seamount is a relatively coherent block of serpentinized peridotite that has been faulted and uplifted by vertical tectonic deformation.

Multichannel seismic reflection studies of Torishima forearc seamount indicate a relatively thin (less than 1 km thick) sequence that lacks internal reflectors mantling the edifice [*Horine et al.,* 1990]. This sequence is underlain by a seismically reflective region, and gravity data indicate that a low-density zone, presumably serpentinized peridotite, underlies the Torishima forearc seamount [*Horine et al.,* 1990]. Torishima forearc seamount, thus, is most likely a composite edifice comprising an uplifted block of low-density, presumably hydrated, forearc mantle through which have erupted serpentine mud flows with entrained clasts. The gravity model proposed by *Horine et al.* [1990] implies that the low density material extends to the decollement, consistent with earlier theoretical calculations [*Fryer and Fryer,* 1987]. *Fryer and Fryer* [1987] suggested that the outer forearc ridge along the inner wall of the Izu-Bonin Trench is related to hydration and associated uplift of the outer portion of the forearc wedge. They suggested that regions of greater hydration might account for the localized highs (seamounts) that are situated along the crest of the Izu-Bonin outer forearc ridge.

3.3. *Fluids and Chimney Materials*

The first carbonate chimney structures, indicative of unusual vent fluids recovered from the Mariana forearc, were retrieved in a dredge haul from about 3800 m on the eastern mid-flank region of the serpentine seamount at 13°45'N. *Haggerty* [1987] noted that compared with fracture-zone carbonates thought to have a dominantly seawater origin, for example, aragonite from the chimney fragments is depleted in Sr (Mariana samples = 7,000-9,400 ppm; fracture zone aragonite = 9,500-11,600 ppm), is enriched in Mg (Mariana = 750-6,300 ppm, fracture zone <300 ppm), and has a significantly lighter carbon isotopic signature (Mariana = -1.2 to -21.2‰ fracture zone = +0.03 to +1.12‰) and a heavier oxygen isotopic signature (Mariana = +5.1 to +7.6‰, fracture zone = +3.16 to +4.87‰. At least two origins are consistent with these values: seawater interactions with ultramafics and dewatering of subducted oceanic lithosphere and/or from the mantle underlying the forearc [*Haggerty,* 1987, *Haggerty,* 1991]. Currently we favor the latter. The presence of the chimney structures on the flank of the seamount implies that as deformation of the flank takes place, pathways are initiated that provide egress for slab-related fluids.

The first observations of venting fluids and of active carbonate chimneys were from Conical seamount. *Alvin* dive observations on the southwest side of the summit of Conical seamount (Figure 8) show fluids actively seeping from carbonate and silicate chimneys (Figures 9 and 10) and related structures [*Fryer et al.,* 1987; *Fryer et al.,* 1990; *Haggerty,* 1991]. Recent *Shinkai 6500* submersible observations on the seamount indicate that carbonate chimney structures are still forming at the summit [pers. comm., *K. Fujioka,* based on dive observations, *Shinkai 6500,* dive 179]. At the depths observed (~3100 m), the carbonate chimney structures should dissolve rapidly unless they were actively being formed [*Haggerty,* 1987; 1991]. The chimney materials from Conical seamount are identical to those recovered from the seamount at 13°45'N [*Haggerty,* 1987; 1991] in mineralogy, although the carbon and oxygen isotope compositions of these samples suggest a slightly different fluid composition (Figure 11). The composition of fluids collected with *Alvin* from a chimney structure on Conical seamount is extremely unusual; high pH (9.28 vs. 7.72 for ambient seawater), high alkalinity (5.53 vs. 2.41 meq/L) and enrichment in CH_4 (1000 vs. 2.1 nM), SiO_2 (0.75 vs. 0.12 mM), SO_4 (30.4 vs. 28.6 mM), and H_2S (2.1 μM vs. not detected) [*Fryer et al.,* 1990]. The CH_4 and H_2S detected in the vent fluids could not have been derived biogenically. The outer Mariana forearc has only a small volume of sedimentary cover [*Hussong and Uyeda,* 1981], with organic carbon contents typically less than 0.3% [*Schorno,* 1981]. Thus, the methane and sulfide in the Mariana vent fluids are unlikely to have formed by the biogenic processes associated with accretionary complexes, such as those described by *Kulm et al.* [1986]. Rather, they were probably generated during serpentinization reactions [*Fryer et al.,* 1990].

Of the three sites drilled on Conical seamount, the one at the summit has the most unusual pore fluid compositions

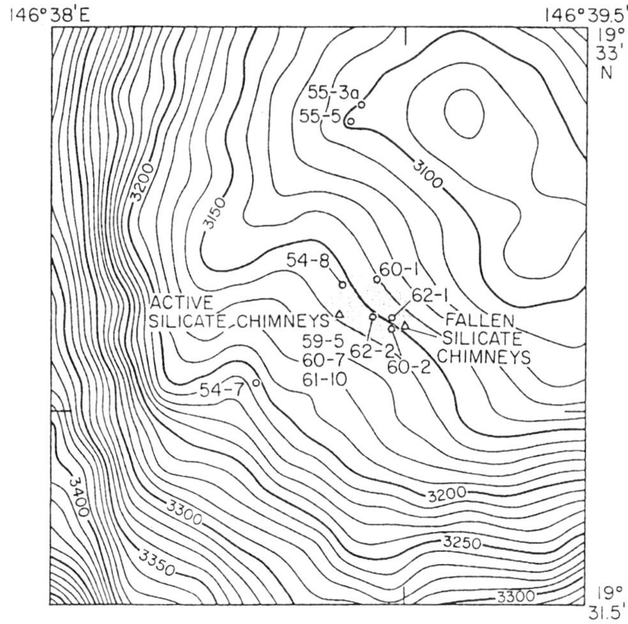

Fig. 8. Bathymetry of the summit region of Conical seamount showing the locations of various chimney structures (open circles and triangles), and the vent region (shaded) inferred from the presence of the chimney structures [after *Fryer et al.*, 1990].

[*Mottl,* 1992]. The pore fluids from the summit drill samples have less than one-half the chloride and bromide of seawater. They also have the highest pH ever measured in samples of pore waters from the deep ocean (maximum value 12.6). Concentrations of H_2S in these samples reach 2 mmol/kg and of ammonia reach 270 μmol/kg. The pore fluids are enriched relative to seawater in alkalinity (×26), sulfate (×1.7), K (×1.5), Rb (×5.6), and B (×10). They are highly depleted in Li, Mg, Ca, Sr, and ^{34}S, and less so in Si, Ba, and Mn. They are depleted in Na, but the Na/Cl ratio increases to nearly twice the seawater value. These compositions are detected in the upper few meters of the serpentine muds. The large gradients in these pore fluid compositions in the surficial sediments are probably maintained by upflow with velocities ranging from a few millimeters to 10 cm per year through the central conduit of the seamount [*Mottl,* 1992]. The considerable variation in composition of the pore fluids at a given depth among these four holes (drilled within 150 m of each other) indicates that the distribution of the flow is variable. Temperature measurements in the summit holes yielded a value of 13.5°C at 58 mbsf in hole 780D. Extrapolation of discrete measurements of temperature in hole 780C yields a value of 3.7°C at the same depth. This suggests that warmer water from the bottom of the deeper hole, hole 780D, welled up into the hole after it was drilled [*Fryer, Pearce, Stokking, et al.,* 1990], supporting the interpretation that a summit conduit supplies the unusual fluids. This upflow of fluids is

Fig. 9. *Alvin* external camera photograph of carbonate chimneys near the summit of Conical seamount, Note the corroded appearance of the surfaces's structure. The large chimney in the left foreground is approximately 1.5 m tall [*Fryer et al.*, 1990].

Fig. 10. *Alvin* external camera photograph of a silicate chimney near the summit of Conical seamount. The structure is approximately 2 m tall [*Fryer et al.*, 1990].

also likely responsible for the persistence of aragonite in the surficial serpentine muds.

Pore fluids having low chlorinity have been recovered from other convergent margins during DSDP and ODP efforts [*Mottl,* 1992]. The low chlorinity of these fluids has been attributed to decomposition of hydrocarbon gas hydrates, membrane filtration, dewatering of clays and opal, and the possible presence of fossil freshwater aquifers. The decrease in chlorinity of the pore fluids in hole 780D is larger than any observed in sediment pore fluids in convergent margins. Mottl [1992] suggests the source for the pore fluids at the summit site are dehydration reactions at depth within the subducted Pacific plate.

The $^{87/86}Sr$ isotopic ratios [*Haggerty and Chadhuri,* 1992] of interstitial waters from this site average 0.70768 (± 0.000011). This value is 0.001569 to 0.001531 lower than the range of values for seawater and also indicates a non-seawater source for the pore fluids.

High levels of methane and the presence of ethane, propane, acetate, and organic acids, were measured in the samples [*Haggerty and Fisher,* 1992; *Mottl,* 1992]. Analyses of the Conical seamount pore waters for organic acids by *Haggerty and Fisher* [1992] indicate that an organic component may be present in the fluids or in the solid materials entrained within the rising serpentine. Presence of these species indicates the existence of more complex organic precursors and, of the pore fluid samples analyzed from leg 125, only those associated with serpentine muds contain organic acids [*Haggerty and Fisher,* 1992]. The ultimate origin of the precursor material cannot be determined unambiguously from the analytical data. That the highest concentrations of organic acids are found at the summit site and that metasediment clasts and slightly higher organic carbon contents also occur in the summit samples suggests that the production of the organic acids results from organic precursors entrained within the rising serpentine muds. The data, however, are also consistent with both the precursory compounds being derived from the down-going slab and with production of organic acids through alkaline hydrolysis of ester linkages within the serpentine muds. The organic acids may have a thermal origin from organic precursors in the downgoing slab or entrained within the serpentine muds, or they may be produced chemically in the rising fluids within serpentine muds of Conical seamount.

By comparison with the pore fluids of the summit site 780, those of the flank sites 778 and 779 show an admixture of a deep source component, possibly from the subducted slab, with one that is derived through the interaction of seawater with peridotite [*Mottl,* 1992]. The pore fluids from site 779 are richer in the slab component.

Although Pacman seamount lacks chimney structures at its summit, there are small carbonate chimneys forming along fractures on its flanks and evidence of fluid migration through fault zones that cut through the seamount. A recent *Shinkai 6500* submersible dive (178) by Fryer discovered carbonate chimney structures on an extension of the fault scarp approximately 8 km east of the *Alvin* dive site (Figure 12). At this location, on the southeastern arm of the seamount, there are two slight depressions in the center of the overlapping high backscatter regions on the up-thrown side of the fault which represent sites from which unconsolidated serpentine flows emanated. The *Shinkai 6500* submersible dive on this feature revealed a highly bioturbated sediment cover of about 5 cm over a terrane composed of serpentine flows and large clasts. The line of small carbonate chimney structures lies near the apparent

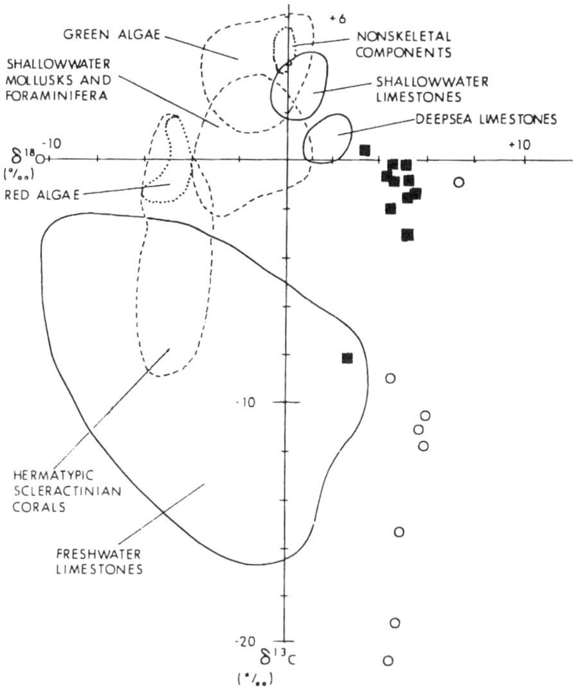

Fig. 11. Comparison of carbonate isotopic compositions of chimney samples from Conical seamount (solid circles) and from a serpentine seamount at 13°45'N on the Mariana forearc [*Haggerty*, 1991].

center of the westernmost depression, on a series of cracks in the seafloor. Preliminary results of analysis of pore fluids from sediments recovered by *Shinkai 6500* push cores from near the chimney sites show compositional tendencies similar to those of Conical seamount summit fluids (pers. comm. *Mottl and Wheat,* 1994). Analyses on these fluids and on the associated sediments are underway. The secondary carbonates deposited in the brecciated samples of the fault zone on the southeast side of Pacman seamount indicate that fluids escaped along the fault zone during brecciation of the bastite. Some of the samples contain acicular radiating carbonate crystals centered on vugs.

The composition of pore fluids from the flanks sites on Torishima forearc seamount, sites 783 (north flank) and 784 (west flank), are similar to seawater with respect to Br, Na, and Na/Cl. Relative to seawater, the pore fluids have low alkalinity, sulfate, K, and B, and slightly lower Rb and Li. They have low Mg and low Si relative to seawater, but are enriched in Ca, Sr, Ba, and ^{34}S. The pH is less than 10. Methane content is low, and no higher hydrocarbons were detected. Thus, if Torishima forearc seamount was constructed in part by serpentine mud flows containing fluids having a deep source, the pore water data of *Mottl* [1992] and *Mottl and Alt* [1992] indicate that those deep fluids have since been supplanted by seawater. However, a

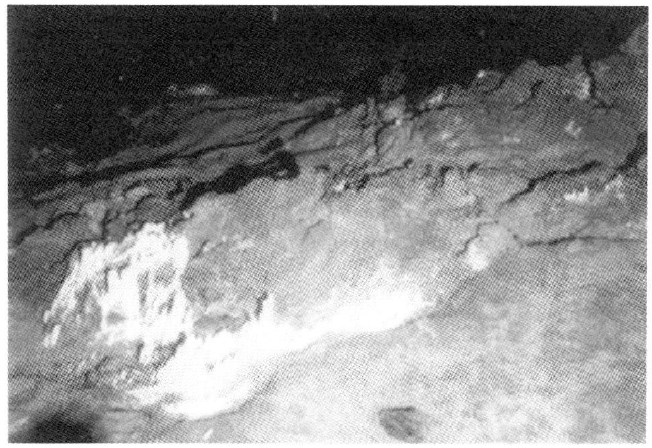

Fig. 12. *Shinkai 6500* external camera photograph of small carbonate chimneys forming along a series of small cracks in the seafloor on the southeastern arm of Pacman seamount.

possible remnant of a slab-related fluid component exists in isotopic composition of the pore fluids. *Haggerty and Chadhuri* [1992] determined that the average $^{87/86}Sr$ isotopic ratio of the site 784 pore fluids is 0.70769 (± 0.000010). This value is 0.001551 to 0.001521 lower than the range of values for seawater and suggests the possibility of multiple sources for these interstitial waters [*Haggerty and Chadhuri,* 1992]. *Haggerty and Chadhuri* [1992] note that it is difficult to distinguish between the specific igneous source or sources that cause the offset from present-day seawater, but they suggest that it is possible that the Torishima forearc seamount fluids may involve a subduction-related component. The seamount is draped by arc sediments that are at least 10 million years old and *Shinkai 6500* dives on the summit region of the seamount showed no recent serpentine flows and no evidence of current vent sites [pers. comm., *K. Fujioka,* 1993].

3.4. *Rheology of Serpentine Matrix*

Tests of the stress-strain behavior of serpentinite muds from hole 780D were performed by *Phipps and Ballotti* [1992] using a Wykeham Farrance torsion-vane (torvane) apparatus onboard the *Joides Resolution.* The results reveal that the serpentinite muds are highly nonideal weak plastic materials having ultimate strengths ranging from 1.3 to 37.5 kilopascals (kPa), and averaging 11.3 kPa (Figure 13). Most serpentinite muds continued to deform at a constant rate at maximum (ultimate) strength. Although no direct determinations of yield strength were made, observations of the behavior of these materials led *Phipps and Ballotti* [1992] to suggest several conclusions: (1) Ultimate strengths of the materials suggest that they could transport blocks of serpentinized peridotite (density approximately 2.6-2.7 g cm^{-3}) as large as 20 m upward against the force of

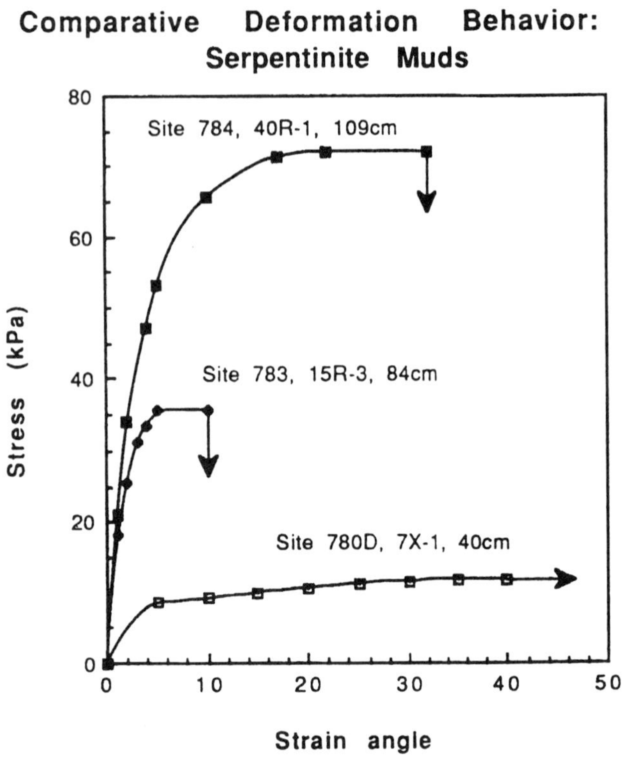

Fig. 13. Rheological behavior of serpentine muds at ODP sites 789, 783, and 784.

gravity. (2) The low yield strengths of the material suggests that unless it continues to well upward from within the conduit of the seamount, the blocks will fall back downward into the conduit. (3) The low strengths are consistent with the interpretation that the seamount is rheologically more like a mud volcano than a salt diapir (salt has an ultimate strength of approximately 10^4 to 10^5 kPa). These findings are also consistent with the observations of shear textures observed in the matrix (described above) that suggest continuing gravitational deformation of the seamount, flattening of the summit and the development of pressure ridges on the flanks.

Rheologically, the sheared, phacoidal serpentinite matrix of Torishima forearc seamount is weaker than the overlying pelagic sediments, but significantly stronger than the serpentine at site 780, and the mode of strain is different (brittle failure vs reaching an ultimate strength)[*Phipps and Ballotti,* 1992]. This rheology reflects the greater dewatering and compaction of the Torishima forearc seamount serpentine matrix.

3.5. *Matrix Lithology*

The lithology of the materials recovered on Conical seamount during leg 125 drilling traces a complex history of episodic protrusion of serpentine flows. The flows tap a variety of sources all of which have been influenced by the presence of deep-seated fluids [*Fryer,* 1992; *Fryer and Mottl,* 1992]. At site 780, on the summit of Conical seamount, faintly layered unconsolidated fine-grained serpentine with occasional clasts of serpentinized ultramafics were recovered to a maximum depth of 163.5 mbsf. Most of the serpentine also contains delicate authigenic aragonite needles. Filamentous opaque debris interpreted as remnants of bacterial mats are found in smear slides throughout the serpentine. Intercalations of biogenic sediments indicate periods of multiple deposition at the seafloor, suggesting periodicity of serpentine protrusion events. The maximum age of these sediments is middle Pleistocene(?), based on fossil assemblages.

Mineralogical differences between the serpentine mud samples from the summit holes, determined by shore-based analysis [*Fryer and Mottl,* 1992], and minor differences between pore fluid compositions [*Mottl,* 1992] from the summit holes suggest a subtle spatial variation near the conduit. An increase in the diversity of the lithology of the muds, based on the shipboard descriptions of the smear slides, begins in core 125-780C-2R at a depth of about 100 mbsf. This change in lithology marks a boundary between two compositional units within this summit hole [*Fryer and Mottl,* 1992]. Members of the rare sjogrenite group (mostly hydroxide carbonate hydrates) and certain other rare secondary phases are present in the muds from the summit holes [*Fryer and Mottl,* 1992; *Heling and Schwarz,* 1992], as well as from the flank sites. The presence of sjogrenite group minerals implies that the alteration of the serpentine muds to form the minerals must be taking place *in situ* [*Fryer and Mottl,* 1992]. The unusual pore waters, which have been suggested to have derived from a deep source [*Fryer et al.,* 1990; *Mottl,* 1992; *Haggerty and Chadhuri,* 1992; *Haggerty and Fisher,* 1992; *Mottl and Alt,* 1992] associated with the rising serpentine muds, likely maintain the stability conditions of these minerals within the mud flows. Iowaite ($Mg_4Fe(OH)_8OCl \bullet xH_2O$), one of the sjogrenite group minerals, requires interaction with seawater to provide the chloride necessary for its formation [*Heling and Schwarz,* 1992]. It was hypothesized that formation of iowaite might explain the low chlorinity of the summit fluids [*Heling and Schwarz,* 1992]. However, iowaite is present in such low concentrations and so sporadically that it is unlikely that this can be the principal explanation for the low chlorinity of the fluids.

Two sites were drilled on the flanks of Conical seamount, sites 778 and 779 (Figure 4). Two pebbles from site 778 contain biogenic components and yielded ages from Neogene to lower Pleistocene(?). The ages estimated for samples from site 779 were lower Pleistocene(?) to lower Pliocene(?). The maximum depth drilled at site 778 was 107.6 mbsf and at site 779 was 303.0 mbsf. The unconsolidated fine-grained serpentine and metamorphic rock clasts recovered there are very similar to those of the summit holes. There are three principal differences in the material recovered from these holes and that from the summit site. The degree of dessication of the unconsolidated serpentine matrix increases with depth on the flank sites. The composition of the matrix [*Fryer and Mottl,* 1992; *Heling and Schwarz,* 1992; *Lagabrielle et al.,* 1992] shows subtle variability suggesting different sources of the serpentine and possibly a periodicity of protrusion events [*Fryer,* 1992]. There were at least 3 events at site 778 and 6 events at 779 [*Fryer and Mottl,* 1992].

Two sites were drilled on the north and west flanks of Torishima forearc seamount during ODP leg 125 (Figure 6). The maximum depth penetrated was 425.3 mbsf at site 784, the western flank site. A matrix of unconsolidated to semiconsolidated serpentine enclosing clasts of serpentinized ultramafic rocks, very similar to that recovered on Conical seamount, underlies a thick sequence of pelagic sediments (maximum age - lower Pliocene). The principal differences between the matrix of Conical seamount and Torishima forearc seamount is the greater degree of dessication and a more complex veining.

The serpentine recovered was primarily phacoidal, sheared serpentine matrix and blocks of serpentinized harzburgite, similar to the recovery from Conical seamount on the Mariana forearc. However, the matrix structures indicate syn- and post-emplacement deformation. Vein textures are well-developed in some places within the serpentinite matrix. This is important because it provides unambiguous evidence that serpentinization occurred in the unlithified serpentine matrix. In several intervals serpentinite clasts are cut by intricate, anastomosing networks of veins that pass out into narrow channels of finely phacoidal serpentinite matrix (e.g., Figure 14). These intervals apparently preserve a late stage of the fragmentation of large clasts by serpentine veins, continuing hydration and softening of the large clasts, and shearing and incorporation of the clasts into the phacoidal matrix. This texture is less prevalent at Conical seamount, although some clasts of ultramafic rock in cores from Conical seamount do show secondary veining that is coherent with veining in the matrix. It is possible that some of the deposits drilled at Torishima forearc seamount are massive blocks of serpentinite that were emplaced by faulting and subsequently subject to further serpentinization *in situ*. This interpretation is consistent with Torishima forearc seamount being a composite of a fault block partially covered by later serpentine flows.

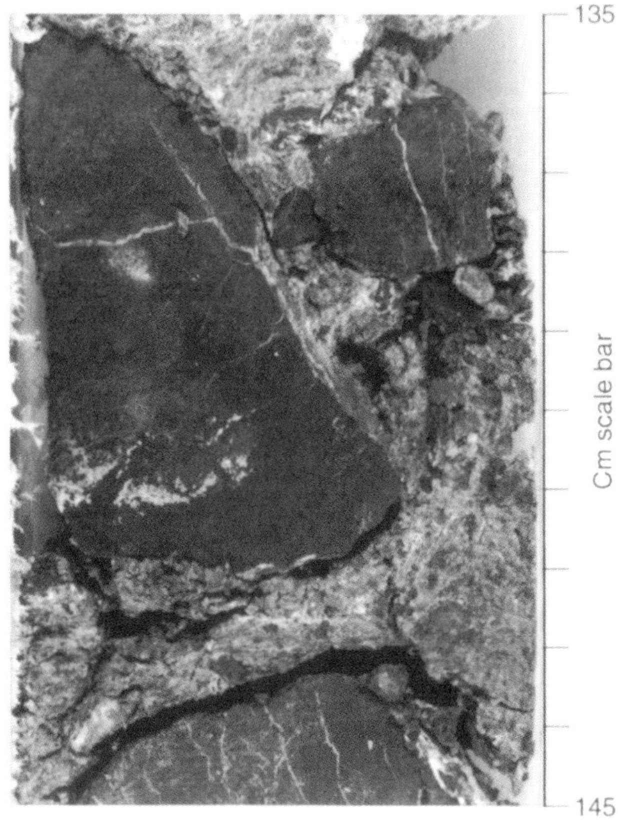

Fig. 14. Core photograph from interval 125-784A-38R-1, 135-145 cm showing the clast in matrix texture of the material. Note the white veins of chrysotile in the dark triangular clast in the middle of the core interval. This vein is continuous into the matrix surrounding the clast. Serpentinization of the clast and the matrix took place through this interval.

3.6 *Metamorphic Rock Clasts*

The composition of the ultramafic samples is dominantly harzburgite and minor dunite [*Ishii et al.,* 1992; *Saboda et al.,* 1992]. Clast types from the flank of the seamount are more diverse in composition than are those from the summit. In general, the compositions of the clasts included in the serpentine flows suggest the flows tap several sources regions, consistent with the interpretation of the variability in the matrix serpentine. Enrichment in light rare-earth elements suggests that a subsolidus enrichment after melting has occurred [*Parkinson et al.,* 1992a,b]. This pattern of enrichment is similar to that observed in ophiolites generated above subduction zones [*Parkinson et al.,* 1992a]. Microstructures in the peridotites from Conical seamount indicate a multistage history of formation [*Girardeau and Lagabrielle,* 1992]. We discuss this history in detail later, as it bears on the origin and emplacement mechanism of the seamount.

Mafic clasts recovered in the serpentine muds of Conical seamount are less than 9% by volume of the total rock recovered [*Johnson,* 1992; *Maekawa et al.,* 1992]. These mafic clasts are dominantly metabasalt or metadiabase, but also include boninite and cumulate gabbro. Based on trace element concentrations, the mafic rocks comprise several geochemical groups, island-arc tholeiite, boninite, mid-ocean ridge basalt, basalt intermediate between island arc tholeiite, and mid-ocean ridge basalt [*Johnson,* 1992]. The presence of basalts having mid-ocean-ridge, trace-element signatures requires that fragments of oceanic plate were accreted to the outer portion of the Mariana forearc [*Johnson and Fryer,* 1990]. Metamorphism of the mafic samples took place principally under low-temperature and low-pressure conditions [*Maekawa et al.,* 1992]. Some of the mafic rocks from site 778, however, contain blueschist-facies metamorphic minerals and are considered to have been derived from a deep (12-18 km) portion of the forearc [*Maekawa et al.,* 1992, 1993, this volume].

Rocks dredged and collected with the *Alvin* submersible from Pacman seamount are similar to the ultramafic rocks collected from Conical seamount. To date, no mafic clasts have been recovered from Pacman seamount.

The rocks retrieved from Torishima forearc seamount are primarily harzburgite and dunite [*Ishii et al.,* 1992] that differ little in composition from those of the Mariana forearc samples recovered at Conical seamount [*Fryer et al.,* 1990; *Saboda et al.,* 1992]. Minor amounts of metamorphosed basalt were recovered, and one sample of a metavolcaniclastic rock was recovered from each of the holes [*Maekawa et al.,* 1992, 1993]. The grade of metamorphism of the Torishima forearc seamount samples is prehnite-actinolite facies which requires a low-temperature/ low-pressure regime (200°-300°C and 0-6 km deep) [*Maekawa et al.,* 1992, 1993].

4. COMPARISON TO SERPENTINE BODIES ON LAND

Lockwood [1971] was the first to recognize that numerous serpentinite bodies exposed on land had originated by a protrusion mechanism prior to tectonic emplacement as exotic terranes. He suggested that the bedded serpentinite "sandstone and shale", similar in texture to the flows on Conical seamount, and the "chaotic breccia (olistostromes)", similar to the deposits recovered on Torishima forearc seamount and perhaps to the oval-shaped flow on Pacman seamount, were formed by "sedimentary" processes. He recognized that the deposits were emplaced during what is now recognized as the early phases of subduction at convergent plate margins ("early in the eugeosynclinal cycle"), and cited evidence for rapid submarine landslides, mudflows, or turbidity currents. He was the first to suggest that sedimentary serpentinites are much more abundant in peridotite bodies than previously thought and that the protrusions responsible for their emplacement were coeval with, instead of younger than, their enclosing sedimentary or metasedimentary rocks. This hypothesis was a marked departure from the previous interpretations of syn- or post-emplacement deformation and serpentinization mechanisms then suggested for the formation of most serpentinites in alpine peridotites. *Lockwood* [1971] examined numerous serpentinite exposures and highlighted the detrital deposits in each. He suggested that the existing interpretations requiring igneous intrusions or extrusions, and syn- or post-emplacement serpentinization were not consistent with the observations. *Lockwood* [1972] suggested that subduction-related fluids were associated with emplacement of the serpentine deposits. His suggestion that the fluids rising with the serpentine were derived by dehydration reactions are borne out by analysis of the pore fluids from Conical seamount. Although his interpretation that the serpentine bodies represent fragments of the subducted plate is inconsistent with the more detailed recent petrologic studies of Mariana and Izu-Bonin protolith [*Bloomer,* 1982; *Bloomer and Hawkins,* 1983; *Parkinson et al.,* 1992b; *Ishii et al.,* 1992], his idea that emplacement occurred during subduction is supported by the observations of serpentine deposits on the both the Mariana and Izu-Bonin forearcs. A reevaluation of several serpentinite bodies on land, particularly those in California, strengthens the interpretations of Lockwood and provides additional evidence for the existence of serpentine mud volcanoes in at least one of these serpentine bodies.

Probably the best analog to the Conical seamount occurrence is the Coalinga asbestos deposit in the New Idria serpentinite of California. The southern end of the San Benito mountains in California is composed of serpentinized ultramafics intruded by later volcanics and is known as the New Idria serpentinite body. A detailed description of the principal asbestos ore deposit, located near the southern boundary of the New Idria serpentinite, approximately 56 km northwest of Coalinga, is presented in *Mumpton and Thompson* [1975]. This deposit, by contrast with most asbestos ores world-wide, is not a cross-fiber or slip-fiber vein deposit in massive serpentinite, but rather consists of friable masses of matted chrysotile surrounding clasts of serpentinitzed ultramafics. The "highly sheared and pulverized" condition of the ore is similar to the texture of the deposits recovered at the Conical seamount on ODP leg 125. The interpretation of the genesis of the mass according to *Mumpton and Thompson* [1975] was by syn- or post-emplacement shear deformation of the serpentine body. *Coleman* [1957] had suggested that the body was serpentinized at depth and rose diapirically by "cold" injection into the overlying Franciscan formation. *Mumpton and Thompson* [1975] suggested that the mass was extensively sheared and serpentinite blocks were broken and crushed against one another to form a thick paste that was pulverized and smeared out. They note that the Coalinga deposit is predominantly chrysotile (over 80% chrysotile vs less than 10% in Alpine-type serpentinite) and note that the predominance of chrysotile and the abundance of brucite (7-8%) are particular to the Coalinga ore. Other portions of the New Idria serpentinite that are also highly sheared lack abundant brucite. *Mumpton and Thompson* [1975] attribute the presence of the chrysotile and brucite to groundwater interaction with the pulverized serpentinite and suggest that the formation of the ore was as a consequence of dissolution of lizardite and antigorite presumed present in the original pulverized serpentinite and reprecipitation as chrysotile. They describe several other serpentine ore deposits and suggest a similar mechanism might explain the origin of all ore deposits composed largely chrysotile. These authors clearly recognized the genetic link between the tectonism and fluid flux through the serpentinite. They were unaware at that time that the formation of chrysotile can take place in active forearcs. If we substitute subduction-related fluid for the groundwater, invoke serpentinization of the peridotite in the original active forearc region in which the protolith was generated, and attribute the deformation of the serpentinite body to tectonic deformation in the active forearc, the mechanism could explain the formation of deposits associated with the serpentine seamounts on the Mariana and Izu-Bonin forearcs.

The whole of the New Idria serpentine body outcrops over an 80 km^2 area; however, the Coalinga open pit asbestos mine is yielding unconsolidated, friable chrysotile from a semicircular body that is approximately 2 km in diameter. The highly disaggregated nature of the ore at this location, the outcrop pattern, and the fact that the body extends to several kilometers depth suggest a remarkable similarity to the interpreted internal structure of the mud-volcano Conical seamount. The descriptions of the ore given in *Mumptom and Thompson* [1975] are consistent with the ore having formed as rising unconsolidated serpentine mud entraining clasts of serpentinized ultramafics. The ore contains acicular calcite crystals similar in size and distribution to the aragonite crystals observed distributed throughout the muds of the summit holes on Conical seamount. These crystals probably formed by precipitation from pore fluids. We suggest that the ore represents the throat of a serpentine mud volcano through which rose unconsolidated serpentine muds, similar to those of Conical seamount, and through which rose fluids that prompted precipitation of carbonates. In further support of this interpretation, the deposit contains various sjogrenite minerals which form as a consequence of interaction between "fresh" water and brucite or serpentine. These minerals are present in the upper zone of surface weathering, but have also been discovered as deep as 30 m below the surface. *Mumpton and Thompson* [1975] interpreted the presence of these minerals well below the normal weathering zone to indicate that such materials oxidized as a consequence of "surface conditions being brought beneath the surface along faults or large fracture zones." The highly friable serpentine deposit probably cannot support brittle fracture. The presence of these minerals in the cores from the summit of Conical seamount indicates that they form during active fluid flow through the serpentine muds rising in the throat of the mud volcano.

Analogous serpentinite deposits to those of the Mariana and Izu-Bonin seamounts also occur at Wilbur Springs, California [*Carlson,* 1984]. In general the serpentinite of this region is thought to be the metamorphosed basal peridòtite of the Coast Range ophiolite upon which the sediments of the Great Valley Sequence were deposited [*Bezore,* 1969; *Bailey et al.,* 1970; *McLaughlin,* 1974; *Evarts,* 1977; *Hopson and Frano,* 1977; *Hopson et al.*, 1981; *Carlson,* 1984; *Phipps,* 1984]. Lens-shaped bodies of serpentinite debris entirely enclosed within sediments of the Great Valley sequences were interpreted in a variety of ways [*Taliaferro,* 1943; *Lawton,* 1956; *Blake, et al.,* 1969; *Rich,* 1968; *Moiseyev,* 1970; *Maxwell,* 1977; *Phipps et al.,* 1979]. Field and petrographic evidence suggested to *Carlson* [1984] that these serpentinite deposits originated as serpentinite flows generated in the forearc environment. Citing the discovery of the serpentine flows [*Hussong and Fryer,* 1985] on the Mariana forearc serpentine seamounts, *Carlson* [1984] suggested a similar origin for the serpentine deposits at Wilbur Springs; however, her interpretation required a diapiric emplacement of the serpentinite on the outer forearc ridge of the proto-Coast Range ophiolite body and restricted the protrusion of serpentine flows to a relatively short period of 10 million years. As we discuss below, the "diapiric" origin of the serpentinite seamounts must be viewed with caution. We also note that the presence of serpentine in the Mariana and Izu-Bonin forearcs appears to have been continuous from the initiation of subduction through the present day.

5. ORIGIN AND EMPLACEMENT OF FOREARC SERPENTINE SEAMOUNTS

The history of the peridotite protolith that gave rise to the serpentine seamounts is best delineated by the combined data from compositional analyses and microstructural studies. We have already discussed composition of the peridotite and noted the characteristics of a supra-subduction zone region. We present a summary of the microstructural studies of the peridotites drilled from Conical seamount described by *Girardeau and Lagabrielle* [1992] and discussed in the drilling synthesis of *Fryer* [1992]. Events recorded in the microstructures and textures of the peridotites from Conical seamount are presented in terms of a sequence of metamorphic, anatectic, and structural phenomena: (1) initial formation of the peridotite under high-temperature asthenospheric conditions, such as are prevalent in the upper mantle of most ophiolite complexes (either at mid-ocean ridges or in island-arc regimes), followed by remelting and possible contamination of the peridotite with a fluid phase accompanied by a period of high-temperature, low-stress deformation of the peridotite, either precursory to initiation of convergence, or during proto-arc formation; (2) development of ductile shear zones coeval with a protracted period of retrograde metamorphism; and (3) brittle failure of the metamorphosed peridotite. These events are consistent with models of evolution of forearc regimes in terms of metamorphic and thermal history [*Fryer*, 1992].

All of the evidence from the vent and pore-fluid studies of Conical seamount indicate that the hydration of the forearc wedge is facilitated by the escape of fluids from the subducting Pacific plate. Thermal modeling suggests that large regions beneath the Mariana forearc develop the stability conditions for the greenschist and blueschist stability fields as subduction proceeds [*Fryer and Fryer,* 1987; *Fryer et al.,* 1989; *Peacock,* 1990]. Timing of the metamorphism has not been established by any of the studies of the forearc samples thus far, except in a relative sense [*Girardeau and Lagabrielle,* 1992; *Maekawa et al.,* 1993]. However, in theory [*Fryer and Fryer,* 1987; *Fryer et al.,* 1989; *Peacock,* 1990], the stability conditions for the type of metamorphism that took place in event 1 develop very early in the history of the convergence process. Virtually the whole of the forearc could be metamorphosed within 1 m.y. after initiation of subduction. There is serpentine in sediments of Middle Eocene age just above the basement on the outer Mariana forearc DSDP site 459B [*Desprairies,* 1981]. Serpentine debris flows in sediments from deposits of this age on the Mariana forearc suggest that formation of some serpentine seamounts, or at least exposure of serpentine masses, and the generation of debris flows from their flanks could have occurred very early in the history of the Mariana forearc. The presence of palygorskite in the same sediments and in the basement beneath suggests an early hydrothermal metamorphic event involving slab-related fluids [*Natland and Mahoney*, 1981]. *Stern and Bloomer* [1992] and *Bloomer et al.* [this volume] present an argument for a rifting event in the forearc at the inception of trench formation. They propose that during the initial phase of development of the forearcs of the Mariana and Izu-Bonin systems, subsidence of oceanic plate initiated tensional stresses in what was to become the forearc region. This triggered a forearc rifting event. Coeval dewatering of the subsiding edge of the plate provided fluids necessary to trigger both magma genesis and metamorphism of the lithosphere in what was to become the supra-subduction zone region. This hypothesis explains the large volumes of supra-subduction zone lavas present in the forearc of the system and could explain the presence of serpentinite masses exposed by rifting in the forearc early in its history. Such a scenario could explain the observations of *Girardeau and Lagabrielle* [1992] noted for the earliest stage of formation of the peridotites (Event 1).

The general lowering of temperature in the forearc, as the gradual overturn of isotherms proceeds during subduction, is well-documented by both theoretical modeling and by the prevalence of blueschist-facies metamorphism in convergent margin regimes. A protracted period of retrograde metamorphism (during Event 2) is implicit in the mineralogic and structural characteristics of the peridotites. *Girardeau and Lagabrielle* [1992] suggest this represents an event that began at a high temperature and ended at a low one and that the temperature decrease might indicate decreasing depth. Alternatively, the simple models for decreasing temperatures with depth in evolving convergent margins can easily explain the retrograde metamorphism that *Girardeau and Lagabrielle* [1992] observe without demanding a change in depth of the peridotite involved [*Fryer,* 1992]. In fact, the presence of blueschist minerals in some of the rocks from Conical seamount [*Maekawa et al.,* 1992; 1993] implies that some of the samples from Conical seamount were metamorphosed at considerable depth. Ductile deformation of the peridotite (also part of Event 2) during the retrograde metamorphism period may be related to any or all of several possible forearc tectonic processes: subduction-related convection processes, tectonic erosion processes, small-scale plate adjustments within forearc blocks, vertical tectonic motions related to subduction of ocean plate seamounts, diapirism of serpentinite blocks, or uprise and protrusion of unconsolidated serpentinite masses [*Fryer,* 1992].

The diapiric emplacement of magmatic peridotite and subsequent serpentinization as suggested by *Girardeau and Lagabrielle* [1992] and *Ishii et al.* [1992] is similar to the emplacement mechanism of alpine serpentine espoused by *Benson* [1918] and by *Hess* [1939, 1955] in his early papers. More recently, field relationships in Columbia [*Hess,* 1966] and particularly in the Coast Ranges and Franciscan Formation of California [*Coleman,* 1957; *Dickinson,* 1966; *Lockwood,* 1971, 1972; *Carlson,* 1984] provide evidence that the mechanism of cold, solid emplacement must account for many of the serpentinites and peridotites of former convergent margins previously interpreted as emplaced magmatically.

Brittle fracture of the metamorphosed peridotite (Event 3) was suggested by *Girardeau and Lagabrielle* [1992] to record diapiric rise of a massive body of serpentinite. Equally, this type of deformation may take place during faulting at relatively shallow depths (a few kilometers) in the forearc lithosphere beneath the seamount, as triggering events leading to eruption of fluid-charged serpentine fault gouge that entrains fragments of the adjacent metamorphosed country rock [*Fryer,* 1992]. *Lockwood* [1972] suggested a similar mechanism by noting that in an area of subduction faults in the overriding plates may provide channels for protrusion of serpentine. *Lockwood* [1972] further suggested that, volatiles derived from sedimentary rocks undergoing metamorphic dehydration in the subducting plate are likely to be concentrated along faults and may also aid protrusion. *Lockwood* [1972] likened the rise of the serpentinite to that of salt diapirism, stressing the density contrast between serpentinite, especially sheared serpentinite such as that we observe in the drilled seamounts.

The problem of sorting out the various physical and theoretical parameters pertaining to the mechanisms by which the serpentine is emplaced hinges on the concept of diapirism in forearc regions. This is so because the term diapirism, with respect to the formation of serpentine seamounts on the Mariana and Izu-Bonin forearcs, has been used in many different ways by various investigators of these seamounts [*Fryer,* 1992]. If we use the term "diapirism" in its broadest sense, the rise of any gravitationally unstable, mobile material, it rapidly becomes necessary to apply several qualifiers in discussions of the Mariana and Izu-Bonin serpentine seamounts We recognize that the term "diapir" is becoming ingrained in discussions of these bodies in the literature, but hope that the danger of this practice will become apparent. We would prefer to drop the term and to describe the mechanisms of emplacement as they apply to each of the seamounts studied.

Fryer and Fryer [1987] suggested that the ridge of serpentinized peridotite exposed in the outer Izu-Bonin forearc results from metamorphic expansion of peridotite at the inner trench slope as a consequence of regional metamorphism (not strictly "diapirism"). They suggested that localized higher degrees of hydration of the forearc peridotite as a consequence of faulting could result in preferential uplift of certain parts of the ridge, thus forming individual serpentinite seamounts [*Fryer and Fryer,* 1987]. *Horine et al.* [1990], by contrast, noted that although the seamounts are composed of low-density material, the intervening sedimented basins do not rest on low-density material. This suggests that the low-density (serpentinized?) zones must be limited to the seamounts. Torishima forearc seamount and, by inference, others along the ridge might form by a multistage process. This process involves initial generation of a "conduit" for the emplacement and protrusion of a small serpentine body ("diapirism"), followed by local expansion of the country rock surrounding the conduit as a consequence of continued hydration and serpentinization by fluids emanating from a deep source. *Horine et al.* [1990] suggested that such a block of uplifted forearc constitutes the core of Torishima forearc seamount. This core, they suggest, was subsequently buried under several hundred meters of serpentine debris and possibly mud flows.

Hussong and Fryer [1981] noted the prevalence of serpentinized peridotite in many of the seamounts on the outer Mariana forearc and suggested "diapirism" as a mechanism of formation, but did not specify the nature of the diapirism. *Bloomer* [1982] studied the petrology and morphology of the seamounts in detail and suggested that the serpentinization of the peridotite occurred at depth, and as tectonic erosion of the Mariana forearc proceeded, large masses of serpentinite were mobilized along deep faults in the forearc and emplaced as diapiric bodies. *Fryer and Fryer* [1987] showed that some serpentine seamounts may form as mud volcanoes erupting serpentine muds from a central conduit and that others may be formed as composites of horst blocks and serpentine debris flows. *Fryer* [1992] suggested that the mud volcano type of serpentine seamount requires formation of a localized conduit by fault movement and implies periodic protrusions over the lifetime of activity on the fault. Certainly this is not "diapirism" any more than a volcanic eruption is diapirism. This mechanism would permit the source materials of various flow units to vary corresponding to the region of the fault along which the bulk of the movement takes place [*Fryer,* 1992]. The observed variability of the mineralogy and lithology of the flows would reflect heterogeneities within the forearc lithosphere.

Although the mud volcano mechanism is a viable one for the formation of serpentine deposits on land, we encourage caution in interpretation of serpentine deposits similar to those described from Conical seamount. Although rising fluids may mobilize fault gouge in any forearc fracture, there is currently evidence of only one active serpentine mud volcano on the Mariana forearc [*Fryer et al.,* 1990]. The oval-shaped serpentinite debris deposit on Pacman seamount is indistinguishable by submersible observations from the older sediment-draped serpentine mud flows of Conical seamount. However, on a gross morphologic scale it is quite different [*Fryer and Fryer,* 1987] and is unlikely to have formed as a highly fluid-charged debris flows such as those of Conical seamount. If Torishima forearc seamount in the Izu-Bonin forearc has a carapace of serpentine debris flows overlying a more coherent horst block, then it is possible that fault-channeled serpentine debris flows may be generated within horst blocks after uplift. Likely there are serpentine seamounts that represent a variety of degrees of combination of the diapir, horst, and mud volcano mechanisms.

The alternative to the mud volcano origin of these seamounts requires large scale faulting in the forearc. Some of the serpentine seamounts are very large. The region south of Pacman seamount is composed of several, large, interconnected serpentine seamounts of highly irregular shape (Figure 2). This region is apparently an uplifted block of serpentinite, approximately 150 km long and over 70 km wide, that has been internally faulted to create discrete edifices. Dredge hauls from the tops of these seamounts show predominantly ultramafic rock types [*Bloomer,* 1982; *Bloomer and Hawkins,* 1983; *Fryer and Fryer,* 1987]. If the seamounts that form this region rose diapirically from depth there is a problem either of unroofing or of continual separation of the bounding faults along which the large serpentinite bodies rose. The crustal materials through which the diapirs rose would have been shed off into the surrounding forearc basins. Drilling to the south of this region on DSDP leg 60 (sites 458 and 459) showed no indication of such shed-off material. It is more likely that the serpentine is exposed by faulting. Certainly, faulting is common in the Mariana forearc [*Mrozowski et al.,* 1981; *Bloomer and Hawkins,* 1983; *Fryer and Fryer,* 1987; *Newsom and Fryer,* 1987; *Johnson and Fryer,* 1990; *Fryer et al.,* 1990; *Marlow et al.,* 1990, *Johnson et al.,* 1991; *Newsom,* 1992; *Fryer,* 1993]. In order to expose a region as large as that of Pacman seamount and adjacent serpentine seamounts to the south, by emplacement along a forearc fault or fault system, it would be necessary to expand the fault, or fault system, by approximately 70 km (the width of the block of serpentine seamounts). Tectonic erosion is one mechanism by which do this. In fact tectonic erosion of this order of magnitude (50-100 km) is proposed for much of the Mariana and Izu-Bonin forearcs [e.g., *Bloomer,* 1982; *Stern and Bloomer,* 1992]. Tectonic erosion may not be necessary, however. The mechanism proposed by *Fryer and Fryer* [1987] modified by that of *Horine et al.* [1990] for the Izu-Bonin outer forearc terrace could also be applied locally to the Mariana forearc. That is, exposure of segments of serpentinized upper mantle in the inner slope of the trench by faulting during the development of the subduction zone would begin the process. Continued serpentinization of the exposures in response to contact with fluids from the downgoing slab would permit local expansion of the serpentinite and creation of an outer forearc high. This would not be "diapirism" strictly, but rather metamorphic expansion. Secondary isolated highs would be formed by a combination of additional local fault-facilitated serpentinization and by protrusion of unconsolidated serpentine flows from depth along faults within the edifices.

6. CONCLUSIONS

Until relatively recently, serpentine deposits exposed in exotic terranes on land have been ascribed to syn- or post-obduction metamorphism. The Mariana and Izu-Bonin forearc serpentine seamounts produce deposits that are indistinguishable from several of these. It is likely that many of the "sedimentary" serpentine bodies in exotic terranes on land have formed during active convergence processes like those of the Mariana and Izu-Bonin seamounts. Emplacement of large serpentinite seamounts is a common occurrence in the outer half of the Mariana forearc and on the ridge that forms the inner trench slope of the Izu-Bonin forearc. There are at least two end-member mechanisms and various combinations of the two by which these serpentine seamounts form. Mud volcanoes composed of unconsolidated serpentine flows make up one end-member type. Horst blocks, probably emplaced by metamorphic expansion of exposed ultramafics are the other end-member type. The bulk of the serpentine seamounts is probably formed by some combination of these mechanisms depending on the local tectonic conditions. Tectonic processes also are probably the major controlling factor influencing the periodicity of the protrusion of serpentine mud flows on seamounts associated with venting of fluids. The common element in emplacement of all types of serpentine seamounts is the presence of large scale faulting.

The potential for metamorphism leading to serpentine formation is ubiquitous in supra-subduction zone regions, and may take place early on, yet none show development of serpentinite seamounts on the scale of that in the Mariana and Izu-Bonin systems. One reason for the broad distribution of serpentine seamounts in the Mariana forearc may relate to the extensive vertical tectonic deformation of this forearc. The increasing curvature of the arc with time, the subduction of numerous plate seamounts, and small-scale rotations of blocks within the forearc region are probably all responsible for the extensive deformation of this forearc. The restriction of serpentine seamounts to the outer forearc in both the Mariana and Izu-Bonin systems may be linked to exposure of peridotites in the inner trench region by faulting early in the development of the subduction zone. Subsequent hydration and metamorphic expansion of the outer toe of the overriding plate and a combination of vertical tectonic processes and eruption of serpentine muds completes the process.

The capacity for incorporation of large volumes of fluids from the downgoing slab into the metamorphosed supra-subduction zone region and the venting of these fluids through forearc faults provides a possible solution to the problem of mass balance of subduction constituents in convergence zones. Unfortunately, the proportion of fluids involved in metamorphic processes versus fault-related venting is unknown, and the exact nature of fluid transport within the supra-subduction zone region is poorly understood. We do know that fluids are venting at several types of sites on serpentine seamounts or along faults through them. Unfortunately the data are limited and the recycling of subduction-related fluids through the forearcs as a major component of global mass balance cannot be determined without more detailed studies. Fluid flux through the forearc probably also plays a critical role in the evolution of the thermal structure. The degree of flux channeled by fractures within the outer forearc may be great enough to create local perturbations of the thermal structure. A better understanding of the distribution of active venting sites will be necessary before accurate models of thermal structure of the forearcs can be constructed. Most of the work on fluid flux in convergent margins has been on accretionary forearcs. Examining the composition of slab-related fluids is difficult enough in nonaccretionary forearcs where channelized fluids have the least opportunity for interaction with the country rock through which they migrate. Limiting studies of fluid flux to accretionary prisms is like placing another filter on the process, the addition of the sedimentary wedge. We would argue that the most direct approach to the study of fluid flux in convergent margins is to study fluids venting along fault zones in nonaccretionary forearcs and the metamorphosed rock in contact with them.

Despite the remaining unanswered questions, the study of serpentine seamounts in the Mariana and Izu-Bonin forearcs has expanded our understanding of degree and timing of metamorphism in both active forearcs and fragments of them emplaced on land. As such bodies accrete to continental masses, such as the serpentinite bodies of the Franciscan Formation in California, they may remain little changed from their original character. These studies have highlighted the importance of uptake of subduction-related fluids in metamorphism of supra-subduction zone regions and of fluid venting as components in understanding mass balance at convergent margins.

Acknowledgments. The interpretation of the synthesized data has benefited particularly from discussions with J. Lockwood, M. Kastner and G. Fryer. We are very grateful to D. Cowan and J. Natland for their helpful reviews of this manuscript. This is SOEST contribution no. 3612 and HIGP contribution no. 777.

REFERENCES

Bailey, E. H., M. C. Blake, Jr., and D. L. Jones, On-land Mesozoic oceanic crust in California Coast Ranges, *U..S Geol Surv. Prof. Paper, 700-C,* 70-81, 1970.

Benson, W. N., The origin of serpentine, a historical and comparitive study, *Am. J. Sci. 4th ser., 46,* 439-460, 1918.

Bezore, S. P., The Mount Saint Helen ultramafic complex of the northern Coast Ranges (abstract), *Geol. Soc. Am., 1(3),* 5, 1969.

Blake, M. C., W. P. Irwin, and R. G. Coleman, Blueschist-facies metamorphism related to regional thrust faulting, *Tectonophysics, 8(3),* 237-246, 1969.

Bloomer, S., Structure and evolution of the Mariana Trench, Petrologic and geochemical studies, Ph.D. Dissertation, Univ. Calif, San Diego, 1982.

Bloomer, S. H. and J. W. Hawkins, Gabbroic and ultramafic rocks from the Mariana Trench: An island arc ophiolite, in *The Tectonic and Geologic Evolution of Southeast Asian Seas and Islands,* Geophys, *Monogr. Ser., 27,* edited by D. E. Hayes, pp. 294-317, AGU, Washington, D.C., 1983.

Carlson, C., Stratigraphic and structural sugnificance of foliate serpentine breccias, Wilbur Springs. in *Soc. Econ. Paleontol. Mineral. Field Trip Guidebook, no. 3,* 108-112, 1984.

Coleman, R. G., Mineralogy and petrology of the New Idria district, California, Ph.D. Dissertation, Stanford University, Stanford, 1957.

Desprairies, A., Authigenic minerals in volcanigenic sediments cored during Deep Sea Drilling Project Leg 60, in *Init. Repts. DSDP, 60,* D. M. Hussong, S. Uyeda, et al., pp. 437-453, U.S. Govt. Printing Office, Washington, D.C., 1981.

Dickinson, W. R., Table Mountain serpentinite extrusion in California Coast Ranges, *Geol Soc. Am. Bull., 77(5),* 451-472, 1966.

Evarts, R. C., The geology and petrology of the Del Puerto ophiolite, Diablo Range, central California Coast Ranges, in Coleman, R.G., and Irwin, W.P. (eds.), *North Americal Ophiolites: Oregon Department of Geology and Mineral Industries Bull., 95,* 121-140, 1977.

Fryer, P., A synthesis of Leg 125 drilling of serpentine seamounts on the Mariana and Izu-Bonin forearcs, in *Proc. ODP, Sci. Results, 125,* P. Fryer, J. A. Pearce, L. B. Stokking, et al., pp. 593-614, Ocean Drilling Program, College Station, TX, 1992.

Fryer, P., The relationship between tectonic deformation, volcanism, and fluid venting in the southeastern Mariana convergent plaate margin, *Proc. JAMSTEC, Symp. Deep Sea Res., 9,* 161-179, 1993.

Fryer, P., and M. J. Mottl, Lithology, mineralogy and origin of Serpentine muds drilled at Conical seamount and Torishima forearc seamount, in *Proc. ODP, Sci. Results, 125,* P. Fryer, J. A. Pearce, L. B. Stokking, et al., pp. 343-362, Ocean Drilling Program, College Station, TX, 1992.

Fryer, G. J., P. Fryer, and K. L. Saboda, Thermal and metamorphic evolution of the Mariana forearc (abstract), *Eos, Trans AGU, 70(43)* ,1381, 1989.

Fryer, P., J. A. Pearce, L. B. Stokking, et al. *Proc. ODP, Init Repts., 125,* Ocean Drilling Program, College Station, TX, 1990.

Fryer, P., J. Haggerty, B. Tilbrook, P. Sedwick, L. E. Johnson, K. L. Saboda, S. Y. Newsom, D. E. Karig, S. Uyeda, and T. Ishii, Results of studies of Mariana forearc serpentinite diapirism (abstract), *Eos, Trans. AGU, 68 (44),* 1534, 1987.

Fryer, P. and G. J. Fryer, Origins of non-volcanic seamounts in a forearc environment, in *Seamounts Islands and Atolls,* edited by B. H. Keating, P. Fryer, R. Batiza, and G. W. Boehlert, *Geophys. Monogr. Ser., 43,* 61-72, Washington, AGU, 1987.

Fryer, P., Ambos, E. L., and Hussong, D. M., Origin and emplacement of Mariana forearc seamounts, *Geology,13,* 774-777, 1985.

Fryer, P., Saboda, K. L., Johnson, L. E., MacKay, M. E., Moore, G. F., and Stoffers, P., Conical seamount: SeaMARC II, *Alvin* submersible, and seismic reflection studies, in *Proc. ODP, Init. Repts.* 125, P. Fryer, J. A. Pearce, L. B. Stokking, et al., pp. 69-80, Ocean Drilling Program, College Station, TX, 1990.

Girardeau J., and Y. Lagabrielle, Deformation history of peridotites from Mariana forearc, Conical seamount, Leg 125, in *Proc. ODP, Sci. Results, 125,* P. Fryer, J. A. Pearce, L. B. Stokking, et al., pp. 519-534, Ocean Drilling Program, College Station, TX, 1992.

Haggerty, J. A., Petrology and geochemistry of Neogene sedimentary rocks from Mariana forearc seamounts: implications for emplacement of the seamounts, in *Seamounts Islands and Atolls,* edited by B. H. Keating, P. Fryer, R. Batiza, and G. W. Boehlert, *Geophys. Monogr. Ser., 43,* 175-185, Washington, AGU, 1987.

Haggerty, J. A., Evidence from fluid seeps atop serpentine seamounts in the Mariana Forearc: Clues for emplacement of the seamounts and their relationship to forearc tectonics, *Mar. Geol., 102,* 293-309, 1991.

Haggerty, J. A., and S. Chaudhuri, Strontium isotopic composition of the interstital waters from Leg 125: Mariana and Bonin forearcs, in *Proc. ODP, Sci. Results, 125,* P. Fryer, J. A. Pearce, L. B. Stokking, et al., pp. 397-400, Ocean Drilling Program, College Station, TX, 1992.

Haggerty, J. A., and J. B. Fisher, Short-chain organic acids in interstitial waters from Nariana and Bonin forearc serpentines, in *Proc. ODP, Sci. Results, 125,* P. Fryer, J. A. Pearce, L. B. Stokking, et al., pp. 387-395, Ocean Drilling Program, College Station, TX, 1992.

Heling, D., and A. Schwarz, Iowaite in serpentinite muds at sites 778,779, 780, and 784: a possible cause for the low chlorinity of pore waters, in *Proc. ODP, Sci. Results, 125,* P. Fryer, J. A. Pearce, L. B. Stokking, et al., pp. 313-324, Ocean Drilling Program, College Station, TX, 1992.

Hess, H. H., Island arcs, gravity anomalies and serpentinite intrusions, *Intl. Geol. Cong. Moscow, 1937, Comptes Rendus 17,* 2:263-283, 1939.

Hess, H. H., Serpentines, orogeny and epeirogeny, *Geol. Soc. Am. Spec. Paper, 62,* 391-409, 1955.

Hess, H. H., Caribbean Research Project, 1965, and bathymeytric chart, *Geol. Soc. Am. Mem., 98,* 1-10, 1966.

Honza, E., and Tamaki, K., The Bonin arc, in *The Ocean Basins and Margins, Volume 7, The Pacific Ocean.,* edited by A. E. M. Nairn and S. Uyeda, pp. 459-502, Plenum Press, New York, 1985.

Hopson, C. A. and C. J. Frano, Igneous history of the point Sal ophiolite, southern California, in Coleman, R. G., and W. P. Irwin (eds.), *North Americal Ophiolites: Oregon Department of Geology and Mineral Industries Bull., 95,* 161-183, 1977.

Hopson, C. A., J. M. Mattinson, and E. A. Pessagno, Jr., Coast

Range ophiolite, western California, in Ernst W. G. (ed.), *Geotectonic Development of California, Ruby Vol. 1*, 706 pp, Prentice Hall, Inc., Englewood Cliffs, N. J., 1981.

Horine, R. L, G. R. Moore, and B. Taylor, Structure of the outer Izu-Bonin forearc from seismic-reflection profiling and gravity modeling, in *Proc. ODP, Init. Repts.* 125, P. Fryer, J. A. Pearce, L. B. Stokking, et al., pp. 81-94, Ocean Drilling Program, College Station, TX, 1990.

Hussong, D. M. and P. Fryer, Structure and tectonics of the Mariana arc and fore-arc: Drill site selection surveys, in *Init. Repts. DSDP, 60*:, D. M. Hussong, S. Uyeda, et al., pp. 33-44, U.S. Govt. Printing Office, Washington, D.C., 1981.

Hussong, D. M., and P. Fryer, Forearc tectonics in the northern Mariana arc, in *Formation of Active Ocean Margins*, edited by N. Nasu, pp. 273-290, TERRA Publishers, Tokyo, 1985.

Hussong, D. M., and S. Uyeda, Tectonic processes and the history of the Mariana Arc, A synthesis of the results of Deep Sea Drilling Leg 60, in *Init. Repts. DSDP, 60*:, D. M. Hussong, S. Uyeda, et al., pp. 909-929, U.S. Govt. Printing Office, Washington, D.C., 1981.

Ishii, T., Dredged samples from the Ogasawara forearc seamount or "Paleoland" --"Forearc ophiolite", in *Formation of Active Ocean Margins*, edited by N. Nasu, pp. 307-342, TERRA Publishers, Tokyo, pp. 307-342, 1985.

Ishii, T., K. Kobayashi, T. Shibata, J. Naka, K. Johnson, K. Ikehara, M. Iguchi, K. Konishi, H. Wakita, F. Zhang, Y. Nakamura, and H. Kayane, Description of samples from Ogasawara forearc, Ogasawara Plateau and Mariana Trough during KH84-1 Cruise, in Kobayashi, K (Ed.), *Preliminary report of the Hakuho Maru Cruise KH84-1*: Tokyo (Ocean Res. Inst., Univ. of Tokyo), 86-167, 1985.

Ishii, T., P. T. Robinson, H. Maekawa, and R. Fiske, Petrological studies of peridotites from diapiric serpentinite seamounts in the Izu-Ogasawara-Mariana forearc; in *Proc. ODP, Sci. Results, 125,* P. Fryer, J. A. Pearce, L. B. Stokking, et al., pp. 445-486, Ocean Drilling Program, College Station, TX, 1992.

Johnson, L. E., Mafic clasts in serpentinite seamounts: petrology and geochemistry of a diverse crustal suite from the outer Mariana forearc, in *Proc. ODP, Sci. Results, 125,* P. Fryer, J. A. Pearce, L. B. Stokking, et al., pp. 401-414, Ocean Drilling Program, College Station, TX, 1992.

Johnson, L. E, and P. Fryer, Petrography, geochemistry and petrogenesis of igneous rocks from the outer Mariana forearc, *Earth Planet. Sci. Lett., 100(1/3)*, 304-316, 1990.

Johnson, L. E., P. Fryer, B. Taylor, M. Silk, D. L. Jones, W. V. Sliter, T. Itaya, and T. Ishii, New evidence for crustal accretion in the outer Mariana forearc: Cretaceous Radiolarian Cherts and mid-ocean ridge basalt-like lavas. *Geology , 19*, 811-814, 1991.

Karig, D. E., Structural history of the Mariana island arc system. *Geol. Soc. Am. Bull.*, 83: 323-344, 1971.

Kulm, L. D. et al., Oregon subduction zone: venting, fauna, and carbonates, *Science, 231*, 561-566, 1986.

Lagabrielle, Y., A.-M. Karpoff, and J. Cotten, Mineralogical and geochemical analyses of sedimentary serpentinites from Conical seamount (hole 778A): implications for the evolution of serpentine seamounts, in *Proc. ODP, Sci. Results, 125,* P. Fryer, J. A. Pearce, L. B. Stokking, et al., pp. 325-342, Ocean Drilling Program, College Station, TX, 1992.

Lawton, J. E., Geology of the north half of the Morgan Valley quadrangle and the south half of the Wilbur Springs quadrangle, California, PhD Dissertation, Stanford University, Stanford, California, 226 pp., 1956.

Lockwood, J. P., Sedimentary and gravity slide emplacement of serpentinite, *Geol. Soc. Am. Bull.*, 82, 919-936, 1971.

Lockwood, J. P., Possible mechanisms for the emplacement of alpine-type serpentinite, *Geol. Soc. Am. Mem., 132*, 273-287, 1972.

Maekawa, H., M. Shozui, T. Ishii, K. L. Saboda, and Y. Ogawa, Metamorphic rocks from the serpentinite seamounts in the Mariana and Izu-Ogasawara forearcs, in *Proc. ODP, Sci. Results, 125,* P. Fryer, J. A. Pearce, L. B. Stokking, et al., pp. 415-430, Ocean Drilling Program, College Station, TX, 1992.

Maekawa, H., M. Shozui, T. Ishii, P. Fryer, and J. A. Pearce, Blueschist metamorphism in an active subduction zone, *Nature, 364(6437)*, 520-523, 1993.

Marlow M. S., D. L. Merril, et al., Underway Geophysics, in *Proc. ODP, Init. Repts., 125,* P. Fryer, J. A. Pearce, L. B. Stokking, et al., pp. 41-69, Ocean Drilling Program, College Station, TX, 1990.

Marlow, M. S., L. E. Johnson, J. A. Pearce, P. Fryer, L. G. Pickthorn, B. J. Murton, Pleistocene volcanic rocks in the Mariana forearc revealed by drilling at ODP site 781, in *Proc. ODP, Sci. Results, 125,* P. Fryer, J. A. Pearce, L. B. Stokking, et al., pp. 293-310, Ocean Drilling Program, College Station, TX, 1992.

Maxwell, J. C., Depositional environment of Jurassic-lower Cretaceous sediments, western margin Sacrimento Valley (abstract), *Geol. Soc. Am. Abstr w/ Programs, 9,* 461-462, 1977.

McLaughlin, J., Preliminary geologic map of the Geysers-Stream field and vicinity, Sonoma County, *California, US Geol. Surv. Open-file Map*, 74-238, 1974.

Moiseyev, A. A., Late serpentinite movements in the California coast ranges: new evidence and its implications, *Geol. Soc. Am. Bull., 81,* 1721-1732, 1970.

Mottl, M. J., Pore waters from serpentine seamounts in the Mariana and Izu-Bonin forearcs, Leg 125: evidence for volatiles from the subducting slab, in *Proc. ODP, Sci. Results, 125,* P. Fryer, J. A. Pearce, L. B. Stokking, et al., pp. 373-386, Ocean Drilling Program, College Station, TX, 1992.

Mottl, M. J. and J. C. Alt, Data Report: Minor and trace element and sulfur isotopic composition of pore watersfrom sites 778 through 786, in *Proc. ODP, Sci. Results, 125,* P. Fryer, J. A. Pearce, L. B. Stokking, et al., pp. 683-690, Ocean Drilling Program, College Station, TX, 1992.

Mrozowski, C. L., D. E. Hayes, and B. Taylor, Multichannel seismic reflection surveys of Leg 60 sites, Deep Sea Drilling Project, in *Init. Repts. DSDP, 60*, D. M. Hussong, S. Uyeda, et al., pp. 57-71, U.S. Govt. Printing Office, Washington, D.C., 1981.

Mumpton, F. A., and C. S. Thompson, Mineralogy and origin of the Coalinga asbestos deposit, *Clays Clay Mineral., 23*, 131-143, 1975.

Natland, J. H., and J. J. Mahoney, Alteration of igneous rocks at Deep Sea Drilling Project sites 458 and 459, Mariana fore-arc region: Relationship to basement structure, in *Init.*

Repts. DSDP, 60, D. M. Hussong, S. Uyeda, et al., pp. 769-788, U.S. Govt. Printing Office, Washington, D.C., 1981.

Newsom, S. Y., and P. Fryer, Three-dimensional gravity modeling of serpentinite seamounts in the Mariana forearc (abstract), *Eos, Trans. AGU, 68(44)*, 1534, 1987.

Newsom S. Y., Gravity interpretation of the Mariana forearc between 19° and 20°N, Masters Thesis, University of Hawaii, Honolulu, 66 pp., 1992.

Parkinson, I. J., J. A. Pearce, M. F. Thirwald, K. T. M. Johnson, and G. Ingram, Trace element geochemistry of peridotites from the Izu-Bonin-Mariana forearc, Leg 125, in *Proc. ODP, Sci. Results Leg, 125,* P. Fryer, J. A. Pearce, L. B. Stokking, et al., pp. 487-506, Ocean Drilling Program, College Station, TX, 1992a.

Parkinson, I. J., G. E. M. Hall, and J. A. Pearce, Palladium, platinum, and gold distribution in serpentinite seamounts in the Mariana and Izu-Bonin forearcs: evidence from Leg 125 fluids and serpentinites, in *Proc. ODP, Sci. Results Leg, 125,* P. Fryer, J. A. Pearce, L. B. Stokking, et al., pp. 507-518, Ocean Drilling Program, College Station, TX, 1992b.

Peacock, S. M., Fluid processes in subduction zones, *Science, 248*, 329-337, 1990.

Phipps, S. P., Mesozoic ophiolitic olistostromes and Cenozoic imbricate thrust faulting in the northern California Coast Ranges; geology of the Mysterious Valley Area, Napa County, PhD Dissertation, Princeton Univ., Princeton, N. J., 669 p, 1984.

Phipps S. P., K. L. Kleinspehn, and J. E. Suppe, Ophiolitic olistostromes in the basal Great Valley Sequence Napa County California coast ranges (abstract), *Geol. Soc. Am. Abstr. w/ Prog. 11,* 122, 1979.

Phipps, S., and D. Ballotti, Rheology of serpentinite muds in the Mariana-Izu-Bonin forearc, in *Proc. ODP, Sci. Results Leg, 125,* P. Fryer, J. A. Pearce, L. B. Stokking, et al., pp. 363-372, Ocean Drilling Program, College Station, TX, 1992.

Rich, E. I., Geology of the Wilbur Springs quadrangle Colusa and Lake Counties, California, PhD Dissertation, Stanford University, Stanford, California, 101 pp., 1968.

Saboda, K. L., P. Fryer, and H. Maekawa, Metamorphism of ultramafic clasts from Conical seamount: sites 778, 779, and 780, *Proc. ODP, Sci. Results , 125,* P. Fryer, J. A. Pearce, L. B. Stokking, et al., pp. 431-444, Ocean Drilling Program, College Station, TX, 1992.

Schorno, K. S., Geochemistry of carbon: International Phase of Ocean Drilling Project Leg 60, in *Init. Repts. DSDP, 60,* D. M. Hussong, S. Uyeda, et al., pp. 501-504, U.S. Govt. Printing Office, Washington, D.C., 1981.

Stern, R. J. and S. H. Bloomer, Subduction zone infancy: Examples from the Eocene Izu-Bonin-Mariana and Jurassic California arcs, *Geol. Soc. Am. Bull., 104,* 1621-1636, 1992.

Stokking L. B., D. L. Merrill, R. B. Haston, J. R. Ali, and K. L. Saboda, Rock magnetic studies of serpentinite seamount in the Mariana and Izu-Bonin regions, in *Proc. ODP, Sci. Results, 125,* P. Fryer, J. A. Pearce, L. B. Stokking, et al., pp. 561-580, Ocean Drilling Program, College Station, TX, 1992.

Taylor, B., Rifting and the volcanic-tectonic evolution of the Izu-Bonin-Mariana arc, in *Proc. ODP, Sci. Results, 126,* B. Taylor, K. Fujioka, et al., pp. 627-651, Ocean Drilling Program, College Station, TX,1992.

Taliaferro, N. L., Franciscan-Knoxville problem, *Am. Assoc. Petrol. Geol. Bull., 27,* 109-219, 1943.

P. Fryer, Hawaii Institute of Geophysics and Planetology, SOEST, 2525 Correa Rd. Honolulu, HI 96822, USA.

M. Mottl, Department of Oceanography, SOEST, 1000 Pope Rd., Honolulu, HI, 96822, USA.

L. Johnson, Naval Research Laboratory, Code 5110, Washington, D.C. 20375, USA.

J. Haggerty, University of Tulsa, 600 S. College Ave., Tulsa, OK 74104, USA.

S. Phipps, Department of Geology, University of Pennsylvania, Philadelphia, PA 19104-6316, USA.

H. Maekawa, Department Earth Sciences, Faculty of Science, Kobe University, Nada, Kobe 657, Japan.

Incipient Blueschist-Facies Metamorphism in the Active Subduction Zone Beneath the Mariana Forearc

Hirokazu Maekawa

Department of Earth and Planetary Sciences, Faculty of Science, Kobe University, Kobe, Japan

Patricia Fryer

Department of Geology and Geophysics, University of Hawaii at Manoa, Honolulu, Hawaii

Atsuo Ozaki

Setsuryo high school, Ibaraki , Japan

Many seamounts composed of serpentinite and peridotite are recognized in the Mariana forearc region. During Ocean Drilling Program Leg 125, high-pressure, low-temperature metabasite clasts were recovered from one of these seamounts. The diagnostic low-variance assemblage of these clasts is lawsonite + pumpellyite + hematite. Using Schreinemakers' method, the stability field of this assemblage is situated at the lowest-pressure and lowest-temperature field in the blueschist-facies metamorphism; these rocks were probably formed under the incipient blueschist-facies conditions. Only a few incipient blueschist-facies rocks have been reported from ancient subduction complexes exposed on land, and none have been studied in any detail. These rocks, however, are very important for understanding the relationship between tectonic processes and subduction-related metamorphism. They preserve the primary nature of subduction processes within a subduction zone, because recrystallization is commonly incomplete and their originally shallow position in subduction zones may prevent overprinting of effects from later-stage disturbance during uplift.

INTRODUCTION

In the Mariana and Izu-Bonin convergent plate margins, enormous amounts of serpentinite and peridotite have been raised from the mantle wedge to the seafloor to form a zone of seamounts in the forearcs. During Ocean Drilling Program Leg 125, many small clasts of high-pressure, low-temperature metabasites were recovered from one of these serpentinite seamounts [*Maekawa et al.*, 1992; *Maekawa et al.*, 1993; Maekawa, in press]. The seamount is called "Conical Seamount" and is one of the youngest serpentinite seamounts in the Mariana forearc (Figure 1). Sediments recovered during drilling at the flanks of the seamount contain Pleistocene nannofossils (<1.6 Ma) [*Ciampo*, 1992], suggesting the recent uplift of serpentinite materials to the seafloor. The fluids (which converted peridotite to serpentinite) are still actively upwelling from its summit [*Haggerty*, 1989; *Mottl*, 1992].

The metamorphic rocks contain typical blueschist facies minerals, lawsonite, sodic pyroxene, and aragonite, which are clearly the products of blueschist-facies (glaucophane-schist facies) metamorphism. The finding of these rocks provides direct evidence of blueschist-facies metamorphism in the subduction zone. In this paper, we discuss the stability field of mineral parageneses of these rocks using Schreinemakers' method and consider the nature and significance of this metamorphism.

Active Margins and Marginal Basins of the Western Pacific
Geophysical Monograph 88

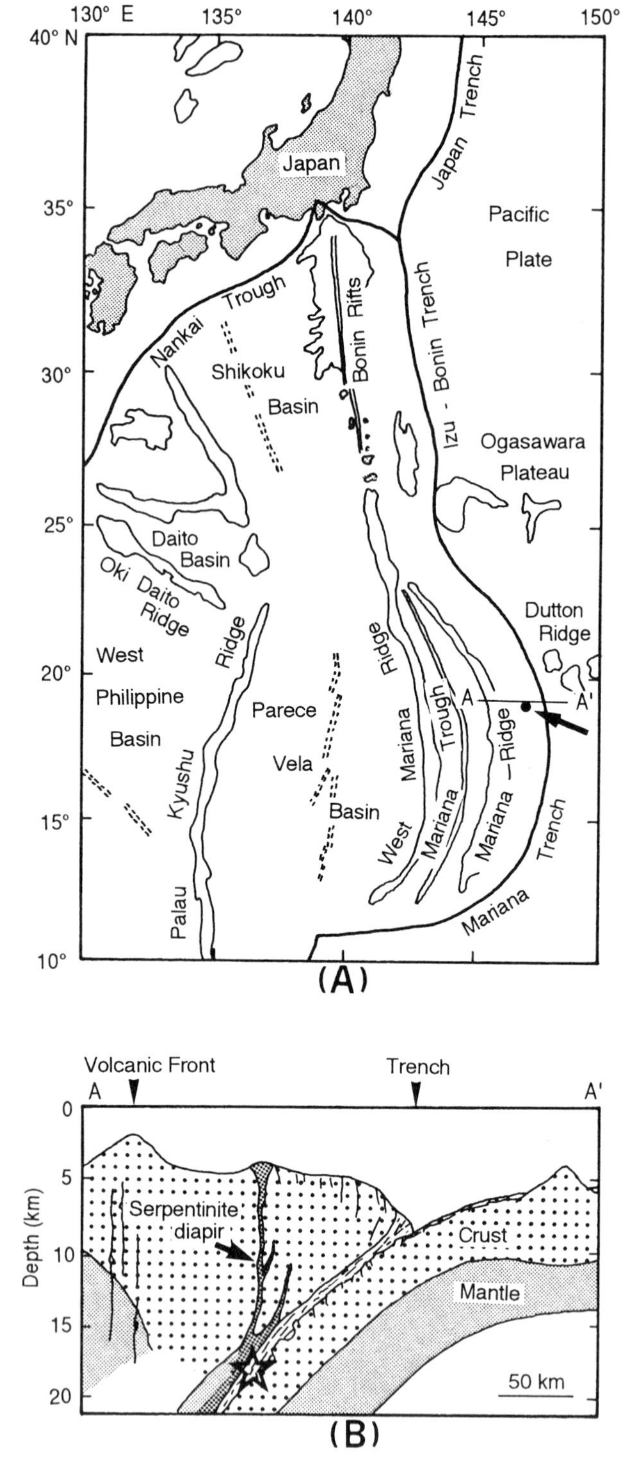

Fig. 1. (A) Regional map of the western Pacific Ocean. An arrow shows the location of Conical Seamount (from *Fryer et al.* [1990]). (B) Schematic cross section showing the tectonic framework of the Mariana arc-trench system (based on the data from *Hussong and Uyeda* [1981] and this study).

METAMORPHIC PETROLOGY

Mineral Descriptions

Approximately 60 metabasites have been recovered from Hole 778A, Site 778, on the southern flank of Conical Seamount. Typical high-pressure minerals, lawsonite, sodic pyroxene, and aragonite, were found in these metabasites. These rocks occur as pebble-size clasts of subrounded shape in a serpentinite matrix. Recrystallization is incomplete, and primary igneous textures are well preserved. Protoliths, identified from their primary textures and mineralogies, are predominantly aphyric to fine-grained basalts and their clastic equivalents. Some protoliths are amphibolitic rocks, and contain relict green hornblende. Geochemical studies indicate the metabasites include rocks having chemical affinity with mid-ocean ridge basalts *[Johnson*, 1992; K. Yamamoto, Kobe University, unpublished data, 1993].

Low-grade metamorphic minerals, such as lawsonite, pumpellyite, sodic pyroxene, and aragonite, are found mainly in veins and cavities and rarely in matrix in metabasites. Fourteen clasts out of twenty-three specimens contain lawsonite, pumpellyite, sodic pyroxene, and/or metamorphic aragonite, and one contains blue amphibole and sodic pyroxene. These minerals were initially identified by optical microscope and confirmed by X-ray diffraction analysis. Lists of mineral associations of the rocks and microprobe analyses of chemical compositions of metamorphic minerals were given in *Maekawa et al.* [1992]. As the modes of occurrence of metamorphic minerals were also described in *Maekawa et al.* [1992], we concentrate on description of mineral associations in this paper.

Lawsonite occurs as an aggregate or radial cluster of fine-grained prismatic or acicular crystals in veins and cavities. In most occurrences, lawsonite is associated intimately with pumpellyite. Microprobe analyses of lawsonite show significant Fe^{3+} substitution for Al (Figure 2). Most analyzed lawsonites usually contain 1.5 to 5.0 wt% Fe_2O_3; however, lawsonite in the hematite-bearing specimen sample 125-778A-6R-1 (piece no. 1, 20-23 cm) contains up to 8 wt% Fe_2O_3 and shows a conspicuous zoning from a Fe^{3+}-rich core (6-8 wt%) to Fe^{3+}-poor rim (2-3 wt%). Fe^{3+}-rich lawsonites with up to 5.8 wt%, 4.3 wt%, and 4.0 wt% have been reported from northwest Turkey [*Okay*, 1982], Baja California [*Sedlock*, 1988], and the Tokoro belt, Japan [*Sakakibara,* 1986], respectively. In the Franciscan terrane, Fe_2O_3 in lawsonite tends to increase with decreasing metamorphic grade and attains a maximum enrichment of 3.2 wt% *[Maruyama and Liou,* 1987]. Fe^{3+} enrichment of lawsonite in the Hole 778A metabasites could be partly due to low-temperature conditions of metamorphism.

Pumpellyite occurs as small spongy aggregates in veins and matrix and commonly coexists with lawsonite, chlorite, quartz, and aragonite. Chemical compositions of

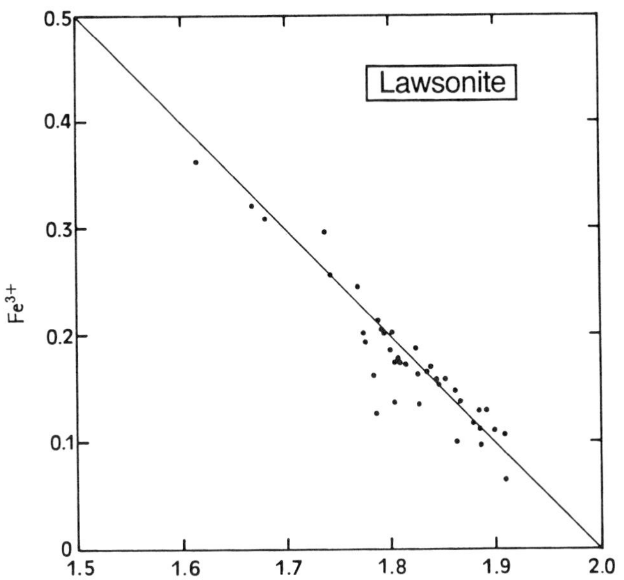

Fig. 2. Chemical compositions of lawsonite on an Al-Fe^{3+} diagram showing significant Fe^{3+} substitution for Al in lawsonite.

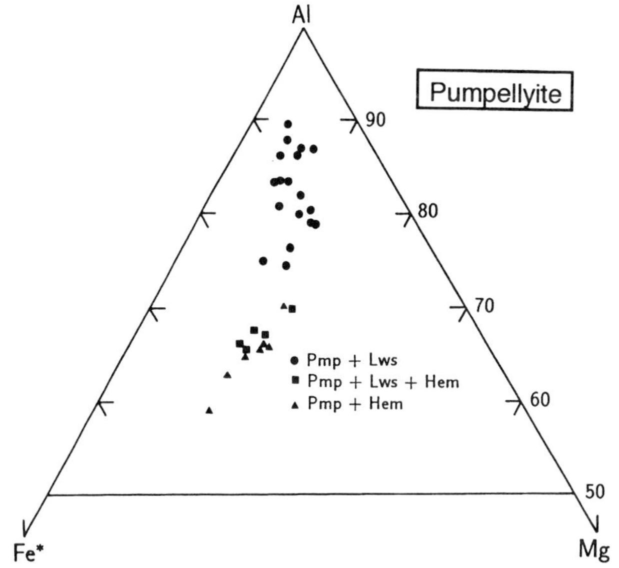

Fig. 3. Chemical compositions of pumpellyite on an Al-Fe*-Mg diagram. Pmp: pumpellyite, Lws: lawsonite, Hem: hematite. Fe*: total Fe as FeO

pumpellyite are apparently controlled by the mineral assemblages: pumpellyites in hematite-bearing samples are richer in Fe^{3+} than those in hematite-free samples (Figure 3). The Fe^{3+}/Al ratios of lawsonite tend to show positive correlation with those of associated pumpellyite, suggesting attainment of equilibrium (Figure 4). Assuming the formula $Ca_4(Fe^{2+}+Mg)(Al,Fe^{3+})_5Si_6O_{23}(OH)_3{\cdot}2H_2O$, the $Fe^{3+}/(Fe^{3+}+Al)$ ratio of pumpellyite that coexists with hematite is estimated from the microprobe analyses to be 0.2.

Sodic pyroxene is found as a tiny spherical aggregate of acicular crystals in plagioclase phenocrysts pseudomorphosed by albite and chlorite [see *Maekawa et al.*, 1993, figure 2], in veins of pumpellyite, and as epitaxial fringes along the cleavage traces of primary clinopyroxene. Spherical aggregates of sodic pyroxene in pumpellyite exhibit irregular shape in comparison with those in albite. Sodic pyroxene usually coexists with albite, chlorite, pumpellyite, and rarely with quartz, but does not coexist with lawsonite. Most sodic pyroxene compositions are scattered around the chloromelanite field and contain 30-35% jadeite molecule.

Aragonite was replaced by calcite along irregular cleavages and was observed as small discrete crystals with identical optical orientations within a calcite pool. Aragonite exhibits a slightly higher refractive index than calcite.

Blue amphibole exhibiting a distinct pleochroism from blue to pale violet can be seen along the rims or cleavage traces of relict hornblendes in sample 125-778A-13R,CC (piece no. 4, 6-10 cm). Blue amphibole is also found as interstitial fillings of some pull-apart relict hornblendes, which suggests syntectonic recrystallization. Microprobe analyses indicate that the blue amphibole is a solid solution between actinolite and magnesioriebeckite, and can be classified as winchite after *Leake* [1978]. Winchite is only found in sample 125-778A-13R,CC (piece no. 4, 6-10 cm), which has a peculiar hematite-bearing, quartz-free composition. As pressure and temperature conditions during metamorphism may have been too low to form winchite in rocks with a wide range of compositions, winchite was formed exclusively in the rocks with highly oxidized peculiar compositions.

Fine-grained aggregates of phengite are found in association with lawsonite as a vein mineral. Chlorite is ubiquitous in varying amounts. The chlorites show a small range in Al/(Fe+Mg) but have an appreciable range in Fe/Mg, reflecting variations in host rock compositions. Quartz occurs mainly as a vein mineral in association with aragonite, pumpellyite, sodic pyroxene, chlorite, and lawsonite. Albite is found as a vein mineral or replacing primary plagioclase. Albite shows a narrow range of composition, with anorthite component of less than 2 % and orthoclase component of less than 1 %.

Sphene is an accessory, but ubiquitous, mineral in all the rocks studied. It is found as dusty aggregates or as tiny anhedral grains scattered in the matrix and in veins.

Opaque minerals include magnetite, ilmenite, hematite, and pyrite. Hematite and magnetite occur mainly as replacements of primary chromian spinel in matrix and as

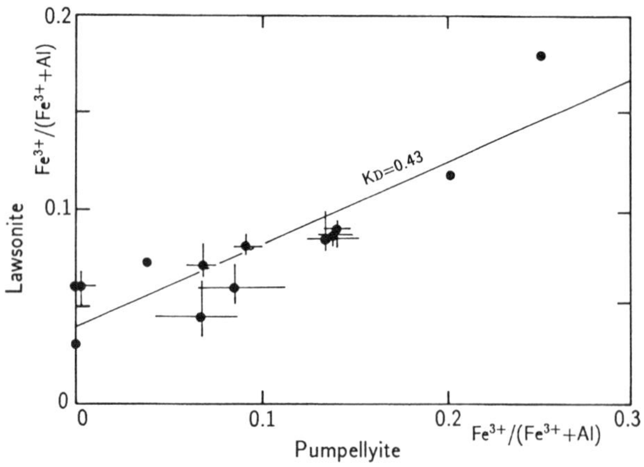

Fig. 4. Fe^{3+}-Al partitioning between lawsonite and pumpellyite. Vertical and horizontal bars represent the ranges of compositional variations in the Fe^{3+}/Al ratio of lawsonite and pumpellyite, respectively.

fine-grained euhedral to subhedral crystals in both matrix and veins. In the former case, these minerals contain variable amounts of Cr_2O_3.

Mineral Parageneses

Since the Hole 778A metabasites are low in metamorphic grade and are very fine-grained, we cannot assume that an entire thin section is in total phase equilibrium. The size of the domain of equilibration must be different from rock to rock, and it should be determined case by case. For the analysis of mineral parageneses, we assume that all phases observed in areas in thin sections no larger than a few millimeters across are in equilibrium with each other if no reaction relation suggesting non-equilibrium is observed. The following discussions may fail to be exact as we have included vein minerals for analyses of mineral parageneses.

The mineral parageneses of the metabasites from Hole 778A were examined by using a pseudoternary system of *Maruyama et al.* [1986] [*Maekawa et al.*, 1992]. In this system, each phase is projected from a constant sub-assemblage: quartz + albite + chlorite + fluids onto the Ca-Fe^{3+}-AFM ($2Al \cdot Fe^{2+}$+Mg) plane, which is normal to the Ca-2Al-(Fe^{2+}+Mg) base and is inclined 80° to the 2Al - Ca join in the Ca-2Al-$2Fe^{3+}$-(Fe^{2+}+Mg) tetrahedron (Figure 5). The projection matrix is calculated using the method described by *Greenwood* [1975]. Winchites from quartz- and albite-free rock were tentatively plotted in Figure 13 of *Maekawa et al.* [1992]. These plots, however, are neglected in Figure 5, because the following discussion is focused on phase relations among Ca-Al silicates and does not deal with the evaluation of these plots. The mineral parageneses observed in the Hole 778A metabasites are as follows.

(1) Lawsonite + pumpellyite + chlorite (12)
(2) Lawsonite + pumpellyite + chlorite + hematite (2)
(3) pumpellyite + sodic pyroxene + chlorite (7)

Parageneses are associated always with quartz and albite and frequently with aragonite. The numbers of thin sections in which each assemblage is observed are in parentheses.

In the Hole 778A metabasites, the characteristic mineral association is lawsonite and pumpellyite. The Fe^{3+} contents in lawsonite show a positive correlation with those of associated pumpellyite, and lawsonite and pumpellyite in hematite-bearing samples contain higher Fe^{3+} contents than those in hematite-free samples. Sodic pyroxene does not occur together with lawsonite. Both minerals are found commonly in a single thin section, but the former always occurs in areas far from the latter; the association lawsonite + sodic pyroxene was probably not stable.

As shown in Figure 5, the important low-variance (three-phase) assemblage is lawsonite + pumpellyite + hematite, and the presence of this assemblage precludes the possible stable association of lawsonite and sodic pyroxene, which is common in the higher-grade blueschists in the Franciscan Complex [*Maruyama et al.*, 1986] and the Kamuikotan metamorphic belt, Japan [*Shibakusa*, 1989].

To clarify paragenetic relations more precisely, the paragenetic relations of the metabasites are again examined using the four component Fe^{3+}-(Fe^{2+}+Mg)-Ca-Al tetrahedron diagram (Figure 6). This diagram well represents the two significant low-variance assemblages observed in the Mariana metabasites, lawsonite + pumpellyite + chlorite + hematite and pumpellyite + chlorite + sodic pyroxene + hematite. The stability field of these parageneses has been examined by using Schreinemakers' method in the Fe^{3+}-(Fe^{2+}+Mg)-Ca-Al system among pumpellyite, lawsonite, epidote, hematite, chlorite, and sodic pyroxene with excess water, quartz, and albite (Figures 7 and 8). The chemical formulae, molar volumes, and entropies of these minerals used in this analysis are given in Table 1. The six univariant reactions and their dP/dT slopes are given in Table 2. After the manner of *Zen* [1966], the absent phase is identified by the phase-label given in parentheses.

In Figure 8, sectors (i), (v), and (vi) are characterized by the association lawsonite + sodic pyroxene that is common in the Franciscan metabasites, and (iii), (iv), and (v) are characterized by chlorite + epidote association. The sectors (i) and (ii) include assemblages with lawsonite + pumpellyite + hematite association. The diagnostic assemblage of the former is lawsonite + pumpellyite + sodic pyroxene + hematite, whereas that of the latter is lawsonite + pumpellyite + chlorite + hematite and includes an assemblage common to the Mariana metabasites: pumpellyite + chlorite + sodic pyroxene + hematite. As mentioned before, the association of pumpellyite + sodic pyroxene is not stable in the Mariana metabasites.

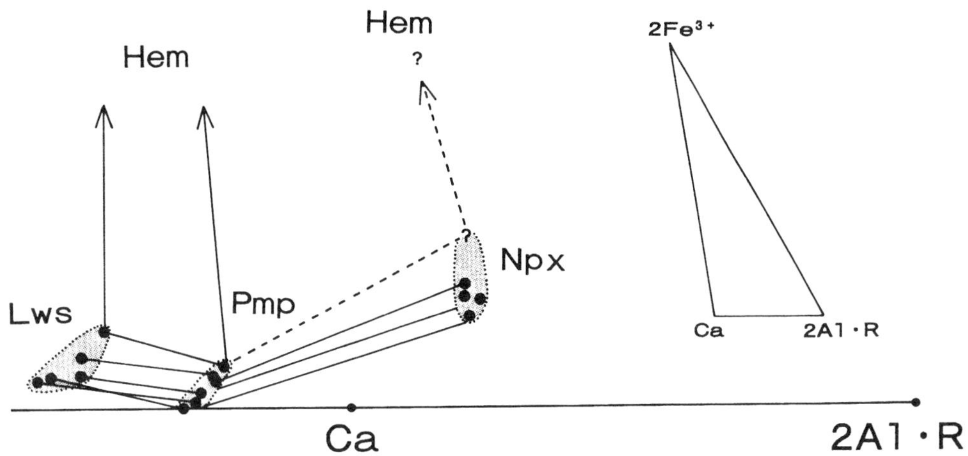

Fig. 5. Paragenetic phase relations of the Hole 778A metabasites graphically represented in the $2Fe^{3+}$ - Ca - AFM (2Al·R) ternary system projected from quartz, albite, chlorite, H_2O, and CO_2. Hem: hematite, Lws: lawsonite, Pmp: pumpellyite, Npx: sodic pyroxene, $R=Mg+Fe^{2+}$.

Therefore, sector (i) is not appropriate, whereas sector (ii) fits the mineral parageneses observed in the Mariana metabasites. Sector (ii) is bounded by the following two reactions:

(E): lawsonite + sodic pyroxene + quartz
= pumpellyite + chlorite + hematite + albite + water
$79CaAl_2Si_2O_7(OH)_2H_2O +$
$110Na_{0.7}Ca_{0.3}R_{0.3}Al_{0.3}Fe_{0.4}Si_2O_6 + 24SiO_2$
$= 28Ca_4RFeAl_4Si_6O_{23}(OH)_3 \cdot 2H_2O +$
$R_5Al_2Si_3O_{10}(OH)_8 + 8Fe_2O_3 + 77NaAlSi_3O_8 +$
$56H_2O$

(N): lawsonite + pumpellyite + hematite + quartz
= epidote + chlorite + water
$2CaAl_2Si_2O_7(OH)_2H_2O +$
$5Ca_4RFeAl_4Si_6O_{23}(OH)_3 \cdot 2H_2O + 3Fe_2O_3 +$
$2SiO_2$
$= 11Ca_2Al_2FeSi_3O_{12}(OH) + R_5Al_2Si_3O_{10}(OH)_8 +$
$12H_2O$

The reaction (E) defines the high-pressure stability limit of the sector (ii). Above the reaction line, the tie-line between lawsonite + sodic pyroxene cuts the plane defined by the stable association of pumpellyite + chlorite + hematite in the sector (ii). As mentioned before, the association, pumpellyite + chlorite + hematite is common in the Hole 778A metabasites, and lawsonite + sodic pyroxene is unstable. The reaction (E) also represents the high-pressure stability limit of the parageneses of the Mariana metabasites.

On the other hand, the reaction (N) defines the high-temperature stability limit of sector (ii). Below the reaction line, the tie-line between epidote + chlorite cuts the stable association of pumpellyite + lawsonite + hematite in sector (ii). The Hole 778A rocks totally lack epidote. According to the analysis by Schreinemakers' method, epidote could occur only in the rocks with a very narrow range of compositions, and it is considered to be unstable in most rocks with ordinary compositions. The presence of lawsonite + pumpellyite + hematite association instead of an epidote-bearing assemblage strongly suggests that the mineral associations observed in the Mariana metabasites represent a lower temperature side of equilibration than those represented by the stable association of epidote + chlorite.

Nakajima et al. [1977] studied the Sanbagawa metamorphism and divided pumpellyite-actinolite facies into two subzones based on this reaction: Fe^{3+}-bearing pumpellyite + hematite + quartz = chlorite + epidote + actinolite + water. The higher-grade subzone is characterized by epidote + pumpellyite association in addition to actinolite and chlorite, and the lower-grade subzone by hematite + pumpellyite association. The Mariana blueschist facies clasts commonly contain metamorphic aragonite. Therefore, as suggested by *Nakajima et al.* [1977], the stable association of lawsonite + pumpellyite + hematite is considered to occur under higher-pressure conditions than that of pumpellyite + hematite of the pumpellyite-actinolite facies in the Sanbagawa metamorphism, in which calcite is ubiquitously stable instead of aragonite.

It is concluded that the stable association of lawsonite + pumpellyite + hematite in the Mariana metabasites only occurs under relative low-temperature and low-pressure conditions of blueschist-facies metamorphism. We strongly suggest that this assemblage is important in that this probably characterizes the lowest-grade incipient blueschist-facies metamorphism in general.

Maekawa et al. [1992] estimated the approximate metamorphic conditions for the Mariana blueschist-facies clasts at about 150 - 250°C and 5 - 6 kb (see Maekawa et al. [1992], figure 14).

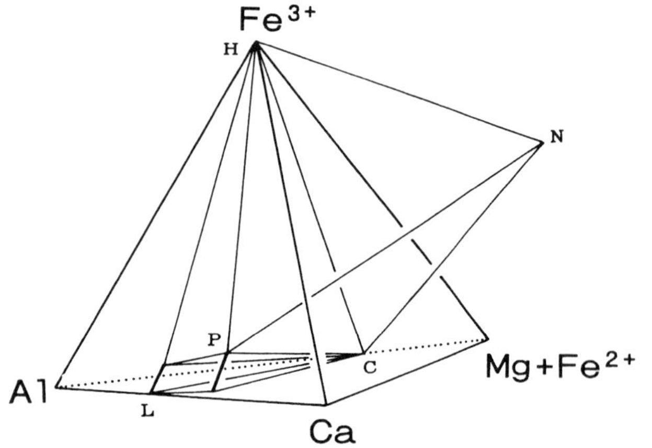

Fig. 6. Mineral parageneses of the Hole 778A metabasites in the Fe^{3+}-(Fe^{2+}+Mg)-Ca-Al tetrahedron. L: lawsonite, P: pumpellyite, N: sodic pyroxene, H: hematite.

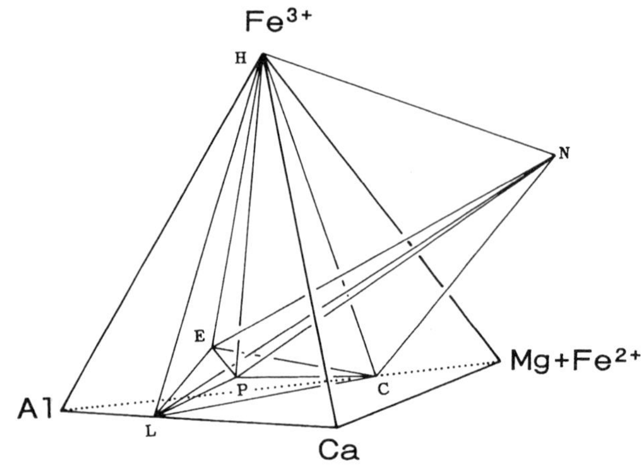

Fig. 7. Plots of six phases used in the phase analyses in the Fe^{3+}-(Fe^{2+}+Mg)-Ca-Al tetrahedron. E: epidote. Other abbreviations are the same as in Fig. 6.

SIGNIFICANCE OF THE INCIPIENT BLUESCHIST FACIES METAMORPHISM IN THE MARIANA FOREARC

Since *Eskola* [1929, 1939] first advocated the glaucophane-schist facies as an independent metamorphic facies, many scientists have engaged in geological, petrological, and geophysical studies of rocks of this facies. The works of *de Roeve* [1955a, b] and *Miyashiro and Banno* [1958] established the idea of the "glaucophane-schist facies" as one of the main metamorphic facies. Although some had asserted the idea of soda metasomatism instead of regional glaucophane-schist facies metamorphism [e.g., *Taliaferro*, 1943], this meta-morphic facies is now widely accepted.

Miyashiro [1961] proposed the idea of paired metamorphic belts in which a metamorphic belt of low- to moderate-pressure type and another metamorphic belt of high-pressure type occur side by side along the convergent plate margins of the circum-Pacific region. He pointed out that the distribution of the latter belt is always limited to the Pacific Ocean side.

On the basis of heat-flow measurements, *Takeuchi and Uyeda* [1965] suggested that high-pressure, low-temperature metamorphism could occur at the continental margin where low heat flow prevails because of the subducting cold oceanic plate. *Oxburgh and Turcotte* [1970] and *Hasebe et al.* [1970] have computed thermal structures of island arcs and indicated a low geothermal gradient in the subduction zones. Since these pioneering works, many scientists have performed heat flow measurements and constructed thermal models to clarify the thermal structures of subduction zones [see *Fryer and Fryer*, 1987; *Peacock,* 1989]. Now, it is widely accepted that the high-pressure blueschist facies metamorphism occurs in the subduction zone where the descent of a cold slab leads to the occurrence of unusually low temperatures at mantle pressures [*Miyashiro*, 1967; *Dewey and Bird*, 1970; *Ernst,* 1971, 1973; *Coleman*, 1972]. Thermal modeling also shows that broad regions underneath the outer forearc are subjected to blueschist-facies metamorphism very early (within 100,000 yrs of initiation of subduction) in the evolution of convergent margins such as the Mariana system [*Fryer and Fryer*, 1987].

Until now, however, the tectonic process of deep-seated blueschist-facies rocks has not been well-clarified. The advantage to studying blueschist-facies metamorphism in the Mariana forearc is that blueschist-facies rocks formed at depths have been brought to the seafloor by upwelling serpentinite materials along trench-parallel fault planes. Since Eocene time, the Pacific Plate has been subducting along the Mariana Trench below the Philippine Sea Plate [*Uyeda and Ben-Avraham*,1972; *Seno and Maruyama*, 1984]. The mantle peridotite above the subducting slab has reacted with water to form serpentinite, and the resulting low-density serpentinite has risen to form a zone of seamounts between 50 and 150 km from the trench [*Fryer et al.,* 1985]. Serpentinite materials, which come up from depths to the seafloor, must have trapped oceanic mantle and crust materials situated within the pathway and entrained them in a fluidized melange rising to the seafloor. Thus, it is reasonable to assume that the blueschist-facies clasts recovered from Hole 778A formed at about 16 to 20 km below the seafloor after initial subduction and were then entrained by uprising fluidized serpentinite materials. Recently we also found sodic pyroxene from a metabasite dredged from a forearc seamount in the Izu-Bonin forearc. The metabasite probably formed under similar conditions to blueschist-facies clasts from the Mariana forearc [Maekawa, in press]. High-pressure blueschist-facies rocks seem to be

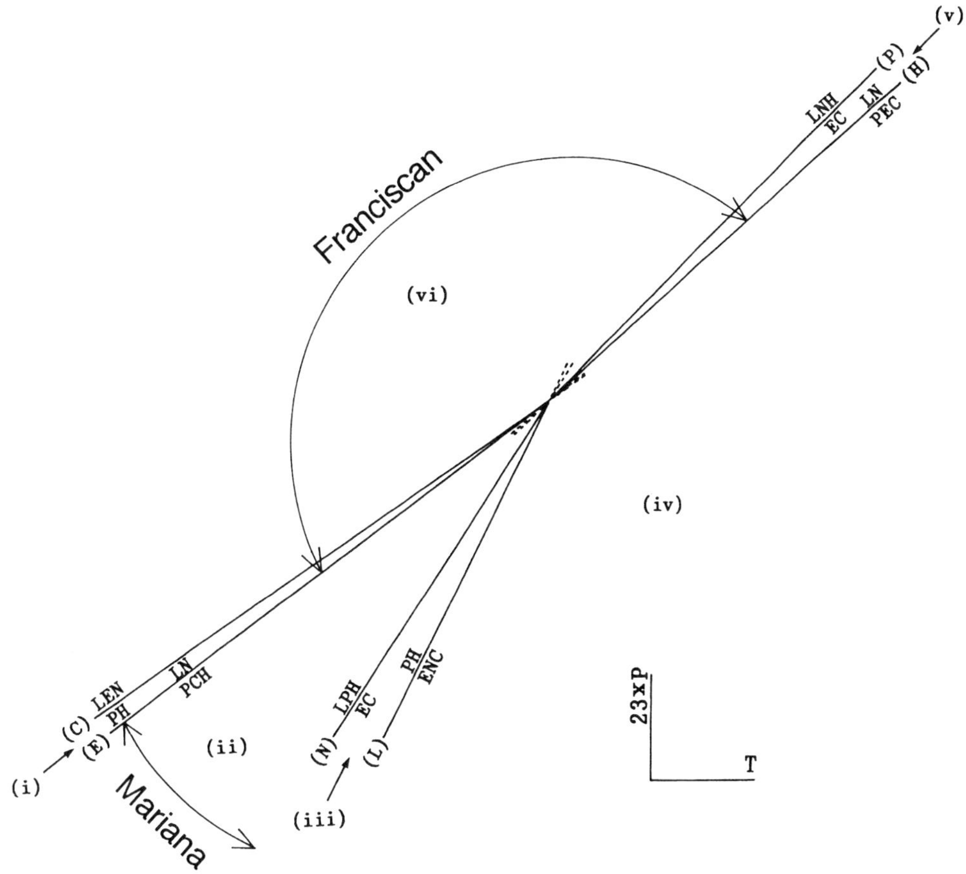

Fig. 8. The Schreinemakers' bundle for the six-phase equilibrium system lawsonite (L), pumpellyite (P), epidote (E), sodic pyroxene (N), chlorite (C), and hematite (H).

TABLE 1. Chemical Formulae, Molar Volumes, and Entropies of Phases used in Schreinemakers' Method.

		Na	Ca	R	Al	Fe^{3+}	Si	H	V(298,1) (cm^3/mol)	S(298,1) (cal/molK)
Lawsonite	(L)	0	1	0	2	0	2	4	101.32(1)	55.8(2)
Pumpellyite	(P)	0	4	1	4	1	6	7	295.6(3)	153.01(3)
Epidote	(E)	0	2	0	2	1	3	1	139.2(2)	75.28(2)
Sodic pyroxene	(N)	0.7	0.3	0.3	0.3	0.4	2	0	60.4(1)	31.9(1)
Chlorite	(C)	0	0	5	2	0	3	8	207.11(2)	111.2(2)
Hematite	(H)	0	0	0	0	2	0	0	30.27(2)	20.94(2)
Albite	(A)	1	0	0	1	0	3	0	100.07(1)	49.51(4)
Quartz	(Q)	0	0	0	0	0	1	0	22.69(2)	9.88(2)
Water	(W)	0	0	0	0	0	0	2	19.8(5)	12(6)

Volume and entropy data for water are of 350°C, 3.5kb. (1) *Robie and Waldbaum* [1968], (2) *Helgeson et al.* [1978], (3) *Schiffman and Liou* [1980], (4) *Openshaw et al.* [1976], (5) *Burnham et al.* [1969], (6) *Fyfe et al.* [1958]. R=Mg+Fe^{2+}

TABLE 2. Six Univariant Reactions and dP/dT Slopes

	Univariant reactions	dP/dT
(H)	23L+30N+8Q=4P+8E+C+21A+24W	18.23
(C)	7L+E+10N+2Q=3P+H+7A+4W	14.51
(N)	2L+5P+3H+2Q=11E+C+12W	32.20
(E)	79L+110N+24Q=28P+C+8H+77A+56W	15.71
(P)	41L+50N+4H+16Q=28E+3C+35A+56W	20.24
(L)	41P+23H+14A+10Q=79E+20N+7C+76W	42.03

not rare in the Izu-Bonin and Mariana forearc seamounts. It is concluded that the presence of blueschist-facies rocks in the forearc seamounts confirms that a blueschist-facies metamorphism actually took place beneath the forearc in the Mariana and Izu-Bonin arc-trench system.

Metamorphic rocks we found from ODP Leg 125 are incipient blueschist-facies rocks. They must have formed at shallower depth than the typical blueschists in the subduction zone. Similar on-land blueschist-facies rocks have been reported from northwest Turkey [*Okay*, 1982], Baja California [*Sedlock*, 1988], and the Cascade Range of the northwest U.S. [*Glassley et al.*, 1976; *Brandon et al.*, 1988; *Brandon,* 1990]. It is difficult and time-consuming to detect incipient blueschist-facies rocks, for metamorphic minerals are very fine-grained. Routine SEM and EPMA analysis are necessary to determine mineral species. Incipient blueschist-facies rocks, however, are very important to the understanding of the interrelationships between tectonic processes and subduction metamorphism. They are sure to preserve the nature of accretion processes, subsequent transportation to the depth, and return path to the surface better than higher-grade rocks, because recrystallization is commonly incomplete. The originally shallower position of these incipient blueschist-facies rocks prevents them from undergoing major disturbance during later uplift.

We obtained only incipient blueschist-facies rocks from ODP Leg 125. However, there is a high probability that higher-grade blueschists and eclogites exist within the serpentinite materials in the Mariana forearc. Further investigation of dredged and drilled samples is necessary in order to understand the interrelationships between metamorphism and subduction tectonics in more detail.

Acknowledgments. We thank H. Shibakusa and T. Nakajima for valuable comments and Y. Tanabe for providing a computer program for the calculation of Schreinemakers' method. In addition we thank S. Maruyama and R. G. Coleman for their careful critiques of the manuscript. This research has been supported by grant from the Japanese Hyogo Science and Technology Association (to HM). This is School of Ocean and Earth Science and Technology contribution no. 3561 and Hawaii Institute of Geophysics and Planetology contribution no. 766.

REFERENCES

Brandon, M. T., High-pressure metamorphism and uplift of the Olympic subduction complex, *Geology, 18,* 1252-1255, 1990.

Brandon, M. T., D. S. Cowan, and J. A. Vance, The Late Cretaceous San Juan thrust system, San Juan Islands, Washington, *Geol. Soc. Am. Spec. Pap. 221,* 81pp., 1988.

Burnham, C. W., J. R. Holloway, and N. F. Davis, Thermodynamic properties of water to 1,000°C and 10,000 bars, *Geol. Soc. Am. Spec. Pap. 132,* 96 pp., 1969.

Ciampo, G., Late Miocene (CN9B subzone) to Holocene calcareous nannofossils from the western and northwestern Pacific Ocean, in *Proc. ODP, Sci. Results, 125,* P. Fryer, J. A. Pearce, L. G. Stokking et al., pp. 15-41, Ocean Drilling Program, College Station, TX , 1992.

Coleman, R. G., Blueschist metamorphism and plate tectonics. *Proc. Int. Geol. Congr., 24th Session,* Section 2, pp. 19-26, 1972.

de Roever, W. P., Genesis of jadeite by low-grade metamorphism, *Am. J. Sci., 253,* 283-298, 1955a.

de Roever, W. P., Some remarks concerning the origin of glaucophane in the North Berkeley Hills, California, *Am. J. Sci., 253,* 240-244, 1955b.

Dewey, J. F., and J. M. Bird, Mountain belt and the new global tectonics. *J. Geophys. Res., 75,* 2625-2647, 1970.

Ernst, W. G., Metamorphic zonations on presumably subducted lithospheric plates from Japan, California and the Alps. *Contrib. Mineral. Petrol., 34,* 43-59, 1971.

Ernst, W. G., Blueschist metamorphism and P-T regimes in active subduction zones, *Tectonophysics, 17,* 255-272, 1973.

Eskola, P., Om mineralfacies, *Geol. Foren. Stockholm Forh., 51,* 157-172, 1929.

Eskola, P., Die metamorphen Gesteine, in *Die Entstehung der Gesteine*, edited by Tom. F. W. Barth, C. W. Correns, u. P. Eskola, pp. 263-407, Julius Springer, Berlin, 1939.

Fryer, P., E. L. Ambos, and D. M. Hussong, Origin and emplacement of Mariana forearc seamounts, *Geology, 13,* 774-777, 1985.

Fryer, P., and G. J. Fryer, Origins of nonvolcanic seamounts in a forearc environment, in *Seamounts, Islands, and Atolls, Geophys. Monogr. Ser., vol. 43,* edited by B. Keating, P. Fryer, R. Batiza, and G. W. Boehlert, pp. 61-69, AGU, Washington, D. C., 1987.

Fryer, P., J. A. Pearce, and L. B. Stokking et al., *Proc. ODP, Init. Repts., 125,* Ocean Drilling Program, College Station, TX, 1990.

Fyfe, W. S., F. J. Turner, and J. Verhoogen, Metamorphic reactions and metamorphic facies, *Geol. Soc. Am. Mem. 75,* 253pp., 1958.

Glassley, W. E., J. T. Whetten, D. S. Cowan, and J. A. Vance, Significance of coexisting lawsonite, prehnite, and aragonite in the San Juan Islands, Washington, *Geology, 4,* 301-302, 1976.

Greenwood, H. J., Thermodynamically valid projections of extensive phase relationships, *Am. Mineral., 60,* 1-8, 1975.

Haggerty, J. A., Fluid inclusion studies of chimneys associated with serpentinite seamounts in the Mariana forearc, PACROFI II, 2, 29, 1989.

Hasebe, K., N. Fujii, and S. Uyeda, Thermal processes under island arcs, *Tectonophysics, 10,* 335-355, 1970.

Helgeson, H. C., J. M. Delany, H. W. Nesbitt, and D. K. Bird, Summary and critique of the thermodynamic properties of rock-forming minerals, *Am. J. Sci., Spec. Vol. 278-A,* pp.1-229, 1978.

Hussong, D. M., and S. Uyeda, Tectonic processes and the history of the Mariana arc: A synthesis of the results of Deep Sea Drilling Project Leg 60, in *Init. Repts. DSDP*, 60, D. M. Hussong, S. Uyeda, et al., pp. 909-929, Ocean Drilling Program, College Station, TX, 1981.

Johnson, L. E., Mafic clasts in serpentine seamounts: petrology and geochemistry of a diverse crustal suite from the outer Mariana forearc, in *Proc. ODP, Sci. Results, 125,* P. Fryer, J. A. Pearce, L. G. Stokking et al., pp. 401-413, Ocean Drilling Program, College Station, TX , 1992.

Leake, B. E., Nomenclature of amphiboles, *Am. Mineral., 63,* 1023-1053, 1978.

Maekawa, H., Metamorphic rocks from serpentinite seamounts in the Mariana and Izu-Bonin forearcs, in *Geology and Geophysics of the Philippine Sea Floor,* edited by S. A. Shcheka and H. Tokuyama, Japanese-Russian-Chinese Monograph (in press).

Maekawa, H., M. Shozui, T. Ishii, P. Fryer, and J. A. Pearce, Blueschist metamorphism in an active subduction zone, *Nature, 364,* 520-523, 1993.

Maekawa, H., M. Shozui, T. Ishii, C. Saboda, and Y. Ogawa, Metamorphic rocks from the serpentinite seamounts in the Mariana and Izu-Ogasawara forearcs, in *Proc. ODP, Sci. Results, 125,* P. Fryer, J. A. Pearce, L. G. Stokking et al., pp. 415-430, Ocean Drilling Program, College Station, TX , 1992.

Maruyama, S., M. Cho, and J. G. Liou, Experimental investigations of blueschist-greenschist transition equilibria: pressure dependence of Al_2O_3 contents in sodic amphiboles- a new geobarometer, in *Blueschists and Related Eclogites,* edited by B. W. Evans and E. H. Brown, *Geol. Soc. Am. Mem. 164,* pp. 1-16, 1986.

Maruyama, S., and J. G. Liou, Petrology of Franciscan metabasites along the jadeite-glaucophane type facies series, Cazadero, California, *J. Petrol., 29,* 1-37, 1987.

Miyashiro, A., Evolution of metamorphic belts. *J. Petrol., 2,* 277-311, 1961.

Miyashiro, A., Orogeny, regional metamorphism, and magmatism in the Japanese Islands, *Medd. Dansk Geol. Foren., 17,* 390-446, 1967.

Miyashiro, A., and S. Banno, Nature of glaucophanitic metamorphism, *Am. J. Sci., 256,* 97-110, 1958.

Mottl, M. J., Pore waters from serpentinite seamounts in the Mariana and Izu-Bonin forearcs, Leg 125: Evidence for volatiles from the subducting slab, in *Proc. ODP, Sci. Results, 125,* P. Fryer, J. A. Pearce, L. G. Stokking et al., pp. 373-385, Ocean Drilling Program, College Station, TX , 1992.

Nakajima, T., Banno, S., and Suzuki, T., Reactions leading to the disappearance of pumpellyite in low-grade metamorphic rocks of the Sanbagawa metamorphic belt in central Shikoku, Japan, *J. Petrology, 18,* 263-284, 1977.

Okay, A. I., Incipient blueschist metamorphism and metasomatism in the Tavsanli Region, Northwest Turkey, *Contrib. Mineral. Petrol., 79,* 361-367, 1982.

Openshaw, R. E., B. S. Hemingway, R. A. Robie, D. R. Waldbaum, and K. M. Krupka, The heat capacities at low temperatures and entropies at 298.15K of low albite, microcline, and high–sanidine, *U. S. Geol. Surv. J. Res. 4,* 195-204, 1976.

Oxburgh, E. R., D. L. Turcotte, Thermal structure of island arcs, *Geol. Soc. Am. Bull., 81,* 1665-1688, 1970.

Peacock, S. M., Thermal modeling of metamorphic pressure-temperature-time paths: a forward approach, in *Short Course in Geology, 7,* edited by F. S. Spear and S. M. Peacock, pp. 57-102, Am. Geophys. Un., Washington, D. C., 1989.

Robie, R. A., and D. R. Waldbaum, Thermodynamic properties of minerals and related substances at 298.15 K (25.0°C) and one atmosphere (1013 bars) pressure and at high temperature, *Geol. Surv. Bull. 1259,* 1968.

Sakakibara, M., A newly discovered high-pressure terrane in eastern Hokkaido, Japan, *J. metamorphic Geol., 4,* 401-408, 1986.

Schiffman, P., and J. G. Liou, Synthesis and stability relations of Mg-Al pumpellyite, $Ca_4Al_5MgSi_6O_{21}(OH)_7$, *J. Petrology 21,* 441-474, 1980.

Sedlock, R. L., Metamorphic petrology of a high-pressure, low-temperature subduction complex in west-central Baja California, Mexico, *J. Metamorphic Geol., 6,* 205-233, 1988.

Seno, T., and Maruyama, S., Paleogeographic reconstruction and origin of the Philippine Sea, *Tectonophysics, 102,* 53-84, 1984.

Shibakusa, H., Lawsonite-pumpellyite-epidote stabilities in glaucophane schists in the Horokanai-Kamietanbetsu area of Kamuikotan zone, Hokkaido, Japan, *Mineral. Petrol., 40,* 241-256, 1989.

Takeuchi, H., and S. Uyeda, A possibility of present-day regional metamorphism, *Tectonophysics, 2,* 59-68, 1965.

Taliaferro, N. L., Franciscan-Knoxville problem, *Bull. Am. Assoc. Petrol. Geol., 27,* 109-219, 1943.

Uyeda, S., and Z. Ben-Avraham, Origin and development of the Philippine Sea, *Nature, 240,* 176-178, 1972.

Zen, E-an, Construction of pressure-temperature diagrams for multicomponent systems after the method of Schreinemakers-a geometric approach, *Bull. U. S. Geol. Surv., 1225,* 56 pp., 1966.

Hirokazu Maekawa, Department of Earth and Planetary Sciences, Faculty of Science, Kobe University, Nada, Kobe 657, Japan

Patricia Fryer, Department of Geology and Geophysics, School of Ocean and Earth Science and Technology, University of Hawaii at Manoa, 2525 Correa Rd, Honolulu, Hawaii 96822, USA

Atsuo Ozaki, Setsuryo high school, Shukunosyo, Ibaraki 567, Japan

Sedimentation in Western Pacific Backarc Basins: New Insights From Recent ODP Drilling

Kathleen M. Marsaglia

Department of Geological Sciences, University of Texas at El Paso, El Paso, Texas

Sam Boggs, Jr.

Department of Geological Sciences, University of Oregon, Eugene, Oregon

Peter Clift

Ocean Drilling Program, College Station, Texas

Abbas Seyedolali

Department of Geological Sciences, University of Oregon, Eugene, Oregon

Randall Smith

Department of Geology, California State University, Fresno, California

During the most recent phase of ocean drilling in the western Pacific, a series of Ocean Drilling Program (ODP) legs explored the history of sedimentation within several types of backarc basins: a nascent intraoceanic backarc basin (Sumisu Rift, leg 126), a young intraoceanic backarc basin (Lau Basin, leg 135), a continental margin backarc basin (Japan Sea, legs 127 and 128), and marginal basins of uncertain origin (e.g., Sulu Sea, leg 124). The Sumisu Rift is a nascent backarc basin characterized by high sediment accumulation rates ($\geq$4 km/yr.) and thick, coarse pumiceous units derived from periodic eruptions of nearby arc calderas and deposited by turbidity-current and mass-flow processes. The mixed volcanic provenance of the Sumisu basin fill can be attributed to syneruptive mixing or mixing during transport into the basin. Sites drilled in a series of sub-basins in the Lau backarc region show similar upward-fining sequences with basal coarse-grained, proximal deposits (volcanic sands and gravels that grade upward into nannofossil-ooze dominated sections with rare air-fall ash beds. The proximal facies record an apparent eastward shift in volcanism across the basin during the Plio-Pleistocene. Geochemical analyses of pyroclastic debris within the coarse basal sections are consistent with local intrabasinal volcanic sources. Lower Miocene volcaniclastic sections recovered in the Sulu Basin support a backarc origin for this basin. These deposits represent the basinal portion of an arc-derived volcaniclastic apron that prograded basinward soon after formation of the underlying backarc-basin crust. Subsequent influx of continental clastics into the Sulu Basin during the middle to late Miocene is attributed to a sea-level lowstand. The Japan Sea is a continental backarc basin that contains Quaternary to Miocene sand with both magmatic-arc and continental provenance. Diagenetic modification of more-deeply buried Miocene sandstones has resulted in some loss of provenance information.

1. INTRODUCTION

Several Deep Sea Drilling Project (DSDP) legs were wholly or partly devoted to understanding backarc basin formation and sedimentation. Although a few sites were

Active Margins and Marginal Basins of the Western Pacific
Geophysical Monograph 88

drilled in the Japan Sea and the Lau Basin, drilling efforts were concentrated in the Philippine Sea. In particular, a transect of sites across the Parece Vela Basin and the Mariana Trough provided the basis for models of intra-oceanic backarc basin evolution and sedimentation (e.g., *Carey and Sigurdsson* [1984] and *Klein* [1985]). However, this drilling left some questions unanswered: the early history of backarc basin formation, including the possible role of forearc rifting; the history of volcanism across the arc during rifting; and the processes of basin formation and sedimentation in continental-margin backarc basins.

These questions were addressed from 1988 to 1991, when a series of ODP legs explored the sedimentation history within the Sulu, Sumisu, Japan, and Lau backarc basins (Figure 1). Results from these legs have greatly improved our knowledge of backarc basin facies relationships and sediment provenance. Within this paper, these results are first summarized, focussing on the depositional styles and provenance of clastic/pyroclastic debris; their implications for current models of backarc-basin sedimentation are then discussed.

2. BACKGROUND

From 1988 to 1991, the Ocean Drilling Program examined several types of backarc basins: nascent and young intra-oceanic-backarc basins (legs 126 and 135), a continental-margin backarc basin (legs 127 and 128), and marginal basins of uncertain origin (leg 124).

The Sumisu Rift is a nascent backarc basin located along the Izu-Bonin arc south of the Japanese islands (Figures 1 and 2); the major objectives of leg 126 included the determination of the time of initiation and development of this basin [*Taylor et al.*, 1990a]. To attain these goals, two sites were drilled on the basin floor and one on the eastern rift-flank uplift (Figures 2 and 3). Despite drilling difficulties, unconsolidated pumiceous sediments were successfully penetrated and the resulting core provides a nearly complete record of rapid sediment accumulation in this basin. Arc volcanoes along this arc segment take the form of submerged felsic calderas, which are thought to have produced the pumice deposits found in the rift [*Nishimura et al.*, 1991;1992].

The Lau Basin has progressed to the seafloor spreading stage of backarc basin formation, a more advanced stage of development than that of the Sumisu Rift. Drilling in the Lau Basin on leg 135 revealed a surprising history of migrating volcanism and sedimentation in a young intra-oceanic backarc basin thought to have developed by forearc rifting close to the volcanic front [*Parson et al.*, 1992]. Several sites were drilled across this backarc region in linear, sediment filled sub-basins (Figures 1, 4 ,and 5).

Continental backarc basin development and sedimentation were the focus of ODP legs 127 and 128 in the Japan Sea (Figures 1 and 6). The sites drilled during these legs are

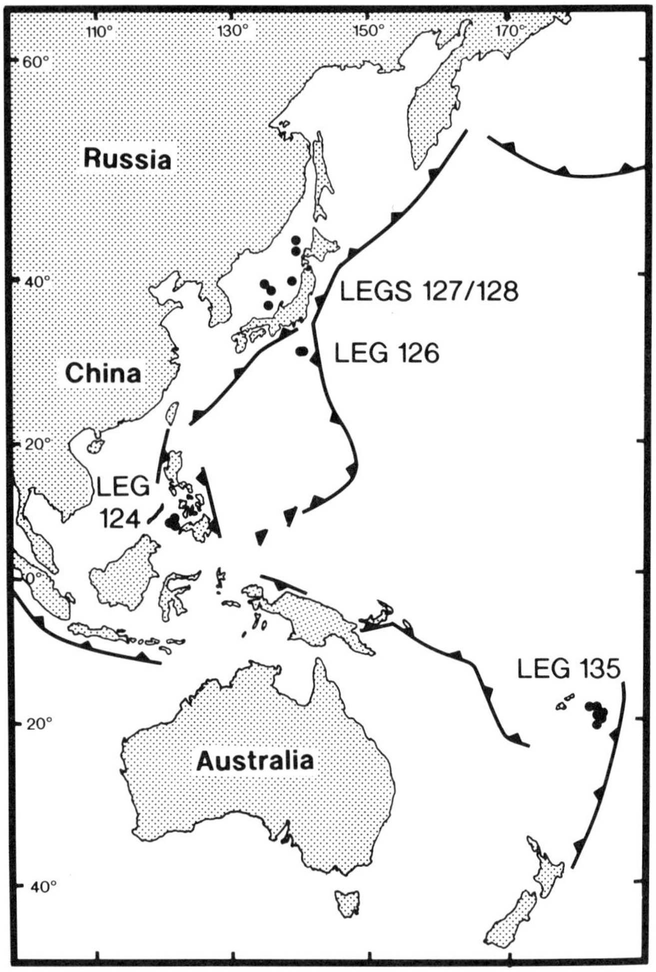

Fig. 1. Map of the western Pacific showing the general location of Ocean Drilling Program Leg sites (filled circles) discussed in this paper. See Figures 2, 4, 6 and 8 for detailed site locations.

located in basinal areas and on high-standing blocks. The nearly complete sedimentary sections and series of volcanic rocks recovered on these legs (e.g., Figure 7) allow for detailed reconstruction of basin evolution and depositional history.

Leg 124 set out to determine the origin of two marginal basins, the Celebes and Sulu Seas, located within a tectonically complex region near the junction of the Pacific, Australian, and Philippine Sea plates [*Rangin et al.*, 1990]. Although information on the nature and age of the basement in these basins was critical to understanding their origin, the sedimentary record also provided significant information. The stratigraphy preserved in the Sulu Basin suggests that it formed as the result of backarc spreading, and so we center our discussion on this basin. Three sites

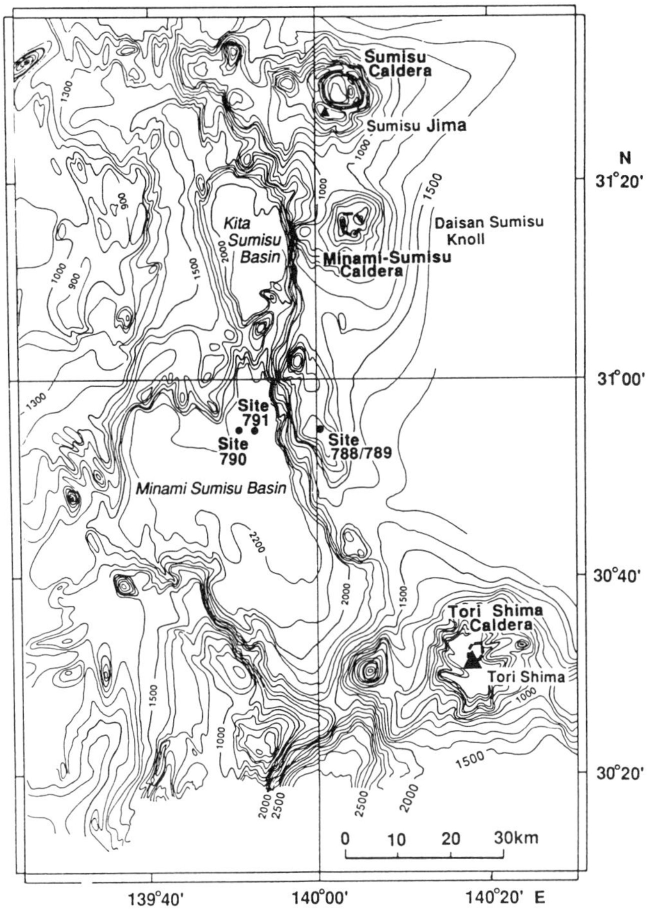

Fig. 2. Detailed bathymetric map of the Sumisu Rift region, Izu-Bonin arc, showing the location of ODP Leg 126 sites and arc calderas. The Sumisu Rift is subdivided into two sub-basins, the Kita (north) Sumisu Basin and the Minami (South) Sumisu Basin by a volcanic ridge. Adapted from *Nishimura et al.* [1992].

were drilled, one on the Cagayan ridge which divides the basin, and two in the SE Sulu Basin (Figures 8 and 9).

3. SUMISU RIFT

3.1. Introduction

The Sumisu Rift is an asymmetrical, normal-fault-bounded basin located just west of the Izu-Bonin arc axis. Basin architecture and surficial sedimentation and volcanism were initially summarized by investigators at the Hawaii Institute of Geophysics and the Geological Survey of Japan prior to ODP drilling [e.g., *Brown and Taylor*, 1988; *Murakami*, 1988; *Nishimura and Murakami*, 1988; *Fryer et al.*, 1990; *Hochstaedter et al.*, 1990; *Taylor et al.*, 1990b]. According to later work by *Klaus et al.* [1992], the basin is subdivided into two depocenters, the Kita and Minami Sumisu basins (Figure 2), by a chain/ridge of submarine volcanoes located on a transfer zone linking opposing border faults. The sub-basin floors are smooth, deep (2100-2275 m) plains that gently shallow to the north and south toward the volcaniclastic aprons that flank adjacent arc volcanoes and calderas [*Klaus et al.*, 1992]. Submarine volcanism within the rift is structurally controlled and concentrated along fault zones [*Klaus et al.*, 1992].

Although this rift has not yet developed into the seafloor spreading phase, fragments of mafic volcanic rock recovered from submarine outcrops by submersible and mafic basement rocks drilled within the basin at sites 790 and 791 are geochemically "backarc basin basalts" [*Gill et al.*, 1992]. Drilling greatly elucidated the nature of the basin fill and the history of sedimentation on the rift-flank, constraining the age of rift inception at approximately 2 Ma [*Taylor*, 1992].

3.2. Volcanism

The Sumisu Rift lies in a volcanically active region (Figure 10) containing over 200 volcanic centers [*Brown and Taylor*, 1988], where volcanic products range from tholeiitic basalt to sodic rhyolite [*Fryer et al.*, 1985]. With rift inception and the resultant change in stress regime along the arc, volcanism became primarily dacitic to rhyolitic and large submarine calderas developed [*Gill et al.*, 1992]. Quaternary tephras within the rift are bimodal in composition and belong to the low-alkali tholeiitic series [*Fujioka et al.*, 1992; *Hiscott and Gill*, 1992]; they are thought to be the differentiation products of subcrustal basalt [*Gill et al.*, 1992]. The composition of pyroxene phenocrysts within the tephras plot within the nonalkali basalt field [*Fujioka and Saito*, 1992], and in general, mineral and whole-rock geochemical analyses indicate source rocks of island arc character [*Fujioka and Saito*, 1992; *Hiscott and Gill*, 1992].

3.3. Sedimentary Facies

As a result of differential subsidence, the basin-fill geometry of the Sumisu Rift is wedge-like, thickening toward the eastern margin (Figure 3). At both basinal sites, vesicular basalt is overlain by a Quaternary sedimentary sequence that increases in thickness from site 790 to site 791 (Figure 3). The basal sedimentary section at each site is composed predominantly of nannofossil mud and mudstone, which is overlain by a coarse pumiceous volcaniclastic sequence. Apparent sediment accumulation rates calculated for the basinal deposits range up to 4 km/m.y. at site 791 (Figure 3). A Pliocene to Quaternary sequence of predominantly coarse-grained pumiceous sediments was recovered at site 788 on the rift flank (Figure 3).

A fine-grained muddy facies dominates the oldest infill of the Sumisu Rift (Figure 3). This facies is nannofossil-rich

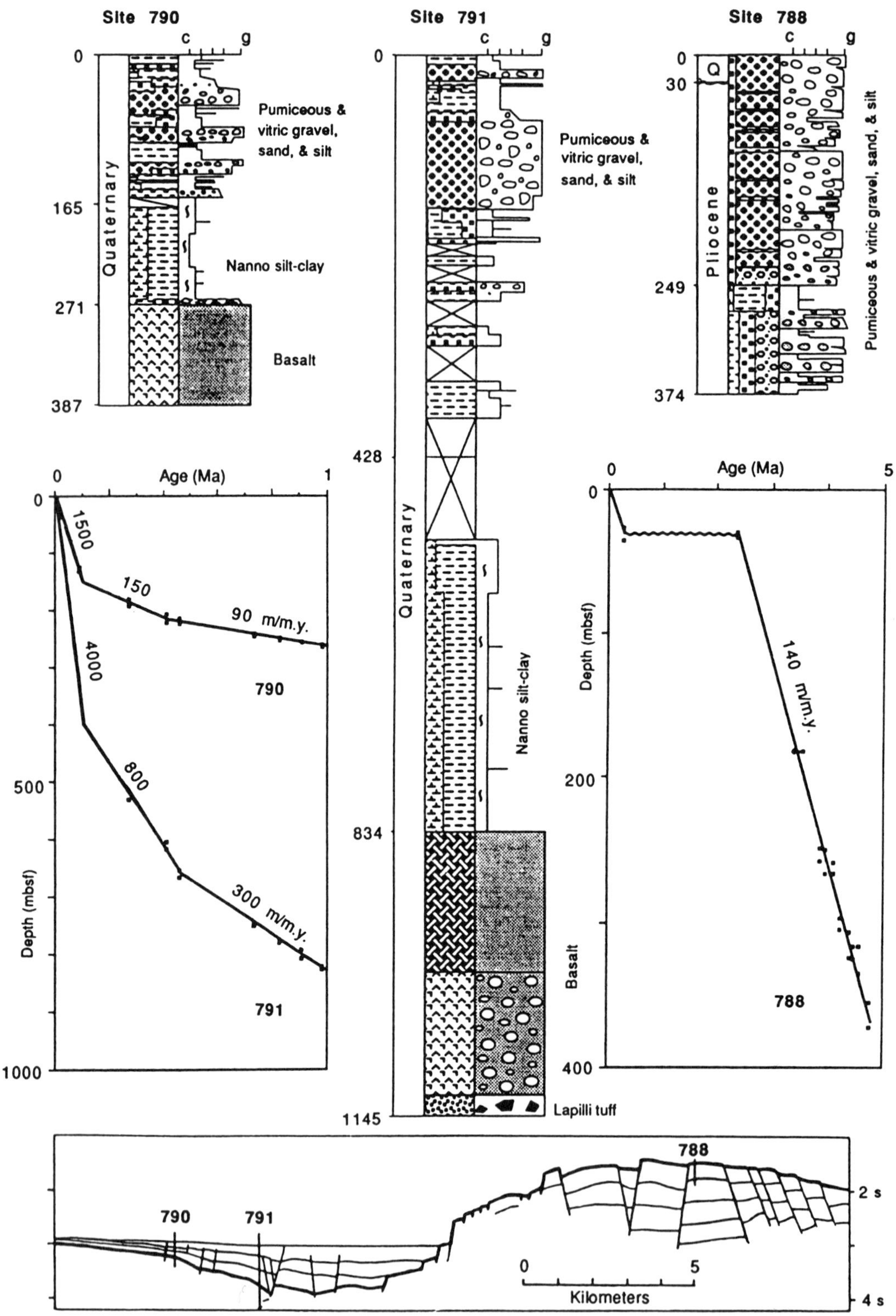

Fig. 3. Diagram summarizing the lithostratigraphy for Leg 126 sites 788, 790, and 791. Units are not pictured but unit boundaries are marked on the left side of the stratigraphic columns in meters below sea floor (mbsf). Relative grain size variations are profiled on the right, from clay (c) to gravel (g). The schematic cross section at the base is simplified from a seismic profile that links the sites. From *Taylor* [1992].

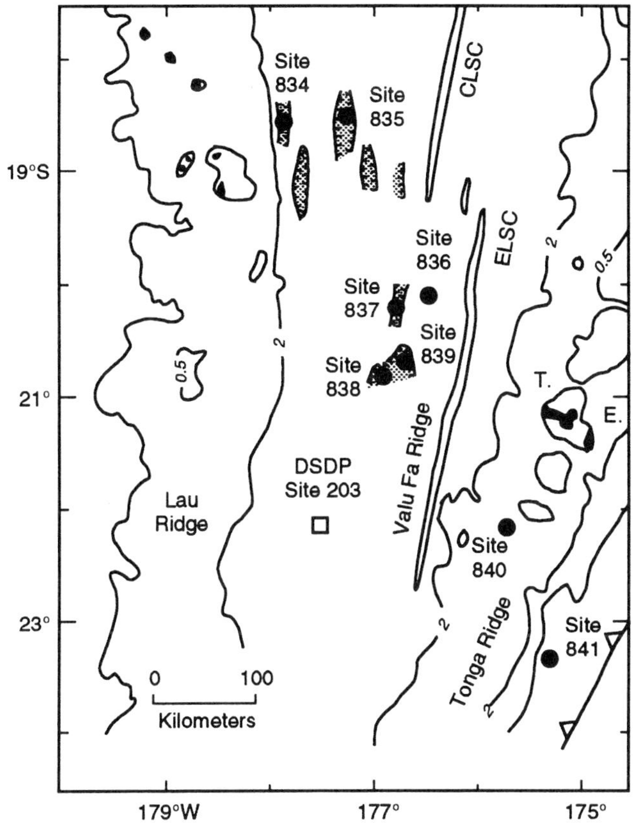

Fig. 4. Map of backarc sites drilled in the Lau Basin during Leg 135 with outline bathymetry (500 and 2000 m isobaths). Contour interval in kilometers. Known sub-basins are stippled. ELSC= East Lau spreading center, CLSC = Central Lau spreading center, T = Tongatapu, and E = 'Eua.

and is characterized by medium to very thick graded beds with local bases of dark silt and sand; the massive appearance of the muddy beds was attributed to intense bioturbation by shipboard scientists [*Taylor et al.*, 1990a], but reinterpreted as a product of mass-flow processes and/or anaerobic conditions by *Colella et al.* [1992].

The layered coarse clastic sequences that comprise the uppermost part of the basin fill are the likely products of turbidity-current and gravity-flow deposition off the flanks of arc calderas and the steep eastern margin of the rift [Figure 11; *Taylor*, 1990a; *Colella et al.*, 1992; *Klaus et al.*, 1992]. These sequences consist of (1) structureless to normally graded beds of pumiceous gravel, volcaniclastic sand and vitric silt; (2) massive to parallel-laminated graded beds of dark volcaniclastic sand that pass upward into mud; (3) graded muddy intervals with isolated volcanic clasts; and (4) thin white to green beds of ash.

Beds of rhyolitic pumiceous gravel range in excess of 20 m in thickness and are thought to have been produced by caldera eruptions along the frontal arc [*Nishimura et al.*, 1992]. The pumiceous gravels are considered to be the coarse basal portions of up to five large-scale, eruption-related, upward-fining sedimentary sequences, that range up to 118 m in thickness [*Nishimura et al.*, 1992]. These pumiceous units exhibit internal layering, grading, and local muddy intraclasts attributed to short-term remobilization and redistribution of material by gravity-flow processes after the initial explosive submarine eruption. They are separated by hemipelagic intervals with intercalated ash layers thought to represent longer quiescent periods between major eruptions [*Nishimura et al.*, 1992]. These eruptions occurred in the last 150,000 years, based on paleontological control [*Taylor et al.*, 1990a]. *Nishimura et al.* [1991,1992] have calculated recurrence intervals for these eruptions, assuming that the pumice deposits were instantaneous events, and extrapolating sediment accumulation rates for the interbedded hemipelagic units (Figure 12). Subtle variations in maximum grain size and percent gravel suggest that the Pliocene pumiceous units recovered at site 788 may also record eruptive cycles [*Nishimura et al.*, 1991, 1992].

3.4. Provenance

The sedimentology of this basin is closely linked to the eruptive history of nearby volcanic centers [*Nishimura et al.*, 1991,1992]. Possible sources of volcaniclastic material deposited within the Sumisu Rift include intrabasinal mafic to dacitic volcanic centers along the ridge that subdivides the basin, and felsic arc calderas that lie at the northeastern and southeastern limits of the rift basin (Figure 10). Other possible sources of clastic debris include reworked and redeposited materials from the eastern rift flank, and less likely, the protoremnant arc to the west.

The pumice clasts from sites 790 and 791 are two-pyroxene rhyolites, most likely derived from local arc volcanoes [*Nishimura et al.*, 1992]. However, the precise volcanic source(s) of the pumiceous deposits cored at sites 790 and 791 remain to be identified. Pumice from the Sumisu Rift ODP sites is geochemically similar to that of Minami-Sumisu Caldera, but the presence of a bathymetric barrier makes transport from the northern to the southern sub-basin unlikely. The most probable source of the pumiceous units at sites 790 and 791 is Tori Shima Caldera, but pumice from this volcano has not been geochemically analyzed [*Nishimura et al.*, 1992].

The sand fraction of the basin fill is predominantly composed of vitric to microcrystalline volcaniclastic debris with phenocrysts of plagioclase, pyroxene and minor olivine [*Fujioka and Saito*, 1992; *Marsaglia*, 1992]. Individual samples are generally inhomogeneous, either as a function of syneruptive mixing or mixing during submarine transport within the basin [*Hiscott and Gill*, 1992; *Marsaglia*, 1992].

On standard plots of sand detrital modes used for tectonic provenance analysis, the volcaniclastic sands from the

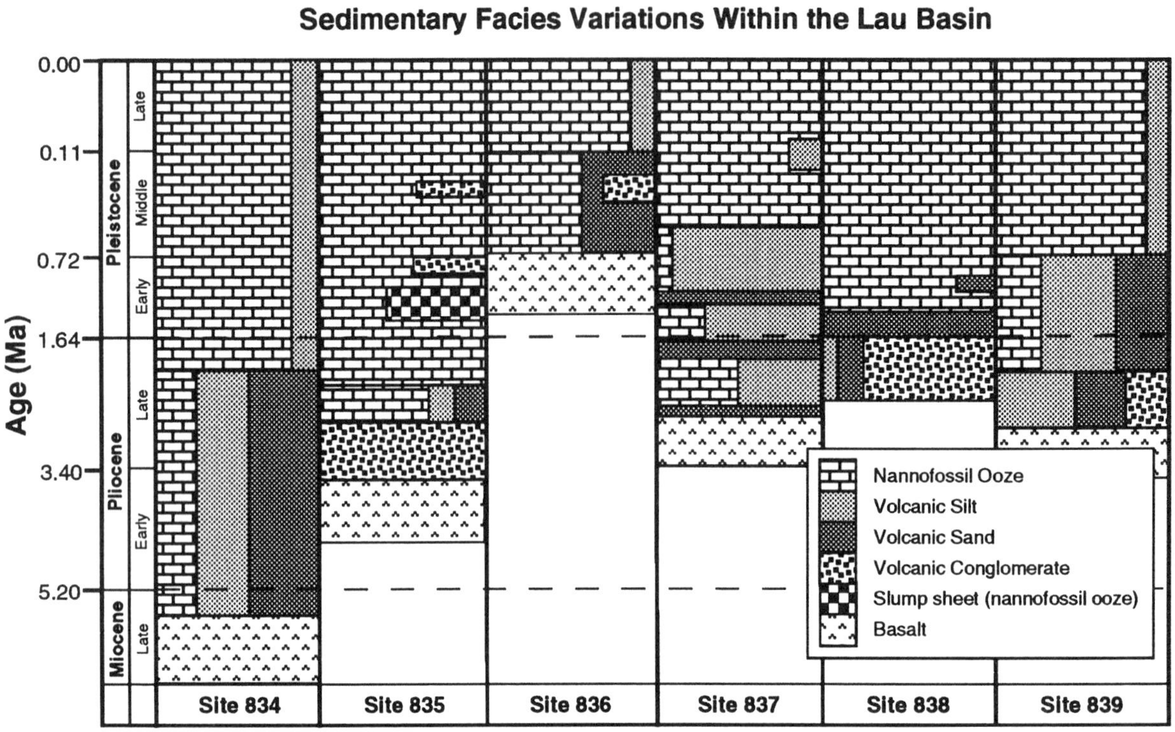

Fig. 5. Time-space distribution of dominant sedimentary facies in the Lau Basin.

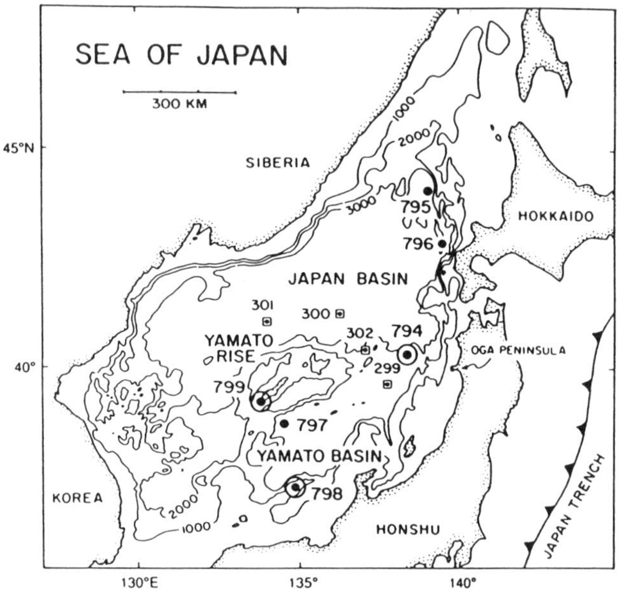

Fig. 6. Location map for the Japan Sea showing drill sites for Leg 127 (dots) and Leg 128 (circled dots). Sites 796, 797, and 799 are the focus of this paper. Boxed dots are DSDP sites drilled in 1975.

Sumisu Rift and the rift flank are fairly uniform in composition (Figure 13); they fall in the "Undissected Arc Field" of *Dickinson et al.* [1983], and would plot at the volcanic apex of a LmLvLs diagram and the plagioclase apex on a QmKP plot [*Marsaglia*, 1992]. Quaternary samples from the basinal sites are slightly enriched in quartz with respect to the rift-flank Plio-Pleistocene sequence, but in general, QFL detrital modes for these sands are similar to those of Izu-Bonin forearc sites and those determined by *Packer and Ingersoll* [1986] across the Mariana arc (Figure 13). The basinal sites are also slightly enriched in the felsitic volcanic component with respect to microlitic and lathwork fractions (Figure 14).

Although the Sumisu sands appear somewhat homogeneous, the proportions of colorless (felsic), brown (intermediate to mafic) and black or tachylitic (mafic) vitric volcanic lithic fragments show great diversity. Some basinal samples are enriched in intermediate to mafic brown and black glassy volcanic lithic fragments with respect to the rift flank site (Figure 15). Bulk geochemical analyses of these same sand samples suggest more intermediate (dacitic) compositions, and do not reflect the full range of felsic to mafic volcanic components present [*Hiscott and Gill*, 1992]. Electron-probe analysis of individual glassy fragments from sites 788, 790 and 791 indicates a

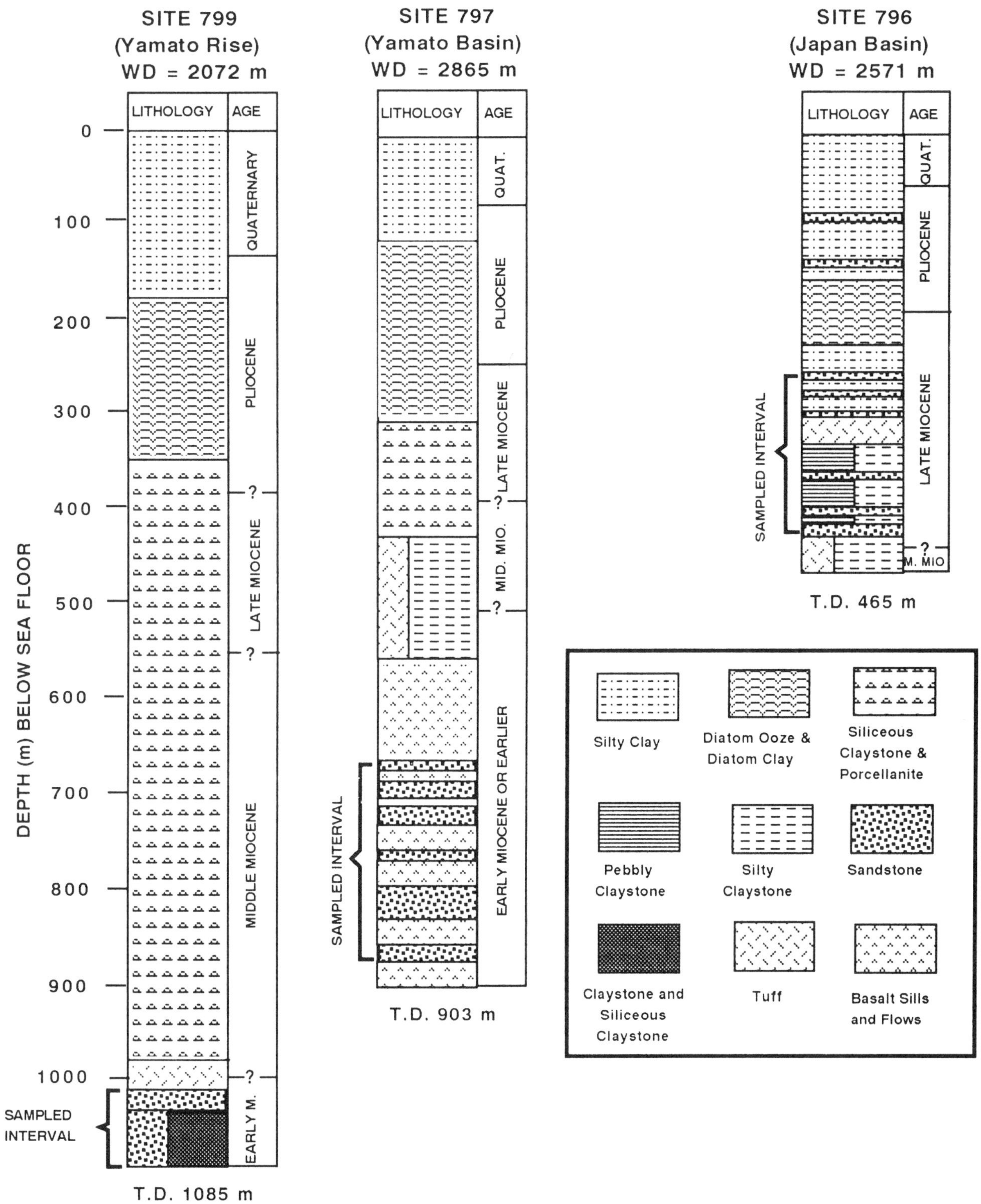

Fig. 7. Generalized lithology of Miocene-Quaternary sediments at ODP sites 796 (northeastern Japan Basin), 797 (Yamato Basin), and 799 (Yamato Rise) in the Japan Sea. Intervals sampled for this study are also shown. WD = water depth; TD = total drilled depth.

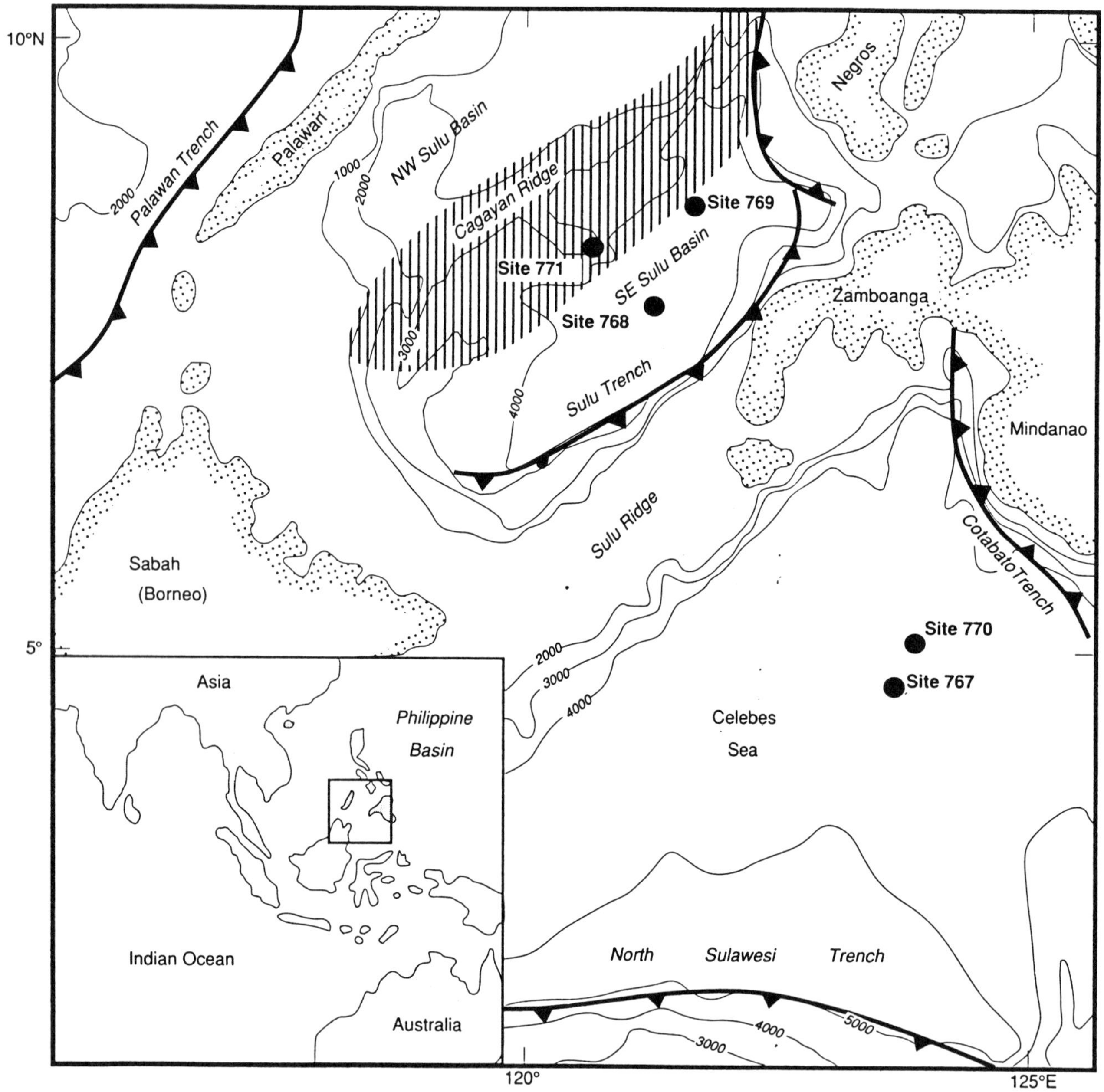

Fig. 8. Location of Leg 124 sites in the Sulu and Celebes seas. Barbed lines indicate subduction zones, and the Cagayan ridge is highlighted with vertical lines. From *Rangin et al.* [1990].

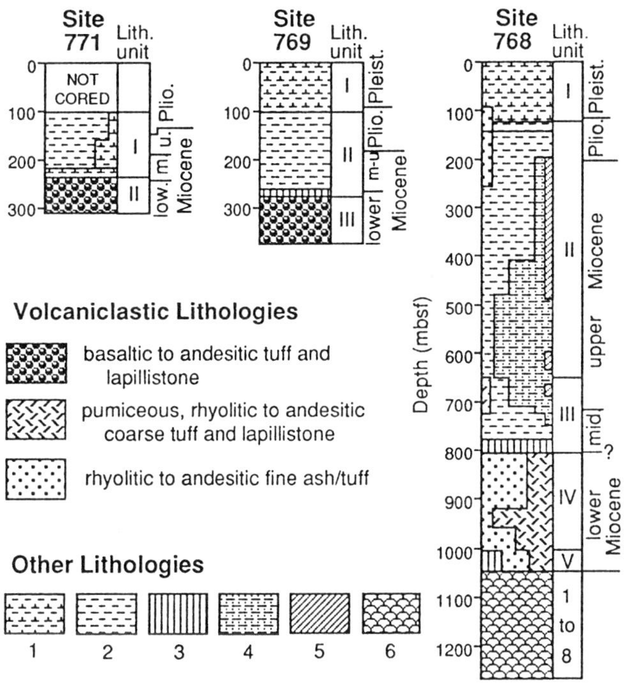

Fig. 9. Simplified lithologic columns for ODP Leg 124 sites in the southeast Sulu Basin and Cagayan ridge, showing relative proportions of interbedded lithologies. Key to symbols used for other lithologies: (1) marl; (2) hemipelagic clay (stone) and silty clay (stone); (3) pelagic brown claystone; (4) terrigenous turbidites; (5) carbonate turbidites; (6) oceanic basement.

dominantly rhyolitic assemblage of clasts with lesser dacite and basaltic andesite, and minor basalt and andesite [*Rodolfo et al.*, 1992].

4. LAU BASIN

4.1. Introduction

The stratigraphy at each of the six sites drilled in the Lau Basin during leg 135 (sites 834-839; Figure 4) shows a similar progression of coarse-grained, proximal volcaniclastic sediments, fining up to a nannofossil ooze-dominated sequence, with rare interbedded ash beds [*Parson et al.*, 1992]. The change from volcanic-dominated to pelagic-dominated sedimentation reflects a fundamental change in the style of extension and volcanism in this backarc basin system.

4.2. Sedimentary Facies

At each backarc site, the basal sedimentary section comprises an assemblage of thick-bedded sands or gravels. These fall into two different types, graded and ungraded. Ungraded sand beds typically have little internal structure and are interpreted to be the products of proximal mass flow processes. In contrast, graded sands commonly have parallel lamination, together with rarer cross lamination or slump folding. Water-escape structures indicate that these sands were rapidly deposited. It seems most likely that the sands were laid down by high-density turbidity currents [cf. *Hiscott and Middleton*, 1979]. Up section, ungraded, mass-flow sands are less common, a trend accompanying a decrease in the thickness and grain size of graded turbidite layers. Typically the upper 50-100 m of sediment at each site are dominantly nannofossil ooze and chalk, representing back-ground sedimentation. Interbedded within the calcareous deposits are thin (5-10 cm), graded, air-fall, ash layers and interspersed pumice fragments. Individual ash layers are sharp-based, and grade up into nannofossil oozes. However, they are devoid of lamination or other sedimentary structures indicative of current deposition. They are the result of air borne ash fall events, or in the case of the pumice may have been rafted in from a number of possible sites from through-out the region (e.g., New Zealand, Fiji, New Hebrides).

4.3. Provenance

Rifting in the Lau Basin separated the Miocene arc from its forearc basin, leaving no arc on the trench side of the basin until 3 Ma, 2.25 m.y. after formation of the oldest backarc basin basalts (BABBs) at site 834 [*Parson et al.*, 1992]. The older Miocene arc is referred to as the Tonga arc, while the new arc created since rifting of the Lau Basin is referred to as the Tofua arc. The sedimentary and topographic evidence point to a Pliocene Lau Basin characterized by proximal sedimentation around intrabasinal seamount volcanoes. The source and distribution of volcaniclastic material at any one place within the Lau backarc basin is dependent on a number of factors including sub-basin topography and proximity to volcanic centers.

Each sub-basin within the Lau backarc region (Figure 4) is typically flanked by significant, usually tectonically controlled, high regions, which form major barriers to sediment transport. The underfilled nature of the sub-basins, coupled with their generally small dimensions (2-5 km across, by 10-15 km long), makes long distance (> 10 km) transport of coarse clastic material by water borne density flows rather unlikely. During formation of the Lau Basin, extension and volcanism migrated east across the basin, and remnants of the original Miocene Tonga arc left on the trench side of the basin were faulted and subdivided into horsts and grabens. The grabens became floored by pillowed and sheet-flow basalts, while volcanic seamounts on the sub-basin margins erupted more explosive, felsic to intermediate lavas. These intrabasinal volcanic centers are the major source of clastic sediment within the backarc region. Some volcanic centers may have been down-faulted into sub-basins as the basin evolved; however, drilling in

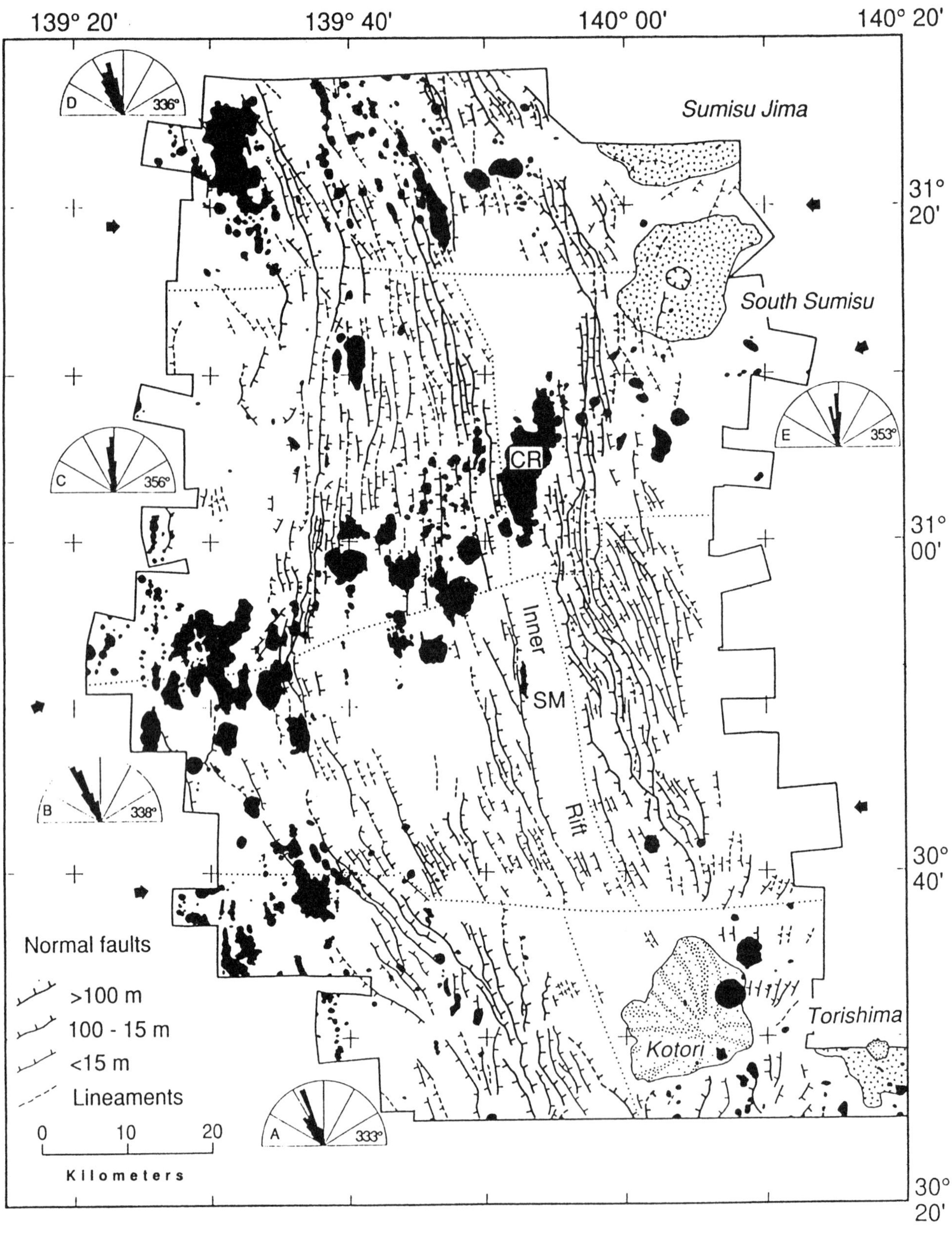

Fig. 10. Structural and volcanic map of the Sumisu Rift. Extrusive volcanic rocks are shaded black, active calderas (Sumisu, South Sumisu, and Torishima) are stippled, and pyroclastic cones (Kotori) are striped and stippled. CR = Central Ridge and SM = Shadow Mountain. From *Klaus et al.* [1992].

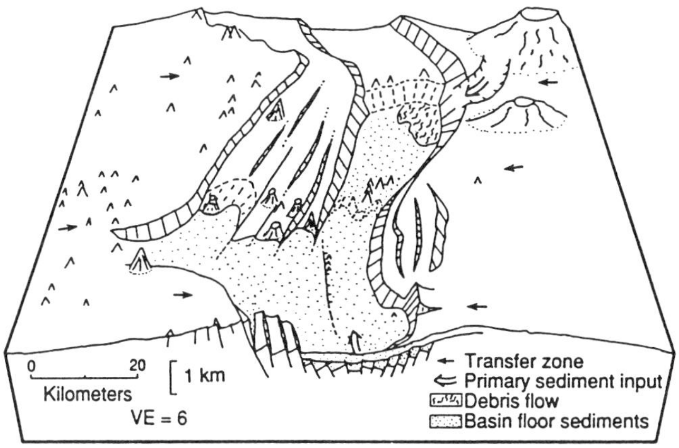

Fig. 11. A block diagram of the Sumisu Rift illustrating the likely direction of sediment input into the rift from arc calderas (primary sediment input). From *Klaus et al.* [1992].

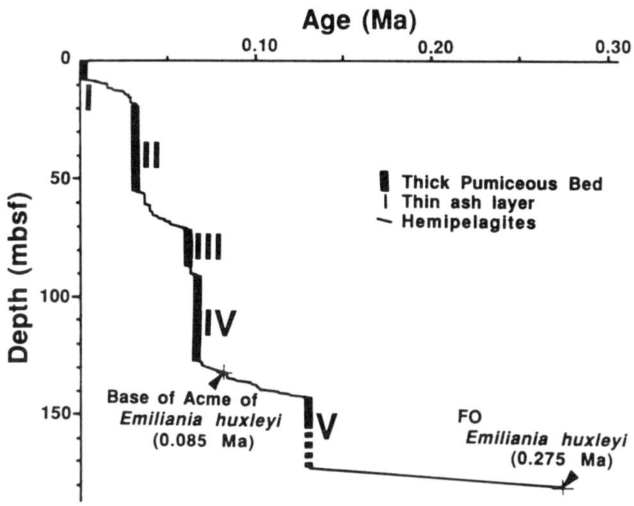

Fig. 12. Age vs. depth plot for site 790 showing apparent periodicity of major pumice-producing eruptions. Thick pumiceous beds are assumed to be instantaneous events, and hemipelagic intervals are assumed to have accumulated at a constant rate. Paleontological data provides age constraints as indicated. From *Nishimura et al.* [1992].

these sub-basins has only recovered basaltic basement rocks. The distribution of basalt could be explained by either post-extensional formation of the seamount volcanoes in any single location or by the presence of a thick carapace of basalt that drilling failed to penetrate.

The simplified sedimentary logs in Figure 5 show the broadly upward-fining nature of the sequence at each of the backarc basin sites. However, the time at which sands or gravels cease to dominate the section differs at each site, being substantially older at sites 834 and 835 compared to sites 837-839. This and the proximal nature of the sediments found at the base of each site imply that the major source of clastics at sites 834-835 was different from the source that supplied clastics to sites 836-839. Proximal sedimentation at site 837 indicates continued volcanism for at least 2 m.y. after eruption of the basement basalts at site 837. This suggests that volcanism and sedimentation during the initial stages of sub-basin formation were localized, with no suggestion of one major sediment source for the entire basin. Despite continued volcanism on the Lau Ridge, the presence of underfilled sub-basins between it and the region of active extension would have prevented it from being a major source after the earliest rift stage (Figure 16).

The delicate morphology [*Bednarz and Schmincke*, 1994] and fresh nature of the pyroclastic debris within the Lau basin fill are incompatible with extensive transport and suggest active local volcanism. The source volcano in each case could not have been part of the Tofua arc, because evidence from site 840 on the Tonga platform [*Shipboard Scientific Party*, 1992; *Tappin et al.*, 1994; *Clift and Dixon*, 1994] suggests a lull in arc volcanism from 5.0 to 3.0 Ma, during the initial stages of basin rifting. However, the Tofua arc is the most likely source of the thin-bedded, fine-grained ash layers at the top of each section (Figure 16). Other regional volcanic centers, such as the Taupo Volcanic Zone of New Zealand [*Shane and Froggatt*, 1991], are known to have spread ash far into the Pacific during the Pleistocene and must remain possible sources for some ash layers. However, the prevailing winds blow west to east [*Kennett*, 1981] and do not favor New Zealand as a provider of large volumes of ash to the Lau Basin. Because Pleistocene deposits in the Lau Basin are not proximal volcanic sediments, the presumed Pliocene seamount sources for volcanic sediment must have become largely extinct. The source of the pumice found throughout the upper parts of the sequence at each site is also difficult to pinpoint. While long distance transport by pumice rafting is possible, the Tofua arc, Valu Fa ridge and submarine seamount volcanoes within the Lau Basin north of the Peggy Ridge are the most likely bulk contributors.

The depositional setting in the Lau Basin appears to be radically different from that proposed by *Carey and Sigurdsson* [1984], in which volcaniclastic sedimentation far into the basin had been established by turbidity currents from the arc shortly after arc break-up. In the central Lau Basin, the rough topography and modest sediment supply have prevented this from happening (Figure 2). In addition, before 3.0 Ma the only arc volcanism was on the Lau Ridge. Continued extension removed the source further and further from each new sub-basin.

4.4. Geochemistry

4.4.1. *Major elements.* Consideration of the geochemistry of individual sediment grains extracted from the sediments in the Lau backarc basin allows the model of

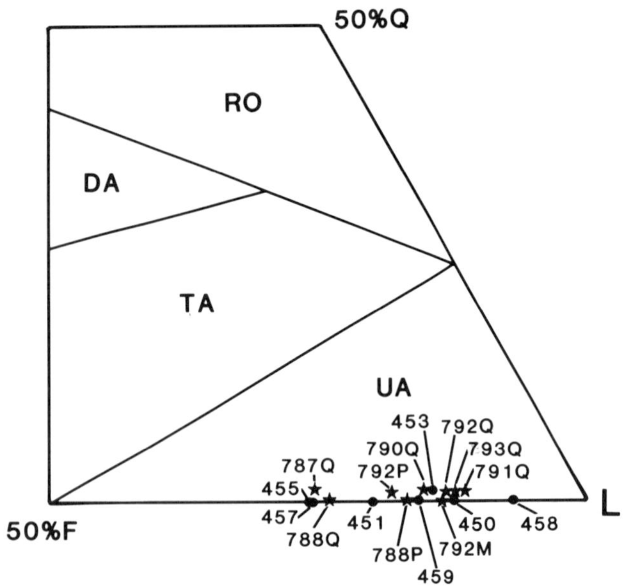

Fig. 13. Lithic corner of a QFL diagram subdivided into provenance fields of *Dickinson et al.* [1983]: RO = recycled orogen, DA = dissected arc, TA = transitional arc, and UA = undissected arc. Site means for Leg 126 forearc (787, 792, 793) and backarc (788, 790, 791) sites have been plotted by age: Q = Quaternary, P = Pliocene, and M= Miocene. Mariana transect site means (450-459) taken from *Packer and Ingersoll* [1986]. From *Marsaglia* [1992].

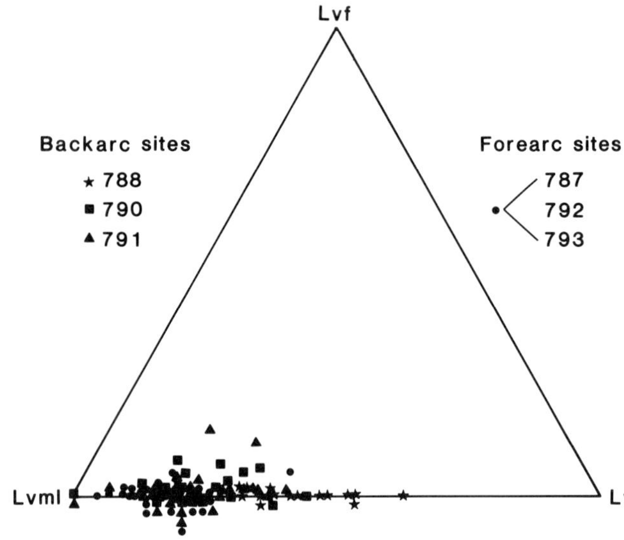

Fig. 14. Ternary plot of microcrystalline volcanic lithic proportions for Leg 126 forearc and backarc sites. Lvf = volcanic lithic with felsitic texture; Lvml = volcanic lithic with microlitic texture; and Lvl = volcanic lithic with lathwork texture. Felsitic textures are characteristic of felsic lavas, microlitic textures of intermediate lavas, and lathwork textures of mafic lavas. Overlapping data along the base of the diagram (0%Lvf) have been plotted in histogram fashion. From *Marsaglia* [1992].

localized sedimentary sources to be tested and extended. Samples were initially analyzed for a suite of 10 major elements using an electron microprobe [*Clift and Dixon*, 1994; *Bednarz and Schmincke*, 1994]. Of these grains, a selected subset of low-silica volcanic glasses was analyzed for a suite of 20 trace and rare-earth elements using a 4f ims Cameca ion microprobe [*Clift and Dixon*, 1994]. For this purpose, low-silica grains were selected to minimize the effect of fractional crystallization. Where possible, aphyric glasses were chosen in preference to phyric glasses, which must have experienced some fractional crystallization, enhancing the amount of petrogenetic information that could be extracted. Some inaccuracy was introduced into the electron probe analyses as a result of volatilization of the Na and K. As well as producing anomalously low alkali contents, this has the effect of boosting the percentage of other major elements. However, this effect is largely negated by the presence of volatiles, mostly H_2O, which reduces the total. *Clift and Dixon* [1994] argue that since high alkali glasses are also the most volatile-rich, a recalculation of major element contents is unnecessary and undesirable, as it introduces further unconstrained errors. Silica values plotted in Figure 17 are therefore the original raw analyses.

The principal aim of this probe work was to determine the dominant types of volcanic glass grains being reworked into the basin and thereby the nature of intrabasinal seamount volcanism. Variability in the total silica contents of glass grains does not seem to change from the base to the top of each site. However, three different patterns are recognized among individual sites (Figure 17). Sites 834 and 835 display a full array of silica values ranging from basaltic andesite to rhyolite. In contrast, sites 837-839 show a strongly bimodal pattern with basaltic andesites and rhyolites dominating. Site 836 has a unimodal glass assemblage, comprising only basaltic andesite glass. The vast majority of glass grains from throughout the basin are of low-K type, similar to glasses analyzed by *Cawood* [1991] from the Tonga forearc. However, a small number of grains from sites 834 and 835 are medium and high-K types. This reinforces the distinction made on the basis of silica contents alone.

Interestingly, the chemical subdivision of the sites mirrors their tectonic positions within the basin [*Clift and ODP leg 135 Scientific Party*, 1994]. Sites 834 and 835 lie some distance from the other sites (Figure 4) and on the basis of the proximal sedimentary facies and rugged surrounding topography would be expected to have a different source. Furthermore, although site 836 is close to sites 837-839, a distinctive geochemical grouping might be expected because it lies within crust generated along the Eastern Lau spreading center. This contrasts with the crust

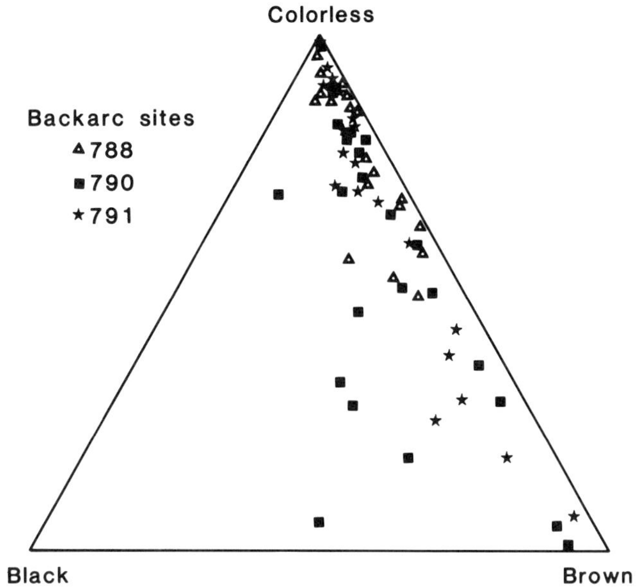

Fig. 15. Ternary plot of colorless, brown and black glassy volcanic lithic proportions for Leg 126 backarc sites. From *Marsaglia* [1992].

of the other sites, which are volcanically floored basins within extended volcanic arc crust [*Parson and Hawkins*, 1994].

4.4.2. *Trace and rare earth elements.* Ion microprobe analyses of low-silica grains from Lau Basin sites [*Clift and Dixon*, 1994] can also be used to test the hypothesis that sediment was derived locally from intra-basinal seamounts. The assumption is that the trace and rare-earth element characteristics of the volcaniclastic sediments should be similar to those of the underlying igneous basement rocks, because the BABBs in each graben presumably share the same source mantle as the seamount volcanoes on the basin flanks. Thus we compare the grain analyses with whole rock analyses of the basement basalts at each site [data from *Ewart et al.*, 1994].

Figure 18 plots Ba/Nb against La/Nb. Ba represents large ion lithophile elements (LILEs), mostly derived from the subducting slab, and Nb represents high-field-strength elements (HFSEs), which are believed to be extracted solely from the mantle wedge. The horizontal axis plots La, a rare-earth element against Nb, a HFSE. The relationship of La to petrogenesis is slightly obscure. To a large degree the concentration of this element is controlled by the degree of partial melting of the source and its degree of depletion. There is some evidence to suggest that LREEs may be water mobile to a limited degree [*Hole*, 1984], and consequently may show some enrichment with increasing slab flux. Therefore, scatter in data points across the

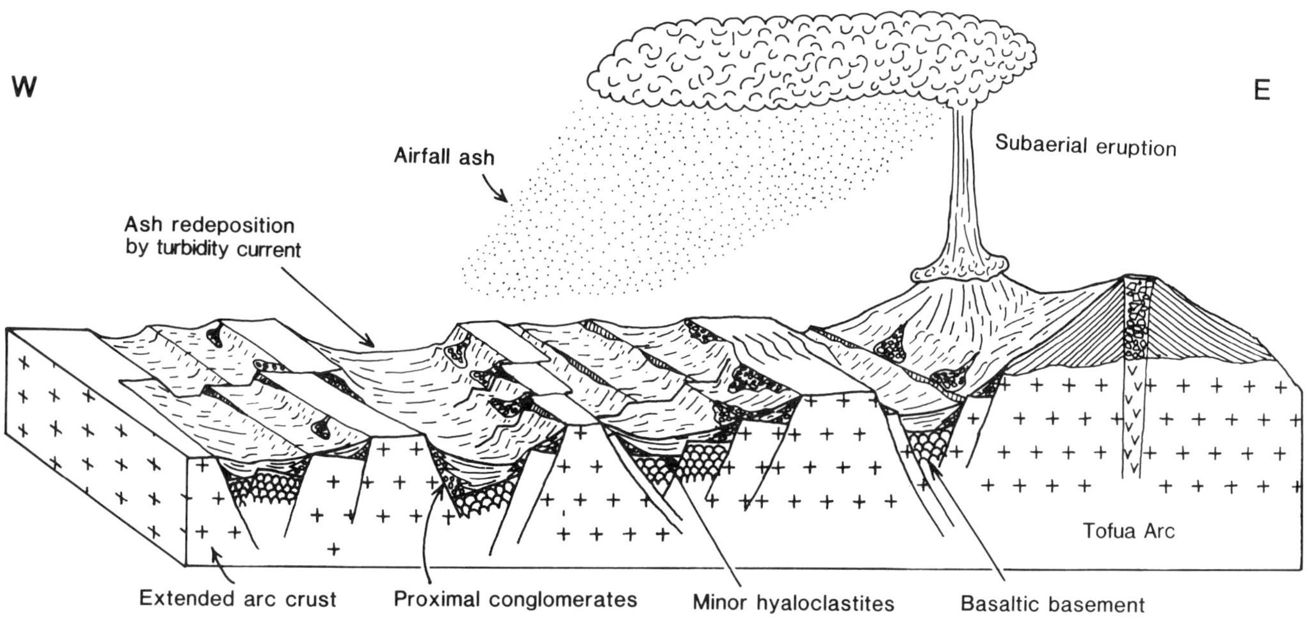

Fig. 16. Schematic representation of the modern Lau Basin, showing that volcaniclastic debris from the arc volcanic front is ponded close to the arc, except in the case of widespread airfall ash material.

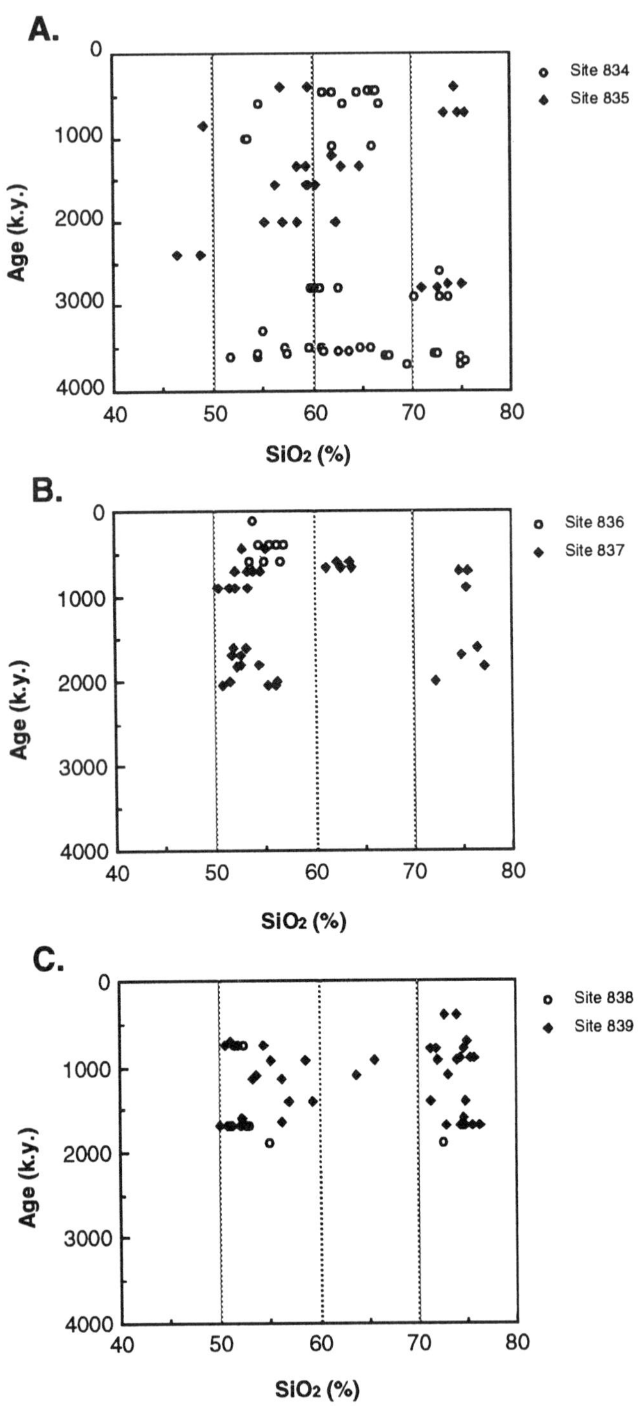

Fig. 17. Age versus %SiO_2 content of glass from Lau Basin sites. Silica values are raw counts uncorrected for volatile loss.

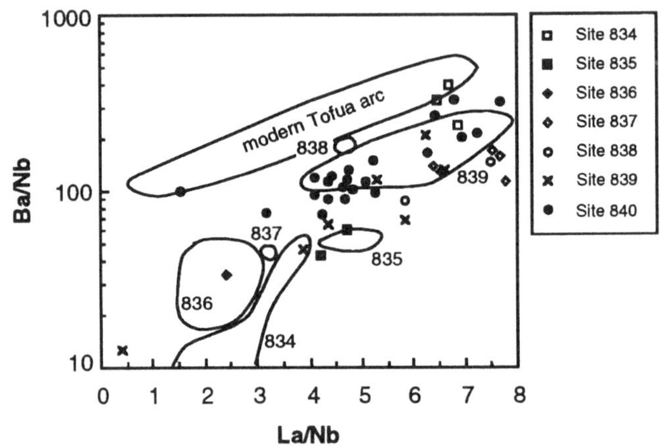

Fig. 18. Discrimination diagram showing the variation in Ba/Nb versus La/Nb for basaltic glass grains (MgO > 4.0%) from sites 834-840. Fields show the distributions of analyses of basement basalts at sites 834-839, together with analyses from the modern Tofua arc [data from *Ewart et al.*, 1993].

discrimination plot in Figure 18 reflects differing degrees of source mantle depletion and degrees of partial melting between different basalts, as well as variations in the fluid flux from the subducting slab.

The fields in Figure 18 show the nature of the BABBs at each backarc site, as well as the arc volcanic rocks from the modern Tofua arc. Data points from site 840 represent the pre-rift Miocene Tonga arc and the oldest parts of the new Tofua arc. These points show La/Nb values similar to the modern arc, but lower Ba/Nb, possibly due to lower slab-derived fluid flux in the past. Analyses of site 834 glasses plot between the modern arc and values from site 840 and are markedly different from the BABBs analyzed at this site in that the site 834 glasses show high Ba/Nb and La/Nb values. Their arc-like composition probably reflects derivation of the sediment at site 834 from the remnant arc (Lau Ridge) early in the rifting process rather than from new intrabasinal seamount volcanoes. In contrast, the similarity of glass grain and BABB compositions at sites 835 and 836 is compatible with the local derivation of sediments at these sites. Site 837 shows little similarity between volcanic ash and local BABB chemistry; the glass grains at site 837 have higher La/Nb and slightly higher Ba/Nb than the underlying BABBs. According to these data, the ashes from sites 837 and 838 were erupted from a volcano distinct in character from their basaltic basements. Based on the overlap of values in Figure 18, the ashes at sites 837-839 were derived from similar sources, possibly the same volcano in view of their relative proximity to one another (Figure 4); overlap of these values with the site 839 BABB field suggests that the source volcano was proximal to that site, but at great enough distance to be influenced by other mantle domains.

5. SULU BASIN

5.1. Introduction

The Sulu Sea is a small yet complex marginal basin lying between northern Borneo and the Philippines (Figure 8). The submerged, northeast-trending Cagayan ridge divides the Sulu Sea into two distinct basins, of which only the south-east Sulu Basin has an oceanic basement. This basin is 4.5 to 5 km deep, with a sedimentary fill 1 to 2 km thick, and is presently subducting southeastward beneath the Sulu Ridge, an active volcanic arc. ODP leg 124 drilled site 768 near the center of the southeast Sulu Basin, and sites 769 and 771 on the southeast flank of the Cagayan ridge. The results of leg 124 [*Rangin et al.*, 1990] have answered several first-order questions concerning the age and stratigraphic framework of the southeast Sulu Basin.

5.2. Age and Mode of Formation of the Southeast Sulu Basin

Drilling results from leg 124 established an early Miocene minimum age of opening for the southeast Sulu Basin. The basal pelagic and volcaniclastic deposits that overly oceanic crust at site 768 have a minimum age of 18.6 Ma by correlation of poorly preserved radiolarian assemblages with magnetic polarity Chrons C5C through C5E [*Shyu et al.*, 1991]. *Roeser* [1991] mapped very weak magnetic anomalies in the northern part of the basin, which he tentatively identified as spanning 30 to 10 Ma (late Oligocene to late Miocene). His model predicts a crustal age at site 768 that is 2 m.y. younger than the age suggested by core data, and an extremely slow half-spreading rate (0.6 cm/yr) that is questioned on morphotectonic grounds by *Rangin and Silver* [1991]. Studies prior to leg 124 interpreted the southeast Sulu Basin as a trapped fragment of oceanic crust [*Lee and McCabe*, 1986] or as an extensional backarc basin opened behind either the Sulu Ridge [*Mitchell et al.*, 1986; *Rangin*, 1989] or the Cagayan ridge [*Holloway*, 1982]. Oceanic basement drilled at site 768 consists of basalt lavas and sills with compositions transitional between mid-ocean ridge basalts and island-arc tholeiites, suggesting generation from subduction-modified intraoceanic mantle sources [*Spadea et al.*, 1991]. Thick sequences of lower Miocene or older pyroclastic deposits were encountered at each of the leg 124 Sulu Sea sites (Figure 8), further supporting a backarc origin. At site 768 in the southeast Sulu Basin, thick rhyolitic to andesitic pyroclastic deposits occur near the base of the sediment column, overlying oceanic crust. The two Cagayan ridge sites (769 and 771) were drilled on the southeast flank of the ridge, which consists of normal-fault blocks stepping downward to the southeast Sulu Basin [*Mascle and Biscarrat*, 1978]. The deepest rocks penetrated at the Cagayan ridge sites are andesitic to basaltic coarse tuff and lapillistone, with compositions comparable to tholeiitic and calc-alkaline island-arc magmas [*Spadea et al.*, 1991; *Smith et al.*, 1991]. Sparse microfossils in the sedimentary sections overlying the pyroclastic deposits at all three sites indicate cessation of volcanism by 16 Ma. K-Ar ages on pyroclasts at the Cagayan ridge sites are as old as 20 Ma, overlapping the oceanic basement age at site 768. Basin formation was thus concurrent with rifting and explosive volcanism on the flank of the Cagayan ridge.

In many western Pacific backarc basins, the polarity of subduction during basin formation is indicated by the contrast between the thick volcaniclastic apron developed behind the frontal arc and the much thinner accumulation on the front flank of the remnant arc [*Karig and Moore*, 1975]. This criterion is difficult to apply to the southeast Sulu Basin, because the volcaniclastic strata are buried beneath 250 to 800 m of younger deposits, and the southeast flank of the basin has been modified by subduction of the basin floor beneath the Sulu Ridge. Arc volcanic rocks correlative with the lower Miocene Cagayan ridge sequence have not been found on the Sulu Ridge, and the structure of its southeast flank is indicative of Cenozoic extension rather than convergence [*Rangin and Silver*, 1991]. Leg 124 drilling results are best explained by opening of the southeast Sulu Basin as a backarc basin behind an active Cagayan arc, with volcanism ceasing in early middle Miocene time, concurrent with collision between the Cagayan arc and the continental Reed Bank-Palawan block [*Silver and Rangin*, 1991].

5.3. Sedimentation in the Southeast Sulu Basin

The stratigraphy of the southeast Sulu Basin records varying sediment input from volcanic, continental, and pelagic sources, related to changes in the tectonic setting of the basin and eustatic sea level effects. Four stages can be recognized in the depositional history of the basin.

5.3.1. *Early Miocene (18.5 to 17 Ma).* During the early Miocene, pyroclastic deposits produced by submarine explosive eruptions accumulated on the Cagayan ridge and in the southeast Sulu Basin. The andesitic to basaltic Cagayan ridge deposits are unstratified coarse tuff and lapillistone, with intercalated thin lava flows. Clasts in these near-vent deposits are moderately to highly vesicular, and they clearly formed by the explosive fragmentation of a vesiculating magma. The pyroclastic deposits are overlain by and locally interbedded with pelagic clay and nannofossil marl, indicating eruption depths of several hundred meters to several thousand meters. *Cashman and Fiske* [1991] have catalogued recent evidence for explosive underwater eruptions as deep as 1800 m in back-arc basins.

Early Miocene rhyolitic to andesitic volcaniclastic sand and sandstone (tuff and lapillistone) overlying oceanic crust at site 768 in the southeast Sulu Basin are 240 meters thick, and are intercalated with brown pelagic claystone in their lower part. The coarser deposits are highly pumiceous, occurring in very thick, massive to laminated graded units interbedded with thinner and finer grained sediments (tuffs). The sequence was deposited by a series of submarine gravity flows and turbidity currents fed by voluminous explosive eruptions, with rates of deposition as high as 160 m/m.y. [*Shipboard Scientific Party*, 1990a]. Co-deposition of pumice and denser lithic particles indicates that the bulk of the pyroclastic material was produced by submarine eruptions [*Cashman and Fiske*, 1991]. These deposits represent the basinal portion of an arc-derived volcaniclastic apron which prograded basinward soon after formation of the underlying back-arc basin crust.

5.3.2. *Middle Miocene (17 to 10.4 Ma).* During middle Miocene time, deposition in the southeast Sulu Basin and Cagayan ridge was dominated by the slow accumulation of pelagic and hemipelagic clay. Admixed pelagic biogenic carbonate is present at site 771 but absent at the deeper sites 769 and 768, so the deeper ridge flank and basin floor lay below the CCD. Deposits are thin at all sites, with sparse microfossils and consequent poor age control. At site 768 sedimentation rates (uncorrected for compaction) may have been as low as 1 m/m.y., and there is some evidence to suggest that a hiatus exists in this interval [*Shyu et al.*, 1991]. Continental and volcanic arc sources began to contribute minor amounts of coarse sediment to the southeast Sulu Basin in the later part of this interval, in the form of rare quartzose and volcaniclastic turbidite beds.

5.3.3. *Late Miocene (10.4 to 5.2 Ma).* Sedimentation in the southeast Sulu Basin during Late Miocene time was dominated by rapid deposition of fine-grained quartzose turbidites and hemipelagic clay, producing a sequence that is 515 m thick at site 768. The shallower Cagayan ridge sites were unaffected by turbidite deposition, and accumulated only pelagic and hemipelagic deposits.

The turbidites are thin- to thick-bedded but predominantly fine-grained, consisting primarily of clayey silt to clay, with less common sandy layers. The large volume of fine-grained detritus and abundance of plant debris suggest that sediment was supplied from coastal plain deltas fringing the basin. A likely source was northern Borneo, where the quartz-rich strata of the Crocker Formation have been exposed since early middle Miocene time [*Rangin and Silver*, 1991]. The peak of turbidite deposition occurred between 10 and 9 Ma in both the Sulu and Celebes Basins. The beginning of this influx was synchronous with a major sea-level lowstand that defines the base of the TB3 global sea-level supercycle [*Haq et al.*, 1988], and subdivisions of the turbidite sequence correlate with third-order eustatic cycles, indicating an important eustatic control on turbidite sedimentation in the basin [*Betzler et al.*, 1991].

Several very thick graded beds of silicic vitric tuff were deposited in the basin near the beginning of this interval. The beds contain plant debris and clay rip-up clasts and are interpreted as reworked subaerially erupted ash redeposited by submarine gravity flows. Their likely source was the Sulu Ridge, where arc volcanism was initiated by middle Miocene time with the beginning of southeastward subduction of the basin [*Rangin and Silver*, 1991].

5.3.4. *Pliocene to Recent (5.2 to 0 Ma).* The Pliocene to Recent sedimentary record at all Sulu Sea sites is dominated by hemipelagic and pelagic deposits. At the deeper sites 768 and 769, biogenic carbonate accumulation began to supplement clay deposition at 2.4 Ma, and Pleistocene sequences at both sites are marls averaging 40% $CaCO_3$. Simultaneous initiation of carbonate sedimentation at both sites resulted from abrupt deepening of the CCD, synchronous with the onset of northern hemisphere glaciation [*Linsley*, 1991].

Beginning about 7 Ma, explosive volcanic activity resumed in the region, and continued into the Pleistocene. This episode is recorded in the southeast Sulu Basin by intermittent deposition of thin, graded, structureless ash layers, interpreted as airfall deposits. These layers probably have multiple sources in the late Neogene arcs of the southern Philippines [*Pouclet et al.*, 1991; *Pubellier et al.*, 1991].

6. JAPAN SEA

6.1. Stratigraphic Characteristics

The basins that make up the Japan Sea backarc are underlain by Quaternary sediments and Tertiary sedimentary rocks that may exceed 2 km in thickness [*Ludwig et al.*, 1975; *Honza*, 1979; *Tamaki et al.*, 1981]. The stratigraphic section in the Japan Sea has been penetrated by ten DSDP and ODP deep holes (Figure 6), as well as numerous piston cores, that encountered at some sites sediments as old as early Miocene. The Quaternary to Miocene stratigraphic section throughout the Japan backarc is composed mainly of fine-grained sediment consisting of silty clay and claystone, diatom ooze and diatom clay, and siliceous claystone. The characteristics of fine-grained sediments at ODP sites 796, 797, and 799 illustrated in Figure 7 are fairly typical. Minor interbeds of volcanic ash or tuff, siltstone, and sandstone occur within these fine-grained sediments at many sites. Quaternary turbidite sands were recovered at two DSDP sites (299, 301) in the Japan Basin and from many piston core sites [*Boggs*, 1984]. Very minor Pliocene-age sands are present at ODP site 796 (northeastern Japan Basin) and moderately well developed turbidite sandstones of Miocene age have been cored at ODP sites 796, 797 (Yamato Basin), and 799 (Yamato Rise) [*Shipboard Scientific Party*, 1990b,c,d]. These sands and sandstones provide important clues to the provenance and diagenetic history of the Japan Sea backarc basin.

6.2. Petrographic Characteristics of Japan Backarc Sands and Sandstones

6.2.1. *Quaternary.* The distribution of Quaternary sands in Japan Sea sediments is discussed by *Boggs* [1984], who suggests that these sands are largely turbidites. The petrographic characteristics of Quaternary sands from

TABLE 1. Average Percent Framework Composition of Sandstone from Sites 796, 797, and 799.

Framework Components	Site 796	Site 797	Site 799
Monocrystalline quartz	0.5 ± 0.6	7.4 ± 2.4	29.6 ± 4.4
Polycrystalline quartz and metaquartzite	0.2 ± 0.3	2.3 ± 1.7	6.1 ± 1.6
Chert	0.3 ± 0.4	1.6 ± 1.5	0.4 ± 0.5
Sanidine	0.1 ± 0.2	6.0 ± 4.0	0.9 ± 1.1
Orthoclase and microcline	0.0	0.3 ± 0.5	18.3 ± 4.6
Plagioclase	28.0 ± 17.8	15.6 ± 9.4	23.7 ± 4.2
Volcanic rock fragments and glass	61.5 ± 17.9	46.1 ± 20.5	2.9 ± 1.7
Plutonic rock fragments	0.0	0.3 ± 0.6	6.3 ± 4.7
Metamorphic rock fragments	0.1 ± 0.2	0.6 ± 0.7	0.4 ± 0.5
Sedimentary rock fragments	0.0	0.5 ± 0.8	4.1 ± 3.0
Indeterminate unstable rock fragments	0.6 ± 0.7	4.0 ± 5.5	2.0 ± 1.2
Ripup clasts (intraclasts)	6.9 ± 9.2	11.1 ± 17.8	3.6 ± 2.8
Micas	0.1 ± 0.2	3.4 ± 2.3	1.4 ± 1.4
Heavy minerals	1.6 ± 1.9	1.0 ± 0.8	0.3 ± 0.4
Q as % QFL	1.1	13.3	38.1
F as % QFL	30.8	25.8	45.3
L as % QFL	68.1	60.9	16.5
L_v as % total unstable rock fragments	98.9	89.5	18.5
L_p as % total unstable rock fragments	0.0	0.6	40.1
L_m as % total unstable rock fragments	0.2	1.2	2.5
L_s as % total unstable rock fragments	0.0	1.0	26.1
L_i as % total unstable rock fragments	1.0	7.8	12.7
Plagioclase/total feldspar ratio	0.99	0.71	0.55
Total Samples	14	43	18

Q = Total quartz + chert; F = total feldspars; L = total unstable rock fragments (does not include intraclasts);
L_v = volcanic rock fragments and glass; L_p = plutonic rock fragments; L_m = metamorphic rock fragments, excluding metaqurtzite;
L_s = sedimentary rock fragments, exclusing chert; L_i = indeterminate unstable rock fragments;
± indicates one standard deviation ().

DSDP sites 299 and 301 are described in detail by *Packer and Ingersoll* [1986] and discussed by *Marsaglia* [1991] and *Marsaglia et al.* [1992]. *Packer and Ingersoll* [1986] report that site 299 (Yamato Basin) sands contain an average of 13 QF% quartz, 24% feldspar (plagioclase to total feldspar ratio of 0.70), 63% unstable lithic fragments (of which volcanic rock fragments make up 78%), and 3% framework mica. Site 301 (Japan Basin) Quaternary sandstones contain an average of 24 QFL% quartz, 53% Feldspar (P/F = 0.54), 23% unstable lithic fragments (of which 58% are volcanic), and 1% framework mica.

6.2.2. *Pliocene.* The only Pliocene sands recovered (to our knowledge) from the backarc are some very thin, unconsolidated deposits at site 796 (Figure 7). These sands have not been studied in detail but appear to be primarily volcanic sands (ashes).

6.2.3. *Miocene. Boggs and Seyedolali* [1992, 1993] studied the composition of Miocene sandstones at ODP sites 796, 797, and 799 (Figure 7). (Miocene sediments at DSDP sites 299 and 301 contain few sandstones, none of which have been studied petrographically.) The results are summarized in Table 1. This table shows that Miocene sandstones from site 796 (northeast Japan Basin) are composed predominantly of feldspar (plagioclase/total feldspar ratio = 0.99) and volcanic rock fragments with only about 1 percent quartz. Sandstones at site 797 (Yamato Basin) contain considerably more quartz (~13%), a much lower ratio of plagioclase to total feldspar (0.71), and abundant volcanic rock fragments. Sandstones from site 799 (Yamato Rise) show a compositional signature very different from that at site 796 and 797. Quartz content is much higher (~38%), volcanic rock fragments are few, and K-feldspars make up nearly half of the total feldspar.

6.3. Provenance Significance of Framework Components

6.3.1. *Quaternary.* The average QFL content (13-24-63) of Quaternary sands at DSDP site 299 reported by *Packer and Ingersoll* [1986] indicates that these sands were derived primarily from a volcanic source, which the authors suggest was western Honshu. The reported QFL content of DSDP site 301 sands (24-53-23) shows a much reduced volcanic influence. Packer and Ingersoll suggest that the sands at this site may have been derived, in part, from the Honshu arc and in part from the Asian mainland. This interpretation agrees with the provenance conclusion of *Sibley and Pentony* [1978], which was based on a chemical and petrographic study of feldspar and quartz in the 34-74-micron fraction of turbidite sediment from the Japan Sea.

6.3.2. *Miocene.* Miocene sandstones from ODP sites 796, 797, and 799 display marked differences in framework composition that suggest that these sandstones were derived from different source rocks and different tectonic provenances. Site 796 (northeast Japan Basin) sandstones are composed almost entirely of volcanic detritus, which was probably from a pyroclastic source located on or near western Hokkaido. The fresh, unweathered nature of this detritus suggests that it may have erupted subaqueously within the Japan Sea.

The sandstones at site 797 (Yamato Basin) also have a strong volcanic signature; however, they also contain epiclastic framework components derived from sedimentary, metamorphic, and possibly granitic rocks. These epiclastic components indicate derivation from a partially dissected volcanic arc, likely located on western Honshu. The framework composition of site 799 sandstones, which approaches that of a classical arkose, is so different from that of site 797 sandstones that the site 799 sandstones must have been derived from a quite different source, even though the two sites are located close together (Figure 19). site 799 sandstones were derived from a dissected-arc or basement-uplift provenance that contained felsic plutonic rocks. The most likely source area is Kita-Yamato Bank (Figure 19), which lies to the northwest on Yamato Rise and is known to contain granite and granodiorite [*Tamaki*, 1988]. Some of the sandy detritus at site 799 may have been derived from the Asian mainland (Korea), which is also host to felsic plutonic rocks; however, the mainland is a less likely source. It lies much farther away and is host also to abundant metamorphic rocks, few fragments of which are present in site 799 sandstones.

The average QFL framework percentages for site 797 (Yamato Basin) Miocene sandstones (~13-26-61) are remarkably similar to values reported by *Packer and Ingersoll* [1986] for DSDP site 299 (Yamato Basin) Quaternary sands (13-24-63). Also, the average QFL composition of site 799 (Yamato Rise) Miocene sandstones (~38-45-17) is reasonably close to that of the Quaternary sands at DSDP site 301 (Japan Basin) (24-53-23) reported by Packer and Ingersoll. These similarities may be pure coincidence. On the other hand, they may indicate that the source rocks that separately furnished sediment to the Japan and Yamato basins may have changed relatively little between Miocene and Quaternary time.

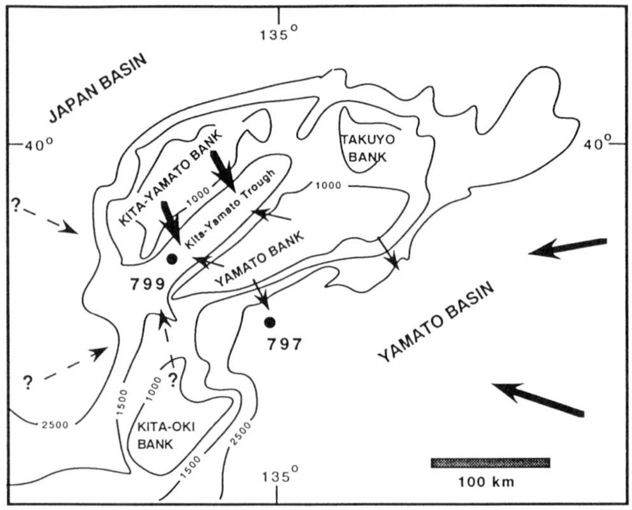

Fig. 19. Bathymetric map of Yamato Rise (see Fig. 1) and adjacent areas showing the locations of sites 797 and 799. The arrows show the inferred general direction of sediment transport from the source areas. The heavy arrows indicate the most important sources. Bathymetric contours generalized from *Tamaki* [1988].

6.4. Diagenesis of Miocene Sandstones

The diagenetic history of backarc sandstones in the Japan Sea has been established only for the Miocene sandstones at sites 796, 797, and 799 [*Boggs and Seyedolali*, 1992]. These sandstones have undergone various degrees of diagenetic alteration: albitization of feldspars, partial replacement of plagioclase feldspars by authigenic K-feldspar, formation of authigenic zeolites, and cementation and replacement by carbonates, clay minerals, sulfides and sulfates. Because these changes have altered the mineral and chemical composition of the sandstones to some extent, they have had some effects on provenance interpretation.

Of particular significance is the change in chemical composition of plagioclase feldspar owing to albitization. The chemical compositions of more than 1000 feldspar grains in 40 sandstone samples from the three Japan Sea ODP sites have been studied by using the electronprobe microanalyzer for chemical analysis and elemental mapping, the SEM in the secondary and back-scattered mode, and the cathodoluminescence scope. Site 796 sandstones contain very little K-feldspar. Plagioclase feldspars display a wide range of compositions from anorthite to oligoclase; albite is absent. The volcanic rocks from which these feldspars were derived probably ranged in composition from andesite to basalt. By contrast, calcic plagioclase is absent at sites 797 and 799 where a significant proportion of the plagioclase is albite. Also, many K-feldspar grains at site 797 and a few at site 799 have been partially albitized.

Albitization of plagioclase in site 797 and 799 sandstones has altered the chemical signature of the original plagioclase. It can still be recognized petrographically as plagioclase; however, its usefulness in provenance interpretation is diminished because its chemical composition cannot be used as a reliable guide to the chemical composition of its parent rocks, and characteristics such as twinning and zoning may have been obliterated. Also, many plagioclase grains in site 797 sandstones have been partially replaced by authigenic K-feldspar.

Albitization appears to be strongly related to temperature. Feldspars in sandstones at site 796, which were buried to shallow depths and subjected to relatively low diagenetic temperatures (40°-65°C) have not been albitized. The most extensive albitization took place in site 797 sandstones, which were intruded by basaltic sills and subjected to temperatures exceeding 100°C. Feldspars at site 799 were altered at somewhat lower diagenetic temperatures than those at site 797 and show less extensive albitization.

In addition to albitization, replacement of feldspars and other framework grains by zeolites and carbonate minerals is also common. For example, more than 5 percent of the framework grains in a few site 796 sandstones are so completely zeolitized that the identity of the original grains cannot be established. Also, significant amounts, as much as 5-10 percent, of the framework grains in some sandstones, particularly a few site 799 sandstones, have been destroyed by carbonate replacement. Much fine-grained glass (matrix glass) has also been destroyed by alteration to mixed-layer clay minerals or zeolites, and some feldspars may have been removed by dissolution.

7. DISCUSSION AND CONCLUSIONS

7.1. Implications of Recent Drilling for Models of Intraoceanic Backarc Sedimentation

General models for backarc-basin sedimentation have been proposed, based to a large degree, on DSDP drilling results [e.g., *Karig and Moore*, 1975; *Carey and Sigurdsson*, 1984; *Klein*, 1985; Figure 20]. We have discussed one example, that of the Sulu Sea (leg 124), where these models were used to determine the likelihood of a backarc basin origin and to interpret the volcanic/depositional history of a marginal basin. In addition, leg 124 results offer new insight into the influence of sea-level changes on backarc-basin sedimentation. *Klein's* [1985] model for the spatial and temporal distribution of facies in backarc basins emphasizes the tectonic rather than eustatic controls on turbidite sedimentation. In contrast, stratigraphic analysis of the sections penetrated on leg 124 in the Sulu and Celebes Seas strongly suggests that the influx of Miocene quartzose turbidites into these basins was eustatically controlled.

Drilling in the Sumisu Rift and Lau Basin has provided more detailed information on the early evolution of intraoceanic backarc basins. The Sumisu Rift is thought to represent the earliest phase of backarc-basin formation. This stage of development was not investigated during previous DSDP legs, and so, was not emphasized in the models of backarc sedimentation proposed by *Carey and Sigurdsson* [1984; Figure 20] and *Klein* [1985]. Drilling on leg 126 in the Sumisu Rift elucidated the nature and distribution of the early rift fill of this nascent intraoceanic backarc basin. Rapid sediment accumulation rates appear to be characteristic of this phase of development, in association with episodic submarine eruptions of arc calderas; felsic pumice and volcaniclastic debris produced during these eruptions are subsequently distributed throughout the basin by submarine gravity flow processes.

Results from leg 135 in the Lau Basin suggest that the locus of rifting and volcanism and the structural evolution of a backarc basin play major roles in determining basin/sub-basin stratigraphy. Following arc rifting, the Lau backarc region was compartmentalized into a series of sub-basins bounded by normal faults. Sedimentary sequences in these sub-basins fine upward and young toward the modern arc, reflecting an eastward shift in the locus of extension, volcanism and associated volcaniclastic sedimentation through time. This shift has not been previously documented and, therefore, has not been represented in models of backarc basin sedimentation. This shift may be a function of initial rifting in front of, rather than behind, the main arc axis. In the model proposed by *Carey and Sigurdsson* [1984; Figure 20], volcaniclastic turbidity currents travel from the arc far into the basin shortly after arc break-up. In the central Lau Basin, however, the rough topography and modest sediment supply prevent transport of arc-derived materials across the basin (Figure 16). The isolated basins and local volcanic sources that characterized the early history of the Lau Basin (Figure 16) contrast sharply with the smooth basin profile and continuous volcaniclastic apron extending across the backarc depicted by *Carey and Sigurdsson* [1984] in Figure 20.

Because of their volcaniclastic nature, intraoceanic backarc sands from various basins exhibit similar QFL, QmKP, and LmLvLs percentages [*Packer and Ingersoll*, 1986; *Marsaglia and Ingersoll*, 1992]. Although these sediments appear somewhat uniform in composition using traditional ternary plots, they vary widely in terms of their felsic, intermediate, and mafic volcaniclastic proportions. As summarized above, provenance studies of volcaniclastic sand from the Sumisu Rift and Lau Basin include information on the percentages of felsic to mafic volcaniclastic debris, geochemical analyses of individual glass shards and phenocrysts, as well as whole-rock geochemical analyses. These studies have supplied more detailed information on source magma composition, in some instances providing a geochemical fingerprint of the source volcano(es). In the case of the Lau Basin, volcani-

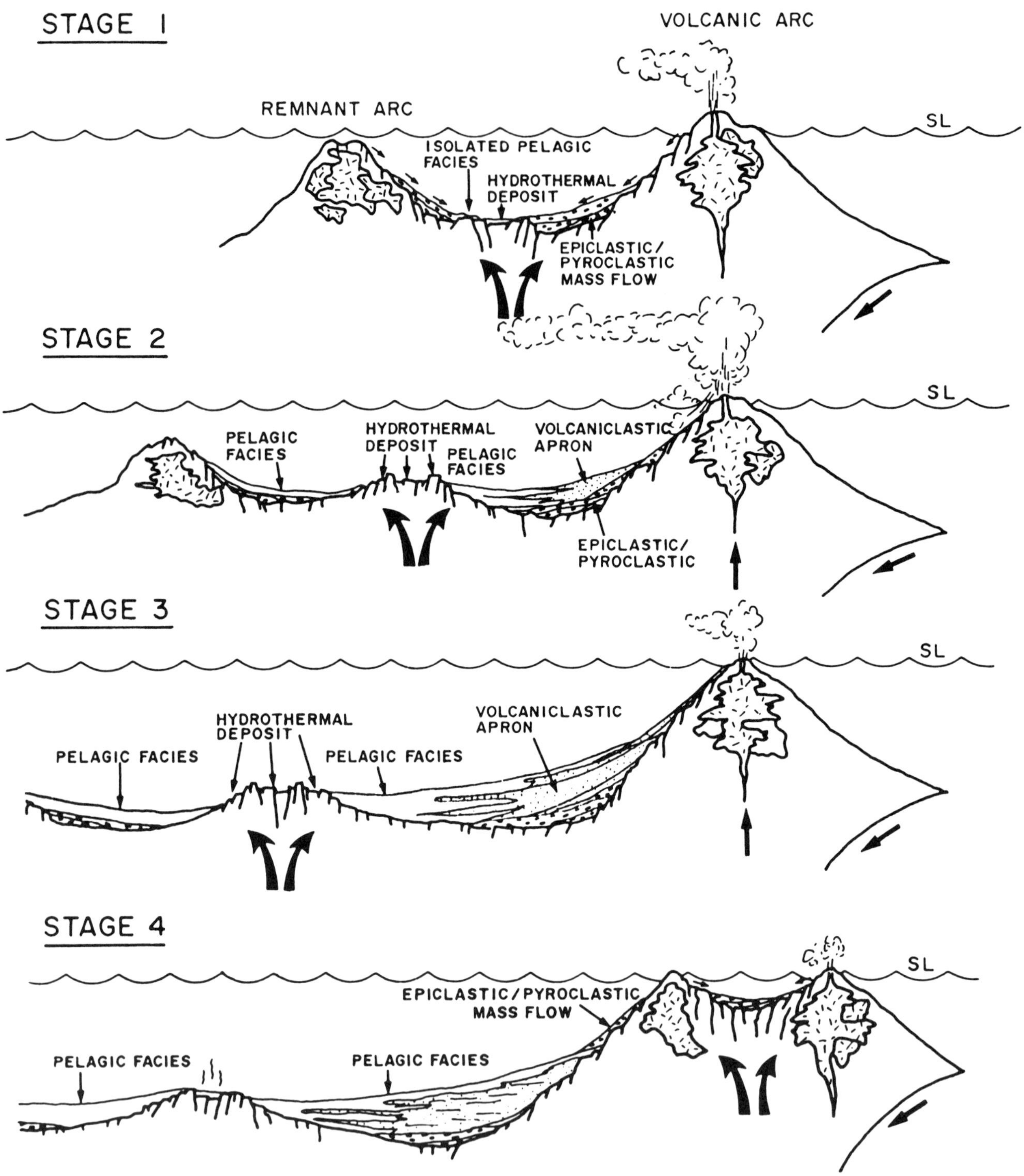

Fig. 20. Model of backarc basin evolution and sedimentation from *Carey and Sigurdsson* [1984].

clastic provenance studies have been used to document basin isolation, whereas in the Sumisu Rift, precise provenance of the sandy units is apparently masked by mixing during eruption and/or transport within the basin.

In summary, compositional variations within volcaniclastic backarc sequences are complex and a function of many variables. The volcaniclastic sequences deposited in backarc (and forearc) basins are now being more closely scrutinized in terms of what their geochemical signatures can tell about the magmatic evolution of the arc.

7.2. *Implications of Recent Drilling for Models of Continental-Margin Backarc Sedimentation*

For the most part, models for backarc-basin sedimentation have emphasized intraoceanic examples. Continental-margin backarc basins, such as the Japan Sea, have not received much attention, due in part to only minor DSDP drilling in this tectonic setting. ODP leg 127 and 128 results [e.g., *Ingle et al.*, 1990] suggest a strong paleoceanographic control on pelagic/hemipelagic sedimentation in the Japan Sea, in part, a function of arc (sill) tectonics. Our discussion has focussed on sand provenance in the Japan Sea, and the dramatic changes in sand composition over short distances in the backarc region. Arkosic, plutoniclastic sands recovered at various DSDP and ODP sites were likely derived from mainland Asia and/or now submerged continental blocks in the Japan Sea. In contrast to western and intrabasinal source terranes, the Japanese magmatic arc to the east sheds volcaniclastic sands into the basin. The degree of arc dissection along Honshu is reflected in the composition of sand deposited at the westernmost DSDP and ODP sites in the Japan Sea. Some provenance information has been lost due to post-depositional alteration.

7.3. *Future Drilling and Models of Backarc Sedimentation*

In order to fully understand how representative the volcaniclastic component of backarc basin sequences is of arc (and intrabasinal) volcanism, dominantly submerged volcanic sources first need to be better characterized. A comprehensive model for the petrologic evolution of intraoceanic arcs has remained elusive because of limited outcrop and core across the time-frame of arc evolution. Initial attempts by the Deep Sea Drilling Project to penetrate unconsolidated volcanic aprons of arc volcanoes failed. On ODP leg 126, similar deposits were successfully penetrated using piston coring techniques. Hole instabilities did finally ensue, however, with eventual loss of equipment. The dearth of information on proximal arc deposits provides a challenge to future drilling efforts and equipment development.

Acknowledgments. We thank Raymond V. Ingersoll and Keith Crook for their helpful and constructive reviews. PC thanks the Royal Society of Edinburgh, British Petroleum, and the Natural Environment Research Council for funding this work.

REFERENCES

Bednarz, U., and H.-U. Schmincke, Composition and provenance of volcaniclastic sediments in the Lau Basin (SW Pacific), ODP Leg 135 (Sites 834 to 839), in *Proc. ODP, Sci. Results, 135,* J. Hawkins, L. Parson, J. Allan, et al., pp. 51-74, College Station, TX, 1994.

Betzler, C., A. J. Nederbragt, and J. G. Nichols, Significance of turbidites at Site 767 (Celebes Sea) and Site 768 (Sulu Sea), *Proc. ODP, Sci. Results, 124,* E. A. Silver, C. Rangin, M. T. von Breymann, et al., pp. 431-446, Ocean Drilling Program, College Station, TX, 1991.

Boggs, S., Jr., Quaternary sedimentation in the Japan arc-trench system, *Geol. Soc. Am. Bull., 95,* 669-685, 1984.

Boggs, S., Jr., and A. Seyedolali, Diagenetic albitization, zeolitization, and replacement in Miocene sandstones, Sites 796, 797, and 799, Japan Sea, *Proc. ODP, Sci. Results, 127/128, Pt. 1,* K. A. Pisciotto, J. C. Ingle, Jr., M. T. von Breymann, et al., pp. 131-151, Ocean Drilling Program, College Station, TX, 1992.

Boggs, S., Jr., and A. Seyedolali, Provenance of Miocene sandstones from ODP sites in the Japan Sea, *J. Sedimentological Soc. Japan, 38,* 5-24, 1993.

Brown, G., and B. Taylor, Sea-floor mapping of the Sumisu Rift, Izu-Ogasawara (Bonin) island arc, *Bull. Geol. Surv. Japan, 39,* 23-38, 1988.

Carey, S., and H. Sigurdsson, A model of volcanogenic sedimentation in marginal basin, in *Marginal Basin Geology, Geol. Soc. London, Spec. Publ., 16,* edited by B. P. Kokelaar, and M. F. Howells, 37-58, Blackwell, London, UK, 1984.

Cashman, K. V., and R. S. Fiske, Fallout of pyroclastic debris from submarine volcanic eruptions, *Science, 253,* 275-280, 1991.

Cawood, P. A., Nature and record of igneous activity in the Tonga arc, SW Pacific, deduced from the phase chemistry of derived detrital grains, in *Developments in Sedimentary Provenance Studies, Geol. Soc., London, Spec. Publ., 57,* edited by A. C. Morton et al., 305-323, Blackwell, London, UK, 1991.

Clift, P. D., and J. E. Dixon, Variations in arc volcanism and sedimentation related to rifting of the Lau Basin, SW Pacific, in *Proc. ODP, Sci. Results, 135,* J. Hawkins, L. Parson, J. Allan, et al., pp. 23-49, Ocean Drilling Program, College Station, TX, 1994.

Clift, P. D., and ODP Leg 135 Scientific Party, Volcanism and sedimentation in a rifting island arc terrain; an example from Tonga, SW Pacific, in *Volcanism Associated with Extension at Consuming Plate Margins,* Geol. Soc. London, Spec. Publ., edited by J. L. Smellie, 1994, in press.

Colella, A., A. D'Alessandro, and R. De Rosa, Deep-water trace fossils and their environmental significance in forearc and backarc Cenozoic successions around Izu-Bonin arc, Leg 126 ODP, *Proc. ODP, Sci. Results, 126,* P. Taylor, K. Fujioka, et al., pp. 209-229, Ocean Drilling Program, College Station, TX, 1992.

Dickinson, W. R., L. S. Beard, G. R. Brakenridge, J. L. Erjavec, R. C. Ferguson, K. F. Inman, R. A. Knepp, F. A. Lindberg, and P. T. Ryberg, Provenance of North American Phanerozoic sandstones in relation to tectonic setting, *Geol. Soc. Am. Bull., 94,* 222-235, 1983.

Ewart, A., W. B. Bryan, B. W. Chappell, and R. L. Rudnick, Regional geochemistry of the Lau-Tonga arc and backarc systems, *Proc. ODP, Sci. Results, 135,* J. Hawkins, L. Parson, J. Allan, et al., pp. 385-425, Ocean Drilling Program, College Station, TX, 1994.

Fryer, P., C. Langmuir, B. Taylor, Y. Zhang, and D. Hussong, Rifting of the Izu Arc, III: Relationship of chemistry to tectonics (abstract), *EOS Transactions AGU,* 66, 421, 1985.

Fryer, P., B. Taylor, C. H. Langmuir, and A. G. Hochstaedter, Petrology and geochemistry of lavas from the Sumisu and Torishima backarc rifts, *Earth Planet. Sci. Lett., 100,* 161-178, 1990.

Fujioka, K., and S. Saito, Composition of heavy minerals from sands and sandstones of the Izu-Bonin arc, Leg 126, *Proc. ODP, Sci. Results, 126,* P. Taylor, K. Fujioka, et al., pp. 155-169, Ocean Drilling Program, College Station, TX, 1992.

Fujioka, K., A. Nishimura, Y. Matsuo, and K. S. Rodolfo, Correlation of Quaternary tephras throughout the Izu-Bonin areas, *Proc. ODP, Sci. Results, 126,* P. Taylor, K. Fujioka, et al., pp. 23-46, Ocean Drilling Program, College Station, TX, 1992.

Gill, J. B, C. Seales, P. Thompson, A. G. Hochstaedter, and C. Dunlap, Petrology and geochemistry of Pliocene-Pleistocene volcanic rocks from the Izu arc, Leg 126, *Proc. ODP, Sci. Results, 126,* P. Taylor, K. Fujioka, et al., pp. 383-404, Ocean Drilling Program, College Station, TX, 1992.

Haq, B. U., J. Hardenbol, and P. R. Vail, Mesozoic and Cenozoic chronostratigraphy and eustatic cycles, in *Sea-Level Changes: an Integrated Approach,* Soc. Econ. Paleontol. Min., Spec. Publ., 23, edited by C. K. Wilgus, B. S. Hastings, C. G. St. C. Kendall, H. W. Posamentier, C. A. Ross, and J. C. Van Wagoner, 71-108, Tulsa, OK, 1988.

Hiscott, R. N., and G. V. Middleton, Depositional mechanics of thick-bedded sandstones at the base of a submarine slope, Tourelle Formation (Lower Ordovician), Quebec, Canada, in *Geology of Continental Slopes,* Soc. Econ. Paleontol. Min., Spec. Publ., 27, edited by L. J. Doyle and O. H. Pilkey, 307-326, Tulsa, Oklahoma, 1979.

Hiscott, R. N., and J. B. Gill, Major and trace element geochemistry of Oligocene to Quaternary volcaniclastic sands and sandstones from the Izu-Bonin arc, *Proc. ODP, Sci. Results, 126,* P. Taylor, K. Fujioka, et al., pp. 467-485, Ocean Drilling Program, College Station, TX, 1992.

Hochstaedter, A. F., J. B. Gill, M. Kusakabe, S. Newman, M. Pringle, B. Taylor, and P. Fryer, Volcanism in the Sumisu Rift, I. Major element, volatiles and stable isotope geochemistry, *Earth Planet. Sci. Lett., 100,* 179-194, 1990.

Hole, M. J., A. D. Saunders, G. F. Marriner, and J. Tarney, Subduction of pelagic sediments: Implications for the origin of Ce-anomalous basalts from the Mariana Islands, *J. Geol. Soc. London,* 141, 453-472, 1984.

Holloway, N. H., North Palawan Block, Philippines—its relation to the Asian mainland and role in evolution of South China Sea, *Am. Assoc. Petrol. Geol. Bull., 66,* 1357-1383, 1982.

Honza, E., (ed.), *Geological investigation of the Japan Sea (GH78-2 Cruise),* Geol. Survey of Japan Cruise Rept. no 13., 1979.

Ingle, J. C., Jr., K. Suyehiro, M. T. von Breymann, et al., *Proc. ODP, Init. Repts., 128,* College Station, TX, 1990.

Karig, D. E., and G. F. Moore, Tectonically controlled sedimentation in marginal basins, *Earth Planet. Sci. Lett., 26,* 233-238, 1975.

Kennett, J. P., Marine tephrachronology, in *The Oceanic Lithosphere. The Sea v. 7,* edited by C. Emiliani, Wiley-Interscience Publications, New York, 1981.

Klaus, A., B. Taylor, G. F. Moore, M. E. MacKay, G. R. Brown, Y. Okamura, and F. Murakami, Structural and stratigraphic evolution of Sumisu Rift, Izu-Bonin arc, *Proc. ODP, Sci. Results, 126,* P. Taylor, K. Fujioka, et al., pp. 555-574, Ocean Drilling Program, College Station, TX, 1992.

Klein, G. deV., The control of depositional depth, tectonic uplift, and volcanism on sedimentation processes in the backarc basins of the western Pacific Ocean, *J. Geol., 93,* 1-25, 1985.

Lee, C. S., and R. McCabe, The Banda Celebes Sulu Basins: a trapped piece of Cretaceous Eocene oceanic crust? *Nature, 322,* 51-54, 1986.

Linsley, B. K., Carbonate sedimentation in the Sulu Sea linked to the onset of Northern Hemisphere glaciation, 2.4 Ma, *Proc. ODP, Sci. Results, 124,* E. A. Silver, C. Rangin, M. T. von Breymann, et al., pp., 375-378, Ocean Drilling Program, College Station, TX, 1991.

Ludwig, W. J., S. Murauchi, and R. E. Houtz, Sediments and structures of the Japan Sea, *Geol. Soc. Am. Bull., 86,* 651-664, 1975.

Marsaglia, K. M., Provenance of sands and sandstones from the Gulf of California, a rifted continental arc, in *Sedimentation in Volcanic Settings,* Soc. Econ. Paleontol. Min., Spec. Publ., 45, edited by R. V. Fisher and G. A. Smith, 237-248, Tulsa, Oklahoma, 1991.

Marsaglia, K. M., Petrography and provenance of volcaniclastic sands recovered from the Izu-Bonin arc, Leg 126, *Proc. ODP, Sci. Results, 126,* P. Taylor, K. Fujioka, et al., pp. 139-154, Ocean Drilling Program, College Station, TX, 1992.

Marsaglia, K. M., and R. V. Ingersoll, Compositional trends in arc-related, deep-marine sand and sandstone: a reassessment of magmatic-arc provenance, *Geol. Soc. Am. Bull., 104,* 1637-1649, 1992.

Marsaglia, K. M., R. V. Ingersoll, and B. M. Packer, Tectonic evolution of the Japanese Islands as reflected in modal compositions of Cenozoic forearc and backarc sand and sandstone, *Tectonics, 11,* 1028-1044, 1992.

Mascle, A., and P. A. Biscarrat, The Sulu Sea: a marginal basin in Southeast Asia, in *Geological and Geophysical Investigations of the Continental Margins,* Am. Assoc. Petrol. Geol. Mem., 29, 373-381, Tulsa, OK, 1978.

Mitchell, A. H. G., F. Hernandez, and A. P. Dela Cruz, Cenozoic evolution of the Philippine Archipelago, *J. Southeast Asian Earth Sci., 1,* 3-22, 1986.

Murakami, F., Structural framework of the Sumisu Rift, Izu-Ogasawara Arc, *Bull. Geol. Surv. Japan, 39,* 1-21, 1988.

Nishimura, A., and F. Murakami, Sedimentation of the Sumisu Rift, Izu-Ogasawara arc, *Bull. Geol. Surv. Japan, 39*, 39-61, 1988.

Nishimura, A., K. M. Marsaglia, K. S. Rodolfo, A. Colella, R. N. Hiscott, K. Tazaki, T. Janecek, J. Firth, M. Isiminger-Kelso, Y. Herman, J. B. Gill, R. N. Taylor, B. Taylor, K. Fujioka, and Leg 126 Shipboard Scientific Party, Pliocene-Quaternary submarine pumice deposits in the Sumisu Rift area, Izu-Bonin arc, in *Sedimentation in Volcanic Settings,* Soc. Econ. Paleontol. Min., Spec. Publ., 45, edited by R. V. Fisher and G. A. Smith, 201-208, Tulsa, Oklahoma, 1991.

Nishimura, A., K. S. Rodolfo, A. Koizumi, J. Gill, and K. Fujioka, Episodic deposition of Pliocene-Pleistocene pumice from the Izu-Bonin arc, Leg 126, *Proc. ODP, Sci. Results, 126,* P. Taylor, K. Fujioka, et al., pp. 3-21, Ocean Drilling Program, College Station, TX, 1992.

Packer, B. M., and R. V. Ingersoll, Provenance and petrology of Deep Sea Drilling Project sands and sandstones from the Japan and Mariana forearc and backarc regions, *Sed. Geol., 51,* 5-28, 1986.

Parson, L., J. Hawkins, J. Allan, et al., *Proc. ODP, Init. Repts., 135,* College Station, TX, 1992.

Parson, L. M., and J. W. Hawkins, Two stage ridge propagation and the geological history of the Lau backarc basin, in *Proc. ODP, Sci. Results, 135,* J. Hawkins, L. Parson, J. Allan, et al., pp. 819-828, Ocean Drilling Program, College Station, TX, 1994.

Pouclet, A., M. Pubellier, and P. Spadea, Volcanic ash from Celebes and Sulu Sea basins off the Philippines (Leg 124): petrography and geochemistry, *Proc. ODP, Sci. Results, 124,* E. A. Silver, C. Rangin, M. T. von Breymann, et al., pp. 467-488, Ocean Drilling Program, College Station, TX, 1991.

Pubellier, M., P. Spadea, A. Pouclet, R. Solidum, A. Desprairies, and H. Cambray, Correlations of tephras in Celebes and Sulu Sea basins: constraints on geodynamics, *Proc. ODP, Sci. Results, 124,* E. A. Silver, C. Rangin, M. T. von Breymann, et al., pp. 459-466, Ocean Drilling Program, College Station, TX, 1991.

Rangin, C., The Sulu Sea, a back arc basin setting within a Neogene collision zone, *Tectonophysics, 161,* 119-141, 1989.

Rangin, C., and E. A. Silver, Neogene tectonic evolution of the Celebes-Sulu basins: new insights from Leg 124 drilling, *Proc. ODP, Sci. Results, 124,* E. A. Silver, C. Rangin, M. T. von Breymann, et al., pp. 39-49, Ocean Drilling Program, College Station, TX, 1991.

Rangin, C., E. Silver, M. T. von Breymann, et al., *Proc. ODP, Init. Repts., 124,* College Station, TX, 1990.

Rodolfo, K. S., R. U. Solidum, A. Nishimura, Y. Matsuo, and K. Fujioka, Major-oxide stratigraphy of glass shards in volcanic ash layers of the Izu-Bonin arc-backarc sites (Sites 788/789 and 790/791), *Proc. ODP, Sci. Results, 126,* P. Taylor, K. Fujioka, et al., pp. 505-517, Ocean Drilling Program, College Station, TX, 1992.

Roeser, H. A., Age of the crust of the southeast Sulu Sea Basin based on magnetic anomalies and age determined at Site 768, *Proc. ODP, Sci. Results, 124,* E. A. Silver, C. Rangin, M. T. von Breymann, et al., pp., 339-343, Ocean Drilling Program, College Station, TX, 1991.

Shane, P. A. R. and P. C. Froggatt, Glass chemistry, paleomagnetism and correlation of middle Pleistocene tuffs in southern North Island, New Zealand and western Pacific, *N. Z. J. Geol. Geophys., 34*, 203-211, 1991.

Shipboard Scientific Party, Site 768, in *Proc. ODP, Init. Repts., 124,* E. A. Silver, C. Rangin, M. T. von Breymann, et al., pp., 195-297, Ocean Drilling Program, College Station, TX, 1990a.

Shipboard Scientific Party, Site 796, *Proc. ODP, Init. Repts., 127,* K. Tamaki, K. Pisciotto, J. Allan, et al., pp. 247-322, Ocean Drilling Program, College Station, TX, 1990b.

Shipboard Scientific Party, Site 797, *Proc. ODP, Init. Repts., 127,* K. Tamaki, K. Pisciotto, J. Allan, et al., pp. 323-421, Ocean Drilling Program, College Station, TX, 1990c.

Shipboard Scientific Party, Site 799, *Proc. ODP, Init. Repts., 128,* J. C. Ingle, Jr., K. Suyehiro, M. T. von Breymann, et al., pp. 237-402, Ocean Drilling Program, College Station, TX, 1990d.

Shipboard Scientific Party, Site 840, *Proc. ODP, Init. Repts., 135,* L. Parson, J. Hawkins, J. Allan, et al., pp. 489-570, Ocean Drilling Program, College Station, TX, 1992.

Shyu, J.-P., D. Merrill, V. Hsu, M. A. Kaminski, C. M. Müller, A. J. Nederbragt, R. P. Scherer, and H. Shibuya, Biostratigraphic and magnetostratigraphic synthesis of the Celebes and Sulu seas, Leg 124, *Proc. ODP, Sci. Results, 124,* E. A. Silver, C. Rangin, M. T. von Breymann, et al., pp. 11-35, Ocean Drilling Program, College Station, TX, 1991.

Sibley, D. F., and K. J. Pentony, Provenance variation in Turbidite sediments, Sea of Japan, *J. Sediment. Petrol., 48,* 1241-1248, 1978.

Silver, E. A., and C. Rangin, Leg 124 tectonic synthesis, *Proc. ODP, Sci. Results, 124,* E. A. Silver, C. Rangin, M. T. von Breymann, et al., pp. 3-9, Ocean Drilling Program, College Station, TX, 1991.

Smith, T. E., C. H. Huang, and F. G. Sajona, Geochemistry and petrogenesis of the volcanic rocks from Holes 768 and 769, Sulu Sea, *Proc. ODP, Sci. Results, 124,* E. A. Silver, C. Rangin, M. T. von Breymann, et al., pp. 297-310, Ocean Drilling Program, College Station, TX, 1991.

Spadea, P., L. Beccaluva, L. Civetta, M. Coltorti, J. Dostal, F. Sajona, G. Serri, C. Vaccaro, and O. Zeda, Petrology of basic igneous rocks from the floor of the Sulu Sea, in *Proc. ODP, Sci. Results, 124,* E. A. Silver, C. Rangin, M. T. von Breymann, et al., pp. 271-296, ODP, College Station, TX, 1991.

Tamaki, K., Geological structure of the Japan Sea and its tectonic implications, *Bull. Geol. Survey Japan, 39,* 269-365, 1988.

Tamaki, K., E. Honza, M. Yuasa, K. Nishimura, and F. Nishimura, Geological map of the central Japan Sea, *Geol. Survey Japan Marine Geology Map Series, 15,* 1981.

Tappin, D. R., T. Bruns, and E. L. Geist, Rifting of the Tonga/Lau Ridge and formation of the Lau backarc basin: evidence from Site 840 on the Tonga Ridge, *Proc. ODP, Sci. Results, 135,* J. Hawkins, L. Parson, J. Allan, et al., pp. 367-371, Ocean Drilling Program, College Station, TX, 1994.

Taylor, B., Rifting and the volcanic-tectonic evolution of the Izu-Bonin-Mariana arc, *Proc. ODP, Sci. Results, 126,* P. Taylor, K. Fujioka, et al., pp. 627-651, Ocean Drilling Program, College Station, TX, 1992

Taylor, B., K. Fujioka, et al., *Proc. ODP, Init. Repts., 126,* College Station, TX, 1990a.

Taylor, B., G. Brown, P. Fryer, J. B. Gill, A. G. Hochstaedter, H. Hotta, C. H. Langmuir, M. Leinen, A. Nishimura, and T. Urabe, ALVIN-Sea Beam studies of the Sumisu Rift, Izu-Bonin arc, *Earth Planet. Sci. Lett., 100,* 127-147, 1990b.

Kathleen M. Marsaglia, Department of Geological Sciences, University of Texas at El Paso, El Paso, TX 79968.

Sam Boggs, Jr., and Abbas Seyedolali, Department of Geological Sciences, University of Oregon, Eugene, OR 97403-1272.

Peter Clift, Ocean Drilling Program, 1000 Discovery Drive, College Station, TX 77845-9547.

Randall Smith, Department of Geology, 2345 Ramon Avenue, California State University, Fresno, Fresno, CA 93740-0024.

Sedimentation in Forearc Basins, Trenches, and Collision Zones of the Western Pacific: A Summary of Results from the Ocean Drilling Program

Michael B. Underwood[1], Peter F. Ballance[2], Peter D. Clift[3], Richard N. Hiscott[4], Kathleen M. Marsaglia[5], Kevin T. Pickering[6], and R. Pamela Reid[7]

The comprehensive strategy pursued by the Ocean Drilling Program in the western Pacific Ocean resulted in significant improvements in our understanding of arc-related geologic processes and products. Many types of depositional environments were examined, including intraoceanic forearc basins, backarc basins, trenches, and ridge-trench collision zones. Most of the ODP sites were chosen to address tectonic, geochemical, or hydrologic objectives, but a wealth of new sedimentologic information was generated at the same time. We now have a better appreciation of how intraoceanic forearcs are affected by diverse depositional processes (e.g., debris flows, turbidity currents, and vertical settling of tephra, hemipelagic mud, and pelagic ooze). Detailed turbidite stratigraphy deduced from FMS borehole imagery shows that the Izu-Bonin forearc basin is characterized by a noncyclic basin-plain style of sedimentation, rather than a channelized submarine fan or a channelized slope apron. Detrital modes for volcaniclastic sands in the Izu-Bonin forearc are consistent with an undissected-arc provenance; subdivisions of volcanic texture and glass type provide additional details and support the idea of extensive subaerial production and mixing of epiclastic detritus. Evolution of the Tonga forearc has been affected by a complicated interplay of tectonic subsidence, eustacy, and arc volcanism. Stratigraphic evolution of the Nankai accretionary prism produced a classic upward-coarsening and upward-thickening mega-cycle representative of transition from abyssal hemipelagites to trench turbidites. Turbidite sands within the outer trench-wedge facies display indicators of northwest-directed paleoflow; those flows probably experienced a phase of upslope movement onto the outer wall and deflection back toward the Nankai trench axis. Sand petrography and clay mineralogy collectively support a mixed detrital provenance for the Nankai trench wedge; the main source area is located in the Izu-Honshu collision zone. Because of high geothermal gradients, clay-mineral diagenesis within the Nankai prism has progressed to advanced stages at relatively shallow depths. The initial occurrence of illite/smectite (I/S) mixed-layer clay, for example, occurs approximately 555 m below the seafloor at a temperature of 65°C. Ridge-trench collision in the Vanuatu subduction zone has affected rates of trench sedimentation, pathways of sediment dispersal, formation and destruction of sediment ponds on the forearc slope, and facies partitioning at the décollement. The collision zone contains a well-defined accretionary prism, and intervals of kaolinite-rich clay serve as preferred sites for propagation of the décollement.

[1]Department of Geological Sciences, University of Missouri, Columbia, Missouri
[2]Department of Geology, University of Auckland, Auckland, New Zealand

Active Margins and Marginal Basins of the Western Pacific
Geophysical Monograph 88

[3]Ocean Drilling Program, College Station, Texas
[4]Earth Sciences Department, Memorial University, St. John's, Canada
[5]Department of Geological Sciences, University of Texas at El Paso, El Paso, Texas
[6]Department of Geological Sciences, University College, London, United Kingdom
[7]Rosenstiel School of Marine and Atmospheric Science, University of Miami, Miami, Florida

1. INTRODUCTION

This paper focuses on several thematic topics associated with ODP (Ocean Drilling Program) activity in the forearc-trench systems of the western Pacific Ocean. Pertinent drilling legs include 125 (Izu-Bonin and Mariana forearcs), 126 (Izu-Bonin forearc), 131 (Nankai Trough accretionary prism), 134 (Vanuatu collision zone), and 135 (Tonga forearc). It is important to realize that most of the western Pacific drilling legs were designed to satisfy objectives in tectonics, structural geology, fluid flow, and geochemistry. The principal targets during Leg 125, for example, were igneous basement rocks, serpentinite seamounts, and dewatering products of the subducting lithosphere. Leg 126 addressed the topics of crustal flexure, the formation of igneous basement within the frontal arc, and the micro-structural and large-scale rotation and translation of the forearc, in addition to sedimentologic objectives such as the dynamics of depositional architecture, changing paleoceanographic conditions, and the influence of arc volcanism as a source of sediment over time. Leg 131 concentrated on how pore fluids affected growth of the Nankai accretionary prism, on how the mechanical state and physical properties of deformed sediments change, and on the fabrics and structural styles of accreted sediments. Leg 134 focused on the style of deformation within a ridge-trench collision zone, the effects of collision on arc evolution, and dewatering processes within the forearc and subducted lithosphere. In addition to backarc targets, two sites in the Tonga forearc (Leg 135) were drilled to identify the lithology and age of basement material, to determine arc-platform and forearc stratigraphy and their relation to arc evolution, to investigate the deformation structures of the forearc slope, and to sample and analyze fluids. Regardless of the pre-cruise objectives, a tremendous wealth of basic sedimentologic information was generated during each of the western Pacific ODP legs.

This summary of sedimentologic themes grew out of discussions among some of the co-authors at the Asilomar workshop on Western Pacific Drilling Results in January 1993. Additional contributions were solicited from several key ODP participants who were unable to attend the workshop. The primary responsibilities for specific segments of the text are as follows: gravity-flow processes (Ballance); borehole imagery (Hiscott); volcaniclastic sands (Marsaglia); forearc subsidence (Clift); Nankai detrital provenance (Marsaglia and Underwood); Nankai stratigraphy (Pickering and Underwood); trench paleocurrents (Pickering and Underwood); clay-mineral diagenesis (Underwood); and ridge-trench collision (Reid). The products of ODP shipboard and shorebased research are important because they allow the sedimentologic community to refine many important ideas and conceptual models of three-dimensional facies architecture and depositional processes along convergent plate boundaries.

2. INTRAOCEANIC FOREARC ENVIRONMENTS

2.1. *Overview of Depositional Processes and Products*

Intraoceanic forearcs are present within all major climatic belts, including low latitudes (Mariana, Solomons, Vanuatu, Tonga, Lesser Antilles), high latitudes (Aleutians, Scotia), and temperate latitudes (Bonin, Kermadec). Only the Mariana and Izu-Bonin forearcs, however, have been drilled in a reasonably comprehensive manner [*Hussong et al.*, 1981; *Fryer et al.*, 1990, 1992; *Taylor et al.*, 1990, 1992]. Limited drilling has taken place in the forearc segments of Tonga [*Parson et al.*, 1992; *Hawkins et al.*, 1994], Vanuatu [*Andrews et al.*, 1975; *Collot et al.*, 1992], the Lesser Antilles [*Biju-Duval et al.*, 1984; *Mascle et al.*, 1988; *J.C. Moore et al.*, 1990], and the Aleutians [*Creager et al.*, 1973].

Oceanic forearc sediments can be considered within a continuum between volcanic and pelagic end-members (Figure 1). The biogenic-pelagic contribution is mostly calcareous at low and temperate latitudes and mostly siliceous at high latitudes. There is almost always a component of clay minerals and scattered volcanic clasts. The source of the clay, however, is not always obvious; much of it probably forms through *in situ* alteration of fine air-fall ash, in addition to a background of fine particles blown by the wind or suspended initially in surface waters. Studies of regional-scale clay mineral distribution [e.g., *Hathon and Underwood*, 1991] show that dispersal patterns can be complicated for hemipelagic muds because of the combined effects of atmospheric circulation, surface currents, thermohaline bottom currents, turbidity currents, and repeated resuspension within the bottom nepheloid layer. Because of this complexity, detrital sources need to be documented carefully rather than assumed.

The steady background of pelagic and hemipelagic sedimentation near active volcanoes invariably is interrupted by sudden incursions of tephra. Most marine tephra layers have sharp bases; many contain internal sedimentary structures of the Bouma sequence and grade upward into muddy background sediment. Cross-laminae and convoluted laminae are particularly well developed in thin, base-missing Bouma sequences, and beds of this type are generally accepted as the depositional products of turbidity currents (Figure 1). Distinguishing between turbidites and

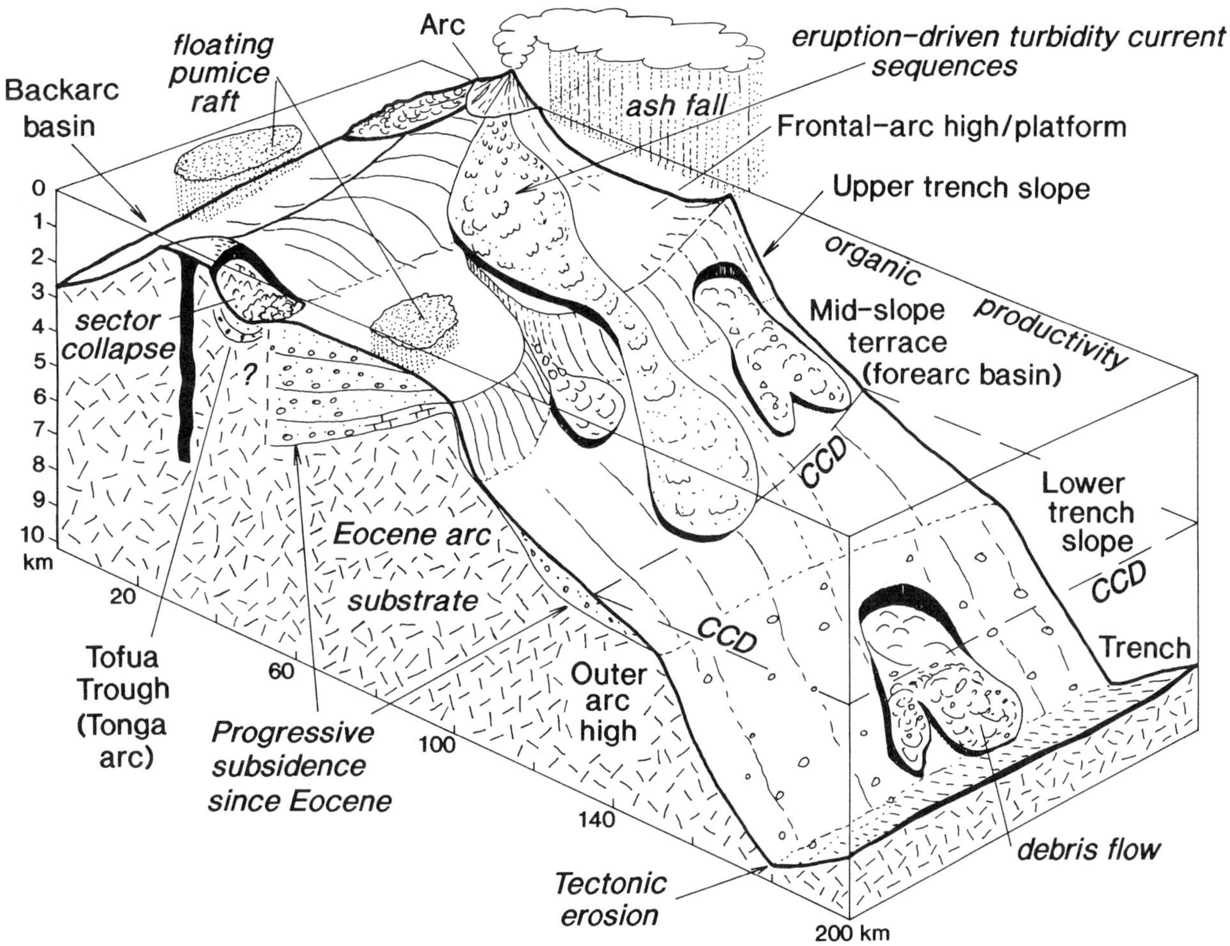

Fig. 1. Block diagram showing major depositional processes in oceanic forearc environments; based mainly on examples from Tonga.

direct ash-fall beds, however, can be problematic because of similarities in their respective sedimentary structures.

Unequivocal ash-fall beds appear to be rare. One prominent example from the Izu-Bonin forearc consists of a 1-cm-thick, normally graded bed of vitric silty ash, overlain by a 10-cm-thick, inversely graded layer with sand-sized to pebble-sized pumice clasts. These deposits could have resulted from a single ash-fall event with delayed settling of floating pumice fragments [*Taylor et al.*, 1990]. The Tonga frontal-arc platform contains a group of seven possible ash-fall layers in near-surface sediments (Site 840) [*Parson et al.*, 1992]. Individual ash beds are up to 10 cm thick with maximum grain diameters of 3 mm; they differ from interbedded turbidites in their lower glass content, and they contain up to 20% by volume of crystals of plagioclase, clinopyroxene, orthopyroxene, and rare olivine.

Unfortunately, the sedimentologic characteristics described above are not unique to one emplacement mechanism. Crystal contents of ash-fall beds, for example, can vary systematically with distance from the explosive source [*Sigurdsson et al.*, 1980]. In addition, crystals derived from volcanic eruptions should be expected in many turbidite parent materials. In a turbidity current, relatively dense constituents in fragmental volcanic material will be among the first particles to settle, so the higher-density crystal fraction may occur only in proximal deposits. In the Izu-Bonin example cited above, a gravity-flow mechanism may have produced normal grading in the relatively dense components, whereas the inverse grading may have resulted from lower density pumice that floated away from the site of initial settling of dense ash. To complicate matters further, ashfalls can occur at the same time as subaerial pyroclastic surges and flows, which themselves can be transformed into subaqueous turbidity currents when they reach the shoreline of a volcanic island [*Sparks et al.*, 1980a, 1980b]. Finally, most forearc deposits are heavily bioturbated, and it is clear from inspection of both outcrops and cores that many thin ash beds (2-3 cm or less) become totally dispersed into the background sediment. This factor obviously hampers the preservation potential and subsequent recognition of discrete ash-fall deposits.

Chemical analyses of glass shards, crystals, and bulk ash samples normally reveal wide ranges of composition within any single depositional interval [*Cawood*, 1985; *Parson et al.*, 1992; *Fujioka et al.*, 1992; *Rodolfo et al.*, 1992a]. Dominant glass compositions can change significantly through time, owing to tectonic cycles and their controls on arc magmatism. Extensive mixing also occurs among materials of different ages and sources. Shallow-water fossils, for example, are incorporated into gravity flows routinely. Thus, in addition to the influence of magmatic processes, the bulk chemical character of deep-water sediments in the intraoceanic depocenters can be affected by residence times and physical mixing processes (waves, currents, etc.) within transitional storage areas along the margins of arc platforms.

Two types of volcanic clasts deserve special consideration during discussions of forearc sedimentation. First, great volumes of pumice erupted during volcanic explosions provide a unique component to the sediment budgets. Submarine eruptions can generate large floating mats of pumice clasts (Figure 1), which sometimes travel incredible distances from their source [*Cas and Wright*, 1987]. Abrasion between floating clasts produces a constant rain of glass shards onto the sea floor. Eventually, however, the pumice becomes waterlogged and sinks. The behavior of pumice clasts in gravity flows is also unusual because the grain densities remain low with respect to crystals and rock fragments, even after they are waterlogged. Pumice fragments, therefore, are usually concentrated higher in the bed and show inverse grading.

A second type of special clast is the accretionary lapillus, which forms as volcanic dust adheres to raindrops or small nuclei on the ground. Most accretionary lapilli are strong enough to behave as regular mineral and rock fragments during marine transport. They can occur by themselves, and in such cases they form the basal interval of a normally graded bed [*Soh et al.*, 1989]. When mixed with denser rock fragments and mineral grains, accretionary lapilli likewise show normal grading, but the lapilli become concentrated higher in the beds.

2.2. *Types of Gravity Flows and Depositional Sequences*

A broad spectrum of gravity-flow deposits has been recognized within intraoceanic forearc environments (Figure 1). Turbidites are reported most frequently, and these deposits range in thickness from a few mm to 4.5 m or more. The thicker beds tend to be coarse-grained and commonly show inverse-to-normal size grading. Basal divisions of alternating basic and felsic laminae (e.g., Site 793) may be expressions of repetitive traction carpets [*Hiscott and Middleton*, 1980; *Lowe*, 1982] or fluctuating flows [*Hiscott*, 1994]. The glass contents of sandy turbidites are consistently greater in the Tonga and Mariana-Bonin systems [e.g., *Marsaglia*, 1992] as compared with the Aleutians, where polycrystalline volcanic-lithic grains are more abundant [*Stewart*, 1978].

Debris-flow deposits appear to be less common than turbidites, but they have been recognized on the basis of large clast sizes, poor sorting, obvious matrix support, and clast heterogeneity [*Vallier et al.*, 1985; *Bloomer and Fisher*, 1987; *Ballance et al.*, 1989; *Ballance*, 1991]. Exposed sequences along the Tonga frontal-arc platform, for example, contain probable debris-flow deposits with clasts up to 1 m or more in diameter, and blocks of reef coral are mixed with volcanic materials. These beds are typically several meters thick and display inverse-to-normal grading; they appear very similar to the Parnell Grit beds of the Waitemata Basin of New Zealand, which are inferred products of sector collapse events on strato-volcanoes [*Ballance and Gregory*, 1991]. Debris-flow deposits probably are preserved preferentially on steep seafloor surfaces such as the lower slopes of sediment-starved trenches or the over-steepened walls of submarine canyons. Dredge hauls from such environments have recovered wide ranges of clast lithologies, clast ages, and clast sizes, including remobilized fragments of vitric mudstone, volcanic sandstone, volcanic breccia, carbonates, and serpentinized ultramafic rock.

In all forearc basins drilled to basement, the basal sedimentary units are conglomeratic, and this common occurrence is probably related to the tectonic trigger of basin formation through rifting of old arc substrate. Sedimen-tation rates in the overlying volcaniclastic turbidite sequences range widely, from less than 15 m/my to more than 800 m/my. Sedimentation rates typically increase as the event-frequency for mass flows increases and the volcanic sediment supply becomes coarser grained. Upward-fining sequences [e.g. Site 840; *Parson et al.*, 1992] have been attributed to increasing distances of transport relative to the arc source. Textural characteristics also respond to waxing and waning of eruptive events at individual volcanoes, however, so the terms 'proximal' and 'distal' require cautious usage as descriptive adjectives. In addition, turbidite successions in basin-plain environments tend to be noncyclic or erratic with respect to vertical changes in bed thickness and texture [*Hiscott et al.*, 1992]. Therefore, a single type of stratigraphic succession cannot be identified which would be diagnostic of all intraoceanic forearcs.

2.3. *Detailed Turbidite Stratigraphy and Borehole Imagery*

Knowledge of three-dimensional facies architecture for turbidite systems within intraoceanic forearc environments was superficial prior to Leg 126. As described above, sandy detritus in such settings is derived entirely from the volcanic arc (either as reworked tephra or as products of subaerial erosion of the exposed volcanoes), and the eruptive material can be resedimented from the flanks of the arc volcanoes into deep water by a variety of sediment gravity flows [see also *Cas and Wright*, 1987]. Potential triggers of gravity flows include sea-level fall, hurricane-scale storm waves, tectonic uplift, and earthquakes.

Sites 787, 792, and 793 were drilled during Leg 126 in the Izu-Bonin forearc basin (Figure 2). Drilling penetrated a thick lower to upper Oligocene turbidite succession that was derived solely from the contemporary Izu-Bonin intraoceanic arc at depositional rates of 100-300 m/my [*Taylor et al.*, 1990]. Multichannel seismic profiles indicate that the Oligocene forearc basin did not contain any large-scale channels or valleys; the depositional environment apparently was a rather featureless, elongate basin plain. Stratigraphic patterns of bed thickness and grain size (upward-thickening, upward-fining, etc.) have long been promoted as important criteria for the recognition of architectural elements produced during the evolution and growth of sandy submarine fans [*Mutti and Ricci Lucchi*, 1972; *Ricci Lucchi*, 1975; *Mutti and Normark*, 1987]. Vertical sequences in basin-plain deposits, however, are supposed to be more irregular [*Stow*, 1986; *Pickering et al.*, 1989]. To help test these ideas, the Formation MicroScanner (FMS) was utilized for the first time by ODP during Leg 126 to generate images of the turbidite successions in the Izu-Bonin forearc [*Lüthi and Banavar*, 1988; *Shipboard Scientific Party*, 1990].

FMS images change in response to grain size, induration and cementation, porosity, mineralogy, and borehole effects such as size, shape, and surface texture (Figure 3). Good quality images of the Oligocene forearc-basin turbidite succession were recorded for depths of 2232-2441 meters below rig floor (mbrf) in Hole 792E and 4023-4330 mbrf in Hole 793B. Distinctive intervals allowed individual sandstone beds to be matched between cores and the FMS records. For parts of the succession where core recovery was incomplete, the FMS images alone provide detailed information on sedimentary facies through construction of bed-by-bed logs (Figure 4).

The total number of sharp-based sandstone beds recognized in the FMS data are 447 for Hole 792E and 807 for Hole 793B. The number of sandstone beds is inversely proportional to bed thickness; that is, there are many thin beds and few thick beds. Some of the very thick beds are debris flow deposits, not turbidites [*Rodolfo et al.*, 1992b]. An appropriate distribution function is a power law, in which the number $N(t)$ of beds whose thickness is greater than t is proportional to t^{-b}, where the exponent b is a positive constant. In log-log space, a power-law distribution plots as a straight line with slope b. For data from both FMS-based sections, b is slightly greater than 1.0 (Figure 5). For beds thinner than 10 cm, the number of beds is smaller than predicted by the power law. *Hiscott et al.* [1992] suggested that this is either because of erosion of an indeterminate number of such thin beds by subsequent sediment gravity flows, or because some beds with thickness near the resolution limit of the FMS tool (2.5 cm) could not be recognized.

In poorly documented depositional environments, such as intraoceanic forearc basins, trends in bed thickness are particularly important because they provide evidence for changes in eruptive style of the arc and changes in relative sea level [*Hiscott et al.*, 1992]. Some parts of each log from the Izu-Bonin forearc consist mainly of very thick beds (Figure 4). Visual inspection, however, reveals no consistent trends in bed thickness across the boundaries of, or within, these intervals. Thinner beds always disrupt any potential asymmetric sequences. Clusters of very thick beds are similar to the so-called "depositional lobes" recognized in submarine-fan deposits [*Mutti*, 1977; *Mutti and Normark*, 1987; *Shanmugam*, 1990]. When viewed in vertical section only, similar groups of beds have also been called "packets" [*Ojakangas*, 1968; *Chipping*, 1972; *Hiscott*, 1980], a term preferred here because there is no three-dimensional evidence for deposition on true submarine fans. The sandstone-rich packets probably accumulated in response to local or regional changes in the volumes or pathways of sediment gravity flows entering the Izu-Bonin forearc.

The data files used to plot the FMS-based sedimentary logs were sorted to determine the number of sandstone beds in each bed-thickness class of *Ingram* [1954], as well as the

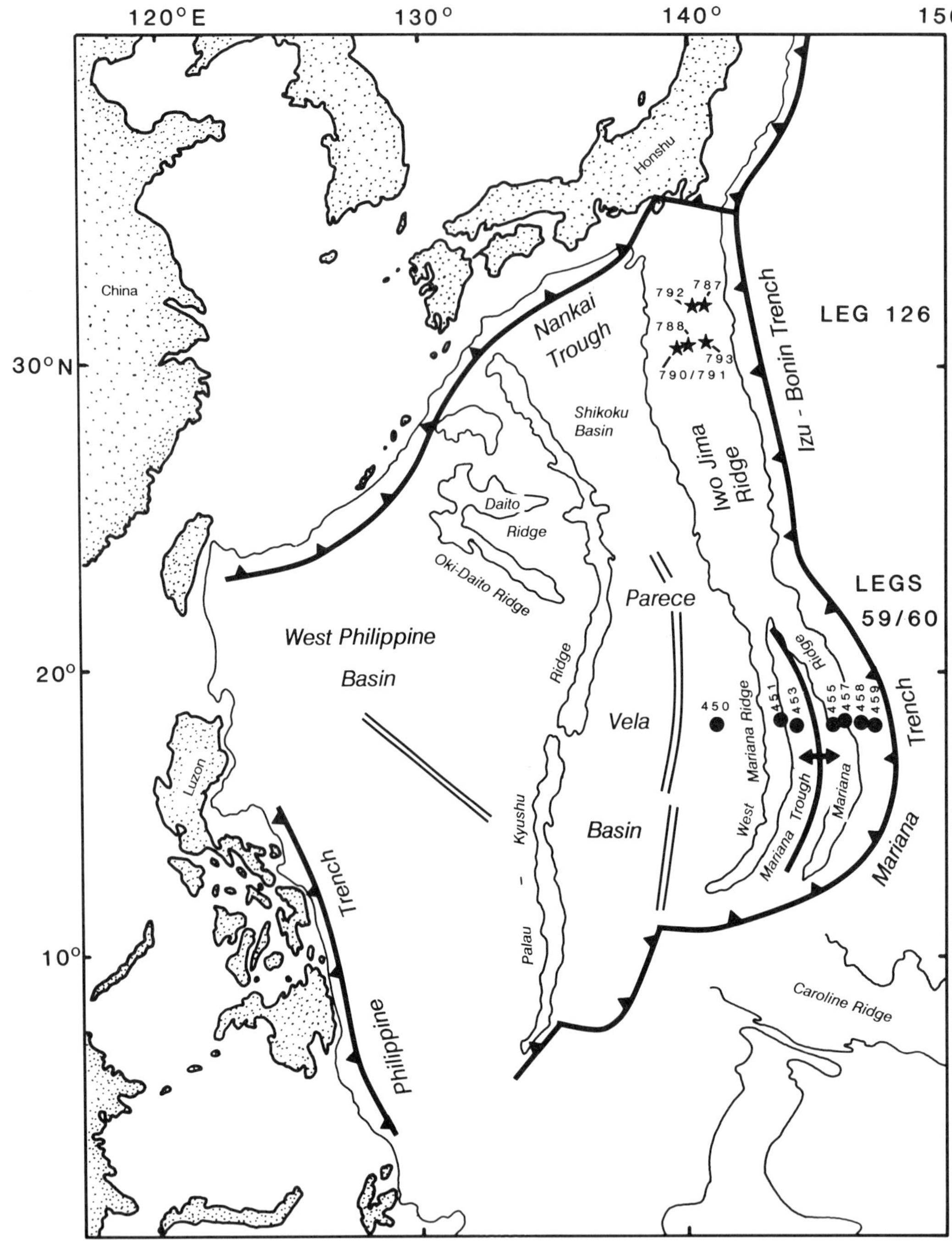

Fig. 2. Map of the Philippine Sea region showing locations of Izu-Bonin forearc sites (ODP Leg 126) and the DSDP Leg 59/60 transect of the Mariana Trench and backarc basin.

number of small to very large sediment gravity flows to reach each site per unit time, assuming no erosion beneath subsequent flows [*Hiscott et al.*, 1994]. A derivative parameter is the recurrence interval for beds of different thicknesses. Except for very thin and very thick beds, for which statistical data are not dependable [*Hiscott et al.*, 1994], recurrence intervals range from about 4,000 yr to 850,000 yr; these values are similar to recurrence intervals reported by *Klein* [1984] for turbidites in the Coral Sea and the Sea of Japan. Because sandstone bed thicknesses are

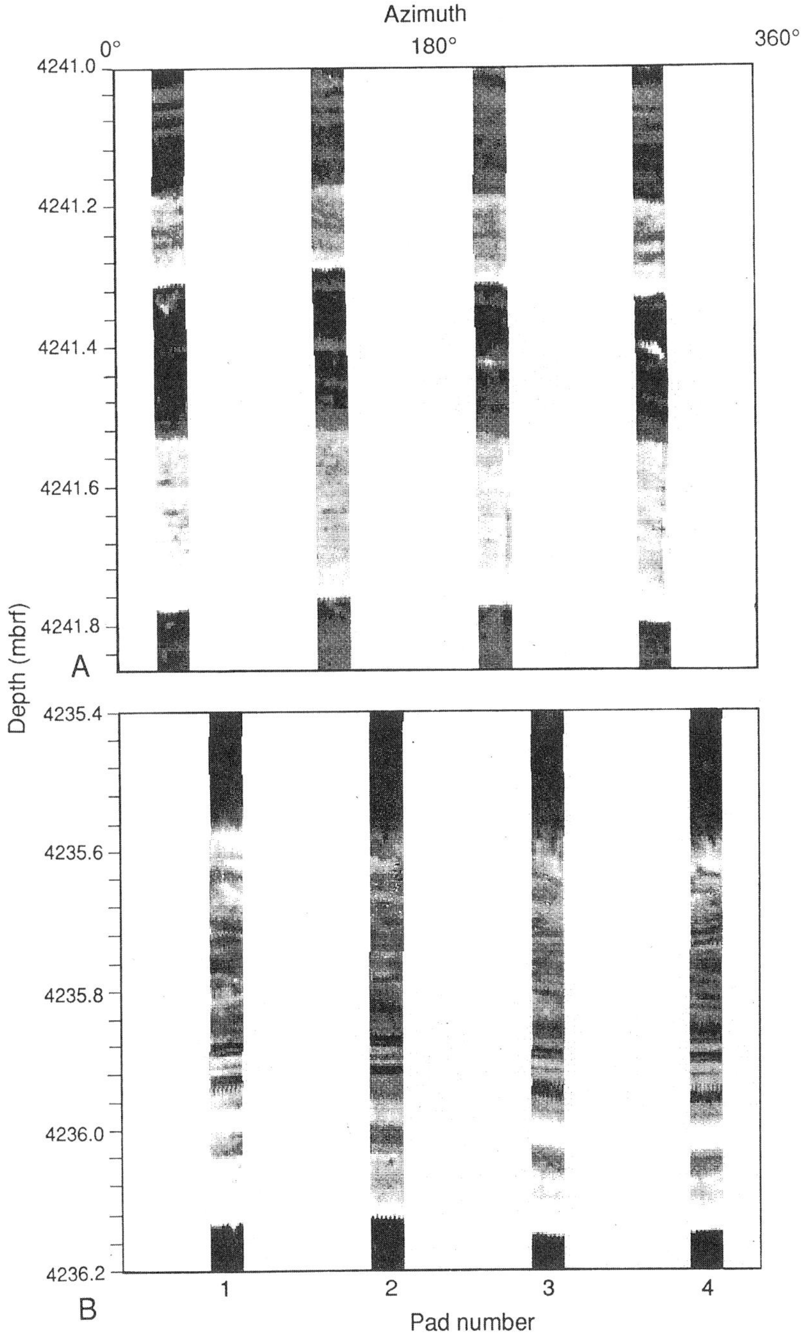

Fig. 3. FMS images of sedimentary structures and grading in medium to thick sandstone beds in Hole 793B. Lighter tones indicate coarser grained sediment. A. Two graded beds, both cross laminated; cross laminae are most visible in the upper bed at 4241.25 mbrf. B. Parallel laminae and wavy laminae in a sharp-based, graded sandstone bed. From *Hiscott et al.* [1992].

distributed according to a power law with a unit exponent, we conclude that the recurrence interval is proportional to bed thickness.

The calculated recurrence intervals are much longer than the recurrence intervals for strong earthquakes along short segments of the Izu-Bonin arc system [*Duda*, 1965; *Mogi*, 1990]. Therefore, the initiation of large turbidity currents must have depended largely on the accumulation of

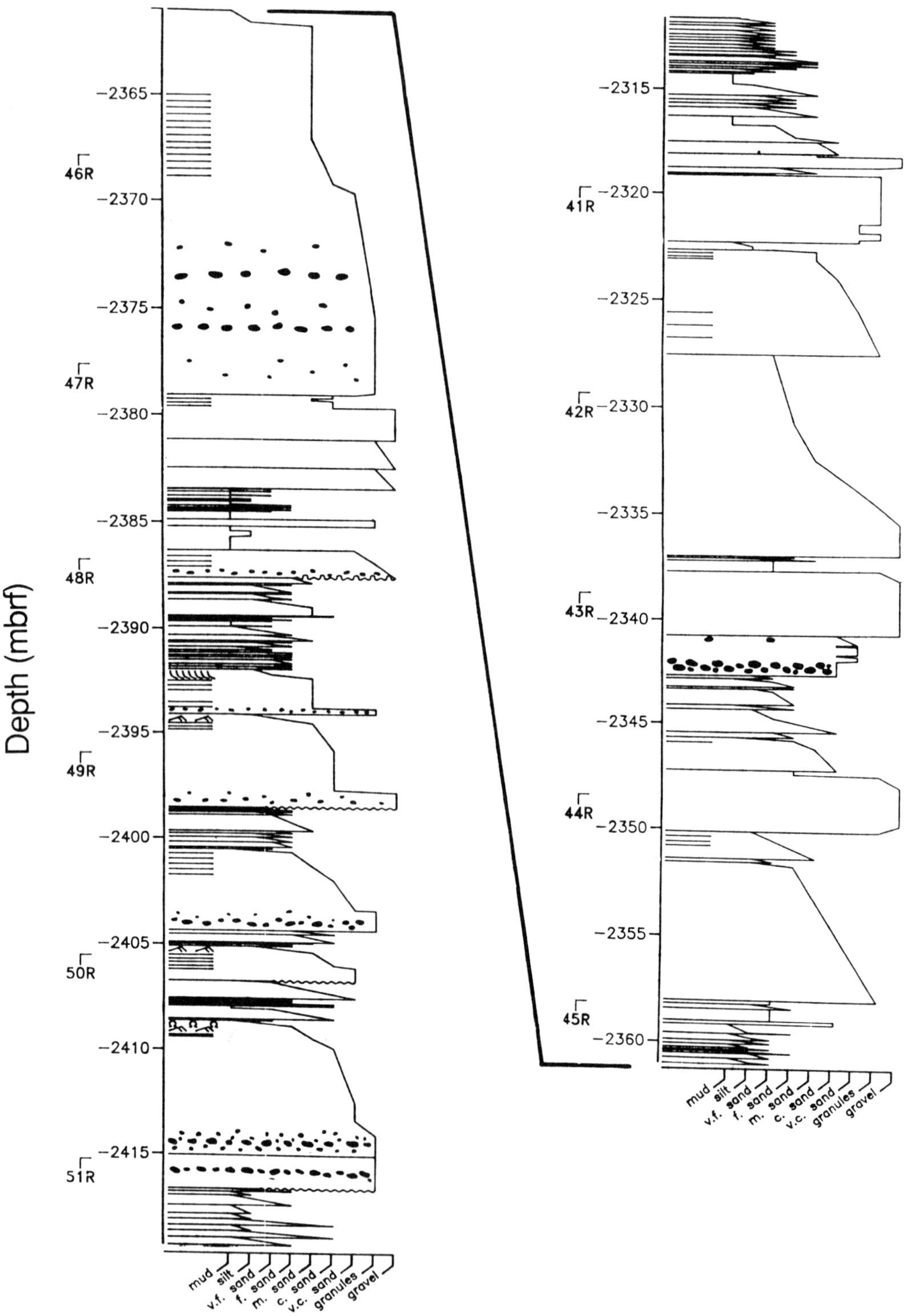

Fig. 4. MS-based log for part of Hole 792E, showing positions of core tops (e.g., 46R), depths in mbrf, estimates of grain size, symbols for sedimentary structures (parallel laminae, cross bedding, ripple laminae, convolution, erosive bases), and positions of fine-grained intraclasts (black). See *Hiscott et al.* [1992] for complete log.

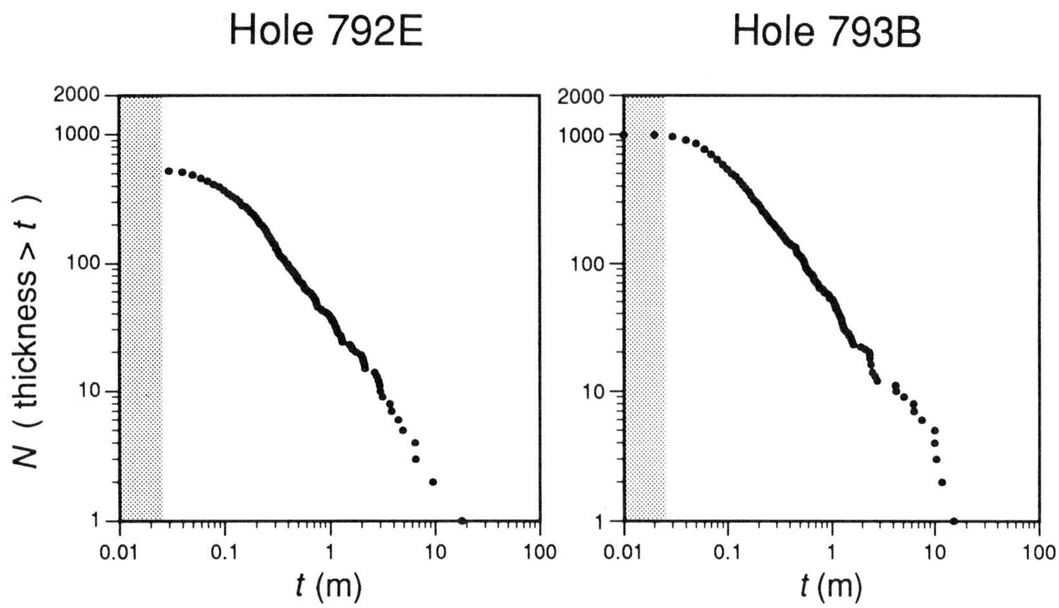

Fig. 5. Logarithmic plot of the number of sandstone-shale couplets of thickness >t versus t in the FMS-based sections. As predicted for a power-law distribution, the plot follows a straight line with a slope that is the characteristic exponent of the distribution (slightly >1 in this case). The stippled areas define the thickness interval below the resolution limit of the FMS tool (i.e., 2.5 cm). From *Hiscott et al.* [1992].

sufficient amounts of unstable sediment near the edge of the arc platform, rather than dynamic loading by an energetic trigger. Prior to resedimentation, temporary storage may have filled depressions on the flanks of volcanoes and resisted mobilization by smaller earthquakes. The time required for accumulation of the volcaniclastic material might be on the order of tens of thousands of years. At other times, smaller amounts of material were mobilized by smaller magnitude (or more distant) tremors. The earthquake magnitudes and associated volumes of liberated material both obey simple power-law distributions.

The sole source for coarse sediment entering the Izu-Bonin forearc was detritus and ejecta from the adjacent chain of diminutive volcanic islands. Oligocene sedimentation rates, nevertheless, were impressively high (100-300 m/my). Comparable rates of 265-425 m/my have been calculated for such on-land forearc-basin examples as the Great Valley Group [*Ingersoll*, 1979], which received sediment supplies from much larger continental drainage basins. On the other hand, the facies relations in the Izu-Bonin intraoceanic forearc basin differ in several important respects from those in well-studied continental-margin analogs: (1) there is a general lack of conglomerate in the intraoceanic forearc; (2) high-relief channel-levee complexes have been undetected to date, in contrast to inferred analogs such as the Cretaceous Nanaimo Group of coastal British Columbia [*England and Hiscott*, 1992]; (3) there is an absence of diversity in submarine-fan facies associations within the Izu-Bonin system, in contrast to more heterogeneous examples such as the Cretaceous Great Valley Group of California [*Ingersoll*, 1978]; and (4) the intraoceanic basin-plain succession contains scattered beds of sandstone reaching dimensions as thick as about 18 m, which conflicts with typical models which show only thin-bedded turbidites in distal depositional environments [e.g., *Mutti and Ricci Lucchi*, 1972].

Leg 126 drilled only two holes to the bottom of the Izu-Bonin forearc basin, so the three-dimensional facies architecture remains poorly documented. Multichannel seismic profiles [*Taylor*, 1992] do not have the vertical resolution to define continuity or geometry of individual sandstone packets. For this reason, the lateral extent of individual sandstone beds, the specific pathways of turbidity currents, and the importance of local erosion all remain unknown. In spite of these limitations, the drilling results clearly document the enormous output of the Oligocene arc volcanoes, estimated at 14 $km^3/10^3$ yr from 29°N to 34°N. In addition, it is obvious that very-large-volume turbidity currents and subaqueous debris flows have the ability to transport sediment at least as far as 40 km from the volcanic front. Small submarine fans with distributary channels, or channelized slope aprons near the Oligocene volcanoes, may have escaped detection, but the apparently random occurrence of beds of different thickness seems more characteristic of a basin-plain style of deposition [*Ricci Lucchi and Valmori*, 1980; *Hiscott et al.*, 1986].

2.4. *Petrography and Provenance of Volcaniclastic Sands and Sandstones*

Most forearc sands are composed largely of volcanic lithic fragments eroded from the adjacent arc, with lesser amounts of feldspar and denser mineral grains [*Packer and Ingersoll*, 1986; *Marsaglia and Ingersoll*, 1992]. Only a few sand-bearing sites were drilled by the Deep Sea Drilling Project (DSDP) in the forearc basins of intraoceanic arcs: Site 186 in Atka basin (central Aleutian forearc), and Sites 458 and 459 in the Mariana forearc [*Marsaglia and Ingersoll*, 1992]. Recent ODP drilling in the Izu-Bonin forearc (Figure 2) resulted in a three-fold increase in petrologic data for this type of tectonic setting.

During Leg 126, the sedimentary fill of the Izu-Bonin forearc was penetrated at three sites (787, 792, 793). *Marsaglia* [1992] found that the compositions of Izu-Bonin forearc sands are similar to sand compositions in other intraoceanic arcs, in terms of detrital modes Q-F-L (total quartz, total feldspar, total lithic fragments), Q_m-K-P (monocrystalline quartz, K-feldspar, plagioclase), and L_m-L_v-L_s (metamorphic, volcanic, sedimentary lithic fragments). The sand samples contain only traces of quartz, no potassium feldspar, and virtually no sedimentary or metamorphic lithic fragments. Their modal values plot on or near the plagioclase apex on a Q_m-K-P diagram and at the L_v apex of an L_m-L_v-L_s diagram. The most variation is observed on a Q-F-L plot (Figure 6), which shows that Quaternary forearc samples are slightly more enriched in total quartz with respect to older units. This enrichment also occurs in the arc-axis and backarc sites, and the increases in quartz probably were caused by felsic volcanism associated with modern submarine calderas along the arc axis. For comparison, detrital modes for Mariana sites [*Packer and Ingersoll*, 1986] plus the tectonic-provenance fields of *Dickinson et al.* [1983] are also shown on Figure 6. As expected, all of these data plot in the undissected-arc field.

Temporal and across-basin trends in sediment composition were recognized through subdivisions of volcanic texture and glass color [*Marsaglia*, 1992]. The textural categories include the vitric, microlitic, and lathwork glassy varieties defined by *Dickinson* [1970]. Glass color is a function of magma composition, so studies

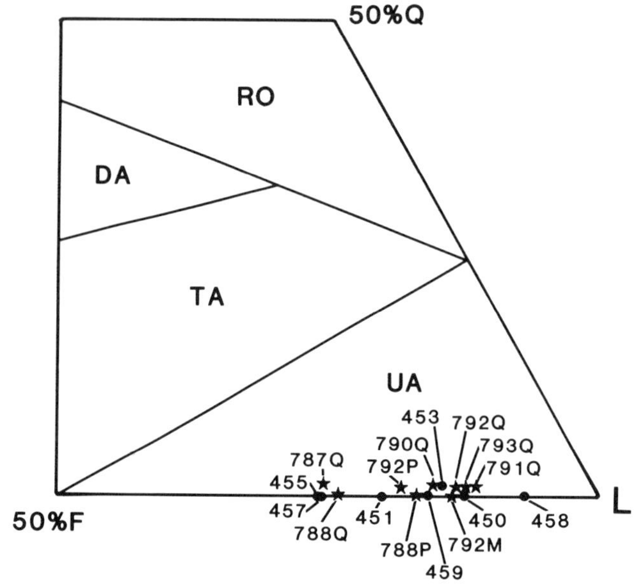

Fig. 6. Lithic corner of a QFL plot showing sand petrography of Izu-Bonin and Mariana forearc and backarc sites; Q = total quartz, F = total feldspar, and L = total lithic proportions. Mean values for Mariana (Sites 450-459) taken from *Packer and Ingersoll* [1986]. Izu-Bonin data (Sites 787-793) taken from *Marsaglia* [1992] and subdivided according to age: Q = Quaternary, P = Pliocene, M = Miocene. Provenance fields are from *Dickinson et al.* [1983], where RO = recycled orogen, DA = dissected arc, TA = transitional arc, and UA = undissected arc. Data acquired using the Gazzi-Dickinson method [*Dickinson*, 1970; *Ingersoll et al.*, 1984] on thin sections stained for feldspar. See *Marsaglia* [1992] and *Packer and Ingersoll* [1986] for further details.

of glass populations yield information regarding the eruptive history of the arc. The relation between glass color and silica content, as determined by refractive index, was documented previously for pyroclastic materials from the Leg 60 Mariana transect [*Schmincke*, 1981] and Site 450 in the Parece Vela Basin [*Rodolfo and Warner*, 1981]. Colorless glass typically is more felsic, and brown glass is more mafic. Sideromelane (brown glass) is produced by rapid quenching of basaltic magma, whereas tachylite (black glass) forms through slower cooling within subaerial or subaqueous settings [*Fisher and Schmincke*, 1984; *Cas and Wright*, 1987]. Tachylite is opaque in transmitted light owing to finegrained Fe and Ti oxides.

Marsaglia [1992] found that the glass proportions of Izu-Bonin samples are extremely heterogeneous (Figure 7a). These results are consistent with data from chemical analyses of the tephras [*Fujioka et al.*, 1992]. In terms of texture, the colorless glassy fragments are predominantly

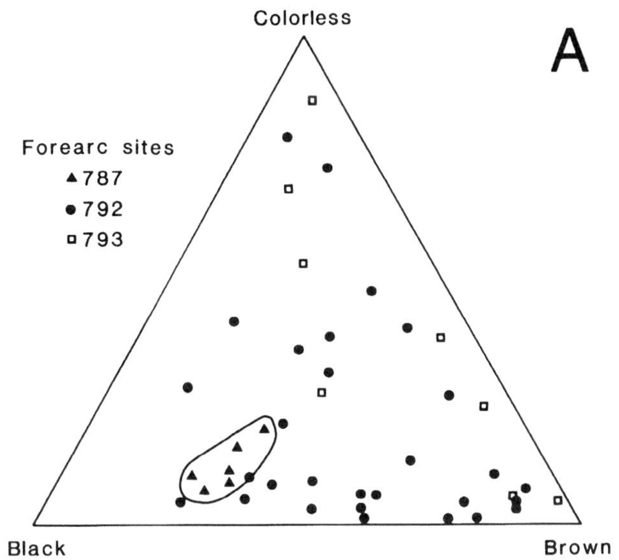

Fig. 7A. Proportions of colorless, black, and brown glass in volcanic sand from Leg 126 forearc sites. After *Marsaglia* [1992].

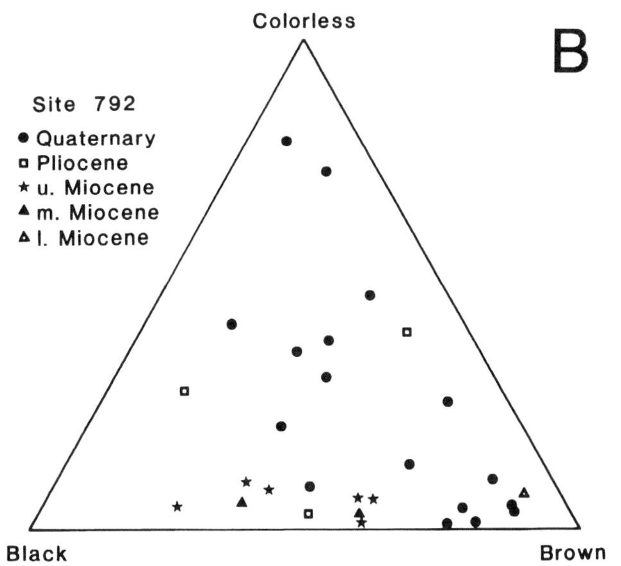

Fig. 7B. Proportions of colorless, black, and brown glass in volcanic sand from Site 792, subdivided by age. From *Marsaglia* [1992].

vitric (holo-hyaline), whereas most textures for brown and black glass are either of microlites in a glassy groundmass or lathworks of fine to medium sand-sized crystals in a glassy or microlitic groundmass. The felsic fractions probably were produced by explosive eruptions associated with submarine calderas. The intermediate to mafic fractions likely formed during slower cooling of arc magmas associated with emergent volcanoes. The documented compositional heterogeneity is explained best by physical mixing in transitional storage areas prior to resedimentation by turbidity currents. Site 787, for example, is located in a submarine channel that begins near arc volcanoes (Figure 2); sands from that site are enriched in tachylite fragments which probably were eroded from emergent arc volcanoes and carried into the adjacent submarine canyon (Figure 2). The abundance of microlitic fragments in all of the glassy fractions is also consistent with extensive subaerial erosion of epiclastic debris.

Temporal trends in volcanic lithic composition also are evident in the Izu-Bonin forearc. At Site 792, most Miocene samples are enriched in black to brown glass, whereas the compositions of Pliocene to Quaternary sand are more heterogeneous (Figure 7b). This change is probably a response to compositional shifts in arc magmatism from intermediate or mafic to more distinctly bimodal (basalt-rhyolite) during the current phase of arc rifting [*Marsaglia*, 1992; *Fujioka et al.*, 1992].

Overall, intraoceanic forearc sands appear to be homogeneous when modal data are plotted on simple ternary formats such as the Q-F-L diagram. Petrographic studies must be completed at a higher level of detail to identify the provenance of volcaniclastic material, through subdivisions of volcanic textures and glass types. Similar methods should be utilized in other types of tectonic settings where volcanic influx is important (e.g., incipient rifts, pull-apart basins, continental forearc basins with mixed provenance).

2.5. *History of Sediment Accumulation and Subsidence in the Tonga Forearc*

Cores recovered at Sites 840 and 841 (Figure 8) record the influence of tectonics on sedimentation along the Tonga intraoceanic forearc. Sediment backstripping techniques [e.g., *Sclater and Christie*, 1980] were used in conjunction with traditional facies analyses to reconstruct the sedimentary and subsidence history. Two critical elements in this approach are the accurate dating of strata and the determination of their paleo-water depths; the latter value is difficult to pinpoint for most marine environments. Estimates were made, however, using microfaunal assemblages, their states of preservation [*Parson et al.*, 1992], and trace fossils [*Ekdale et al.*, 1984].

Clift [1994] demonstrated that the water depth at Site 840 remained close to the present-day depth of 743 m throughout the accumulation of the section cored (i.e., since

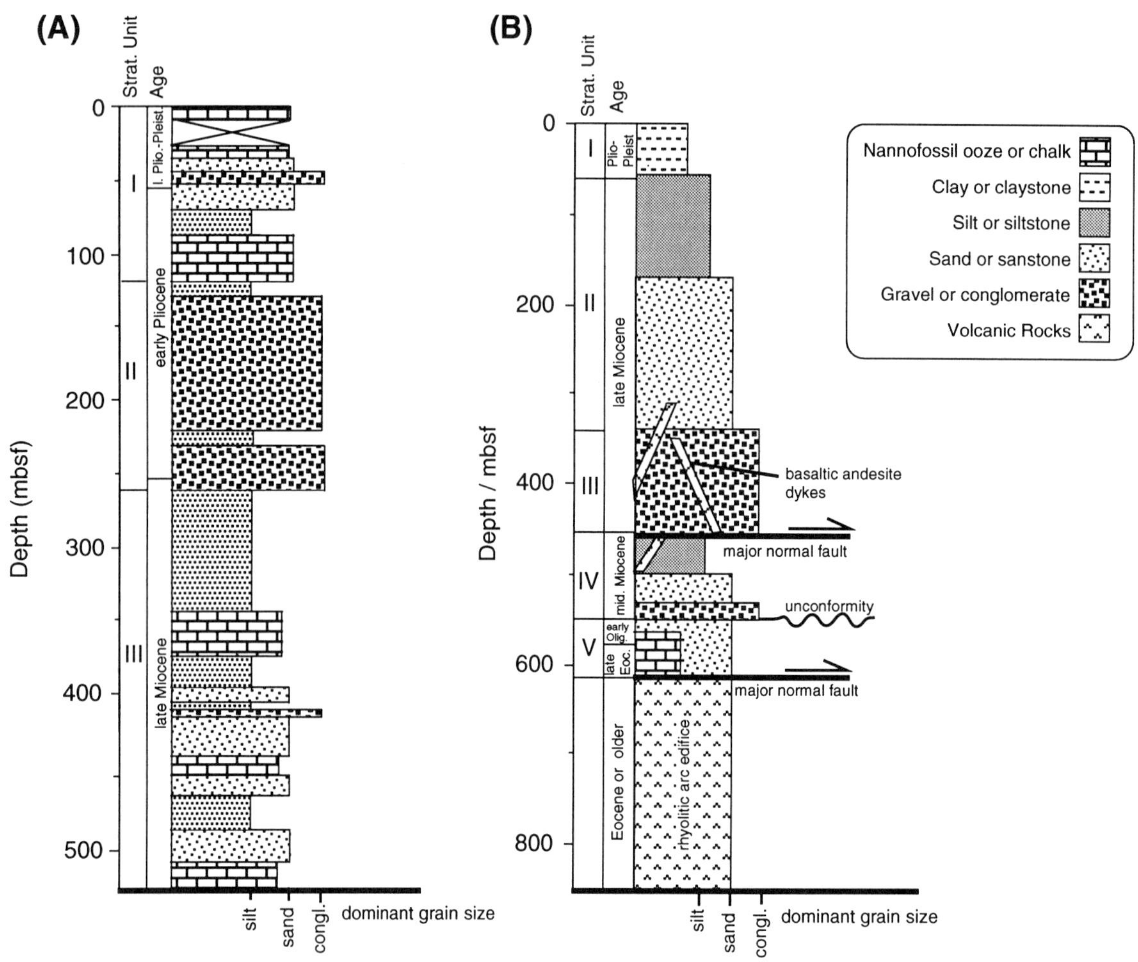

Fig. 8. Stratigraphic columns for (a) Site 840 and (b) Site 841, located on the Tonga Platform and inner trench slope, respectively. Based on recovered cores and FMS logs. From *Clift et al.* [1994].

7.1 Ma). Seismic-reflection data [*Austin et al.*, 1989] and structural information from the cores [*MacLeod*, 1994] point to subsidence of the platform in its present paleogeographic state by the progressive rotation of the Tonga Platform, without dramatic uplift or subsidence events. Site 841 now lies at a depth of 4,850 m, and the sedimentary section there spans a much longer time period of 37 million years. Shallow-water foraminifers place the water depth precisely within the photic zone at 37.0 Ma [*Parson et al.*, 1992]. By 35.0 Ma, however, corrosion of nannofossils and the disappearance of benthic foraminifers indicate subsidence to a position below the lysocline. The record between 35.0 and 16.2 Ma has been lost because of an unconformity, but nannofossils indicate that water depths were above the lysocline immediately after the hiatus. The calcite compensation depth (CCD) and lysocline deepened significantly in the Pacific between the early Oligocene and the middle Miocene [*van Andel*, 1975], so it is debatable whether or not uplift of Site 841 occurred during the hiatus. Structural information [*MacLeod*, 1994] shows a gradual tilting of Site 841 towards the trench since the start of sedimentation. Further water deepening to a position close to the CCD (3,900 m) occurred by 10.0 Ma. The final disappearance of nannofossils took place at 5.3 Ma as Site 841 passed through the CCD (4,200 m).

The subsidence histories for Sites 840 and 841 were modeled by assuming that the crust behaved according to Airy isostasy; this is a likely condition if the forearc had been in a state of extension throughout the period of interest [*Clift*, 1994]. The widespread presence of normal faults in cores supports this assumption, except during a collisional phase between the Louisville Ridge and the Tonga Trench, which occurred directly opposite Site 841 at approximately 2.0 Ma [*Dupont and Herzer*, 1985]. The time interval of collision is brief at Site 841, and collision did not generate a noticeable effect at Site 840. Additional compaction of the sediment may have occurred beyond that calculated for shallow water depths by *Sclater and Christie* [1980], but these effects are probably negligible [*Clift*, 1994].

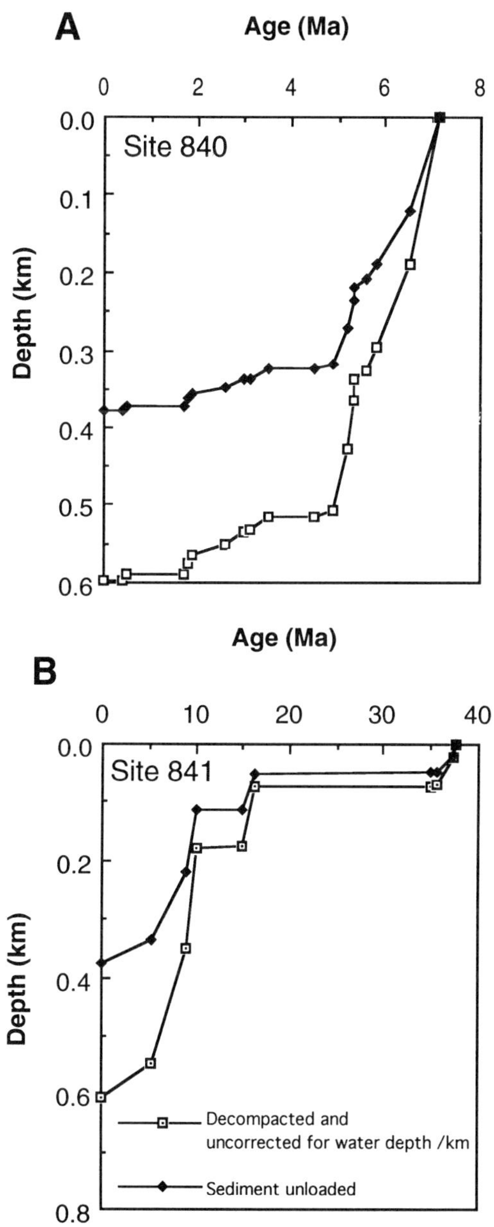

Fig. 9. Decompacted and water-unloaded curves for the sedimentary section drilled at (a) Site 840 and (b) Site 841. Accumulation of sediment above basement relative to Present sea level, with no correction for paleo-water depth. From *Clift* [1994].

Figure 9 illustrates the calculated sediment accumulation rates at both Sites 840 and 841, without corrections for paleo-water depth. In the first of the two curves, a correction was made only for compaction of the sediment; in the second, the loading effect of the sediment on the underlying basement also was considered. The curves clearly show rapid deposition at Site 840 between 7.1 and 5.3 Ma. Following 5.3 Ma, deposition increased for a short time but subsequently slowed to a modest rate after 4.9 Ma. In contrast, Site 841 shows three depositional events, a modest one at 37.0 Ma and larger events at 16.2 and 10.0 Ma (Figure 9).

Figure 10 illustrates the sections corrected for variations in paleo-water depth. Two curves for each site correspond to minimum and maximum depth estimates. At Site 840, the minimum-depth and maximum-depth curves show parallel trends, which indicates that subsidence was balanced by sedimentation. The time period between 7.1 and 5.3 Ma was one of rapid basement subsidence. Subsidence increased again after 5.3 Ma and subsequently slowed to a gradual rate from 4.9 Ma to the present. The apparent rapid subsidence between 1.69 Ma and the present is purely the result of a poor water-depth estimates, since no major deepening is inferred for this time period on the basis of the sedimentary evidence. Data from Site 841, in contrast, show major basement subsidence at 35.5 Ma, but there is a significant discrepancy between the two curves from 35.0 Ma to the present. The maximum water-depth model shows a steady subsidence since 35.0 Ma, whereas the curve based on minimum depth estimates predicts some basement uplift between 35.0 Ma and 16.2 Ma and strong subsidence at 16.2 Ma (Figure 10). As mentioned above, structural data [*MacLeod*, 1994] are consistent with a gradual tilting and continuous subsidence at Site 841, thus favoring the curve based on maximum water-depth estimates.

Subsidence rates in the Tonga forearc clearly changed at 5.3 Ma and 35.5 Ma, with a possible third event at 16.2 Ma. Interestingly, the events at 5.3 Ma and 35.5 Ma correlate well with episodes of arc rifting, prior to the formation of the Lau and South Fiji backarc basins, respectively [*Weissel and Watts*, 1975; *Parson et al.*, 1992]. Subsidence at 16.2 Ma does not correlate well with any known tectonic event in the region. Instead, the event may have been controlled by local extensional faulting in the forearc owing to ongoing subduction erosion [see *von Huene and Scholl*, 1991, for general discussion of this phenomenon]. If a major normal fault were initiated close to Site 841 at 16.2 Ma, movement along it could explain the influx of coarse clastic material, as well as any deepening that might have occurred. Although the influx of coarse clastic material at 10.0 Ma does not seem to match

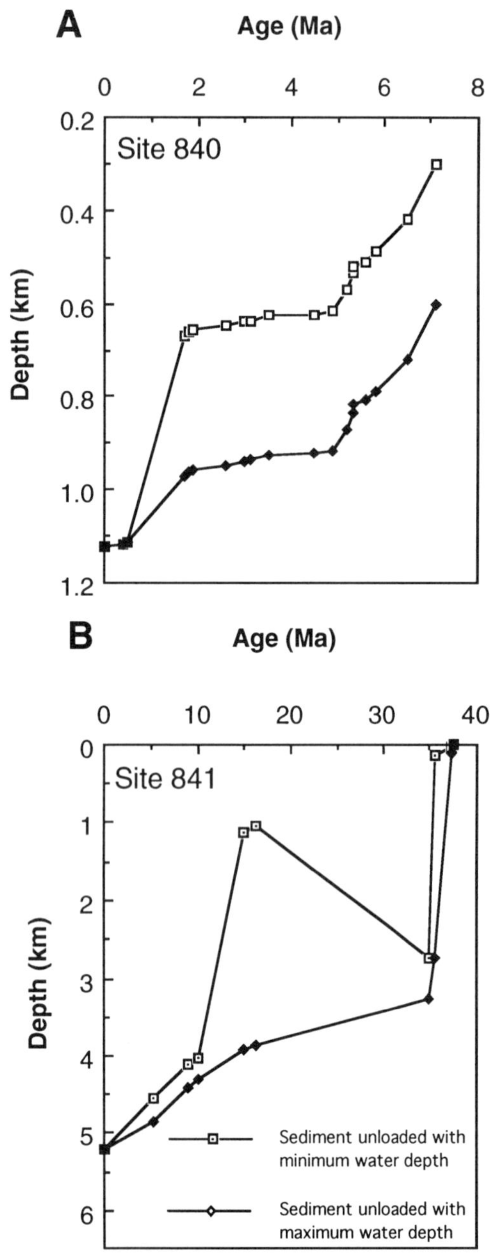

Fig. 10. Decompacted and water-unloaded subsidence curves for (a) Sites 840 and (b) Site 841, with corrections for paleo-water depths. From *Clift* [1994].

tectonically driven subsidence when water depths are accounted for, this inconsistency may merely reflect the difficulty in making accurate water-depth estimates. One reason for suggesting that the forearc did experience a change at that time comes from Site 840. The shallow water depths at that site provide much better evidence for rapid subsidence dating from at least 7.1 Ma and conceivably starting at 10.0 Ma. If this conclusion is correct, then the subsidence may be related to extension and rotation of the Tonga Arc following subduction polarity reversal in the New Hebrides [*Moberly*, 1971].

The Tonga Platform has been positioned next to active volcanic arcs throughout its history. One pulse of coarse volcaniclastic sedimentation preceded and accompanied the breakup of the original (Miocene) Tonga Arc. This type of detritus may provide the record of enhanced high-silica volcanism during the breakup period [e.g., Izu-Bonin; *Nishimura et al.*, 1991]. Slow subsidence of the Tonga Platform after 4.9 Ma was also marked by accumulation of nannofossil chalk, especially during the period 4.9-3.0 Ma. This changeover from a volcaniclastic-dominated sequence in the Miocene to a carbonate-dominated sequence in the Pliocene-Pleistocene can be attributed to subsidence of the original arc edifice (Lau Ridge) below sea level and its separation from the Tonga Platform during the early stages of arc rifting. Intense siliceous volcanism continued within the Lau Basin as it opened, but the submarine eruptions were incapable of transporting significant amounts of volcaniclastic sediment onto the Tonga Platform. The reappearance of volcaniclastic debris within the sedimentary column at Site 840 indicates that island-arc volcanism re-established itself on the trench side of the basin by 3.0 Ma. The relatively small size of the modern Tofua Arc [*Ewart et al.*, 1973] shows that it is a young feature; arc volcanism adjacent to the Tonga Platform probably was not continuous throughout the rifting period.

Electron microprobe analyses of individual volcanic glass shards document the temporal record of arc-magma chemistry [*Clift and Dixon*, 1994; *Bøe*, 1993]. As in the Izu-Bonin system [*Hiscott and Gill*, 1992; *Gill et al.*, 1994], volcaniclastic debris in the Tonga forearc was eroded from active volcanoes and transported by mass flows. In both cases, the time delay between eruption of volcanic material and resedimentation onto the forearc slope was brief. This rapid sedimentologic connection between arc volcanoes and forearc depocenters should be considered during the interpretation of other modern and ancient volcaniclastic sequences.

3. NANKAI SUBDUCTION ZONE

3.1. *Stratigraphic Evolution from Abyssal Plain to Trench*

Facies distributions and stratigraphic models for accretionary prisms have been described by many previous workers [e.g., *Piper et al.*, 1973; *Schweller and Kulm*, 1978; *Dickinson and Seely*, 1979; *Underwood and Bachman*, 1982; *Thornburg and Kulm*, 1987]. Upward-coarsening sequences should form in response to temporal and spatial facies changes as depositional sites migrate from the open ocean to the trench axis. Pelagic sediments on the oceanic plate typically pass upward into hemipelagic muds of the outer trench-slope environment; the trench wedge, at the top of the succession, is characterized by silty and sandy turbidites, debris-flow deposits, and sediment slides. This paradigm is somewhat limited in its application, however, because it focuses on such examples as the Aleutian Trench and the Chile Trench, where normal thicknesses of oceanic crust are being subducted and rates of sedimentation are high. The Nankai Trough southwest of Japan is not a typical example because it involves subduction of marginal-basin lithosphere complete with an extinct backarc spreading center [*Jolivet et al.*, 1989; *Moore et al.*, 1990; *Taira et al.*, 1992].

Site 808 was drilled within the toe of the Nankai accretionary prism, just landward of the surface expression of the frontal thrust (Figure 11). This site is also near the point of intersection between the deformation front, which strikes northeast-southwest, and the north-northwest trending spreading ridge of Shikoku Basin [*Le Pichon et al.*, 1987]. Thus, prior to their accretion, most of the strata penetrated at Site 808 had been deposited on or near the crest of a basement high on the subducting plate, thereby keeping the total sediment thickness at a regional minimum [*Pickering et al.*, 1993a]. The stratigraphy of Site 808 is particularly impressive because it displays the stratigraphic evolution of an accretionary prism in its entirety (Figure 12). Significant structural modifications of the primary depositional architecture have been caused by offsets along the frontal thrust, as well as seaward propagation of the prism's basal décollement (Figure 13) [*Shipboard Scientific Party*, 1991].

The overall stratigraphy recorded at Site 808 mimics most aspects of the upward-coarsening model proposed by *Piper et al.* [1973]. The oldest sedimentary rocks above basaltic basement were deposited at approximately 13.8 Ma [*Shipboard Scientific Party*, 1991]. Miocene, Pliocene, and early Pleistocene deposition within the central part of Shikoku Basin was characterized by slow suspension fall-out of hemipelagic mud and periodic influx of volcanic ash (Figure 12). Much of this phase of sedimentation evidently occurred near or below the CCD, based on the paucity of preserved foraminifera and calcareous nannofossils. Rates of sedimentation within the marginal basin varied from approximately 6 m/my to 215 m/my (Figure 14) [*Shipboard Scientific Party*, 1991]. Significantly, the position of the décollement within the upper Shikoku Basin facies unit does not correspond to any type of fundamental change in lithology; instead, the fault has propagated through a monotonous section of bioturbated hemipelagic mudstone [*Taira et al.*, 1992].

There are three important differences between the Shikoku Basin succession recovered at Site 808 and the stratigraphies of previous DSDP sites located farther to the south (Figure 11). First, layers of volcanic ash are common throughout the sections at Sites 442, 443, and 444 [*White et al.*, 1980], whereas ash and tuff layers at Site 808 are concentrated in the lower part of the Pleistocene section [*Pickering et al.*, 1993a]. Chemical analyses indicate that the ash layers were derived from a wide variety of volcanic sources, and perhaps from more than one of the arcs surrounding the marginal basin [*Masuda et al.*, 1993; *Underwood et al.*, 1993b]. Second, terrigenous turbidites of Pliocene age are present within the lower Shikoku Basin strata only at Site 297 [*Karig et al.*, 1975]. The turbidity currents responsible for these enigmatic sands never reached the ridge crest within the central part of the basin (i.e., the backtracked position of Site 808). The Pliocene turbidite sands at Site 297 probably were eroded from uplifted parts of the Shimanto accretionary complex exposed on the island of Kyushu [*Marsaglia et al.*, 1992]. Third, Miocene deposits immediately above basaltic basement at Site 808 consist of thick rhyolitic tuffs and variegated mudstones. Similar tuff beds also were discovered near the base of Hole 297 [*Karig et al.*, 1975], but not at Sites 442, 443, and 444 [*White et al.*, 1980]. Some of the pyroclastic explosions may have been subaqueous, thereby triggering voluminous ash-bearing turbidity currents which moved onto the crest of the abandoned ridge and into the eastern half of the backarc basin. The pyroclastic deposits are coeval with the widespread tuffs and near-trench granitic plutons of the Outer Zone of Japan [*Oba*, 1977; *Terakado et al.*, 1988; *Hibbard and Karig*, 1990]. Magmatism in the Outer Zone occurred during a fundamental reorganization of plate boundaries at approximately 15 Ma, when very young Shikoku Basin lithosphere first entered the subduction zone [*Hibbard and Karig*, 1990]. The plate reorganization, however, was also contemporaneous with opening of the Sea of Japan [*Jolivet and Tamaki*, 1992]; in addition, seafloor magnetic anomalies within Shikoku Basin indicate that a late phase of unstable and disorganized spreading occurred between 14 and 12 Ma [*Chamot-Rooke et al.*, 1987]. Ties between specific tectonic causes and specific volcanic effects, therefore, remain uncertain.

As the reference position of Site 808 moved close to the Nankai subduction front, Pleistocene sediments were deposited in the trench mainly by turbidity currents, debris flows, and sediment slides. One of the complications in deciphering this part of the stratigraphic history is the repetition of trench-wedge facies units owing to offset along

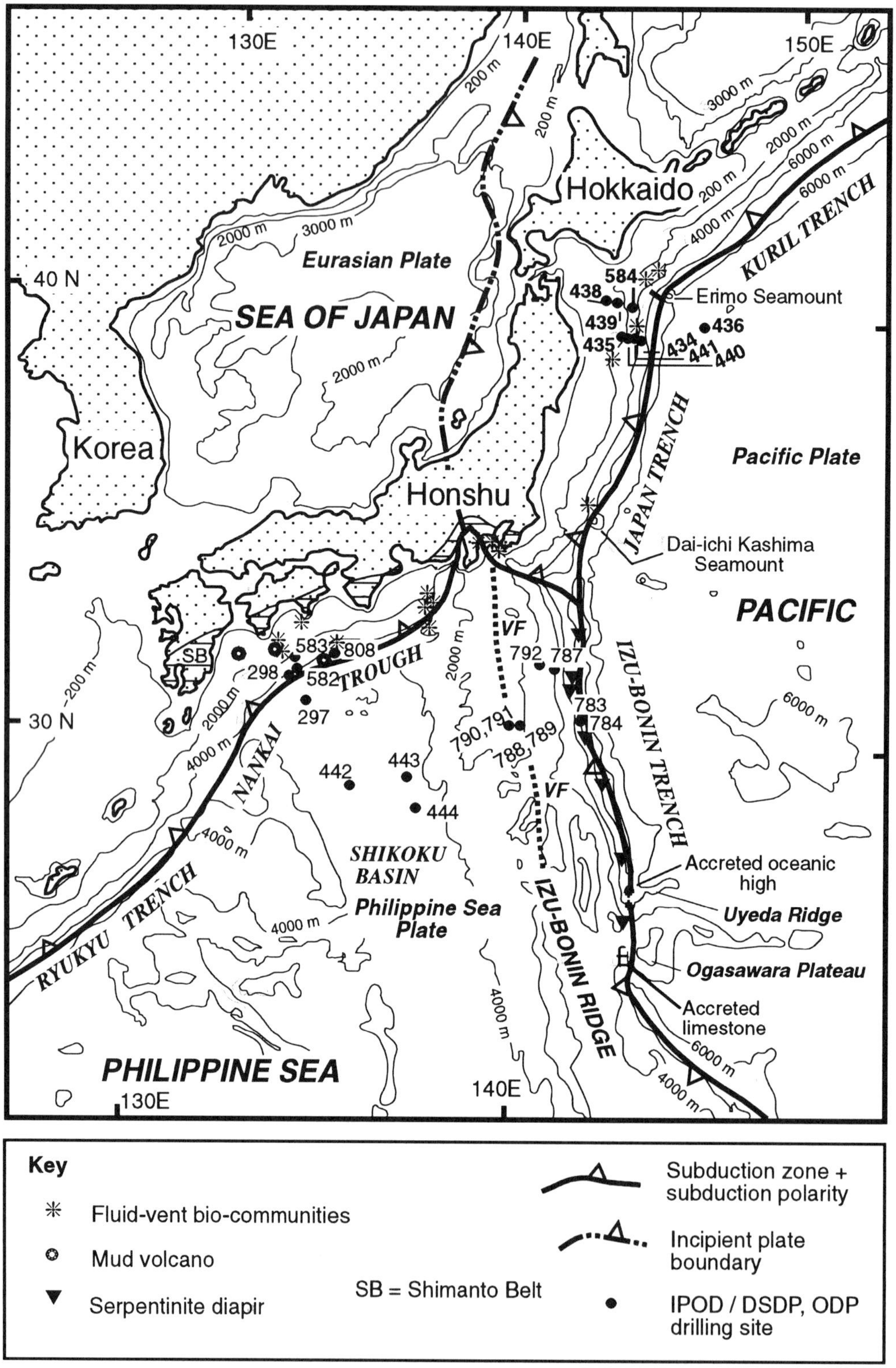

Fig. 11. Map of southwest Japan and vicinity showing major tectonic features, plus locations of DSDP and ODP drill sites. From *Pickering et al.* [1993a].

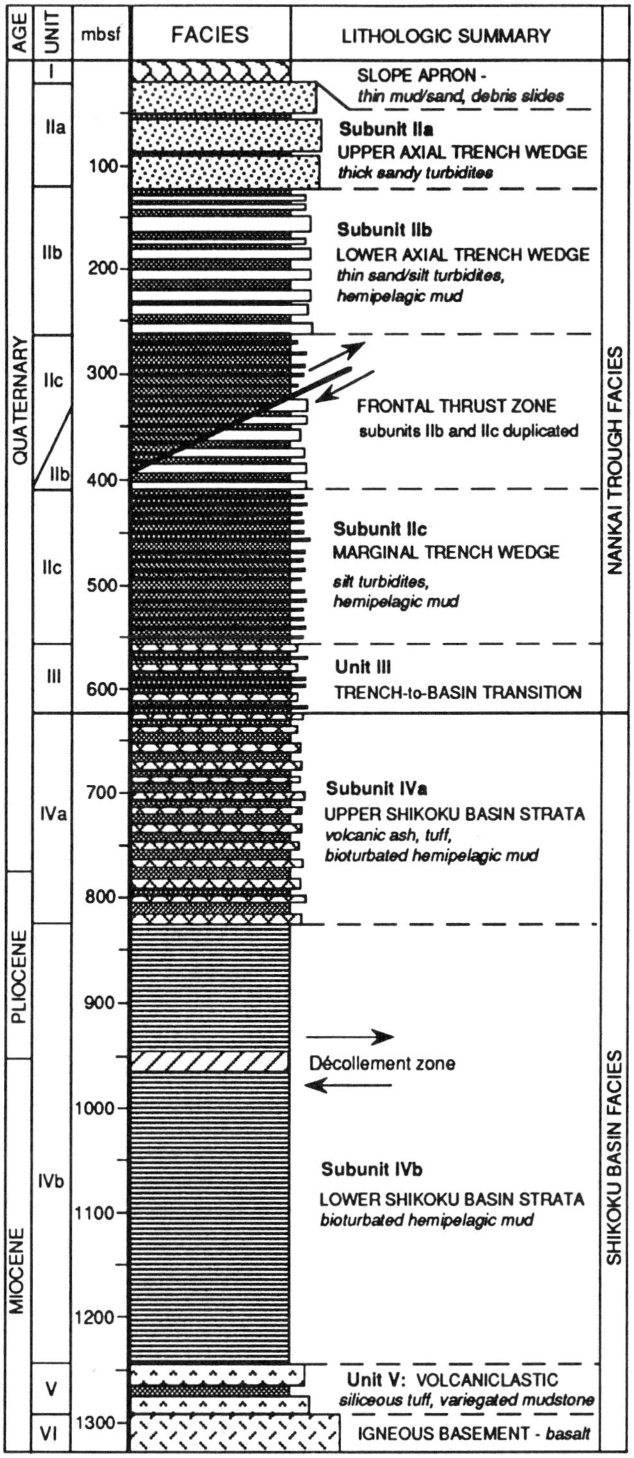

Fig. 12. Summary stratigraphy for Site 808. A classic upward-coarsening sequence passes from pelagic and hemipelagic deposits of the Shikoku Basin to sandy turbidites of the Nankai trench wedge. After *Shipboard Scientific Party* [1991].

the frontal thrust (Figures 12 and 13). A distinctive matrix-supported mudstone-pebble conglomerate and a yellow carbonate layer provide the most reliable marker beds for calculating vertical offset, which amounts to roughly 145 m [*Shipboard Scientific Party*, 1991]. Rates of Quaternary sediment accumulation within the trench wedge were as high as 1,380 m/my (Figure 14) [*Shipboard Scientific Party*, 1991].

Downslope transverse transport clearly is the most important component of sediment dispersal for nearby forearc-basin and slope-apron environments [*Blum and Okamura*, 1992; *Taira and Ashi*, 1993], but is a secondary contribution to the Nankai trench wedge. Sediment slides and debris flows near the prism toe probably have been responsible for some cannibalization of recently accreted sediments [*J.C. Moore and Karig*, 1976; *Taira and Ashi*, 1993]. There are many submarine canyons across the strike length of the forearc, but only the largest canyons (Figure 15) provide efficient, through-going conduits for transverse flux onto the trench floor [*de Rosa et al.*, 1986; *Shimamura*, 1989]. Small lobate sediment accumulations (e.g., Tenryu fan) prove that the transverse delivery systems were active in the recent past [*Shimamura*, 1989; *Soh et al.*, 1991]. Conversely, petrographic data [e.g., *Taira and Niitsuma*, 1986; *Marsaglia et al.*, 1992] indicate that most of the terrigenous debris was supplied by subaerial erosion within the Izu-Honshu Collision Zone to the northeast (Figure 15). Axial flow paths are also substantiated by the presence of a major channel which can be traced from the mouth of Suruga Trough for approximately 400 km [*Shimamura*, 1989]. Most of the sediment transport, therefore, has been parallel to the trench axis.

The offscraped trench sediments are capped by a thin carapace (less than 20 m) of lower-slope deposits (Figure 12). The slope sediments display several effects of mass wasting, including discontinuous lenses of sand, mesoscopic fold hinges, and reversals in the facing direction of thin sand turbidites [*Pickering et al.*, 1993a]. Megascopic submarine slides are also visible on sidescan sonar images of the seafloor near Site 808 [*Taira and Ashi*, 1993]. Some of the mass wasting may have remobilized offscraped trench fill in addition to the overlying slope apron. On the other hand, the drilling site is located only 150 m above the trench floor. It is possible, therefore, that the thin-bedded turbidites were not accreted but deposited initially on the lowermost slope from relatively thick sediment plumes as trench turbidity currents moved to the southwest down the trench axis.

Sedimentation patterns in backarc basins such as the Shikoku Basin are known to be influenced by many factors

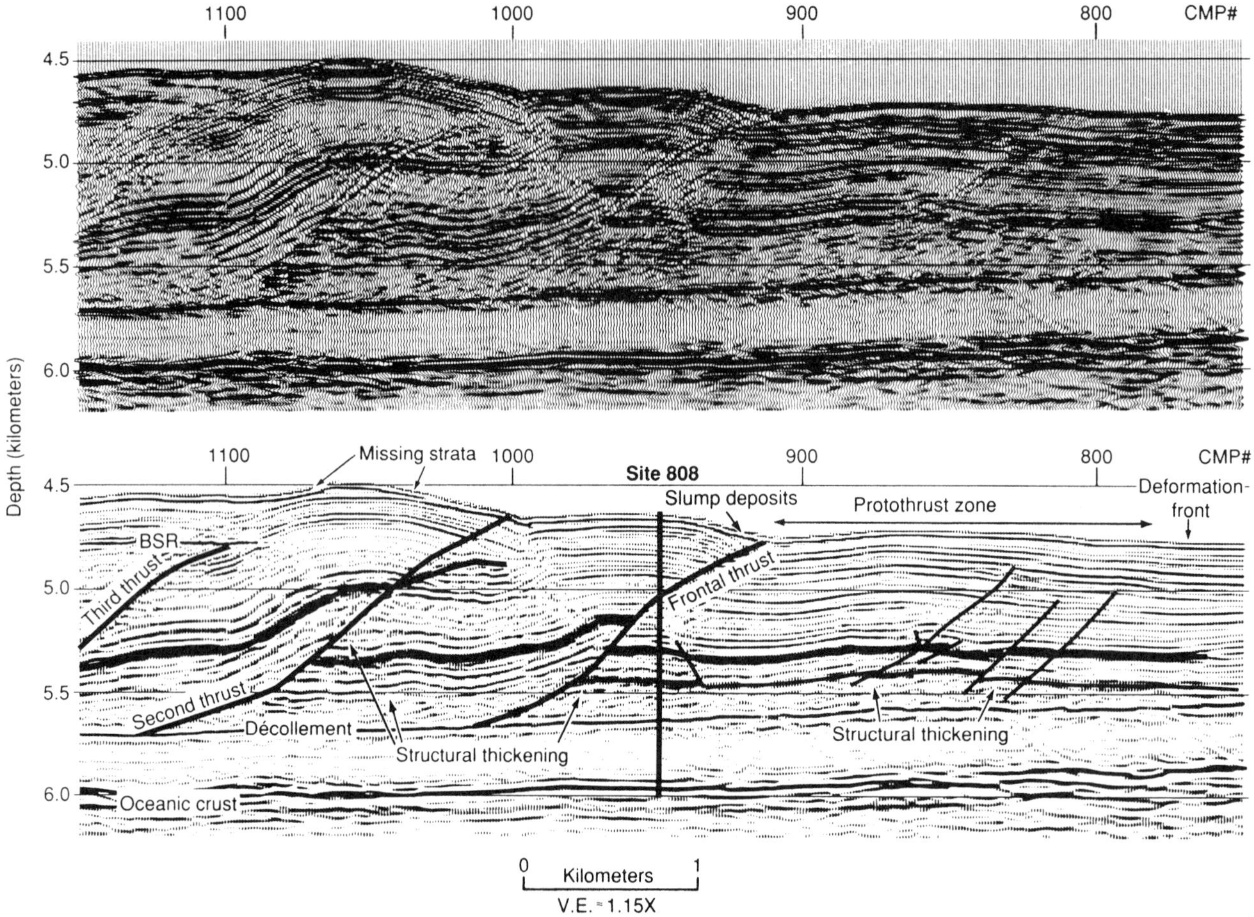

Fig. 13. Depth-migrated multichannel seismic-reflection profile through Site 808; interpretation below highlights top of oceanic basement, décollement, deformation front, folded and thrust-faulted sedimentary rocks of the accretionary prism, and underthrusting of hemipelagites below décollement. From *Moore et al.* [1991].

including biogenic productivity, oceanic circulation, continental sediment yield, climate, regional volcanism, wind patterns, and rates of uplift in nearby source terrains [*Klein*, 1985]. Tectonic processes clearly exerted first-order control over the stratigraphic succession accreted at Site 808. Eustatic fluctuations in sea level [e.g., *Haq et al.*, 1987] undoubtedly affected the region as well, but their influence is beyond the resolution of existing biostratigraphic data or isotope stratigraphy. Instead, major changes in the rates of sediment accumulation (Figure 14) were dictated by proximity to the trench axis and proximity to terrigenous sediment supplies. The Izu-Honshu Collision Zone, for example, experienced rapid rates of uplift and erosion during the Quaternary [*Soh et al.*, 1991], and seafloor relief associated with the extinct spreading ridge of Shikoku Basin also exerted a fundamental control on sedimentation through early-phase bifurcation of the trench bathymetry, particularly along its seaward margin. Because of rapid sedimentation, however, this topographic barrier became completely buried during the late Quaternary, thereby allowing axial turbidity currents to flow from one end of the trench to the other [*Pickering et al.*, 1993a].

3.2. *Paleoflow Patterns for Turbidites in Nankai Trough*

Deflection and reflection of turbidity currents have been cited to explain the characteristics of several unusual deposits in both modern and ancient deep-marine basins [*van Andel and Komar*, 1969; *Ricci Lucchi and Valmori*, 1980; *Pickering and Hiscott*, 1985; *Marjanac*, 1990]. In addition, upslope movement of turbidity currents has been invoked to explain occurrences of anomalously coarse-grained sediments resting on bathymetric highs such as oceanic plateaus or the outer slopes of trenches [*Damuth*

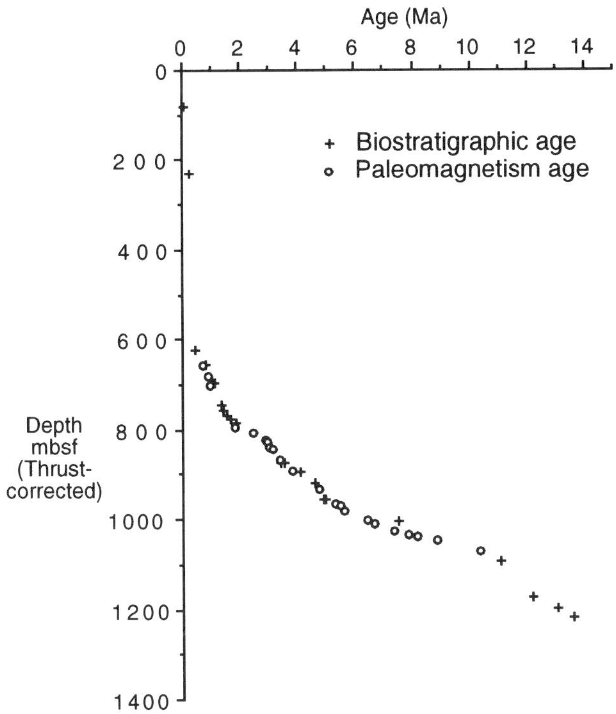

Fig. 14. Stratigraphic age versus depth for Site 808. Rate of sediment supply increased significantly as Shikoku Basin lithosphere approached the deformation front of Nankai Trough. After *Pickering et al.* [1993a].

and Embley, 1979; *Dolan et al.*, 1989; *Underwood*, 1991]. Soliton wave theory provides a reasonable explanation for the physical behavior of reflected turbidity currents [*Pantin and Leeder*, 1987], and scaled laboratory models have been used both to visualize flow trajectories [*Kneller et al.*, 1991] and to determine ideal combinations of boundary conditions for maximum vertical run-up [*Muck and Underwood*, 1990]. Core recoveries from Site 808 are important in this regard because they display the first convincing evidence from internal sedimentary structures of upslope flow and deflection off the seaward slope of a modern trench [*Pickering et al.*, 1992, 1993b].

Mottled hemipelagic mudstone is the most common lithology within the outer marginal-trench facies of Nankai Trough (Figure 12) [*Shipboard Scientific Party*, 1991]. The beds of interest are thin, fine-grained sandstone and coarse-grained siltstone with sharp, erosive bases; they are typically normally graded and display Bouma T_{cde} divisions. These probable turbidites are bipartite, with a lower laminated sand interval overlain by a diffusely laminated to structureless mud cap decimeters in thickness. The absence of burrowing indicates rapid deposition of the mud caps, probably through suspension fallout from the tails of turbidity currents. In contrast, the background hemipelagic sediment above the mud caps is thoroughly bioturbated; typical ichnogenera include *Zoophycos* and *Chondrites*.

Many of the basal sand and silt intervals of the bipartite turbidites show repeated rippled divisions (Bouma T_c), including high-angle climbing-ripple laminae, low-angle to wavy laminae, convolute laminae, and over-steepened lee faces [*Pickering et al.*, 1993b]. Grain size breaks occur between the cross-laminated divisions, with the intervening mudstone partings ranging from cm to mm scale. Stoss-side erosion and lee-side preservation of cross-laminae are typical, and these geometries of internal sedimentary structures are produced by traction-related bedforms as they migrate down current. In the most dramatic cases, ripple cross-laminae at Site 808 are oriented in opposite directions within a single turbidite bed. Collectively, the small-scale depositional cycles, grain-size breaks, and internal sedimentary structures within the thin turbidites indicate rapid deposition from unsteady and nonuniform flows.

Analysis of magnetic fabrics within cores from DSDP Site 582 (Figure 16) shows that most of the turbidity currents responsible for deposition in the western corridor of the Nankai Trough moved parallel to the trench axis [*Taira and Niitsuma*, 1986]. In contrast, paleoflow directions from Site 808 were obtained from dip azimuths of the ripple cross-laminae after the directions had been corrected for core rotation using shipboard paleomagnetic data [*Shipboard Scientific Party*, 1991]. The error associated with this technique is approximately $\pm 10°$ [*Pickering et al.*, 1993b], and the resulting rose diagram (Figure 16) represents the first such plot of corrected paleocurrent azimuths since the inception of the DSDP-ODP program in 1968. The data show three principal flow directions: (1) to the north-northwest (down the maximum slope of outer trench wall); (2) up the outer trench slope (toward the open Shikoku Basin); and (3) away from the Izu-Honshu collision zone toward the west-southwest (parallel to the trench axis).

Pickering et al. [1992, 1993b] suggested that the complicated pattern of paleoflow directions is the result of two closely related, though not necessarily mutually exclusive, processes: (1) deceleration and collapse of thick, bank-full axial flows; according to this idea, without requiring any vertical shift in the flow's center of gravity, the upper body and dilute entrained layer wedges out against the marginal walls of the trench and moves down-gradient during the final stage of flow; (2) radial down-axis expansion and oblique upslope movement onto the seaward margin of the trench; in this case, movement is accompanied by a true vertical shift in the center of gravity, and upslope flow is followed by gradual deflection,

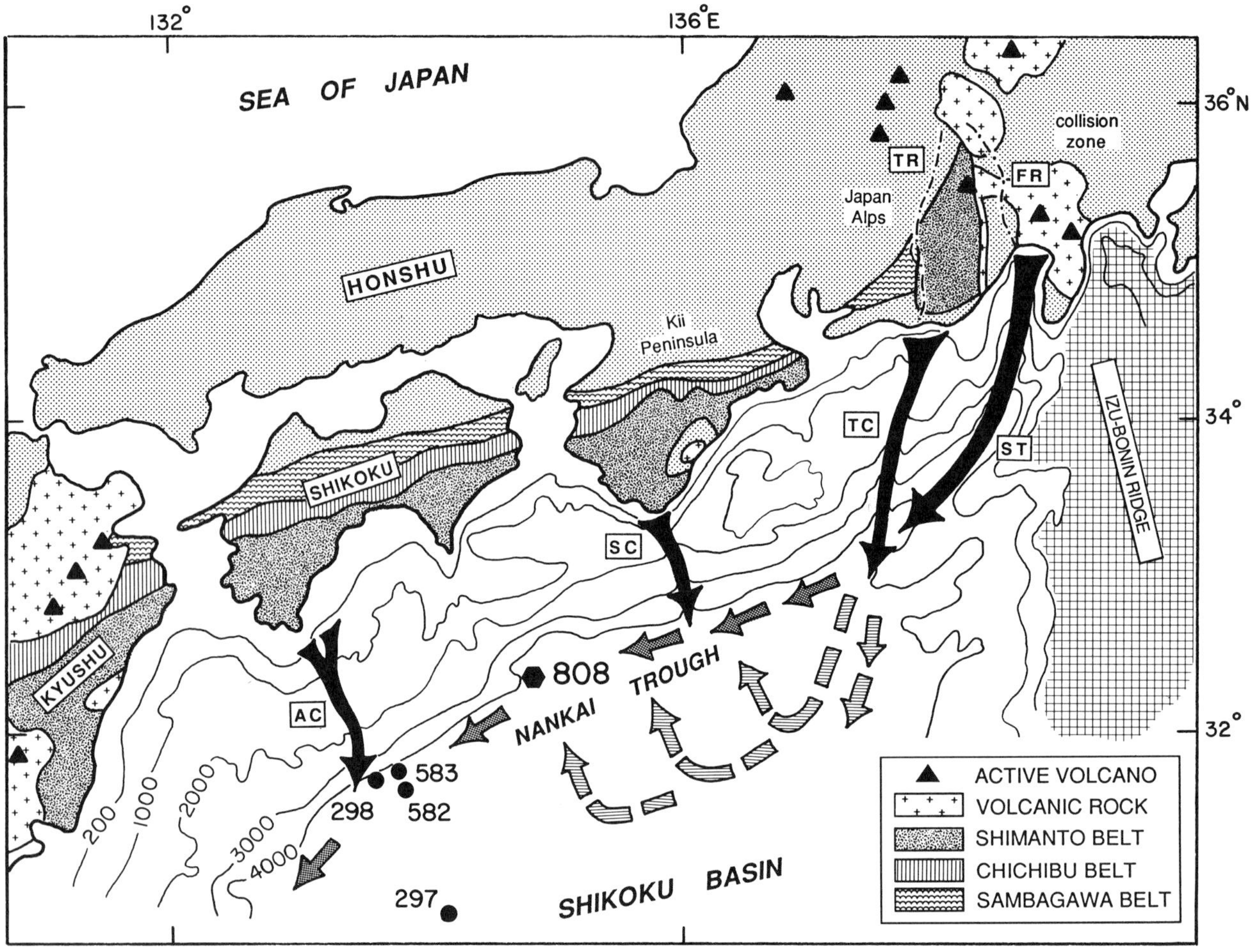

Fig. 15. Location map of Site 808, neighboring DSDP Leg 37 and Leg 87A sites, and major rock units within onland exposures of southwest Japan. Major sediment entry points are as follows: ST = Suruga Trough, TC = Tenryu Canyon, SC = Shionomisaki Canyon, AC = Ashizuri Canyon, FR = Fuji River, TR = Tenryu River. Bathymetric contours are in meters. Regions of volcanic rocks include Neogene-Quaternary lithologies of mafic-intermediate to felsic compositions. From *Underwood et al.* [1993a].

momentum decay, and a final surge down the lines of maximum slope inclination. In addition, it is conceivable that some turbidity currents first move down the forearc slope (i.e., toward the southeast) before undergoing a nearly orthogonal reflection off the opposite wall. One of the more curious aspects of the cores is that the diagnostic cross-laminae preferentially preserve the effects of the collapsing or rebounding turbidity currents, rather than the upslope-directed phase of transport. *Pickering et al.* [1993b] deduced that early-phase ripples form within the region of vertical run-up and tend to be remolded during subsequent flow collapse, thereby generating downslope-facing ripples as the turbidity currents experience their final stages of deceleration.

Figure 17 illustrates the expected pattern of paleocurrents influenced by combinations of oblique deflection and orthogonal reflection. Neither existing petrofacies data nor paleoflow data are sufficient to pinpoint the relative contributions of all possible sources on regional, long-term, three-dimensional sedimentation budgets in Nankai Trough. Nevertheless, the results of paleocurrent analysis have profound implications for the interpretation of analogous deposits, both modern and ancient. Flow directions just prior to deposition can be totally unrelated to the direction of transport out of the detrital source area. Therefore, caution must be exercised when interpreting the provenance of turbidite sands based on paleocurrent data alone.

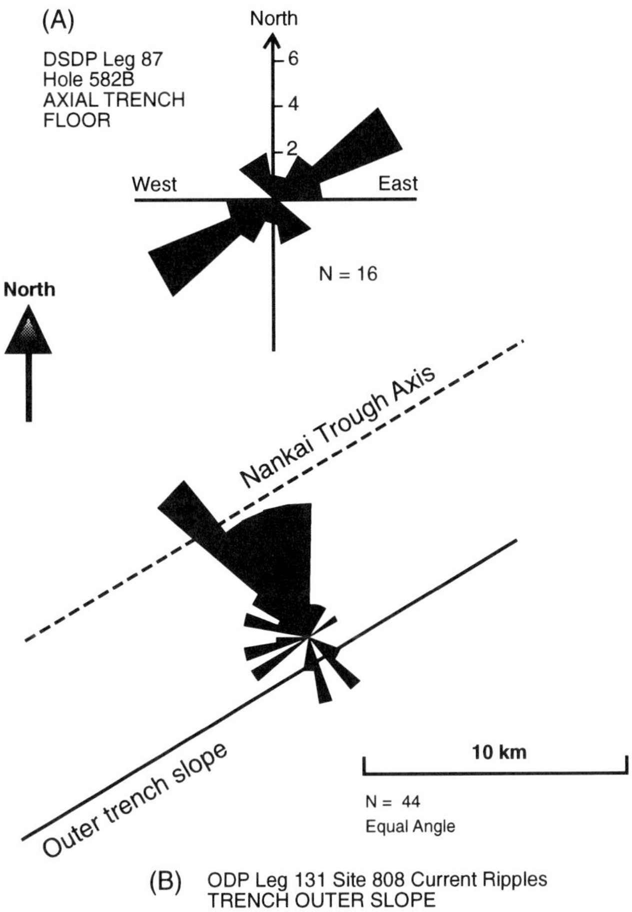

Fig. 16. Rose diagrams of (a) mean preferred grain orientations from trench-axis turbidites at Site 582, determined by magnetic fabric analysis of *Taira and Niitsuma* [1986], and (b) restored paleocurrent directions from sandy and silty turbidites of the outer trench-slope facies at Site 808; these data represent dip azimuths of ripple cross-laminae reoriented using detrital natural remanent magnetism. From *Pickering et al.* [1993b].

3.3. *Provenance of Nankai Trough Sediments*

Most well-developed siliciclastic accretionary complexes have formed where thick sections of sediment accumulate outboard of a deformation front [*Schweller and Kulm*, 1978; *Underwood and Bachman*, 1982; *von Huene and Scholl*, 1991]. In the simplest types of dispersal systems, erosion of the volcanic arc, transport across the forearc, and deposition in the trench and abyssal plain are all confined to common, and fairly narrow, geographic corridors. Some abyssal-plain sands are far travelled, however, because of transport by turbidity currents and post-depositional movement of the oceanic plate [*Dickinson*, 1982; *Underwood*, 1986]. Some trench sediments likewise exhibit anomalous petrographic characteristics because of long-distance axial transport from provenance corridors to the depositional corridor. Dispersal patterns for hemipelagic muds can be even more complicated [e.g., *Hathon and Underwood*, 1991]. Provenance interpretations for a given accreted succession, therefore, may not match the tectonic character of the same corridor within the overriding plate [*Velbel*, 1985; *Marsaglia and Ingersoll*, 1992].

Subduction beneath the southwest margin of Japan probably involved the Pacific Plate when Shikoku Basin lithosphere was forming [*Hibbard and Karig*, 1990; *Jolivet and Tamaki*, 1992]. The margin changed dramatically during the middle Miocene, with opening of the Sea of Japan, rotation of large crustal blocks, initiation of subduction of the Philippine Sea Plate, anomalous near-trench magmatism, and collision of the Izu-Bonin arc with the Honshu arc [*Otofuji et al.*, 1985; *Niitsuma*, 1988; *Taira et al.*, 1989]. The Izu-Honshu Collision Zone started to form about 15 Ma; rapid uplift and erosion resulted in a dramatic increase in the sediment supply, particularly near the collision zone [*Soh et al.*, 1991]. Compositional signals of this tectonic complexity are recorded by Nankai trench sediments.

3.3.1. *Sand provenance.* Several studies of detrital provenance have been attempted using sand samples from the DSDP and ODP sites off southern Japan (Figure 11). Site 297, located on the northern edge of the Shikoku Basin, contains trench overspill sediments of Pliocene age [*Karig et al.*, 1975]. Sites 298 and 583 penetrated the toe of the Nankai accretionary prism, whereas Site 582 was drilled into relatively undeformed trench fill [*Moore and Karig*, 1976; *Coulbourn*, 1986]. As discussed previously, Site 808 is located east of the DSDP sites on trend with the fossil spreading axis of Shikoku Basin [*Pickering et al.*, 1993a].

Unfortunately, petrographic results from earlier DSDP studies [*Harrold and Moore*, 1975; *Taira and Niitsuma*, 1986; *de Rosa et al.*, 1986] are mildly to dramatically inconsistent with one another, probably because of variations in sample preparation and petrographic techniques. In response to this problem, *Marsaglia et al.* (1992) generated a uniform data set for Sites 297, 298, 582 and 583, and *Underwood et al.* (1993a) published compatible petrographic data from Site 808 sands (Figure 18). *Marsaglia et al.* [1992] confirmed that the Pliocene sands from Site 297 (northern Shikoku Basin) are very similar in composition to sand eroded from the Shimanto Belt, and they concurred with *Harrold and Moore* [1975] that the likely Pliocene sources were located on Shikoku and/or Kyushu Island (Figure 15). These deposits contain higher

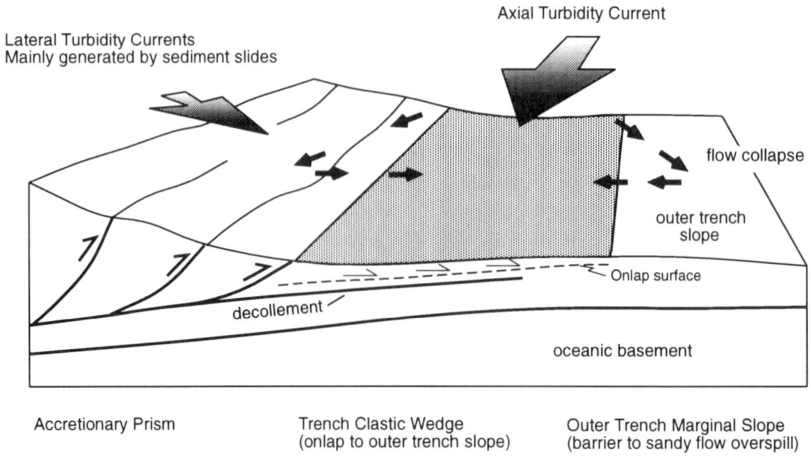

Fig. 17. Conceptual diagram to explain diverse turbidity-current flow directions at Site 808. Lateral (transverse) input of turbidity currents from upslope on Nankai accretionary prism is followed by high-angle reflection off the outer trench slope. Axial flows go through radial expansion, gravitational collapse, and oblique deflection off the outer slope, with the last phase of movement directed back toward the trench axis. From *Pickering et al.* [1993b].

percentages of total quartz and greater proportions of felsic-volcanic rock fragments than the Quaternary trench sediments (Figure 18). The more quartzose Pliocene sands evidently were partitioned to the western side of the Shikoku Basin spreading ridge.

Compositions of Quaternary sands from all the trench-floor and accretionary-prism sites fall on a mixing line between sources in the Shimanto Belt and sources in the Northern Honshu magmatic arc (Figure 19). This mixing effect can be seen most clearly in the polycrystalline lithic populations, which include fresh pyroclastic debris, a wide variety of paleovolcanic rock fragments, sedimentary rock fragments, and clasts of low-grade metasedimentary rock (Figure 18). The most likely source of this polymict material is the Izu-Honshu Collision Zone, where uplift and erosion of the Shimanto Belt and overlying arc and forearc deposits, such as the Miura Group [*Soh et al.*, 1991], allow intermingling of sand from different tectono-stratigraphic units. In addition, neovolcanic material from active volcanoes, such as Mt. Fuji, is mixed with recycled sedimentary and volcaniclastic debris from the accretionary forearc. These sediments are then funneled through Suruga Trough into the axial channel system of Nankai Trough (Figure 15).

As discussed previously, there is convincing morphologic evidence to support the hypothesis of limited transverse influx through some of the larger forearc canyons, particularly Tenryu and Shiono-misaki canyons [*Shimamura*, 1989; *Soh et al.*, 1991]. When compared to such examples as the Aleutian Trench [*Underwood*, 1991], however, along-strike variations in trench-floor petrofacies are subtle. Perhaps the best evidence for transverse influx comes from the Quaternary sands at Site 298, which are slightly enriched in quartz relative to those from Sites 582, 583, and 808 (Figure 15). Because of the proximity of Site 298 to the mouth of Ashizuri Canyon, this sediment probably includes local erosional products from the Hada Peninsula of Shikoku Island, where middle Miocene granitic plutons and rocks of the Shimanto Belt are exposed [*Taira et al.*, 1988].

3.3.2. *Clay mineral provenance*. Relative percentages of clay minerals also can be used to infer sediment provenance, although regional patterns displayed by clay minerals are sometimes quite different than the dispersal patterns for interbedded sands [*Hathon and Underwood*, 1991; *Underwood*, 1991]. *Chamley et al.* [1986] first documented a preponderance of illite and chlorite within the Pleistocene turbidite section of the western Nankai Trough and suggested that most of the muddy sediment was fed axially from a source near the Izu Peninsula of central Japan. Data from Site 808 provide many additional details [*Underwood et al.*, 1993a]. Mineralogic trends below depths of 600 mbsf have been influenced by both changing sediment supplies and clay-mineral diagenesis [*Underwood et al.*, 1993b], so that part of the data set will be discussed separately.

In spite of its proximity to active volcanic sources, which might be expected to produce large amounts of smectite, the dominant clay mineral in the central part of the Nankai Trough is illite, with typical relative percentages of 70-80% (Figure 20). The relative abundances of all three major mineral constituents (illite, chlorite, and smectite) are remarkably uniform throughout the trench wedge, and there are no meaningful differences between the respective clay mineralogies of hemipelagic muds and matrix clay washed from turbidite sands and silts (Figure 21). This mineralogic uniformity is consistent with a common source for both the hemipelagic muds deposited from the bottom nepheloid layer and the interstitial clays that were entrained into trench-floor turbidity currents. A modest amount of mineral partitioning occurs as a function of grain size; specifically, chlorite content increases in the 2-6-μm size fraction and smectite is more abundant in the less than 2-μm size fraction (Figure 20). Illite crystallinity indices, illite/mica b_o lattice dimensions, and relative percentages of mica polytypes are all consistent with source terranes that were exposed to the most advanced stages of diagenesis or lowermost greenschist-facies metamorphic conditions [*Underwood et al.*, 1993a]. There is no evidence for extensive erosion of rocks subjected to high-pressure (blueschist-facies) metamorphism, such as the Sambagawa Belt (Figure 15). As with sand petrography, clay-mineral abundances in Nankai Trough can be accounted for through subaerial erosion within the Fuji River drainage basin (Izu-Honshu Collision Zone) and subaqueous transport through Suruga Trough into the eastern end of Nankai Trough.

3.3.3. *Triple-junction provenance.* The arc-trench collision zone of central Japan, a trench-trench-trench triple junction, has profoundly affected the volume and composition of sediment shed into the Nankai Trough. The modern magmatic arc, however, is poorly developed. Active volcanoes are not present, for example, on the island of Shikoku (Figure 15). Consequently, the expected magmatic-arc provenance signature has been overprinted by the influx of recycled sedimentary material eroded from the uplifted accretionary prism, particularly within the collision zone. This type of petrologic overprint also has been documented near the Chile trench-ridge-trench triple junction [*Marsaglia et al.*, in press]. Several types of triple-junction configurations and migration scenarios have been proposed for the pre-Quaternary Japanese margin [e.g., *Seno and Maruyama*, 1984; *Seno*, 1985; *Niitsuma*, 1988; *Jolivet et al.*, 1989; *Hibbard and Karig*, 1990; *Otsuki*,

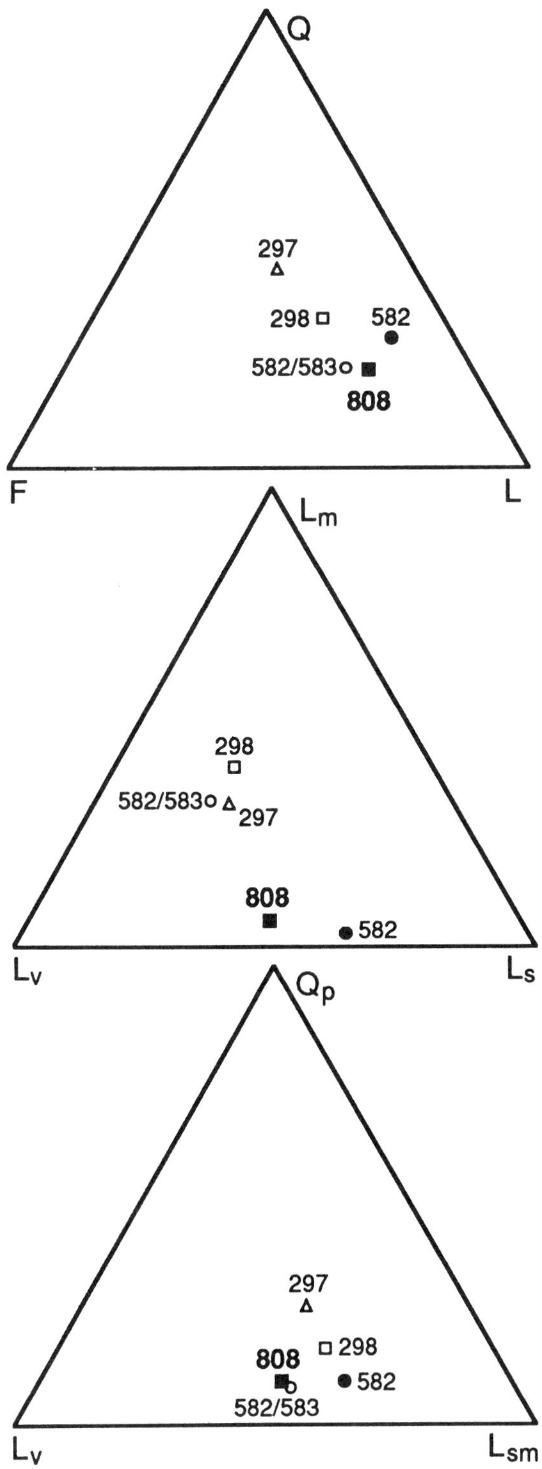

Fig. 18. Detrital modes for sands at DSDP and ODP sites off southern Japan. Q = total quartz; F = total feldspar; L = total lithic fragments; L_m = metamorphic lithic fragments; L_V = volcanic lithic fragments; L_S =sedimentary lithic fragments; Q_P = polycrystalline quartz; and L_{sm} = sedimentary and metamorphic rock fragments. Site means are for data presented by *Shipboard Scientific Party* [1991] for Site 808 (open square), *Taira and Niitsuma* [1986] for Site 582, and *Marsaglia et al.* [1992] for Sites 297, 298, and 582/583 combined (solid dots). From *Underwood et al.* [1993a].

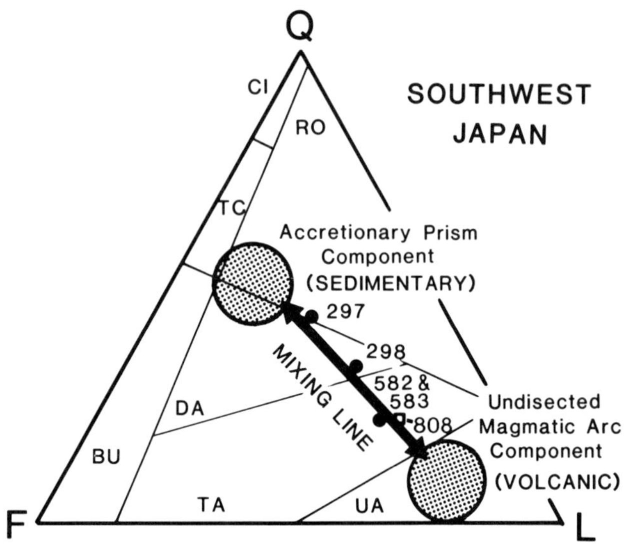

Fig. 19. Average Q-F-L proportions for sands at Sites 297, 298, 582/583 [from *Marsaglia et al.*, 1992] and Site 808 [from *Shipboard Scientific Party*, 1991]. Mixing between two sources (accretionary prism and undissected magmatic arc) indicated by arrowed line. Tectonic provenance fields from *Dickinson et al.* [1983]: UA = undissected arc; TA = transitional arc; DA = dissected arc; RO = recycled orogen; BU = basement uplift; TC = transitional continental; CI = continental interior. After *Marsaglia et al.* [1992].

1990; *Jolivet and Tamaki*, 1992; *Pickering et al.*, 1993a]. Given the gross similarity of sediment compositions at Sites 298, 582, 583, and 808, it seems that overlapping detrital modes and clay-mineral signatures can be produced through any one of several plate-margin geometries. Thus, diagnostic capabilities of standard analytical methods in sedimentary petrology must be improved before we can expect to pinpoint the sedimentary signatures associated with each specific type of triple junction.

3.4. *Clay Mineral Diagenesis within Nankai Prism*

The same basic detrital assemblage is characteristic of the Nankai trench wedge and underlying hemipelagic deposits of Shikoku Basin (Figure 20). One important change is a monotonic increase in smectite content within the upper Shikoku Basin facies. This enrichment can be attributed, in part, to diagenetic alteration of disseminated volcanic ash [*Masuda et al.*, 1993; *Underwood et al.*, 1993b]. The gradual increase in smectite coincides with a facies boundary at approximately 560 mbsf (Unit III), below which volcanic ash becomes more abundant. Local spikes in smectite content probably represent cryptic bentonite layers (Figure 20). Compilation of geochemical data from discrete ash layers (Site 808, plus DSDP Sites 297, 582, and 583) shows that pyroclastic material has been transported into the Nankai trench wedge and the Shikoku Basin from a wide variety of andesitic to rhyolitic sources [*Donnelly*, 1980; *Sugisaki*, 1980; *Minai et al.*, 1986; *Kawahata et al.*, 1986; *Pouclet et al.*, 1986; *Masuda et al.*, 1993; *Underwood et al.*, 1993b]. For the most part, however, dissemination of ash into the background mud deposits did not result in significant departures from the average bulk-mudstone geochemistry [*Underwood et al.*, 1993b].

Smectite decreases relative to other clay minerals beginning at a depth of about 820 mbsf (Figure 20). There probably were lesser amounts of detrital smectite, disseminated glass shards, and discrete layers of ash transported into the Shikoku Basin during its Miocene and early Pliocene history. In addition, however, Site 808 is located within a region of unusually high heat flow [*Yamano et al.*, 1984, 1992; *Kinoshita and Yamano*, 1986; *Ashi and Taira*, 1993]. Measurements of borehole temperature, for example, are consistent with an average linear gradient of approximately 110°C/km to depths of 350 mbsf [*Shipboard Scientific Party*, 1991]. Consequently, burial temperatures almost certainly reached high enough levels within the past 0.5 my to promote the diagenetic transformation of smectite to illite. As supporting evidence, X-ray diffraction analyses show a regular increase in the illite content of I/S mixed-layer clay within Units III and IV (Figure 22). There is also a slight increase in K_2O and Rb within mudstones of the lower Shikoku Basin facies, probably due to the uptake of those cations by illite layers in I/S mixed-layer clays [*Underwood et al.*, 1993b].

The first appearance of diagenetic I/S at Site 808 occurs at a depth of approximately 555 mbsf (Figure 22). If linear extrapolation of the shallow geothermal gradient is valid to that depth, then the present-day temperature associated with the initial formation of random (R = 0) I/S interstratification is ~65°C. At a depth of 900 mbsf (~100°C), the I/S phase generally contains greater than 30% smectite (Figure 22). In typical shales, the transition from random to ordered (R = 1) interlayers occurs with 60-70% illite layers [*Bethke et al.*, 1986]. At Site 808, R = 1 ordering in the less than 2-μm size fraction first occurs at a depth of 1220 mbsf and a temperature of ~135°C. Within the less than 0.2-μm size fraction, R = 1 ordering first appears at a depth of 1060 mbsf and a temperature of ~120°C (Figure 22). This acceleration of illitization within the finer size fraction is also typical [*Hower et al.*, 1976; *Jennings and Thompson*, 1986].

Clay mineral diagenesis is influenced by many factors in addition to burial temperature, such as fluid pressure, pore-

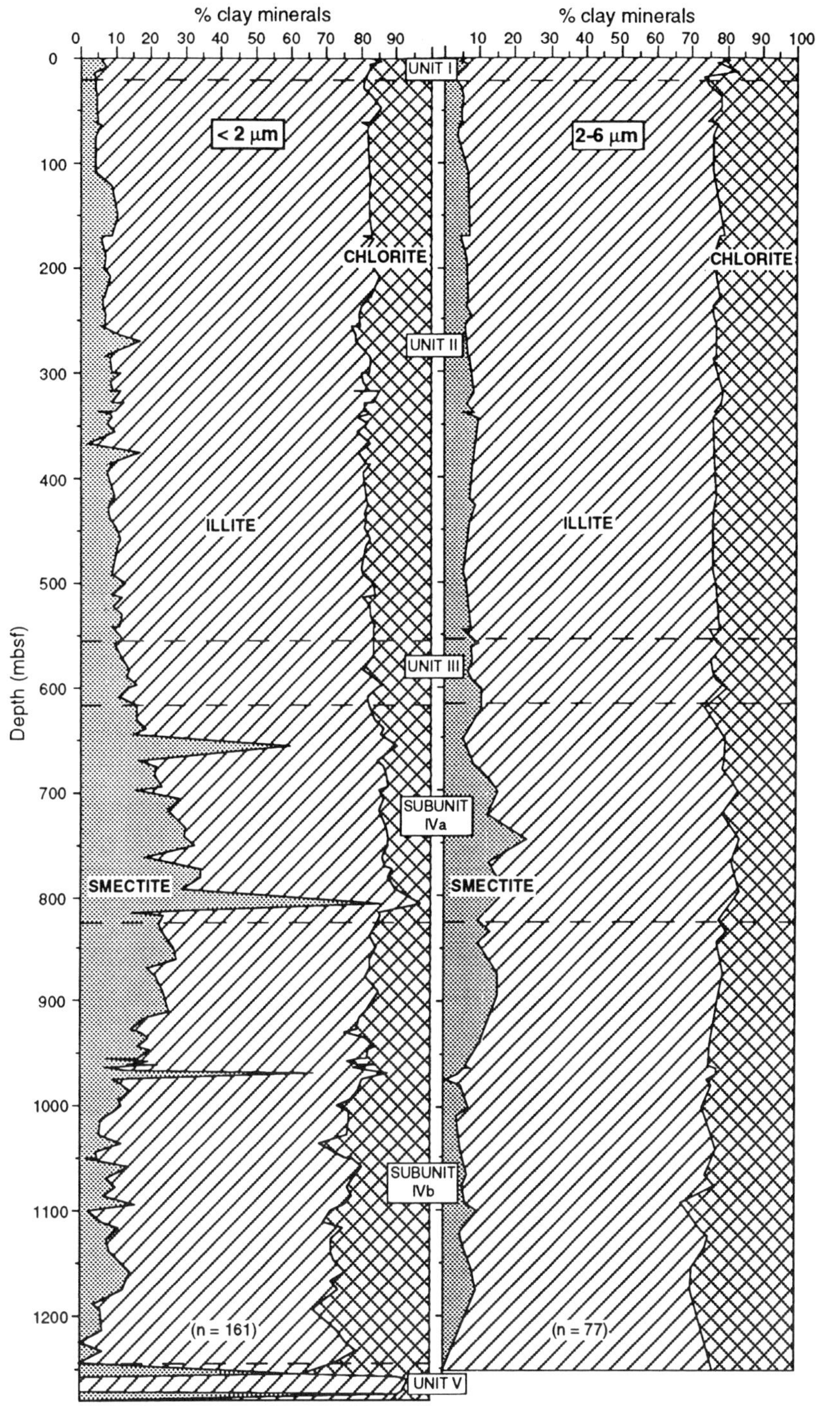

Fig. 20. Relative percentages of smectite, illite, and chlorite for hemipelagic muds at Site 808. Traces of kaolinite are included with chlorite, and mixed-layer clays are not discriminated from discrete illite and smectite. Based on integrated X-ray diffraction peak areas (glycolated, oriented aggregates) and weighting factors established by *Underwood et al.* [1993a]. Separate analyses were completed for less than 2-μm and 2-6-μm size fractions. From *Underwood et al.* [1993b].

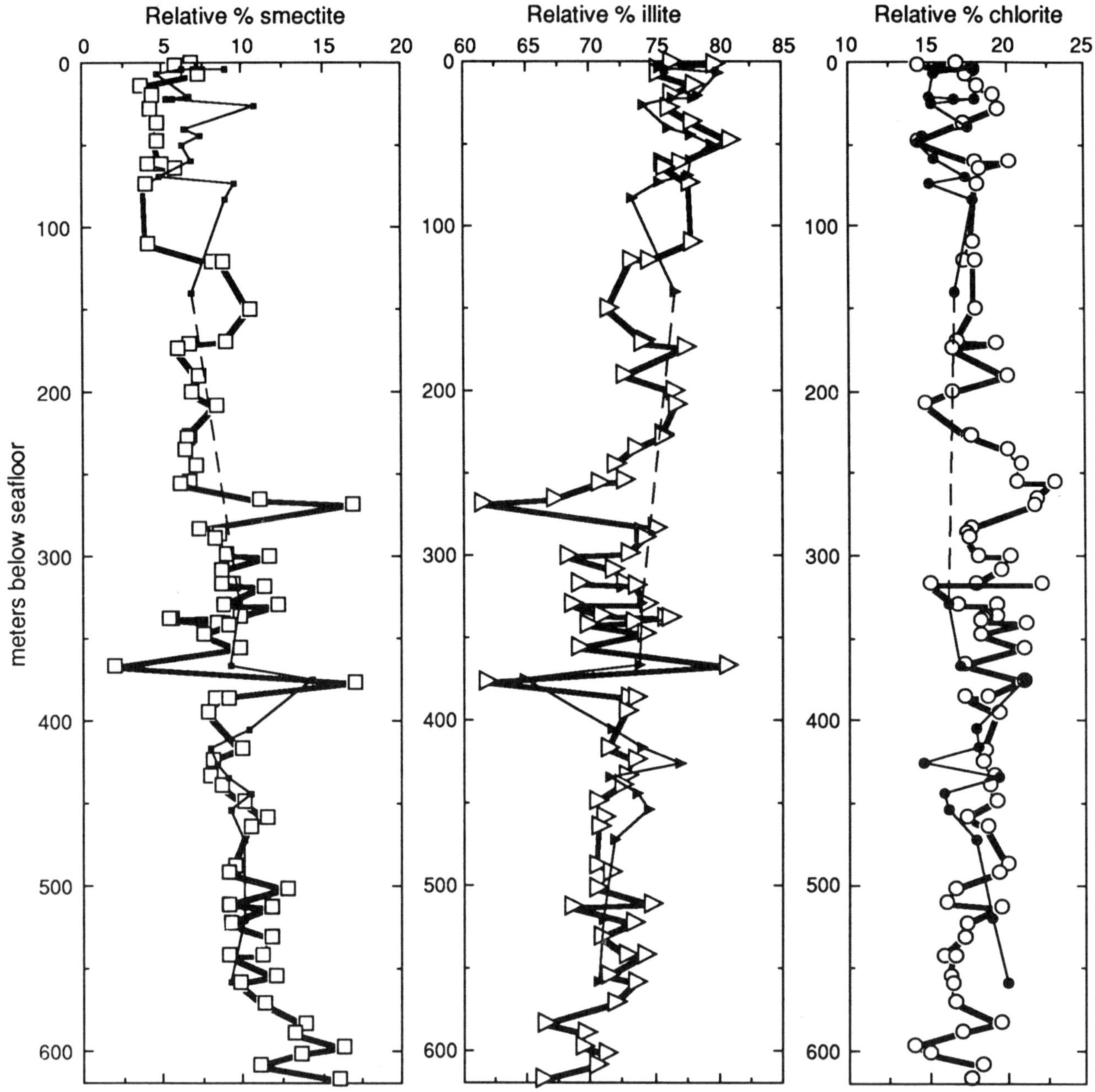

Fig. 21. Comparison of relative clay-mineral percentages at Site 808 (less than 2-μm size fraction) for hemipelagic muds (open symbols, heavy lines) and clay-sized matrix washed from turbidites (closed symbols, narrow lines). From *Underwood et al.* [1993a].

water geochemistry, permeability and bulk mineralogy of the host rocks, and reaction time [*Perry and Hower*, 1970; *Bruce*, 1984; *Bethke and Altaner*, 1986; *Pytte and Reynolds*, 1989]. Consequently, diagenetic gradients vary considerably from one type of basin to another [*Velde et al.*, 1986; *Velde and Iijima*, 1988; *Hansen and Lindgreen*, 1989; *Freed and Peacor*, 1989a]. Active continental geothermal fields provide the best match for results at Site 808 in terms of the temperature and depth of illitization thresholds [*Jennings and Thompson*, 1986; *Walker and Thompson*, 1990]. The geothermal sites prove that I/S reactions can move to completion in sediments that are only 1 to 2 Ma in age; the temperatures required to attain R = 1 ordering under these conditions are between 135°C and 155°C.

Two of the more important aspects of the smectite-to-illite transformation involve the release of interlayer water

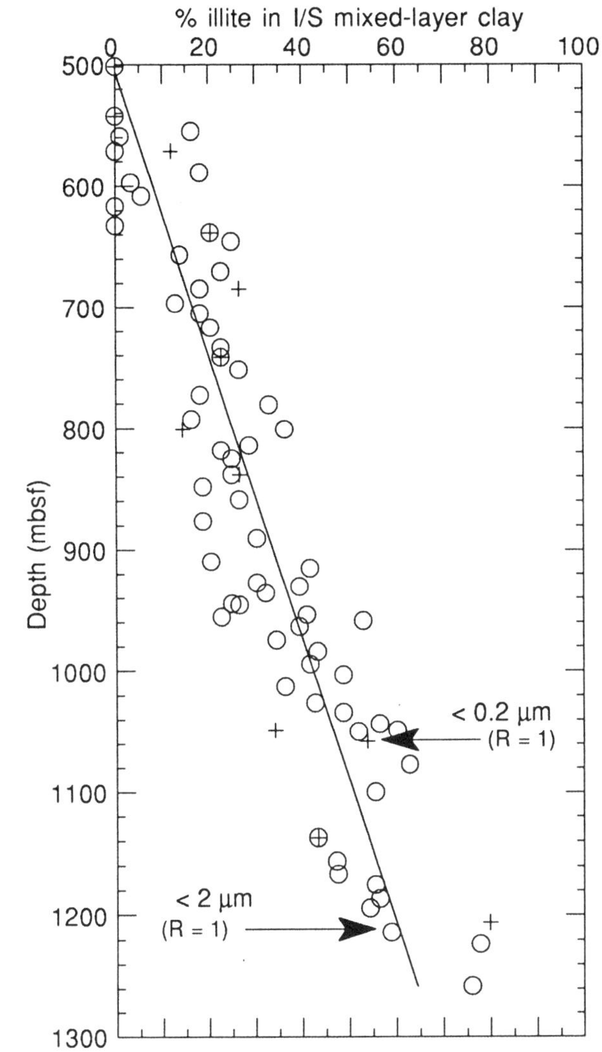

Fig. 22. Percent illite in I/S mixed-layer clays in hemipelagic mudstones as a function of burial depth at Site 808. Symbols denote two size fractions: less than 2-mm (open circles) and less than 0.2-mm (crosses). Based on angular separation (Δ°2q) between composite illite (001) / smectite (002) peak at 9.0-10.3°2q and illite (002) / smectite (003) peak at 15.8-17.4°2q [see *Moore and Reynolds*, 1989]. Initial appearance of R = 1 ordering based on second-order superstructure reflection (002*) at diffraction angle of approximately 6.5°2q. From *Underwood et al.* [1993b].

into the pore system of the host mudrock and the corresponding development of excess pore-fluid pressure if the liberated water cannot dissipate at the rates of production [*Burst*, 1969; *Bruce*, 1984; *Colten-Bradley*, 1987; *Freed and Peacor*, 1989b; *Vrolijk*, 1990]. At Site 808, the progressive increase in illite interlayers in I/S mixed-layer clays occurs over the same depth interval as a documented reduction in pore-water chlorinity [*Kastner et al.*, 1993]. Significantly, however, the mass-balance procedure of *Vrolijk et al.* [1991] shows that the absolute abundance of smectite in the Shikoku Basin section at Site 808 (Unit IV) is much too small to generate the necessary volume of fresh water entirely through *in situ* smectite dehydration. Low-chlorinity pore waters, therefore, probably migrated to Site 808 from other sources. The most likely sources are located deeper in the accretionary prism, where lower porosities or higher smectite percentages would promote greater amounts of pore-water dilution during smectite dehydration [*Kastner et al.*, 1993]. Alternatively, some lateral migration may have occurred within the prism toe from pockets of Shikoku Basin mudstone containing greater percentages of detrital or authigenic smectite [*Underwood et al.*, 1993b].

4. RIDGE-TRENCH COLLISION ZONES

One of the principal objectives of Leg 134 was to investigate the effects of subduction and collision between the forearc slope of the New Hebrides (Vanuatu) island arc, the North d'Entrecasteaux Ridge (NDR), and Bougainville Guyot (Figure 23). In the vicinity of the collision zone, the NDR forms a nearly continuous edifice composed of igneous rocks with both mid-ocean-ridge basalt (MORB) and island-arc tholeiite (IAT) chemical affinities [*Maillet et al.*, 1983; *Collot and Fisher*, 1992]. The NDR can be traced eastward on seismic-reflection profiles for 15 km underneath the toe of the deformation front [*Fisher et al.*, 1991]. Collision has resulted in deformation of a broad area of the lower and upper trench slope, involving as much as 1500-2000 m of uplift [*Collot and Fisher*, 1991, 1992].

ODP Sites 827, 828 and 829 were drilled within the NDR collision zone (Figure 23) to study the many geologic consequences of ridge-trench interaction. Because of vertical and lateral displacements of the NDR, depositional sites have been transferred through a variety of sedimentary environments. The collision, moreover, has influenced the discharge of sediment from the subaerial arc. Dramatic amounts of uplift within the forearc have formed the emergent islands of Espiritu Santo and Malakula (Figure 23). Uplift was especially significant during the Pleistocene [*F.W. Taylor*, 1992], and abundant epiclastic sediment has been distributed preferentially to the west by the prevailing ocean currents and turbidity currents [*Reid et al.*, 1994].

4.1. *Influence of Collision on Trench Sedimentation*

Site 828 was drilled about 2 km seaward of the trace of the deformation front (Figure 23). Abrupt facies changes (Figure 24) are responses to translation of the borehole site

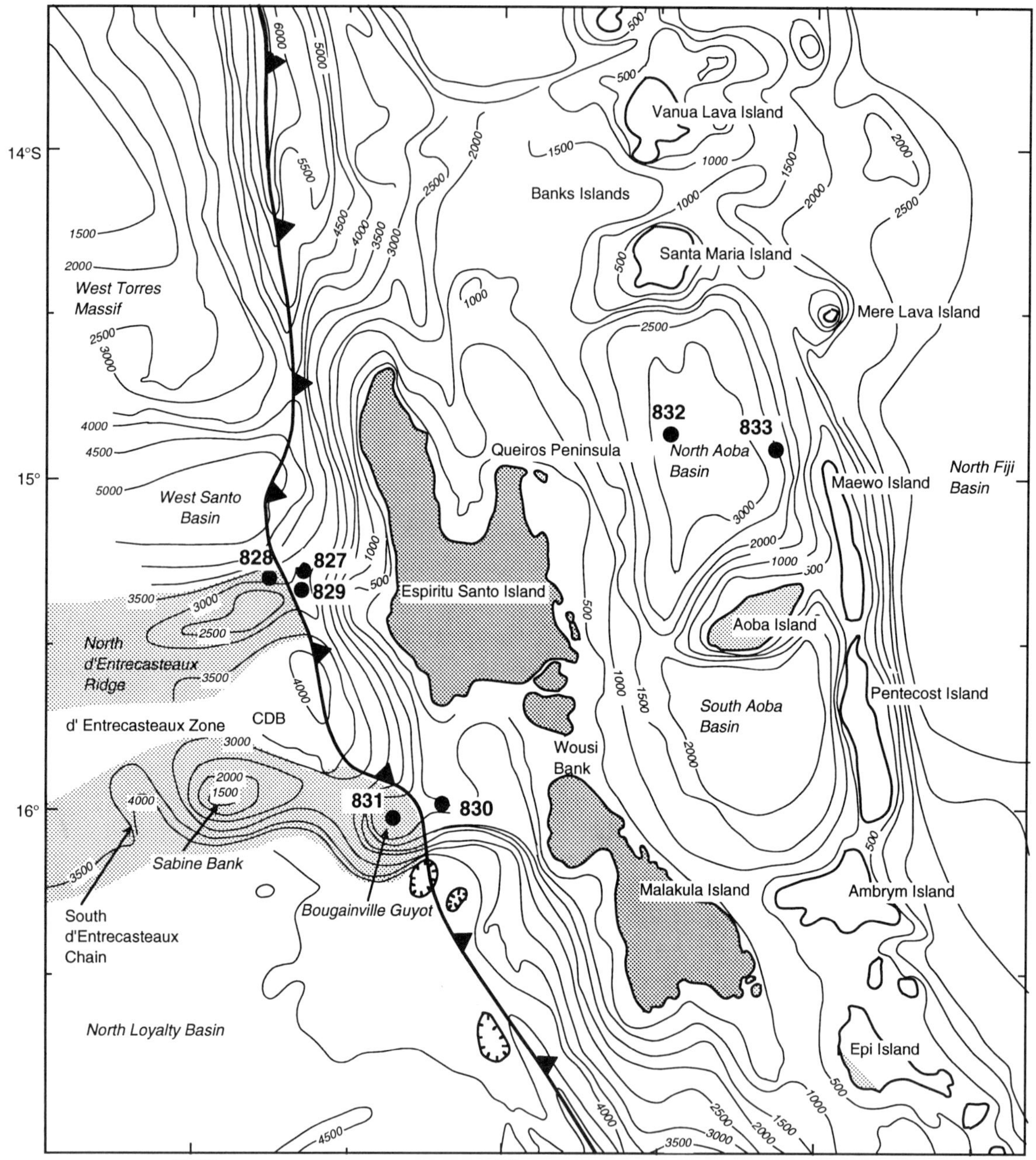

Fig. 23. Locations of Sites 827, 828, and 829 in the collision zone between North d'Entrecasteaux Ridge and New Hebrides (Vanuatu) Island Arc. After *Collot et al.* [1992].

by relative plate movement (13 cm/yr) from the abyssal plain into the zone of trench sedimentation. The igneous rocks recovered at the base of the borehole represent the eroded and weathered detritus of a volcanic basement complex [*Collot et al.*, 1992]. Basement strata consist of immature volcanic breccia, brecciated lava flow deposits, and brecciated aphyric basalt and diabase. Middle Eocene pelagic foraminifers occur locally, and smectite occurs in the fractures of some of the volcanic rock fragments. The sedimentary sequence above the volcanic complex records a gradual transition from pelagic deposition (nannofossil and foraminiferal oozes) during the late Eocene, Oligocene, and early Pliocene, to hemipelagic sedimentation and volcaniclastic input during the middle and late Pleistocene

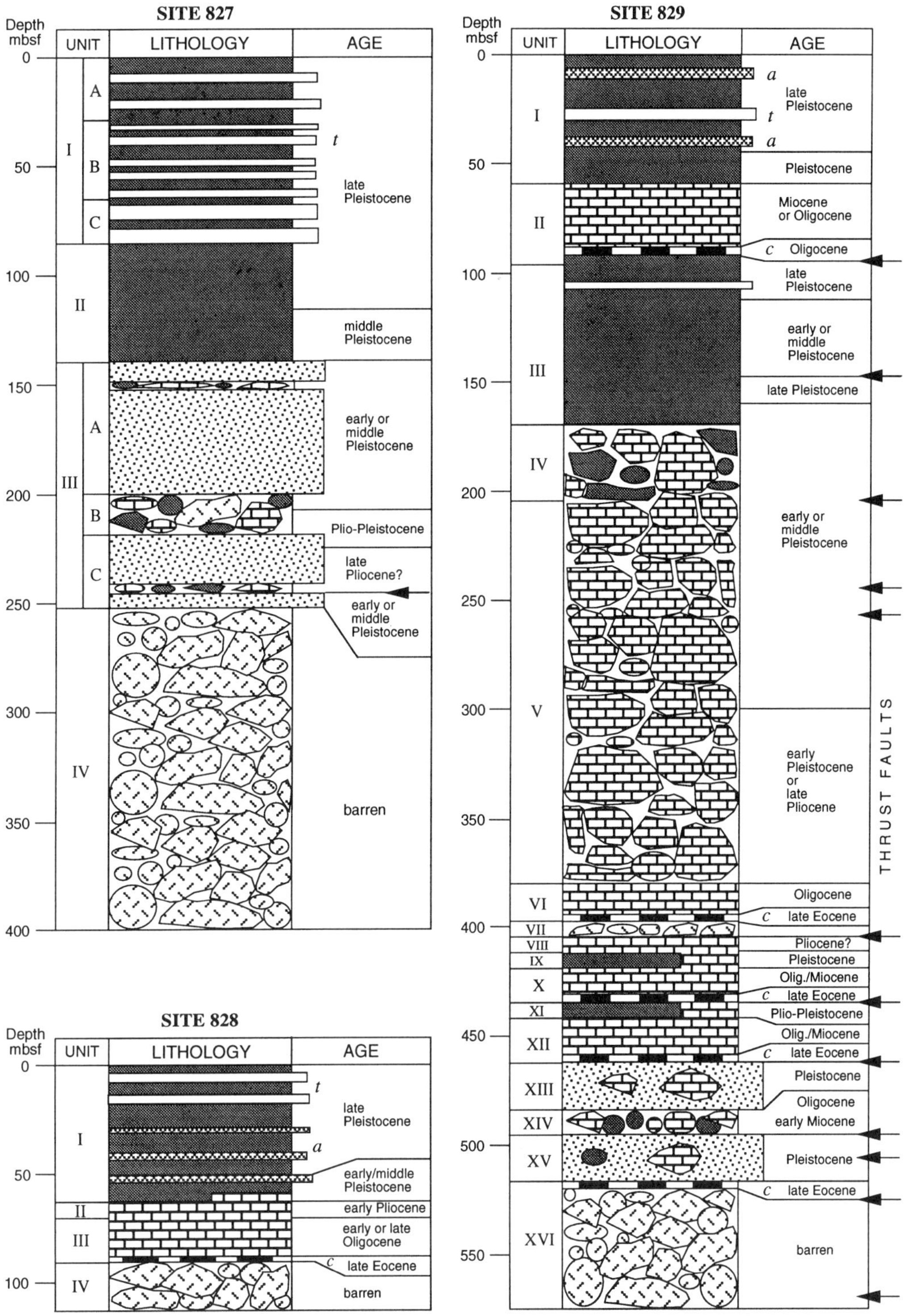

Fig. 24. Revised lithostratigraphy for Sites 827, 828, and 829, after *Reid et al.* [1993]. Arrows indicate positions of thrust faults. Brick pattern = chalk, carbonate ooze, and limestone breccia; igneous pattern = volcanic breccia; gray background pattern = volcanic silt and siltstone; stippled pattern = volcanic sand and sandstone; *c* = kaolinite-rich clay; *t* = turbidite; *a* = volcanic ash.

(Figure 24). Upper Pleistocene volcanic silt at the top of the succession (Unit I) grades downward to middle or lower Pleistocene calcareous silt, mixed sediment, and biogenic ooze. Crystal-rich and lithic-rich turbidites occur in the upper part of Unit I; these beds display normal size grading and are composed primarily of sand and silt-sized volcanic minerals (e.g., plagioclase and pyroxenes), polycrystalline volcanic-rock fragments, glass shards, pumice, and scoria fragments. Vitric-rich interbeds occur in the lower part of Unit I and probably represent volcanic ash-fall deposits transported through the atmosphere and water column from explosive eruptions in the New Hebrides island arc. An upward decrease in carbonate content within Unit I coincides with an increase in sedimentation rate, from about 4.4 cm/1000 yr during the middle Pleistocene to about 18 cm/1000 yr in the late Pleistocene (Figure 25).

During the 0.50 my represented by the Pleistocene section of Site 828, the NDR travelled approximately 65 km towards the subduction zone, based on a convergence rate of about 13 cm/yr [*Collot et al.,* 1992]. Structural blockage of axial flow paths on the trench floor was one obvious outcome of ridge collision. The increase in proximity to a volcaniclastic sediment source also resulted in a decrease in carbonate content, an increase in grain size of the background hemipelagic sediment, and an increase in the frequency of turbidity currents at the position of Site 828 [*Reid et al.*, 1993]. The dominant source of turbidite sediment transported onto the ridge crest was probably Espiritu Santo Island. Arc-derived sediment suspended in the surface waters was preferentially distributed to the west of Espiritu Santo Island by the South Equatorial Current and Trade Drift Convergence. Storm-generated and seismically triggered turbidity currents likely provided a mechanism for transferring material directly downslope toward the zone of ridge-trench intersection, and turbidite activity was enhanced during lowstands of sea level [*Reid et al.*, 1993].

Espiritu Santo Island has experienced repeated episodes of uplift and subsidence; the net geologic result is a combination of carbonate and volcaniclastic rocks which range in age from Oligocene to Pleistocene [*F.W. Taylor*, 1992]. The most recent phase of island uplift probably was caused by the initial collision of the NDR with the New Hebrides forearc. Late Quaternary uplift rates, based on the dating of raised coral terraces, indicate that cumulative uplift moved strata from average trench-slope depths to an subaerial island elevation of 1400 m above sea level in about 2 my [*F.W. Taylor*, 1992]. Emergence exposed the island to vigorous erosion processes, thereby creating a robust source of epiclastic sediment entering the forearc and trench.

The NDR must have remained elevated above the trench axis as it moved progressively towards the front of the collision zone. Site 828 is currently at a water depth 150 m above the bathymetric low that separates the NDR from the inner trench slope [*Collot et al.*, 1992]. Deposition of sands on top of the ridge shows that the arc-derived turbidity currents probably moved upslope. Alternatively, the flows must have reached thicknesses greater than 150 m. Upslope deposition by turbidity currents has been documented in a number of convergent margin settings [*Moore et al.*, 1982; *Dolan et al.*, 1989; *Muck and Underwood*, 1990; *Underwood*, 1991; *Pickering et al.*, 1992]. If upslope deposition were caused by a component of near-orthogonal vertical run-up, following descent of the inner trench slope, then the respective thicknesses still would have to exceed 100 m [*Muck and Underwood*, 1990]. These amounts of upslope transfer or flow thickness are not particularly unusual, however, as plumes of suspended sediment commonly rise 500 to 1000 m above trench floors [*Moore et al.*, 1982]. In extreme cases, turbidity currents probably attain thicknesses in excess of 1500 m [*Dolan et al.*, 1989; *Underwood*, 1991].

4.2. *Influence of Collision on Forearc Sedimentation*

The two most important consequences of the NDR subduction for *in situ* inner-trench-slope sedimentation have been: (1) to initiate a new source of epiclastic sediment through uplift and erosion of Espiritu Santo Island; and (2) to cause vertical movement and horizontal translation of forearc depositional sites. Site 827 is in an ideal position to quantify these processes. Located 4 km east of the subduction front on an elevated terrace that is bounded to the north by a steep scarp, the terrace is aligned with the trace of the NDR where it impinges on the forearc slope from the west (Figure 23). Drilling penetrated 252 m of upper Pliocene to upper Pleistocene slope deposits resting above 148 m of accreted sedimentary-lithic conglomerate of unknown age (Figure 24). The conglomerate contains andesitic clasts and amphibole grains, so debris flows with arc-derived material either slid into the trench or moved farther seaward onto a flat spot of the NDR prior to accretion. Unit III, above the conglomerate, is composed of Pleistocene to upper Pliocene sandstone, sandy siltstone, and conglomerate; these strata are partially lithified and highly sheared. Biostratigraphic inversion near the base of Unit III places upper Pliocene sediment over lower Pleistocene sediment (Figure 24). Accretion of Unit IV and sediment at the base of Unit III probably occurred within the past 1.58 my [*Reid et al.*, 1993].

Initial deposition within the overlying forearc slope apron produced beds of sandy siltstone and conglomerate (Unit III)

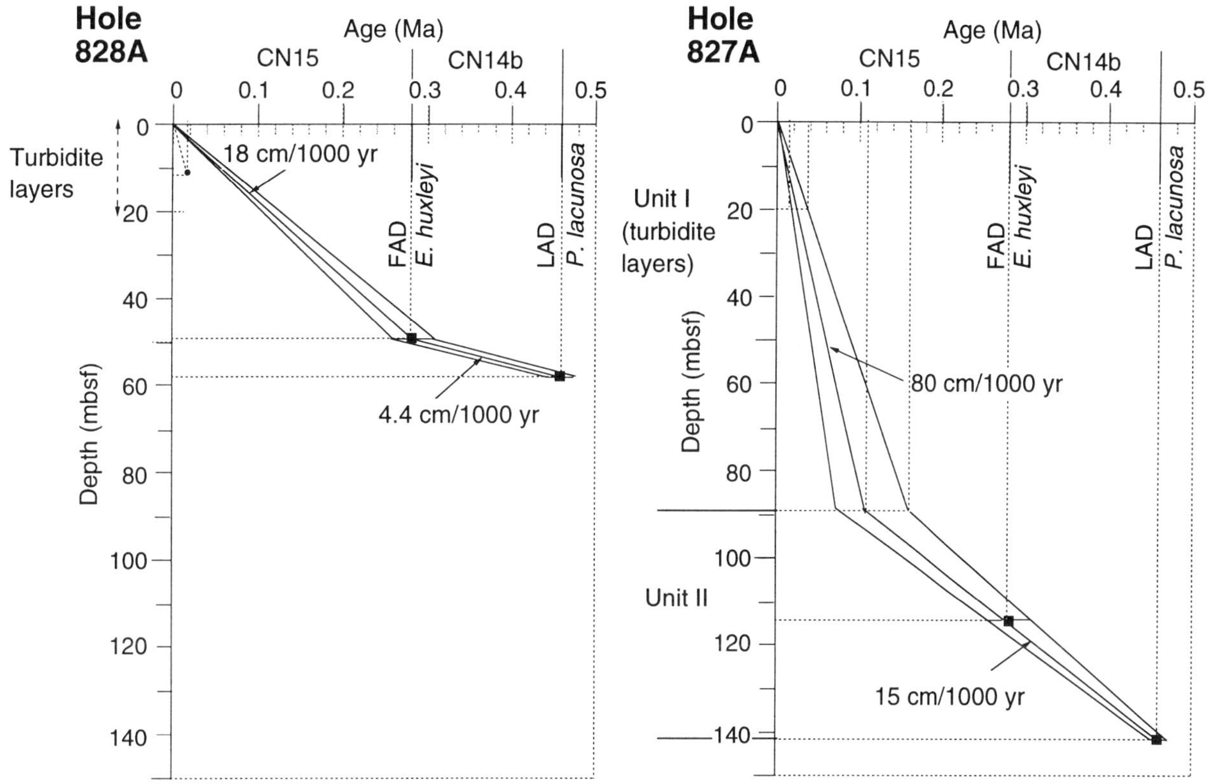

Fig. 25. Sedimentation rates for the past 0.5 my at Sites 827 and 828. After *Reid et al.* [1993].

which, in turn, were buried by upper and middle Pleistocene volcanic silt (Unit II). The onset of turbidite deposition during the late Pleistocene coincides with the base of Unit I (Figure 24). This fundamental change in sedimentation was abrupt and characterized by interlayers of volcanic silt and normally graded silty sand. Extrapolation of a constant sedimentation rate (15 cm/1000 yr) throughout Unit II (Figure 25) indicates that turbidite deposition began sometime between 110,000 and 160,000 yr B.P. The peak in turbidite frequency was between 18,000 and 38,000 yr B.P., and the average accumulation rate was 80 cm/1000 yr (Figure 25). An age of 18,000 yr would correspond to the last glacial lowstand [*Imbrie et al.*, 1984], so eustacy may have played a role in the generation and movement of turbidity currents. *Reid et al.* [1993] argued, however, that fluctuations in sea level were not the sole influence on sedimentation rates. Eustacy, for example, does not provide a convincing explanation for the overall trend in which the proportion of turbidite layers, maximum layer thickness, and grain size all decrease gradually up-section. Two alternative hypotheses could account for these trends. First, the source of turbidite sediment may have become finer grained and less abundant through the depositional history of Unit I. This scenario seems unlikely, however, in light of the uplift history of Espiritu Santa Island, which should have resulted in an increase in sedimentation rates through the late Pleistocene. Alternatively, the dominant flow directions of turbidity currents may have changed as a result of tectonic modification of the forearc substrate. Progressive uplift, for example, may have moved the site from a relatively depressed zone of turbidite ponding to an elevated zone of flow deflection and diminishing accumulation. The initiation of turbidite deposition at the base of Unit I likewise may have been a result of tectonic uplift, coinciding with a phase in the deformation history in which offset of the seafloor created a depositional sink for turbidity currents descending the forearc slope [*Collot et al.*, 1992].

The sedimentary record at Site 827 bears some similarity to other uplifted trench-slope and slope-basin sequences, such as those exposed on Nias Island off the coast of Sumatra [*G.F. Moore and Karig*, 1976]. The Nias beds display a pronounced upward-thickening and upward-coarsening trend that probably resulted from progressive uplift from a zone of initial isolation into a position in which submarine canyons and channels supplied the basin with increasing amounts of coarse-grained turbidites [*Moore et al.*, 1980]. Unlike Nias Island tectonism, however, uplift at Site 827 has progressed into a later phase in which the seafloor once again became relatively isolated from coarse-grained turbidites because of flow deflection around bathymetric highs.

4.3. *Facies Partitioning at the Décollement*

One of the time-honored paradigms of convergent-margin tectonics points out that pelagic abyssal-plain sediments should be selectively subducted and terrigenous trench deposits should be accreted by offscraping at the toes of accretionary prisms [*Karig and Sharman*, 1975; *Moore*, 1975]. Recent studies have emphasized the role of seamount collision and ridge collision in the process of tectonic erosion [*Lallemand and Le Pichon*, 1987; *von Huene and Cullotta*, 1989; *von Huene and Lallemand*, 1990; *von Huene and Scholl*, 1991]. In apparent defiance of these conceptual models, the stratigraphic section at Site 829 records a complicated history of tectonic accretion within a collision zone, with mechanical transfer of both pelagic ridge sediment and trench fill into the insular slope of the Vanuatu forearc [*Collot et al.*, 1992]. The site is located at the toe of the accretionary prism, 2 km east of the trace of the subduction front, at a water depth of 2905 m (Figure 23). Numerous biostratigraphic inversions and zones of structural deformation provide convincing evidence for thrust imbrication (Figure 23); individual shear zones range in thickness from a few cm to 3 m [*Collot et al.*, 1992].

Sixteen lithostratigraphic units have been recognized at Site 829 (Figure 24). Dominant lithologies include interbeds of volcanic silt, sand, and ash, foraminiferal chalk, brown clay, siltstone-chalk breccia, and volcanic breccia. The thrust sheets were not emplaced systematically in their original stratigraphic order (i.e., with the oldest imbricate in the highest structural position). Instead, some of the more recently accreted packets (Units I and II) rest structurally above older thrust slices, a pattern consistent with displacement along landward-verging faults. Accretion occurred at least within the past 1.58 my, and the presence of upper Pleistocene microfossils in the upper part of Unit III (2nd thrust sheet) indicates that Units I to III probably were offscraped during the past 280,000 yr [*Reid et al.*, 1993].

Vrolijk [1990] recognized that décollements in subduction zones tend to occur in smectite-rich horizons; smectite is a common detrital and authigenic constituent of hemipelagic sediments derived from volcanic arcs. The décollement associated with the NDR collision zone, in contrast, appears to have formed along kaolinite-rich, rather than smectite-rich, horizons (Figure 24). Similarities of age and mineralogic composition support a correlation between brown, clay-rich horizons on the NDR (Site 828) and clays found along many of the thrust faults at Site 829 [*Reid et al.*, 1993]. The kaolinite-rich clay evidently provides a preferential zone of weakness along which the pelagic cover becomes delaminated from the top of volcanic basement, as the downgoing plate is thrust under the Vanuatu trench slope.

5. CONCLUSIONS

ODP drilling legs within forearc-trench segments of the western Pacific Ocean added a wealth of definitive sedimentologic information. The geologic community now has a much better understanding of the diverse array of sedimentary processes responsible for transport of volcaniclastic sediment within intraoceanic forearcs. Precise definition of turbidite stratigraphy, using FMS technology, showed that there are significant differences between the Izu-Bonin intraoceanic forearc and well-studied continental-margin forearc basins, such as the Great Valley Group of California; specifically, there is a general lack of conglomerate in the Izu-Bonin forearc, facies associations are limited, and there are no consistent vertical trends in bed thickness. As expected, the detrital modes of intraoceanic forearc sands are consistent with an undissected-arc tectonic provenance; spatial and temporal variations in volcaniclastic input can be defined much more precisely, however, through meticulous subdivision of volcanic textures and glass types. Stratigraphic studies of the Tonga forearc led to a more sophisticated appreciation of the interplay that exists between tectonically driven basin subsidence, rates of sediment accumulation, and arc volcanism.

Cores from the toe of the Nankai accretionary prism represent the first complete record of facies evolution at a single site along a convergent margin. The section begins with basaltic basement of the subducting plate, passes through abyssal-plain hemipelagic deposits, a sandy trench wedge, and a thin slope apron above the accretionary prism. For the first time, dip azimuths for ripple cross-laminae were restored to their true geographic orientations using paleo-magnetic data and then used to define paleocurrent directions within the trench facies. Surprisingly, most paleoflow indicators are oriented back toward the trench axis in response to the reflection or deflection of turbidity currents off the outer trench slope. Analyses of sand petrography and clay mineralogy showed that the Izu-Honshu collision zone served as the principal source of sediment in the Nankai trench wedge. Geothermal gradients with the trench and accretionary prism have been high enough to promote diagenetic reactions. There has been a systematic replacement of detrital and authigenic smectite by I/S mixed-layer clay, and vertical illitization gradients are similar to those documented within active continental geothermal fields. These reactions play an important role in the hydrologic budget of the accretionary prism.

Collision between the New Hebrides (Vanuatu) forearc slope and the North d'Entrecasteaux Ridge exerted profound influence on patterns of trench and forearc sedimentation. Diverse responses to collision include blockage of axial flow paths, increased rates of uplift and sediment influx by turbidity currents, upslope flow of turbidity currents onto the subducting ridge crest, formation and destruction of sediment ponds on the lower forearc slope, and facies partitioning at the décollement. In opposition to long-standing paradigms favoring tectonic erosion during collisional events, the NDR collision zone includes a well-defined accretionary prism with numerous thrust packets, and the accreted strata include both pelagic deposits delaminated from the subducting igneous basement and turbidites offscraped from the trench wedge. Stratigraphic horizons enriched in kaolinite, rather than smectite, apparently provide the zones of weakness favored for propagation of the décollement.

Acknowledgments. We thank the many shipboard and shorebased co-workers, students, and technicians who assisted in our studies of western Pacific sedimentology. Funds for much of the shorebased research were provided by the U.S. Scientific Support Program. The manuscript was reviewed by Ralph Moberly and William Dickinson; we appreciate their insightful comments. Sarah Simpson assisted with editing.

REFERENCES

Andrews, J. E., G. Packham, et al., *Init. Repts. DSDP*, vol. 30, 753 pp., U.S. Govt. Printing Office, Washington, D.C., 1975.

Ashi, J., and A. Taira, Thermal structure of the Nankai accretionary prism as inferred from the distribution of gas hydrate BSRs, in *Thermal Evolution of the Tertiary Shimanto Belt, Southwest Japan: An Example of Ridge-Trench Interaction*, edited by M. B. Underwood, Geol. Soc. Am. Spec. Paper 273, pp. 137-150, 1993.

Austin, J., F. W. Taylor, and C. D. Cagle, Seismic stratigraphy of the central Tonga Ridge, *Mar. Petrol. Geol.*, *6*, 71-92, 1989.

Ballance, P. F., Gravity flows and rock recycling on the Tonga landward trench slope: relation to trench-slope tectonic processes, *J. Geol.*, *99*, 817-827, 1991.

Ballance, P. F., J. A. Barron, C. D. Blome, D. Bukry, P. A. Cawood, G. C. H. Chaproniere, R. Frisch, R. H. Herzer, C. S. Nelson, P. Quinterno, H. Ryan, D. W. Scholl, A. J. Stevenson, D. R. Tappin, and T. L. Vallier, Late Cretaceous pelagic sediments, volcanic ash and biotas from near the Louisville hotspot, Pacific Plate, paleolatitude ~42°S, *Palegeogr. Paleoclimatol. Paleoecol.*, *71*, 281-299, 1989.

Ballance, P. F., and M. R. Gregory, Parnell Grits—large subaqueous volcaniclastic gravity flows with multiple particle support mechanisms, in *Sedimentation in Volcanic Settings*, edited by R. V. Fisher and G. A. Smith, Soc. Econ. Paleontol. Mineral. Spec. Publ. 45, 189-200, 1991.

Bethke, C. M., and S. P. Altaner, Layer-by-layer mechanism of smectite illitization and application to a new rate law, *Clays Clay Miner.*, *34*, 136-145, 1986.

Bethke, C. M., N. Vergo, and S. P. Altaner, S.P., Pathways of smectite illitization, *Clays Clay Miner.*, *34*, 125-135, 1986.

Biju-Duval, B., J. C. Moore, et al., *Init. Repts. DSDP*, vol. 78A, 621 pp., U.S. Govt. Printing Office, Washington, D.C., 1984.

Bloomer, S. H., and R. L. Fisher, Petrology and geochemistry of igneous rocks from the Tonga Trench, a non-accreting plate boundary, *J. Geol.*, *95*, 469-495, 1987.

Blum, P., and Y. Okamura, Pre-Holocene sediment dispersal systems and effects of structural controls and Holocene sea-level rise from acoustic facies analysis: SW Japan forearc, *Mar. Geol.*, *108*, 295-322, 1992.

Bøe, R., Nature and record of late Miocene mass-flow deposits of the Lau-Tonga forearc basin, Tonga Platform (Hole 840B), in *Proc. ODP, Sci. Results*, vol. 135, J. F. Hawkins, L. M. Parson, and J. F. Allan, pp. 87-100, Ocean Drilling Program, College Station, TX, 1994.

Bruce, C. H., Smectite dehydration—its relation to structural development and hydrocarbon accumulation in northern Gulf of Mexico basin, *Bull. Am. Assoc. Petrol. Geol.*, *68*, 673-683, 1984.

Burst, J. F., Diagenesis of Gulf Coast clayey sediments and its possible relation to petroleum migration, *Bull. Am. Assoc. Petrol. Geol.*, *53*, 73-93, 1969.

Cas, R. A. F., and J. V. Wright, *Volcanic Successions, Modern and Ancient*, 528 pp., Unwin Hyman, London, 1987.

Cawood, P. A., Petrography, phase chemistry and provenance of volcanogenic debris from the southern Tonga ridge: Implications for arc history and magmatism, in *Geology and Offshore Resources of Pacific Island Arcs—Tonga Region*, edited by D. W. Scholl and T. L. Vallier, pp. 149-169, Circum-Pacific Council for Energy and Mineral Resources, Earth Science Series, vol. 2, Houston, TX, 1985.

Chamley, H., J.-P. Cadet, and J. Charvet, Nankai Trough and Japan Trench late Cenozoic paleoenvironments deduced from clay mineralogy data, in *Init. Repts. Deep Sea Drilling Project*, vol. 87, H. Kagami, D. E. Karig, W. T. Coulbourn, et al., pp. 633-642, U. S. Govt. Printing Office, Washington, D.C., 1986.

Chamot-Rooke, N., V. Renard, and X. Le Pichon, Magnetic anomalies in the Shikoku Basin: a new interpretation, *Earth Planet. Sci. Lett.*, *83*, 214-228, 1987.

Chipping, D. H., Sedimentary structures and environment of

some thick sandstone beds of turbidite type, *J. Sediment. Petrol., 42*, 587-595, 1972.

Clift, P.D., Controls on the sedimentary and subsidence history of an active plate margin; An example from the Tonga Arc, SW Pacific, in *Proc. ODP, Sci. Results*, vol. 135, J. F. Hawkins, L. M. Parson, J. F. Allan, et al., pp. 173-188, Ocean Drilling Program, College Station, TX, 1994.

Clift, P. D., and J. E. Dixon, Variations in arc volcanism and sedimentation related to rifting of the Lau Basin, SW Pacific, in *Proc. ODP, Sci. Results*, vol. 135, J. F. Hawkins, L. M. Parson, J. F. Allan, et al., pp. 23-49, Ocean Drilling Program, College Station, TX, 1994.

Collot, J.-Y., and M. A. Fisher, The collision zone between the North d'Entrecasteaux Ridge and the New Hebredes island arc, 1. Sea beam morphology and shallow structure, *J. Geophys. Res., 96*, 4457-4478, 1991.

Collot, J.-Y., and M. A. Fisher, The d'Entrecasteaux Zone-New Hebrides Island Arc Collision Zone: an overview, in *Proc. ODP, Init. Repts.*, vol. 134, J.-Y. Collot, H. G. Greene, L. B. Stokking, et al., pp. 19-32, Ocean Drilling Program, College Station, TX, 1992.

Collot, J.-Y., Greene, H.G., Stokking L.B., et al., *Proc. ODP, Init. Repts.*, vol. 134, 1136 pp., Ocean Drilling Program, College Station, TX, 1992.

Colten-Bradley, V. A., Role of pressure in smectite dehydration—effects on geopressure and smectite-to-illite transformation, *Bull. Am. Assoc. Petrol. Geol., 71*, 1414-1427, 1987.

Coulbourn, W. T., Sedimentologic summary, Nankai Trough Sites 582 and 583, and Japan Trench Site 584, in *Init. Repts. DSDP*, vol. 87, H. Kagami, D. E. Karig, W. T. Coulbourn, et al., pp. 909-926, U. S. Govt. Printing Office, Washington, D.C., 1986.

Creager, J. S., D. W. Scholl, et al., *Init. Repts. DSDP*, vol. 19, 913 pp., U.S. Govt. Printing Office, Washington, D.C., 1973.

Damuth, J. E., and R. W. Embley, Upslope flow of turbidity currents on the northwest flank of the Ceara Rise: western equatorial Atlantic, *Sedimentology, 26*, 825-834, 1979.

de Rosa, R., G. G. Zuffa, A. Taira, and J. K. Leggett, Petrography of trench sands from the Nankai Trough, southwest Japan: implications for long-distance turbidite transportation, *Geol. Mag., 123*, 477-486, 1986.

Dickinson, W. R., Interpreting detrital modes of graywacke and arkose, *J. Sediment. Petrol., 40*, 695-707, 1970.

Dickinson, W. R., Composition of sandstones in circum-Pacific subduction complexes and fore-arc basins, *Bull. Am. Assoc. Petrol. Geol., 66*, 121-137, 1982.

Dickinson, W. R., and D. R. Seely, Structure and stratigraphy of forearc regions, *Bull. Am. Assoc. Petrol. Geol., 63*, 2-31, 1979.

Dickinson, W. R., L. S. Beard, G. R. Brakenridge, J. L. Erjavec, R. C. Ferguson, F. K. Inman, R. A. Knepp, F. A. Lindberg, and P. T. Ryberg, Provenance of North American Phanerozoic sandstones in relation to tectonic setting, *Bull. Geol. Soc. Am., 94*, 222-235, 1983.

Dolan, J., C. Beck, and Y. Ogawa, Upslope deposition of extremely distal turbidites: An example from the Tiburon Rise, west-central Atlantic, *Geology, 17*, 990-994, 1989.

Donnelly, T. W., Appendix. Chemical composition of deep sea sediments—Sites 9 through 425, Legs 2 through 54, Deep Sea Drilling Project, in *Init. Repts. DSDP*, vol. 54, B. R. Rosendahl, R. Hekinian, R., et al., pp. 899-949, U.S. Govt. Printing Office, Washington, D.C., 1980.

Duda, S. J., Secular seismic energy release in the circum-Pacific belt, *Tectonophysics, 2*, 409-452, 1965.

Dupont, J., and R. H. Herzer, Effect of subduction of the Louisville Ridge on the structure and morphology of the Tonga arc, in *Geology and Offshore Resources of Pacific Island Arcs—Tonga Region*, edited by D. W. Scholl and T. L. Vallier, pp. 323-332, Circum-Pacific Council for Energy and Mineral Resources, Earth Science Series, vol. 2, Houston, TX, 1985.

Ekdale, A. A., R. G. Bromley, and S. G. Pemberton, *Ichnology. Trace Fossils in Sedimentation and Stratigraphy*, Soc. Econ. Paleontol. Mineral. Short Course 15, 317 pp., 1984.

England, T. D. J., and R. N. Hiscott, Lithostratigraphy and deep-water setting of the upper Nanaimo Group (Upper Cretaceous), outer Gulf Islands of southwestern British Columbia, *Can. J. Earth Sci., 29*, 574-595, 1992.

Ewart, A., W. B. Bryan, and J. B. Gill, Mineralogy and geochemistry of the younger volcanic islands of Tonga, SW Pacific, *J. Petrol., 14*, 429-465, 1973.

Fisher, M. A., J.-Y. Collot, and E. L. Geist, The collision zone between the North d'Entrecasteaux Ridge and the New Hebrides island arc, 2. Structure from multichannel seismic data, *J. Geophys. Res., 96*, 4457-4478, 1991.

Fisher, R. V., and H.-U. Schmincke, *Pyroclastic Rocks*, 528 pp., Springer-Verlag, New York, 1984.

Freed, R. L., and D. R. Peacor, Variability in temperature of the smectite/illite reaction in Gulf Coast sediments, *Clay Miner., 24*, 171-180, 1989a.

Freed, R. L., and D. R. Peacor, Geopressured shale and sealing effect of smectite to illite transition, *Bull. Am. Assoc. Petrol. Geol., 73*, 1223-1232, 1989b.

Fryer, P., J. A. Pearce, et al., *Proc. ODP, Init. Repts.*, vol. 125, 1092 pp., Ocean Drilling Program, College Station, TX, 1990.

Fryer, P., J. A. Pearce, L. B. Stokking, et al., *Proc. ODP, Sci. Results*, vol. 125, 716 pp., Ocean Drilling Program, College Station, TX, 1992.

Fujioka, K., Y. Matsuo, A. Nishimura, M. Koyama, and K. S. Rodolfo, Tephras of the Izu-Bonin forearc (Sites 787, 792, and 793), in *Proc. ODP, Sci. Results*, vol. *126*, B. Taylor, K. Fujioka, et al., pp. 47-74, Ocean Drilling Program, College Station, TX, 1992.

Gill, J. B., R. N. Hiscott, and Ph. Vidal, Turbidite geochemistry and evolution of the Izu-Bonin arc and continents, *Lithos*, in press, 1994.

Hansen, P. L., and H. Lindgreen, Mixed-layer illite/smectite diagenesis in Upper Jurassic claystones from the North Sea and onshore Denmark, *Clay Miner., 24*, 197-213, 1989.

Haq, B. U., J. Hardenbol, and P. R. Vail, Chronology of fluctuating sea levels since the Triassic, *Science*, *235*, 1156-1167, 1987.

Harrold, P. J., and J. C. Moore, Composition of deep-sea sands from marginal basins of the northwestern Pacific, in *Init. Repts. DSDP,* vol. 31, D. E. Karig and J. C. Ingle, Jr., pp. 507-514, U. S. Govt. Printing Office, Washington, D.C., 1975.

Hathon, E. G., and M. B. Underwood, Clay mineralogy and chemistry as indicators of hemipelaic sediment dispersal south of the Aleutian arc, *Mar. Geol.*, *97*, 145-166, 1991.

Hawkins, J. F., L. M. Parson, J. F. Allan, et al., *Proc. ODP, Sci. Results*, v. 135, Ocean Drilling Program, College Station, TX, 1994.

Hibbard, J. P., and D. E. Karig, An alternative plate model for the early Miocene evolution of the SW Japan margin, *Geology*, *18*, 170-174, 1990.

Hiscott, R. N., Depositional framework of sandy mid-fan complexes of Tourelle Formation, Ordovician, Quebec, *Bull. Am. Assoc. Petrol. Geol.*, *64*, 1052-1077, 1980.

Hiscott, R. N., Traction-carpet stratification in turbidites—fact or fiction? *J. Sediment. Res.*, *64*, 204-208, 1994.

Hiscott, R. N., and G. V. Middleton, Fabric of coarse deep-water sandstones, Tourelle Formation, Quebec, Canada, *J. Sedi- ment. Petrol.*, *50*, 703-722, 1980.

Hiscott, R. N., and J. B. Gill, Major and trace element geochemistry of Oligocene to Quaternary volcaniclastic sands and sandstones from the Izu-Bonin Arc, in *Proc. ODP, Sci. Results,* vol. 126, B. Taylor, K. Fujioka, et al., pp. 467-486, Ocean Drilling Program, College Station, TX, 1992.

Hiscott, R. N., A. Colella, P. Pezard, M. A. Lovell, and A. Malinverno, Sedimentology of deep-water volcaniclastics, Oligocene Izu-Bonin forearc basin, based on Formation Micro-Scanner images, in *Proc. ODP, Sci. Results,* vol. 126, B. Taylor, K. Fujioka, et al., pp. 75-96, Ocean Drilling Program, College Station, TX, 1992.

Hiscott, R. N., A. Colella, P. Pezard, M. A. Lovell, and A. Malinverno, Basin plain turbidite succession of the Oligocene Izu-Bonin intraoceanic forearc basin, *Mar. Petrol. Geol.*, *10*, in press, 1994.

Hiscott, R. N., K. T. Pickering, and D. R. Beeden, Progressive filling of a confined Middle Ordovician foreland basin associated with the Taconic Orogeny, Quebec, Canada, in *Foreland Basins*, edited by P. A. Allen and P. Homewood, pp. 309-325, Spec. Publs. Int. Assoc. Sediment., vol. 8, Blackwell Scientific Publications, Oxford, 1986.

Hower, J., E. V. Eslinger, M. E. Hower, and E. A. Perry, Mechanism of burial metamorphism of argillaceous sediment: 1. Mineralogical and chemical evidence, *Geol. Soc. Am. Bull.*, *87*, 725-737, 1976.

Hussong, D. M., S. Uyeda, et al., *Init. Repts. DSDP*, vol. 60, 929 pp., U.S. Govt. Printing Office, Washington, D.C., 1981.

Imbrie, J., J. D. Hays, D. G. Martinson, A. McIntyre, A. C. Mix, J. J. Morley, N. G. Pisias, W. L. Prell, and N. J. Shackleton, The orbital theory of Pleistocene climate: support from a revised chronology of the marine $\delta^{18}O$ record, in *Milankovitch and Climate, Part 1*, edited by A. L. Berger et al., pp. 269-305, D. Reidel, Hingham, Mass., 1984.

Ingersoll, R. V., Submarine fan facies of the Upper Cretaceous Great Valley Sequence, northern and central California, *Sediment. Geol.*, *21*, 205-230, 1978.

Ingersoll, R. V., Evolution of the Late Cretaceous forearc basin, northern and central California, *Bull. Geol. Soc. Am.*, *90*, 813-826, 1979.

Ingersoll, R. V., T. F. Bullard, R. L. Ford, J. P. Grimm, J. D. Pickle, and S. W. Sares, The effect of grain size on detrital modes: a test of the Gazzi-Dickinson point-counting method, *J. Sediment. Petrol.*, *54*, 103-116, 1984.

Ingram, R. L., Terminology for the thickness of stratification units and parting units in sedimentary rocks, *Bull. Geol. Soc. Am.*, *65*, 937-938, 1954.

Jennings, S., and G. R. Thompson, Diagenesis of Plio-Pleistocene sediments of the Colorado River Delta, southern California, *J. Sediment. Petrol.*, *56*, 89-98, 1986.

Jolivet, L., P. Huchon, and C. Rangin, Tectonic setting of western Pacific marginal basins, *Tectonophysics*, *160*, 23-47, 1989.

Jolivet, L., and K. Tamaki, Neogene kinematics in the Japan Sea region and volcanic activity of the Northeast Japan Arc, in *Proc. ODP, Sci. Results*, vol. 127/128, K. Tamaki, K. Suyehiro, et al., pp. 1311-1332, Ocean Drilling Program, College Station, TX, 1992.

Karig, D. E., J. C. Ingle, Jr., et al., *Init. Repts. DSDP*, vol. 31, 927 pp., U.S. Govt. Printing Office, Washington, D.C., 1975.

Karig, D. E., and G. F. Sharman, III., Subduction and accretion at trenches, *Bull. Geol. Soc. Am.*, *86*, 377-389, 1975.

Kastner, M., H. Elderfield, W. J. Jenkins, J. M. Geiskes, and T. Gamo, Geochemical and isotopic evidence for fluid flow in the western Nankai subduction zone, Japan, in *Proc. ODP, Sci. Results*, vol. 131, I. A. Hill., A. Taira, J. Firth, et al., pp. 397-413, Ocean Drilling Program, College Station, TX, 1993.

Kawahata, H., K. Fujioka, and T. Ishizuka, Sediments and interstitialwater at Sites 582 and 584, the Nankai Trough and the Japan Trench landward slope, in *Init. Repts. DSDP*, vol. 87, H. Kagami, D. E. Karig, W. T. Coulbourn, et al., pp. 865-875, U.S. Govt. Printing Office, Washington, D.C., 1986.

Kinoshita, H., and M. Yamano, The heat flow anomaly in the Nankai Trough area, in *Init. Repts. DSDP*, vol. 87, H. Kagami, D. E. Karig, W. T. Coulbourn, et al., pp. 737-743, U.S. Govt. Printing Office, Washington, D.C., 1986.

Klein, G. deVries, Relative rates of tectonic uplift as determined from episodic turbidite deposition in marine basins, *Geology*, *12*, 48-50, 1984.

Klein, G. deVries, Sedimentation patterns in relation to rifting, arc volcanism and tectonic uplift in back-arc basin of the western Pacific Ocean, in *Formation of Active Ocean Margins*, edited by N. Nasu et al., pp. 517-549, Terra Sci. Publ. Co., Tokyo, 1985.

Kneller, B., D. Edwards, W. McCaffrey, and R. Moore, Oblique reflection of turbidity currents, *Geology, 19*, 250-252, 1991.

Lallemand, S., and X. Le Pichon, Coulomb wedge model applied to subduction of seamounts in the Japan Trench, *Geology, 15*, 1065-1069, 1987.

Le Pichon, X., T. Iiyama, H. Chamley, J. Charvet, M. Faure, H. Fujimoto, T. Furuta, Y. Ida, H. Kagami, S. Lallemant, J. Leggett, A. Murata, H. Ogada, C. Rangin, V. Renard, A. Taira, and H. Tokuyama, Nankai Trough and fossil Shikoku Ridge: results from Box 6 Kaiko Survey, *Earth Planet. Sci. Lett., 83*, 186-198, 1987.

Lowe, D. R., Sediment gravity flows: II. Depositional models with special reference to the deposits of high-density turbidity currents, *J. Sediment. Petrol., 52*, 279-297, 1982.

Lüthi, S. M., and J. R. Banavar, Application of borehole images to three-dimensional geometric modeling of eolian sandstone reservoirs, Permian Rotliegende, North Sea, *Bull. Am. Assoc. Petrol. Geol., 72*, 1074-1089, 1988.

MacLeod, C.J., Structure of the outer Tonga forearc at ODP Site 841, in *Proc. ODP, Sci. Results*, vol. 135, J. F. Hawkins, L. M. Parson, J. F. Allan, et al., pp. 313-329, Ocean Drilling Program, College Station, TX, 1994.

Maillet, P., M. Monzier, M. Selo, and D. Storzer, The d'Entre-casteaux zone (southwest Pacific). A petrological and geochronological reappraisal, *Mar. Geol., 53*, 179-197, 1983.

Marjanac, T., Reflected sediment gravity flows and their deposits in flysch of middle Dalmatia, Yugoslavia, *Sedimentology, 37*, 921-929, 1990.

Marsaglia, K. M., Petrography and provenance of volcaniclastic sands recovered from the Izu-Bonin Arc, Leg 126, in *Proc. ODP, Sci. Results*, vol. 126, B. Taylor, K. Fujioka, et al., pp. 139-154, Ocean Drilling Program, College Station, TX, 1992.

Marsaglia, K. M., and R. V. Ingersoll, R.V., Compositional trends in arc-related, deep-marine sand and sandstone: A reassessment of magmatic-arc provenance, *Bull. Geol. Soc. Am., 104*, 1637-1649, 1992.

Marsaglia, K. M., R. V. Ingersoll, and B. M. Packer, Tectonic evolution of the Japanese islands as reflected in modal compositions of Cenozoic forearc and backarc sand and sandstone, *Tectonics, 11*, 1028-1044, 1992.

Marsaglia, K. M., X. Torres, I. Padilla, and K. Rimkus, Provenance of Pleistocene and Pliocene sand and sandstone, ODP Leg 141, Chile Margin, in *Proc. ODP, Sci. Results*, vol. 141, S. Lewis, J. Berhmann, et al., Ocean Drilling Program, College Station, TX, in press, 1994.

Mascle, A., J. C. Moore, et al., *Proc. ODP, Init. Repts.*, vol. 110, 603 pp., Ocean Drilling Program, College Station, TX, 1988.

Masuda, H., H. Tanaka, T. Gamo, W. Soh, and A. Taira, Major-element chemistry and alteration mineralogy of volcanic ash, Site 808 in the Nankai Trough, in *Proc. ODP, Sci. Results*, vol. 131, I. A. Hill, A. Taira, J. V. Firth, et al., pp. 175-183, Ocean Drilling Program, College Station, TX, 1993.

Minai, Y., R. Matsumoto, and T. Tominaga, Geochemistry of deep sea sediments from the Nankai Trough, the Japan Trench, and adjacent regions, in *Init. Repts. DSDP*, vol. 87, H. Kagami, D. E. Karig, W. T. Coulbourn, et al., pp. 643-657, U.S. Govt. Printing Office, Washington, D.C., 1986.

Moberly, R., Youthful oceanic lithosphere of marginal isles, western Pacific, *Proc. Pac. Sci. Cong., 12*, Canberra, 393, 1971.

Mogi, K., Seismicity before and after large shallow earthquakes around the Japanese islands, *Tectonophysics, 175*, 1-34, 1990.

Moore, D. M., and R. C. Reynolds, Jr., *X-ray Diffraction and the Identification and Analysis of Clay Minerals*, Oxford University Press, New York, 332 pp, 1989.

Moore, G. F., and D. E. Karig, Development of sedimentary basins on the lower trench slope, *Geology, 4*, 693-697, 1976.

Moore, G. F., H. G. Billman, P. E. Hehanussa, and D. E. Karig, Sedimentology and paleobathymetry of Neogene trench-slope deposits, Nias Island, Indonesia, *J. Geol., 88*, 161-80, 1980.

Moore, G. F., T. H. Shipley, P. L. Stoffa, D. E. Karig, A. Taira, S. Kuramoto, H. Tokuyama, and K. Suyehiro, Structure of the Nankai Trough accretionary zone from multichannel seismic reflection data, *J. Geophys. Res., 95*, 8753-8765, 1990.

Moore, G. F., D. E. Karig, T. H. Shipley, A. Taira, P. L. Stoffa, and W. T. Wood, Structural framework of the ODP Leg 131 area, Nankai Trough, in *Proc. ODP, Init. Repts.*, vol. *131*, A. Taira, I. Hill, J. Firth, et al., pp. 15-23, Ocean Drilling Program, College Station, TX, 1991.

Moore, J. C., Selective subduction, *Geology, 3*, 530-532, 1975.

Moore, J. C., and D. E. Karig, Sedimentology, structural geology, and tectonics of the Shikoku subduction zone, southwestern Japan, *Bull. Geol. Soc. Am., 87*, 1259-1268, 1976.

Moore, J. C., J. S. Watkins, K. J. Mcmillen, S. B. Bachman, J. K. Leggett, N. Lundberg, T. H. Shipley, J. Stephan, F. W. Beghtel, A. Butt, B. M. Didyk, N. Niitsuma, L. E. Shephard, and H. Stradner, Facies belts of the Middle America Trench and forearc region, southern Mexico: results from Leg 66 DSDP, in *Trench-Forearc Geology*, edited by J. K. Leggett, pp. 77-94, Geol. Soc. London Spec. Publ. 10, 1982.

Moore, J. C., A. Mascle, et al., *Proc. ODP, Sci. Results*, vol. 110, 448 pp., Ocean Drilling Program, College Station, TX, 1990.

Muck, M. T., and M. B. Underwood, Upslope flow of turbidity currents: A comparison among field observations, theory, and laboratory models, *Geology, 18*, 54-57, 1990.

Mutti, E., Distinctive thin-bedded turbidite facies and related depositional environments in the Eocene Hecho Group (south-central Pyrenees, Spain), *Sedimentology, 24*, 107-131, 1977.

Mutti, E., and W. R. Normark, Comparing examples of modern and ancient turbidite systems: problems and concepts, in

Marine Clastic Sedimentology, edited by J. K. Leggett and G. G. Zuffa, pp. 1-38, Graham and Trotman, London, 1987.

Mutti, E., and F. Ricci Lucchi, Le torbiditi dell'Apennino settentrionale: introduzione all'analisi di facies, *Mem. Soc. Geol. Italia, 11*, 161-199, 1972. (English translation by Nilsen, T. H., *Int. Geol. Rev., 20*, 1125-166, 1978.)

Niitsuma, N., Neogene tectonic evolution of southwest Japan, *Modern Geol., 12*, 497-532, 1988.

Nishimura, A., K. M. Marsaglia, K. S. Rodolfo, A. Colella, R. N. Hiscott, K. Tazaki, J. B. Gill, T. Janecek, J. Firth, M. Isiminger-Kelso, Y. Herman, R. N. Taylor, B. Taylor, K. Fujioka, and Leg 126 Scientific Party, Pliocene-Quaternary submarine pumice deposits in the Sumisu Rift area, Izu-Bonin Arc, in *Sedimentation in Volcanic Settings*, edited by R. V. Fisher and G. A. Smith, Soc. Econ. Paleontol. Mineral. Spec. Publ. 45, 201-208, 1991.

Oba, N., Emplacement of granitic rocks in the Outer Zone of southwest Japan and geologic significance, *J. Geol., 85*, 383-393, 1977.

Ojakangas, R. W., Cretaceous sedimentation, Sacramento Valley, California, *Bull. Geol. Soc. Am., 79*, 973-1008, 1968.

Otofuji, Y., A. Hayashida, and M. Torii, When was the Japan Sea opened?: Paleomagnetic evidence from southwest Japan, in *Formation of Active Ocean Margins*, edited by N. Nasu, S. Uyeda, I. Kushiro, K. Kobayashi, and H. Kogami, pp. 551-566, Terra Sci. Publ. Co., Tokyo, 1985.

Otsuki, K., Westward migration of the Izu-Bonin Trench, northward motion of the Philippine Sea Plate, and their relationships to the Cenozoic tectonics of the Japanese island arcs, *Tectonophysics, 180*, 351-367, 1990.

Packer, B. M., and R. V. Ingersoll, Provenance and petrology of Deep Sea Drilling project sands and sandstones from the Japan and Mariana forearc and backarc regions, *Sediment. Geol., 51*, 5-28, 1986.

Pantin, H. M., and M. R. Leeder, Reverse flow in turbidity currents: The role of internal solitons, *Sedimentology, 34*, 1143-1155, 1987.

Parson, L. M., J. W. Hawkins, J. F. Allan, et al., *Proc. ODP, Init. Repts.*, vol. 135, 677 pp., Ocean Drilling Program, College Station, TX, 1992.

Perry, E., and J. Hower, Burial diagenesis in Gulf Coast pelitic sediments, *Clays Clay Miner., 18*, 165-177, 1970.

Pickering, K. T., and R. N. Hiscott, Contained (reflected) turbidity currents from the Middle Ordovician Cloridorme Formation, Quebec, Canada: An alternative to the antidune hypothesis, *Sedimentology, 32*, 373-394, 1985.

Pickering, K. T., R. N. Hiscott, and F. J. Hein, *Deep Marine Environments, Clastic Sedimentation and Tectonics*, Unwin Hyman, London, 416 pp., 1989.

Pickering, K. T., M. B. Underwood, and A. Taira, Open ocean to trench turbidity current flow in the Nankai Trough: Flow collapse and reflection, *Geology, 20*, 1099-1102, 1992.

Pickering, K. T., M. B. Underwood, and A. Taira, Stratigraphic synthesis of DSDP-ODP sites in the Shikoku Basin, Nankai Trough, and accretionary prism, in *Proc. ODP, Sci. Results*, vol. 131, I. A. Hill, A. Taira, J. V. Firth, et al., pp. 313-330, Ocean Drilling Program, College Station, TX, 1993a.

Pickering, K. T., M. B. Underwood, and A. Taira, Open-ocean to trench turbidity-current flow in the Nankai Trough: Flow collapse and flow reflection, in *Proc. ODP, Sci. Results*, vol. 131, by I. A. Hill, A. Taira, J. V. Firth, et al., pp. 35-43, Ocean Drilling Program, College Station, TX, 1993b.

Piper, D. J. W., R. von Huene, and J. R. Duncan, Late Quaternary sedimentation in the active eastern Aleutian Trench, *Geology, 1*, 19-22, 1973.

Pouclet, A., K. Fujioka, J. Charvet, J., and J.-P. Cadet, Petrography and geochemistry of volcanic ash layers from Leg 87A, Nankai Trough (south Japan), in *Init. Repts. DSDP*, vol. 87, H. Kagami, D. E. Karig, W. C. Coulbourn, et al., pp. 695-701, U.S. Govt. Printing Office, Washington, D.C., 1986.

Pytte, A. M., and R. C. Reynolds, The thermal transformation of smectite to illite, in *Thermal History of Sedimentary Basins*, edited by N. D. Naeser and T. H. McCulloh, pp. 133-140, Springer-Verlag, New York, 1989.

Reid, R. P., S. N. Carey, T. S. Staerker, and M. Ask, Sedimentology of the collision zone between the North d'Entrecasteaux Ridge and New Hebrides island arc (Sites 827, 828, and 829), in *Proc. ODP, Sci. Results*, vol. 134, H. G. Greene, J.-Y. Collot, L. B. Stokking, et al., Ocean Drilling Program, College Station, TX, in press, 1994.

Ricci Lucchi, F., Depositional cycles in two turbidite formations of northern Apennines, *J. Sediment. Petrol., 45*, 1-43, 1975.

Ricci Lucchi, F., and E. Valmori, Basin-wide turbidites in a Miocene, over-supplied deep-sea plain: A geometrical analysis, *Sedimentology, 27*, 241-270, 1980.

Rodolfo, K. S., and R. J. Warner, Tectonic, volcanic, and sedimentologic significance of volcanic glasses from Site 450 in the eastern Parece Vela Basin, Deep Sea Drilling Project Leg 59, in *Init. Repts. DSDP*, vol. 59, L. W. Kroenke, R. W. Scott, et al., pp. 603-607, U. S. Govt. Printing Office, Washington, D.C., 1981.

Rodolfo, K. S., R. U. Solidum, A. Nishimura, Y. Matsuo, and K. Fujioka, Major-oxide stratigraphy of glass shards in volcanic ash layers of the Izu-Bonin arc-backarc sites (Sites 788/789 and 790/791), in *Proc. ODP, Sci. Results*, vol. 126, B. Taylor, K. Fujioka, et al., pp. 505-518, Ocean Drilling Program, College Station, TX, 1992a.

Rodolfo, K. S., A. Arguden, and R. U. Solidum, Textural characteristics of Oligocene submarine debris-flow deposits, Sites 792 and 793, in *Proc. ODP, Sci. Results*, vol. 126, B. Taylor, K. Fujioka, et al., pp. 97-99, Ocean Drilling Program, College Station, TX, 1992b.

Schmincke, H.-U., Ash from vitric muds in deep sea cores from the Mariana Trough and fore-arc region (South Philippine Sea) (Sites 453, 454, 455, 458, 459, and SP), Deep Sea Drilling Project Leg 60, in *Init. Repts. DSDP*, vol. 60, D. M. Hussong, S. Uyeda, et al., pp. 473-481, U.S. Govt. Printing Office, Washington, D.C., 1981.

Schweller, W. J., and L. D. Kulm, Depositional patterns and

channelized sedimentation in active eastern Pacific trenches, in *Sedimentation in Submarine Canyons, Fans, and Trenches*, edited by D. J. Stanley and G. Kelling, pp. 323-350, Dowden, Hutchinson, and Ross, Stroudsburg, PA, 1978.

Sclater, J. G., and P. A. F. Christie, Continental stretching: An example of the post-mid-Cretaceous subsidence of the central North Sea basin, *J. Geophys. Res.*, *85*, 3711-3739, 1980.

Seno, T., Age of subducting lithosphere and back-arc basin formation in the Western Pacific since the middle Tertiary, in *Formation of Active Ocean Margins*, edited by N. Nasu, S. Yueda, I. Kushiro, K. Kobayashi, and H. Kagami, pp. 469-481, Terra Sci. Publ. Co., Tokyo, 1985.

Seno, T., and S. Maruyama, Paleogeographic reconstruction and origin of the Philippine Sea, *Tectonophysics*, *102*, 53-84, 1984.

Shanmugam, G., Deep-marine facies models and the inter-relationship of depositional components in time and space, in *Deep Marine Sedimentation: Depositional Models and Case Histories in Hydrocarbon Exploration and Development*, edited by G. C. Brown, D. S. Gorsline and W. J. Schweller, pp. 199-246, Pacific Sect. Soc. Econ. Paleontol. Mineral., Spec. Publ. 66, 1990.

Shimamura, K., Topography and sedimentary facies of the Nankai deep sea channel, in *Sedimentary Facies in the Active Plate Margin*, edited by A. Taira and F. Masuda, pp. 529-556, Terra Sci. Publ. Co., Tokyo, 1989.

Shipboard Scientific Party, Explanatory notes, in *Proc. ODP, Init. Repts.*, vol. 126, B. Taylor, K. Fujioka, et al., pp. 51-60, Ocean Drilling Program, College Station, TX, 1990.

Shipboard Scientific Party, Site 808, in *Proc. ODP, Init. Repts.*, vol. 131, A. Taira, I. Hill. J. Firth, et al., pp. 71-269, Ocean Drilling Program, College Station, TX, 1991.

Sigurdsson, H., R. S. J. Sparks, S. N. Carey, and T. C. Huang, Volcanogenic sedimentation in the Lesser Antilles arc, *J. Geol.*, *88*, 523-540, 1980.

Soh, W., A. Taira, Y. Ogawa, H. Taniguchi, K. T. Pickering, and D.A.V. Stow, Submarine depositional processes for volcaniclastic sediments in the Mio-Pliocene Misaki Formation, Miura Group, central Japan, in *Sedimentary Facies in the Active Plate Margin*, edited by A. Taira and F. Masada, pp. 619-630, Terra Sci. Publ. Co., Tokyo, 1989.

Soh, W., K. T. Pickering, A. Taira, and H. Tokuyama, Basin evolution in the arc-arc Izu Collision Zone, Mio-Pliocene Miura Group, central Japan, *J. Geol. Soc. London*, *148*, 317-330, 1991.

Sparks, R. S. J., H. Sigurdsson, and S. N. Carey, The entrance of pyroclastic flows into the sea, I. oceanographic and geologic evidence from Dominica, Lesser Antilles, *J. Volcanol. Geotherm. Res.*, *7*, 87-96, 1980a.

Sparks, R. S. J., H. Sigurdsson, and S. N. Carey, The entrance of pyroclastic flows into the sea, II. theoretical considerations on subaqueous emplacement and welding, *J. Volcanol. Geotherm. Res.*, *7*, 97-105, 1980b.

Stewart, R. J., Neogene volcaniclastic sediments from Atka Basin, Aleutian Ridge, *Bull. Am. Assoc. Petrol. Geol.*, *62*, 87-97, 1978.

Stow, D. A. V., Deep clastic seas, in *Sedimentary Environments and Facies*, second edition, edited by H. G. Reading, pp. 399-444, Blackwell Scientific, Oxford, 1986.

Sugisaki, R., Major-element chemistry of argillaceous sediments at DSDP Sites 442, 443, and 444, Shikoku Basin, in *Init. Repts. DSDP*, vol. 58, G. deV. Klein, K. Kobayashi, et al., pp. 719-735, U.S. Govt. Printing Office, Washington, D.C., 1980.

Taira, A., and J. Ashi, Sedimentary facies evolution of the Nankai forearc and its implications for the growth of the Shimanto accretionary prism, in *Proc. ODP, Sci. Results*, vol. 131, I. A. Hill, A. Taira, J. V. Firth, et al., pp. 331-341, Ocean Drilling Program, College Station, TX, 1993.

Taira, A., and N. Niitsuma, Turbidite sedimentation in the Nankai Trough as interpreted from magnetic fabric, grain size, and detrital modal analysis, in *Init. Repts. DSDP*, vol. 87, H. Kagami, D. E. Karig, W. C. Coulbourn, et al., pp. 611-632, U. S. Govt. Printing Office, Washington, D.C., 1986.

Taira, A., J. Katto, M. Tashiro, M. Okamura, and K. Kodama, The Shimanto Belt in Shikoku, Japan—evolution of Cretaceous to Miocene accretionary prism, *Modern Geol.*, *12*, 5-46, 1988.

Taira, A., H. Tokuyama, and W. Soh, Accretion tectonics and evolution of Japan, in *The Evolution of the Pacific Ocean Margins*, edited by Z. Ben-Avraham, pp. 100-123, Oxford University Press, Oxford, 1989.

Taira, A., I. Hill, J. Firth, and Shipboard Scientific Party, Sediment deformation and hydrogeology of the Nankai Trough accretionary prism: synthesis of shipboard results of ODP Leg 131, *Earth Planet. Sci. Lett.*, *109*, 431-450, 1992.

Taylor, B., Rifting and the volcanic-tectonic evolution of the Izu-Bonin-Mariana Arc, in *Proc. ODP, Sci. Results*, vol. 126, B. Taylor, K. Fujioka, et al., pp. 627-651, Ocean Drilling Program, College Station, TX, 1992.

Taylor, B., K. Fujioka, et al., *Proc. ODP, Init. Repts.*, vol. 126, 1002 pp., Ocean Drilling Program, College Station, TX, 1990.

Taylor, B., K. Fujioka, et al., *Proc. ODP, Sci. Results*, vol. 126, 709 pp., Ocean Drilling Program, College Station, TX, 1992.

Taylor, F. W., Quaternary vertical tectonics of the central New Hebrides island arc, in *Proc. ODP, Init. Results*, vol. 134, J.-Y. Collot, H. G. Greene, L. B. Stokking, et al., pp. 33-42, Ocean Drilling Program, College Station, TX, 1992.

Terakado, Y., H. Shimizu, and A. Masuda, Nd and Sr isotopic variations in acidic rocks formed under a peculiar tectonic environment in Miocene southwest Japan, *Contrib. Mineral. Petrol.*, *99*, 1-10, 1988.

Thornburg, T. M., and L. D. Kulm, Sedimentation in the Chile Trench: Depositional morphologies, lithofacies and stratigraphy, *Bull. Geol. Soc. Am.*, *98*, 33-52, 1987.

Underwood, M. B., Sediment provenance within subduction complexes—an example from the Aleutian forearc, *Sediment. Geol.*, *51*, 57-73, 1986.

Underwood, M. B., Submarine canyons, unconfined turbidity

currents, and sedimentary bypassing of forearc regions, *Rev. Aquatic Sci.*, *4*, 149-200, 1991.

Underwood, M. B., and S. B. Bachman, Sedimentary facies associations within subduction complexes, in *Trench-Forearc Geology*, edited by J. K. Leggett, Geol. Soc. London Spec. Publ. 10, 537-550, 1982.

Underwood, M. B., K. T. Pickering, and A. Taira, Provenance and dispersal patterns for sediments in the turbidite wedge of Nankai Trough, in *Proc. ODP, Sci. Results*, vol. 131, I. A. Hill, A. Taira, J. V. Firth, et al., pp. 15-34, Ocean Drilling Program, College Station, TX, 1993a.

Underwood, M. B., K. Pickering, J. M. Gieskes, M. Kastner, and R. Orr, Sedimentary geochemistry, clay mineralogy, and diagenesis: A synthesis of data from Leg 131 Nankai Trough, in *Proc. ODP, Sci. Results*, vol. 131, I. A. Hill, A. Taira, J. V. Firth, et al., pp. 343-363, Ocean Drilling Program, College Station, TX, 1993b.

Vallier, T. L., R. M. O'Connor, D. W. Scholl, A. J. Stevenson, and P. J. Quinterno, Petrography of rocks dredged from the landward slope of the Tonga Trench: implications for middle Miocene volcanism and subsidence of the Tonga Ridge, in *Geology and Offshore Resources of Pacific Island Arcs—Tonga Region*, edited by D. W. Scholl and T. L. Vallier, pp. 109-120, Circum-Pacific Council for Energy and Mineral Resources, Earth Science Series, vol. 2, Houston, TX, 1985.

van Andel, T. H., Mesozoic/Cenozoic compensation depth and the global distribution of calcareous sediments, *Earth Planet. Sci. Lett.*, *26*, 187-194, 1975.

van Andel, T. H., and P. D. Komar, Ponded sediments of the Mid-Atlantic Ridge between 22 and 23 north latitude, *Bull. Geol. Soc. Am.*, *80*, 1163-1190, 1969.

Velbel, M. A., Mineralogically mature sandstones in accretionary prisms, *J. Sediment. Petrol.*, *55*, 685-690, 1985.

Velde, B., and A. Iijima, Comparison of clay and zeolite mineral occurrences in Neogene age sediments from several deep wells, *Clays Clay Miner.*, *36*, 337-342, 1988.

Velde, B., T. Suzuki, and E. Nicot, Pressure-temperature-composition of illite/smectite mixed-layer minerals: Niger Delta mudstones and other examples, *Clays Clay Miner.*, *34*, 435-441, 1986.

von Huene, R., and R. Cullotta, Tectonic erosion at the front of the Japan convergent margin, *Tectonophysics*, *160*, 75-90, 1989.

von Huene, R., and S. Lallemand, Tectonic erosion along the Japan and Peru convergent margins, *Bull. Geol. Soc. Am.*, *102*, 704-720, 1990.

von Huene, R., and D. W. Scholl, Observations at convergent margins concerning sediment subduction, subduction erosion, and the growth of continental crust, *Rev. Geophys.*, *29*, 279-316, 1991.

Vrolijk, P., On the mechanical role of smectite in subduction zones, *Geology*, *18*, 703-707, 1990.

Vrolijk, P., A. Fisher, and J. Gieskes, J., Geochemical and geothermal evidence for fluid migration in the Barbados accretionary prism (ODP Leg 110), *Geophys. Res. Lett.*, *18*, 947-950, 1991.

Walker, J. R., and G. R. Thompson, Structural variations in chlorite and illite in a diagenetic sequence from the Imperial Valley, California, *Clays Clay Miner.*, *38*, 315-321, 1990.

Weissel, J.K., and A. B. Watts, Tectonic complexities in the south Fiji marginal basin, *Earth Planet. Sci. Lett.*, *28*, 121-126, 1975.

White, S. M., H. Chamley, D. Curtis, G. deV. Klein, A. Mizuno, and D. M. Fountain, Sediment synthesis: Deep Sea Drilling Project Leg 58, Philippine Sea, in *Init. Repts. Deep Sea Drilling Project*, vol. 58, G. deV. Klein, K. Kobayashi, et al., pp. 963-1013, U.S. Govt. Printing Office, Washington, D.C., 1980.

Yamano, M., S. Honda, and S. Uyeda, Nankai Trough: A hot trench?, *Mar. Geophys. Res.*, *6*, 187-203, 1984.

Yamano, M., J.-P. Foucher, M. Kinoshita, A. Fisher, R. D. Hyndman, and ODP Leg 131 Shipboard Scientific Party, Heat flow and fluid flow regime in the western Nankai accretionary prism, *Earth Planet. Sci. Lett.*, *109*, 451-462, 1992.

M. B. Underwood, Department of Geological Sciences, 101 Geology Building, University of Missouri, Columbia, MO, 65211, U.S.A.

P. F. Ballance, Department of Geology, Private Bag 92019, University of Auckland, Auckland, New Zealand.

P. D. Clift, Ocean Drilling Program, 1000 Discovery Drive, Texas A&M University Research Park, College Station, TX, 77845-9547, U.S.A.

R. N. Hiscott, Earth Sciences Department, Memorial University, St. John's, Newfoundland, A1B 3X5, Canada.

K. M. Marsaglia, Department of Geological Sciences, University of Texas at El Paso, El Paso, TX, 79968, U.S.A.

K. T. Pickering, Department of Geological Sciences, University College London, Gower Street, London, WC1E 6BT, United Kingdom.

R. P. Reid, Rosenstiel School of Marine and Atmospheric Science, Departmenmt of Marine Geology and Geophysics, 4600 Rickenbacker Causeway, University of Miami, Miami, FL, 33149, U.S.A.

Paleomagnetic Rotations and the Japan Sea Opening

Laurent Jolivet

Laboratoire de Géologie, Ecole Normale Supérieure, Paris, France

Hidetoshi Shibuya

Department of Earth Sciences, University of Osaka Prefecture, Japan

Marc Fournier

Laboratoire de Géologie, Ecole Normale Supérieure, Paris, France

The apparent incompatibility of the classical "bar door" opening model of the Japan Sea based on paleomagnetic studies and the pull-apart geometry based on the observation of shear zones in and around Japan is symptomatic of a lack of understanding of the tectonic history of the Japan Sea. After a critical review of paleomagnetic data and a discussion on the possible error bars we present information concerning the internal strain of the Japan arc during the opening. We show that it is possible to integrate both sets of data in a single model, provided that some of the paleomagnetic rotations are due to distributed deformation of SW and NE Japan and not to a rigid body rotation of 600 km long blocks. Rotations occur at all scales and the 50° of clockwise rotation of southwest Japan is the sum of 30° of rotation of southwest Japan as a whole and of 20° due to internal deformation. The amount of paleomagnetic rotation is then compatible with the kinematics based on the internal structure of the Japan Sea and the geometry of major dextral shear zones. The rigid body rotation of southwest Japan is accommodated by extension and left-lateral motion along the Median Tectonic Line which is a second order fault between the two major dextral shear zones that bound the Japan Sea to the east and to the west. The problem of the apparent instantaneous rotation of southwest Japan as opposed to the progressive opening of the Japan Sea probably results from the way paleomagnetic data and radiometric dates are averaged.

PROBLEM SETTING

The past ten years have seen the development of a debate on the opening mechanism of the Japan Sea (Figure 1). An important breakthrough was made by Otofuji et al. [*Otofuji and Matsuda*, 1983, 1984, 1987; *Otofuji et al.*, 1985, 1991, 1994] with their "bar door" opening model based on the observation that SW and NE Japan show consistent easterly and westerly deflected paleomagnetic declinations. The rotation interval was assigned to a very short period, less than 2 m.y. long, about 15 Ma. The model is that of a double rotation about two nearby poles and a double fan-shaped opening of the Japan Sea. The rigid behavior of the two continental blocks, each of them being more than 500 km long, implies that oceanic crust formation in the Japan Sea was contemporaneous with the rotation. A rapidly opening model was then postulated in concert with rapid rates of motion of the rotating blocks.

Based on different data, *Lallemand and Jolivet* [1985] postulated that major dextral strike slip faults guided the opening of the Japan Sea in a pull-apart manner, in opposition with the fan-shaped kinematics of *Otofuji et al.* Later *Jolivet et al.* [1989, 1991] and *Jolivet and Tamaki* [1992] introduced block rotations in the pull-apart model, describing the whole region as a wide dextral shear zone (Figure 2).

Active Margins and Marginal Basins of the Western Pacific
Geophysical Monograph 88

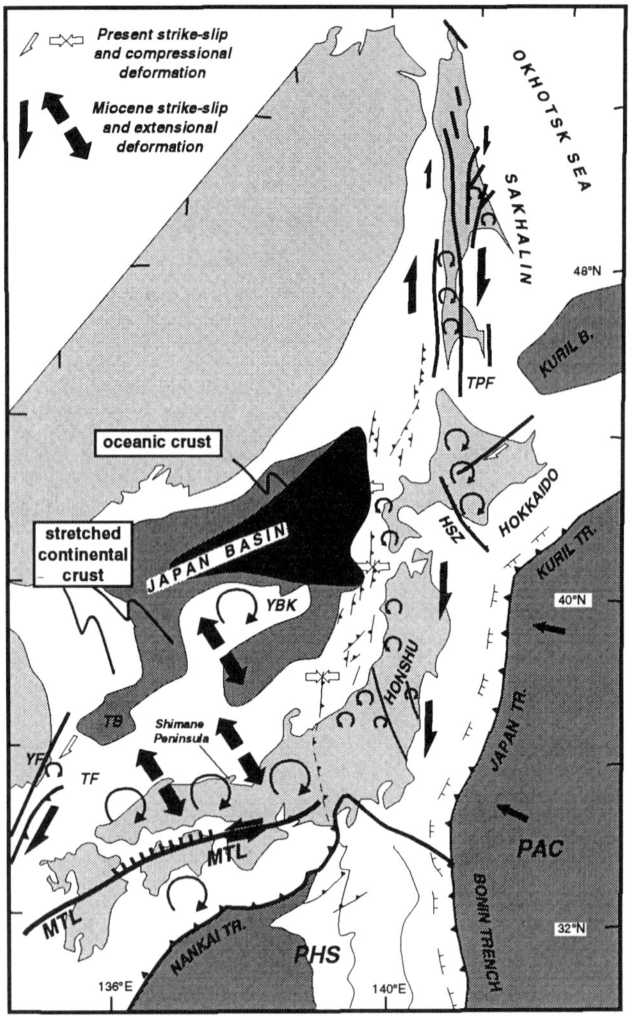

Fig. 1. Tectonic map of the Japan Sea area after *Jolivet and Tamaki* [1992]. Two successive stress fields are shown (Miocene and Present). Paleomagnetic rotations are indicated. YBK: Yamato Bank, TPF: Tym Poronaisk fault, TB: Tsushima basin, TF: Tsushima fault, YF: Yangsan fault, MTL: Median Tectonic Line, HSZ: Hidaka shear zone, PAC: Pacific Plate, PHS: Philippine Sea Plate. Oblique Mercator projection.

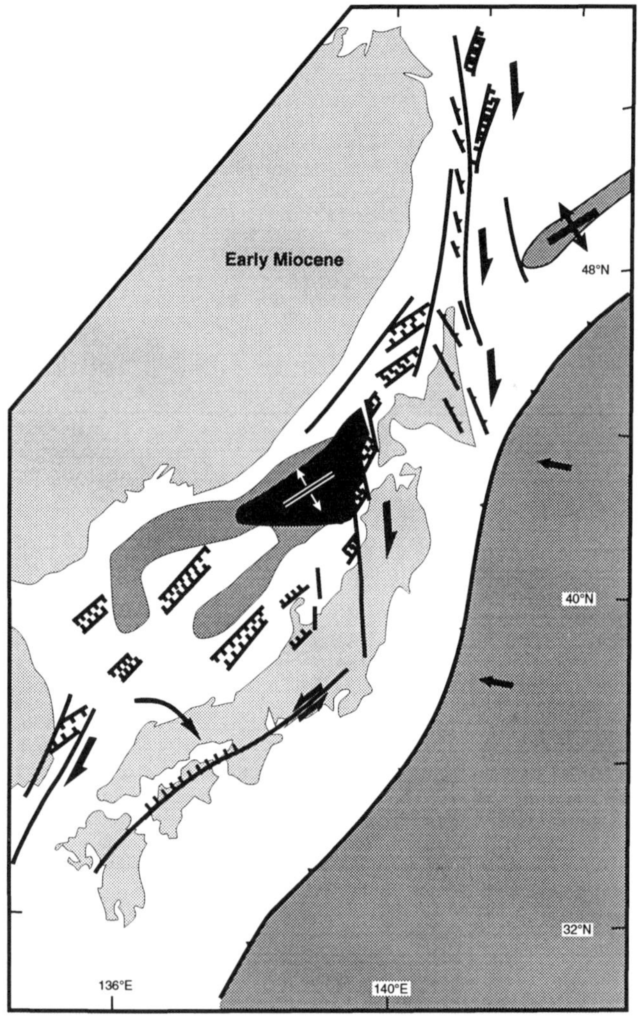

Fig. 2. Possible reconstruction of the Japan Sea in the early Miocene during the opening after *Jolivet et al.* [1992].

Even with these recent adaptations the strike-slip model remains apparently incompatible with paleomagnetic data because it suggests a progressive opening between 25 and 12 Ma (Figure 3), with progressive block rotations instead of the instantaneous ones in the fan-shaped model. Most geological and geophysical data, acquired on land as well as offshore, show that the Japan Sea did not open instantaneously and that it was already widely opened by 15 Ma [*Tamaki*, 1988; *Tamaki et al.*, 1992]. Recent deep sea drilling at several sites in the Japanese waters of the Japan Sea reached the volcanic basement, which is everywhere older than 15 Ma, and subsidence curves show a maximum during the early Miocene [*Ingle* 1992]. Previous geophysical investigations had also shown that heat flow and depth distribution in the basin are compatible with an early Miocene opening [*Tamaki*, 1986, 1988].

We start this paper with the assumption that both paleomagnetic data and other geological data are relevant and that the apparent inconsistency is symptomatic of a lack of understanding of the geometry and timing of opening, and we try to reconcile both sets of data in a single model. We first present a critical review of published paleomagnetic data putting the emphasis on the length of error bars. We then present observations on the way the Japan arc deformed during the opening of the Japan Sea and on the geometrical relationships between this

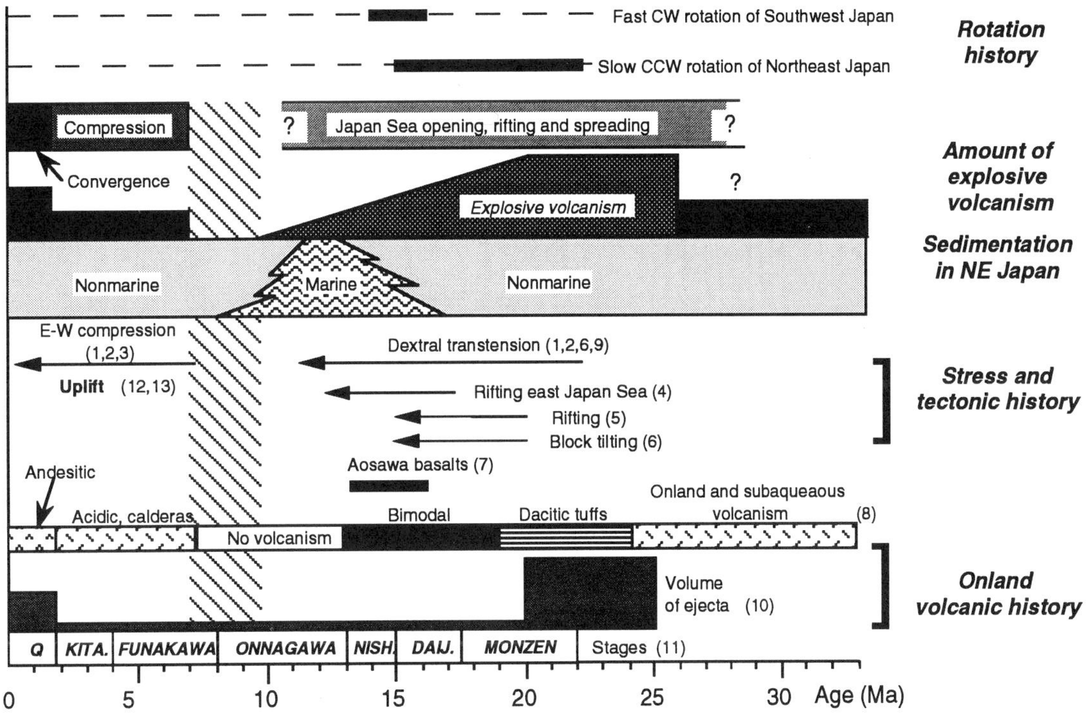

Fig. 3. Synthesis of the tectonic and volcanic events in northeast Honshu after *Jolivet and Tamaki* [1992] compared to the timing of large scale paleomagnetic rotations of NE and SW Japan discussed in text: (1) *Jolivet and Huchon* [1989], (2) *Jolivet et al.* [1991], (3) *Nakamura and Uyeda* [1980], (4) *Suzuki* [1989], (5) *Yamaji* [1989, 1990], (6) *Amano and Sato* [1989], (7) *Tsuchiya* [1989, 1990], (8) *Usuta* [1989], (9) *Otsuki* [1989], (10) *Sugimura et al.* [1963], (11) *Fujioka* [1986], (12) *Sugi et al.* [1983], and (13) *Iijima and Tada* [1990].

deformation and internal structures of the Japan Sea. We emphasize the fact that deformation occurred at various scales, and that distributed strain of the SW Japan and NE Japan blocks might partly explain the observed rotations. We finally propose a geometry of opening which partly reconciles both data sets, and we discuss points that do not fit in this scheme and possible explanations for these remaining discrepancies.

PALEOMAGNETIC CONSTRAINTS

The counterclockwise rotation of NE Japan and the clockwise rotation of SW Japan (Figure 4) were discovered in the early works on paleomagnetism in Japan [*Kawai et al.*, 1961, 1969]. Kawai and his colleagues measured remanent magnetization of granitic rocks in NE and SW Japan. Their material would be considered inappropriate by today's standards, but these were the only rocks systematically dated in those days with a strong enough magnetization for the magnetometer they used.

This work, which was summarized in *Kawai et al.* [1971], suggested that the Japanese islands had been bent between 120 and 80 Ma. No link was proposed with the opening of the Japan Sea. These results were not easily accepted because the material was not suitable for paleomagnetic tests. Because all samples were Cretaceous in age they all showed normal polarity. Their magnetization was not usually strong and stable enough for the demagnetization procedures in use in the sixties.

Yaskawa [1979] first proposed a possible correlation between the observed rotations and the opening of the Japan Sea, but the data he gathered did not unambiguously constrain the age of rotation.

Although Ito and his colleagues [*Ito and Tokieda*, 1980; *Ito et al.*, 1980] also considered a possible link between the observed rotations and the Japan Sea opening, it was not until the work of Otofuji and colleagues that the rotation of SW Japan and the opening of the Japan Sea were linked in the same mechanism and dated as middle Miocene. The first paper appeared in 1983 [*Otofuji and Matsuda*, 1983]. A thermal demagnetization technique was applied to acidic volcanic rocks which were not suitable for alternating field demagnetization because of a slight metasomatism. They also dated the samples and found that the clockwise

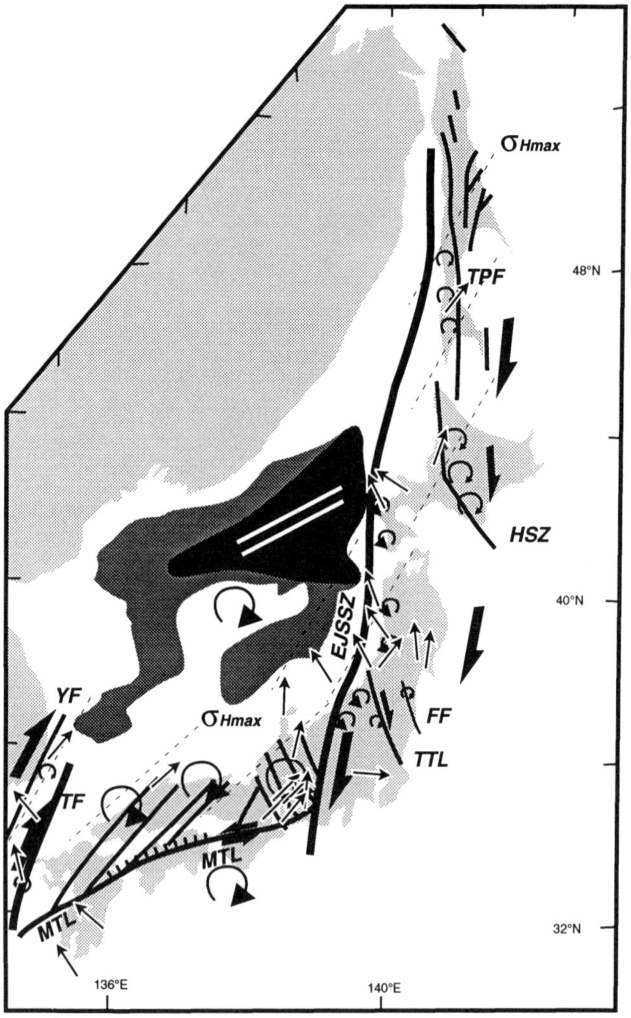

Fig. 4. Major faults and sense of block rotations around the Japan Sea. Most rotations are deduced from paleomagnetic studies (see text for references) except for those of NE Sakhalin which are suspected by *Fournier et al.* [1994] but not demonstrated and of the Yamato Bank which is deduced from *Jolivet et al.* [1991] kinematics. Also shown is the distribution of crustal blocks suspected by *Kanaori* [1990] in SW Japan and the direction of the maximum horizontal stress after *Jolivet et al.* [1991, 1992] and *Charvet et al.* [1992]. Paleomagnetic remanent directions were compiled from *Kodama et al.* [1991], *Kodama and Nakayama* [1993], *Ishikawa and Tagami* [1991], *Ishikawa et al.* [1989], *Kim et al.* [1986], *Otofuji et al.* [1985, 1991, 1994], *Hayashida and Ito* [1984], *Hayashida* [1986], *Hayashida et al.* [1991], *Itoh* [1988], *Hirooka et al.* [1986], *Hyodo and Niitsuma* [1986], *Yamazaki* [1989], *Tosha and Hamano* [1988], *Kodama et al.* [1993], *Takeuchi et al.* [1992]. Counterclockwise rotations in south Kyushu are post late Miocene.

rotation was as young as 20 Ma. Subsequent studies [*Otofuji and Matsuda*, 1984; *Hayashida et al.,* 1991] on sediments precisely dated by bio- and magnetostratigraphy suggested that the rotation occurred within a short period around 15 Ma. *Otofuji et al.* [1985] later found that the counterclockwise rotation of NE Japan was as young as that of SW Japan. Later works sought to constrain more precisely the timing and amount of rotation of NE and SW Japan.

Fast Rotation of SW Japan at 15 Ma

The study that most precisely constrains the timing of rotation was published by *Hayashida et al.* [1991] and *Hayashida and Ito* [1984]. They studied sediments of the Setouchi Miocene basins, distributed on the northern side of the Median Tectonic Line (MTL), which are dated with biostratigraphic and magnetostratigraphic techniques. Their data indicate that the rotation occurred after 16 Ma. Slightly younger sediments of the Morozaki area, in Blow's N9 foraminiferal zone (younger than 15 Ma) have smaller clockwise deflections, though the normal polarity leaves open the possibility that they were recently remagnetized. A similar chronology is found in the northern Chubu district (Yatsuo area) where *Itoh* [1988] also supports a smaller rotation within the *Denticulopsis Lauta* zone (younger than 15.7 Ma). Both studies found no significant rotation between 20 and 15 Ma. K-Ar dating by *Otofuji and Matsuda* [1983] suggested that most rotation occurred around 15 Ma within a period shorter than can be presently resolved by the available dating techniques.

Some regions of SW Japan nevertheless show different amounts of rotation. The Hokuriku and Chichibu regions are located on both sides of the MTL, which makes a sharp bend from an ENE trend to a more northerly direction west of the collision zone of central Japan. The bend is often attributed to the collision of the Izu arc, assuming that the MTL was straight before the collision. Paleomagnetic data in this region are consistent with this interpretation [*Itoh*, 1988; *Hyodo and Niitsuma*, 1986]. The change in rotation angle is consistent with this model, which assumes that the whole of SW Japan was first rotated clockwise uniformly, and that later its eastern part was rotated back CCW during the indentation of central Japan by the Izu arc. The rotation angle is calculated for the main part of SW Japan by *Otofuji et al.* [1985]. They selected reliable paleomagnetic data with the following criteria: (1) good age assignment, (2) demagnetization with both thermal and alternating field techniques, (3) small 95% confidence circle (<30°), and (4) data correction for tilt. Fitting a function on the selected data gave a finite rotation of 47°. However the estimation

TABLE 1. Paleomagnetic Dpata

Formation	Lat	Long	Age	D	Dcorr	I	N	a	Ref.
SW Japan									
Yoboshiyama	34.6	131.9	33	46.4	36.7	34	2	8.2	1
Hirefuriyama	34.7	131.9	33	49.1	39.4	47	4	21.2	1
Okami	34.8	131.9	33	64.8	55.1	61	3	16.5	1
Hamada	34.9	132.1	32	78.5	68.8	38	5	16.6	1
Harifuku	35	132.2	35	86.3	76.6	46	4	38.1	1
Kawamoto	35	132.5	33	61.6	51.9	53	7	6.4	1
Kawauchi	35	132.5	28	52.2	52.2	34	6	15.3	1
Hata	35.1	132.5	21	69.9	64.3	50	6	14.5	1
Ichishi	34.7	136.4	16.4	44.6	44.6	47	11	11.9	2
Morozaki	34.7	136.8	16	38.2	38.2	56	6	12.7	3
Kani	35.4	137	16.3	-22.2	49.9	53	6	9.7	4
N. Shidara	35.1	137.6	16	47	47	47	4	14	5
NE Japan									
H. Ohtorigawa	38.5	139.8	21	-27.5	-27.5	51	1	...	6
Okatsugawa	39	140.4	25	-40.5	-40.5	54	5	11.6	6
Yunosawagawa	39.1	140.4	33	-43.5	-53.3	63	3	21.4	6
Hatamura	39.2	140.3	21.9	-43.4	-43.4	65	4	24.2	6
Daijima	39.9	139.8	20.8	-45	-45	51	6	9.8	7
Monzen	39.9	139.7	35	-38.3	-48.1	54	6	9.7	6
Nisatai	40.3	141.3	21.8	-78.8	-78.8	43	2	41.2	6
Gongenzaki	41.1	140.3	E.Mio	-64.9	-64.9	56	3	27.7	6
Kunitomi	43	140.9	27.1	-64.2	-64.2	65	2	53.5	6
Fukuyama	46.4	140.2	23	-87.1	-87.1	43	2	2.6	6, 7

Table 1: Paleomagnetic data used in this paper to calculate the mean value of rotation of SW and NE Japan. Lat: Latitude, Long: Longitude, D: declination, Dcorr: declination corrected for the polar wander after *Otofuji et al.* (1985). 1: *Otofuji and Matsuda* [1983, 1984], 2: *Hayashida and Ito* [1984], 3: *Hayashida* [1986], 4: *Hayashida et al.* [1991], 5: *Torii* [1983], 6: *Otofuji et al.* [1985, 1994], 7: *Tosha and Hamano* [1988].

of uncertainties is difficult with this method so we used simple means for paleomagnetic data older than 16 Ma, including some recent data [*Hayashida et al.*, 1991; *Otofuji et al.*, 1991] (Table 1). We obtain a rotation angle of 50.9 ± 7.3° (95% confidence limit).

In contrast to NE Japan the rigid rotation of SW Japan as a single block has been widely accepted because of the limited variations of the paleomagnetic directions and mainly because the geological provinces arranged along the arc south of the MTL are straight and do not show an obvious division into smaller blocks. The only regions with a different trend of magnetization are located near the collision zone as discussed above. The whole of SW Japan has been sampled for paleomagnetic measurement from the forearc region of the Shimanto belt [*Tagami*, 1982] to the backarc side near the Japan Sea. It is difficult to divide SW Japan into smaller blocks without violating the apparent simple arrangement of the area.

Ambiguous Timing of the Rotation of NE Japan Between 22 and 15 Ma

Fewer systematic studies of the paleomagnetism of NE Japan are available in the literature than for SW Japan [*Otofuji et al.*, 1985; *Tosha and Hamano*, 1986, 1988]. These works indicated that counterclockwise rotation occurred sometime between 22 and 15 Ma. *Tosha and Hamano* [1986] and *Tanaka et al.* [1991] showed that the rotation had been completed before 15 Ma. *Yamazaki* [1989] collected samples from the east coast of Honshu and showed that the Matsushima formation (older than 15.7 Ma) has not been rotated. However, despite the observation of both normal and reverse polarities, a secondary remagnetization is still possible. Confusing paleomagnetic results are obtained in the Gongenyama formation 100 km to the northeast [*Momose et al.*, 1990]. The authors concluded that the rotation was not finished before N9 and N10 foraminiferal zones. The result is ascertained through the presence of both reverse and normal polarities and a slight improvement of clustering of data after tilt correction. The remanent magnetic direction close to the present field however leaves the possibility that the demagnetization has not removed all secondary magnetizations. Several other works also suggest that the rotation occurred before 15 Ma and after 22 Ma [*Tosha and Hamano*, 1988]. This makes the rotation older than that postulated by *Otofuji and Matsuda* [1984] for SW Japan. In a more recent paper *Otofuji et al.* [1994] proposed that NE Japan (NE Honshu and SW Hokkaido) rotated counterclockwise as a single block with a "climax" at 15 Ma.

The rotation angle is more difficult to determine by comparison to SW Japan. If we take the mean of data with ages between 20 and 40 Ma from *Otofuji et al.* [1985] and *Tosha and Hamano* [1988] (Table 1), the declination would be -56.1 ± 10.6° (95% confidence limit).

The Double Rotation Model and Japan Sea Opening

Despite a less precisely constrained timing of rotation in NE Japan, *Otofuji et al.* [1985] favored a model with rapid opening of the Japan Sea by rotation of NE and SW Japan about two poles located near the northern and southern ends of the Japan arc in a very short period about 15 Ma. The almost instantaneous rotation leads to implausibly rapid rates of spreading near the hinge between SW and NE Japan (more than 60 cm/year).

Until ODP legs 127 and 128 [*Tamaki et al.*, 1992], there had been no precise dating of oceanic spreading in the Japan Sea. The interval of rifting was rather well constrained by various studies on shore and offshore along the eastern margin in NE Japan but the oceanic basement had never been directly dated. Only in the northern part of the Japan Sea had poorly defined magnetic anomalies been recently discovered [*Kobayashi et al.*, 1988], which suggest an early Miocene spreading older than the proposed fast rotation of SW Japan. The volcanic basement was penetrated in four sites during the two drilling legs, and all results suggest that the Japan Sea was widely opened before 15 Ma. There is an apparent contradiction between the observed age of opening and paleomagnetic data that suggest rapid rigid rotations of two large blocks.

As we shall see later, it is possible to reconcile the amount and sense of paleomagnetic rotations with the observed deformation of the Japan arc, but not the timing of rotation. Did the rotation actually occur in this short time span, or is it possible that the error on the rotation of SW Japan is larger than usually considered?

Otofuji et al. [1991] give error estimates on the radiometric dating they performed in San'in district. According to them the rotation occurred within a short period of 1.8 ± 1.5 m.y., dated by the non-rotated Omori formation above and the rotated Kawai formation below. The maximum interval for the rotation is 3.3 m.y., which is still shorter than the duration of opening. Ages given for the two formations are means calculated with 8 and 15 different samples coming from various places in each formation. Is it reasonable to calculate a mean age when the individual ages of samples vary widely (from 13.7 ± 0.3 to 18.3 ± 0.4 Ma for the Kawai formation)? If the analytical error is small, there is no reason to deny the obtained ages and to prefer a mean value. Assuming that the mean value represents the age of the Kawai formation is equivalent to supposing that this formation was deposited instantaneously, which is very unlikely. Lithologic features of the Kawai formation are quite variable and its deposition could have lasted longer than supposed by the authors. If the rotation actually occurred in the time interval defined by the two formations, this interval should be defined by the youngest age found in the Kawai formation and the oldest age found in the Omori formation, and not between two mean ages. Furthermore some of the ages obtained in the underlying Kawai formation are younger (13 Ma) than the oldest obtained in the overlying Omori formation. Because some of these ages violate stratigraphic relations, why use them to calculate a mean value?

The authors also calculate average virtual geomagnetic poles (VGP) for each formation and give a 95% confidence interval. If block rotations occurred at a small scale, there could be significant local differences between the rotations of individual sites. As suggested by *Kanaori* [1990] on the basis of field observation in Shimane Peninsula and discussed in a later section of this paper, rotation of smaller scale blocks occurred at the same time as the fast rotation of southwest Japan. What is the meaning of an average VGP in this context? Calculating it, one assumes that the block has rotated rigidly, which is not proven a priori. In their recent paper *Otofuji et al.* [1994] compare the 95% confidence intervals of NE Honshu and SW Hokkaido (considered a priori as two rigid blocks) and conclude that the rotations are statistically identical and thus that the two regions rotated as a single rigid block. These statistics in fact show that the data do not allow separation of the two sets of data, but the inference that they rotated as a single rigid block is only one possible interpretation.

There could thus be an answer to our problem in the way data are treated statistically. If so, many more samples and a much denser sampling procedure is the only way to answer the question of timing and amount of rotation of a complex region; the sampling must also be done in connection with a good structural map where shear zones and other block boundaries are identified. A "statistically rigid" block might be so only because there is no other simple way to consider the numerical data without a priori information on the structure of the concerned region.

An alternative solution is to consider that the observed rotations partly correspond to the internal deformation of the Japan arc. In a classical domino system with a simple geometry crustal blocks rotate by the same amount and it might be impossible to see a difference in the distribution of paleomagnetic rotations between a single rigid block and several dominoes bounded by parallel strike-slip faults. *Kanaori* [1990] suggested that some of the rotation of SW Japan might be accounted for by rotation of large-scale dominoes (Figure 4).

We shall first see that some paleomagnetic data already show that local rotations occur along major shear zones, and then identify these and the strain regime they imply for the Japan arc.

Local Paleomagnetic Rotations along Shear Zones

NE Japan. There are many reports of paleomagnetic directions anomalous with respect to the rigid rotation model of NE Japan. *Otofuji et al.* [1985] themselves reported easterly deflected directions from eastern Asahi, which they attributed to the influence of the Tanakura tectonic line (TTL). *Oda et al.* [1989] observed a 20-30° clockwise rotation south of Sendai city, in the Yanagawa and Takadate area, dated at 15 Ma. They attributed these rotations to drag along the Futaba fault which runs parallel to the TTL in the Abukuma massif. Regional variations of the declination were already discussed based on data from granitic rocks [*Ito and Tokieda*, 1986; *Kawai et al.*, 1971]. It is thus possible that the rotation did not occur with a single rigid block. The basement of NE Japan is also less continuous than in SW Japan. However the data available do not define the size and boundaries of rotating blocks.

Tsushima Strait and Korea peninsula. Two sets of paleomagnetic data show local Miocene rotations. One is the clockwise rotation (45°) of the Guryongpo area on the east coast of the Korean peninsula [*Kim et al.*, 1986] and the second is a counterclockwise rotation of Tsushima island (30°) [*Ishikawa et al.*, 1989]. Both regions are located next to large shear zones and the authors attribute these anomalous direction to local strain. It is noteworthy that the two areas rotated in opposite sense despite the fact that they are located on opposite sides of the same shear zone.

Sakhalin and Hokkaido. Kodama et al. [1993] and *Takeuchi et al.* [1992] analyzed samples from Hokkaido and Sakhalin along the major dextral shear zone that bounds the Japan Sea to the east (east Japan Sea shear zone, EJSSZ). In Sakhalin they obtained reliable measurements on the west side of the Tym-Poronaisk fault, which is a major dextral dislocation that runs N-S for more than 600 km [*Rozhdestvenskiy*, 1982]. Clockwise rotations are observed consistently in the whole area, which the authors attribute to distributed deformation with a simple domino model. There do not seem to be discrete shear zones dividing crustal blocks west of the Tym-Poronaisk fault but rather en echelon folds whose axes are curved clockwise near the fault [*Fournier et al.*, 1994]. The dextral rotation can then be explained as well with a more continuous deformation model.

In a recent paper *Kodama et al.* [1993] describe the results of a sampling campaign in Hokkaido and they show very similar results with dextral rotations, which they attribute to the dextral motion along the axial zone of Hokkaido. The dextral rotations seem to last until the early late Miocene which is later than the end of the Japan Sea opening.

During the opening of the Japan Sea, major strike-slip shear zones were active such as those in central Hokkaido and Sakhalin, and older strike-slip faults such as the MTL or TTL were reactivated. We describe in the following the geometry of deformation of the arc during the opening seen from the view point of the structural geologist and see how it is related to the geometry of opening.

DEFORMATION OF THE JAPAN ARC DURING THE JAPAN SEA OPENING

Oceanic crust occurs in only a restricted area in the northern part of the Japan Sea (Figure 1). Magnetic anomalies identified by *Kobayashi et al.* [1988] suggest that it dates back to the early Miocene. The geometry of anomalies also suggests that the spreading center had been propagating westward during the opening [*Tamaki et al.*, 1992; *Jolivet and Tamaki*, 1992]. This inference is supported by the general triangular shape of the area where oceanic crust is distributed with a wide base along the EJSSZ and a tip further west in the Russian waters. In other regions of the Japan Sea, geophysical surveys have shown that no true oceanic crust is present even in deep basins. The seismic velocity structure of the Tsushima and Yamato basins suggest that they are floored by a thinned continental crust intruded by numerous basaltic sills [*Tokuyama et al.*, 1987]. True continental crust, similar to that of the Japan arc, is found under the Yamato Bank, which has been stranded during the southeastward drift of SW Japan. The Japan Sea can thus be divided into two very different regions. The northern part is characterized by the presence of oceanic crust and a very narrow margin on the Japanese side. Little is known on the Siberian side, but bathymetry also suggests a narrow margin. On the other hand the western part of the Japan Sea was formed by distributed intra-continental extension.

Major Dextral Strike-Slip Shear Zones

As described in several earlier papers [*Lallemand and Jolivet*, 1985; *Jolivet et al.*, 1991; *Jolivet et al.*, 1992; *Fournier et al.*, 1994] the eastern margin of the Japan Sea is paralleled by a major dextral shear zone which trends N-S from the northern tip of Sakhalin to central Japan (Figures 1 and 2). The most obvious zones of strain localization within this shear zone are the Hidaka shear zone (central Hokkaido) and the Tym-Poronaisk fault (central Sakhalin). The Hidaka shear zone runs along 200 km in southern Hokkaido and shows a complete crustal section from brittle to ductile structures compatible with a dextral shear with a component of westward thrusting. Farther north and west, en echelon folds and thrusts [*Kimura et al.*, 1983] form part of the same dextral shear. Radiometric dating of the high temperature metamorphism contemporaneous with the dextral motion, as well as paleostress tensor analysis in the external domains where biostratigraphy allows dating of the successive tectonic events, suggests that the dextral motion has been active from the late Oligocene to the middle Miocene [*Jolivet and Huchon,* 1989]. As discussed above, recent paleomagnetic data [*Kodama et al.*, 1993] in central Hokkaido suggest that the dextral motion might have lasted until the late Miocene though different interpretations of the same data are still probably possible (see discussion in *Jolivet*, 1994). The dextral strike-slip motion stopped sometime in the late Miocene, and the Hidaka shear zone was reworked as a thrust fault, in the hanging wall of which the Hidaka mountains were recently uplifted.

The Tym-Poronaisk fault [*Rozhdestvenskiy*, 1982] divides the island of Sakhalin into two narrow stripes. The western side shows en echelon folds in Cretaceous and early Cenozoic sediments similar to those of Hokkaido which suggest a component of dextral shear. Tectonic analysis of the Tym-Poronaisk fault itself shows that it is made of N-S trending strike-slip segments alternating with NW-SE trending thrust faults. Paleostress analysis near the fault suggests that a transpressional regime has been active during the Miocene until recently; the abrupt change in stress regime seen in Hokkaido is not obvious here though the dextral shear seems less active now [*Jolivet et al.*, 1992, *Fournier et al.*, 1994]. The East Sakhalin mountains are cut by several NE-trending dextral strike-slip faults which branch from the Tym-Poronaisk fault like R-Riedel shears. Counterclockwise rotation of the stripes in between are inferred by *Fournier et al.* [1994], but no paleomagnetic data are yet available to test this hypothesis. A major N-S trending strike-slip fault runs offshore along the western margin of Sakhalin and it led to the formation of second order pull-apart basins in the Tatar Strait.

The transpressional dextral shear zone of Sakhalin and Hokkaido is more than 1000 km long. It is relayed at the latitude of SW Hokkaido by a narrow zone of transtensional dextral shear parallel to the margin of the Japan Sea. En echelon grabens and dextral strike-slip faults are seen offshore. Paleostress analysis in this region from Oga Peninsula to Sado island and Noto peninsula [*Jolivet et al.*, 1991, *Fournier et al.*, 1994] shows that the direction of the extensional principal stress has been oblique on the margin with a NW-SE trend. It is seen in rocks older than the late Miocene. It is then replaced by the E-W compression that is active today, as attested to by the frequent compressional tsunamigenic earthquakes on this side of the Japan arc [*Nakamura,* 1983; *Fukao and Furumoto,* 1975; *DeMets*, 1992]. This zone of transtensional dextral shear divides the NE Japan arc from the oceanic crust of the northern Japan Sea and has been active during its spreading. The Japan Sea is thus bordered on its eastern side by a major dextral shear zone, more than 2000 km long including the transpressional and the transtensional domains, which was active from the late Oligocene to the middle or late Miocene and is still active in the north.

A second dextral strike-slip shear zone is known along the southeastern coast of Korea and in the Tsushima Strait [*Otsuki and Ehiro*, 1978]. The NE-trending Yangsan fault offsets the Cretaceous Bulgugsa granites by several tens of kilometers. It also offsets metallogenic belts of Korea, which are then offset again dextrally on the Japanese side of the Tsushima Strait [*Sillitoe*, 1977]. The direction of early Miocene basaltic dykes intruding the Miocene Pohang basin is compatible with dextral motion along the Yangsan fault [*Lee*, 1988]. Paleostress analysis in the Cretaceous Gyongsan basin and the Miocene Pohang Basin shows two successive stages of dextral motion during the late Oligocene and early Miocene [*Hwang*, 1992]. This shear zone was thus active while the Japan Sea was opening.

The geology of Tsushima island is puzzling in this context. It clearly shows en echelon folds and thrusts formed inside a left-lateral shear zone parallel to the dextral one previously described. Recent observations by *Fabbri and Charvet* [1994] suggest that this compressional deformation occurred more recently than the Japan Sea opening (from 15 Ma onward) and was preceded by a phase of dextral shear. What is more puzzling is that focal mechanisms of earthquakes in the western part of the Japan Sea [*Jun*, 1990] are compatible with a mixture of shortening perpendicular to the Yangsan fault and dextral strike-slip. Similar mechanisms are seen further inland in the gulf of Bohai [*Chen and Nabelek*, 1988]. The left-lateral event is thus bracketed between two episodes of right-lateral motion.

Reactivated Shear Zones

Formed initially as a left-lateral strike-slip fault in the Late Cretaceous and early Cenozoic the Tanakura tectonic line was reactivated as a dextral shear zone during the Miocene [*Otsuki and Ehiro*, 1978] though the Miocene offset is probably small. Parallel faults cut through the basement of the Abukuma massif such as the Futaba fault.

The Median Tectonic Line is a very large shear zone which runs E-W between the inner and outer zones of SW Japan [*Ichikawa*, 1980]. It was most active during the Late Cretaceous and early Cenozoic as a left-lateral fault. Ductile deformation is observed on its northern side in the granitoids of the Cretaceous Ryoke belt. It is active today as a dextral strike-slip shear zone that accommodates the dextral component of the oblique subduction of the Philippine Sea Plate under SW Japan.

Very little is known on the behavior of the MTL during the Miocene. *Jolivet et al.* [1989] suggested that it had already reversed its motion to a dextral shear because of the apparent dextral torsion of paleomagnetic directions near the fault. Recent observations of trend of Miocene dikes by *Takeshita* (oral communication, 1990) on the contrary suggest that it was left-lateral around 15 Ma. *Fournier*'s [1994] recent observations confirm Takeshita's findings. Not only do the trends of Miocene basaltic dikes imply a left-lateral component of motion, but also the fault set analysis we performed in Miocene basins distributed along the fault is compatible with this sense of motion with an additional component of extension perpendicular to the fault. Observation in the west (Shikoku) shows the MTL as a normal fault while in the east (Shidara basin) it has a significant left-lateral component. No evidence is known for the amount of relative left-lateral displacement during the Miocene

Distributed Deformation in SW Japan

The rotations of dominoes can produce an apparently homogenous strain field if only rotations are taken into account and the dominoes are not identified. The linearity of the MTL and other boundaries between the belts of SW Japan precludes major internal strain in a first approach. Linearity is, however, a characteristic of strike-slip shear zones and, because outcrops of SW Japan are far from continuous, some important shear zones bounding dominoes might have been missed by geologists. Already *Kanaori* [1990] recognized large-scale fault-bounded blocks in SW Japan.

To approach this problem we surveyed the San'in district (central SW Japan on the backarc side) where *Otofuji et al.* [1991] have recognized a fast clockwise rotation at about 15 Ma. The deformation is very different depending upon the observed stratigraphic level. Miocene strata older than 15 Ma show both normal and strike-slip faults and the inferred direction of extension varies from WNW-ESE to NW-SE. Shimane peninsula (Figures 1 and 5) is the area where *Otofuji et al.* [1991] obtained some of the measurements indicating a fast rotation of SW Japan at 15 Ma. *Fournier* [1994] has performed paleostress analysis using fault sets with *Angelier*'s [1984] method. The paleostress tensors are shown on Figure 5. A mixture of strike-slip and extensional tensors is observed plus some additional compressional ones that are related to the recent compression. The mean direction of extension (Figure 5) is NW-SE, which is compatible with the general dextral pull-apart geometry. Figure 6 shows a simplified block diagram of a measured site on the western tip of the Shimane peninsula where pure N150°E dextral strike-slip faults and N50°E left-lateral and normal faults were observed. The computed stress tensor [*Angelier*, 1984] (Figure 7) implies a strike-slip regime with σ_1 and σ_3 horizontal. This outcrop is a good example of how the brittle deformation is distributed in the older Miocene strata. Here N150°E dextral strike-slip faults are regularly spaced every two meters. These are large fault planes, several tens of meters long. Oblique to this first set is a second one that trends N50°E on average and which corresponds to oblique slip faults with a component of left-lateral slip and a component of normal slip. These planes are also regularly spaced every meter or so and are never longer than 2 or 3 m. They show normal offsets of approximately 30 cm. The offsets induced by the first set of faults are not seen in the outcrop. Inside the blocks defined by these two sets of faults a third set (N20°E) with very small offsets are observed. These are dextral faults but no striae could be seen. The two largest sets of faults define regularly spaced dominoes and the geometry implies a clockwise rotation (Figure 6). Similar outcrops were observed along the coastline of the San'in district with the same regular spacing of faults and several sets with smaller offsets on smaller faults [*Fournier*, 1994]. This deformation is characteristic of the Kawai-Kuri and older formations. The Kawai-Kuri formation is dated by *Otofuji et al.* [1991] as 16.0 ± 1.4 Ma.

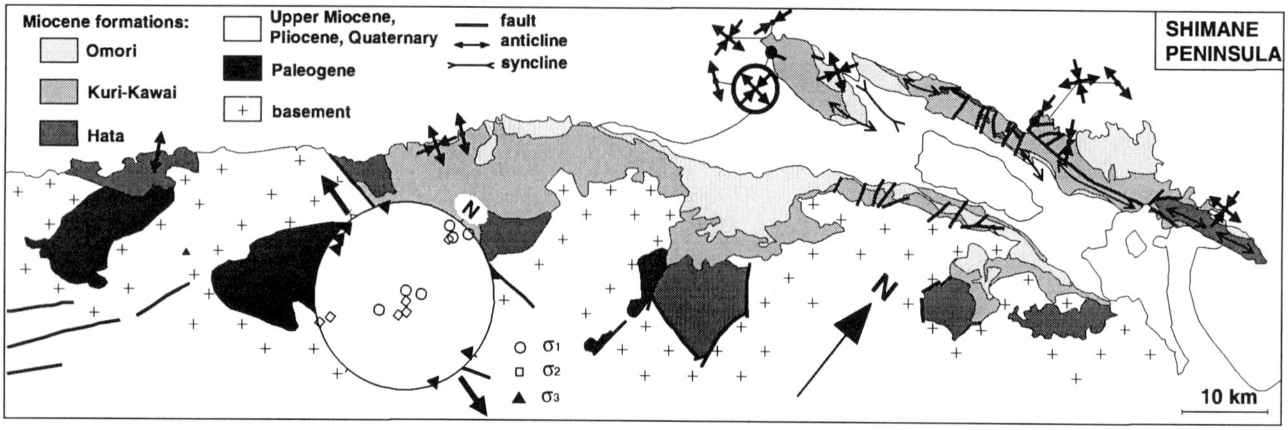

Fig. 5. Simplified geological map of Shimane peninsula (see Figure 1 for location) and paleostress tensors deduced from inversion of fault kinematic data [*Angelier*, 1984]. The site discussed detailed in Figures 6 and 7 is located by a black dot and the computed tensor is circled. Insert: 6 stress tensors with σ_3 horizontal showing a mean direction of extension ENE-WSW [after *Fournier*, 1994].

Younger formations such as the Omori formation (14.2 ± 0.6 Ma) do not show this deformation. Most of the outcrops are little deformed and show only tilting and long wavelength folding which can be attributed to very recent deformation. Only one outcrop in the Omori formation showed a set of faults that indicates a NE-trending compression, which could be compatible with the NW-trending extension described above. It could also be the local result of folding and thus have no relation with the older deformation. Except for this outcrop, all formations younger than the Kawai-Kuri formation are little deformed and never show the fault-bounded domino geometry seen below.

According to *Otofuji et al.* [1991], the Omori formation did not suffer the clockwise rotation observed in the underlying rocks; this allowed them to constrain the timing of rotational motion of SW Japan to a very short period about 15 Ma. We see from this example that the rotation was locally accompanied by a strong internal deformation with a domino geometry. Because the outcrops are not continuous enough to draw a complete map of these structures it is impossible to quantify the amount of finite internal deformation of SW Japan. Furthermore the blocks we have seen might be only second or third order blocks within larger ones which are more difficult to see. We can only conclude that some internal deformation of SW Japan occurred during the paleomagnetic rotation and thus that the 50° rotation is only partly rigid. *Kanaori* [1990] proposed a similar geometry at the scale of central Japan.

STRIKE-SLIP SHEAR ZONES AND PALEOMAGNETIC ROTATIONS

Jolivet et al. [1991] used the internal structure of the Japan Sea to close it back to its pre-opening position. One of the constraints was the shape of the northern side of the Yamato Bank, which fits reasonably with the 2000 m isobath of a portion of the Siberian coast. This allowed them to calculate an approximate pole of rotation which falls in the Tsushima Strait. The same authors used the same geometry to rotate SW Japan back to its position before opening and obtained a finite rotation about the same pole of 30° clockwise. This was based upon the assumption that the Yamato basin was floored with oceanic crust, which has since been proved wrong. [*Tamaki et al.*, 1992]. The 30° of rotation completely closes the Yamato basin and the western part of the Japan basin. The presence of highly stretched continental crust instead of oceanic crust is an argument to reduce the finite rotation. However the thickness of the Japan arc crust before the Japan Sea opening is not known precisely enough and it is then impossible to calculate the thinning factor. We shall then consider that 30° of clockwise rigid rotation of SW Japan is a maximum. The remaining 20° must be found in the internal strain we have described above.

The rigid clockwise rotation of SW Japan is compatible with a left-lateral motion along the MTL in a simple domino model. SW Japan inner zones and outer zones can be considered as 600 km long dominoes, bounded by left-

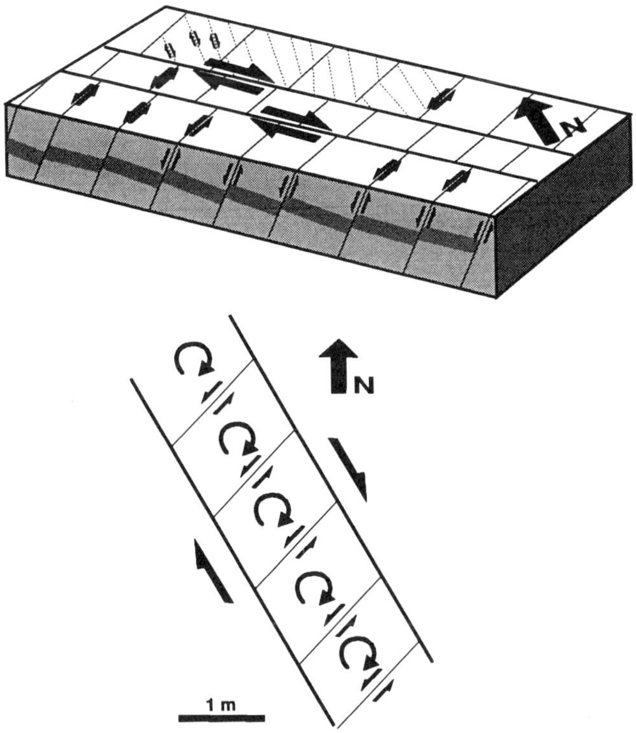

Fig. 6. Schematic representation of an outcrop in the Kawai-Kuri formation in Shimane peninsula showing the distribution of strain along the boundaries of small blocks and implying dextral rotation.

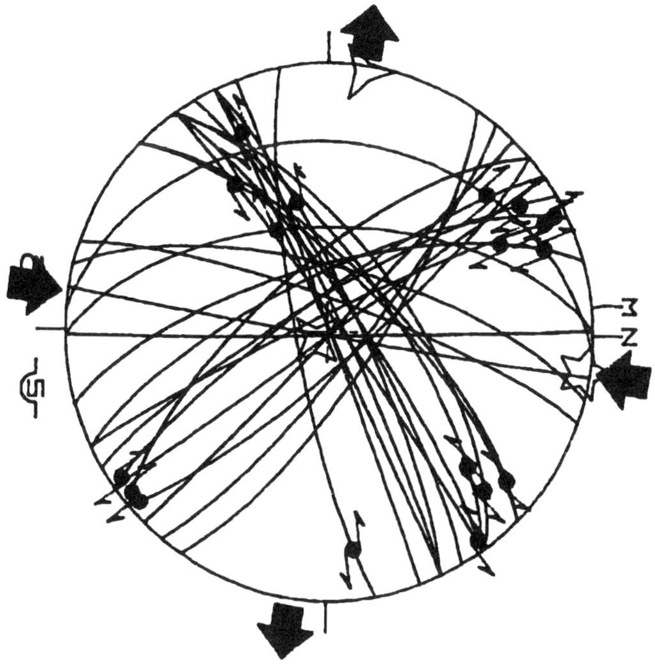

Fig. 7. Stress tensor inferred from the faults measured on the outcrop shown in Figure 6 with *Angelier*'s method [1984].

lateral E-W cross faults (MTL) and N-S dextral master faults.

To properly close the northern part of the Japan Sea, a 20° counterclockwise rotation was used in *Jolivet et al.* [1991]. This number is similar to what is given by paleomagnetic data for this region.

We can already conclude that if one accepts that a part of the clockwise rotation of SW Japan is due to internal strain actually observed at the time of paleomagnetic rotations, there is no major conflict between paleomagnetic data and the dextral pull-apart model.

Paleomagnetic studies along the major shear zones have shown that local rotation occurs and can be related to the observed deformation in Hokkaido and Sakhalin. Local rotation effects are also observed along less important shear zones such as the Tanakura tectonic line.

RECONSTRUCTIONS

The reconstructions shown here (Figure 8) are simplified from those described in *Jolivet and Tamaki* [1992]. The entire Japan Sea is considered as a wide dextral pull-apart zone between two major dextral strike-slip shear zones. The largest one is along the eastern margin of the Japan Sea and in Hokkaido and Sakhalin. A second one divides the Korean peninsula from SW Japan. The 30° of clockwise rotation of SW Japan accommodates the difference in the amount of offset on these two shear zones. In the reconstructions 400 km of dextral offset are imposed on the eastern shear zone and only 150 km on the western one. Extension is distributed in the domain of relay between these two shear zones; it eventually localizes at the southern tip of the eastern transpressional domain and finally leads to a complete rupture of the crust and formation of oceanic crust in a triangular zone. The geometry of the oceanic domain is that of a crustal-scale tension crack propagating westward near the end of a strike-slip shear zone. Crustal thinning continues even after the initiation of the spreading center, in the southern part of the Japan Sea. On the reconstructions only the rigid rotations are shown. Local rotations along the major shear zones occur in the meantime. Extension, spreading and strike-slip motion last until the middle Miocene, with strike-slip motion alone until recent time in the north (Sakhalin). After 10 Ma, a compressional stress field forms in NE Japan and the deformation progressively localizes along the eastern margin where compressional earthquakes

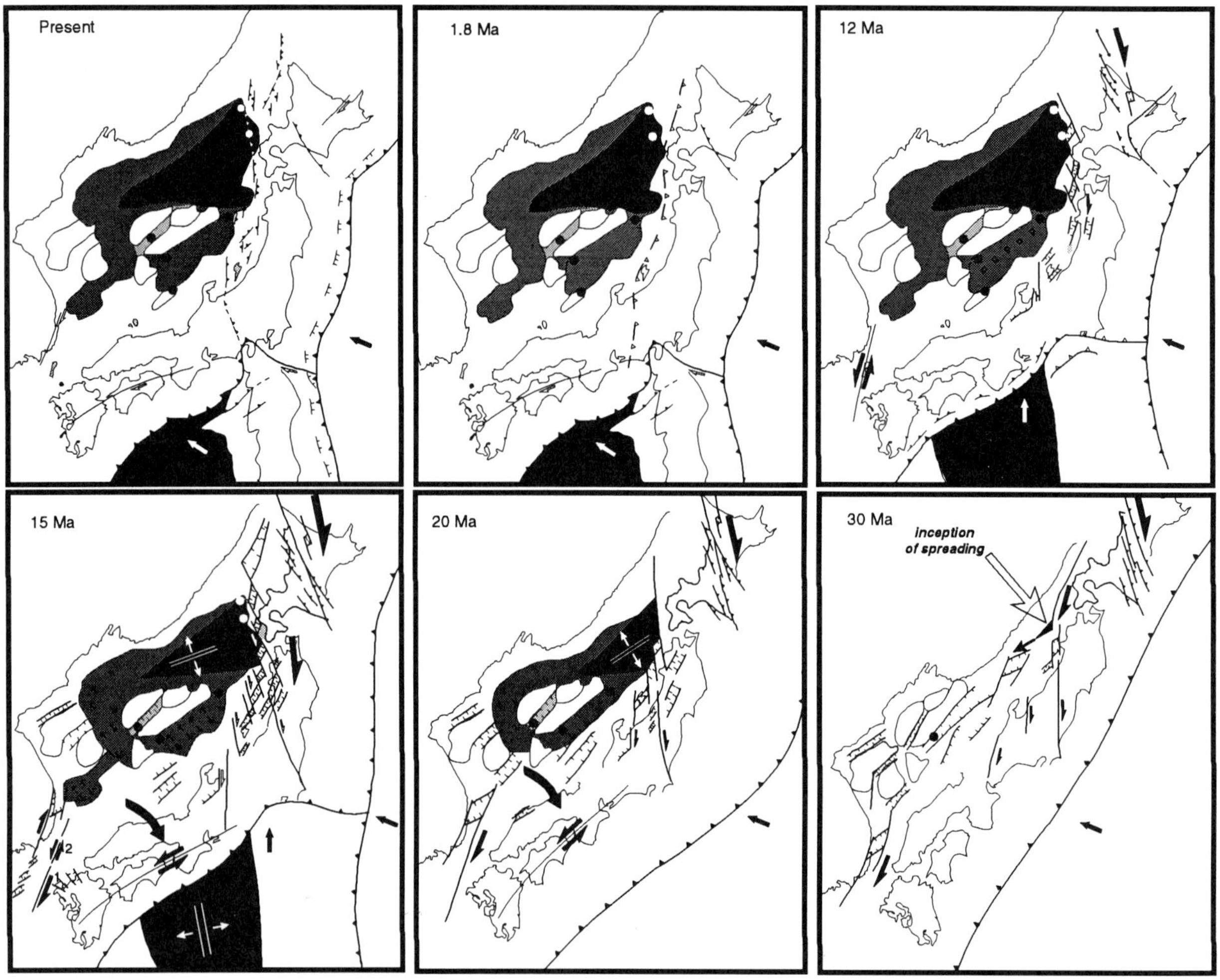

Fig. 8. Reconstructions of the Japan Sea evolution from 30 Ma to the present. Oceanic crust is shown in black and stretched continental crust in gray. Black and white dots represents ODP legs 127 and 128 drilling sites.

are recorded nowadays. This zone has been interpreted as a future subduction zone by *Nakamura* [1983].

CONCLUSION

This paper shows that the pull-apart model and the amounts of paleomagnetic rotations of SW and NE Japan can be compatible. The left-lateral behavior of the MTL during the Miocene accommodated the clockwise rotation of SW Japan between the two major dextral strike-slip faults, one along Sakhalin, Hokkaido, and NE Japan, and one between the Korean peninsula and SW Japan. One has, however, to accept that a part of the 50° of clockwise rotation of SW Japan is due to internal deformation of the rotating blocks with second order dominoes. The observation of such distributed brittle deformation in San'in district is in favor of this conclusion.

Local dextral rotations are seen along the major dextral shear zones in Sakhalin and Hokkaido, as well as along the Tanakura tectonic line and the Futaba fault. This geometry shows that rotations involve blocks of different sizes, major blocks such as SW Japan, smaller blocks along the major shear zones, or inside the major blocks as we have seen in San'in district. It is thus most important that the paleomagnetic sampling is made in connection with a precise identification of block boundaries.

The timing of rotation remains difficult to reconcile with other evidence for the age of opening. It is inescapable that the clockwise rotation of a 600 km block must leave a hole of the size of the Japan Sea, and thus the rotation and the basin formation should have the same age. The evidence for rifting and spreading before the given age for the rotation cannot be ignored. There must then be a problem in the timing of rotation. The discussion of radiometric dating in San'in suggest that the error bar might be much larger than suspected by *Otofuji et al.* [1991]. We suggest that the data be reconsidered without averaging the ages nor the declinations, and that measurements be made on the basis of a structural map showing the possible tectonic boundaries in order to differentiate between first and second order rotations. This job might be difficult to do in Japan because of poor outcrop conditions, but this difficulty must be taken into account when giving uncertainties on the values and ages of rotation.

Acknowledgments. This short discussion comes after many years of research in Japan in the field as well as onboard oceanographic vessels. Laurent Jolivet is indebted to many of his colleagues in Japan, among them M. Nakagawa, N. Kito, T. Bamba, S. Miyashita and his family, G. Kimura. Special thanks to K. Tamaki for his efforts to promote ODP drilling in the Japan Sea and his help to understand the geometry of opening of the Japan Sea. Thanks also to J. P. Cadet, J. Charvet, X. le Pichon, P. Huchon, P. Davy, J. P. Brun and C. Rangin. Thanks to V. Rozhdestvenskiy, K. Sergeyev, O. Melnikov, R. Kovalenko and H. Gnibidenko for the help they provided in the field in Sakhalin. We benefited also from discussions with T. Ohguchi, K. Kaneko, Y. I. Otofuji, R. Tada and A. Yamaji. S. Uyeda and J. F. Dewey have also encouraged this work. Y. Otofuji kindly suggested to us papers to read and provided us with data before these were published.

REFERENCES

Amano, K., and H. Sato, Neogene tectonics of the central part of the Northeast Honshu arc, *Mem. Soc. Geol. Jpn.*, 32, 81-96, 1989.

Angelier, J., Tectonic analysis of faultslip data sets, *J. Geophys. Res.*, 89, 5835-5848, 1984.

Charvet, J., K. Grimm, J. Griffin, L. Jolivet, and A. Pouclet, Structural features in ODP leg 128 cores: relationship with the tectonic evolution of the Japan Sea, in *Proc. ODP, Sci. Results, 127-128*, part 2, edited by K. Tamaki, K. Suyehiro, J. Allan and M. McWilliams, pp. 1175-1193, Ocean Drilling Program, College Station, TX, 1992.

Chen, W. P., and J. Nabelek, Seismogenic strike-slip faulting and the development of the North China basin, *Tectonics*, *7*, 975-989, 1988.

De Mets, C., A test of present-day plate geometries for northeast Asia and Japan, *J. Geophys. Res.*, *12*, 17,627-17,636, 1992.

Fabbri, O., and J. Charvet, Paleocontraintes et mouvements decrochants dans le détroit de Corée d'après l'exemple des îles Tsushima (Japon Sud-Ouest), *C. R. Acad. Sci. Paris*, *318*, 397-404, 1994.

Fournier, M., Ouverture de bassins marginaux et déformation continentale: l'exemple de la Mer du Japon, Thèse Université Pierre et Marie Curie, Paris, 1994.

Fournier, M., L. Jolivet, P. Huchon, V. S. Rozhdestvensky, K. F. Sergeyev, and L. Oscorbin, Dextral motion in Sakhalin, and the Japan Sea opening, *J. Geophys. Res.*, *99*, 2701-2725, 1994.

Fujioka, K., Synthesis of Neogene explosive volcanism of the Tohoku arc, deduced from the marine tephra drilled around the Japan trench region, Deep Sea Drilling Project Legs 56, 57 and 87B, in *Init. Repts. DSDP*, 87, H. Kagami, D. E. Karig, W. T. Colbourn, et al., pp. 703-721, U.S. Govt. Printing Office, Washington D.C., 1986.

Fukao, Y., and M. Furumoto, Mechanisms of large earthquakes along the eastern margin of the Japan sea, *Tectonophysics*, *25*, 247-266, 1975.

Hayashida, A., Timing of rotational motion of southwest Japan inferred from paleomagnetism of the Setouchi Miocene series, *J. Geomagn. Geoelectr.*, *38*, 295-310, 1986.

Hayashida, A., and I. Y. Ito, Paleoposition of southwest Japan at 16 Ma: implication from magnetism of the Miocene Ichishi group, *Earth Planet. Sci. Lett.*, *68*, 335-342, 1984.

Hayashida, A., T. Fukui, and M. Torii, Paleomagnetism of the early Miocene Kano Group in Southwest Japan and its implication for the opening of the Japan Sea, *Geophys. Res. Lett.*, *18*, 1095-1098, 1991.

Hirooka, K., H. Sakai, T. Takahashi, H. Kinoto, and A. Takeuchi, Tertiary tectonic movement of central Japan inferred from paleomagnetic studies, *J. Geomagn. Geoelectr.*, *38*, 311-323, 1986.

Hyodo, H., and N. Niitsuma, Tectonic rotation of the Kanto mountains, related with the opening of the Japan Sea and collision of the Tanzawa block since Middle Miocene, *J. Geomagn. Geoelectr.*, *38*, 335-348, 1986.

Hwang, J. H., L'évolution tectonique post-Jurassique de la Corée: apport des reconstitutions des paleocontraintes dans les bassins Crétacé méridionaux, Thèse université Pierre et Marie Curie, Paris, 1992.

Ichikawa, K., Geohistory of the Median Tectonic Line of Southwest Japan, *Mem. Geol. Soc. Jpn.*, *18*, 187-212, 1980.

Iijima, A., and R. Tada, Evolution of Tertiary sedimentary basins of Japan in reference to opening of the Japan Sea, *J. Fac. Sci., Tokyo Univ.*, 22, 121-171, 1990.

Ingle, C. J., Subsidence of the Japan Sea: stratigraphic evidence from ODP sites and onshore sections, in *Proc. ODP, Sci. Results, 127/128*, K. Tamaki, K. Suyehiro, J. Allan, M. McWilliams, et al., pp. 1197-1218, Ocean Drilling Program, College Station, TX, 1992.

Ishikawa, N., and T. Tagami, Paleomagnetism and fission-track geochronology on the Goto and Tsushima islands in the Tsushima Strait area: implications for the opening mode of the Japan Sea, *J. Geomagn. Geoelectr.*, *43*, 229-253, 1991.

Ishikawa, N., M. Torii, and K. Koga, Paleomagnetic study of the Tsushima Islands, southern margin of the Japan Sea, *J. Geomagn. Geoelectr.*, *41*, 797-811, 1989.

Ito, H., and K. Tokieda, An interpretation of paleomagnetic results from Cretaceous granites in South Korea, *J. Geomagn. Geoelectr.*, *32*, 275-284, 1980.

Ito, H., and K. Tokieda, K., Tilting movements of the Japanese islands inferred from Cretaceous and Early Tertiary paleomagnetic data, *J. Geomagn. Geoelectr.*, *38*, 361-386, 1986.

Ito, H., K. Tokieda, and Y. Notsu, A paleomagnetic approach to possible tilting movements of Kitakami mountains and Oshima peninsula of Hokkaido, *Rock Mag. Paleogeophys., 7*, 113-117, 1980.

Itoh, Y., Differential rotation of the eastern part of southwest Japan inferred from paleomagnetism of Cretaceous and Neogene rocks, *J. Geophys. Res., 93*, 3401-3411, 1988.

Jolivet, L., Comment on "Clockwise tectonic rotation of Tertiary sedimentary basins in central Hokkaido, northern Japan," by K. Kodama, T. Takeuchi and T. Ozawa, *Geology*, 21, 431-434, 1993, *Geology, 22*, 94-95, 1994.

Jolivet, L., and Ph. Huchon, Crustal scale strike-slip deformation in Hokkaido, Northeast Japan, *J. Struct. Geol., 11*, 509-522, 1989.

Jolivet, L., and K. Tamaki, Neogene kinematics in the Japan Sea region and the volcanic activity of the Northeast Japan arc, in *Proc. ODP, Sci. Results, 127/128*, K. Tamaki, K. Suyehiro, J. Allan, M. McWilliams, et al., pp. 1311-1331, Ocean Drilling Program, College Station, TX, 1992.

Jolivet, L., P. Huchon, and C. Rangin, Tectonic setting of Western Pacific marginal basins, *Tectonophysics, 160*, 23-47, 1989.

Jolivet, L., Ph. Huchon, J. P. Brun, N. Chamot-Rooke, X. Le Pichon, and J. C. Thomas, Arc deformation and marginal basin opening, Japan Sea as a case study, *J. Geophys. Res., 96*, 4367-4384, 1991.

Jolivet, L., M. Fournier, P. Huchon, V. S. Rozhdestvenskiy, K. F. Sergeyev, and L. S. Oscorbin, Cenozoic intracontinental dextral motion in the Okhotsk-Japan Sea region, *Tectonics, 11*, 968-977, 1992.

Jun, M. S., Source parameters of shallow intraplate earthquakes in and around the Korean peninsula and their tectonic implication, *Acta Universitatis Upsaliensis, Comprehensive Summaries of Uppsala Dissertation from the Faculty of Science, 285*, Uppsala, 1990.

Kanaori, Y., Late Mesozoic - Cenozoic strike-slip and block rotation in the inner belt of Southwest Japan, *Tectonophysics, 177*, 381-399, 1990.

Kawai, N., H. Ito, and S. Kume, Deformation of the Japanese islands as inferred from rock magnetism, *Geophys. J. R. Astron. Soc., 6*, 124-130, 1961.

Kawai, N., K. Hirooka, and T. Nakajima, Paleomagnetic and potassium-argon age information supporting Cretaceous-Tertiary hypothetic bend of the main island, Japan, *Palaeogeogr. Palaeoclim. Palaeoecol., 6*, 277-282, 1969.

Kawai, N., T. Nakajima, and K. Hirooka, The evolution of the island arc of Japan and the formation of granites in the circum Pacific belt, *J. Geomagn. Geoelectr., 23*, 267-293, 1971.

Kim, K. H., J. K. Won, J. Matsuda, K. Nagao, and M. W. Lee, Paleomagnetism and K-Ar age of volcanic rocks from Guryongpo area, Korea, *J. Korean Inst. Mining Geol.*, 19, 231-237, 1986.

Kimura G., S. Miyashita, and S. Miyasaka, Collision tectonics in Hokkaido and Sakhalin, in *Accretion Tectonics in the Circum Pacific Regions*, edited by M. Hashimoto and S. Uyeda, pp. 117-128, Terrapub, Tokyo, 1983.

Kobayashi, K., K. Tamaki, M. Nakanishi, N. Isezaki, and K. Sayanagi, Complex pseudofault pattern of the Japan Sea - results of detailed geomagnetic survey, *EOS Trans. AGU, 67*, 1227, 1988.

Kodama, K., and K. Nakayama, Paleomagnetic evidence for post-Late Miocene intra-arc rotation of south Kyushu, Japan, *Tectonics, 12*, 35-47, 1993.

Kodama, K., T. Ozawa, K. Inoue, Y. Maeda, and T. Tohru, Paleomagnetism and post-Middle Miocene counter-clockwise rotation of Tanegashima island off Kyushu, Japan, *J. Geomagn. Geoelectr., 43*, 721-740, 1991.

Kodama, K., T. Takeuchi, and T. Ozawa, Clockwise tectonic rotation of Tertiary sedimentary basins in central Hokkaido, northern Japan, *Geology, 21*, 431-434, 1993.

Lallemand, S., and L. Jolivet, Japan Sea: a pull apart basin. *Earth Planet. Sci. Lett., 76*, 375-389, 1985.

Lee, J. S., Pétrologie et relations structurales des volcanites Crétacé à Cénozoïques de la Corée du Sud: implications géodynamiques sur la marge est-eurasiatique, Thèse Université d'Orléans, 1988.

Momose, H., M. Torii, and H. Tanaka, Counterclockwise paleomagnetic direction from the Gongenyama Formation (N9-N10) on the coastal area of Northeast Japan, *Rock. Mag. Paleogeophys., 17*, 47-50, 1990.

Nakamura, K., Possible nascent trench along the eastern Japan sea as the convergent boundary between Eurasia and North American plates (in Japanese), *Bull. Earthquake Res. Inst., 58*, 721-732, 1983.

Nakamura, K., and S. Uyeda, Stress gradient in back arc regions and plate subduction, *J. Geophys. Res., 85*, 6419-6428, 1980.

Oda, H., M. Torii, and A. Hayashida, Paleomagnetic study and fission-track dating in Yanagawa and Takadate area, Northeast Japan, *Rock Mag. Paleogeophys., 19*, 231-237, 1989.

Otofuji, Y., and T. Matsuda, Paleomagnetic evidence for the clockwise rotation of Southwest Japan, *Earth. Planet. Sci. Lett., 62*, 349-359, 1983.

Otofuji, Y., and T. Matsuda, Timing of rotational motion of Southwest Japan inferred from paleomagnetism, *Earth Planet. Sci. Lett., 70*, 373-382, 1984.

Otofuji ,Y., and T. Matsuda, Amount of clockwise rotation of Southwest Japan - fan shape opening of the southwestern part of the Japan Sea, *Earth Planet. Sci. Lett., 85*, 289-301, 1987.

Otofuji, Y., T. Matsuda, and S. Nohda, Paleomagnetic evidences for the Miocene counterclockwise rotation of northeast Japan - rifting process of the Japan arc, *Earth Planet. Sci. Lett., 75*, 265-277, 1985.

Otofuji, Y., T. Itaya, and T. Matsuda, Rapid rotation of southwest Japan - paleomagnetism and K-Ar ages of Miocene volcanic rocks of southwest Japan, *Geophys. J. Int., 105*, 397-405, 1991.

Otofuji, Y., A. Kambara, T. Matsuda, and S. Nohda, Counterclockwise rotation of northeast Japan: paleomagnetic evidence for regional extent and timing of rotation, *Earth Planet. Sci. Lett., 121*, 503-518, 1994.

Otsuki, K., Reconstruction of Neogene tectonic stress field of Northeast Honshu arc from metalliferous veins, *Mem. Soc. Geol. Jpn., 32*, 281-304, 1989.

Otsuki K., and M. Ehiro, Major strike slip faults and their bearing on the spreading of the Japan Sea, *J. Phys. Earth, supplement issue, 26*, 537-555, 1978.

Rozhdestvenskiy, V. S., The role of wrench faults in the structure of Sakhalin, *Geotectonics, 16*, 323-332, 1982.

Sillitoe, R. H., Metallogeny of an Andean type continental margin in South Korea, implications for opening of the Japan Sea, in *Island Arcs, Deep Sea Trenches and Back Arc Basins, Maurice Ewing Ser.*, vol. 1, edited by M. Talwani and W. C. Pitman III, pp. 303-310, AGU, Washington, D. C., 1977.

Sugi, N., K. Chinzei, and S. Uyeda, Vertical crustal movements of northeast Japan since middle Miocene, in *Geodynamics of the Western Pacific and Indonesian Region, Geodynamic Ser.*, vol. 11, edited by W. C. T. Hilde and S. Uyeda, pp. 317-329, AGU, Washington, D.C., 1983.

Sugimura, A., T. Matsuda, K. Chinzei, and K. Nakamura, Quantitative distribution of late Cenozoic volcanic material in Japan, *Bull. Volcanol.*, *26*, 125-140, 1963.

Suzuki, K., On the Late Cenozoic history in the southern part of northeast Honshu in Japan, *Mem. Geol. Soc. Jpn.*, *32*, 97-112, 1989.

Tagami, T., Paleomagnetism and fission-track ages of the Kumano acidic rocks, *News of Osaka Micropaleontologists, 9*, 23-32, 1982.

Takeuchi, T., T. Ozawa, and K. Kodama, Paleomagnetism of the Late Cretaceous to Neogene deposits in central Hokkaido, Japan and south Sakhalin of the USSR: multiple rotations caused by transcurrent faults (abstract), 29th International Geological Congress, Kyoto, 24 August-3 September 1992, 108, 1992.

Tamaki, K., Age estimation of the Japan sea on the basis of stratigraphy, basement depth and heat flow data, *J. Geomagn. Geoelectr.*, *38*, 427-446, 1986.

Tamaki, K., Geological structure of the Japan sea and its tectonic implications, *Bull. Geol. Surv. Jpn.*, *39*, 269-365, 1988.

Tamaki, K., K. Suyehiro, J. Allan, J. C. Ingle, and K. Pisciotto, Tectonic synthesis and implications of Japan Sea ODP drilling, in *Proc. ODP, Sci. Results, 127/128*, K. Tamaki, K. Suyehiro, J. Allan, M. McWilliams, et al., pp. 1333-1350, Ocean Drilling Program, College Station, TX, 1992.

Tokuyama, H., M. Suyemasu, K. Tamaki, E. Nishiyama, S. Kuramoto, K. Suyehiro, H. Kinoshita, and A. Taira, Report on DELP cruises in the Japan Sea, Part III: seismic reflection studies in the Yamato Basin and the Yamato Rise area, *Bull. Earthq. Res. Inst., Univ. Tokyo, 62*, 367-390, 1987.

Torii, M., Paleomagnetism of Miocene rocks in the Setouchi province: evidence for rapid clockwise rotation of Southwest Japan in middle Miocene time, Ph. D. Thesis, Kyoto Univ., 1983.

Tosha, T., and Y. Hamano, Paleomagnetic study on the dyke swarm in the Oga peninsula, Northeast Honshu island, *J. Geomagn. Geoelectr., 38*, 349-360, 1986.

Tosha, T., and Y. Hamano, Paleomagnetism of Tertiary rocks from the Oga peninsula and the rotation of northeast Japan, *Tectonics*, *7*, 653-662, 1988.

Tsuchiya, N., Submarine basalt volcanism of Miocene Aosawa formation in the Akita-Yamagata oil filed basin, back-arc region of northeast Japan, *Mem. Geol. Soc. Jpn.*, *32*, 399-408, 1989.

Tsuchiya, N., Middle Miocene back-arc rift magmatism of basalt in the Northeast Japan arc, *Bull. Geol. Surv. Jpn.*, *41*, 473-505, 1990.

Usuta, M., Geotectonic history of the southern part of Akita prefecture, Northeast Japan, *Mem. Geol. Soc. Jpn.*, *32*, 57-80, 1989.

Yamaji, A., Geology of Atsumi area and early Miocene rifting in the Uetsu district, northeast Japan, *Mem. Geol. Soc. Jpn., 32*, 305-320, 1989.

Yamaji, A., Rapid intra-arc rifting in Miocene northeast Japan, *Tectonics*, *9*, 365-378, 1990.

Yamazaki, T., Paleomagnetism of Miocene sedimentary rocks around Matsushima Bay, Northeast Japan and its implication for the time of the rotation of Northeast Japan, *J. Geomagn. Geoelectr., 41*, 533-548, 1989.

Yaskawa, K., On the paleomagnetic approach to the origin of the Japan Sea, *Nihon Kai (Japan Sea)*, *10*, 120-126, 1979, in Japanese.

L. Jolivet, and M. Fournier, Laboratoire de Géologie, Ecole Normale Supérieure, 24 rue Lhomond, 75231 Paris cedex 05, France

H. Shibuya, Department of Earth Sciences, University of Osaka Prefecture, Sakai 593, Japan

The Philippine Sea Plate: Magnetism and Reconstructions

Robert Hall

Department of Geological Sciences, University College London, U.K.

Michael Fuller

Department of Geological Sciences, University of California, Santa Barbara, California

Jason R. Ali

Oceanography Department, Southampton University, U.K.

Charles D. Anderson

Department of Geological Sciences, University of California, Santa Barbara, California

Paleomagnetic results from ocean drilling and from land on the Philippine Sea Plate indicate progressive northward translation of the entire plate during the Tertiary, although different sites suggest different ages of equatorial crossing. The difficulty of orienting ocean drilling cores has recently been remedied and Ocean Drilling Program leg 126 provided the first Philippine Sea Plate declination data from sea bottom sites. These showed large clockwise declination shifts of up to ~90° since the early Oligocene. Similar large declination shifts have been reported from land sites at the eastern margin of the plate; these record similar changes in inclinations with age as ocean drilling sites. The cause of the declination deflections is controversial and they have been explained as the result of entire plate rotation, marginal basin opening, and/or local tectonic deformation at the plate edge. Because most declination data have been obtained from the eastern margin of the plate, it has proved impossible to distinguish between these three possible causes. However, new paleomagnetic data from land sites in the southern part of the Philippine Sea Plate in eastern Indonesia define two sub-areas: a region within the Sorong fault system that records variable rotations attributed to deformation at the plate edge, and a region north of the Sorong fault that records a large Tertiary, but discontinuous, clockwise rotation. Both areas record northward paleolatitude shifts with decreasing age since the beginning of the Neogene but inclinations of lower Eocene sites suggest southward movement during the Paleogene. We consider the area north of the Sorong fault to form one of the oldest parts of the plate and assume that the declination shifts record the Tertiary rotation of the western part of the plate. Using this assumption it is possible to calculate rotation poles and reconstruct the Philippine Sea Plate after calculating the rotational effects of opening the marginal basins. The model predictions closely fit all paleomagnetic data and also satisfy constraints on rotation inferred from magnetic anomaly skewness and seamount magnetization studies. We therefore propose that the plate has rotated clockwise in a discontinuous manner since the early Tertiary: about the present Eurasia-Philippine Sea Plate pole of 48.2°N, 157°E by 5.45° between 0 and 5 Ma, and about best fit rotation poles of 15°N, 160°E by 34° between 5 and 25 Ma, and about 10°N, 150°E by 50° between 40 and 50 Ma. Changes in rates of rotation have been driven by large scale tectonic changes, including the change in Pacific motion at ~43 Ma and arc-continent collision at the north Australian margin at ~ 22 Ma. Some eastern plate margin sites record rotations that differ from those predicted by the model and are interpreted to reflect local tectonic deformation.

INTRODUCTION

The Philippine Sea (Figure 1) was among the earliest targets of the Deep Sea Drilling Project (DSDP) since its origin and history were enigmatic and even today remain

Active Margins and Marginal Basins of the Western Pacific
Geophysical Monograph 88

controversial. The Philippine Sea Plate is surrounded by subduction zones that separate it from the oceanic ridge system and consequently its present motion with respect to other major plates is difficult to determine. Its origin remains uncertain partly because it is the one major plate whose history of motion remains poorly known.

Early suggestions of large northward motions and rotation of the entire plate were based on magnetic anomaly studies and inclination data from ocean drilling [*Louden*, 1976, 1977]. Later paleomagnetic studies carried out during ocean drilling [*Kinoshita*, 1980; *Keating*, 1980; *Keating and Herrero*, 1980; *Bleil*, 1981; *Haston et al.*, 1992; *Koyama et al.*, 1992] provided further support for northward translation of the plate, and this aspect of its motion history is now generally accepted. However, the rotational history of the plate has proved more controversial. Anomaly skewness studies in the West Philippine Basin [*Louden*, 1977; *Shih*, 1980] were interpreted to indicate substantial clockwise rotation of the plate during the Tertiary but appear to conflict with seamount magnetization studies [*Vacquier and Uyeda*, 1967] and anomaly skewness studies [*Chamot-Rooke et al.*, 1987] in the Shikoku Basin.

The small islands at the eastern edge of the plate have been the site of a substantial paleomagnetic effort [*Kobayashi et al.*, 1971; *Aoki*, 1972; *Larson et al.*, 1975; *Fuller et al.*, 1980; *Kodama*, 1981; *Kodama et al.*, 1983; *Keating et al.*, 1983; *Keating and Helsley*, 1985; *Adachi et al.*, 1987; *Haston et al.*, 1988; *Haston and Fuller*, 1991]. All reported large clockwise rotations, but the interpretation of these results has been questioned because local rotations about a vertical axis provide a possible explanation, in which case plate wide rotations would not be needed. Until recently it was impossible to orient ocean drill cores and therefore all DSDP paleomagnetic studies lack declination data. Some progress has been made toward overcoming this difficulty and declination results were obtained from Ocean Drilling Program (ODP) leg 126 [*Koyama et al.*, 1992] that suggest large clockwise rotations since the mid-Tertiary. However, the several phases of backarc spreading that have led to opening of marginal basins within the plate are potentially the cause of rotations of some sites. Models invoking local deformation have also been proposed [*Larson et al.*, 1975; *Kodama et al.*, 1983; *McCabe and Uyeda*, 1983] to account for declination deflections. Distinguishing these effects from plate-wide movements has therefore proved difficult. All of the models are reviewed by *Koyama et al.* [1992].

The motion history of the plate is important since it is a key element in the tectonic evolution of the western Pacific. It has proved difficult, using chemical methods, to distinguish between the origin of the Philippine Sea by trapping of Pacific ocean crust or by spreading in a marginal basin whereas different models for its origin do make specific predictions about the orientation of the West Philippine Sea spreading center. Many of our ideas of island arc development and magmatic processes have been based on interpretations of the tectonics of the Philippine Sea Plate, most recently those concerning the origin of boninites. The history of plate motions in the western Pacific is also critical to understanding the origin of marginal basins at the eastern Eurasia margin. Although no widely accepted solution has yet emerged there seems to be no way of determining the motion history of the Philippine Sea Plate except by using magnetic data.

In this paper we summarize magnetic data from the Philippine Sea Plate and then outline a model for its Tertiary motion based upon new paleomagnetic data from the southern part of the Philippine Sea Plate. Rotation poles for the plate were determined using a forward modelling approach and provide the basis for a tectonic model that was then refined to include the effects of marginal basin opening within the plate. The predictions of the model are compared with all the magnetic data from the plate, including the paleomagnetism of DSDP and ODP cores, of the islands, and of seamounts, and the phase shifting of magnetic anomalies. We propose that the majority of the data is consistent with clockwise rotation of the entire plate by ~90° since 50 Ma with additional local rotations of a few eastern margin sites.

PALEOMAGNETIC RESULTS FROM THE PHILIPPINE SEA PLATE

Paleomagnetism of DSDP and ODP cores

Six DSDP/ODP legs (6, 31, 58, 59, 60, 125 and 126) have drilled the Philippine Sea Plate. The majority of the drill sites are from the northern half of the plate, with most of these in the Oligocene and younger backarc and forearc basins on the eastern half of the plate. The pioneering DSDP work was by *Louden* [1977], who recovered reliable directions of magnetization from sites 290, 292, and 294 (Figure 1). Site 292 yielded particularly consistent results from uniform calcareous ooze with a sedimentation rate of 1 cm/10^3 years. The intensity of magnetization was relatively strong and the AF demagnetization characteristics excellent. The results gave a well defined trend in inclination with age (Table 1) and were interpreted as a linear change in paleolatitude with age from south to north, with an equatorial crossing at approximately 35 Ma. Results from sites 290 and 294 were compared on the assumption that they were on the same plate and shown to be self consistent.

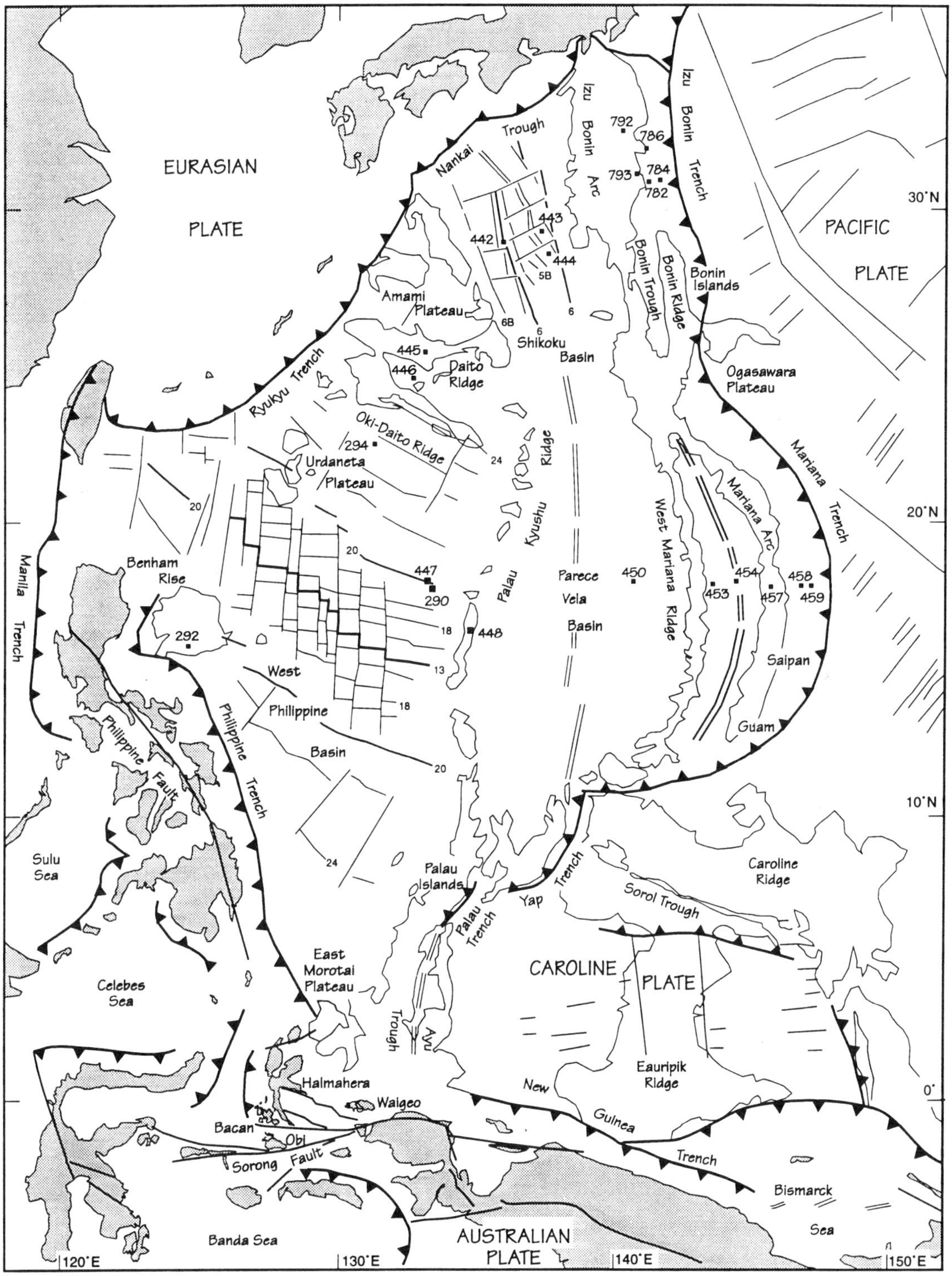

Fig. 1. Principal tectonic, bathymetric and magnetic features of the Philippine Sea region. Solid squares indicate DSDP and ODP sites; important magnetic anomalies are numbered. West Philippine Basin anomalies from *Hilde and Lee* [1984]. Shikoku Basin anomalies from *Chamot-Rooke et al.* [1987]. Caroline Plate anomalies from *Weissel and Anderson* [1978].

TABLE 1. Latitudes calculated from inclination data for DSDP and ODP Sites on the Philippine Sea Plate

Location	Age	Palaeolatitude	Change in Latitude	Source
290	35.0 ± 1.0	26.7 ± 9.7	9.1 ± 9.7	*Louden* [1977]
290	15.0 ± 2.0	4.2 ± 6.0	-13.4 ± 6.0	*Louden* [1977]
292	41.0 ± 2.0	-4.9 ± 2.2	-20.5 ± 2.2	*Louden* [1977]
292	37.5 ± 1.5	0.8 ± 1.2	-14.8 ± 1.2	*Louden* [1977]
292	33.0 ± 1.0	1.7 ± 1.5	-13.9 ± 1.5	*Louden* [1977]
292	28.0 ± 1.0	5.0 ± 2.8	-10.6 ± 2.8	*Louden* [1977]
292	24.0 ± 1.5	7.4 ± 2.1	-8.2 ± 2.1	*Louden* [1977]
292	8.5 ± 1.5	8.8 ± 2.9	-6.8 ± 2.9	*Louden* [1977]
294	50.0 ± 5.0	-13.2 ± 6.4	-35.7 ± 6.4	*Louden* [1977]
294	24.0 ± 2.0	7.8 ± 8.5	-14.7 ± 8.5	*Louden* [1977]
442-4	17.0 ± 2.0	18.0 ± 12.0	-11.0 ± 12.0	*Kinoshita et al.* [1980]
442-4	3.0 ± 2.0	21.0 ± 17.0	-8.0 ± 17.0	*Kinoshita et al.* [1980]
445	44.0 ± 7.0	4.5 ± 2.9	-21.0 ± 2.9	*Kinoshita et al.* [1980]
445	30.0 ± 7.0	10.8 ± 4.4	-14.7 ± 4.4	*Kinoshita et al.* [1980]
445	14.0 ± 9.0	17.4 ± 5.1	-8.1 ± 5.1	*Kinoshita et al.* [1980]
445	2.5 ± 2.5	18.8 ± 12.0	-6.7 ± 12.0	*Kinoshita et al.* [1980]
446	44.0 ± 8.0	6.2 ± 4.5	-20.4 ± 4.5	*Kinoshita et al.* [1980]
446A	45.0 ± 8.0	4.3 ± 3.0	-20.3 ± 3.0	*Kinoshita et al.* [1980]
447	33.0 ± 2.0	6.0 ± 13.1	-12.0 ± 13.1	*Keating and Herrero* [1980]
448	33.0 ± 2.0	7.8 ± 10.0	-8.5 ± 10.0	*Keating and Herrero* [1980]
448	32.5 ± 5.0	4.7 ± 5.4	-11.6 ± 5.4	*Keating* [1980]
450	15.0 ± 5.0	7.8 ± 10.8	-10.2 ± 10.8	*Keating* [1980]
453	3.7 ± 1.0	11.2 ± 1.2	-6.8 ± 1.2	*Bleil* [1981]
454	1.2 ± 0.3	12.7 ± 1.7	-5.3 ± 1.7	*Bleil* [1981]
456	2.2 ± 0.3	19.6 ± 3.1	1.6 ± 3.1	*Bleil* [1981]
458	32.0 ± 2.0	4.5 ± 1.5	-13.5 ± 1.5	*Bleil* [1981]
458	32.0 ± 2.0	9.1 ± 0.8	-8.9 ± 0.8	*Bleil* [1981]
459	39.0 ± 2.0	6.3 ± 1.2	-11.7 ± 1.2	*Bleil* [1981]
459	39.0 ± 2.0	0.0 ± 3.0	-18.0 ± 3.0	*Bleil* [1981]
459	25.5 ± 4.5	9.2 ± 0.8	-8.8 ± 0.8	*Bleil* [1981]
459	17.5 ± 6.5	12.2 ± 0.8	-5.8 ± 0.8	*Bleil* [1981]
782A	13.4 ± 3.0	23.3 ± 5.2	-7.5 ± 5.2	*Haston et al.* [1992]
782A	7.8 ± 2.6	24.0 ± 4.6	-6.9 ± 4.6	*Haston et al.* [1992]
782A	0.5 ± 0.5	27.2 ± 6.0	-3.7 ± 6.0	*Haston et al.* [1992]
784A	3.4 ± 1.8	4.3 ± 7.1	-26.6 ± 7.1	*Haston et al.* [1992]
786A	4.3 ± 0.9	22.8 ± 4.4	-9.1 ± 4.4	*Haston et al.* [1992]
786A	2.5 ± 0.9	32.0 ± 8.7	0.1 ± 8.7	*Haston et al.* [1992]
787	27.9 ± 0.3	21.0 ± 5.3	-11.4 ± 5.3	*Koyama et al.* [1992][a]
787	27.4 ± 0.6	16.7 ± 6.9	-15.7 ± 6.9	*Koyama et al.* [1992][a]
787	26.1 ± 1.1	26.6 ± 8.2	-5.8 ± 8.2	*Koyama et al.* [1992][a]
792	33.0 ± 1.2	20.0 ± 15.0	-12.4 ± 15.0	*Koyama et al.* [1992][a]
792	28.7 ± 0.3	18.3 ± 5.6	-14.1 ± 5.6	*Koyama et al.* [1992][a]
792	27.9 ± 0.3	21.3 ± 5.0	-11.1 ± 5.0	*Koyama et al.* [1992][a]
792	11.0 ± 1.2	24.2 ± 6.1	-8.2 ± 6.1	*Koyama et al.* [1992][a]
792	6.5 ± 0.8	27.4 ± 18.5	-5.0 ± 18.5	*Koyama et al.* [1992][a]
793	31.0 ± 0.3	12.6 ± 5.1	-18.6 ± 5.1	*Koyama et al.* [1992][a]
793	28.7 ± 0.6	16.1 ± 4.9	-15.0 ± 4.9	*Koyama et al.* [1992][a]
793	27.5 ± 0.7	19.3 ± 4.0	-11.8 ± 4.0	*Koyama et al.* [1992][a]
793	22.0 ± 1.0	14.9 ± 6.0	-16.2 ± 6.0	*Koyama et al.* [1992][a]
793	15.8 ± 0.8	25.0 ± 6.3	-6.1 ± 6.3	*Koyama et al.* [1992][a]
Leg 126 mean	29.8 ± 6.1	18.4 ± 2.1	-13.3 ± 2.1	*Koyama et al.* [1992]
Leg 126 mean	19.8 ± 3.5	21.3 ± 5.1	-10.4 ± 5.1	*Koyama et al.* [1992]
Leg 126 mean	10.8 ± 5.6	25.4 ± 6.0	-6.3 ± 6.0	*Koyama et al.* [1992]

[a] Discrete sample inclinations obtained from their Figure 6

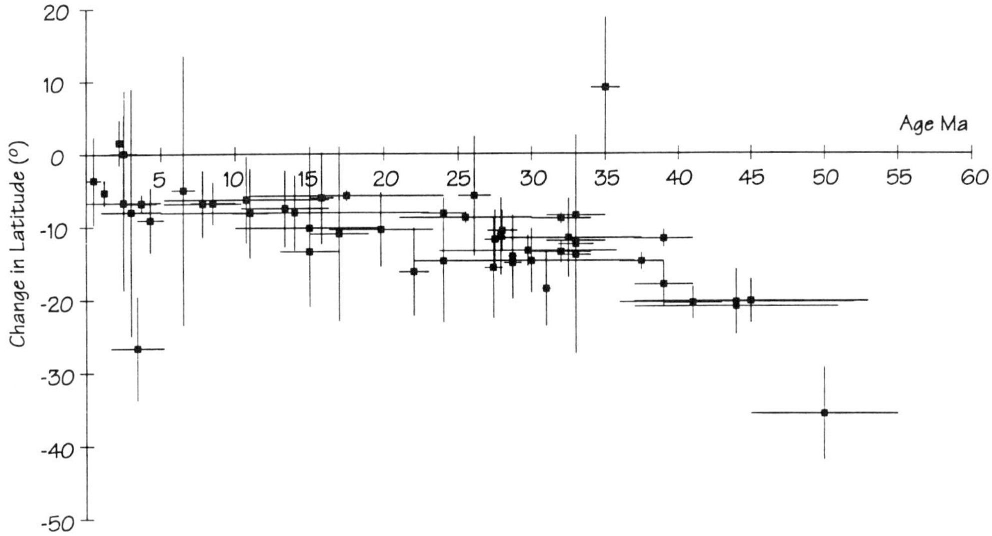

Fig. 2. Age versus change in latitude calculated from inclination data from all DSDP and ODP sites in the Philippine Sea Plate. Since most sites lack declination results this plot assumes a simple linear change in paleolatitude with time and many sites older than ~ 20 Ma could be interpreted as northern rather than southern hemisphere. Table 1 gives data sources.

Paleomagnetic studies have subsequently come from numerous Philippine Sea Plate sites [*Kinoshita,* 1980; *Keating,* 1980; *Keating and Herrero,* 1980; *Bleil,* 1981; *Haston et al.,* 1992; *Koyama et al.,* 1992]. All report northward motion of the Philippine Sea Plate with estimates of changes of about 20° in latitude since the Oligocene. In such a scenario, the holes studied by *Louden* [1977] crossed the equator in latest Eocene. The data are summarized in Table 1 and Figure 2.

There is, however, an ambiguity in the interpretation of these data because only inclination is known. The data also permit an approach to the equator from the north and a return to higher northern latitudes. All published interpretations of inclination data have therefore followed *Louden* [1977] in assuming a simple linear change in paleolatitude with time as suggested by Figure 2. This interpretation is reasonable for younger sites at higher northerly latitudes, but for older sites with inclinations that typically indicate sub-equatorial latitudes more complex motion histories are possible.

The only declination data reported for the Philippine Sea Plate from DSDP and ODP sites come from the Izu-Bonin forearc region [*Koyama et al.,* 1992]. At sites 787, 792 and 793 of leg 126, paleomagnetic declinations were recovered for the first time using the formation microscanner to orient the cores (Table 2). The bedding orientation is given by the downhole microresistivity measurement, and so the bedding plane observed in the specimen can be oriented with respect to the downhole observation. The results show clockwise rotated declinations (Figure 3). The mean deflections increase with age from 34° ± 20° for the middle Miocene, to 66° ± 25° for the late Oligocene and 85° ± 17° for the early Oligocene, but the scatter is large, so that progressively older α_{95} confidence limits overlap. The middle Miocene direction is, however, distinguishable from the early Oligocene direction. Secondary viscous magnetization was also used to orient the cores, but this method [*Fuller,* 1969] relies on faithful recording of the present field direction by the soft magnetization, which is not always the case in the drill cores. The scatter in results reported for samples oriented by this method was larger than for those oriented with the microscanner but broadly consistent in giving substantial clockwise rotations. The DSDP and ODP results suggest that the various sites have moved northward by as much as 20° since the early Oligocene, and the sites in the Bonin forearc indicate that these have also rotated clockwise as they moved north.

Paleomagnetic Studies of the Islands Within the Philippine Sea Plate

Paleomagnetic surveys from the islands on the Philippine Sea Plate can, in principle, provide both inclination and declination from which paleolatitudes and rotation about vertical axes can be recovered. Paleomagnetic studies have been made in the Palau islands, on Guam and Saipan in the

TABLE 2. Declination data from sites on the Philippine Sea Plate

Location	Age	I	D	α_{95}	Source
ODP SITES					
793	30.5 ± 0.5		85.0	13.0	*Koyama et al.* [1992]
792	28.0 ± 1.0		66.0	21.0	*Koyama et al.* [1992]
792	11.0 ± 1.0		34.0	17.0	*Koyama et al.* [1992]
LAND SITES					
Bonins, Anijima, Nishijima	41.0 ± 3.0	9.9	105.6	11.6	*Haston and Fuller* [1991][a]
Bonins, Hahajima	41.0 ± 3.0	4.7	211.1	26.4	*Haston and Fuller* [1991][a]
Bonins, Mukojima	41.0 ± 3.0	-25.0	203.7	9.9	*Haston and Fuller* [1991][a]
Bonins, Mukojima	41.0 ± 3.0	14.1	209.2	2.6	*Haston and Fuller* [1991][a]
Guam	35.0 ± 5.0	15.1	66.1	11.1	*Haston and Fuller* [1991][a]
Palau	32.3 ± 1.1	10.2	281.8	20.1	*Haston and Fuller* [1991][a]
Palau	30.3 ± 0.9	5.3	68.4	12.8	*Haston and Fuller* [1991][a]
Palau	20.1 ± 0.5	3.5	54.5	17.4	*Haston and Fuller* [1991][a]
Saipan	41.4 ± 0.9	-12.0	43.0	12.5	*Haston and Fuller* [1991][a]
Saipan	12.9 ± 0.3	-30.7	208.1	8.4	*Haston and Fuller* [1991][a]
Saipan	2.0 ± 1.5	8.6	350.1	22.5	*Haston and Fuller* [1991][a]
Kasiruta	33.3 ± 0.3	33.1	231.2	5.0	*Hall et al.* [1994a]
Kasiruta	33.3 ± 0.3	23.9	217.9	6.1	*Hall et al.* [1994a]
Kasiruta	32.3 ± 2.8	18.9	221.1	5.3	*Hall et al.* [1994a]
Kasiruta	32.3 ± 2.8	24.2	218.9	12.8	*Hall et al.* [1994a]
Waigeo	29.0 ± 6.0	11.2	214.2	7.2	*Hall et al.* [1994a]
Waigeo	29.0 ± 6.0	18.1	229.9	13.9	*Hall et al.* [1994a]
Halmahera	47.2 ± 2.9	-3.7	264.3	15.9	*Hall et al.* [1994a]
Halmahera	40.0 ± 2.0	-24.1	39.1	19.5	*Hall et al.* [1994a]
Halmahera	29.4 ± 6.1	-33.1	41.6	4.1	*Hall et al.* [1994a]
Halmahera	26.7 ± 1.9	27.5	222.3	4.6	*Hall et al.* [1994a]
Halmahera	7.0 ± 2.0	35.7	192.2	3.2	*Hall et al.* [1994a]
Halmahera	7.0 ± 2.0	38.4	192.2	9.0	*Hall et al.* [1994a]
Halmahera	5.0 ± 1.0	0.5	10.1	7.3	*Hall et al.* [1994a]
Halmahera	5.0 ± 1.0	-2.5	7.1	4.2	*Hall et al.* [1994a]
Halmahera	2.2 ± 0.7	27.0	176.7	7.4	*Hall et al.* [1994a]
Halmahera	2.2 ± 0.7	26.9	189.8	1.9	*Hall et al.* [1994a]

[a] Sources from which these data are compiled are cited in text and in *Haston and Fuller* [1991]

Marianas, and in the Bonin islands [*Kobayashi et al.*, 1971; *Aoki*, 1972; *Larson et al.*, 1975; *Fuller et al.*, 1980; *Kodama*, 1981; *Kodama et al.*, 1983; *Keating et al.*, 1983; *Keating and Helsley*, 1985; *Adachi et al.*, 1987; *Haston et al.*, 1988, *Haston and Fuller*, 1991]. *Haston and Fuller* [1991] reviewed all paleomagnetic results then available from land sites on the plate and selected reliable data. They essentially included all results with adequate site structural information that yield stable directions with small values of α_{95}. We have used their selected data set, with the addition of some Bonin islands sites accepted by *Koyama et al.* [1992], as summarized below. A further contribution to the paleomagnetism of the islands of the Philippine Sea Plate has come from Halmahera and nearby islands in the southernmost part of the plate [*Hall et al.*, 1994a; *Ali and Hall*, 1994]. These results are also reviewed. The data are summarized in Tables 2 and 3.

The Northern Islands of the Philippine Sea Plate

The Mariana and Bonin islands have a middle-upper Eocene volcanic basement that was originally contiguous with the Palau-Kyushu Ridge (Figure 1). During the late Oligocene and early Miocene (30-17 Ma) the Parece Vela Basin opened, and in so doing it carried the West Mariana Ridge and the Mariana arc east. In the northern part of Philippine Sea Plate the opening of the Shikoku Basin (25-15 Ma) had a similar effect, carrying the Izu-Bonin arc and Bonin Ridge east. The development of the Mariana Trough since the late Miocene has resulted in separation of the West

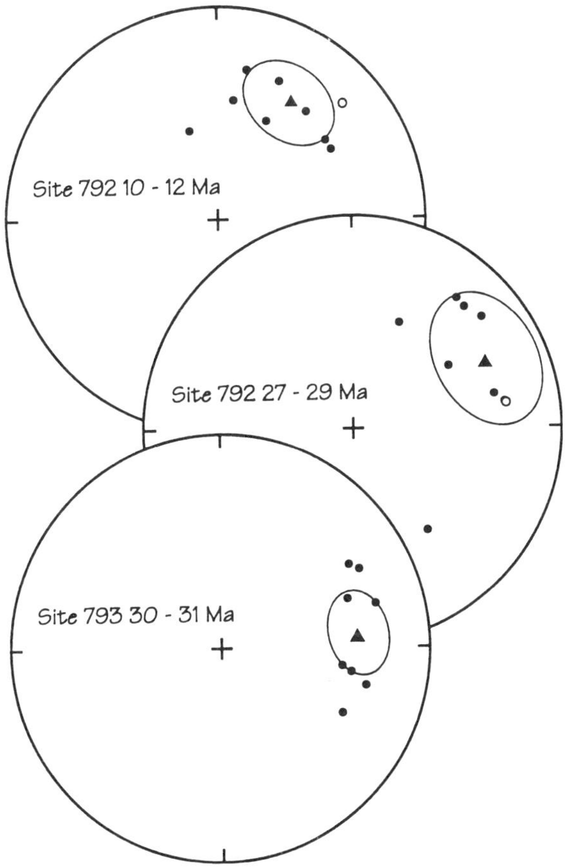

Fig. 3. Declination plots for sites 792 and 793 (leg 126) in the Izu-Bonin forearc region after *Koyama et al.* [1992]. Mean vector directions are shown as triangles and α_{95} circles of confidence are also plotted.

Mariana Ridge and the Mariana arc. Today the Izu-Bonin arc is in the early stage of extension. The Palau islands have not suffered eastward transport by the two phases of extension described above, and so paleomagnetic data could provide a direct record of West Philippine Sea Plate motion.

In addition to the middle and late Tertiary backarc basin growth of the eastern Philippine Sea Plate, there has been subduction at the eastern boundary for much of the Tertiary. There have been numerous suggestions, following the original ideas of *Vogt* [1973] and *Vogt et al.* [1976] that the arcuate shape of the Marianas is the result of deformation of this eastern boundary of the plate by subduction-related processes. All of the islands at the eastern edge of the plate are situated close to active subduction zones. It is therefore unfortunate that one of the places where we are able to carry out standard paleomagnetic surveys is a region which may have suffered tectonic deformation.

Palau islands. The Palau islands are composed of island-arc subaerial and submarine volcanic rocks and reef limestones ranging in age from late Eocene to Holocene [*Mason et al.,* 1956]. Pliocene-Pleistocene reef limestones and Miocene-Pliocene clays overlie a volcanic section that consists of the Oligocene to Miocene Ngermelengui formation including the Medorm and Arakabesan members, the lower Oligocene Aimeliik formation, and the upper Eocene Babeldaup formation. Ages are based on K-Ar determinations [*Meijer et al.,* 1983; *Haston et al.,* 1988].

Aoki [1972] reported results from the lower Miocene Arakabesan member which show clockwise deflected directions. *Adachi et al.* [1987] presented data from eleven sites in Eocene and Oligocene rocks; eight sites show clockwise deflected directions. *Haston et al.* [1988] presented data from twelve sites located in the volcanogenic rocks (Figure 4). Five sites from the lower Miocene Arakabesan member gave an average direction that is clockwise deflected ($I = 3.5°$, $D = 54.5°$, $\alpha_{95} = 17.4°$, $K = 20.3$). Two sites from the middle Oligocene Medorm member show larger clockwise deflected directions and two sites from the upper Eocene Babeldaup formation with similar directions were interpreted as indicating middle Oligocene remagnetization based on rock magnetic evidence [*Haston,* 1988]. These four sites give an average middle Oligocene direction of $I = 5.3°$, $D = 68.4°$, $\alpha_{95} = 12.8°$, $K = 52.4$. Three sites from the lower Oligocene Aimeliik formation are strongly (>90°) clockwise deflected and were initially interpreted as local rotation associated with the opening of the Parece Vela Basin [*Haston et al.,* 1988]. However, remagnetization and poor magnetic stability in many of the samples make the data unreliable. A consistent picture of clockwise rotation emerges from all results from the Palau islands from at least the middle Oligocene onwards. Minor northward translation is permitted within the error limits of the inclination data, but is not required (Figure 4).

Guam. Tracey et al. [1964] described Tertiary limestones unconformably resting on the volcanogenic Umatac and Alutom formations. They assigned a Miocene age to the Umatac formation and its Facpi member, and an Eocene to Oligocene age to the Alutom formation. Subsequent mapping [*Reagan and Meijer,* 1984] and radiometric age dating [*Meijer et al.,* 1983] indicate a late Eocene-early Oligocene age for the Alutom formation and a late middle Eocene age for the Facpi formation.

Keating and Helsley [1985] presented graphical plots of declination data from nineteen sites taken from the various formations that, in general, show clockwise deflected directions. *Haston and Fuller* [1991] reported data from Pliocene to Eocene rocks on Guam (Figure 4). Three sites

TABLE 3. Latitudes calculated from inclination data for land sites on the Philippine Sea Plate

Location	Age	Palaeolatitude	Change in Latitude	Source
Bonins, Chichijima	41.0 ± 3.0	-0.4 ± 8.5	-27.5 ± 8.5	*Haston and Fuller* [1991]
Bonins, Anijima, Nishijima	41.0 ± 3.0	5.0 ± 5.9	-27.1 ± 5.9	*Haston and Fuller* [1991]
Bonins, Hahajima	41.0 ± 3.0	2.4 ± 13.3	-24.8 ± 13.3	*Haston and Fuller* [1991]
Bonins, Mukojima	41.0 ± 3.0	13.1 ± 5.7	-14.0 ± 5.7	*Haston and Fuller* [1991]
Bonins, Mukojima	41.0 ± 3.0	7.2 ± 1.4	-19.9 ± 1.4	*Haston and Fuller* [1991]
Guam	35.0 ± 5.0	7.7 ± 5.8	-5.8 ± 5.8	*Haston and Fuller* [1991]
Palau	32.3 ± 1.1	-5.1 ± 10.3	-12.6 ± 10.3	*Haston and Fuller* [1991]
Palau	30.3 ± 0.9	2.7 ± 6.4	-4.8 ± 6.4	*Haston and Fuller* [1991]
Palau	20.1 ± 0.5	1.8 ± 8.7	-5.7 ± 8.7	*Haston and Fuller* [1991]
Saipan	41.4 ± 0.9	-6.1 ± 6.5	-21.3 ± 6.5	*Haston and Fuller* [1991]
Saipan	12.9 ± 0.3	16.5 ± 5.2	1.3 ± 5.2	*Haston and Fuller* [1991]
Saipan	2.0 ± 1.5	4.3 ± 11.4	-10.9 ± 11.4	*Haston and Fuller* [1991]
Izu	16.5 ± 5.0	14.5 ± 4.9	-20.5 ± 4.9	*Hirooka et al.* [1985]
Izu	16.0 ± 5.0	21.0 ± 12.1	-14.0 ± 12.1	*Hirooka et al.* [1985]
Izu	11.0 ± 1.0	21.6 ± 2.4	-20.5 ± 2.4	*Hirooka et al.* [1985]
Izu	5.6 ± 0.5	25.3 ± 2.0	-14.0 ± 2.0	*Hirooka et al.* [1985]
Izu	4.0 ± 0.0	34.5 ± 4.4	-13.4 ± 4.4	*Hirooka et al.* [1985]
Kasiruta	33.3 ± 0.3	-18.1 ± 3.2	-17.7 ± 3.2	*Hall et al.* [1994a]
Kasiruta	33.3 ± 0.3	-12.5 ± 3.5	-12.2 ± 3.5	*Hall et al.* [1994a]
Kasiruta	32.3 ± 2.8	-9.7 ± 2.9	-9.3 ± 2.9	*Hall et al.* [1994a]
Kasiruta	32.3 ± 2.8	-12.7 ± 7.3	-12.3 ± 7.3	*Hall et al.* [1994a]
Waigeo	29.0 ± 6.0	-5.7 ± 3.7	-5.4 ± 3.7	*Hall et al.* [1994a]
Waigeo	29.0 ± 6.0	-9.3 ± 7.5	-9.0 ± 7.5	*Hall et al.* [1994a]
Halmahera	47.2 ± 2.9	1.9 ± 8.0	0.6 ± 8.0	*Hall et al.* [1994a]
Halmahera	40.0 ± 2.0	-12.6 ± 11.2	-13.2 ± 11.2	*Hall et al.* [1994a]
Halmahera	29.4 ± 6.1	-18.1 ± 2.7	-19.5 ± 2.7	*Hall et al.* [1994a]
Halmahera	26.7 ± 1.9	-14.6 ± 2.7	-14.6 ± 2.7	*Hall et al.* [1994a]
Halmahera	7.0 ± 2.0	-19.8 ± 2.2	-19.8 ± 2.2	*Hall et al.* [1994a]
Halmahera	7.0 ± 2.0	-21.6 ± 6.4	-23.3 ± 6.4	*Hall et al.* [1994a]
Halmahera	5.0 ± 1.0	-0.3 ± 3.7	-1.5 ± 3.7	*Hall et al.* [1994a]
Halmahera	5.0 ± 1.0	1.3 ± 2.1	-2.5 ± 2.1	*Hall et al.* [1994a]
Halmahera	2.2 ± 0.7	-14.3 ± 4.4	-16.0 ± 4.4	*Hall et al.* [1994a]
Halmahera	2.2 ± 0.7	-14.2 ± 1.1	-15.9 ± 1.1	*Hall et al.* [1994a]
Obi	30.5 ± 3.0	-12.3 ± 5.0	-11.0 ± 5.0	*Ali and Hall* [1994]
Obi	12.5 ± 2.5	-7.0 ± 5.6	-5.4 ± 5.6	*Ali and Hall* [1994]
Obi	11.3 ± 3.0	-13.8 ± 2.4	-12.2 ± 2.4	*Ali and Hall* [1994]
Obi	11.0 ± 1.0	-24.1 ± 2.1	-22.5 ± 2.1	*Ali and Hall* [1994]

from the Miocene Alifan Limestone yielded stable results that are indistinguishable from the present reversed field direction; NRM intensities suggest that the magnetization is not primary. Twelve sites collected from the Alutom formation yielded stable results. Three of these sites show directions close to the present field, but slightly counterclockwise deflected. The remaining nine sites show strongly clockwise deflected directions. The magnetization is considered to predate folding and is probably primary. The mean of the corrected directions is I = 15.1°, D = 66.1°, α_{95} = 11.1°, K = 20.1. Three samples collected by *Kobayashi et al.* [1971] from the Facpi formation show a clockwise deflected direction of about 80°; eight sites reported by *Larson et al.* [1975] have directions that are scattered and clockwise deflected between 12° and 87°. *Haston and Fuller* [1991] reported results from ten sites in dikes and flows of the Facpi formation but observe that structural corrections are not very reliable. In both *in situ* and corrected coordinates, all of the sites show clockwise deflected directions that range from 18° to 92°. The inclinations of the sites are also widely scattered and range between -32° and 52°. There is no correlation between geographic location or rock type with direction, indicating that the scatter in the data cannot be solely caused by lack

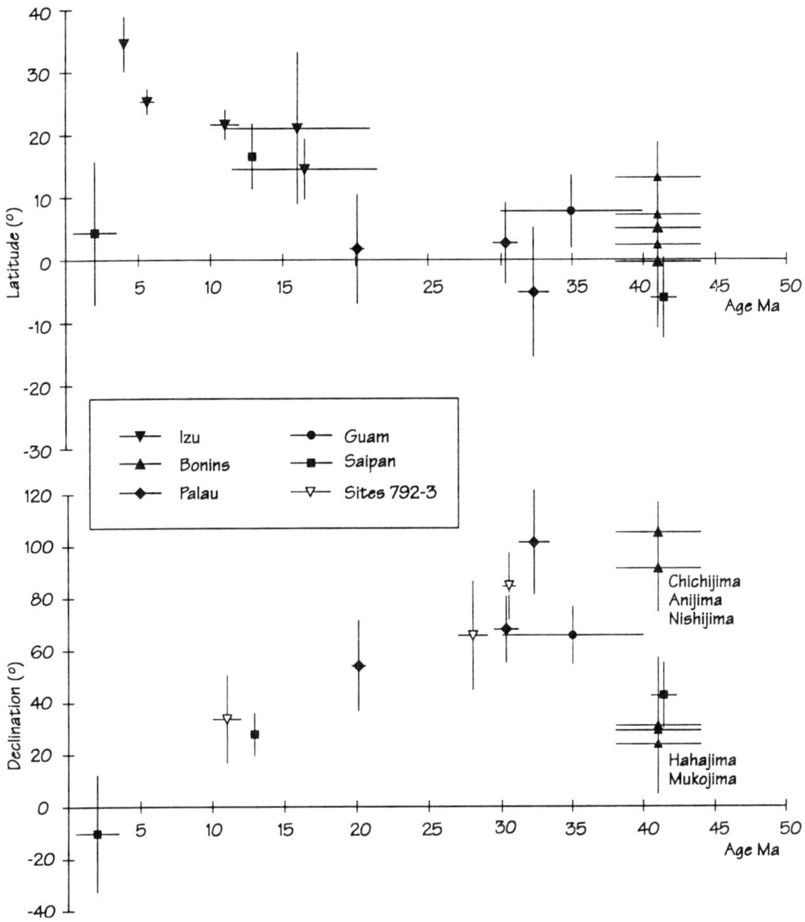

Fig. 4. Paleolatitude and declination versus age for land sites at the eastern margin of the Philippine Sea Plate; declination plot includes results from ODP sites 792-3. Tables 2 and 3 give data sources.

of structural control or different age units, although rock magnetic studies suggest that the magnetization is primary. It is not clear why the results show such extreme scatter, but it may be due to a combination of secular variation, rotation after magnetization and poor structural control. This phenomenon of good individual site determinations but wide between-site scatter is a recurrent puzzle of work in these islands, which requires further investigation. The results from the remagnetized Alifan limestone and Alutom formation permit a minor recent episode of counterclockwise rotation, but from upper Eocene and lower Oligocene sites strong clockwise rotations are found by all workers.

Saipan. Cloud et al. [1956] describe Saipan as a volcanic core enveloped by reef limestones. The oldest rocks are dacitic volcanic flows and tuffs of the upper Eocene Sankakuyama formation [*Meijer et al.*, 1983], unconformably overlain by andesitic pyroclastic rocks and water-laid volcanogenic sediments of the Hagman formation, which yielded a K-Ar age of 35.7 ± 0.5 Ma [*Meijer et al.*, 1983] and late Eocene fossil ages [*Cloud et al.*, 1956]. Overlying the older volcanic rocks are reef limestones of the Oligocene-Miocene Tagpachau limestone formation [*Cloud et al.*, 1956] and basaltic flows and pyroclastics of the Fina Sisu formation which yielded a mid-Miocene K-Ar date of 12.9 ± 0.3 Ma [*Meijer et al.*, 1983]. The youngest rocks are Pliocene-Pleistocene reef limestones of the Mariana limestone formation.

Fuller et al. [1980] first reported paleomagnetic data from Saipan and suggested that Saipan has undergone significant clockwise rotation since the late Eocene. *Keating and Helsley* [1985] also reported paleomagnetic data suggesting that Saipan has undergone progressive clockwise rotation since the Miocene. *Haston and Fuller* [1991] reported results from rocks ranging from Pliocene-Pleistocene to Eocene in age (Figure 4). Four sites from the Pliocene-

Pleistocene Mariana limestone formation yielded stable results that have a mean direction of I = 8.6°, D = 350.1°, α_{95} = 22.5°, K = 17.5, which is slightly counterclockwise deflected, although the large α_{95} makes the result indistinguishable from the present dipole field. Four sites collected from basaltic flows of the middle Miocene Fina Sisu formation yielded stable magnetic directions although their flat-lying nature and extreme weathering means that structural corrections are uncertain. These were combined with one site from the Oligocene-Miocene Tagpachau limestone formation, which yielded a similar result to give a mean middle Miocene direction of I = -30.7°, D = 208.1°, α_{95} = 8.4°, K = 64.6. Results from the Hagman formation [*Fuller et al.*, 1980; *Keating and Helsley*, 1985; *Haston and Fuller*, 1991] suggest around 40-50° of clockwise deflection, but the quality of the results is very poor. It appears that these rocks are flow breccias, brecciated after the magnetization was acquired, and thus the results cannot be used in tectonic analysis. *Fuller et al.* [1980] reported clockwise-deflected declinations between 9° and 46° and negative inclinations from four sites in dacite flows of the upper Eocene Sankakuyama formation. *Haston and Fuller* [1991] reported additional sites which indicate a mean *in situ* direction significantly clockwise-deflected from the present dipole field position of I = -10.3°, D = 40.9°, α_{95} = 18.5°, K = 9.8. Determination of bedding attitude in many flows is problematic, but there is little difference between the mean-corrected and the *in situ* directions and the magnetization is interpreted to be pre-folding.

The Pliocene-Pleistocene Mariana limestone from Saipan is counterclockwise rotated, but indistinguishable at the 95% confidence level from the present field, or the corresponding reversed direction. However, all older rocks show clockwise rotations, with the upper Eocene Sankakuyama formation revealing substantial northward motion. The combination of the inclination data from the Sankakuyama formation from Saipan and the Alutom formation from Guam suggests an equatorial crossing between the age of magnetization of these two formations. This places the equatorial crossing as latest Eocene (Figure 4), similar to that interpreted from ODP leg 31 sites [*Louden*, 1977] but younger than that suggested by ODP leg 58 sites [*Kinoshita*, 1980]. The combination of the inclination data from the two islands is of course not dependent upon the assumption of common tectonic history. The declinations revealing progressive clockwise rotation of the Saipan and Guam sites are indistinguishable, but the large α_{95} limits do not exclude somewhat different rotation histories. Although they both form part of the East Mariana Ridge they may have suffered local rotations as a part of the two phases of backarc extension or during deformation of the Mariana arc associated with subduction of the Pacific Plate. Indeed the model of *Karig et al.* [1978] predicts that both islands will have suffered counterclockwise rotation in the past 5 m.y. associated with opening of the Mariana Trough. Prior to this there is a clear record of clockwise rotation observed by all workers.

Bonin islands. The Bonin islands, located on the Bonin Ridge, consist of three main groups of islands that are approximately 50 km apart. The islands are composed predominantly of volcanic rocks thought to have formed in an active arc prior to the opening of the Parece Vela and Shikoku Basins and are the type locality for the high MgO boninites.

Mukojima and Yomejima are the northernmost of the Bonin islands and consist of boninitic and andesitic pillow lavas, hyaloclastites, and dikes [*Kodama*, 1981]. Two K-Ar dates of 42 and 22 Ma have been obtained from these islands [*Tsunakawa*, 1983] with the likely age of the rocks being late Eocene. Only three sites have been reported from these islands [*Kodama*, 1981], all of which are clockwise-deflected between 23° and 52°, although one site has an α_{95} of 37°.

The central group of islands consists predominantly of andesitic and boninitic pillow lavas, dikes and hyaloclastites. One K-Ar age determination from Ototojima yielded a 22 Ma age [*Tsunakawa*, 1983]. Lower Eocene radiolaria and nannofossils are reported from northern Chichijima and a wide range of K-Ar ages have been obtained [*Tsunakawa*, 1983], with the oldest being late Eocene and younger ages attributed to a thermal disturbance around 5-10 Ma. Geological similarities of the islands suggests that all the rocks are of late Eocene age. Results from twenty sites from Anijima, two from Nishijima, and six from Ototojima were reported by *Keating et al.* [1983]. Although there is a large amount of scatter, the results predominantly cluster around 090°. A mean *in situ* direction of I = -0.8°, D = 91.7°, α_{95} = 16.9°, K = 4.9 was obtained from the sites on Anijima and Nishijima. The six sites from Ototojima are widely scattered. Twenty-seven sites from Chichijima yield stable magnetic directions [*Kodama*, 1981; *Kodama et al.*, 1983; *Keating et al.*, 1983]. The majority of the sites show strongly clockwise-deflected directions, and the mean *in situ* direction is I = 9.5°, D = 105.3°, α_{95} = 12.6°, K = 6.1. *Haston and Fuller* [1991] estimated a structurally corrected direction by averaging known corrected directions with *in situ* directions from the sites where there is no tectonic information. There is only a small difference between the mean *in situ* and corrected directions and since the strata are only gently dipping they suggest that the *in situ* direction is a late Eocene direction.

Hahajima at the south end of the Bonin islands comprises

andesitic lavas and volcanic breccias together with a number of dikes. Most of the strata are gently dipping, usually less than 20° [*Kodama et al.*, 1983]. Marine sediments contain fossils of late middle Eocene age [*Saito*, 1962; *Ujiie and Matsumaru*, 1977]. K-Ar dates from the island have a wide scatter and range from 40 Ma to 9 Ma [*Kaneoka et al.*, 1970; *Tsunakawa*, 1983]. The similarity of the geology to the other islands suggests that these rocks are also upper Eocene. Only four sites with stable magnetic directions have been reported from Hahajima [*Kodama*, 1981]; the directions are scattered, but all show clockwise deflections.

The paleomagnetic data from the Bonin islands are reminiscent of that from the Facpi formation from Guam: individual sites give internally consistent results with low α_{95} limits, but between-site scatter is very large. This pattern of results cannot be explained simply by local rotations about a vertical axis, for example, by block rotation in the vicinity of strike-slip faults, but requires additional rotation about non-vertical axes. While the directions are predominantly clockwise rotated, both the inclination and declination vary so much that neither the amount of rotation nor the magnitude of the latitudinal shift can be accurately determined. Nevertheless, *Haston and Fuller* [1991] used the results from Chichijima and those from Anijima and Nishijima (Figure 4) because so many sites were available that even though the dispersion measured by K was only 6, the α_{95} was 11. The results from Hahajima, Mukojima, Otootojima, and Yomejima were not used by them in tectonic analysis because of the relatively small number of sites. In contrast, *Koyama et al.* [1992] accepted the smaller rotation of Hahajima, Yomejima, and Mukojima and excluded only the very badly scattered data from Otootojima from their tectonic analysis. We have followed *Koyama et al.* [1992] and excluded data from Otoojima but have also excluded the single site from Yomejima with an α_{95} of 37.

Japan: the Izu Peninsula

The Izu peninsula is the northern tip of the Izu-Bonin arc which has become accreted to Japan during subduction of the northern Philippine Sea in the last 12 m.y. [*Taira et al.*, 1989]. *Hirooka et al.* [1985] describe paleomagnetic results from lower Miocene sediments of the Nishina formation, lower Miocene volcanics of the Kadono formation and Miocene-Pliocene sediments of the Shirahama group. Nineteen sites were described, summarized in Table 3, among which four Pliocene sites have inclinations indicating present-day latitudes. Sites in older rocks all show inclinations indicating lower paleolatitudes and *Hirooka et al.* [1985] suggested that the peninsula moved north at a rate of approximately 1°/m.y. between the early Miocene and accretion to Japan at about 3-5 Ma. They attributed this movement to rotation of the Philippine Sea Plate although declinations from all sites are scattered.

The Southern Islands of the Philippine Sea Plate

The southern part of the Philippine Sea Plate has been ignored in most tectonic syntheses. Recent studies of seismicity and marine geology [*McCaffrey*, 1982; *Nichols et al.*, 1990] indicate that the islands between Halmahera and Waigeo form part of the Philippine Sea Plate with the southern boundary of the plate being the Sorong fault system (Figure 1). Interpretation of marine magnetic anomalies in the West Philippine Basin [*Mrozowski et al.*, 1982; *Hilde and Lee*, 1984] indicate that the early Tertiary opening of the central West Philippine Sea has separated the Daito Ridge province of the northern part of the basin from the southern part of the plate. The similarities of the pre-Neogene geology of the Halmahera-Waigeo region [*Hall and Nichols*, 1990; *Hall et al.*, 1994b] to the Daito Ridge province record a similar early arc history and indicate that the northern and southern regions of the West Philippine Basin include the oldest rocks of the plate.

Halmahera and Waigeo group of islands. The Indonesian islands between Halmahera and Waigeo constitute the largest area of land within the Philippine Sea Plate: ~5 x 10^4 km^2 spread over ~25 x 10^4 km^2. There is a basement of ophiolitic and arc rocks [*Hall et al.*, 1988; 1991] overlain by, and imbricated with, arc volcanic and sedimentary rocks of late Cretaceous-Eocene age; arc plutonic rocks intrude the ophiolites. A late Eocene regional unconformity corresponds to the change in Pacific Plate motion at ~43 Ma [*Clague and Jarrard*, 1973]. The ophiolitic and arc basement rocks are overlain by shallow water upper Eocene limestones and a deep water Oligocene extensional sequence including basaltic pillow lavas and volcaniclastic turbidites [*Hall et al.*, 1994b]. Arc volcanic activity ceased in the region in the latest Oligocene to earliest Miocene. A second regional unconformity separates older rocks from early-middle Miocene limestones and crustal fragments of Australian and Philippine Sea Plate origin have a common stratigraphic history after the early Miocene. Eastward subduction of the Molucca Sea Plate at the western edge of the Philippine Sea Plate resulted in late Neogene volcanic activity producing the Halmahera arc.

254 paleomagnetic sites have been drilled in these islands in rocks of Jurassic to late Neogene age and 63 of these sites have yielded stable results. All sites north of the Sorong fault were drilled in rocks which are currently part of the

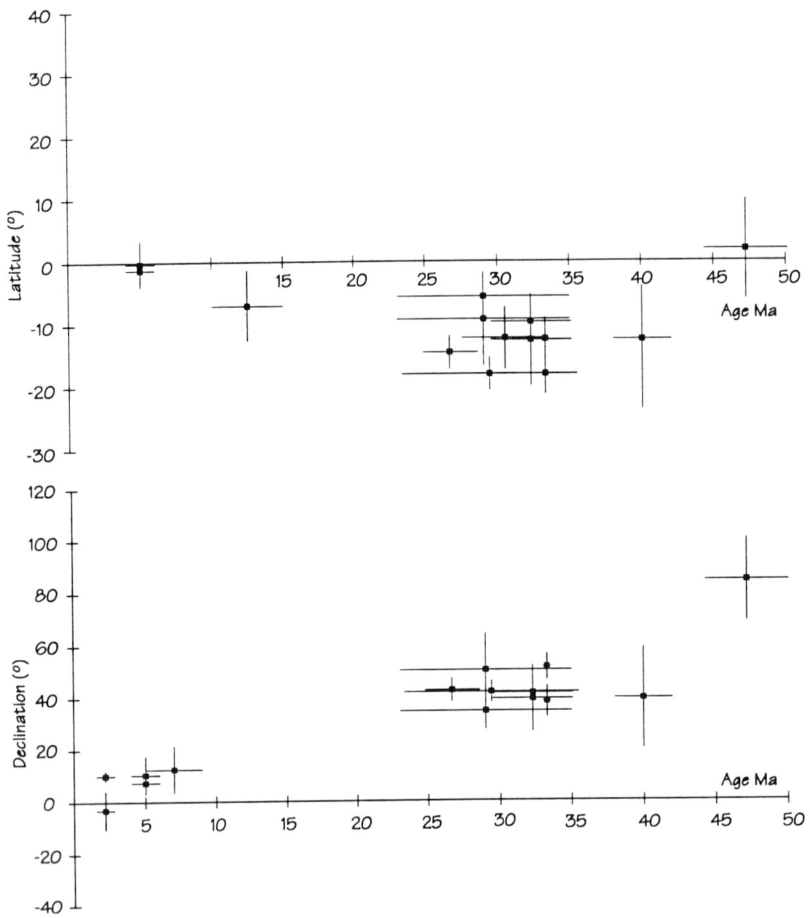

Fig. 5. Paleolatitude and declination versus age for land sites within the southern part of the Philippine Sea Plate in east Indonesia. Tables 2 and 3 give data sources.

Philippine Sea Plate [*Hall et al.,* 1994a]. Other sites drilled in rocks of Philippine Sea Plate origin are parts of crustal blocks now within the Sorong fault system [*Ali and Hall,* 1994]. The results from sites in Tertiary rocks in the area including Halmahera, Bacan, Obi and Waigeo are summarized below.

Declination data define two sub-areas: an area within the Sorong fault system and an area north of the Sorong fault. The area within the fault zone includes crustal fragments of Australian and Philippine Sea Plate origin. Within this area Oligocene and Neogene rocks of Philippine Sea Plate origin record both counterclockwise rotations and clockwise rotations interpreted as block movements within the left-lateral fault system resulting from movement on the Sorong fault [*Ali and Hall,* 1994].

The area north of the Sorong fault is currently part of the Philippine Sea Plate and includes rocks of Philippine Sea Plate origin only. In this area there is a record of long-term, but discontinuous, clockwise rotation (Figure 5). Six sites in upper Neogene volcanic and sedimentary rocks record small clockwise declination deflections [*Hall et al.,* 1994a]. The rotations are consistent with those expected from angular velocities and rotation poles calculated for the Philippine Sea Plate for the present-day [*Ranken et al.,* 1984; *Huchon,* 1986; *Seno et al.,* 1987a, b; 1993]. However, because the rocks are all very young ($<7 \pm 2$ Ma) only the two oldest sites in volcanic rocks show clear differences from the present field direction with structurally-corrected directions of $D = 192.2°$, $I = 38.4°$, $\alpha_{95} = 9.0$, $K = 239$ and $D = 192.2°$, $I = 35.7°$, $\alpha_{95} = 3.2$, $K = 256$. The inclination values suggest that the effects of secular variation have not been fully averaged-out in these sites. No stable results were obtained from early-middle Miocene limestones because of their very weak magnetism and poor magnetic behaviour.

All rocks of ages between ~25 and ~40 Ma north of the Sorong fault show clockwise declination deflections of ~40°. Twenty-two sites in rocks of various formations of Oligocene to late Eocene age on different islands within this extensive area have similar inclinations and declinations. These include pillow basalts with interbedded pelagic sediments and deep water volcaniclastic sediments. The magnetization in these rocks is undoubtedly primary. This is indicated by magneto-mineralogical studies, positive fold tests, and localities at which sites of both normal and reversed polarities have been identified. The consistent declination shifts observed throughout the area north of the Sorong fault indicate that the region has behaved as a coherent unit since at least the end of the Eocene.

Results from five lower Eocene sites drilled in folded limestones have declinations indicating ~90° of rotation. Application of a structural correction produces a mean direction of D = 264.3°, I = -3.7°, α_{95} = 15.9°, K = 24. The tilt-corrected data pass both the *McElhinny* [1964] and the *McFadden* [1990] tests at the 95% confidence level indicating a pre-folding magnetization. The declinations of the lower Eocene rocks are therefore interpreted to indicate ~50° of clockwise rotation between ~40 and 50 Ma.

The inclinations of these lower Eocene rocks indicate sub-equatorial latitudes (2 ± 8°) interpreted to be of northern hemisphere origin (Figure 5) by assuming a progressive clockwise rotation north of the Sorong fault. All other Tertiary rocks north of, and within, the Sorong fault system formed at low southerly latitudes. These data therefore indicate southward movement of the present southern part of the plate between ~50 and ~40 Ma. Within the Sorong fault system, rocks from twenty-five sites of Oligocene to Miocene age have yielded stable results. Although they have different rotation histories, rocks of similar ages in the two sub-areas north of, and within, the Sorong fault system have similar inclinations. The inclination data are interpreted to indicate no significant latitudinal shift between ~40 and ~25 Ma, the interval of no rotation. After ~25 Ma, there was northward movement of ~10-15° during the Neogene, accompanying the ~40° clockwise rotation. The Neogene northward change in latitude is similar to that recorded by DSDP and ODP sites elsewhere in the plate. Young Neogene lavas at four sites within the Sorong fault system and four sites north of the fault record similar declinations but significantly higher inclinations than associated sediments, although they also indicate northward motion with decreasing age. Demagnetization experiments indicate that the remanence of the lavas is primary and these inclination data are attributed to a failure to average geomagnetic secular variation.

OTHER MAGNETIC DATA

Phase Shifting of Magnetic Anomalies

The shape of sea floor magnetic anomalies depends, among other factors, upon the orientation of the local geomagnetic field. This means that, in principle, the direction of the geomagnetic field can be recovered from what has become known as the skewness, or phase shift, of the magnetic anomalies. This makes possible the interpretation of plate motions from this phase shifting of the anomalies [*Schouten and McCamy,* 1972; *Schouten and Cande,* 1976].

These techniques have been applied to the West Philippine Basin anomalies by *Louden* [1976], *Watts et al.* [1977], *Bowin et al.* [1978] and *Shih* [1980]. Their analyses were reviewed in detail by *Jarrard and Sasajima* [1980]. As these authors noted there are conflicting results. Both *Louden* [1976] and *Shih* [1980] report pole positions which imply about 60° of clockwise rotation for the West Philippine Basin since anomaly 20 (44 Ma), whereas *Bowin et al.* [1978] reported a pole implying about 50° of counterclockwise rotation. This latter pole was questioned by *Jarrard and Sasajima* [1980], on the grounds that the set of anomalies analyzed by *Bowin et al.* [1978] were poorly defined. On the other hand, the analyses of *Louden* [1976] and *Shih* [1980] were not corrected for anomalous skewness.

Anomalous skewness can be demonstrated when conjugate anomalies on either side of a ridge are compared, or where the skewness can be determined because the ridge geometry during anomaly formation is known [e.g., *Harrison,* 1987]. It can be caused by non-vertical magnetic source boundaries, or differential rotations subsequent to magnetization, which both give systematically different skewness on either side of a spreading ridge. Anomalous skewness is maximized if the ridge is oriented east-west, and thus the anomalies compared are north and south of the ridge. The analysis of *Watts et al.* [1977] did include a correction for anomalous skewness that reduced the estimate of northward motion. For this reason *Jarrard and Sasajima* [1980] preferred that analysis. *Chamot-Rooke et al.* [1989] report a mean skewness value indicating no significant rotation of the Shikoku Basin since anomaly 6 (20 Ma).

Seamount Paleomagnetism

With the general paucity of paleomagnetic data for the oceanic regions in the early 1960's, an ingenious technique was developed by *Vacquier* [1962] and others to calculate the direction of magnetization of seamounts by treating the whole seamount effectively as a single paleomagnetic sample. To do this it is assumed that (1) the magnetization is uniform, (2) that the structure is flat-bottomed, so that the topography gives the magnetized volume, and finally (3) that the Königsberger ratio of the magnetic material is sufficiently large, that the induced magnetization in the present field can be ignored and the magnetization of the seamount be treated as if it is entirely due to the remanent moment. With the availability of a magnetic survey over the seamount, one then seeks to find the direction of magnetization for the seamount which best fits, in a least squares sense, the observed magnetic field.

This technique was applied to seamounts in the Western Pacific, three seamounts in the Shikoku Basin, and to volcanoes in Japan by *Vacquier and Uyeda* [1967]. The group of seamounts in the Shikoku Basin gave a pole position of 76°N, 44°W with an α_{95} of 18° and it was noted that this result was significantly different from the results for three seamounts east of the Izu-Bonin ridge in the Pacific which gave a pole of 56°N, 66°W. A Cretaceous age was obtained for the Pacific Ocean seamounts by K-Ar determinations on dredged samples, but the authors were careful to note that the samples were altered and that the age was best regarded as a minimum age. No ages were quoted for the Shikoku Basin seamounts, but we now know that two of the three seamounts are on young seafloor close to the spreading axis and so their ages cannot be greater than about 15 Ma [*Chamot-Rooke et al.*, 1987]. Their calculated paleolatitudes are significantly different from the present latitudes, but their declinations are indistinguishable from the present field (Table 4). Because of the short length of intervals of constant polarity in the Neogene, it is likely that these seamounts are not of uniform polarity, which violates assumptions of a uniform direction of magnetization. The situation is made worse by the large errors associated with determinations of the directions of magnetization by inversion of seamount data [e.g., *Harrison*, 1987]. The most reliable seamount poles have precision estimates of ± 10-20° [*Harrison et al.*, 1975].

RECONSTRUCTION OF THE PHILIPPINE SEA PLATE

There have been several previous approaches to reconstructing the Philippine Sea Plate. One of the earliest was by *Keating and Helsley* [1985] who produced a schematic reconstruction about a pole of rotation to the northeast of the plate. This was based on paleomagnetic data and was consistent with the observed clockwise rotation and northward motion. They also noted the possible role of 'ball-bearing' tectonics at the eastern boundary of the plate. *Seno and Maruyama* [1984] based reconstructions on land geological data from Japan and paleomagnetic data from DSDP cores in the West Philippine Basin, which they interpreted as indicating a relatively simple northward motion of the plate since ~50 Ma. They argued that paleomagnetic data from Shikoku Basin seamounts indicated no significant rotation since 17 Ma and closed the Mariana Trough assuming no rotation of the plate. For the interval before 17 Ma they produced alternative reconstructions assuming an 'anchored slab' and a 'retreating trench' at the eastern margin of the plate and closing the older marginal basins. They suggested that rotations inferred from paleomagnetic data in the Bonin, Guam, and Saipan islands could be adequately accounted for by rotation caused by trench retreat before 17 Ma. Rotations inferred from magnetic anomaly skewness data in the West Philippine Basin [*Louden*, 1977; *Shih*, 1980] were suggested to be too large and not adequately substantiated. They concluded that the Philippine Sea Plate originated by backarc spreading. *Jolivet et al.* [1989] also used the method of *Seno and Maruyama* [1984] to constrain the kinematics of the Philippine Sea Plate when reconstructing western Pacific marginal basins.

TABLE 4. Seamount Computations of *Vaquier and Uyeda* [1967]

Seamount	Latitude	Longitude	D	I	Palaeolatitude
IV-5	27.7°N	140.4°E	5	39	22°N
IV-6	29.6°N	137.1°E	-1	31	17°N
IV-7	30.2°N	136.7°E	-1	7	4°N

Byrne and DiTullio [1992] used geological studies from Japan and paleomagnetic data to make partial reconstructions of the Philippine Sea Plate and concluded that the plate originated as trapped Kula-Pacific oceanic crust. We consider that their innovative approach is flawed since it fails to account for the early arc history recorded by drilling and dredging of the Daito Ridge province and the geology of the eastern Indonesian islands of the southern Philippine Sea Plate.

Haston and Fuller [1991] produced alternative models based essentially on paleomagnetic data. They closed the marginal basins within the plate using visual fit by two methods, minimizing overlap and minimizing open space, and then corrected the observed declinations of eastern Philippine sites for the rotations required to close the basins. They argued that the linear trend of declination with age for both models indicated continuous Tertiary rotation of the entire plate with approximately 70° of clockwise rotation since 30 Ma and suggested that the pole of rotation was situated near to the southeast boundary of the plate.

Packham [1990] argued that the models of *Seno and Maruyama* [1984] and *Jolivet et al.* [1989] require too much subduction at the Nankai Trough-Ryukyu Trench. He accepted the latitudinal shift and rotation amount for the plate suggested by *Louden* [1977] and estimated the rotation pole position as 16°N and 152°E, east of Guam. *Koyama et al.* [1992] reviewed all earlier tectonic models proposed for the plate and also favored plate-wide rotation, although they suggested that the rotation pole was situated near Palau (7.5°N and 134.5°E).

We adopt a different approach. As noted above, the area north of the Sorong fault is currently part of the Philippine Sea Plate. Interpretations of marine magnetic anomalies in the West Philippine Sea show that this region includes some of the oldest rocks in the Philippine Sea Plate. We consider that the region has belonged on the southern part of the Philippine Sea Plate since the early Tertiary and this is supported by the geological evidence summarized by *Hall et al.* [1994b]. The area close to the present-day southern boundary of the Philippine Sea Plate, the Sorong fault, includes sites with local rotation histories different from those north of the fault. As discussed by *Ali and Hall* [1994] we interpret data from these sites to reflect deformation at the plate edge and suggest that in eastern Indonesia, unlike areas at the present eastern edge of the plate, the distinction between local and plate-wide rotations is clear. We therefore assumed that the paleomagnetic data from the area north of the Sorong fault system in eastern Indonesia record the Tertiary motion history of the oldest and most stable part of the Philippine Sea Plate. This assumption offers a means of incorporating the Philippine Sea Plate into a global kinematic model. The assumption was used, as described below, to determine rotation poles for the plate using a forward modelling approach. We then estimated the effects of closing the marginal basins by two tectonic models and compared the results. Model 1 uses three fragments, rotating the Izu-Bonin arc back to the Palau Kyushu Ridge to close the Shikoku Basin, and rotating the West Mariana and East Mariana Ridges as separate fragments to close the Parece Vela Basin and Mariana Trough. Model 2 cuts these fragments into smaller segments (Figure 6) each of which was rotated separately to simulate spreading center propagation during opening of the basins.

Calculating Rotation Poles for the Philippine Sea Plate

Haston and Fuller [1991] summarized existing Philippine Sea Plate paleomagnetic data from ODP sites, the islands between Palau and the Bonins, and the Izu peninsula. We used this data set and added to it new data from ODP legs 125 and 126 [*Koyama et al.*, 1992; *Haston et al.*, 1992] and data from land sites in eastern Indonesia [*Ali and Hall*, 1994; *Hall et al.*, 1994a] that are within the Philippine Sea Plate. In calculating rotation poles for the Philippine Sea Plate our approach was to start with a simple rotation about a single pole of rotation for the plate and test predictions of latitude changes against data from the plate. The rotation pole is therefore effectively a Philippine Sea Plate-magnetic north Euler pole. We then searched for rotation poles for the plate that could account for the observed latitude shifts. The model was modified and made more complex, as described below, in a series of stages.

There is a large set of inclination data from the plate indicating northward changes of latitude with decreasing age for almost all sites, with the exception of the east Indonesian sites which record a southward change of latitude between 50 and 40 Ma. It is also clear that similar changes in latitude have been experienced by different sites. In our initial modelling we therefore divided the inclination data into subsets based on the present day positions of the sites from which they were obtained. The purpose of this approach was to attempt to model the paleolatitudes for the entire range of latitudes (0° and 35°N) over which the plate currently extends. Because there are insufficient data from some sites to obtain a long record of paleolatitudes we initially considered inclination data as subgroups and compared the observed and predicted paleolatitudes from only the following areas: Izu peninsula (modelled as a point at 35°N, 139°E), ODP sites 442-6 (modelled as 27°N, 133°E), ODP sites 450-9 (modelled as 18°N, 146°E), ODP sites 290-4 (modelled as 16°N, 133°E), and eastern Indonesia (modelled as 0°, 127°E).

There is much less declination information from the plate, and the interpretation of the results from sites in the eastern part of the plate is controversial. The amounts of rotation chosen for the modelling were therefore those indicated by the east Indonesia declination data, which were assumed to record the rotation history of the oldest part of the plate and indicate a total rotation of ~90° clockwise since ~50 Ma.

Present Day Rotation Pole

The first step was to determine the effect of rotating about the present Eurasia (EU)-Philippine Sea Plate (PH) pole (Table 5) which is situated somewhere near, or north of, the northern edge of the plate. The paleomagnetic data cannot be accounted for by rotation about any of these poles. Those which are further south [*Ranken et al.*, 1984; *Huchon*, 1986] could account for the declination shifts but result in southward motion with decreasing age for all sites. Poles which are further north [*Seno et al.*, 1987a, b, 1993] all result in small changes in latitude for most sites using the

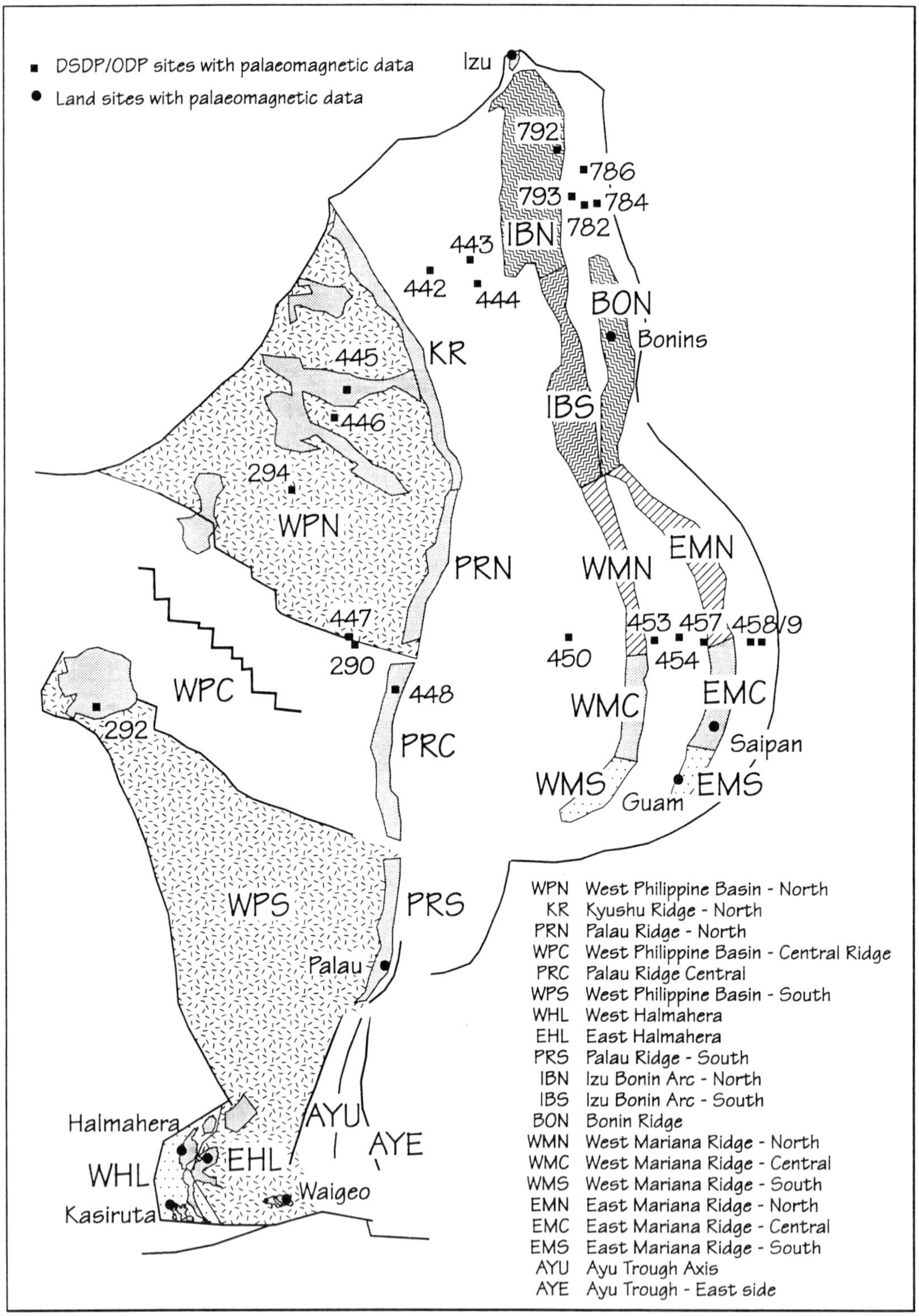

Fig. 6. Fragments used in reconstructing the Philippine Sea Plate. Areas within the plate left unshaded represent crust formed during episodes of marginal basin formation since the early Tertiary.

TABLE 5. Philippine Sea Plate-Eurasia poles

	Latitude	Longitude	ω (deg/m.y.)
Seno et al. [1993]	48.2°N	157.0°E	1.09
Seno et al. [1987a]	49.3°N	159.1°E	1.08
Seno et al. [1987b]	47.3°N	154.4°E	1.06
Huchon [1986]	38.0°N	142.0°E	1.95
Ranken et al. [1984]	37.0°N	141.0°E	1.60

present-day rotation rates (Figure 7) and rotation rates are also too small to produce the observed declination shifts. Using the *Seno et al.* [1987a, b, 1993] poles but increasing the rotation rate results in southward motions of all sites similar to those using the poles of *Ranken et al.* and *Huchon*. According to *Seno and Maruyama* [1984], *Huchon* [1986], and *Jolivet et al.* [1989] the present motion of the plate was established between 3 and 5 Ma. We use the EU-PH pole and rotation rate of *Seno et al.* [1993] for the period 0-5 Ma only. Using the other EU-PH poles for this interval makes no significant difference to the model.

Pre-5 Ma Pole

As a second step we then searched for a pole that could account for the data set for the interval 5-50 Ma. *Koyama et al.* [1992] suggested that during the Neogene the rotation pole for the plate was situated near Palau, since there is little change in inclinations recorded by sites on Palau. However, a pole near Palau does not satisfy the inclination data from elsewhere in the plate (Figure 7). We found that rotation poles that can account for the paleomagnetic data for the interval 5-25 Ma lie on a path trending broadly northeast from Guam. Poles on this path also account for most of the data from the plate for the interval 25-50 Ma, although they do not produce the southward motion of the east Indonesian sites for the early part of this interval. However, this method only produces a zone within which poles can lie. For a given rotation, the farther the pole is from the Philippine Sea Plate, the greater the distance moved by a point at the southern part of the plate. Poles that are in this set but very far from the plate are therefore geologically and kinematically unreasonable since they would result in the plate originating very far east in the Pacific basin, but there is no way from the paleomagnetic data alone to estimate the position of the rotation pole.

Choice of the 5-25 Ma Pole Position

In order to provide a limit on possible pole positions we introduced one geological condition into the model. We propose, as explained by *Hall et al.* [1994b], that the early Miocene regional unconformity in eastern Indonesia represents collision of the Philippine Sea Plate arc with Australia and any pole must permit the Halmahera-Waigeo region of east Indonesia to arrive at the Australian margin at ~25 Ma. As a third step we therefore searched for poles to satisfy this geological condition, and best fit the paleomagnetic data for the interval 5-25 Ma. The position of the north Australian margin is based on the stage poles for India-Antarctica motion from *Royer and Sandwell* [1989]. This condition resulted in a rotation pole within a very small area (approximately ± 5° for both latitude and longitude) with the best fit pole near to 15°N and 160°E (Figure 7). The Neogene pole of 16°N and 152°E suggested by *Packham* [1990] is quite near to our estimate but leaves the Halmahera-Waigeo region about 5°N of the Australian margin at 25 Ma.

Pre-25 Ma Pole

Our data indicate no rotation of the plate between ~25-40 Ma, consistent with the geological evidence for a convergent plate boundary between the Philippine Sea Plate and Australia during this interval [*Hall et al.*, 1994b]. The Halmahera-Waigeo region moved south at some time between 40 and 50 Ma whereas paleomagnetic investigations elsewhere in the Philippine Sea Plate report northward motions. The rotation pole at 15°N and 160°E cannot account for this difference, but a movement of the pole from its Neogene position could. As a fourth step we therefore searched for poles for the interval 40-50 Ma that would allow northward motion of the northern part of the plate and southward motion for the southern part of the plate. Only a small shift from the Neogene position was required and a best fit pole position for the interval 40-50 Ma is 10°N and 150°E (with an error of approximately ± 5° for both latitude and longitude). However, it should be noted that there are few inclination data for the interval 40-50 Ma and that modelling the paleolatitude trends for this interval requires a compromise solution (Figure 7). Poles that completely satisfy the inclination data from eastern Indonesia cause significant southward, instead of the observed northward, motion of other parts of the plate. Furthermore, for the older sites the initial modelling does not take into account the effects of opening of the marginal basins within the plate, which could have important effects on latitude changes of some sites. These effects are considered below. Additional results from Eocene and older rocks are needed and may help to locate the pole more precisely.

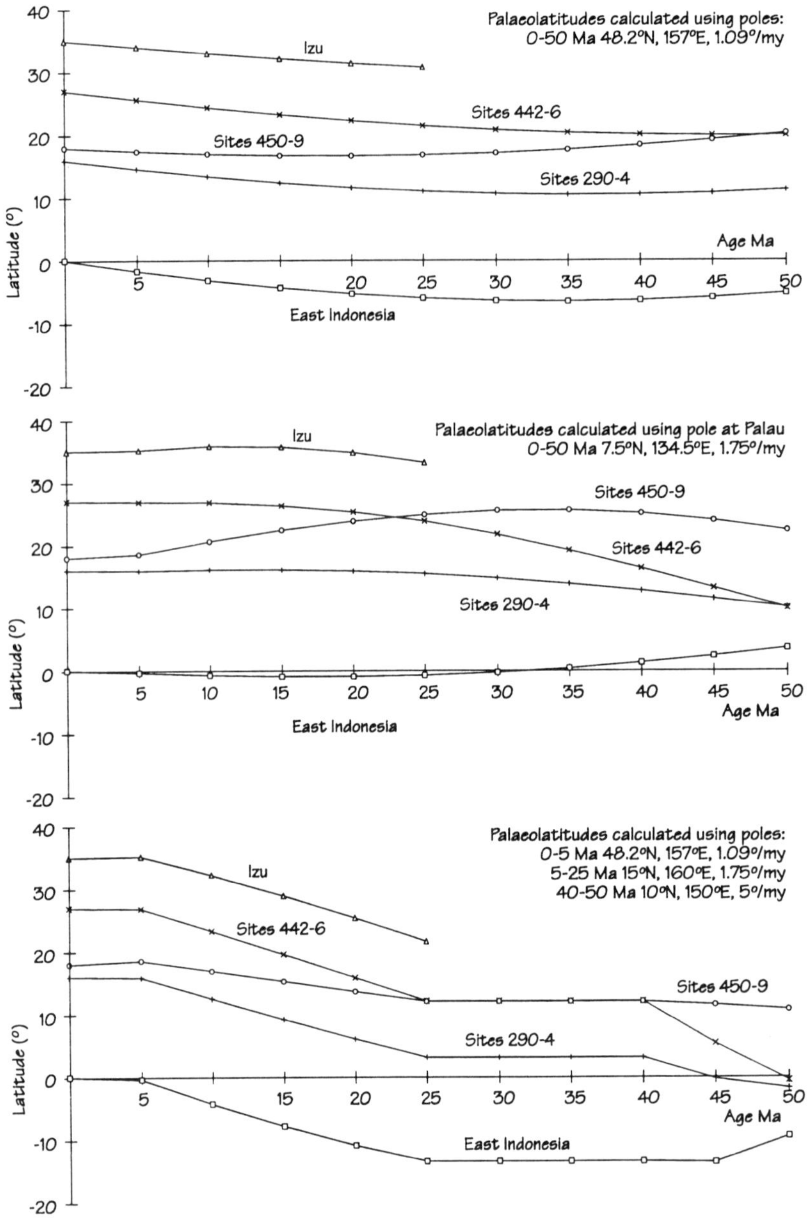

Fig. 7. Paleolatitudes calculated for different sites in the Philippine Sea Plate for different rotation poles. Top shows paths calculated using the present day pole of *Seno et al.* [1993]. Center shows estimates based on the pole near Palau suggested by *Koyama et al.* [1992]. The lowermost plot shows the results using the poles calculated in this study. See text for discussion.

Opening of Philippine Sea Plate Marginal Basins

To calculate rotation poles about which the Tertiary marginal basins opened we used published maps and assigned ages to marine magnetic anomalies using the time scale of *Harland et al.* [1990]. We have defined fragments within the plate separated by marginal basins and considered them as rigid blocks in a similar manner to *Seno and Maruyama* [1984] and *Haston and Fuller* [1991] and made conservative reconstructions by allowing no overlap of fragments. We first attempted to calculate opening poles for marginal basins using marine magnetic anomalies but these are poorly defined or absent in the Parece Vela Basin and Mariana Trough. We later also rotated fragments using the ATLAS [*Cambridge Paleomap Services,* 1993] computer program to achieve a visual fit and constrained the choice of poles by a number of criteria: visual best fit of the fragment edges, fit of declination data from paleomagnetic sites, smooth basin opening along the length of the marginal basins, and a minimum number of rotations, in most cases one, but in some cases two. This method means there is no numerical quantification of goodness of fit but, as shown by *Scotese et al.* [1988], interactive computer graphical techniques can achieve results closely comparable to mathematical methods. A qualitative assessment of the models is possible by comparison of the model predictions and magnetic data.

Because there is no way of relating the movement of the Philippine Sea Plate to any other plate, the model uses paleomagnetic data to calculate the position of the plate with respect to present day magnetic north. A reference position within the plate is therefore required, to which the movement of all other fragments can be referred. Because we have based the model on the east Indonesian paleomagnetic data we chose the southern part of the West Philippine Basin as the reference. The fragments used in the reconstructions are shown in Figure 6. The basic model is therefore based upon the poles estimated by the methods described above and declination data from the east Indonesia sites. The motion of the southern West Philippine Basin (Figure 6, WPS) is described by the PH-EU pole of *Seno et al.* [1993] for the interval 0-5 Ma, followed by 35° rotation between 5 and 25 Ma about a pole at 15°N, 160°E, no rotation between 25 and 40 Ma, and 50° rotation about a pole at 10°N, 150°E between 40 and 50 Ma (Table 6). The movement of Eurasia in the paleomagnetic reference frame is small for the whole of the Tertiary and the paleomagnetic, mean lithosphere and hotspot reference frames differ by only a few degrees for the Cenozoic [*Gordon and Jurdy,* 1986].

In order to account for the effects of opening of the marginal basins in the plate (West Philippine, Parece Vela, Shikoku, Mariana Trough) opening poles were calculated for each of the marginal basins. These were then used with the poles calculated for the whole plate to make reconstructions using the ATLAS [*Cambridge Paleomap Services,* 1993] computer program. As described below the first model was then modified to allow a more complex opening of the Parece Vela Basin. The final preferred kinematic model for the whole plate is listed in Table 6. Negative signs correspond to counterclockwise rotations about the Euler pole during opening. Reconstructions based on these poles are shown in Figures 8 to 10.

West Philippine Basin. The central West Philippine Basin was closed between 44 and 34 Ma, using anomaly identifications and poles calculated from Figure 4 of *Hilde and Lee* [1984] and the revised anomaly ages of *Harland et al.* [1990]. Poles were calculated from transform orientations and anomaly intersections. For the period between 54 and 44 Ma the estimates of the poles are based on fitting anomalies 24 in the northern and southern West Philippine Basin. The results of these estimates are that for the interval between 34 and 54 Ma the southern and northern parts of the West Philippine Basin move symmetrically with respect to the central basin ridge, they each move about 8°, and they remain parallel to the ridge. Since 34 Ma the whole of the West Philippine Basin has been a single rigid fragment and all the fragments forming the West Philippine Basin have the same motion history. Models 1 and 2 are identical for the West Philippine Basin.

Shikoku Basin. The Shikoku Basin was first closed using data from *Chamot-Rooke et al.* [1987] and considering the whole of the Izu-Bonin arc north of about 23°N (Figure 6, IBN and IBS) as a single fragment. The oldest anomaly in the Shikoku Basin is ?7 dated as ~25 Ma. There is a change in anomaly orientation at ~ anomaly 6 (20 Ma) and spreading ceased at anomaly 5B (15 Ma). Two poles have been calculated from Figure 5 of *Chamot-Rooke et al.* [1987] to rotate the Izu-Bonin arc with respect to the Kyushu Ridge. A first pole requiring a rotation of -1.6° closes the central Shikoku Basin between 15-20 Ma; a second pole with a rotation of 5° closes the remainder of the Shikoku Basin between 20-25 Ma. This reconstruction (Model 1) leaves a triangular gap widening northwards at the north end of the Shikoku Basin (Figure 9).

To better close the Shikoku Basin the Izu-Bonin arc was then considered as two fragments (Figure 6, IBN and IBS), which were rotated separately in Model 2. The northern fragment (IBN) was rotated back to the Kyushu Ridge to fit the boundary of rifted blocks of Oligocene arc crust shown as Figure 11 of *Taylor* [1992]. For the opening phase between 15-20 Ma we used the same pole as Model 1 for IBN and IBS but then searched for poles that close the basin fully. A number of poles can achieve a good fit; they have similar longitudes (~138-142°E) but vary in latitude between ~24°N and 10°S. In our preferred reconstruction (Model 2) we chose a pole near the south end of the Shikoku Basin (24°N, 138°E). This pole maximizes the amount of rotation of the Izu-Bonin arc relative to the Kyushu Ridge in an attempt to fit declination data from sites in the Bonin forearc. The southern fragment (IBS) was then fitted as closely as possible to the Kyushu Ridge by rotation between 20-25 Ma while minimizing overlap of IBS and IBN. Again we searched for poles that maximize rotations relative to the Kyushu Ridge to best fit declination data from sites in the Bonin islands. Poles that achieve a good fit

TABLE 6. Rotation poles used in Philippine Sea Plate reconstruction (Model 2)

Fragment[a]		Start Ma	End Ma	Euler Pole Lat.	Euler Pole Long.	Total Rotation	Comment
WPS	EUR	0.00	5.00	48.2	157.0	5.45	*Eurasia-Philippine Sea Plate pole of Seno et al.* [1993]
WPS		5.00	25.00	15.0	160.0	34.0	*S WPSP wrt present N; 40° total CW rotation 0-25 Ma*
WPS		25.00	40.00				*No rotation*
WPS		40.00	50.00	10.0	150.0	50.0	*50° further rotation: 90° total rotation 0-50 Ma*
WPC	WPS	0.00	34.26				*Fix Central Basin Axis to S WPSP*
WPC	WPS	34.26	43.13	0.16	39.09	-3.3	*Close WPSP (S side)(C13-C20): fit anom 20*
WPC	WPS	43.13	53.39	1.79	28.0	-5.0	*WPSP closure to C24: fit anom 24*
WPN	WPC	0.00	34.26				*Fix N WPSP to Central Basin Axis*
WPN	WPC	34.26	43.13	0.09	41.4	-3.0	*Close W PSP (N side)(C13-C20): fit anom 20*
WPN	WPC	43.13	53.39	1.79	28.0	-5.0	*WPSP closure to C24: fit anom 24*
WMN	WPN	0.00	17.00				*Fix N end of W Mariana Ridge to PK Ridge*
WMN	WPN	17.00	20.00	28.0	137.0	-9.0	*Close N end of Parece Vela Basin: stage 1*
WMN	WPN	20.00	30.00	28.0	137.0	-27.0	*Close N end of Parece Vela Basin: stage 2*
WMC	WPC	0.00	17.00				*Fix central part of W Mariana Ridge to PK Ridge*
WMC	WPC	17.00	30.00	60.0	120.0	-10.0	*Close Parece Vela Basin central part*
WMS	WPC	0.00	17.00				*Fix S part of W Mariana Ridge to PK Ridge*
WMS	WPC	17.00	20.00	2.0	147.0	8.0	*Close S end of Parece Vela Basin*
WMS	WPC	20.00	25.00	3.0	142.5	18.0	*Three stages to keep*
WMS	WPC	25.00	30.00	2.0	140.0	14.0	*East edge of plate smooth*
EMN	WMN	0.00	6.00	23.0	142.0	-20.0	*Close N Mariana Trough: fit N EMR to WMR*
EMC	WMN	0.00	5.00	32.0	136.0	-6.0	*Close central Mariana Trough: fit central EMR to WMR*
EMC	WMN	5.00	6.00	14.6	144.8	6.0	*Two stages for smoothness*
EMS	EMC	0.00	5.00				*Fix south EMR to EMR*
EMS	EMC	5.00	6.00	14.6	144.8	-20.0	*Close S Mariana Trough: fit S EMR to WMR*
KR	WPN	0.00	50.00				*Move Kyushu Ridge with N WPSP*
IBN	KR	0.00	14.98				*Fix Izu-Bonin Arc to Kyushu Ridge*
IBN	KR	14.98	19.26	1.2	7.8	-1.6	*Close N Shikoku Basin (C5B.1-C6)*
IBN	KR	19.26	25.01	24.0	138.0	30.0	*Fit N end of Izu-Bonin Arc to PK Ridge (C6-C7)*
IBS	KR	0.00	14.98				*Fix Izu-Bonin Arc to Kyushu Ridge*
IBS	KR	14.98	19.26	1.2	7.8	-1.6	*Close central Shikoku Basin(C5B.1-C6)*
IBS	KR	19.26	25.01	10.0	140.0	10.0	*Fit S end of Izu-Bonin Arc (C6-C7)*
BON	IBS	0.00	25.00				*Fix Bonin Ridge to Izu-Bonin Arc*
BON	IBS	25.00	35.00	22.94	141.86	6.0	*Close Bonin Trough*
AYU	WPS	0.00	5.00				*Move Ayu Trough axis with south WPSP*
AYU	WPS	5.00	15.00	6.0	133.5	-10.0	*Close Ayu Trough: west side*
AYE	AYU	0.00	5.00				*Move Ayu Trough E side with south WPSP*
AYE	AYU	5.00	15.00	6.0	133.5	-10.0	*Close Ayu Trough: east side*
EHL	WPS	0.00	50.00				*Fix E Halmahera to WPSP S side*
WHL	EHL	0.00	3.00				*Move W Halmahera with E Halmahera*
WHL	EHL	3.00	5.00	90.0	0.0	-0.6	*W Halmahera moves E wrt E Halmahera*

[a] Fragment in column 1 rotated with respect to fragment in column 2; fragment codes are given in Figure 6

have similar longitudes (~140°E) and latitudes between 10°N and 10°S; the northernmost pole maximizes the rotation (10°) but the amount is reduced only to about 5° for a pole at 10°S.

Parece Vela Basin. Poles that close the Parece Vela Basin between 17-30 Ma were calculated from topographic elements and magnetic lineations shown on Figure 4 of *Mrozowski and Hayes* [1979] for Model 1. All calculated poles lie along a narrow zone of longitude in the range 139-141°E but no calculated latitude fits the full data set. If the north part of the basin is closed completely about a suitable pole there is a significant overlap for the ridges at the edge of the south part of the basin and vice versa. To eliminate overlap and maintain the calculated longitude for the pole the best fit was found at 35°N, 140°E with a rotation of -15°. This is close to the pole used by *Haston and Fuller*

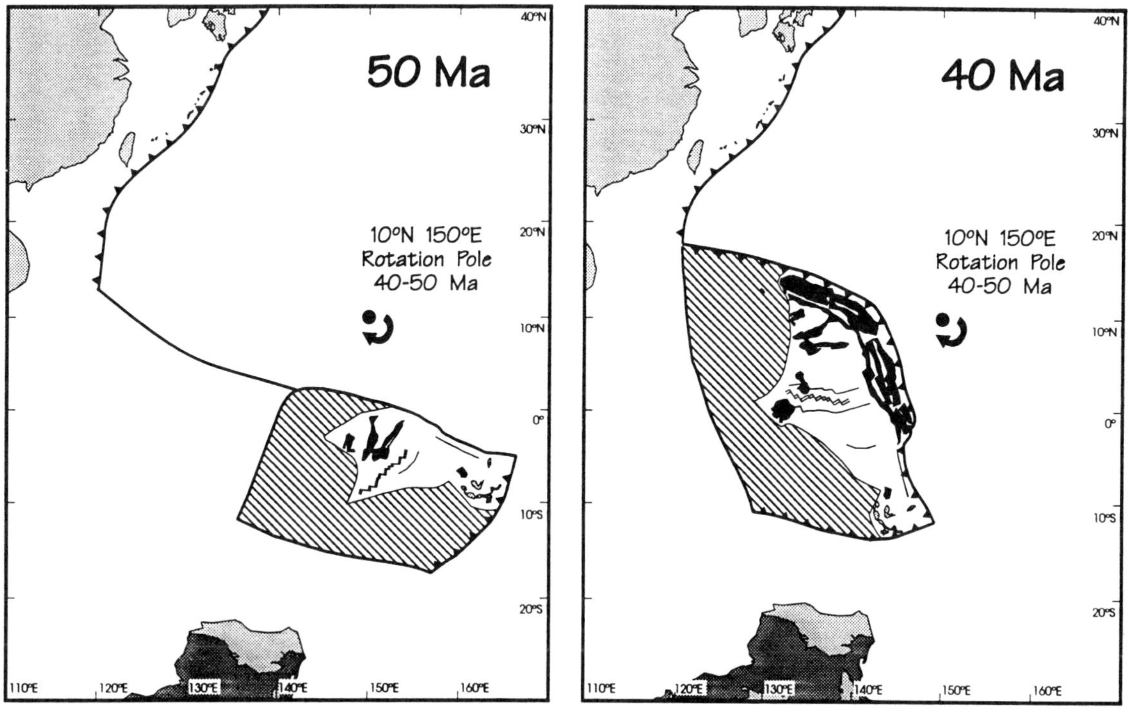

Fig. 8. Reconstructions of the Philippine Sea Plate at 50 and 40 Ma based on Model 2 described in the text. The shaded area shows the approximate area of the plate inferred to have been subducted. The western edge of the plate is schematic and does not portray the complex development of the Philippine islands. Japan has been reconstructed using the model of *Jolivet et al.* [1991]. The position of the north Australian margin is based on *Royer and Sandwell* [1989].

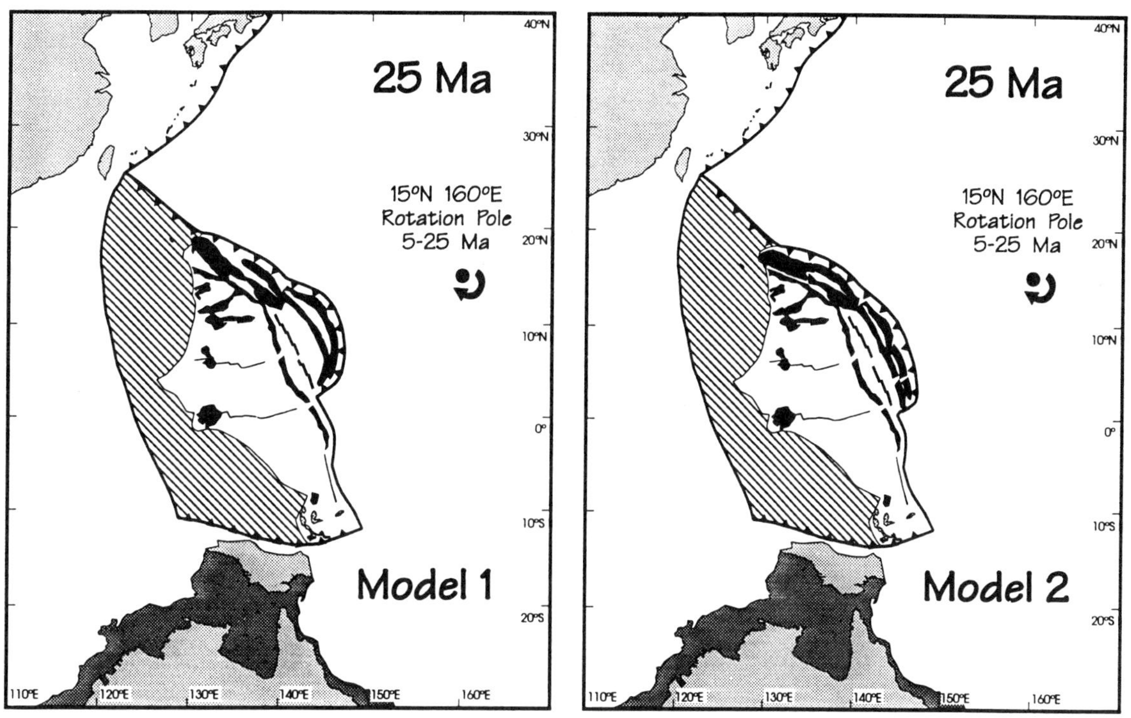

Fig. 9. Alternative reconstructions of the Philippine Sea Plate at 25 Ma based on the two models for marginal basin opening described in the text. See Figure 8 for other details.

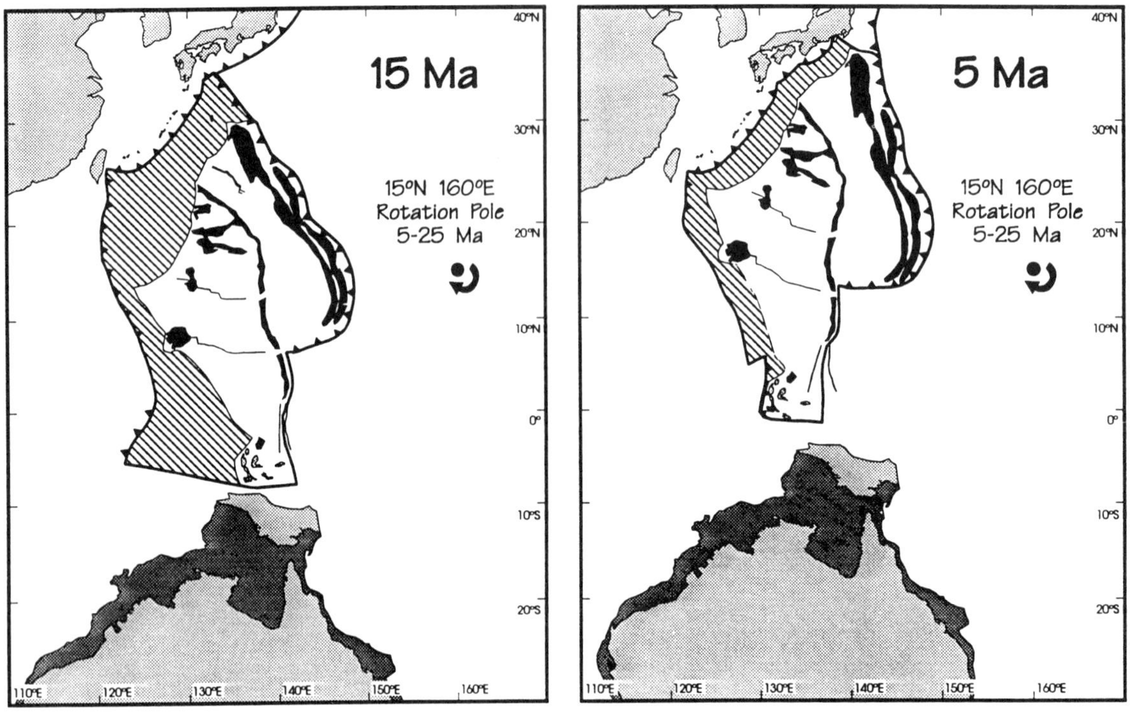

Fig. 10. Reconstructions of the Philippine Sea Plate at 15 and 5 Ma based on Model 2 described in the text. See Figure 8 for other details.

[1991] in their minimum overlap model although with less rotation. Similar effects can be obtained by moving the pole northward, which reduces further the amount of rotation, for example, a pole at 45°N, 140°E requires only -10° rotation to close the basin.

Model 1 has some obvious deficiencies which can be seen in alternative reconstructions for 25 Ma (Figure 9). First, because no overlap has been permitted, neither the Parece Vela Basin nor the Mariana Trough can be fully closed and there are large gaps in their central parts. Second, because the Parece Vela Basin began opening earlier than the Shikoku Basin the poles calculated using Model 1 cause a significant discontinuity, implying a transform offset between the two marginal basins, in the front of the forearc between 30 and 15 Ma. *Chamot-Rooke et al.* [1987] and *Taylor* [1992] suggest that spreading started in the central part of the Parece Vela Basin and propagated north and south. To test this possibility the West Mariana Ridge was divided into three segments, a northern, central and southern section (WMN, WMC and WMS of Figure 6) and these segments were fitted separately to the Palau-Kyushu Ridge in Model 2. The best visual fits, based on rotating different segments of the West Mariana Ridge back to the Palau-Kyushu Ridge were found using different poles and rotations for different segments (Table 6). Poles were also chosen to avoid or minimize overlap of the West Mariana fragments along strike and of WMN with IBS.

Model 2 smooths the opening of the Parece Vela and Shikoku Basins (Figure 9). It implies differential rotations through the West Mariana Ridge (and hence the East Mariana Ridge) between 30 and 17 Ma, with large counterclockwise rotations in the north (-36°), smaller counterclockwise rotations in the center (-10°), and clockwise rotations (40°) at the south end. *Mrozowski and Hayes* [1979] show that at the south end of the basin the rift valley is asymmetrically situated and runs into the arc feature on the west side of the Yap Trench. Thus the rigid opening model constructed here is likely to be an over-simplification. Nonetheless, as shown below, the Model 2 does reduce the discrepancies between the predicted latitudes and paleolatitudes calculated from inclination data (Figures 11 to 14), in particular those from the sites drilled on leg 60, as well as discrepancies in predicted and observed rotations for Guam and Saipan (see below). This suggests that there was curvature of the West Mariana Ridge between 30 and 17 Ma as a result of spreading.

Mariana Trough. The Mariana Trough opened between 0-6 Ma. The *Le Pichon et al.* [1975] Euler pole of 21.4°N, 138.1°E with 29.5° rotation results in a large overlap of the East and West Mariana Ridges but the overlap can be minimized if rotation about this pole is reduced to ~15°. A practically identical result is obtained by closing the Mariana Trough about a pole at its north end at the junction of the East and West Mariana Ridges and is the solution adopted in Model 1 using a rotation of -12°. Model 1 leaves an arcuate gap narrowing north and south in the center of the Mariana Trough.

In Model 2 the East Mariana Ridge was separated into north, central, and south segments (EMN, EMC, and EMS of Figure 6) at similar latitudes to the segments of the West Mariana Ridge. The different rotations thus simulate spreading propagating from the center of the Mariana Trough and permit a much closer fit of the East and West Mariana Ridges by additional rotations of segments of the East Mariana Ridge relative to the West Mariana Ridge. The best fit poles were chosen to fit the segments of the East and West Mariana Ridges as closely as possible, to avoid overlap of segments of the East Mariana Ridge, and to smooth opening of the Mariana Trough. For EMN Model 2 uses the same rotation pole as Model 1 for the interval 0-6 Ma but to fully close the gap in the Mariana Trough there is a total rotation of -20°. For the EMC two rotations relative to WMN were found to be necessary in order to prevent overlap and smooth Mariana Trough opening. Between 0-5 Ma there was rotation of -6° followed by a further rotation of 6° between 5 and 6 Ma. For the interval 0-5 Ma EMS rotated with EMC and then rotated -20° relative to the EMC between 5 and 6 Ma. Model 2 eliminates discrepancies in predicted and observed rotations for Guam and Saipan (Figure 15). This suggests that curvature of the East Mariana Ridge developed between 30 and 17 Ma and between 6 and 0 Ma as a result of backarc spreading.

Bonin Trough. Rifting in the Izu-Bonin arc, for example in the Sumisu Rift, has occurred since 2 Ma but extension of the Bonin Trough is mid-Oligocene in age [*Taylor*, 1992]. The amount of Quaternary extension is small (<5 km, B. Taylor, pers. comm., 1994) but in the Oligocene there may have been some rotation of the Bonin Ridge relative to the Izu-Bonin arc if the pole of rotation is close to the trough. The amount of rotation cannot have exceeded ~6°.

Ayu Trough. Spreading in the Ayu Trough may be continuing at the present day as the Caroline-Philippine Sea Plate rotation pole is situated close to Palau [*Ranken et al.*, 1984; *Seno et al.*, 1993]. *Weissel and Anderson* [1978] suggested that the main phase of spreading was during the middle and late Miocene, approximately 5 to 15 Ma. This is incorporated in the model by the addition of crust to the Philippine Sea Plate west of the Ayu Trough, symmetrical addition of crust to the Caroline Plate east of the Ayu Trough, and a rotation of the Ayu Trough relative to the southern part of the Philippine Sea Plate of -10° about the present pole of rotation near Palau. Because the Ayu Trough rotated away from the former edge of the plate it will have no effect on sites within the plate.

Halmahera. East and West Halmahera are separated by a mid-Pliocene fold and thrust zone due to about 60 km of relative movement [*Hall and Nichols,* 1990] on strands of the Sorong fault system. This is unlikely to have caused relative rotation of the major parts of the island and is included in the model by a rotation of <1° between 5 and 3 Ma.

The Final Solution

Finite rotation poles that map any site on the plate from a past position to its present position have been calculated for 5 Ma intervals between 0 and 50 Ma. Using the finite rotation poles allows a complete comparison of predicted and observed rotations and latitudes for different sites within the plate with time. We consider that these show that the new and the pre-existing paleomagnetic data are part of a single set consistent with an origin in a single plate.

Inclination data. Figures 11 to 14 show the predicted latitudes from the two models compared to paleolatitudes calculated from inclination data for different parts of the plate. The earliest paleomagnetic study of the Philippine Sea Plate using inclination data was by *Louden* [1977] based on DSDP sites 290, 292, and 294. The most complete record is from site 292 and *Louden* inferred a broadly linear northward motion of the plate with decreasing age although he recognized that each mean paleolatitude could have originated on either side of the equator. Figure 11 shows *Louden's* interpretation of inclination data from site 292 but also plots the alternative southern hemisphere paleolatitudes calculated from 24 to 41 Ma sediments. This shows that there is a better fit to the model if all of the pre-Neogene sediments are considered to have originated in the southern hemisphere. In general, there is an excellent correlation of predicted latitudes and paleolatitudes calculated from inclination data for both ocean drilling (Figures 11 to 13) and land sites (Figure 14) in the plate, with the most deviation for the oldest sites (>40 Ma) for which there are the least data. The calculated latitudes are not greatly sensitive to small movement of either the poles of rotation for the whole plate or opening poles for marginal basins. A few points that fall significantly off the calculated curves are clearly problems with the data: the 35 Ma site 290 point is the only result from ocean drilling that indicates a significant southward latitude shift, and one result from site 784A suggests an improbable northward shift of the site by 27° since the early Pliocene. The east Indonesia data

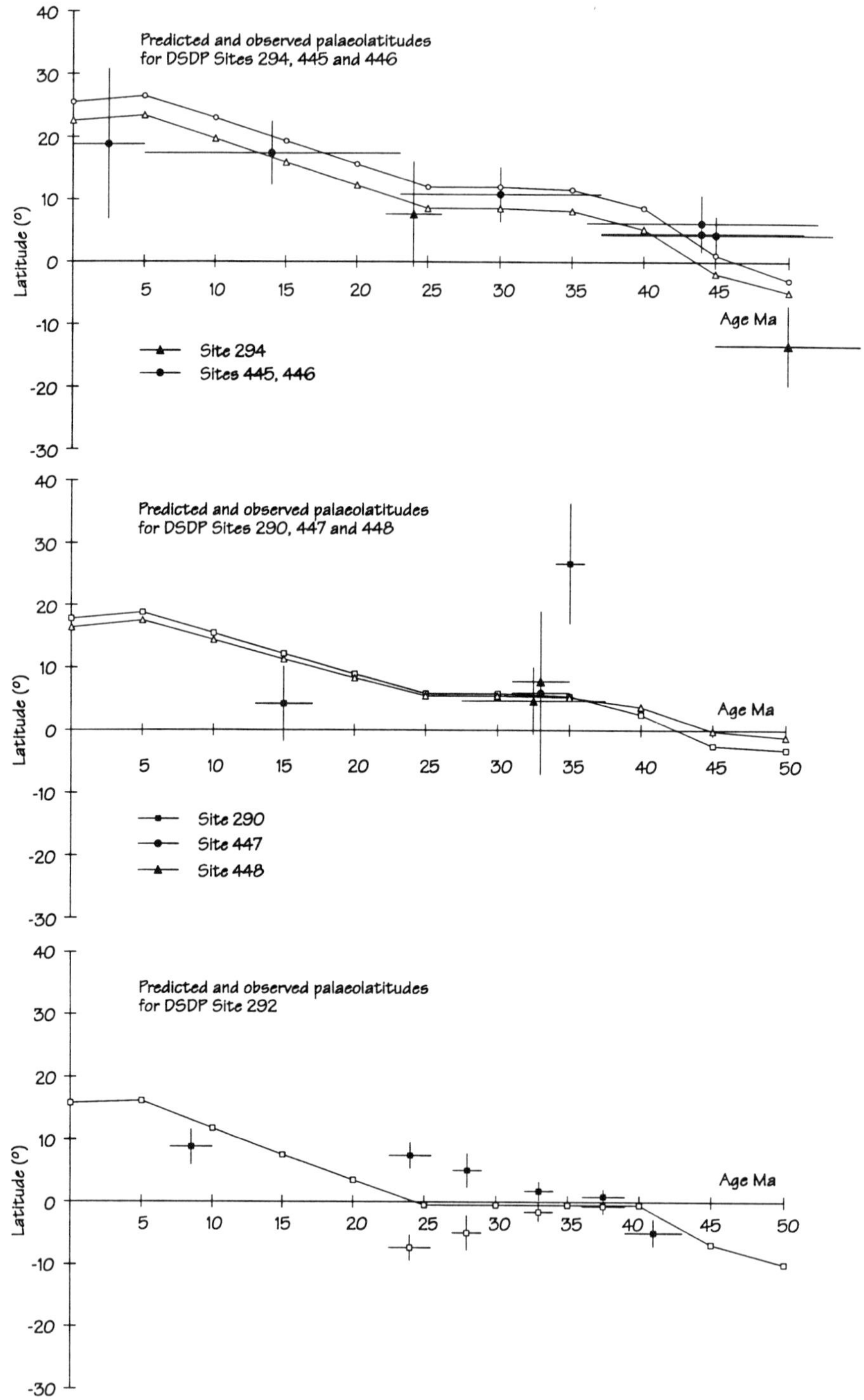

Fig. 11. Comparison of paleolatitudes predicted from Model 2 and calculated from inclination data for sites in the north (top), central (center) and south (bottom) West Philippine Basin. Model 1 predicts identical latitudes for these areas. Sites 445 and 446 yield almost identical results. Sites 290 and 447 yield almost identical results. For site 292 the plot shows paleolatitudes inferred by *Louden* [1977] to be northern hemisphere (solid squares) and equally feasible southern hemisphere paleolatitudes (open squares).

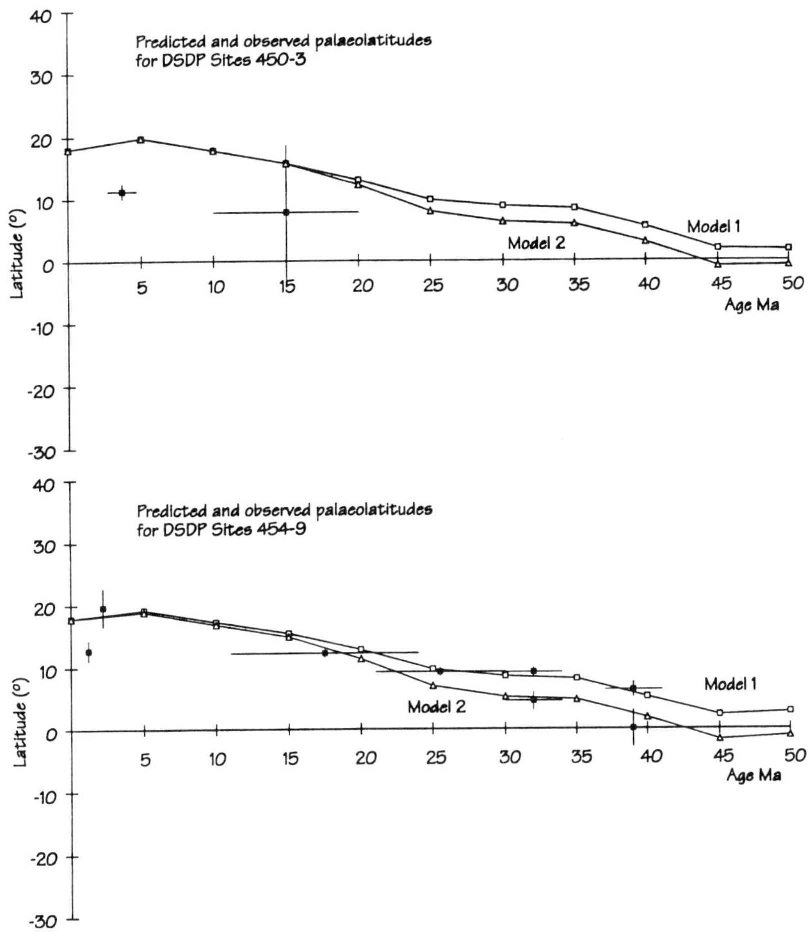

Fig. 12. Comparison of paleolatitudes predicted from Models 1 and 2 and calculated from inclination data for sites in the Parece Vela Basin (above) and Marianas (below). The latitudes calculated are for site 451 and for site 457.

indicate a 50° rotation of the entire plate between 40 and 50 Ma but do not constrain the rotation time very closely. Thus, deviations of older sites from the model curves could be accounted for by postulating that all the rotation occurred in a much shorter interval, for example between 50 and 45 Ma, if rotation of the plate was linked in some way to the sudden change in Pacific Plate motion at ~43 Ma. The Bonin islands results (Figure 14) could be accounted for by relatively small difference in site ages if the variations in inclination values are real; none of the sites has been separately dated and the late Eocene age assigned is poorly controlled since many of the volcanic rocks have yielded much younger isotopic ages. The most significant discrepancies are for Neogene sites in the Izu peninsula, which is now off the Philippine Sea Plate. Because the apparent Neogene northward movement of these sites is so much faster than any other sites on the Philippine Sea Plate possible explanations are that the Izu peninsula moved separately from the plate or that the tilt corrections applied do not adequately describe the complex structural deformation of the peninsula. The latter suggestion seems plausible in view of the scatter in corresponding declinations of these sites [*Hirooka et al.*, 1985].

Declination data. Information on rotations is much more limited. Since the declination data from the east Indonesian sites provide the basis for the models there is an excellent fit of observed and predicted rotations (Figure 15). The only other declination data come from sites in the eastern margin of the plate where ODP leg 126 results [*Koyama et al.*, 1992] and land studies consistently imply large clockwise rotations. However, interpretation of these results has been disputed because all sampling sites have been at the eastern

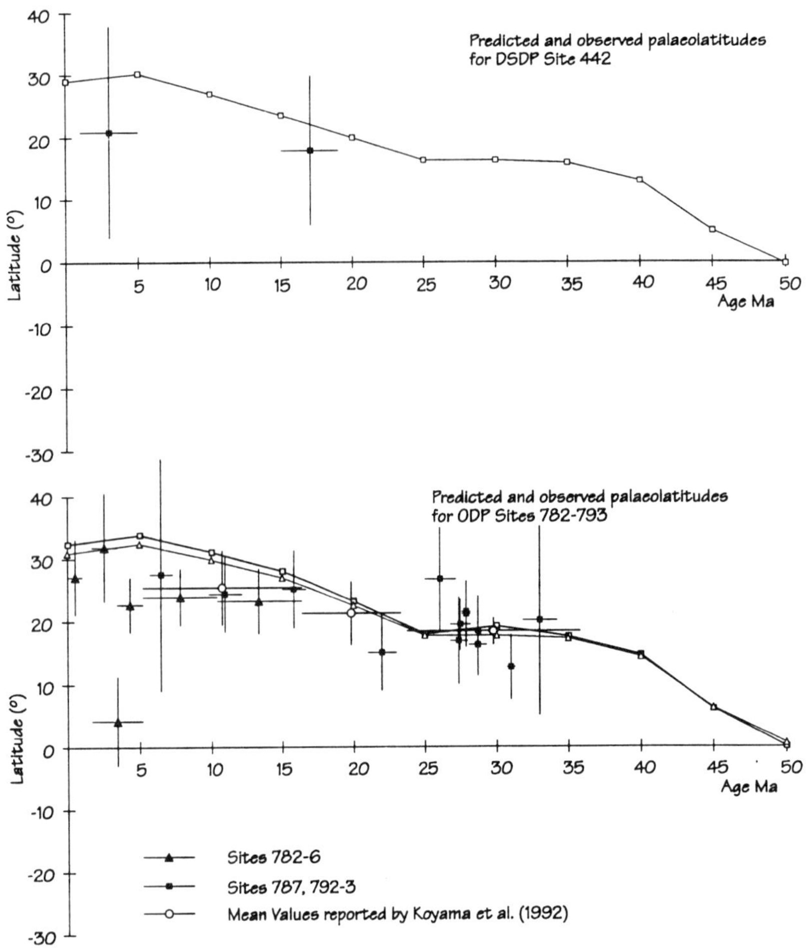

Fig. 13. Comparison of paleolatitudes predicted from Model 2 and calculated from inclination data for sites in the Shikoku Basin (above) and Bonin forearc (below). Model 1 predicts identical latitudes for these areas. The Bonin forearc latitudes calculated are for site 782A and for site 792. Leg 126 results (sites 787, 792 and 793) are based on discrete sample inclinations shown by *Koyama et al.* [1992] on their Figure 6; the mean values reported for these samples are also shown.

edge of the plate where local tectonic rotations are a potential explanation. Model 2 achieves a good match of predicted and observed rotations for most of the eastern margin sites. Fitting the northern Izu-Bonin arc fragment to the Kyushu Ridge at the boundary of Oligocene arc crust increases the predicted rotation of leg 126 sites and improves the fit to observed rotation (Figure 15). Differences, particularly the middle Miocene result, could reflect problems discussed above with orientation of the cores because the rotation estimates of *Koyama et al.* [1992] are averages of a rather scattered data set (Figure 3). Alternatively, if Neogene rotation of the Philippine Sea Plate started later than assumed in the model (e.g., 20 Ma instead of 25 Ma) the predicted declination curve would be steeper than that shown (Figure 15) and the fit would improve. This suggestion would also apply to the Miocene result from Saipan. The Bonin islands results fall off the predicted curves. The sites with lower declinations are those from Hahajima and Mukojima rejected by *Haston and Fuller* [1991] but accepted by *Koyama et al.* [1992]. The sites with higher inclinations are from Chichijima, Anijima, and Nishijima and were accepted by both sets of authors. A possible explanation is that the sites with smaller declinations, which have the higher inclinations, have been paleomagnetically and isotopically reset consistent with the early Neogene K-Ar ages recorded by some of the volcanic rocks. This explanation would require the sites with larger declinations to be older than the poorly constrained late Eocene age shown. A simpler explanation is local tectonic rotation during forearc extension. Of the other eastern

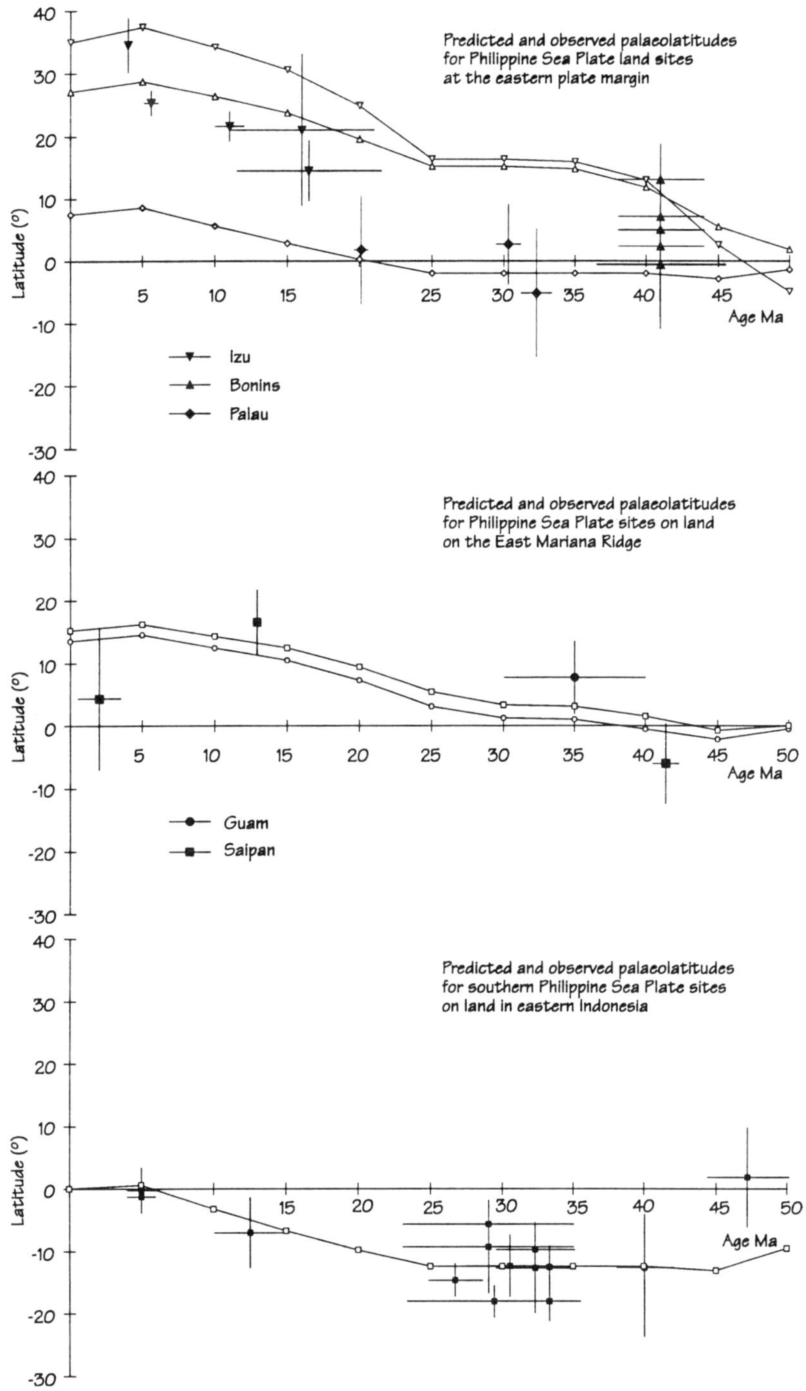

Fig. 14. Comparison of paleolatitudes predicted from Model 2 and calculated from inclination data for sites in the eastern plate margin (top and center) and the southern Philippine Sea Plate (bottom). Model 1 predicts very similar latitudes for sites in the eastern plate margin.

margin sites, those on Palau have significantly different declinations (Figure 15) than the model although inclinations fit well (Figure 14). Here, rotations about vertical axes related to deformation are the obvious explanation and the arrival of the Caroline Ridge at the trench may have caused additional clockwise rotation as suggested by *McCabe and Uyeda* [1983].

Phase shifted magnetic anomalies. As summarized above,

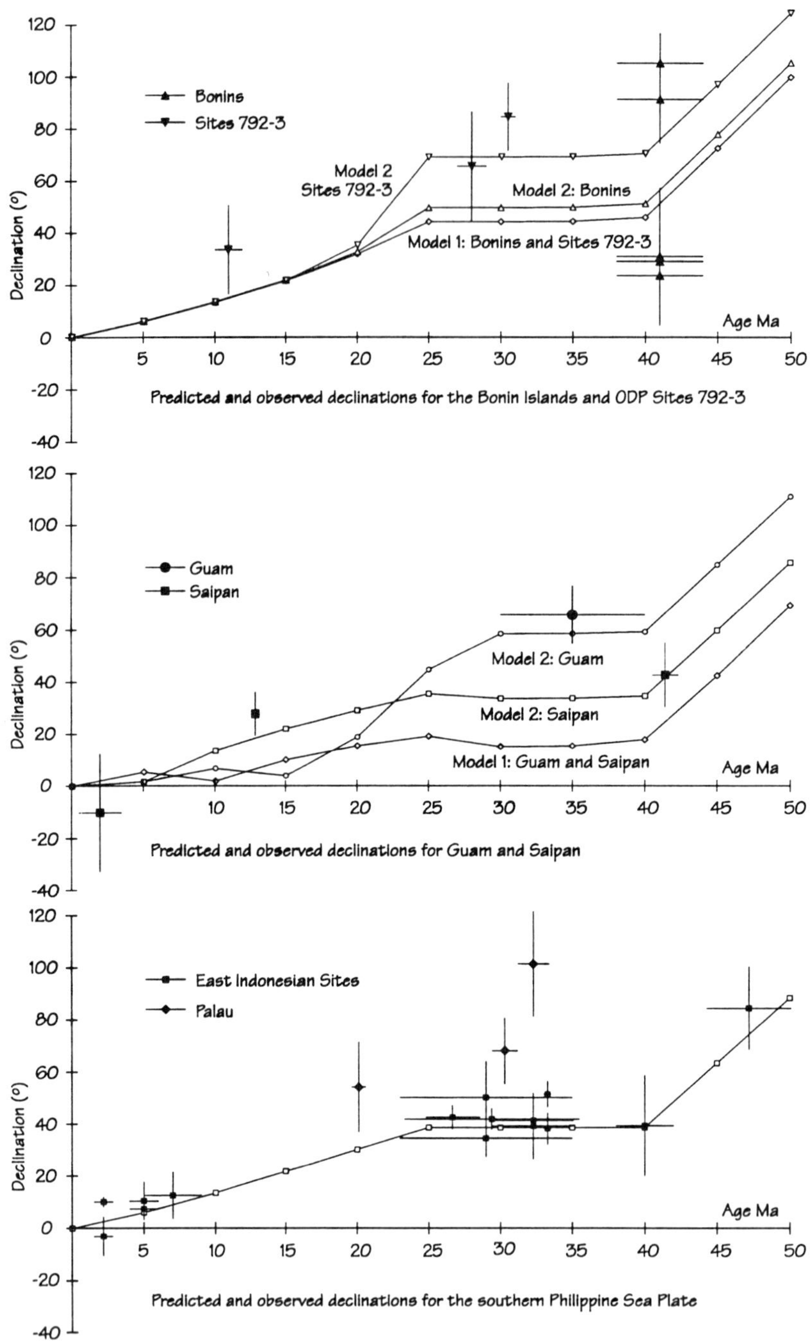

Fig. 15. Comparison of declinations predicted from Model 2 and observed for sites in the eastern plate margin (top and center) and the southern Philippine Sea Plate (bottom). Model 1 predicts different latitudes for sites in the Mariana region and these are also shown. The predicted declinations of Palau and eastern Indonesian sites are identical.

on the basis of magnetic anomaly skewness, both *Louden* [1976] and *Shih* [1980] suggested about 60° of clockwise rotation for the West Philippine Basin since anomaly 20 (44 Ma), whereas *Bowin et al.* [1978] inferred about 50° of counterclockwise rotation. Our model predicts between 57° and 60° clockwise rotation for West Philippine Basin sites since 44 Ma. For the Shikoku Basin the abstract of *Chamot-Rooke et al.* [1989] suggested no significant rotation since anomaly 6 (20 Ma) which *Koyama et al.* [1992] cite to indicate <20° rotation. Our model predicts 30° clockwise rotation for sites in the Shikoku Basin since 20 Ma.

Seamount paleomagnetism. The results from the Shikoku

Basin seamounts have been used to argue that there has been no significant rotation of the Philippine Sea Plate since 17 Ma [*Seno and Maruyama,* 1984; *Koyama et al.*, 1992]. However, if the computational method of *Vacquier and Uyeda* [1967] is applicable to the Shikoku Basin seamounts it should predict inclination as well as declination, but it is clear that the inclinations they calculated for two of the seamounts are improbable (Table 4). These seamounts are located on the former spreading center of the Shikoku Basin with suggested ages of 14-12 Ma [*Chamot-Rooke et al.*, 1989] and should therefore record paleolatitudes that are less than 10° south of their present positions based on the inclination data from ocean drilling sites on the plate (Figure 2). In fact, seamount IV-6 records a latitude shift of 13° and IV-7 a latitude shift of 26°. These are the seamounts that record no apparent rotation (-1°). In contrast, seamount IV-5 has an inclination value consistent with its likely age indicating a latitude shift of 6°; this seamount records a rotation of 5°. Because of the problems with the seamount computational methods summarized above, and likely errors, we consider that these data do not provide a critical test of Philippine Sea Plate motion, although they weaken arguments for rotations in excess of about 20° in the past 15 Ma. For seamount ages of 14 Ma our model predicts clockwise rotations of 20° and for ages of 12 Ma rotations of 16°.

Local Tectonic Rotations

The evidence of rotation of the Philippine Sea Plate has previously been explained in three ways: as a consequence of plate-wide rotation, as the result of opening of marginal basins, or as a result of local tectonic causes. We consider that there is now little doubt that the entire Philippine Sea Plate has rotated clockwise by about 90° since 50 Ma. The model developed using this estimate of whole plate rotation also incorporates the likely effects of marginal basin opening. We have shown that this model provides an excellent match between model latitudes and inclination results. We have also shown that the model predicts large rotations for the land and ODP sites in the eastern part of the plate, which have previously proved so controversial. It provides an excellent fit to rotations of the West Philippine Basin estimated from magnetic anomaly skewness by *Louden* [1976] and *Shih* [1980]. For the Shikoku Basin we consider that there is no conflict with studies of seamount magnetization although our model predicts a larger rotation during the last 20 Ma than that suggested by skewness analysis. We can see no obvious explanation of the latter result, which conflicts with all Neogene paleomagnetic data from the plate, most notably the nearby site 792 result from leg 126 [*Koyama et al.*, 1992].

Our modelling shows that a discontinuous rotation is more likely than the progressive regular rotation of the plate suggested by *Haston and Fuller* [1991], but the declination data set from the plate remains very small. After the effects of whole plate rotation and marginal basin opening have been taken into account there remain discrepancies between the rotations inferred from declination data and the model predictions for a few sites near the eastern margin of the plate. The Palau rotations appear to be due to deformation at the plate margin but discrepancies for other sites could equally be inadequacies of the model. More paleomagnetic data are needed, particularly from lower-middle Eocene and Miocene sites, to determine the timing of changes in rotation. Equally, a better understanding of the opening mechanisms of the Philippine Sea marginal basins is needed to improve modelling of their effects. The Parece Vela Basin is particularly poorly known.

IMPLICATIONS FOR THE PHILIPPINE SEA PLATE BOUNDARIES

The reconstructions of the Philippine Sea Plate by *Seno and Maruyama* [1984] were based on arguments concerning the location of the trench triple junction near the Japanese islands. Although they ignored the southern part of the plate the reconstructions (Figures 8 to 10) here are compatible with their interpretation of the plate boundary history around Japan, which gives some confidence in our new model. Their favored retreating trench model for the subsequent development of the eastern boundary is also consistent with our reconstructions.

The western boundary of the plate in the Philippines has not been modeled in our reconstructions. However, the Eurasian margin and the present-day extent of the South China Sea provide a western limit for the plate. Subduction at the Philippine Trench is known to be very young [*Cardwell et al.*, 1980] and much of the eastern Philippines formed part of the Philippine Sea Plate probably until the end of the Miocene. In contrast, subduction at the Manila Trench began between the late Oligocene and middle Miocene [*Hayes and Lewis,* 1984], consistent with our suggestion of a change in motion of the plate at ~25 Ma. However, there is very little space in the reconstructions for major eastward subduction of the South China Sea suggesting that the Neogene Philippine islands were similar to those of the present, associated with intermittent and limited subduction and important strike-slip faulting.

At ~45 Ma the Izu-Bonin-Mariana forearc was the site of boninite generation over a distance of several hundreds of kilometers. Some models [e.g., *Casey and Dewey*, 1984; *Stern and Bloomer*, 1992] have suggested that the unusually high heat flow required for boninite formation was provided by the initiation of subduction at a former N-S transform fault within the Pacific ocean basin [*Uyeda and Ben-Avraham*, 1972; *Hilde et al.*, 1977] that became the Palau-Kyushu Ridge after the change in Pacific plate motion at ~43 Ma. However, since the ridge-transform boundary is considered to have been orthogonal this model can only account for boninite magmatism in a single area. Our reconstructions predict that at this time the Palau-Kyushu Ridge was oriented approximately WNW-ESE, also indicating that the N-S transform model for boninite genesis should be modified or rejected. High heat flow along an extended zone could result from ridge subduction if an ocean spreading center were subparallel to the Palau-Kyushu Ridge but the early Tertiary distribution of plates and plate boundaries in the western Pacific basin is still very poorly understood. Figure 16 shows a speculative cartoon based upon the suggestion of *Seno and Maruyuma* [1984] that there was a North New Guinea Plate south of the Pacific Plate in the early Tertiary, subsequently removed by subduction with a Pacific-North New Guinea spreading center configuration suggested by *Osozawa* [1992]. This configuration would account for early-middle Eocene subduction recorded in the Halmahera-Waigeo region [*Hall et al.*, 1994b], causing the West Philippine Basin to open as a backarc basin separating an Eocene Halmahera-Waigeo arc from the Daito Ridge province remnant arc. Boninite magmatism would follow from subduction at the Palau-Kyushu Ridge and the rapidly increased length of the subducted boundary at its southern edge could account for the change in motion of the Pacific Plate soon after the ridge was subducted. The change in Pacific Plate motion is temporally linked, and may be mechanically associated, with the rapid rotation of the Philippine Sea Plate recorded paleomagnetically between 50 and 40 Ma.

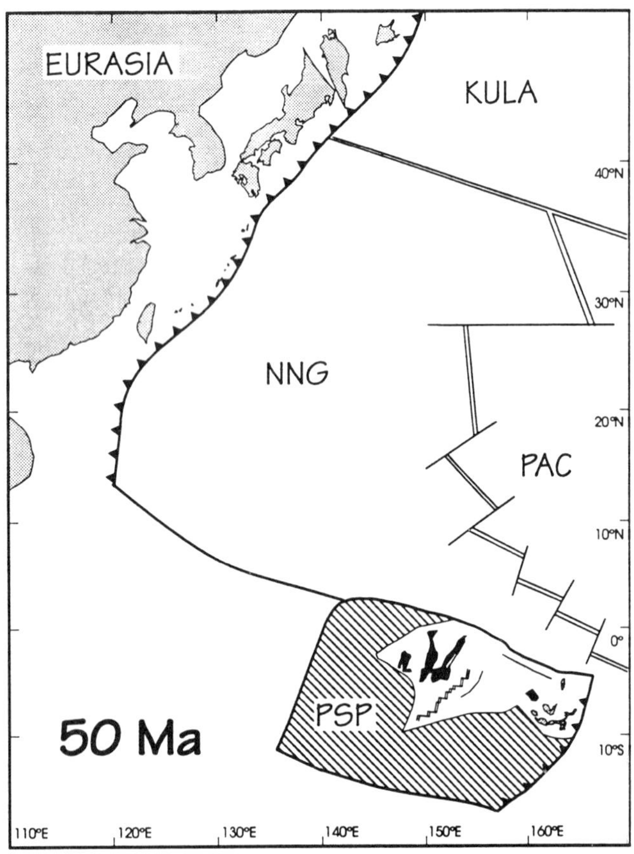

Fig. 16. Speculative configuration of plates in the Western Pacific at 50 Ma based on the reconstructions presented here, and the suggestion of *Seno and Maruyuma* [1984] that there was a North New Guinea Plate (NNG) south of the Pacific Plate (PAC) in the early Tertiary with the spreading center configuration suggested by *Osozawa* [1992]. Japan has been reconstructed using the model of *Jolivet et al.* [1991].

The model has some implications for the Pacific-Caroline-Philippine Sea Plate boundaries. The Ayu Trough opened during the middle and late Miocene [*Weissel and Anderson*, 1978] and spreading may be continuing at the present day. There has been no subduction at the Caroline-Philippine Sea Plate boundary and little convergence at the Caroline-Pacific boundary since ~25 Ma. We have modelled these boundaries using the Caroline-Philippine Sea Plate rotation pole and rate of *Seno et al.* [1993] for the period 0-5 Ma and then moving the Caroline Plate with the Ayu Trough east side (AYE of Figure 6) before this. The consequences for Caroline-Pacific boundary motion are that there has been predominantly left-lateral motion since 25 Ma (approximately 1500 km), minor oblique convergence between 15 and 5 Ma, and minor oblique extension between 5 and 0 Ma. The timing of convergence and extension are partly dependent on the model since the period of 15 and 5 Ma was chosen as the interval of the main opening of the Ayu Trough (see above). However, these conclusions are consistent with the evidence of young oblique extension in the Sorol Trough [*Weissel and Anderson*, 1978]. Models involving less rotation of the Philippine Sea Plate will require subduction at either the Pacific-Caroline boundary or the Caroline-Philippine Sea Plate boundary during the Neogene.

One important consequence of our model is that motion since 25 Ma results in the movement of the southern part of the plate along the northern margin of Australia. This implies that the Philippine Sea Plate-Australia plate boundary has been largely a strike-slip boundary since the beginning of the Miocene. This is a conclusion indicated by geological arguments, principally to account for the lack of evidence for subduction between Australia and the Philippine Sea Plate during this interval. However, it is by no means a necessary consequence of the model and gives some confidence that the combination of paleomagnetic and geological data, as used here, is useful. Arc volcanic rocks of Oligocene age mark the subduction of Australian-Indian oceanic crust beneath the Philippine Sea Plate margin before the Miocene. *Ali and Hall* [1994] suggest that the change from convergent to strike-slip boundary in the early Miocene resulted in accretion to the north Australian margin of Philippine Sea Plate arc fragments that were subsequently displaced and dismembered during the Neogene, producing the numerous 'terranes' of the New Guinea orogenic belt [*Pigram and Davies,* 1987]. The collision event may be the cause of the Neogene rotation of the plate.

The Philippine Sea Plate is one key to understanding the tectonic and magmatic development of the western Pacific and paleomagnetism has a major role to play in deciphering its history. To test and improve the models and speculations presented here there is a need for deep (old) oriented core. As technology improves this could be achieved by a new round of ODP drilling in the region. More declination data is badly needed from the western part of the plate to test the rotation model and in areas farther from the plate boundaries. Other important targets are the small basins west of the Philippines. Drilling the center of these basins (instead of looking at their exposed edges) is more likely to provide information about their overall motion and about their response to movements of the adjacent major plates. The only paleomagnetic data from any of these basins are from a single site in the Celebes Sea [*Shibuya et al.*, 1991] and clearly more sites are required before the information can be considered as representing plate-wide movements.

Acknowledgments. This work was supported by NERC award GR3/7149, grants from the Royal Society, the University of London SE Asia Geological Research Group, the University of London Central Research Fund and NSF grant EAR 84 08354. We thank S. J. Baker, F. T. Banner, T. R. Charlton, E. M. Finch, E. A. Hailwood, H. Y. Ling, J. A. Malaihollo, G. J. Nichols and S. J. Roberts for discussions and their contributions to the Sorong Fault Project; S. Cisowski and J. R. Dunn for advice and assistance with paleomagnetic work; and C. C. Rundle and D. C. Rex for guidance and help with isotopic dating. Fieldwork in Indonesia was facilitated by GRDC, Bandung and the Director, R. Sukamto with field support by D. A. Agustiyanto, S. Atmawinata, A. Haryono, Kusnama, T. Padmawijadja and S. Pandjaitan. This paper is publication #49 of the Birkbeck College and UCL Research School.

REFERENCES

Adachi, Y., H. Inokuchi, Y. Otofuji, N. Isezaki, and K. Yaskawa, Rotation of the Philippine Sea Plate inferred from paleomagnetism of the Palau and the Yap Islands, *Rock Magnet. Paleogeophys., 14,* 72-75, 1987.

Ali, J. R., and R. Hall, Evolution of the boundary between the Philippine Sea Plate and Australia: palaeomagnetic evidence from eastern Indonesia, *Tectonophysics,* in press, 1994.

Aoki, Y., A post depositional magnetization mechanism of sediments and its application to geophysical problems - paleosecular variation in Japan, Ph.D. thesis, 101 pp., Univ. of Tokyo, Tokyo, 1972.

Bleil, U., Paleomagnetism of DSDP Leg 60 sediments and igneous rocks from the Mariana region, in *Init. Repts DSDP*, 60, D. M. Hussong and S. Uyeda et al., pp. 855-876, U.S. Govt. Printing Office, Washington, D.C., 1981.

Bowin, C., R. S. Lu, C. Lee, and H. Schouten, Plate convergence and accretion in the Taiwan-Luzon region, *Am. Assoc. Pet. Geol. Bull., 62,* 1645-1672, 1978.

Byrne, T., and L. DiTullio, Evidence for changing plate motions in southwest Japan and reconstructions of the Philippine Sea plate, *The Island Arc, 1,* 148-165, 1992.

Cambridge Paleomap Services, ATLAS version 3.3. Cambridge Paleomap Services, P.O. Box 246, Cambridge, U.K., 1993.

Cardwell, R. K., B. L. Isacks, B. L., and D. E. Karig, The spatial distribution of earthquakes, focal mechanism solutions, and subducted lithosphere in the Philippine and northeastern Indonesian islands, in *The Tectonic and Geologic Evolution of Southeast Asian Seas and Islands, Geophys. Monogr. Ser.,* vol. 23, edited by D. E. Hayes, pp. 1-35, AGU, Washington, D.C., 1980.

Casey, J. F., and J. F. Dewey, Initiation of subduction zones along transform and accreting plate boundaries, triple junction evolution and forearc spreading centres - implications for ophiolitic geology and obduction. *Spec. Publ. Geol. Soc. London, 13*, 269-290, 1984.

Chamot-Rooke, N., V. Renard, and X. Le Pichon, Magnetic anomalies in the Shikoku Basin: a new interpretation, *Earth Planet. Sci. Lett., 83*, 214-228, 1987.

Chamot-Rooke, N., K. Tamaki, and K. Kobayashi, Deskewed magnetic profiles of the Shikoku Basin and the past kinematics of the Philippine Sea Plate, *Eos*, 70, 1365-1366, 1989.

Clague, D. A., and R. D. Jarrard, Tertiary Pacific plate motion deduced from the Hawaiian-Emperor Chain, *Geol. Soc. Am. Bull., 84*, 1135-1154, 1973.

Cloud, P. E., R. G. Schmidt, and H. W. Burke, Geology of Saipan, Mariana Islands, 1, General Geology, *U.S. Geol. Surv. Prof. Pap., 280-A*, 126 pp., 1956.

Fuller, M., Magnetic orientation of borehole cores, *Geophysics, 34,* 772-774, 1969.

Fuller, M., J. R. Dunn, G. Green, Jin-Lu Lin, R. McCabe, K. Toney, and I Williams, Paleomagnetism of Truk Islands, eastern Carolines and of Saipan, Marianas, in *The Tectonic and Geologic Evolution of Southeast Asian Seas and Islands, Geophys. Monogr. Ser.,* vol. 23, edited by D. E. Hayes, pp. 235-246, AGU, Washington, D.C., 1980.

Gordon, R. G., and D. M. Jurdy, Cenozoic plate motions, *J. Geophys. Res., 91*, 12,389-12,406, 1986.

Hall, R., and G. J. Nichols, Terrane amalgamation in the Philippine Sea margin, *Tectonophysics*, *181*, 207-222, 1990.

Hall, R., M. G. Audley-Charles, F. T. Banner, S. Hidayat and S. L. Tobing, The basement rocks of the Halmahera region, east Indonesia: a Late Cretaceous-Early Tertiary forearc, *J. Geol. Soc. London, 145*, 65-84, 1988.

Hall, R., G. J. Nichols, P. D. Ballantyne, T. Charlton, and J. Ali, The character and significance of basement rocks of the southern Molucca Sea region, *J. SE Asian Earth Sci., 6*, 249-258, 1991.

Hall, R., J. R. Ali, and C. D. Anderson, Tertiary motion history of the Philippine Sea Plate: paleomagnetic evidence from eastern Indonesia, *Earth Planet. Sci. Lett.*, in press, 1994a.

Hall, R., J. R. Ali, C. D. Anderson, and S. J. Baker, Origin and motion history of the Philippine Sea Plate, *Tectonophysics,* in press, 1994b.

Harland, W. B., R. L. Armstrong, A. V. Cox, L. E. Craig, A. G. Smith, and D. G. Smith, *A Geologic Time Scale 1989*, Cambridge University Press, 263 pp., 1990.

Harrison, C. G. A., R. D. Jarrard, V. Vaquier, and R. L. Larson, Palaeomagnetism of Cretaceous Pacific seamounts, *Geophys. J. R. Astron. Soc., 42*, 859-882, 1975.

Harrison, C. G. A., Marine magnetic anomalies - the origin of the stripes, *Ann. Rev. Earth Planet. Sci., 15*, 505-543, 1987.

Haston, R., *Paleomagnetic results from Palau, West Caroline Islands: A constraint on Philippine Sea plate motion,* M.A. thesis, Univ. of Calif., Santa Barbara, 98 pp., 1988.

Haston, R., M. Fuller, and E. Schmidtke, Paleomagnetic results from Palau, West Caroline Islands: A constraint on Philippine Sea plate motion, *Geology, 16*, 654-657, 1988.

Haston, R. B., and M. Fuller, Paleomagnetic data from the Philippine Sea Plate and their tectonic significance, *J. Geophys. Res., 96*, 6073-98, 1991.

Haston, R. B., L. B. Stocking, and J. R. Ali, Palaeomagnetic data from holes 782A, 784A and 786A, Leg 125, in *Proc. ODP Sci. Results*, 125, P. Fryer, J. A. Pearce, L. B. Stocking et al., pp. 535-545, Ocean Drilling Program, College Station, TX, 1992.

Hayes, D. E., and S. D. Lewis, A geophysical study of the Manila Trench, Luzon, Philippines. 1. Crustal structure, gravity and regional tectonic evolution, *J. Geophys. Res., 89,* 9171-9195, 1984.

Hilde, T. W. C., and C-S. Lee, Origin and evolution of the West Philippine Basin: a new interpretation, *Tectonophysics, 102*, 85-104, 1984.

Hilde, T., Uyeda, S., and Kroenke, L., Evolution of the western Pacific and its margin, *Tectonophysics, 38,* 145-165, 1977.

Hirooka, K., T. Takahasi, H. Sakai, and T. Nakajima, Paleomagnetic evidence of the northward drift of the Izu peninsula, central Japan, in *Formation of Active Ocean Margins,* edited by N. Nasu, pp. 775-787, Terra Scientific, Tokyo, 1985.

Huchon, P., Comment on "Kinematics of the Philippine Sea Plate", *Tectonics, 5*, 1355-1383, 1986.

Jarrard, R. D., and S. Sasajima, Paleomagnetic synthesis for Southeast Asia: constraints on plate motions, in *The Tectonic and Geologic Evolution of Southeast Asian Seas and Islands, Geophys. Monogr. Ser.*, vol. 23, edited by D. E. Hayes, pp. 293-316, AGU, Washington, D.C., 1980.

Jolivet, L., P. Huchon, J. P. Brun, X. Le Pichon, N. Chamot-Rooke, and J. C. Thomas, Arc deformation and marginal basin opening: Japan Sea as a case study, *J. Geophys. Res., 96, B3*, 4367-4384, 1991.

Jolivet, L., P. Huchon, and C. Rangin, Tectonic setting of the western Pacific marginal basins, *Tectonophysics, 160,* 23-47, 1989.

Kaneoka, I., N. Iaahiki, and A. Zashu, K-Ar ages of the Izu-Bonin Islands, *Geochem. J., 4,* 53-60, 1970.

Karig, D. E., R. N. Anderson, and L. D. Bibee, Characteristics of back arc spreading in the Mariana Trough, *J. Geophys. Res., 83,* 1213-1226, 1978.

Keating, B., Paleomagnetic study of sediments from Deep Sea Drilling Leg 59, in *Init. Repts DSDP*, 59, L. W. Kroenke and R. B. Scott et al., pp. 523-532, U.S. Govt. Printing Office, Washington, D.C., 1980.

Keating, B. H., and C. E. Helsley, Implication of island arc rotations to the studies of marginal terranes, *J. Geodynamics, 2,* 159-181, 1985.

Keating, B. H., and E. Herrero, Paleomagnetic studies of basalts and andesites from DSDP Leg 59, in *Init. Repts DSDP*, 59, L. W. Kroenke and R. B. Scott et al., pp. 533-544, U.S. Govt. Printing Office, Washington, D.C., 1980.

Keating, B. H., K. Kodama, and C. E. Helsley, Paleomagnetic Studies of the Bonin and Mariana Island Arcs, in *Arc Volcanism: Physics and Tectonics,* edited by D. Shimozuru and I. Yokoyama, pp. 243-259, Terra Scientific Publishing Co., Tokyo, Japan, 1983.

Kinoshita, H., Paleomagnetism of sediment cores from DSDP Leg 58, Philippine Sea, in *Init. Repts DSDP*, 58, G. deVries Klein, and K. Kobayashi et al., pp. 765-768, U.S. Govt. Printing Office, Washington, D.C., 1980.

Kobayashi, K., M. Hoshino, and H. Aoki, *Reconnaissance paleomagnetic and rock magnetic study of Guam, Mariana and related sites (in Izu Peninsula)*, Tokai University Press, Japan, pp. 385-390, 1971.

Kodama, K., A palaeomagnetic reconnaissance of the Bonin Islands, *Bull. Earthquake Res. Inst. Univ. Tokyo, 56,* 347-365, 1981.

Kodama, K., B. H. Keating, and C. E. Helsley, Paleomagnetism of the Bonin Islands and its tectonic significance, *Tectonophysics, 95*, 25-42, 1983.

Koyama, M., S. M. Cisowski, and P. Pezard, Paleomagnetic evidence for northward drift and clockwise rotation of the Izu-Bonin forearc since the early Oligocene, in *Proc. ODP Sci. Results*, 126, B. Taylor and K. Fuijioka et al., pp. 353-370, Ocean Drilling Program, College Station, TX, 1992.

Larson, E. E., R. L. Reynolds, M. Ozima, Y. Aoki, H. Kinoshita, S. Zasshu, N. Kawai, T. Nakajima, K. Hirooka, R. Merril, and S. Levi, Paleomagnetism of Miocene volcanic rocks of Guam and the curvature of the southern Mariana island arc, *Geol. Soc. Am. Bull., 86*, 346-350, 1975.

Le Pichon, X., J. Francheteau, and G. F. Sharman, Rigid plate accretion in an inter-arc basin: Mariana Trough, *J. Phys. Earth, 23*, 251-256, 1975.

Louden, K., Magnetic anomalies in the West Philippine Basin, in

The Geophysics of the Pacific Ocean Basin and its Margin, edited by G. H. Sutton, M. H. Manghnani and R. Moberly, *Geophys. Monogr. Ser., vol. 19*, 253-267, AGU Washington, D.C., 1976.

Louden, K. E., Paleomagnetism of DSDP sediments, phase shifting of magnetic anomalies and rotations of the West Philippine Basin, *J. Geophys. Res., 82*, 2989-3002, 1977.

McCabe, R., and S. Uyeda, Hypothetical model for the bending of the Mariana arc, in *The Tectonic and Geologic Evolution of Southeast Asian Seas and Islands, Part 2, Geophys. Monogr. Ser.*, vol. 27, edited by D. E. Hayes, pp. 281-293, AGU, Washington, D.C., 1983.

McCaffrey, R., Lithospheric deformation within the Molucca Sea arc-arc collision: evidence from shallow and intermediate earthquake activity, *J. Geophys. Res., 87*, 3663-3678, 1982.

McElhinny, M. W., Statistical significance of the fold test in palaeomagnetism, *Geophys. J. R. Astron. Soc., 8*, 338-340, 1964.

McFadden, P. L., A new fold test for paleomagnetic studies, *Geophys. J. Int., 103*, 163-169, 1990.

Mason, A. C., G. Corwin, C. L. Rodgers, O. Elmquist, A. J. Vessel, and R. J. McCracken, Military geology of the Palau Islands, Caroline Islands, report, *Intel. Div. Off. Eng. Headquarter, U.S. Army (rear), U.S. Geol. Survey*, Washington, D.C., 1956.

Meijer, A., M. Reagan, H. Ellis, M. Shafiqullah, J. Sutter, P. Damon, and S. Kling, Chronology of volcanic events in the eastern Philippine Sea, in *The Tectonic and Geologic Evolution of Southeast Asian Seas and Islands, Part 2, Geophys. Monogr. Ser.*, vol. 27, edited by D. E. Hayes, pp. 360-372, AGU, Washington, D.C., 1983.

Mrozowski, C. L., and D. E. Hayes, The evolution of the Parece Vela Basin, eastern Philippine Sea, *Earth Planet. Sci. Lett., 46*, 49-67, 1979.

Mrozowski, C. L., S. D. Lewis, and D. E., Hayes, Complexities in the tectonic evolution of the West Philippine Basin, *Tectonophysics, 82*, 1-24, 1982.

Nichols, G. J., R. Hall, J. S. Milsom, D. Masson, L. Parson, N. Sikumbang, B. Dwiyanto, and H. Kallagher, The southern termination of the Philippine Trench, *Tectonophysics, 183*, 289-303, 1990.

Osozawa, S., Double ridge subduction recorded in the Shimanto accretionary complex, Japan, and plate reconstruction, *Geology, 20*, 939-942, 1992.

Packham, G. H., Plate motions and Southeast Asia: some tectonic consequences for basin development, *SEAPEX (SE Asia Pet. Expl.) Proc., 9*, 55-68, 1990.

Pigram, C. J., and H. L. Davies, Terranes and the accretion history of the New Guinea orogen, *BMR J. Aust. Geol. Geophys., 10*, 193-211, 1987.

Ranken, B., R. K. Cardwell, and D. E. Karig, Kinematics of the Philippine Sea Plate, *Tectonics, 3*, 555-575, 1984.

Reagan, M. K., and A. J. Meijer, Geology and geochemistry of early arc volcanic rocks from Guam, *Geol. Soc. Am. Bull., 95*, 701-713, 1984.

Royer, J-Y., and D. T. Sandwell, Evolution of the eastern Indian Ocean since the Late Cretaceous: constraints from Geosat altimetry, *J. Geophys. Res., 94*, 13, 755-13, 782, 1989.

Saito, T., Eocene planktonic Foraminifera from Hahajima (Hillsborough Island), *Trans. Proc. Paleontol. Soc. Jpn., N. W., 65*, 209-225, 1962.

Schouten, H., and S. C. Cande, Palaeomagnetic poles from marine magnetic anomalies, *Geophys. J. R. astr. Soc., 44*, 567-575, 1976.

Schouten, H., and K. McCamy, Filtering marine magnetic anomalies, *J. Geophys. Res., 77*, 7089-7099, 1972.

Scotese, C. R., L. M. Gahagan, and R. L. Larson, Plate tectonic reconstructions of the Cretaceous and Cenozoic ocean basins, *Tectonophysics, 155*, 27-48, 1988.

Seno, T., and S. Maruyama, Paleogeographic reconstruction and origin of the Philippine Sea, *Tectonophysics, 102*, 53-84, 1984.

Seno, T., T. Moriyama, S. A. Stein, D. F. Woods, and C. R. Demets, Redetermination of Philippine Sea Plate motion, *Seism. Soc. Japan Program with Abstr., 1*, 86, 1987a.

Seno, T., T. Moriyama, S. Stein, D. F. Woods, C. Demets, D. Argus, and R. Gordon, Redetermination of Philippine Sea Plate motion, *Eos Trans. AGU*, 68, 1474, 1987b.

Seno, T., S. A. Stein, and A. E. Gripp, A model for the motion of the Philippine Sea plate consistent with NUVEL-1 and geological data, *J. Geophys. Res.*, 98, 17,941-17,948, 1993.

Shibuya, H., D. L. Merril, V. Hsu, and Leg 124 Shipboard Scientific Party, Paleogene counterclockwise rotation of the Celebes Sea - orientation of ODP cores utilizing the secondary magnetization, in *Proc. ODP Sci. Results*, 124, E. A. Silver and C. Rangin et al., pp. 519-523, Ocean Drilling Program, College Station, TX, 1991.

Shih, T-C., Marine magnetic anomalies from the western Philippine Sea: implications for the evolution of marginal basins, in *The Tectonic and Geologic Evolution of Southeast Asian Seas and Islands, Geophys. Monogr. Ser.*, vol. 23, edited by D. E. Hayes, pp. 49-75, AGU, Washington, D.C., 1980.

Stern, R. J., and S. H. Bloomer, Subduction zone infancy: examples from the Eocene Izu-Bonin-Mariana and Jurassic California arcs, *Geol. Soc. Am. Bull., 104*, 1621-1636, 1992.

Taira, A., H. Tokuyama, and W. Soh, Accretion tectonics and evolution of Japan, in *The Evolution of Pacific Ocean Margins*, edited by Z. Ben-Avraham, pp. 100-123, Oxford University Press, New York, 1989.

Taylor, B., Rifting and the volcano-tectonic evolution of the Izu-Bonin-Mariana arc, in *Proc. ODP Sci. Results*, 126, B. Taylor and K. Fuijioka et al., pp. 627-651, Ocean Drilling Program, College Station, TX, 1992.

Tracey, J. L., S. O. Schlanger, J. T. Stark, D. B. Doan, and H. G. May, General geology of Guam, *U.S. Geol. Surv. Prof. Pap., 403-A*, 104 pp., 1964.

Tsunakawa, H., K-Ar dating on volcanic rocks on the Bonin Islands and its tectonic implication, *Tectonophysics, 95*, 221-232, 1983.

Ujiie, H., and K. Matsumaru, Stratigraphic outline of Hahajima (Hillsborough Island) Bonin Islands, *Mem. Natl. Sci. Mus., 10*, 5-18, 1977.

Uyeda, S., and Z. Ben-Avraham, Origin and development of the Philippine Sea, *Nature (London) Phys. Sci., 240*, 176-178, 1972.

Vacquier, V., A machine method for computing the magnetization of a uniformly magnetized body from its shape and a magnetic survey, Proc. Benendum Earth Magnetism Symp., Pittsburgh, pp. 123-137, 1962.

Vaquier, V., and S. Uyeda, Palaeomagnetism of nine seamounts in the western Pacific and of three volcanoes in Japan, *Bull. Earthquake Res. Inst., Tokyo, 45*, 815-848, 1967.

Vogt, P. R., Subduction and aseismic ridges, *Nature, 241,* 189-191, 1973.

Vogt, P. R., A. Lowrie, D. R. Bracey, and R. N. Hey, Subduction of aseismic ocean ridges: Effects on shape, seismicity, and other characteristics of consuming plate boundaries, *Spec. Pap. Geol. Soc. Am., 172,* 59 pp., 1976.

Watts, A. B., J. K. Weissel, and R. L. Larson, Seafloor spreading in marginal basins of the western Pacific, *Tectonophysics, 37,* 167-181, 1977.

Weissel, J. K., and R. N. Anderson, Is there a Caroline Plate?, *Earth Planet. Sci. Lett., 41,* 143-158, 1978.

R. Hall, Department of Geological Sciences, University College London, Gower Street, London WC1E 6BT, U.K.

M. Fuller, and C. D. Anderson, Department of Geological Sciences, University of California, Santa Barbara, CA 93106.

J. R. Ali, Oceanography Department, Southampton University, Southampton SO9 5NH, U.K.

Closure of Western Pacific Marginal Basins: Rupture of the Oceanic Crust and the Emplacement of Ophiolites

Claude Rangin

Département de Géologie Structurale, Université Pierre et Marie Curie, Paris, France

Eli A. Silver

Department of Earth Sciences, University of California Santa Cruz, CA

Kensaku Tamaki

Ocean Research Institute, University of Tokyo, Japan

Most of the marginal basins of the western Pacific region opened in Cenozoic time and many of these are presently closing. Late Cenozoic basin closing has been documented in the recently drilled Celebes and Sulu seas (Ocean Drilling Program leg 124) and in the Japan Sea (leg 127). The oceanic floors of these young marginal basins were deformed locally before being actively consumed along young subduction zones, with parts of their sedimentary section and crust incorporated into accretionary wedges. The youngest parts of these basins are the first to be deformed, to subduct, or to be incorporated in these accretionary wedges. The initial flexural stage affecting the crust before rupture appears to be a local process.

INTRODUCTION

The way marginal basins are deformed in response to collision at their margins is a matter of debate. Compression of a basin can induce flexure and rupture of the oceanic crust. Commonly a new subduction zone forms along one of the basin margins. The crust can also break in the basin itself or along the axial ridge, inducing intra oceanic crust deformation. Creation of such slivers of oceanic lithosphere can facilitate later incorporation of these marginal basin fragments into accretionary wedges or obduction as ophiolites onto adjacent continental margins.

Here we explore several questions related to closure of marginal basins:

• Is there a relationship between the age and thickness of the marginal basin lithosphere and the preferred site of rupture in the basin ?

• Are the slivers of oceanic lithosphere that are created during rupture incorporated into accretionary wedges?

Answers to these questions can help us understand the process by which dense pieces of oceanic crust become incorporated into accretionary wedges. This process has been described on land in the Shimanto complex [*Taira et al.*, 1988], the Nias complex of western Indonesia [*Moore and Karig*, 1980], the Papuan Peninsula of Papua New Guinea [*Davies*, 1971], and the east arm of Sulawesi [*Silver et al.*, 1983a,b]. The process was inferred to occur in a submarine accretionary wedge south of New Britain island, on the basis of multichannel seismic, and ancillary magnetics, data [*Bernstein et al.*, 1992a,b]. However, such an occurrence has not been tested by drilling.

The location of the crustal rupture in the basin is important for constraining the thermal and mechanical regime of intra-oceanic thrusting, an initial process of some ophiolite emplacement. Modeling of intra-oceanic thrusting followed by emplacement of ophiolites onto a continental margin [*Boudier et al.*, 1985] implies that many ophiolites may be tectonically decoupled and obducted as slices of hot and thin lithosphere close to spreading centers undergoing

Active Margins and Marginal Basins of the Western Pacific
Geophysical Monograph 88

compression. A fast spreading rate seems to be the most influential parameter favoring the production of thin lithosphere at oceanic ridges. The limited thickness of ophiolites [*Coleman*, 1977; *Nicolas*, 1989] and the small age difference between ophiolite igneous crystallization and metamorphic sole formation [*Coleman*, 1977] are basic constraints for locating the initial stage of intra-oceanic deformation along ocean ridges; these same constraints imply that marginal basins are good candidates for future ophiolite formation.

Marginal basins, with their relatively thin crust and rapid opening, have been suspected for some time as possible sources of ophiolitic slabs [*Amstrong and Dick*, 1974]. Rapid closure of these small basins, only a few million years after opening, was recently confirmed by the Ocean Drilling Program (ODP) leg 124 in the Sulu Sea basin and leg 127 in the Japan Sea (Figure 1).

In this paper we review data supporting onset of compression in these basins. We examine the deformation of the oceanic floor in the Celebes-Sulu and Japan Seas drilled during legs 124, 127, and 128; we locate the ruptures in the oceanic crust of these basins; and we investigate how oceanic fragments were eventually incorporated into the new accretionary wedges along the margins of these small basins.

THE JAPAN SEA

Results of ODP drilling in the Japan Sea showed that active spreading in the Yamato and Japan Basins, as well as dextral transtension along the eastern margin (Figure 2), began during the early Miocene and continued for 5 to 10 m.y. [see *Jolivet et al.*, this volume].

East-west compression was superimposed on this dextral transtension and is recorded in the late Miocene sediments as well as in reverse faulting and thrusting along the eastern margin of the Yamato and Japan Basins (Figure 2a). This intraplate shortening probably started in late Neogene time.

Subsidence in the Japan Sea: A Matter of Debate?

Flexural strain within the Japan Sea is marked by uplift of NW Japan [*Jolivet and Tamaki*, 1992] and by varying subsidence rates in and around the Japan Sea. Paleobathymetric evidence and radiometric and biostratigraphic ages established at the base of the sediment columns at ODP sites 794, 795, and 797 [*Ingle*, 1992] have established that the Yamato and Japan Basins of the Japan Sea had subsided to near their present water depth by 20-18 Ma, i.e., at the end of the opening stage. Thus all of the major deep basins of the Japan Sea were in place and with their present basin plains at minimum depths of 2 to 3 km by that time. According to Ingle, the fundamental change from an extensional to a compressional tectonic regime in the Japan backarc region at 11-10 Ma is not marked by accelerated subsidence but instead by widespread uplift and deformation of Miocene basin-fill sequences along the southern margin of the Japan Sea. In his interpretation, flexure does not affect the Japan Sea in response to the compressional regime established in this backarc area at 11-10 Ma. In contrast, *Jolivet and Tamaki* [1992], following the model of *Cloething et al.* [1985], interpret further subsidence at site 794, located at the front of the compressive deformation zone, to suggest that while the margins of the basin were uplifted, the center was subsiding. This subsidence may mark the onset of compression in the Japan Sea.

The best way to reconcile these two interpretations is to consider that subsidence under horizontal compression was limited to the eastern Japan Sea, just in front of the nascent thrust zone.

Rupture of the Crust

A thrust zone in the eastern Japan Sea (Figure 2) was identified by *Tamaki and Honza* [1985] on the basis of seismic reflection profiles, gravity data, and earthquake focal mechanism solutions. Numerous ridges are clearly formed by thrust movements associated with uplift of the edge of the overthrust plate. Incipient subduction is suggested along the eastward dipping thrust faults. The active thrusting outlined by uplift of Okoshuri ridge at site 796 is dated 1.8 Ma [*Tamaki et al.*, 1990] and is marked by east or west verging thrust faults (Figure 2a). *Jolivet and Tamaki* [1992] used ash layers recovered in drill holes from the Japan arc to link this late compressional evolution of the Japan Sea with the magmatic evolution of the arc. This study allows precise determination of the period, 10 to 7 Ma, when the stress field from dextral transtension related to the Japan Sea opening changed to compression leading to intraoceanic plate deformation (Figure 2b). Thus a relatively long period of local flexure of the oceanic plate preceded rupture.

Geodynamic Explanation for the Stress Variation

The change of stress direction between 10 and 7 Ma was interpreted in various ways. *Tamaki and Honza* [1985] inferred that the origin of the convergent stress was due to intraplate or inter-microplate relative motion (extrusion of the Amurian plate toward the east) in response to the India-Eurasia collision, following the model of *Tapponnier et al.*

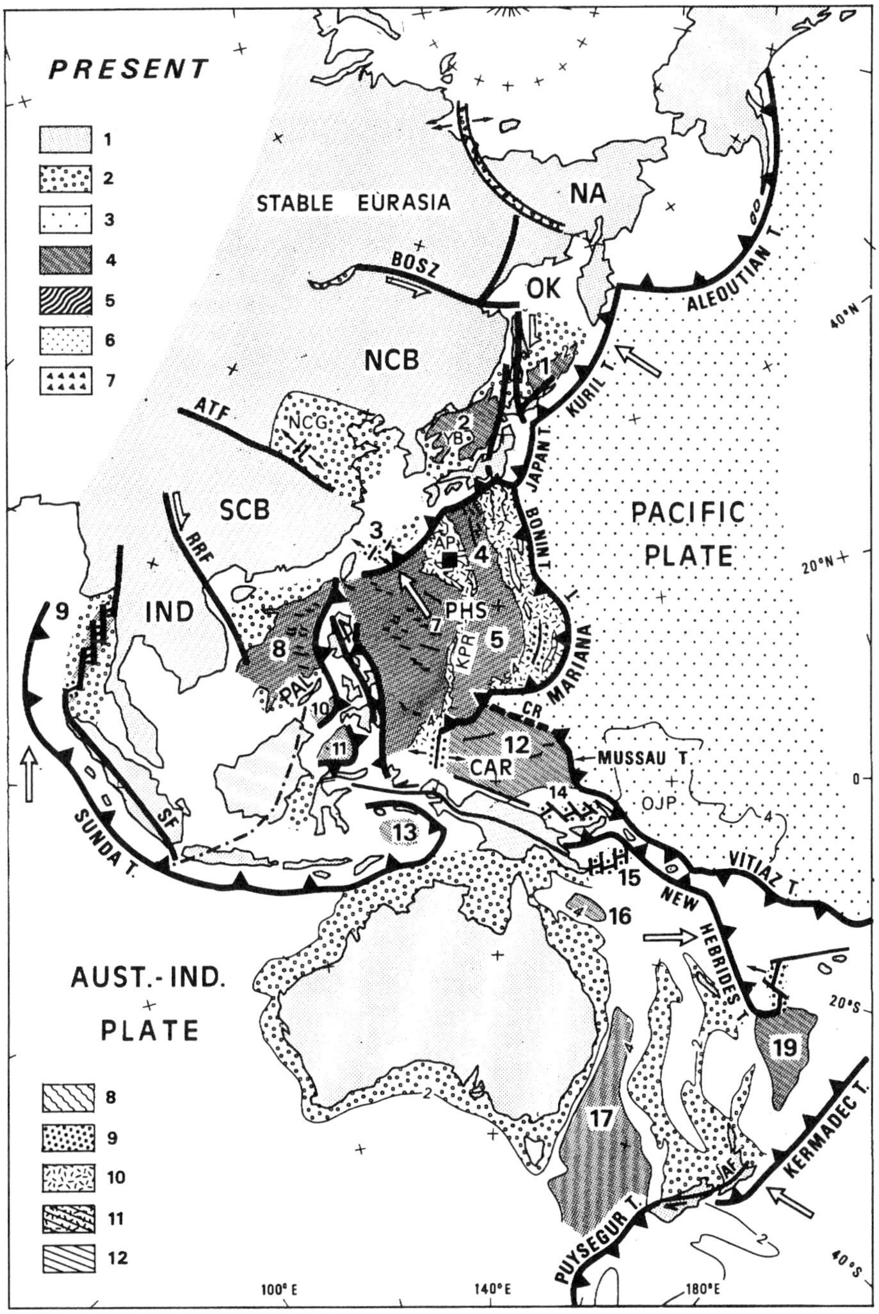

Fig. 1. Geodynamics of the Western Pacific Region [from *Jolivet et al.*, 1989] showing the distribution of the marginal basins. 1 - Kuril Basin, 2 - Sea of Japan, 3 - Okinawa basin, 4 - Shikoku Basin, 5 - Parece Vela Basin, 6 - Mariana Basin, 7 - West Philippine Basin, 8 - South China Sea, 9 - Andaman Sea, 10 - Sulu Sea, 11 - Celebes Basin, 12 - Caroline Basin, 13 - Banda Sea, - 14 - Bismarck rift, 15 - Woodlark rift, 16 - Coral Sea, 17 - Tasman Sea, 18 - North Fiji Basin. OK - Okhotsk plate, NCB - north China block, SCB - south China block, IND - Indochina block, AP - Amami Plateau, KPR - Kyushu-Palau Ridge, CR - north Caroline Ridge, OJP - Ontong Java Plateau, BOSZ - Baikal-Okhotsk shear zone, HTSZ - Hidaka-Tartary shear zone, ATF - Altyn-Tagh fault, RRF - Red River fault, SF - Sumatra fault, AF - Alpine fault. Black squares in the PHS plate are DSDP sites (445 - north; 292 - south). Patterns 1 to 12: 1 = emerged continental crust, 2 = stretched continental crust, 3 = Pacific plate, 4 = marginal basins oceanic crust (extinct), 5 = Molucca Sea oceanic crust (only on the 12 Ma stage), 6 = marginal basins oceanic crust (active spreading, 7 = AUS-ANT plate oceanic crust), 8 = ANT plate oceanic crust, 9 = New Guinea plate oceanic crust, 10 = crust of the Bonin arc and Kyushu-Palau Ridge, 11 = same as 10, now subducted, 12 = subducted part of the Shikoku Basin.

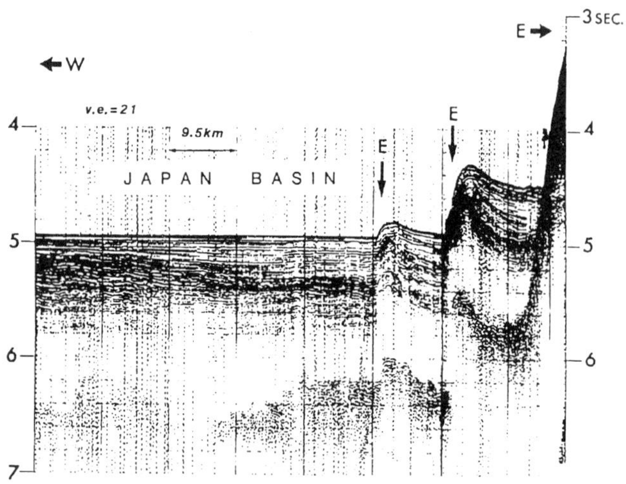

Fig. 2a. Japan Sea: Change in stress regime and deformation of the basin floor. Examples of thrust faults at the foot of the Oshima plateau, from *Tamaki and Honza* [1985].

[1982]. *Jolivet and Tamaki* [1992] interpreted this change as the local effect of a change from transtension to transpression along the N-S Sakhalin-Oga shear zone. They also noted that this event is coeval with more regional plate reorganizations, as marked by onset of the present subduction stage along the Nankai Trench, initiated at 8 Ma and documented by Deep Sea Drilling Program (DSDP) results [*Kagami et al.*, 1983]. The onset of new subduction was also explained by *Huchon* [1985] as the response to detachment of the Philippine Sea slab following the hypothesis of *Niitsima* [1982] for the Pacific Plate, which suffered the same process at 9 Ma. According to Niitsima this slab detachment and the onset of a new subduction zone induced a compressive stress field across the convergent zone that could have induced flexure within the Japan Sea.

An additional effect could be the collision of the Izu-Bonin arc with Honshu, marked by strong uplift in the Kanto Mountains at 8 Ma. This uplift is dated by sudden

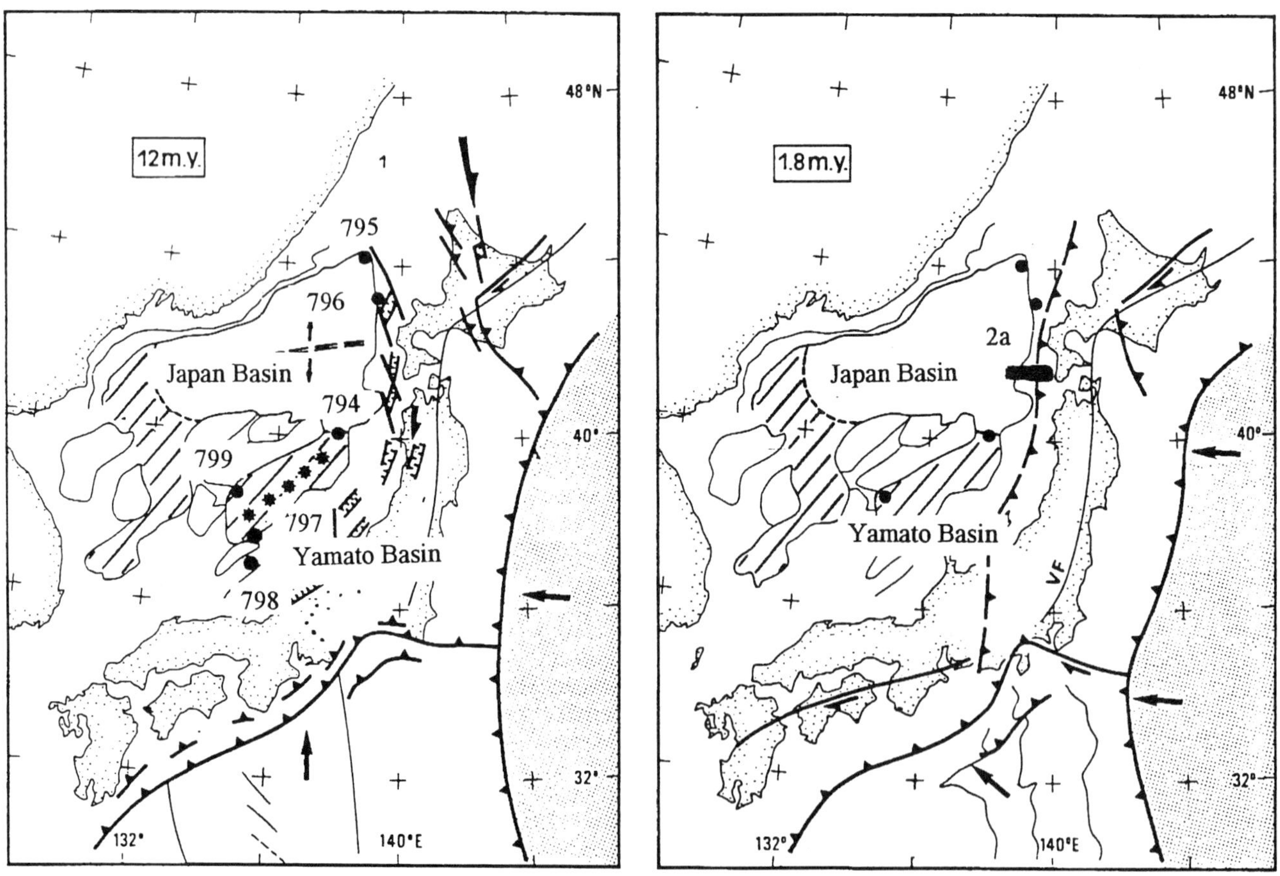

Fig. 2b. Reconstruction of the Japan Sea, slightly modified from *Jolivet and Tamaki* [1992]. Time for onset of compression in the basin and subsidence of basinal parts close to the margin is 12 Ma. Time of initiation of rupture of the Japan Sea crust along the eastern margin of the basin is 1.8 Ma. Location of seismic profile is marked by black bar.

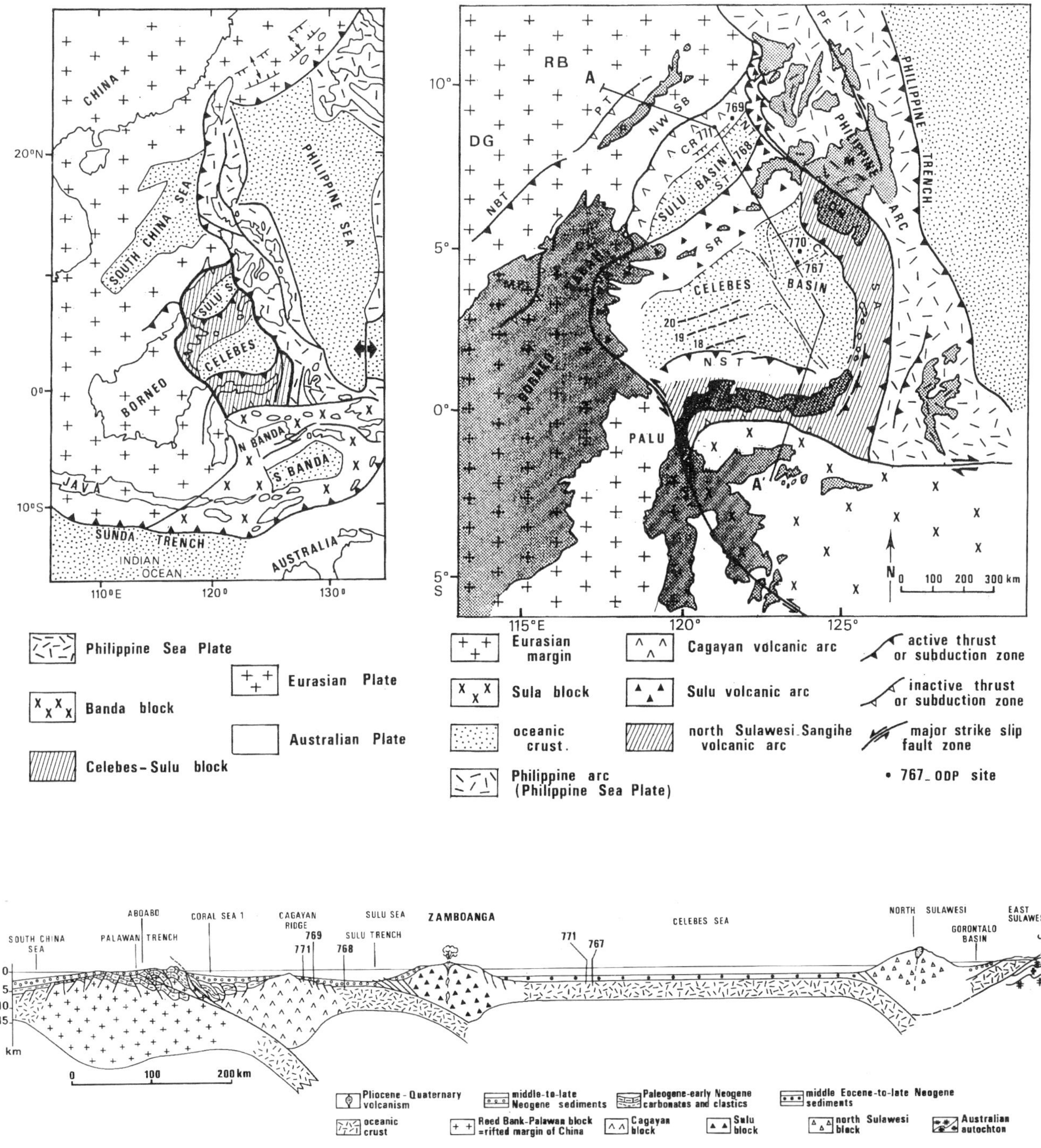

Fig. 3. The Celebes and Sulu Marginal basins [from *Rangin and Silver*, 1991]. NST, North Sulawesi Trench; SR, Sulu Ridge; CR, Cagayan Ridge; NW SB, NW Sulu Basin; PT, Palawan Trench; NBT, North Borneo Trench; DG, Dangerous Grounds; RB, Reed Bank; SA, Sangihe Arc; M, Mindanao; CT, Cotobato Trench; PF, Philippine Fault; AA', trace of cross section.

accumulation of conglomerates (Kawaino formation) in the Fuji River and south of the Kanto Mountains (Iwatono San formation). No clear evidence exists for collision of the Izu-Bonin arc with Honshu prior to 8 Ma. Thus incipient flexure of part of the Japan Sea crust appears coeval with regional tectonic processes such as onset of new subduction along the Nankai Trench and collision of the Izu-Bonin ridge with Honshu.

THE SOUTHEAST ASIAN MARGINAL BASINS

The Sulu and Celebes Seas are located at the convergent triple junction of three major plates: The Eurasian, Philippine Sea, and Australian plates (Figure 3). Therefore the sources of potential intraplate stress are diverse. In the east, these basins are currently being overridden by the Philippine mobile belt, which is colliding against the Palawan, Cagayan, and Sulu ridges [*Rangin et al.*, 1990]. In the south, a block that originated in Australia (the Sula platform, Figure 3) is colliding with Sulawesi forming the southern margin of the Molucca and Celebes Sea [*Silver et al.*, 1983a,b].

Subsidence Recorded in the Southeast Asia Marginal Basins

On the basis of chronostratigraphy and porosity of the sediments recovered from ODP sites 767 and 768 in the Celebes and Sulu Seas, *Huang et al.* [1991] draw conclusions regarding sedimentation and basement subsidence rates. Among the tectonic events recognized with this method there is a pronounced increase in decompacted sedimentation rate in the Celebes Sea at 23.4 Ma. This is approximately the age for suspected incipient closure of the Proto South China Sea in the model of *Rangin et al.* [1990]; these workers believe that all of the SE Asia marginal basins enter compression at that time. This age is also the oldest for volcanic rocks collected in the Cagayan ridge [*Bellon and Rangin*, 1991]. However, the basement subsidence rate curve presented by *Huang et al.* [1991] shows no increase in subsidence rate that could indicate flexure of the crust at that time.

More significant is the increase in basement subsidence rate recorded in both basins between 10.8 and 8.9 Ma (Figure 4), interpreted by *Huang et al.* [1991] as the initiation of the Sulu and Cotobato troughs. These trenches are directly related to collision of the Philippine arc with the Celebes and Sulu margins (Sulu and Palawan block respectively) as discussed later in this paper. Volcanism related to these two newborn subduction zones (Negros and Cotobato) is less than 5 m.y. old [*Pubellier et al.*, 1991a; *Sajona et al.*, 1994].

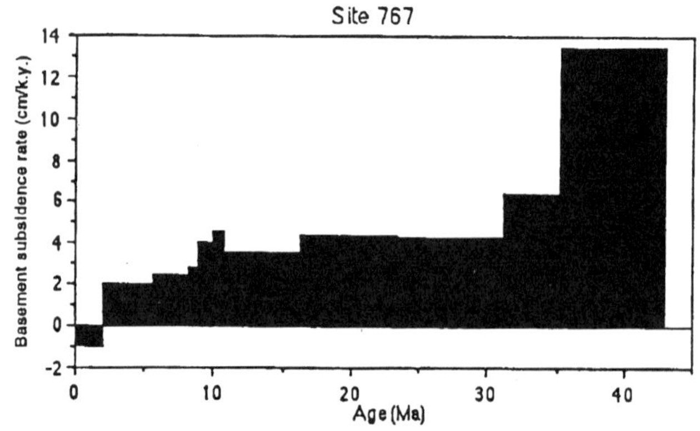

Basement subsidence rates (cm/k.y.) at Site 767, Celebes Sea.

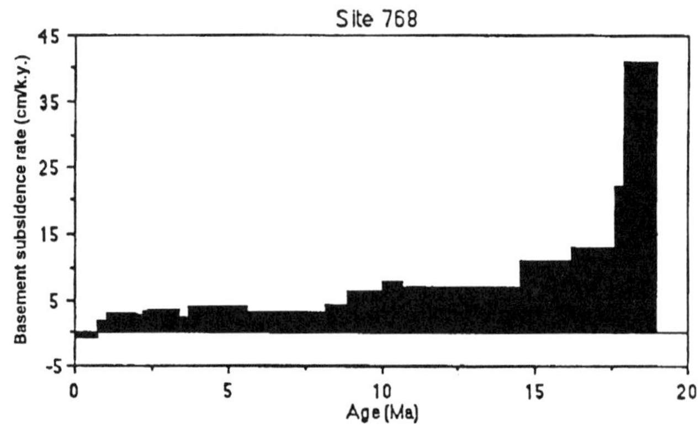

Basement subsidence rates (cm/k.y.) at Site 768, Sulu Sea.

Fig. 4. Basement subsidence rates in the Celebes and Sulu Basins at sites 767 and 768 respectively [from *Huang et al.*, 1991].

Uplift in the basement since 2.0 Ma, first in the Celebes Sea and then in the Sulu Sea (around 0.7 Ma), is interpreted by Huang et al. as the effect of collision in the Molucca Sea. These observed negative subsidence rates are coincident with a new increase of decompacted sedimentation rates in both basins. Recent paleomagnetic data collected by *Surmont et al.* [1994] in the north arm of Sulawesi indicate that rotation and the related subduction began after 4 Ma. The question remains whether this sudden increase of basement subsidence in the Celebes Sea could be explained by proximity to the Cotobato trench. We do not have sufficient evidence to suppose the observed subsidence at site 767 and 768 is representative of the whole Celebes and Sulu Basins.

Deformation of the Southeast Asia Marginal Basins Oceanic Crust

Deformation occurs mainly along the margin of the basins; the "oceanic floor" remains quite rigid. New subduction zones were created in the Late Cenozoic along the Sulu archipelago and the southern margin of the Celebes Basin (Figure 5). The axis of the South China Sea could accommodate part of the regional stress, as evidenced by a compressive focal mechanism at the axis of the basin [*Wang et al.*, 1979]. However, a fine-scale multibeam survey with single channel seismics and 3.5 kHz seismic data recorded during the Nanhai cruise [*Pautot et al.*, 1986] revealed no clear evidence for deformation at the axis of the South China Sea.

In the eastern part of the Celebes Sea, seismic data for ODP leg 124 site surveys [*Hinz and Block*, 1990; *Lewis* 1991] show that thick sediments and underlying oceanic crust remain undeformed and plunge eastward in the Cotobato trench. The same situation is observable along the nascent North Celebes trench [*Silver et al.*, 1983b]. So no clear evidence for thickening and intraplate thrusting has been found in the Celebes Sea, even along nascent trenches. The situation is different in the Sulu Sea. Accretion of oceanic fragments along the inner wall of the Negros and Sulu trenches [*Hinz and Block*, 1990] is probably the most convincing piece of evidence for thrusting within the Sulu Sea basement and incorporation of oceanic crust fragments to the trench wedge (Figure 6). Here multichannel seismic data suggest the presence of a splinter of oceanic crust forming the backstop of the present wedge. On land, the Zamboanga peninsula is considerably uplifted and pre-Cenozoic metamorphic basement is exposed.

Neogene evidence for similar accretion of oceanic crust fragments in wedges is found around the South China Sea. Along its eastern margin this basin subducts in the Manila Trench. To the south, this trench connects with the Mindoro collision zone. Here the Luzon arc overrides the Palawan block, a continental fragment that drifted from China during opening of the South China Sea between 32 and 15 Ma [*Taylor and Hayes*, 1983]. In Mindoro, the accretionary wedge is exposed on land and shows the presence of middle Oligocene ophiolites [*Rangin et al.*, 1985]. These ophiolites, dated by nannoplankton found in the interpillow deep sea sediments, are coeval with the oldest magnetic anomalies of the South China Sea basin, identified immediately to the west by *Taylor and Hayes* [1980]. These ophiolites could be interpreted as fragments of South China Sea oceanic crust jammed between Palawan and Luzon (Figure 7). Imbrication of these ophiolites occurred during early middle Miocene time, the age of the sediments in which the ophiolites are engulfed in Mindoro island. Early middle Miocene is the age for the initiation of the youngest arc volcanism event related to initial subduction along the Manila Trench [*Rangin et al.*, 1990].

Similar observations have been made at the northern tip

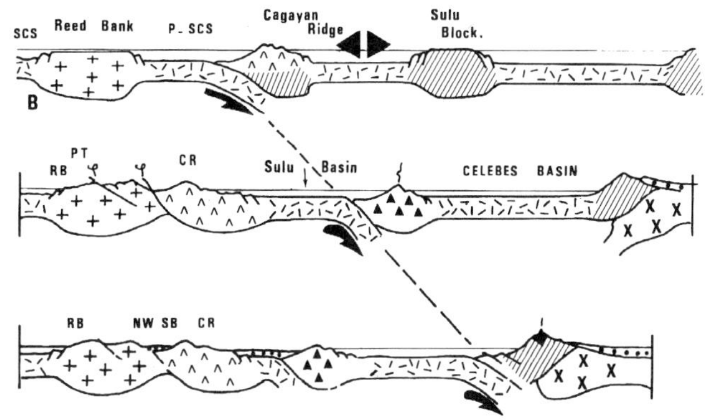

Fig. 5. Interpretative diagram for the evolution of the Celebes and Sulu Basins during their late stage compressional episode [from *Rangin and Silver*, 1991].

of this arc system in Taiwan. Here the Luzon arc has collided with the continental margin of China. At the southern tip of Taiwan, ultramafics and early Miocene pillow lavas are part of the collision wedge [*Stephan et al.*, 1986] and were also interpreted as scraped-off oceanic fragments of the South China Sea. Middle Oligocene ophiolites are reworked into early middle Miocene olistostromes and melanges of the wedge. In the South China and Sulu Seas less than 15 m.y. separate the age of the ophiolites and the time of their incorporation into an accretionary wedge. For the Celebes Sea, 30 to 40 m.y. separate these two events. In the first case the lithosphere is still hot and can be easily deformed, for the Celebes Sea it is more rigid and easily subducted.

Geodynamic Explanations for Stress Variations in the Southeast Asia Marginal Basins

On the basis of ODP data (leg 124) in the Celebes-Sulu seas and geological data collected on land around these basins, *Rangin and Silver* [1991] have proposed a simple scenario for the closure of the SE Asia marginal basins (Figures 5 and 7). In early Miocene time the southern margin of the Celebes-Sulu block collided with the Banda block in central Sulawési. This collision induced rapid subduction and closing of the Proto South China Sea along the northern margin of the Celebes-Sulu block, and the Cagayan volcanic arc was rapidly built. Ages older than 22 Ma cannot be demonstrated for this subduction, suggesting that the Proto South China Sea was subducted in a few million years [*Rangin et al.*, 1990]. This timing indicates that the Proto South China Sea was smaller and narrower than originally proposed by *Taylor and Hayes* [1983]. The significant change in the decompacted sedimentation rate

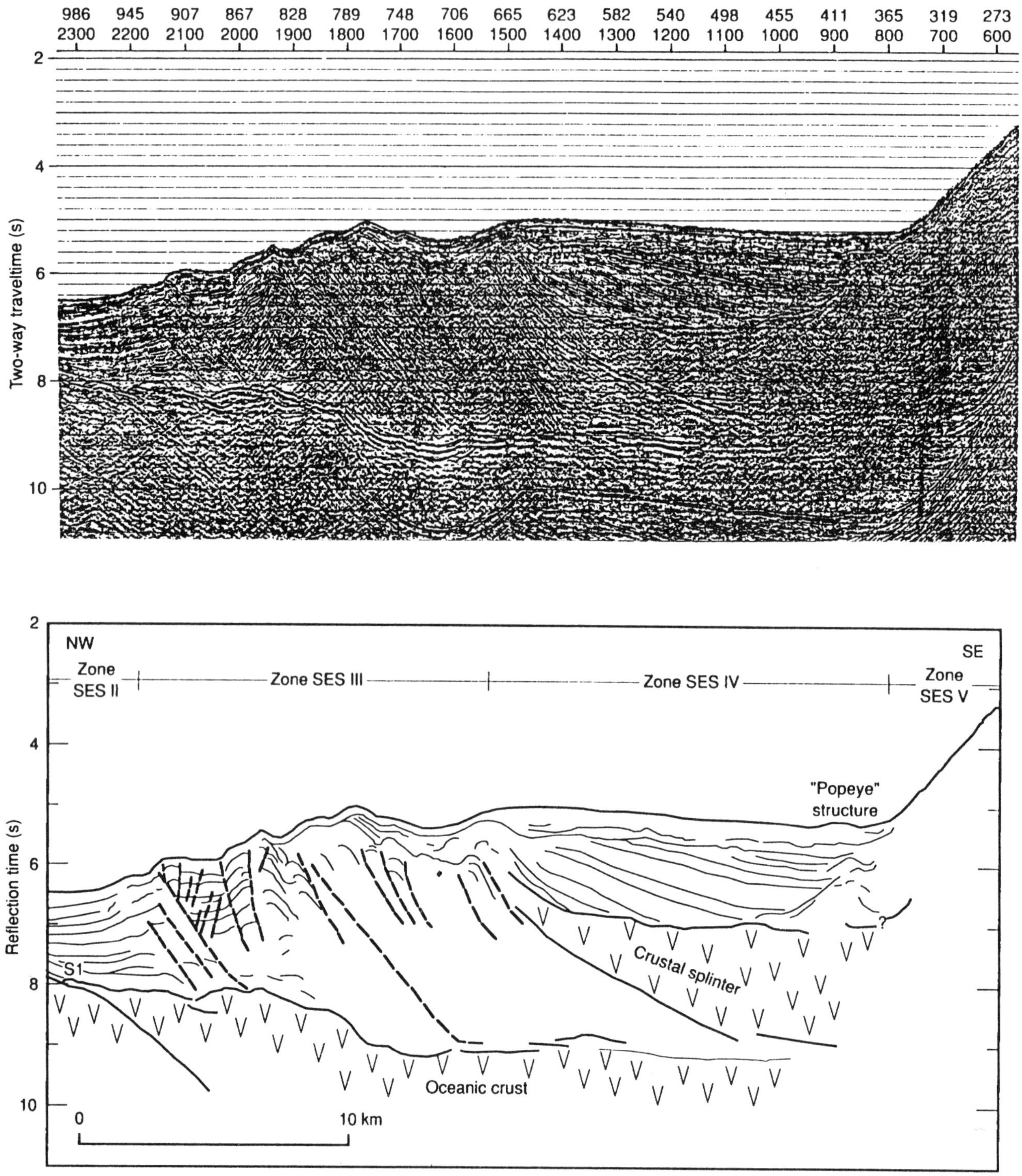

Fig. 6. Sulu crust splinter incorporated to the accretionary wedge of the Negros trench [from *Hinz and Block*, 1990].

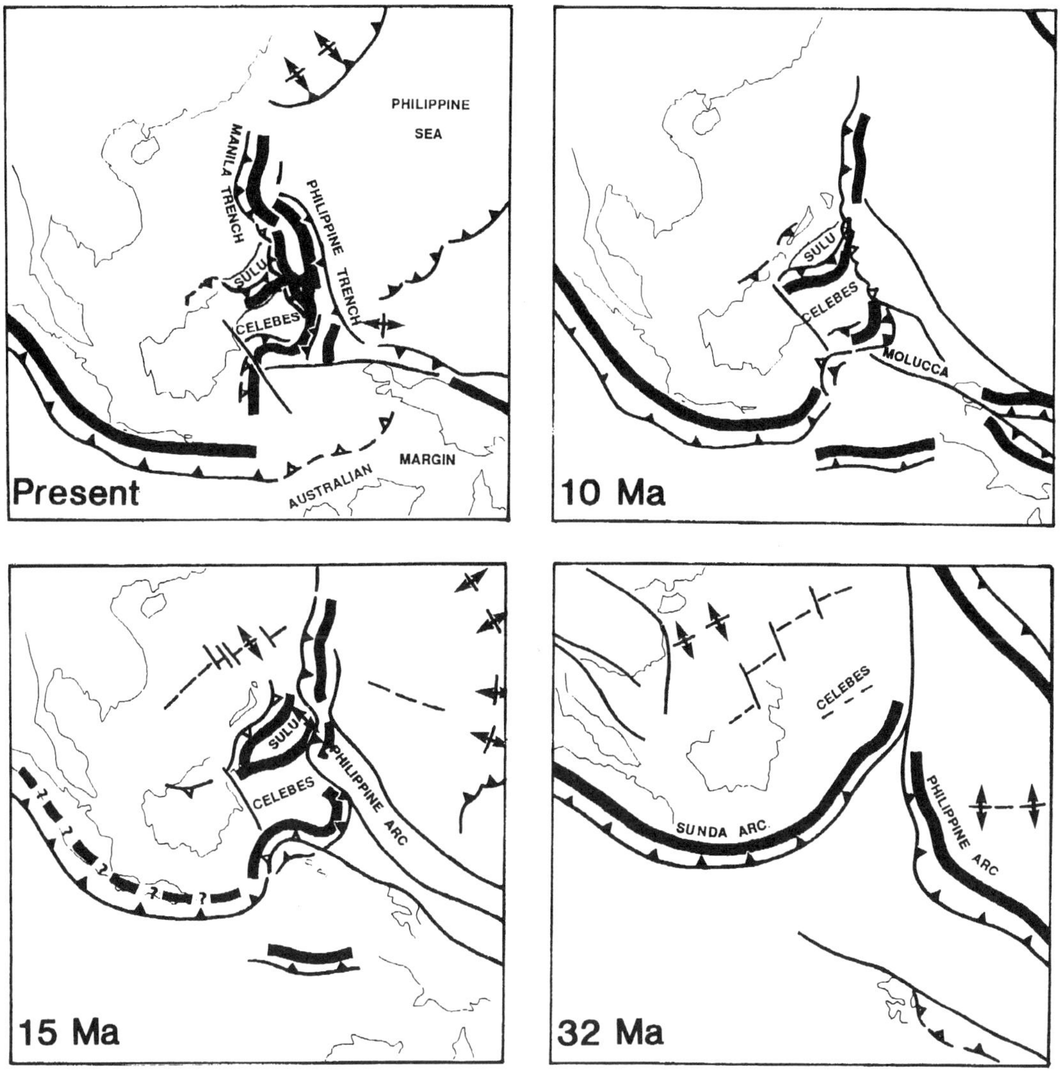

Fig. 7. Simplified palinspatic reconstructions of the convergent Philippine Sea Plate, Australian Plate, Eurasian plate in SE Asia in Neogene time. [*Pubellier et al.*, 1991a, simplified and modified from *Rangin et al.* 1990].

curve in the Celebes Sea at site 767 around 23 Ma is also interpreted by *Huang et al.* [1991] as evidence for basin transformation from a larger, more open oceanic setting to a marginal basin.

Subduction of the Proto South China Sea probably induced backarc opening of the Sulu marginal basin (Figure 7). Termination of this subduction and incipient collision of the Cagayan arc with the rifted fragments from China could be responsible for the obduction of Proto South China Sea Ophiolites on Palawan. These ophiolites, exposed in central and south Palawan island, overthrust the North Palawan continental microblock [*Mitchell et al.*, 1986]. They are not dated in Palawan but extend into NE Borneo [*Müller*, 1991] where this complex is late Cretaceous in age. These ophiolites are much older than the South China Sea magnetic anomalies and can be interpreted as remnant pieces of the former Proto South China Sea, obducted onto the North Palawan rifted microcontinent after complete closure of the Proto South China Sea. These ophiolites are not really incorporated in the Palawan wedge but are thrust above this accretionary complex. After locking of the collision at 15 Ma, the Sulu Sea subducted southward along the Sulu arc, up to late Miocene. A new subduction started in the Celebes Sea with fast rotation of the north arm of Sulawesi after 4 Ma [*Surmont et al.*, 1994].

The closure of these marginal basins also could have been affected by the oblique collision of the Philippine arc (Figure 7) with the basin margins (Palawan and Sulu arcs)

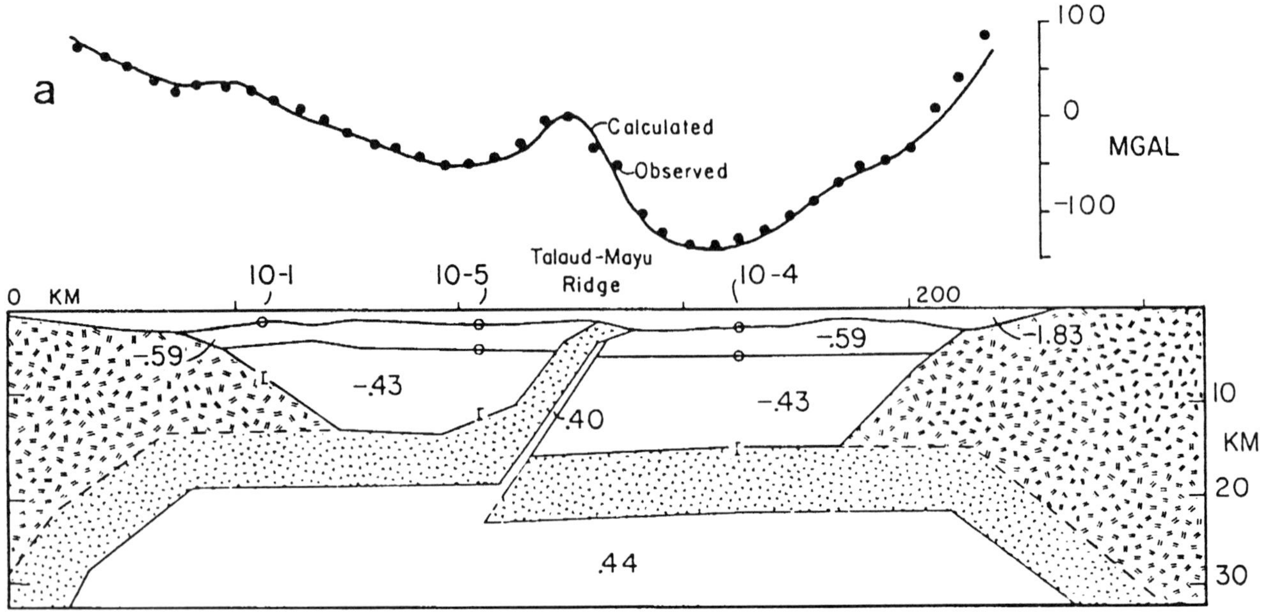

Fig. 8. Molucca Sea oceanic floor affected by thrusting in the Sangihe-Halmahera collision zone. Gravity model [from *Silver and Moore*, 1978] is across central Molucca Sea.

because the Philippine Sea Plate has rotated clockwise since the middle Eocene 90 to 100° [*Haston and Fuller*, 1991]. The Philippine mobile belt extending southward into the Molucca Sea, is in fact the product of a collision of two facing arcs and of the effects of subduction of the Molucca oceanic plate [*Silver and Moore*, 1978] separating the Philippine Sea plate from the Eurasian Plate [*Rangin et al.*, 1990]. The Halmahera arc in the east is still in frontal collision with the Sangihe arc, which limits the Celebes Sea to the west. This collision zone is marked by thickening of the crust of the Molucca Sea by intraoceanic imbrication (Figure 8). The ophiolites exposed in the Talaud Islands are Eocene in age [*Moore et al.*, 1981] and the thrusting is still ongoing as indicated by active seismicity in this area. It is reasonable to suggest that after complete closure of this basin, Molucca sea oceanic crust will be overthrust onto one of the colliding arcs. This effect can be verified in Mindanao because the oblique motion of the Philippine Plate with respect to Eurasia permitted a diachronic closure of the Molucca Basin that could have extended largely to the north in Mindanao and the Visayas (Central Philippines), whereas the Sangihe arc can be traced towards the Daguma Range in the southwest of the island, and the Halmahera arc can be traced in the central and Pacific cordilleras of Mindanao [*Pubellier et al.*, 1991a]. During this oblique collision, the Halmahera arc could have been displaced left laterally with respect to the Sangihe arc along NW-SE strike slip faults, and such faults are seen in the field. Progressive north to south locking of this oblique collision can be dated by cessation of volcanism in the colliding arcs and subsequent induced flip of subduction zones. The Eurasian/Philippine Sea plate is currently being absorbed along two new subduction zones in backarc position, the Philippine trench on the east and the Cotobato-Negros trenches on the west.

Cessation of volcanism along the Sangihe arc from north to south has been discussed by *Marchadier and Rangin* [1990], *Pubellier et al.* [1991a], *Bellon and Rangin* [1991], *Pubellier et al.* [1991b], and *Sajona et al.* [1994]. The most significant event related to this collision in the central Philippines (Tablas and Panay islands) is recorded around 10 Ma by cessation of volcanism in the Sangihe arc and massive pyroclastic discharge in the Sulu Basin [*Rangin et al.*, 1990]. This collision could explain the rapid increase in basement subsidence [*Huang et al.*, 1991] in the Sulu and Celebes Basins near the Negros and Cotobato trenches respectively.

The age of related volcanism for the Philippine, Negros, and Cotobato trench system is quite young. The oldest arc volcanism was recently dated between 2.5 and 2.3 Ma in East Mindanao and Leyte [*Sajona et al.*, 1993]. These ages are in good agreement with the short length of the slab [*Cardwell et al.*, 1980]. Also the oldest age of the Philippine fault is less than 4 Ma [*Aurelio et al.*, 1991]. This fault is mechanically related to the Philippine Trench by shear partitioning [*Fitch*, 1972], which implies that the trench is very young. The Negros trench is the source for recent arc volcanism on Negros island, which is seated on early Pliocene sediments typical of a middle to upper bathyal environment [*Müller et al.*, 1989]. In western Mindanao the Plio-Pleistocene volcanic arc is only marked by recent calderas [*Pubellier et al.*, 1991a] and the slab is also very short [*Cardwell et al.*, 1980].

Consequently, in the Celebes Sea, subsidence of the basement at 10 Ma could have been a response to arc collision on both sides of the Molucca Sea. Rupture of the crust marked by nascent Cotobato-Negros subduction zones less than 3 m.y. old occurred 6 to 7 m.y. after flexure effects in the basin. This timing is quite similar to that observed in the Japan Sea [*Jolivet and Tamaki*, 1992].

CONCLUSIONS

In both the Japan and southeast Asia marginal basins, rupture of the crust is preferentially developed along one of the basin margins and may evolve rapidly into new subduction zones. In these basins the youngest part of oceanic crust was initially subducted. The Sulu Sea began to subduct immediately after opening of the basin, and the same scenario is observed in the Japan Sea. The close age relationship between the formation of the oceanic crust and the age of its incipient deformation makes these basins good candidates for possible subsequent development of ophiolites. This is also verified for the ophiolites in Mindoro and southern Taiwan.

The oceanic crust of marginal basins appears to behave rigidly during closure of the basin when the age of the crust is more than 30 m.y. older. Subsidence is generally located near nascent trenches, and no clear evidence of flexure of the whole basin is recorded.

Obduction of oceanic crust is only really demonstrated in very specific and local geodynamic settings like Mindoro, Taiwan, or eventually Palawan, where an island arc is colliding with a continental margin. The collision of two converging volcanic arcs could induce intraplate thrusting in the subducting oceanic slab as seen in the Molucca Sea by seismic refraction observations, gravity modeling, and earthquake focal mechanisms [*McCaffrey et al.*, 1980]. Thus the last step of closure for a marginal basin could induce thrusting into the oceanic crust and later incorporation in the arc-arc collision complex as suspected in Mindanao.

REFERENCES

Amstrong, R. L., and H. Dick, A model for the development of thin overthrust sheets of crystalline rocks, *Geology*, *2*, 35-40, 1974.

Aurelio, M., E. Barrier, C. Rangin, and C. Müller, The Philippine fault in the late Cenozoic evolution of the Bondoc-Masbate-N Leyte area, Philippines, in *Orogenesis in Action*, edited by R. Hall, G. Nichols, and C. Rangin, *J. SE Asian Earth Sci.*, Special Issue, pp. 221-238, 1991.

Bellon, H., and C. Rangin, Geochemistry and isotopic dating of Cenozoic volcanic arc sequences around the Celebes and Sulu Sea, in *Proc. ODP, Sci. Results, 124*, E. A. Silver, C. Rangin, and M. T. Von Breymann, et al., pp. 321-338, Ocean Drilling Program, College Station, TX, 1991.

Bernstein-Taylor, B. L., K. M. Brown, E. A. Silver, and K. S. Kirshoff-Stein, Basement slivers within the New Britain accretionary wedge: Implications for the emplacement of some ophiolites slivers, *Tectonics*, *11*, 753-765, 1992a.

Bernstein-Taylor, B. L., K. S. Kirshoff-Stein, E. A. Silver, D. L. Reed, and M. Mackay, Large scale dupleixes within the New Britain accretionary wedge: A possible example of accreted ophiolite slivers, *Tectonics*, *11*, 732-752, 1992b.

Boudier, F., J. L. Bouchez, A. Nicolas, M. Cannat, G. Ceuleneer, M. Misseri, and R. Montigny, Kinematics of oceanic thrusting in the Oman ophiolite: Model of plate convergence, *Earth Planet. Sci. Lett.*, *75*, 215-222, 1985.

Cardwell, R. K., B. L. Isacks, and D. E. Karig, The spatial distribution of earthquakes, focal mechanism solutions and subducted lithosphere in the Philippines and northern Indonesia regions, in *The Tectonic and Geologic Evolution of Southeast Asian Seas and Islands*, *Geophys. Monogr. Ser.*, vol. 23, edited by D. Hayes, pp. 1-35, AGU, Washington, D.C., 1980.

Cloething, H. S., H. McQueen, and K. Lambeck, On a tectonic mechanism for fluctuations of relative sea level, *Geology*, 14, 617-620, 1985.

Coleman, R. G., *Ophiolites*, Springer-Verlag, New York, 119 pp, 1977.

Davies, H. L., Peridotite-gabbro-basalt complex in eastern Papua: An over-thrust plate of oceanic mantle and crust, *Aust. Bur. Mineral. Resour. Bull.*, *128*, 48 pp, 1971.

Fitch, T. J., Plate convergence, transcurrent faults and internal deformation adjacent to Southeast Asia and the western Pacific, *J. Geophys. Res.*, *77*, 4432-4460, 1972.

Haston, R. B., and M. Fuller, Paleomagnetic data from the Philippine Sea Plate and their tectonic significance, *J. Geophys. Res.*, *96*, 6073-6078, 1991.

Hinz, K., and M. Block, Summary of geophysical data from the Sulu and Celebes seas, in *Proc. ODP, Init. Repts., 124*, C. Rangin, E. A. Silver, and M. T. Von Breymann, et al., pp. 87-92, Ocean Drilling Program, College Station, TX, 1990.

Huang, Z., F. M. Gradstein, and K. E. Louden, Subsidence and sedimentation analysis of marginal basins: Celebes Sea and Sulu Sea, leg 124 sites 767 and 768, in *Proc. ODP, Sci. Results*, *124*, E. A. Silver, C. Rangin, M. T. Von Breymann, et al., pp. 399-407, Ocean Drilling Program, College Station, TX, 1991.

Huchon, P., Géodynamique de la zone de collision d'Izu, et de la jonction triple du Japon Central, Thèse Université Pierre et Marie Curie, 1985.

Ingle Jr., J. C., Subsidence of the Japan Sea: Stratigraphic evidence from ODP sites and onshore sections, in *Proc. ODP, Sci. Results*, *127/128 Pt. 2*, K. Tamaki, K. Suyehiro, J. Allen, and M. McWilliams, et al., pp. 1197-1218, Ocean Drilling Program, College Station, TX, 1992.

Jolivet, L., P. Huchon, and C. Rangin, Tectonic setting of western Pacific marginal basins, *Tectonophysics*, *160*, 23-47, 1989.

Jolivet, L., and K. Tamaki, Neogene Kinematics in the Japan Sea Region and volcanic activity of the Northeast Japan arc, in *Proc. ODP, Sci. Results, 127/128 Pt. 2*, K. Tamaki, K. Suyehiro, J. Allen, and M. McWilliams, et al., pp. 1311-1331, Ocean Drilling Program, College Station, TX, 1992.

Kagami, H., S. Shiono, and A. Taira, Plate subduction and formation of accretionary bodies in the Nankai Trough, *Kagasu (Science)*, *53*, 429-438, 1983.

Lewis, S. D., Geophysical setting of the Sulu and Celebes seas, in *Proc. ODP, Sci. Results*, *124*, E. A. Silver, C. Rangin, and M. T. Von Breymann, et al., pp. 65-74, Ocean Drilling Program, College Station, TX, 1991.

Marchadier, Y., and C. Rangin, Polyphased tectonics at the southern tip of the Manila Trench, Mindoro-Tabalas Islands Philippines, *Tectonophysics*, *183*, 273-288, 1990.

McCaffrey, R., E. A. Silver, R. W. Raitt, Crustal structure of the Molucca Sea collision zone, Indonesia, in *The Tectonic and Geologic Evolution of Southeast Asian Seas and Islands*, *Geophys. Monogr. Ser.*, vol. 23, edited by D. E. Hayes, pp. 161-177, AGU, Washington D.C., 1980.

Mitchell, A. H. G., F. Hernandez, and A. P. de la Cruz, Cenozoic evolution of the Philippine archipelago, *J. SE Asian Earth Sci.*, *1*, 3-22, 1986.

Moore, G. F., and D. E. Karig, Structural geology of Nias island, Indonesia: Implications for subduction zone and tectonics, *Am. J. Sci.*, *280*, 193-223, 1980.

Moore, G. F., D. Kadarisman, C. A. Evans, and J. W. Hawkins, Geology of the Talaud Island, Molucca Sea collision zone, Northwest Indonesia, *J. Struct. Geol.*, *3*, 467-475, 1981.

Müller, C., Biostratigraphy and geological evolution of the Sulu Sea and surrounding area, in *Proc. ODP, Sci. Results*, *124*, E. A. Silver, C. Rangin, M. T. Von Breymann, et al., pp. 121-132, Ocean Drilling Program, College Station, TX, 1991.

Müller, C., C. H. Von Daniels, P. Cepek, F. Gramann, G. J. G. Bausa, and M. M. De Leon, Biostratigraphy and paleoenvironmental studies in the Tertiary of the Visayan Basin, Philippines, in *On the Geology and Hydrocarbon Prospects of the Visayan Basin, Philippines*, edited by H. Porth and C. H. Von Daniels, *Geol. Jahrbuch Reihe B*, *70*, 303-316, 1989.

Nicolas, A., *Structure of Ophiolites and Dynamics of Oceanic Lithosphere*, Kluwer Acad. Publi., Dordrecht, 367 pp, 1989.

Niitsima, N., Touchstone for plate tectonics, Southern Fossa Magna, *The Earth Monthly*, *4*, 326-333, 1982.

Pautot, G., C. Rangin, A. Briais, P. Tapponier, P. Beuzart, G. Lericolais, X. Mathieu, J. Wu, S. Han, H. Li, Y. Lu, and J. Zhao, Spreading direction in the central South China Sea, *Nature*, *321*, 150-154, 1986.

Pubellier, M., R. Quebral, C. Rangin, B. Deffontaines, C. Müller, J. Buttrelin, and J. Manzano, The Mindanao collision zone: A soft collision event within a Neogene strike slip setting, *J. SE Asian Earth Sci.*, *6*, 239-248, 1991a.

Pubellier, M., P. Spadea, A. Pouclet, R. Solidum, A. Desprairies, and H. Cambray, Correlations of tephras in Celebes and Sulu sea basins: Constraints on geodynamics, in *Proc. ODP, Sci. Results*, *124*, E. A. Silver, C. Rangin, M. T. Von Breymann, et al., pp. 459-465, Ocean Drilling Program, College Station, TX, 1991b.

Rangin, C., J. F. Stephan, and C. Müller, Middle Oligocene oceanic crust of South China Sea jammed into Mindoro collision zone, *Geology*, *13*, 425-428, 1985.

Rangin, C., M. Jolivet, M. Pubellier, and the Tethys Pacific working group, A simple model for the tectonic evolution of Southeast Asia and Indonesia regions for the past 43 Ma, *Bull. Soc. Geol. Fr.*, *VI*, 889-905, 1990.

Rangin, C., E. A. Silver, and M. T. Von Breymann, *Proc. ODP, Init. Repts.*, *124*, Ocean Drilling Program, College Station, TX, 1990b.

Rangin, C., and E.A. Silver, Neogene tectonic evolution of the Celebes-Sulu basins: New insights from Leg 124 drilling, in *Proc. ODP, Sci. Results*, *124*, E. A. Silver, C. Rangin, M. T. Von Breymann, et al., pp. 51-64, Ocean Drilling Program, College Station, TX, 1991.

Sajona, D., H. Bellon, R. Maury, M. Pubellier, J. Cotten, and C. Rangin, Magmatic response to abrupt changes in geodynamic settings: Pliocene-Quaternary lavas from Mindanao (Philippines), submitted to *Tectonophysics*.

Sajona, F. G., R. C. Maury, H. Bellon, J. Cotten, M. J. Defant, and M. Pubellier, Initiation of subduction and the generation of slab melts in western and eastern Minadano, *Geology*, *21*, 1007-1010, 1993.

Silver, E. A., R. McCaffrey, Y. Joyodiwiriyo, and S. Stevens, Ophiolite emplacement by collision between the Sula Platform and the Sulawesi Island arc, Indonesia, *J. Geophys. Res.*, *88*, 9419-9435, 1983a.

Silver, E., R. McCaffrey, R. B. Smith, et al., Collision, rotation and the initiation of subduction in the evolution of Sulawesi, Indonesia, *J. Geophys. Res.*, *88*, 9407-9418, 1983b.

Silver, E.A., and J. C. Moore, The Molucca Sea collision zone, Indonesia., J. Geophys. Res., *83*, 1681-1691, 1978.

Snyder, D. B., J. H. McBride, and the BIRPS core group, Marine deep seismic reflection signatures of continental convergent zones: Examples from the Archean-Proterozoic to the active (abstract), *Eos Trans. AGU*, *74*, 430-431, 1993.

Stephan, J. F., R. Blancjet, C. Rangin, B. Pelletier, J. Letouzey, and C. Müller, Geodynamic evolution of the Taiwan-Luzon-Mindoro belt since the Late Eocene, *Tectonophysics*, *125*, 245-268, 1986.

Surmont J., C. Laj, C. Kissel, C. Rangin, H. Bellon, and B. Priadi, New paleomagnetic constraints on the Cenozoic tectonic evolution of the north arm of Sulawesi, Indonesia, *Earth Planet. Sci. Lett.*, *121*, 629-638, 1994.

Taira, A., et al., The Shimanto Belt in Shikoku, Japan - evolution of Cretaceous to Miocene accretionary prism, *Mod. Geol.*, *12*, 5-46, 1988.

Taira, A, H. Tokuyama, and W. Soh, Accretion Tectonics and Evolution of Japan Sea, in *The Evolution of the Pacific Ocean Margin*, edited by Z. Ben-Avraham, pp. 100-123, Oxford Univ. Press, Oxford, UK, 1989.

Tamaki, K., K. Pisciotto, and J. Allan, *Proc. ODP, Init. Repts.*, *127*, Ocean Drilling Program, College Station, TX, 1990.

Tamaki, K., and E. Honza, Incipient subduction and obduction along the eastern margin of the Japan Sea, *Tectonophysics*, *119*, 381-406, 1985.

Tapponnier, P., G. Peltzer, A. Y. Le Dain, R. Armijo, and P. Cobbold, Propagating extrusion tectonics in Asia: New insights from simple experiments with plasticine, *Geology*, *10*, 611-616, 1982.

Taylor, B., and D. E. Hayes, The tectonic evolution of the South China Sea Basin, in *The Tectonic and Geologic Evolution of Southeast Asian Seas and Islands*, part 1, *Geophys. Monogr.*

Ser., vol. 23, edited by D. E. Hayes, pp. 89-104, AGU, Washington D.C., 1980.

Taylor, B., and D. E. Hayes, The tectonic evolution of the South China Sea Basin, in *The Tectonic and Geologic Evolution of Southeast Asian Seas and Islands*, part 2, *Geophys. Monogr. Ser.*, vol. 27, edited by D. E. Hayes, pp. 23-56, AGU, Washington D.C., 1983.

C. Rangin, Département de Géologie Structurale T26-0 Université Pierre et Marie Curie, Boite 129, 4 Place Jussieu, 75252, Paris, France.

E. A. Silver, Department of Earth Science, University of California Santa Cruz, Santa Cruz, CA 95064, USA.

K. Tamaki, Ocean Research Institute, University of Tokyo, 1-15-1 Minami Dai, Nakano Ku, Tokyo, Japan.